ABSOLUTE VALUE

$$|a| = \begin{cases} a & \text{if } a \geq 0 \\ -a & \text{if } a < 0 \end{cases}$$

$$|a + b| \leq |a| + |b|$$

$$|ab| = |a||b|$$

$|x| < c$ if and only if $-c < x < c$

QUADRATIC FORMULA

The solutions of $ax^2 + bx + c = 0$ are given by

$$x = \frac{-b \pm \sqrt{b^2 - 4ac}}{2a}$$

LAWS OF EXPONENTS

$$a^x a^y = a^{x+y}$$

$$(a^x)^y = a^{xy}$$

$$(ab)^x = a^x b^x$$

$$\left(\frac{a}{b}\right)^x = \frac{a^x}{b^x}$$

$$\frac{a^x}{a^y} = a^{x-y}$$

$$a^{-x} = \frac{1}{a^x}$$

$$a^0 = 1$$

$$a^1 = a$$

$$1^x = 1$$

LAWS OF LOGARITHMS

$$\log_a (xy) = \log_a x + \log_a y$$

$$\log_a \frac{1}{x} = -\log_a x$$

$$\log_a \frac{x}{y} = \log_a x - \log_a y$$

$$\log_a (x^c) = c \log_a x$$

$$\log_b x = \frac{\log_a x}{\log_a b}$$

$$\log_a 1 = 0$$

$$\log_a a = 1$$

$$a^{\log_a x} = x$$

$$\log_a (a^x) = x$$

PRECALCULUS

FOURTH EDITION

PRECALCULUS

FOURTH EDITION

Robert Ellis ▪ Denny Gulick

University of Maryland at College Park

Saunders College Publishing

Harcourt Brace Jovanovich College Publishers

Fort Worth Philadelphia San Diego New York Orlando Austin San Antonio
Toronto Montreal London Sydney Tokyo

Associate Editor: Pamela Whiting
Manuscript Editor: Cheryl Hauser
Production Editor: Judi McClellan
Designers: Ann Smith, Cathy Reynolds
Art Editor: Judith Frazier
Production Manager: Mandy Van Dusen

TO OUR MOTHERS AND THE MEMORY OF OUR FATHERS

Cover: "Ad Parnassum" © estate of Paul Klee/VAGA, New York, 1991

0 © Photo Edit. **58, 84** © Dan Helms/Duomo Photography. **73** Delia Flynn/Stock Boston. **132, 161** © Barry L. Runk/Grant Heilman. **171** © Tiers, Monkmeyer Press Photo Service. **194** © Grant Heilman. **239** Official U.S. Navy Photograph. **239** The Port Authority of New York and New Jersey. **263** National Park Service. **264** AP/Wide World Photos. **289** UPI Photo. **295** The Metropolitan Museum of Art, Museum excavations, 1922–1923; Rogers Fund, 1923. (26.3.11–12). **297** Courtesy Biblical Archaeology Society. **306, 327** © Helen Faye. **326** Courtesy Daytona International Speedway. **351** © Harry W. Rinehart. **362** © From *Sounds of Music*, By Charles Taylor, BBC, London. **364** HBJ Photo Library. **391** © From *Sounds of Music*, By Charles Taylor, BBC, London. **400** © B. Christensen/Stock Boston. **435** National Park Service. **438, 446** © Robin Risque. **508** Courtesy Hewlett Packard. **530** © Culver Pictures. **570** © Melanie Carr/Zephyr Pictures, 1988. **573** © Ken Karp Photography. **591** Courtesy of the Oriental Institute, University of Chicago.

ISBN: 0-15-571067-2

Library of Congress Catalog Card Number: 91-72315

Printed in the United States of America

PREFACE

Precalculus is designed for students with the equivalent of two years of high school algebra. It contains more than enough material for a one-semester or two-quarter course, enabling an instructor to choose topics in accordance with the background of the students and the objectives of the course.

The preliminary Chapter 0 contains a review of basic algebraic concepts and can be covered at a pace dictated by the ability of the student. Chapters 1 through 4 present the main topics of college algebra and related topics from analytic geometry. These include solutions of equations and inequalities, functions and equations and their graphs, and the exponential and logarithmic functions. Chapters 5 through 7 deal with trigonometry and trigonometric functions, trigonometric identities, and the solving of triangles. The remainder of the book examines a variety of other precalculus topics: systems of equations and matrices, complex numbers, roots of polynomials, and topics in discrete algebra (including mathematical induction, sequences, series, permutations and combinations, and the binomial theorem).

In writing this textbook, which is based on the authors' third edition of *College Algebra and Trigonometry*, we have attempted to place significant emphasis on functions and on approximation techniques that can be implemented on a scientific or graphics calculator. The main differences between the two texts follow:

1. Preliminary topics have been placed in Chapter 0.
2. A new section deals with approximating zeros of functions by the bisection method and by graphics calculator, and approximating solutions of inequalities by graphics calculator.
3. A new section on iterates of functions has been added, including a graphical technique for constructing iterates.
4. We discuss the uses of graphics calculators, both for graphing functions and for approximating zeros of functions as well as solutions of inequalities.

5. Many new calculator exercises for scientific as well as graphics calculators have been added.

One of our major goals has been to provide a readable book on precalculus mathematics. As a result, we have tried to present each new topic carefully, with plenty of explanation and illustrations, supporting examples, and cautionary remarks (which are set off from the text to enhance their impact). The examples thoroughly reinforce new concepts and provide computational practice as they show the many diverse fields to which mathematics can be applied. Furthermore, we include those theorems and proofs of theorems that provide additional insight into the fundamental ideas presented. The symbol □ signals the completion of an example.

The exercise sets have been designed to afford ample practice on the topics appearing in the text. The exercises are graded, from routine to more sophisticated; the most difficult exercises are accompanied by an asterisk (*) for easy recognition. Generally the exercises are paired, odd and even, in order to give additional practice and because answers to the odd-numbered exercises are included in the back of the book. Where appropriate we have provided calculator exercises; these are clearly identified with the symbol **C**. In keeping with current trends, both text and exercises include numerous applications to such diverse areas as chemistry, geology, acoustics, motion, demography, forestry, economics, archeology, architecture, engineering, and games. Both the metric and the English systems of measurement are employed in applied exercises. Each chapter concludes with a list of key terms and formulas and a set of review exercises that give practice on the major topics of the chapter. A set of cumulative review exercises for Chapters 0–3 follows Chapter 3, and a set for Chapters 4–7 follows Chapter 7. For convenient reference, the major formulas and laws from algebra and trigonometry are also printed on the inside covers. An Instructor's Manual containing answers to the even-numbered exercises is also available to instructors.

We wish to thank the reviewers of all editions of this book. They include Murray Cantor (Shell Development Company, Houston, Texas), Edward Chapin (University of Maryland, Eastern Shore), C. Patrick Collier (University of Wisconsin, Oshkosh), Joan Dykes (Edison Community College), Mary Farantonello (Radio Shack Computer Center, Houston, Texas), Daniel Mosenkis (City College of the City University of New York), Nancy Jim Poxon (California State University, Sacramento), Helen Salzberg (Rhode Island College), Helen Santiz (University of Michigan, Dearborn), Erik Schreiner (Western Michigan University), Ron Smith (Edison Community College), William Smith (University of North Carolina, Chapel Hill), Shirley Sorensen (University of Maryland, College Park), F. Eugene Tidmore (Baylor University), Marvel Townsend (University of Florida), and Cynthin Wilson (Mississippi State University).

We very much appreciate the help, encouragement, and vision of two fine editors: Marilyn Davis, who worked with us on the first edition, and

Richard Wallis, who worked with us on the second and third editions. In addition, we appreciate the assistance of the following staff members at Harcourt Brace Jovanovich for their endeavors on behalf of the Fourth Edition: Judi McClellan, Cathy Reynolds, Judith Frazier, Mandy Van Dusen, and Pamela Whiting.

ROBERT ELLIS
DENNY GULICK

CONTENTS

3 Polynomial and Rational Functions

4 Exponential and Logarithmic Functions

5 The Trigonometric Functions

6 Trigonometric Identities and Equations

7 Trigonometry and Its Applications

RENT ME
RENT ME
RENT ME
RENT ME

0 REVIEW OF BASIC ALGEBRA

Chapter 0 is a preliminary chapter that contains a review of basic algebra. If you are thoroughly acquainted with basic algebra, you should be able to scan this chapter and dwell only on those topics with which you are less familiar.

We begin with a description of the various kinds of real numbers and the basic properties of the arithmetic operations of addition, subtraction, multiplication, and division. Next we show how to associate the real numbers with the points on a line. Then we branch out and discuss exponents, radicals, polynomials, and rational expressions, all of which are derived in one way or another from the four basic arithmetic operations.

The results of this chapter form the foundation for the many subjects that will be developed in the remaining chapters. Moreover, the results and laws we discuss in this chapter can help in solving everyday problems such as the following one.

> An agency normally rents mid-size cars at $30 per day. It gives a 5% discount to students but raises the rental price 3% on holiday weekends. For such weekends, would the final cost be affected if the agent first took the discount and then added the 3% surcharge, rather than adding the 3% surcharge first and then taking the discount?

We will solve this problem in Example 4 of Section 0.1.

0.1 REAL NUMBERS

Real numbers appear everywhere in our lives. Indeed, we encounter them anywhere from a grocery store to a doctor's office to a sports page. Let us begin by describing the various kinds of real numbers.

Kinds of Real Numbers

There are several basic collections of real numbers:

Natural numbers (or ***counting numbers***): 1, 2, 3, 4, . . .

Even integers: integers that are multiples of 2, such as 0, 2, -2, 4, -4, 6, -6, . . .

Odd integers: integers that are not even, such as 1, -1, 3, -3, 5, -5, . . .

Rational numbers: numbers that can be written in the form a/b, where a and b are integers with $b \neq 0$; examples are $22/7$, $\dfrac{123}{49{,}265}$, and -11.6 $\left(\text{which equals } \dfrac{-116}{10}\right)$

Irrational numbers: any number that is not rational, that is, that cannot be written as a/b with a and b integers.

Examples of irrational numbers are π (which is the ratio of the circumference of a circle to its diameter) and $\sqrt{2}$ (which is the ratio of the length of the diagonal of a square to the length of a side).

We remark that every integer a is a rational number, since a can be written as a/b with $b = 1$. For example,

$$13 = \frac{13}{1} \quad \text{and} \quad -7 = \frac{-7}{1}$$

Every real number has a decimal expansion. The decimal expansion of a rational number is either terminating or repeating. For example,

$$\frac{5}{4} = 1.\overbrace{25}^{\text{terminating}}, \quad \frac{2}{3} = \overbrace{.66666}^{\text{repeating}}\ldots, \quad \text{and} \quad \frac{39}{11} = 3.\overbrace{545454}^{\text{repeating}}\ldots$$

By contrast, the decimal expansion of any irrational number neither terminates nor repeats. For example, the decimal expansion of π begins

$$3.14159265358979\ldots$$

with no known pattern to the digits appearing in the expansion. However, the digits in the expansion of an irrational number *can* have an easily recognizable pattern (see Exercise 39).

Operations and Conventions

There are four fundamental operations of arithmetic. For two real numbers a and b, we denote the operations as follows:

Arithmetic Operations
Sum: $a + b$
Difference: $a - b$
Product: ab, or $a \cdot b$, or $(a)(b)$
Quotient: $\frac{a}{b}$, or a/b, or $a \div b$, with $b \neq 0$

The numbers a and b are ***factors*** of the product ab, and in the quotient a/b, a is the ***numerator*** and b the ***denominator***. Frequently the quotient a/b of two real numbers is called a ***fraction***.

CAUTION: You will notice that a/b is not defined for $b = 0$. Thus dividing by 0 is not permitted and the quotient $a/0$ is not defined. In particular, $0/0$ and $1/0$ are not defined.

It will be important to keep in mind two conventions that are universally applied in evaluating arithmetic expressions. First, in evaluating an expression such as $a + bc$, multiplication and division take precedence over addition and subtraction. Thus

$$3 + 4 \cdot 5 = 3 + 20 = 23$$

and

$$3 - 2 \div 6 = 3 - \frac{1}{3} = \frac{8}{3}$$

Second, we perform operations inside parentheses before those outside the parentheses. As a result,

$$4(-3 + 2 \cdot 5) = 4(7) = 28$$

More generally, computation progresses from the inside expressions outward.

EXAMPLE 1. Simplify $-2[4 - 3 \cdot 2 + (-1 + 7)2]$.

Solution. Working from the inside out, we find that

$$\begin{aligned} -2[4 - 3 \cdot 2 + (-1 + 7)2] &= -2(4 - 3 \cdot 2 + 6 \cdot 2) = -2(4 - 6 + 12) \\ &= -2(10) = -20 \quad \square \end{aligned}$$

Properties of the Arithmetic Operations

We now list the most important properties of addition and multiplication. In the list a, b, and c denote arbitrary real numbers except where otherwise indicated.

Properties of Addition and Multiplication

Addition	*Multiplication*	*Name*
$a + b = b + a$	$ab = ba$	Commutative properties
$(a + b) + c = a + (b + c)$	$(ab)c = a(bc)$	Associative properties
$a + 0 = a = 0 + a$	$a \cdot 1 = a = 1 \cdot a$	Identity properties
For any number a there is a number $-a$ (called the ***additive inverse*** of a) such that $a + (-a) = 0 = (-a) + a$	For any number $a \neq 0$ there is a number $1/a$ or a^{-1} (called the ***multiplicative inverse*** of a) such that $a \cdot \frac{1}{a} = 1 = \frac{1}{a} \cdot a$	Inverse properties
$a(b + c) = ab + ac$ $(a + b)c = ac + bc$		Distributive properties

The associative property says that $(a + b) + c = a + (b + c)$, and as a result we can eliminate the parentheses and write $a + b + c$ for either $(a + b) + c$ or $a + (b + c)$. Thus by definition,

$$a + b + c = (a + b) + c = a + (b + c)$$

For example, the sum $6 + 4 + 1$ can be computed either as $(6 + 4) + 1$ or as $6 + (4 + 1)$, each of which equals 11.

In the same vein, we define abc by

$$abc = (ab)c = a(bc)$$

If we apply the distributive law several times, we find that

$$(a + b)(c + d) = a(c + d) + b(c + d) = ac + ad + bc + bd$$

Consequently

$$\boxed{(a + b)(c + d) = ac + ad + bc + bd} \qquad (1)$$

This result is obtained by multiplying the first terms (F), the outer terms (O), the inner terms (I), and the last terms (L) according to the following scheme:

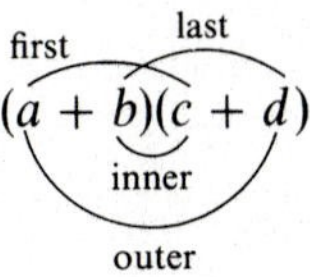

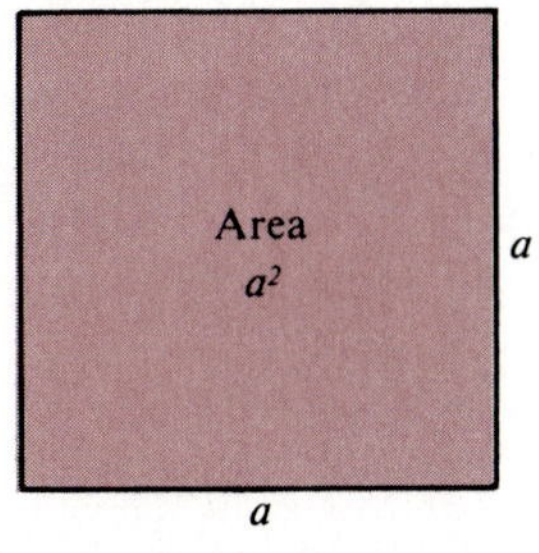

FIGURE 0.1

Thus the formula in (1) is sometimes associated with the word FOIL.

The use of formula (1) is illustrated in Example 2, in which we use the fact that the square a^2 of any number a is defined by

$$a^2 = a \cdot a$$

The number a^2 is called the square of a because if $a > 0$, then a^2 is the area of a square with sides of length a (Figure 0.1).

EXAMPLE 2. Find the product $(3a + 2)(4a + 5)$.

Solution. By (1),

$$\begin{aligned}(3a + 2)(4a + 5) &= (3a)(4a) + (3a)(5) + (2)(4a) + (2)(5) \\ &= 12a \cdot a + 15a + 8a + 10 \\ &= 12a^2 + 23a + 10 \quad \square\end{aligned}$$

A special case of (1) arises if $c = a$ and $d = b$:

$$(a + b)(a + b) = a \cdot a + a \cdot b + b \cdot a + b \cdot b$$

or more succinctly.

$$\boxed{(a + b)^2 = a^2 + 2ab + b^2} \tag{2}$$

Similarly, other special cases of (1) are

$$\boxed{(a - b)^2 = a^2 - 2ab + b^2} \tag{3}$$

and

$$\boxed{(a + b)(a - b) = a^2 - b^2} \tag{4}$$

EXAMPLE 3. Using the fact that $(\sqrt{2})^2 = 2$, verify that $(3 + \sqrt{2})^2 = 11 + 6\sqrt{2}$.

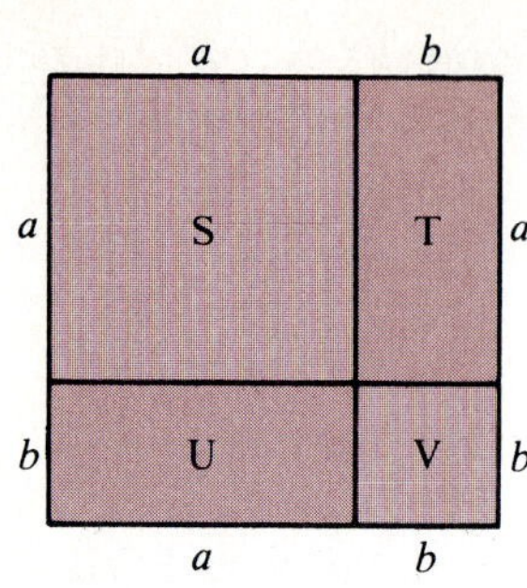

FIGURE 0.2

Solution. By (2),

$$(3 + \sqrt{2})^2 = 3^2 + 2(3)(\sqrt{2}) + (\sqrt{2})^2$$
$$= 9 + 6\sqrt{2} + 2 = 11 + 6\sqrt{2} \quad \square$$

If a and b are positive numbers, then (2) can be interpreted geometrically in terms of area. As portrayed in Figure 0.2, a square R with sides of length $a + b$ can be decomposed into a square S with sides of length a, two rectangles T and U with sides of length a and b, and a square V with sides of length b. Equation (2) relates the area of R, S, T, U, and V as follows:

$$\overbrace{(a+b)^2}^{\text{area of R}} = \overbrace{a^2}^{\text{area of S}} + \overbrace{2ab}^{\text{areas of T and U}} + \overbrace{b^2}^{\text{area of V}} \tag{5}$$

We will use the following property of multiplication in solving equations.

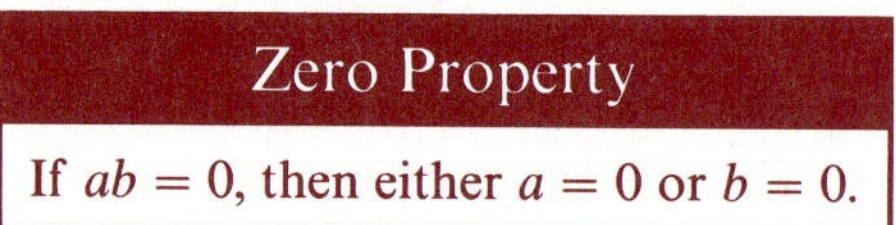

Zero Property

If $ab = 0$, then either $a = 0$ or $b = 0$.

Before we move on to subtraction and division, let us see how the laws already presented in this section can be used in solving everyday problems. Example 4 repeats the problem posed at the outset of this chapter.

EXAMPLE 4. An agency normally rents mid-size cars at \$30 per day. It gives a 5% discount to students but raises the rental price 3% on holiday weekends. For such weekends, would the final cost be affected if the agent first took the discount and then added the surcharge, rather than adding the surcharge first and then taking the discount?

Solution. We observe that giving a 5% discount on \$30 is equivalent to multiplying \$30 by .95; attaching a 3% surcharge at this point amounts to multiplying (.95)(30) dollars by 1.03, yielding (1.03)[(.95)(30)] dollars. In contrast, if we first add the 3% surcharge on \$30, we obtain (1.03)(30) dollars; discounting that amount by 5% yields (.95)[(1.03)(30)] dollars. But by the commutativity and associativity laws for multiplication,

$$(1.03)[(.95)(30)] = (.95)[(1.03)(30)]$$

so it does not matter which order the agent uses in computing the rental price to the student on a holiday weekend. $\square$

Properties of Subtraction and Division

The basic properties involving subtraction and division of real numbers are listed next.

$a - b = a + (-b)$	$-(ab) = (-a)b = a(-b)$
$-(-a) = a$	$(-a)(-b) = ab$
$-a = (-1)a$	$a(b - c) = ab - ac$
$\dfrac{a}{b} = \dfrac{c}{d}$ if and only if $ad = bc$	$\dfrac{a}{b} + \dfrac{c}{b} = \dfrac{a + c}{b}$
$\dfrac{a}{b} = a\left(\dfrac{1}{b}\right)$	$\dfrac{a}{b} \cdot \dfrac{c}{d} = \dfrac{ac}{bd}$ $\dfrac{ac}{bc} = \dfrac{a}{b}$
$-\left(\dfrac{a}{b}\right) = \dfrac{-a}{b} = \dfrac{a}{-b}$	$\dfrac{1}{1/a} = a$
$\dfrac{a}{b} + \dfrac{c}{d} = \dfrac{ad + bc}{bd}$	$\dfrac{a/b}{c/d} = \dfrac{a}{b} \cdot \dfrac{d}{c} = \dfrac{ad}{bc}$

If a and b are integers with $b \neq 0$, then the fraction a/b is in ***lowest terms*** if a and b have no common integer factors besides 1 and -1. Thus $\frac{4}{7}$ is in lowest terms, whereas $\frac{6}{8}$ is not, because 2 is an integer factor of both 6 and 8. The law

$$\frac{ac}{bc} = \frac{a}{b}$$

assists us in reducing a fraction to lowest terms. When we wish to reduce a fraction to lowest terms we first determine those integers that are factors of both numerator and denominator, and then cancel them from both numerator and denominator. For example,

$$\frac{198}{54} = \frac{9 \cdot 11 \cdot 2}{9 \cdot 3 \cdot 2} = \frac{11}{3}$$

If we add $\frac{3}{8}$ and $\frac{7}{12}$ by using the formula

$$\frac{a}{b} + \frac{c}{d} = \frac{ad + bc}{bd}$$

we obtain

$$\frac{3}{8} + \frac{7}{12} = \frac{3 \cdot 12 + 8 \cdot 7}{8 \cdot 12} = \frac{92}{96}$$

(which still needs to be reduced to lowest terms). An equivalent way of obtaining the same result is to write $\frac{3}{8}$ and $\frac{7}{12}$ with the same denominator, called a ***common denominator***. One common denominator is the product, 96, of the denominators 8 and 12. We would obtain

$$\frac{3}{8} = \frac{3 \cdot 12}{8 \cdot 12} = \frac{36}{96} \quad \text{and} \quad \frac{7}{12} = \frac{7 \cdot 8}{12 \cdot 8} = \frac{56}{96}$$

and then

$$\frac{3}{8} + \frac{7}{12} = \frac{36}{96} + \frac{56}{96} = \frac{36 + 56}{96} = \frac{92}{96} \tag{6}$$

The common denominator 96 was obtained by taking the product of the denominators 8 and 12 of the fractions $\frac{3}{8}$ and $\frac{7}{12}$. However, 24 is also a common denominator of $\frac{3}{8}$ and $\frac{7}{12}$, because

$$\frac{3}{8} = \frac{3 \cdot 3}{8 \cdot 3} = \frac{9}{24} \quad \text{and} \quad \frac{7}{12} = \frac{7 \cdot 2}{12 \cdot 2} = \frac{14}{24}$$

Since the common denominator 24 is less than 96, the calculations required to add $\frac{3}{8}$ and $\frac{7}{12}$ are simpler if we use 24 instead of 96:

$$\frac{3}{8} + \frac{7}{12} = \frac{9}{24} + \frac{14}{24} = \frac{23}{24} \tag{7}$$

Of course $\frac{23}{24} = \frac{92}{96}$, so that the results of (6) and (7) are the same. In general the calculations are simplest if we use the smallest possible positive common denominator when adding two fractions. That common denominator is called the ***least common denominator*** (often abbreviated l.c.d.) of the two fractions. One way of obtaining the least common denominator of a/b and c/d is to determine the smallest multiple of b that is also a multiple of d.

EXAMPLE 5. Find the least common denominator of $\frac{3}{16}$ and $\frac{5}{12}$; then write $\frac{3}{16} + \frac{5}{12}$ as one fraction with that denominator.

Solution. Since the smallest multiple of 16 that is also a multiple of 12 is 48, the least common denominator is 48. Consequently

$$\frac{3}{16} + \frac{5}{12} = \frac{9}{48} + \frac{20}{48} = \frac{29}{48} \quad \square$$

EXAMPLE 6. Write the sum $\dfrac{b}{a-b} + \dfrac{a}{a+b}$ as one fraction.

Solution. Using the properties of fractions, we have

$$\frac{b}{a-b}+\frac{a}{a+b}=\frac{b(a+b)+(a-b)a}{(a-b)(a+b)}=\frac{ba+b^2+a^2-ba}{a^2-b^2}$$
$$=\frac{a^2+b^2}{a^2-b^2} \quad \square$$

Care in Using Rules of Algebra

When the rules of algebra are used correctly, they lead to valid conclusions. However, it is easy to be careless in using the rules or to manufacture rules that may look plausible but unfortunately are not valid. Below we list several incorrect formulas; beside each we state the corresponding correct formula:

Incorrect	*Correct*
$(a+b)^2 \overset{?}{=} a^2+b^2$	$(a+b)^2 = a^2+2ab+b^2$
$a(b+c) \overset{?}{=} ab+c$	$a(b+c) = ab+ac$
$a-(b-c) \overset{?}{=} a-b-c$	$a-(b-c) = a-b+c$
$\frac{-a}{-b} \overset{?}{=} -\frac{a}{b}$	$\frac{-a}{-b} = \frac{a}{b}$
$\frac{a}{b+c} \overset{?}{=} \frac{a}{b}+\frac{a}{c}$	$\frac{a}{b+c}$ remains $\frac{a}{b+c}$
$a/(b+c) \overset{?}{=} a/b+c$	$a/(b+c)$ remains $a/(b+c)$
$\sqrt{a+b} \overset{?}{=} \sqrt{a}+\sqrt{b}$	$\sqrt{a+b}$ remains $\sqrt{a+b}$
$\frac{a}{c}\cdot\frac{b}{c} \overset{?}{=} \frac{ab}{c}$	$\frac{a}{c}\cdot\frac{b}{c} = \frac{ab}{c^2}$
$\frac{a}{b}+\frac{b}{c} \overset{?}{=} \frac{a}{c}$	$\frac{a}{b}+\frac{b}{c} = \frac{ac+b^2}{bc}$

Computations by Calculator

Hand-held calculators perform addition, multiplication, and division incredibly rapidly and accurately. Thus they vastly expand our ability to obtain quick numerical results.

Because calculators can store and manipulate only 10 (or perhaps 15 digits) at a time, there are limitations to the results we can obtain from calculators. For instance, if we key 2/3 into the calculator, and if the calculator can display only 10 places, then it would likely represent 2/3 as .6666666667 (that is, rounded up) or .6666666666 (that is, truncated). Since neither of these decimal numbers is exactly 2/3, using either of these decimal numbers, rather than 2/3, in calculator computations will normally result in answers that are only approximations.

We must be mindful of the fact that calculators frequently yield only approximate answers. Consequently, in writing results that we know to be approximate, we will use the symbol $\approx$ in place of $=$. For example, our calculator displays .0531443299 when we key in $\frac{3.141 - 2.11}{61.9 - 42.5}$. Not knowing whether .0531443299 is the exact, or only an approximate, answer for the fraction, we would write

$$\frac{3.141 - 2.11}{61.9 - 42.5} \approx .0531443299 \tag{8}$$

A digit appearing in a decimal approximation is ***significant*** unless it is a zero that indicates the location of the decimal point. Thus 2.01034 has six significant digits, whereas 0.123 and .0000399 each have three significant digits. Both significant and nonsignificant zeros may occur in the same number. Thus

$$.\overbrace{00}^{\text{nonsignificant}}\underbrace{12004}_{\text{significant}}$$

has five significant digits, since the first two zeros indicate where the decimal point is. Similarly, the number 12500 has three significant digits because the two zeros tell us where the decimal point of the number 12500 is.

We will follow two general rules in writing down answers obtained by calculator computations:

1. If all numbers keyed into the calculator are known to be accurate, we will normally use the answer that the calculator displays (since calculators round off before displaying their answers). Thus in (8) we would write our answer as

$$\frac{3.141 - 2.11}{61.9 - 42.5} \approx 0.0531443299$$

2. If the numbers keyed into the calculator are known to be merely approximate, we will round off the answer to the least number of significant digits of any number we key in. For example, if all the numbers on the left side of (8) are known to be approximate, then since 2.11, 61.9, and 42.5 have three significant digits each, and since 3.141 has four significant digits, we would round the answer to three significant digits:

$$\frac{3.141 - 2.11}{61.9 - 42.5} \approx .0531$$

Had the calculator display shown .05316 . . . , then the rounded off three-significant digit answer would have been .0532.

On some occasions, common sense or the purpose for which an answer is designed will determine the number of significant digits given in an answer. For example, suppose we wished to determine the (approximate) average precipitation per year in Chicago. If the calculator responded by displaying the number 33.34251872, we would likely round off to the nearest hundredth (or to four significant digits). In a similar fashion, if a student scored a total of 141 points on 16 homework assignments, then the student's homework average would be

$$\frac{141}{16} = 8.8125$$

Even though 8.8125 is the exact average, an instructor would probably round off to 8.8 or to 9 in the gradebook.

EXERCISES 0.1

In Exercises 1–6, find the given product or power.

1. $(2a - 1)(3a + 4)$

2. $(-a + 1)(a + 1)$

3. $(\frac{1}{2}a + 4)(a - \frac{1}{2})$

4. $(2 + \sqrt{2})(5 - 3\sqrt{2})$

5. $(5 + \sqrt{2})^2$

6. $(\frac{1}{2} - \frac{2}{3}\sqrt{3})^2$

In Exercises 7–8, write the expression without parentheses.

7. $(2a + 3b)^2$

8. $\left(a - \frac{1}{a}\right)^2$

In Exercises 9–10, reduce the given fraction to lowest terms.

9. $\frac{60}{42}$

10. $-\frac{286}{520}$

In Exercises 11–14, carry out the indicated operation. Then reduce your answer to lowest terms.

11. $\frac{21}{25} \cdot \frac{5}{9}$

12. $\left(-\frac{4}{9}\right) \cdot \left(-\frac{27}{10}\right)$

13. $\dfrac{-\frac{1}{6}}{\frac{1}{12}}$

14. $\dfrac{-\frac{2}{15}}{-\frac{7}{75}}$

In Exercises 15–20, find the least common denominator of the two fractions and use it to perform the indicated operation on the two fractions.

15. $\frac{1}{6} + \frac{1}{8}$

16. $\frac{5}{12} - \frac{7}{9}$

17. $\frac{5}{9} + \frac{7}{15}$

18. $-\frac{11}{24} + \frac{13}{36}$

19. $\frac{23}{30} - \frac{29}{36}$

20. $\frac{2}{35} + \frac{3}{49}$

In Exercises 21–24, write the given expression as one fraction.

21. $\frac{a-1}{a+1} - \frac{a+1}{a-1}$

22. $\frac{1}{a+b} + \frac{b}{a^2-b^2}$

23. $\frac{2}{a} - \frac{3}{b} + \frac{4}{ab}$

24. $\frac{1}{a+2} + \frac{4}{a-2} - \frac{2a}{a^2-4}$

In Exercises 25–32, correct the given incorrect formula.

25. $(a + 1)(b + 1) \stackrel{?}{=} ab + 1$

26. $\frac{1/a}{1/b} \stackrel{?}{=} \frac{1}{ab}$

27. $a - (b + c) \stackrel{?}{=} a - b + c$

28. $(-a)(-b) \stackrel{?}{=} -ab$

29. $(a + b)^3 \stackrel{?}{=} a^3 + b^3$

30. $(-a)^2 \stackrel{?}{=} -a^2$

31. $\dfrac{1}{a+b} \stackrel{?}{=} \dfrac{1}{a} + \dfrac{1}{b}$

32. $\dfrac{a}{b} + \dfrac{c}{d} \stackrel{?}{=} \dfrac{a+c}{b+d}$

33. Use (2) to calculate $(9.1)^2$.

34. Use (3) to calculate $(.99)^2$.

In Exercises 35–38, use a calculator to approximate the given expression.

35. $\dfrac{23}{247} - \dfrac{59}{1001}$

36. $\dfrac{(3.0107)(16.38)^2}{49.07}$

37. $\dfrac{3.487 - 2.3496}{48.63 + 3.012}$

38. $\dfrac{\pi - \sqrt{2}}{\pi + \sqrt{2}}$

39. Explain why the number 0.10110111011110 . . . , whose decimal expansion has increasingly long strings of 1's, is irrational.

40. Use formula (1) to prove formulas (3) and (4).

41. Show that $(2a)^2 + (a^2 - 1)^2 = (a^2 + 1)^2$ for any real number a.

42. Show that

$$n = \left(\frac{n+1}{2}\right)^2 - \left(\frac{n-1}{2}\right)^2$$

for any real number n. Conclude that every odd integer is the difference of the squares of two integers.

43. a. Show that subtraction is neither commutative nor associative by finding numbers, a, b, and c such that $a - b \neq b - a$ and $a - (b - c) \neq (a - b) - c$.
b. Show that division is neither commutative nor associative by finding numbers a, b, and c such that $a/b \neq b/a$ and $a/(b/c) \neq (a/b)/c$.

44. Show that $a \cdot 0 = 0$ for any real number a. (*Hint:* First use the fact that $a \cdot 0 = a(0 + 0)$. Then use the distributive property. Finally, subtract $a \cdot 0$ from both sides of the resulting equation.)

45. Prove the Zero Property, that is, show that if a and b are real numbers with $ab = 0$, then either $a = 0$ or $b = 0$. (*Hint:* If $a \neq 0$, multiply both sides of the equation $ab = 0$ by a^{-1}.)

46. Show that if a is a real number with $a^2 = a$, then $a = 0$ or $a = 1$. (*Hint:* Rewrite the equation $a^2 = a$ first as $a^2 - a = 0$ and then as $a(a - 1) = 0$, and use Exercise 45.)

47. Let a be a real number. Show that $a = -a$ if and only if $a = 0$.

48. Show that if a and b are nonzero real numbers, then $ab(a^{-1} + b^{-1}) = a + b$.

49. Let a, b, and c be real numbers. Show that if $a + b = a + c$, then $b = c$.

50. Let a, b, and c be real numbers. Show that if $ab = ac$ and $a \neq 0$, then $b = c$.

51. a. Show that $(a + b + c)^2 = a^2 + b^2 + c^2 + 2ab + 2bc + 2ca$ for any real numbers a, b, and c.
b. Use Figure 0.3 to give a geometric interpretation of the equation in part (a) in case a, b, and c are positive numbers. Pattern your answer after (5).

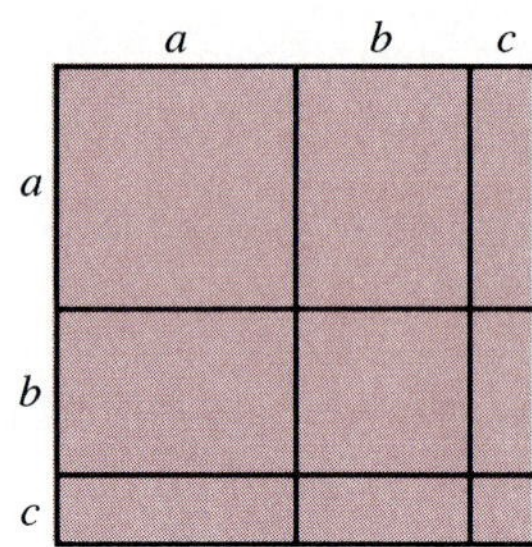

FIGURE 0.3

52. The area A of a circle with radius r is given by $A = \pi r^2$. Approximate the area of a circle with radius approximately equal to 6.132.

53. The surface area S of a cube of side length s is given by $S = 6s^2$. Approximate the surface area of a cube whose side length is approximately 2.1.

0.2 THE REAL LINE

It is possible to associate the real numbers with the points on a given line l so that every real number is associated with one and only one point on l, and so that, conversely, every point on l is associated with one and only one real number. To establish such an association we first select a particular line l, usually drawn horizontally as in Figure 0.4, and then choose an arbitrary point 0 on l to associate with the number 0. The point 0 is called the ***origin***. Next we select the points to associate with the positive integers $1, 2, 3, \ldots$, by marking off line segments of equal length to the right of 0 as in Figure 0.4, and points to associate with the negative integers $-1, -2, -3, \ldots$, by marking off similar line segments to the left of 0. To determine the points on l associated with the rational numbers, we subdivide appropriate portions of l into smaller line segments of equal length. For example, to determine the points to associate with $\frac{1}{3}$ and $\frac{2}{3}$ we subdivide the line segment determined by 0 and 1 into three segments of equal length (Figure 0.4). All points on l that are not associated with rational numbers are associated with irrational numbers. The points associated with the irrational numbers π and $\sqrt{2}$ are exhibited in Figure 0.4.

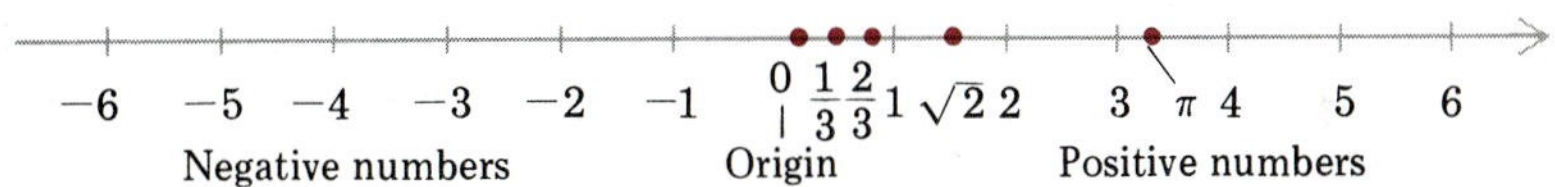

FIGURE 0.4

The ***positive direction*** on l, pointing from left to right, is indicated by an arrow on l (Figure 0.4). Those numbers corresponding to points to the right of 0 are positive numbers, and those numbers corresponding to points to the left of 0 are the negative numbers.

The number that is associated with an arbitrary point A on l is called the ***coordinate*** of A, and the association of the points on l with real numbers is frequently called a ***coordinate system*** for l. The line l with a coordinate system is often referred to as a ***real number line***, or ***real line***.

Inequalities

Let a and b be real numbers. If $a - b$ is positive, we say that a is ***greater than*** b and write $a > b$; or alternatively, we say that b is ***less than*** a and write $b < a$. Geometrically, $a > b$ means that the point corresponding to a on the real line in Figure 0.5(a) lies to the right of the point corresponding to b. For example, $-3 > -5$, and -3 lies to the right of -5 on the real line (Figure 0.5(b)). Using this geometric interpretation and the fact that one of two distinct points on the real line must lie to the right of the other, we conclude that if a and b are real numbers, then exactly one of the following three possibilities is true:

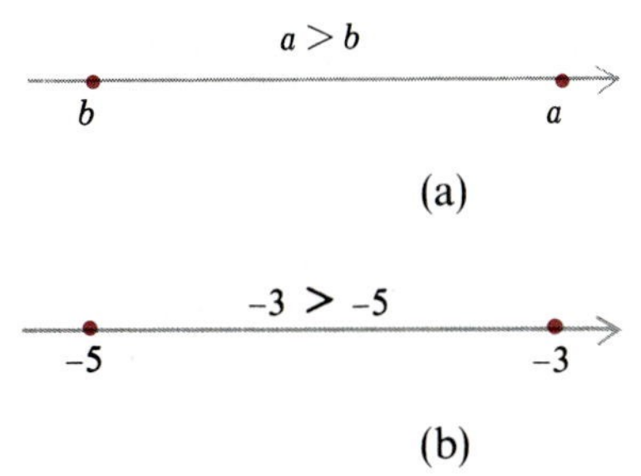

FIGURE 0.5

$$a < b, \qquad a = b, \quad \text{or} \quad a > b$$

This result is called the ***trichotomy law***.* Taking $b = 0$ in the trichotomy law, we see that any real number a satisfies exactly one of the following:

$$a < 0, \qquad a = 0, \quad \text{or} \quad a > 0$$

* The word *trichotomy* comes from a Greek word meaning "threefold division."

The relations $a < b$ and $a > b$ are called ***inequalities***, and the symbols $<$ and $>$ are ***inequality signs***. Although we will return to the general laws governing inequalities in Sections 1.6 and 1.7, we will now present a few basic properties of inequalities.

To simplify the statements of the properties, we say that two nonzero numbers a and b have the ***same sign*** if $a > 0$ and $b > 0$, or if $a < 0$ and $b < 0$. If $a > 0$ and $b < 0$, or if $a < 0$ and $b > 0$, then we say that a and b have ***opposite signs***. We are thus ready to give the properties:

$$ab > 0 \text{ if and only if } a \text{ and } b \text{ have the same sign.} \quad (1)$$

$$ab < 0 \text{ if and only if } a \text{ and } b \text{ have opposite signs.} \quad (2)$$

$$a > 0 \text{ if and only if } -a < 0, \text{ and } a < 0 \text{ if and only if } -a > 0. \quad (3)$$

CAUTION: It is sometimes tempting to regard $-a$ as a negative number simply because the expression $-a$ contains a minus sign. However, as the second half of (3) indicates, if a is negative, then $-a$ is actually positive. For instance, if $a = -5$, then $-a = -(-5) = 5$, a positive number.

EXAMPLE 1. Let $a \neq 0$. Show that a and $1/a$ have the same sign.

Solution. Recall that

$$a \cdot \frac{1}{a} = 1 > 0$$

Since the product of a and $1/a$ is positive, a and $1/a$ must have the same sign by (1). □

We write $a \geq b$ to mean that either $a > b$ or $a = b$, and express this by saying that a is ***greater than or equal to*** b. Alternatively, we can write $b \leq a$ and say that b is ***less than or equal to*** a. Thus the statement "$x + 1$ is less than or equal to -1" can be written in mathematical notation as $x + 1 \leq -1$. The symbols $\geq$ and $\leq$ are also called ***inequality signs***.

If $a \geq 0$, then a is not negative, so we say that a is ***nonnegative***. For example, the number of miles a person travels during a given day is a nonnegative number.

If a, b, and c are three numbers, then the compound inequality $a < c < b$ means that $a < c$ and $c < b$, and we say that ***c is between a and b***. For example, $2 < 2.13 < \frac{5}{2}$. Thus, 2.13 is between 2 and $\frac{5}{2}$. Other compound inequalities, such as $a < c \leq b$ and $a \leq c \leq b$, are defined analogously.

CAUTION: In any compound inequality, the inequality signs must all point in the same direction. We never write a compound inequality such as $3 > x \leq 5$, in which the inequality signs point in opposite directions.

Absolute Value

We define the ***distance*** $d(a, b)$ between two numbers a and b (or equivalently, between the corresponding points on a line l) to be either $a - b$ or $b - a$, whichever is nonnegative (Figure 0.6). Thus the distance $d(2, -5)$ between 2 and -5 is $2 - (-5) = 7$, and the distance $d(3, 9)$ between 3 and 9 is $9 - 3 = 6$ (Figure 0.7).

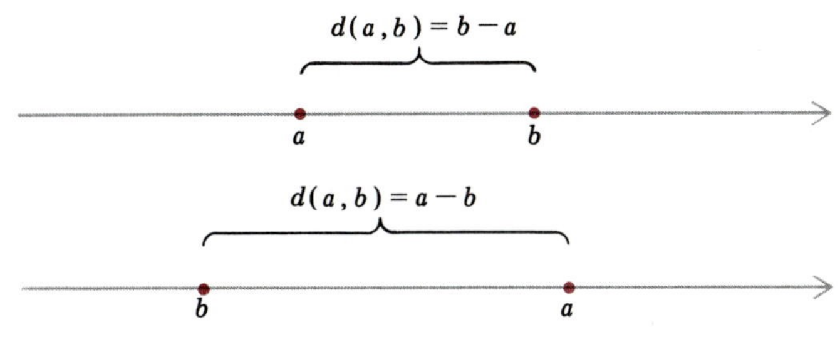

FIGURE 0.6

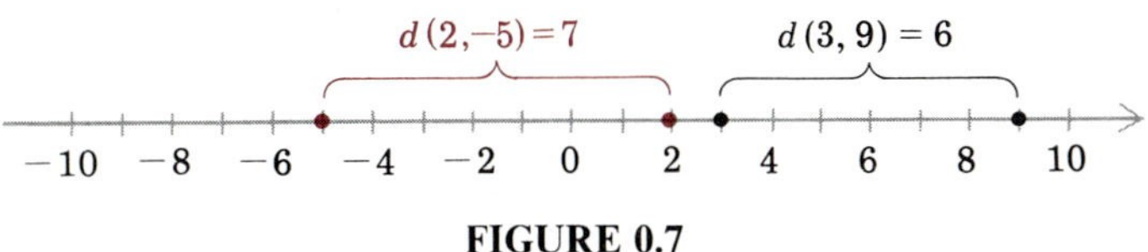

FIGURE 0.7

By our definition, the distance $d(0, a)$ between 0 and a is either a or $-a$, whichever is nonnegative. Of the two numbers a and $-a$, the one that is nonnegative is very important in mathematics and is called the ***absolute value*** $|a|$ of a:

$$|a| = \begin{cases} a & \text{if} \quad a \geq 0 \\ -a & \text{if} \quad a < 0 \end{cases}$$

In the absolute value notation, the distance between 0 and a is given by

$$d(0, a) = |a|$$

Since $b - a = -(a - b)$, it follows from the definition of absolute value that $|a - b|$ is the nonnegative number of the two numbers $a - b$ and $b - a$, so that the distance between a and b can be written in the succinct form

$$d(a, b) = |b - a|$$

EXAMPLE 2. Write each of the numbers $|5|$, $|-5|$, $|0|$, $|\pi - 3|$, and $|\sqrt{2} - 7|$ without absolute value signs.

Solution. Since 5, 0, and $\pi - 3$ are nonnegative, whereas -5 and $\sqrt{2} - 7$ are negative, we have

$$|5| = 5 \qquad |-5| = -(-5) = 5 \qquad |0| = 0$$

$$|\pi - 3| = \pi - 3 \qquad |\sqrt{2} - 7| = -(\sqrt{2} - 7) = 7 - \sqrt{2} \quad \square$$

EXAMPLE 3. Write the expression $|-a^2|$ without absolute values.

Solution. Since a^2 is nonnegative for each value of a, it follows that $-a^2$ is nonpositive. Therefore $|-a^2| = -(-a^2) = a^2$. $\square$

EXAMPLE 4. Find the distances between the following pairs of numbers on the real line.

a. $a = -9, b = 0$
b. $a = 2, b = 6$
c. $a = 7, b = -1$
d. $a = -2, b = -5$

Solution.

a. $d(a, b) = d(-9, 0) = |0 - (-9)| = |9| = 9$
b. $d(a, b) = d(2, 6) = |6 - 2| = |4| = 4$
c. $d(a, b) = d(7, -1) = |-1 - 7| = |-8| = 8$
d. $d(a, b) = d(-2, -5) = |-5 - (-2)| = |-3| = 3$ $\square$

We complete the section with a list of several special properties of absolute values:

$\|a\| = 0$ if and only if $a = 0$	$\|a + b\| \leq \|a\| + \|b\|$
$\|a\| \geq 0$	$\|ab\| = \|a\|\|b\|$
$\|a\| = \|-a\|$	$\left\|\frac{a}{b}\right\| = \frac{\|a\|}{\|b\|}$
$\|a - b\| = \|b - a\|$	

From the property $|a + b| \leq |a| + |b|$ it follows that either $|a + b| = |a| + |b|$ or $|a + b| < |a| + |b|$. See if you can find values of a and b for which $|a + b| = |a| + |b|$, and values for which $|a + b| < |a| + |b|$. Exercise 50 discusses exactly when equality holds.

EXERCISES 0.2

1. Draw a line, set up a coordinate system on the line, and locate the points corresponding to $-1, 2, -2, 3, -3, \frac{5}{2}, -\frac{5}{2}, \frac{7}{4}$, and $-\frac{7}{4}$.

In Exercises 2–12, write out the given statement using the symbols for inequalities.

2. x is less than 0.
3. x is greater than or equal to $\sqrt{2}$.
4. c is less than or equal to $\frac{1}{10}$.
5. $4x$ is greater than or equal to 8.

6. a is between 1 and 2.

7. $6 - r$ is between -1 and 1.

8. y is positive.

9. z is negative.

10. $x + 1$ is nonnegative.

11. $|x - 2|$ is less than 0.01.

12. $|x - 2|$ is greater than or equal to d.

In Exercises 13–18, write out the pairs of numbers in the form $a = b$, $a < b$, or $a > b$, whichever is correct.

13. $\sqrt{2}, 1$ **14.** $|-4|, 4$ **15.** $(-2)^2, 3$

16. $0, \pi - 3$ **17.** $\frac{22}{7}, \pi$ **18.** $\frac{5}{7}, 0.7$

In Exercises 19–24, find the distance between a and b.

19. $a = 0, b = -1$ **20.** $a = 5, b = 5$

21. $a = 6, b = 0$ **22.** $a = \frac{1}{2}, b = \frac{1}{4}$

23. $a = 9.6, b = 1.1$ **24.** $a = \pi, b = \sqrt{2}$

In Exercises 25–32, write the numbers without using absolute values.

25. $|-7| + |-9|$ **26.** $|-5| + |5|$

27. $|-5| - |5|$ **28.** $-2 - |-2|$

29. $|4 - \sqrt{2}| - 5$ **30.** $|3 - \pi| + 3$

31. $\frac{1}{3}|4 - 10|$ **32.** $\dfrac{-|-4|}{|12|}$

In Exercises 33–42, write the expression without absolute values.

33. $|x^2|$ **34.** $|(-4 - x)^2|$

35. $|x^2 + 1|$ **36.** $|-2 - y^2|$

37. $|a - 4|$ if $a \geq 4$ **38.** $|a - 4|$ if $a < 4$

39. $|a - b|$ if $a \geq b$ **40.** $|a - b|$ if $a < b$

41. $|a - b| - |b - a|$ **42.** $\dfrac{|a - b|}{|b - a|}$ if $a \neq b$

43. Archimedes calculated that π lies between $\frac{22}{7}$ and $\frac{223}{71}$. Determine which is the larger and which is the smaller of these two fractions.

44. In the fifth century, the Chinese scientist Tsu Ch'ung-chih found the approximate value $\frac{355}{113}$ for π. Determine whether $\frac{355}{113}$ is larger or smaller than π.

45. Determine the values of a for which $|-a| = a$.

46. Show that $|-a| = |a|$.

47. Show that $|a^2| = |a|^2$.

48. Show that $-|a| \leq a \leq |a|$.

49. Show that $a^2 \geq 0$.

***50.** Show that $|a + b| = |a| + |b|$ if and only if one of the following conditions is satisfied:
a. $a = 0$
b. $b = 0$
c. a and b have the same sign

0.3 INTEGRAL EXPONENTS

Expressions of the form a^n, where a is a number and n an integer, appear in many calculations. For example, if \$1 is deposited into a savings account that pays 7% interest compounded annually, then $(1.07)^4$ is the amount of money in the account at the end of four years. For another example, suppose that when a superball is dropped and bounces, it attains $\frac{6}{7}$ of the height from which it was dropped. After bouncing n times it attains $(\frac{6}{7})^n$ of the height from which it was originally dropped.

In Section 0.1 we noted that

$$a^2 = a \cdot a$$

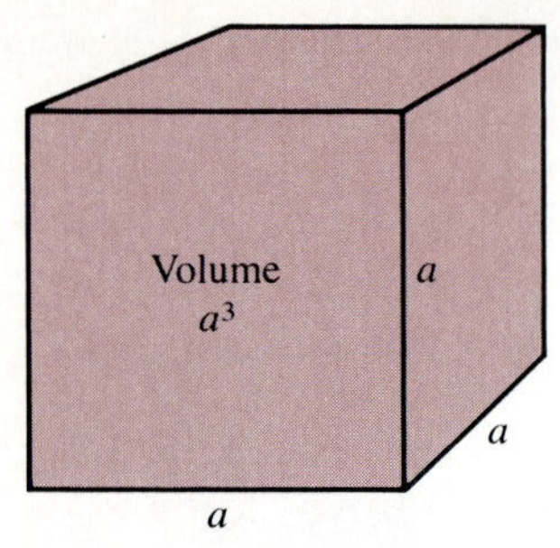

FIGURE 0.8

Similarly, the cube a^3 is defined by the formula

$$a^3 = a \cdot a \cdot a$$

Geometrically, if $a > 0$ then a^3 represents the volume of a cube with sides of length a (Figure 0.8).

In general, if n is any positive integer, the expression a^n stands for the product of n factors of a:

$$\boxed{a^n = \overbrace{a \cdot a \cdot a \cdots a}^{n \text{ factors}}}$$

with the understanding that $a^1 = a$. The expression a^n is read "a to the nth power," or "the nth power of a." Thus

$$2^5 = 2 \cdot 2 \cdot 2 \cdot 2 \cdot 2 = 32, \qquad \left(\frac{1}{3}\right)^2 = \frac{1}{3} \cdot \frac{1}{3} = \frac{1}{9}$$

and $$(1.07)^4 = (1.07)(1.07)(1.07)(1.07) = 1.31079601$$

Next we define negative powers of numbers. Let n be a positive integer. We define a^{-n} by the formula

$$a^{-n} = \frac{1}{a^n}, \quad \text{for} \quad a \neq 0 \tag{1}$$

Thus $$2^{-3} = \frac{1}{2^3} = \frac{1}{8}$$

and $$\left(\frac{1}{3}\right)^{-4} = \frac{1}{\left(\frac{1}{3}\right)^4} = \frac{1}{\frac{1}{81}} = 81$$

To complete the definition of integral powers of numbers, we define

$$a^0 = 1, \quad \text{for} \quad a \neq 0 \tag{2}$$

For example,

$$(1.7)^0 = 1 \quad \text{and} \quad \left(-\frac{1}{6}\right)^0 = 1$$

CAUTION: In (1) we did not define a^{-n} for $a = 0$, since the expression $\frac{1}{0}$ is meaningless. In (2) we did not define a^0 for $a = 0$ because no definition would be reasonable for all the various ways one might wish to assign a value to 0^0. Thus 0^n is not defined for any integer $n \le 0$.

As a result of (1) and (2), a^n is defined for any integer n, with the restriction that $a \neq 0$ if $n \le 0$. In the expression a^n, a is called the ***base*** and n the ***exponent*** or ***power***. Thus a^n has the form

$$(\text{base})^{\text{exponent}}$$

Laws of Exponents

There are many ways of combining powers of numbers. First of all, let us consider the product $2^4 \cdot 2^3$, rearranged as follows:

$$2^4 \cdot 2^3 = (\overbrace{2 \cdot 2 \cdot 2 \cdot 2}^{4\text{ factors}})(\overbrace{2 \cdot 2 \cdot 2}^{3\text{ factors}}) = (\overbrace{2 \cdot 2 \cdot 2 \cdot 2 \cdot 2 \cdot 2 \cdot 2}^{(4+3)\text{ factors}}) = 2^{4+3}$$

Thus
$$2^4 \cdot 2^3 = 2^{4+3}$$

More generally, if a is any number and m and n are positive integers, then

$$a^m a^n = (\overbrace{a \cdot a \cdot a \cdots a}^{m\text{ factors}})(\overbrace{a \cdot a \cdot a \cdots a}^{n\text{ factors}}) = \overbrace{a \cdot a \cdot a \cdots a}^{(m+n)\text{ factors}} = a^{m+n}$$

Therefore
$$a^m a^n = a^{m+n}$$

Actually, this formula is valid even if we remove the restriction that m and n be positive. This and several other formulas involving exponents are listed together below.

Laws of Integral Exponents

Let a and b be real numbers and m and n integers. Each of the following formulas is valid for all values of a and b for which both sides of the equation are defined.

i. $a^m a^n = a^{m+n}$ ii. $(a^m)^n = a^{mn} = (a^n)^m$ iii. $(ab)^n = a^n b^n$

iv. $\left(\dfrac{a}{b}\right)^n = \dfrac{a^n}{b^n} = a^n b^{-n}$ v. $a^{-n} = \dfrac{1}{a^n}$ vi. $\dfrac{a^m}{a^n} = a^{m-n}$

Recall that division by 0 is meaningless, as is raising 0 to a power that is 0 or negative. This implies, for example, that (iv) does not hold if $b = 0$, or if $a = 0$ and $n \le 0$.

EXAMPLE 1. Simplify the following expressions.

a. $5^7 5^{-4}$ b. $[(-3)^3]^2$ c. $\left(\dfrac{2}{5}\right)^3\left(\dfrac{5}{4}\right)^3$ d. $\dfrac{2^3 3^{-1}}{2 \cdot 3^2}$

Solution.

a. By (i),

$$5^7 5^{-4} = 5^{7-4} = 5^3 = 125$$

b. By (ii),

$$[(-3)^3]^2 = (-3)^{3 \cdot 2} = (-3)^6 = 729$$

c. By (iii),

$$\left(\frac{2}{5}\right)^3 \left(\frac{5}{4}\right)^3 = \left(\frac{2}{5} \cdot \frac{5}{4}\right)^3 = \left(\frac{1}{2}\right)^3 = \frac{1}{8}$$

d. By (v) and (vi),

$$\frac{2^3 3^{-1}}{2 \cdot 3^2} = \frac{2^3}{2^1} \frac{3^{-1}}{3^2} = 2^{3-1} 3^{-1-2} = 2^2 3^{-3} = 4\left(\frac{1}{27}\right) = \frac{4}{27} \quad \square$$

EXAMPLE 2. Simplify the following expressions.

a. $\dfrac{(ab)^{-2}}{a^{-3}b^4}$

b. $(a^{-1} + b^{-1})^{-2}$

Solution.

a. Using the Laws of Integral Exponents, we have

$$\frac{(ab)^{-2}}{a^{-3}b^4} = \frac{a^{-2}b^{-2}}{a^{-3}b^4} = \frac{a^{-2}}{a^{-3}} \cdot \frac{b^{-2}}{b^4} = a^{-2-(-3)}b^{-2-4} = ab^{-6}$$

b. Using the Laws of Integral Exponents, we have

$$(a^{-1} + b^{-1})^{-2} = \left(\frac{1}{a} + \frac{1}{b}\right)^{-2} = \left(\frac{b + a}{ab}\right)^{-2} = \left(\frac{ab}{b + a}\right)^2$$

$$= \frac{a^2b^2}{(b + a)^2} \quad \square$$

The exponential laws can be extended to include products of more than two numbers. For example,

$$a^m a^n a^p = a^{m+n+p}$$

$$(abc)^n = a^n b^n c^n$$

EXAMPLE 3. Simplify the expression $\dfrac{(2r^3s^{-2})^4}{(rs^{-2}t^{-3})^2}$ and write it with only positive exponents.

Solution.

$$\frac{(2r^3s^{-2})^4}{(rs^{-2}t^{-3})^2} = \frac{16r^{12}s^{-8}}{r^2s^{-4}t^{-6}} = 16r^{12-2}s^{-8-(-4)}t^{-(-6)}$$

$$= 16r^{10}s^{-4}t^6 = \frac{16r^{10}t^6}{s^4} \qquad \square$$

Let a be any real number. It follows from (1) in Section 0.2 that

$$a^2 \geq 0 \tag{3}$$

In contrast,

$$a^3 \quad \text{has the same sign as} \quad a \tag{4}$$

The statements in (3) and (4) and special cases of the following general results:

> a^n is nonnegative if n is an even integer.
>
> a^n has the same sign as a if n is an odd integer.

EXAMPLE 4. Determine whether the given number is positive or negative.

a. $(-2.17)^3(-4.63)^{-2}$ b. $(-1)^5(-2)^{-3}(\frac{3}{4})^{-14}$

Solution.

a. Notice that

$(-2.17)^3$ is negative because 3 is odd and -2.17 is negative

$(-4.63)^{-2}$ is positive because -2 is even

Therefore the product $(-2.17)^3(-4.63)^{-2}$, being the product of a negative and a positive number, is negative.

b. Observe that

$(-1)^5$ is negative because 5 is odd and -1 is negative

$(-2)^{-3}$ is negative because -3 is odd and -2 is negative

$(\frac{3}{4})^{-14}$ is positive because -14 is even

Consequently the number in (b) is the product of two negative numbers and a positive number and is therefore positive. $\square$

Scientific Notation

Quantities in the physical world come in all sizes, from the microscopic to the astronomical. For instance, the mass of an electron is approximately 0.000000000000000000000000000000911 kilograms, and the mass of the sun is approximately 1,987,000,000,000,000,000,000,000,000,000 kilograms. Many, perhaps most, quantities that arise in the physical sciences are either very small or very large. To make writing such numbers more convenient, scientists have adopted the standard practice of writing any quantity, regardless of its size, as a product $b \times 10^n$, where $1 \le b < 10$ and n is an integer. This notation for the number is called the ***scientific notation*** for the number. In scientific notation the mass of the electron mentioned above is approximately 9.11×10^{-31} kilograms, and the mass of the sun is approximately 1.987×10^{30} kilograms. These numbers are obviously easier to write and remember when given in scientific notation.

Let a given positive number be equal to $b \times 10^n$ in scientific notation. On the one hand, if the number is greater than or equal to 1, then n is the number of places the decimal point must be moved to the *left* in order to make the decimal expansion of the resulting number lie between 1 and 9.9999 . . . . For example, $n = 3$ for the number 2341, since 2.341 lies between 1 and 9.9999 . . . . Thus

$$2341 = 2.341 \times 10^3$$

On the other hand, if the number is less than 1, then n is the negative of the number of places the decimal point must be moved to the *right* in order to make the decimal expansion of the resulting number lie between 1 and 9.9999 . . . . For example, $n = -4$ for the number 0.00073, since 7.3 lies between 1 and 9.9999 . . . . Thus

$$0.00073 = 7.3 \times 10^{-4}$$

EXAMPLE 5. Write the following numbers in scientific notation.

a. 14,753 b. 0.23 c. 0.00000912 d. 1,000,000

Solution.

a. $14{,}753 = 1.4753 \times 10^4$
b. $0.23 = 2.3 \times 10^{-1}$
c. $0.00000912 = 9.12 \times 10^{-6}$
d. $1{,}000{,}000 = 1 \times 10^6$ (or simply 10^6) □

Since most calculators display ten digits, they normally use scientific notation in displaying very large or very small numbers. For example, in scientific notation, 2^{50} is approximately

$$1.125899907 \times 10^{15}$$

A calculator might display this as

1.125899907   15   or   1.125899907E15

In contrast, 3^{-26} is approximately

$$3.934117957 \times 10^{-13}$$

and a calculator might display this as

$$3.934117957 \quad -13 \quad \text{or} \quad 3.934117957\text{E} - 13$$

(Scientific) calculators have a key for calculating powers. On some the key is marked with the symbol ^. To calculate 5^8, for example, one would press 5, ^, 8 and either = or ENTER, depending on the type of calculator. There is also a special key, usually marked EE, for entering a number in scientific notation. To enter 3×10^{-4}, for example, one would press 3, EE, (−), and 4.

EXERCISES 0.3

In Exercises 1–6, compute the given number.

1. 2^{-2} **2.** $\left(\frac{1}{2}\right)^3$ **3.** $\left(\frac{1}{2}\right)^{-3}$

4. 10^3 **5.** 10^{-3} **6.** $(0.03)^2$

In Exercises 7–18, simplify the given expression.

7. $4^2 \cdot 4^3$ **8.** $(-7)^4(-7)^8$

9. $2^4(-2)^5$ **10.** $3^{-6} \cdot 3^4$

11. $10^9 \cdot 10^{-11}$ **12.** $(3 \cdot 3^3)^2$

13. $4^4 \cdot 4^2 \cdot 4$ **14.** $2^5 \div 2^{-3}$

15. $\dfrac{(-7)^5}{7^6}$ **16.** $\dfrac{\pi^2\pi^5}{\pi^3}$

17. $[(-4)^5]^6$ **18.** $[(-3)^{-3}]^{-3}$

In Exercises 19–32, simplify the expression.

19. a^5a^7 **20.** a^4a^{-2}

21. $y^{-2}y^{-6}$ **22.** $r^{-8}r^8$

23. $\dfrac{b^3}{b^{-5}}$ **24.** $(-c^2)^4$

25. $(-c^2)^5$ **26.** $(xy^2)^3$

27. $(x^2y^3z)^4$ **28.** $(\frac{2}{3}x^4)^{-2}$

29. $rs^2(r^5s^4)^3$ **30.** $(t^{-1})^{-1}$

31. $\dfrac{(st^{-1})^{-1}}{s^{-1}t^{-1}}$ **32.** $\dfrac{\dfrac{1}{s^{-1}} + \dfrac{1}{t^{-1}}}{s^{-1}t^{-1}}$

In Exercises 33–36, write the expression as a quotient involving only positive exponents.

33. $a^{-1}b^{-1}$ **34.** $a^{-1} + b^{-2}$

35. $(a^{-1} + b^{-1})^{-3}$ **36.** $(a + b)^{-1}(a^{-1} + b^{-1})$

In Exercises 37–38, determine whether the given number is positive or negative.

37. $(-1)^3(5.2)^3$ **38.** $(-1.3)^{-2}7^{-2}$

In Exercises 39–44, write the number in scientific notation.

39. 483.2 **40.** 0.791 **41.** 1.009

42. 891,134 **43.** 0.9999 **44.** 0.0000134

In Exercises 45–56, use a calculator to approximate the given value and write the answer in scientific notation.

45. (9876)(2751) **46.** (3715)(0.0015)

47. (0.4646)(0.3801) **48.** (55.55)(6.148)

49. (0.0012)(0.00025)

50. (0.00000007)(0.000009)

51. $(7.9 \times 10^3)(2.3 \times 10^5)$

52. $(4.791 \times 10^7)(9.31 \times 10^{-8})$

53. $9824 \div 112{,}344$

54. $83.74 \div 0.012$

55. $(3.246 \times 10^{-6}) \div (4.158 \times 10^{-4})$

56. $(1.111 \times 10^8) \div (5.876 \times 10^{-9})$

In Exercises 57–61, write the statement as an equation.

57. The area A of a circle is π times the square of the radius r.

58. The area A of a square is the square of a side s.

59. The volume V of a sphere is $\frac{4}{3}\pi$ times the cube of the radius r.

60. The volume V of a cylinder is π times the product of the height h and the square of the radius r of the base. (See Figure 0.9.)

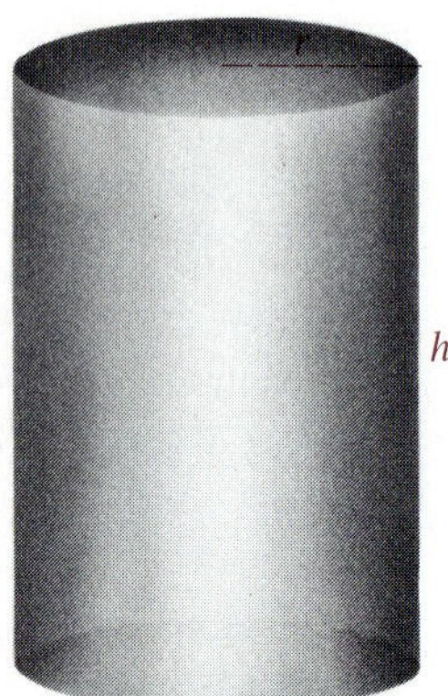

FIGURE 0.9

61. The volume V of a cone is π times $\frac{1}{3}$ the product of the height h and the square of the radius r of the base. (See Figure 0.10.)

62. a. Show that if a and b have the same sign, then $(ab)^n > 0$ for any integer n.
b. Show that if a and b have different signs, then $(ab)^n > 0$ if and only if the integer n is even.

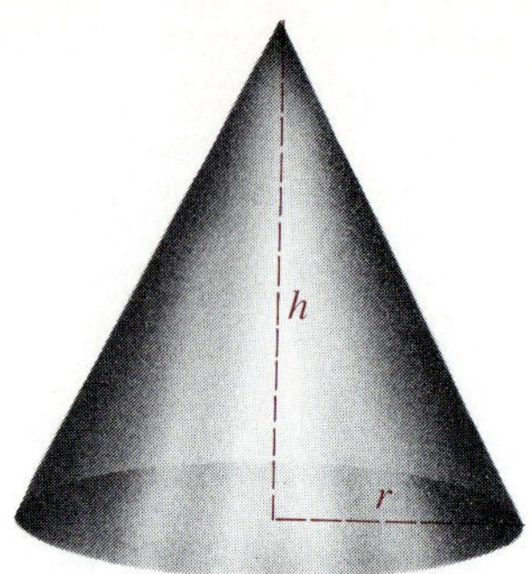

FIGURE 0.10

63. Show that $(a^2+b^2)(c^2+d^2)=(ac+bd)^2+(ad-bc)^2$.

64. Show that $3(2ab)^2 + (a^2 - 3b^2)^2 = (a^2 + 3b^2)^2$.

65. Suppose $2 \cdot 2^2 \cdot 4^3 \cdot 4^4 \cdot 8^5 \cdot 8^6 = (32)^n$. Find the value of n.

***66.** a. For what values of a, m, and n does part (ii) of the Laws of Integral Exponents not hold?
b. For what values of a, b, and n does part (iii) of the Laws of Integral Exponents not hold?

67. The mass of the earth is approximately 5.98×10^{24} kilograms. Write this number in decimal form.

68. The mass of a proton is approximately 1.67×10^{-27} kilograms. Write this number in decimal form.

69. If there are approximately one hundred billion galaxies, each with approximately one hundred billion stars, find the total number of stars in the universe. Write your answer in scientific notation.

70. The atoms in cesium oscillate exactly 9,192,631,770 times per second.
a. Write the number 9,192,631,770 in scientific notation.
b. Determine the number of oscillations a cesium atom would make during a century.

71. The average distance between the earth and the sun is approximately 149,000,000 kilometers. Assuming that 1 kilometer is equal to 0.621 miles, compute the average distance in miles between the earth and the sun. Write your answer in scientific notation.

72. The mass of a proton is approximately 1.673×10^{-24} grams, whereas the mass of an electron is approximately 9.11×10^{-28} grams. How many times more massive than an electron is a proton?

73. A ***light year*** is the distance light in a vacuum travels in one year (approximately $365\frac{1}{4}$ days).
   a. Assuming that light in a vacuum travels 186,000 miles per second, use a calculator to compute the number of miles in one light year. Write your answer in scientific notation.
   b. Sirius is approximately 5.106×10^{13} miles from earth. How many light years away from earth is Sirius?

74. Under ideal conditions, if a superball is dropped from a height of one meter above ground, the nth bounce will have a height of approximately $(\frac{6}{7})^n$ meter.
   a. Compute the height of the 15th bounce.
   b. Would the 15th bounce of the superball be greater than the 6th bounce of a ball whose nth bounce has a height of $(\frac{2}{3})^n$ meter? Explain your answer.

0.4 RADICALS AND RATIONAL EXPONENTS

If the area of a square is 9 square units, then any side of the square has length 3 (Figure 0.11), and we write $\sqrt{9} = 3$. Similarly, for any $a \geq 0$ we define $\sqrt{a}$ to be the nonnegative number b whose square is a. Thus

$$\sqrt{a} = b \quad \text{if and only if} \quad b \geq 0 \quad \text{and} \quad b^2 = a$$

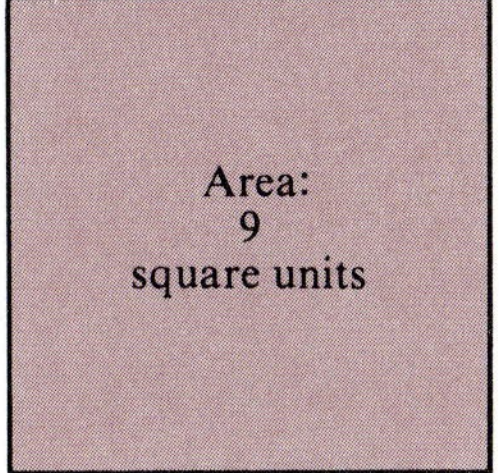

Side length: 3

FIGURE 0.11

For example,

$$\sqrt{1.21} = 1.1 \quad \text{because} \quad 1.1 \geq 0 \quad \text{and} \quad (1.1)^2 = 1.21$$

$$\sqrt{\frac{1}{4}} = \frac{1}{2} \quad \text{because} \quad \frac{1}{2} \geq 0 \quad \text{and} \quad \left(\frac{1}{2}\right)^2 = \frac{1}{4}$$

We call $\sqrt{a}$ the ***square root*** of a, and in the expression $\sqrt{a}$ the a is the ***radicand*** and $\sqrt{\ }$ the ***radical sign***. The expression $\sqrt{a}$ is also called a ***radical***.

By definition, $\sqrt{a} = b$ only if $a = b^2 \geq 0$, so $\sqrt{a}$ is defined only for $a \geq 0$. That the number $\sqrt{a}$ exists for *every* $a \geq 0$ is proved in more advanced books. We observe also that

$$(-\sqrt{a})^2 = (\sqrt{a})^2 = a \quad \text{for} \quad a \geq 0$$

so the squares of $-\sqrt{a}$ and of $\sqrt{a}$ are the same.

Two basic rules for combining radicals pertain to products and quotients:

$$\sqrt{ab} = \sqrt{a}\sqrt{b} \tag{1}$$

and

$$\sqrt{\frac{a}{b}} = \frac{\sqrt{a}}{\sqrt{b}} \tag{2}$$

Frequently we use (1) and (2) to simplify radicals and combinations of radicals.

EXAMPLE 1. Simplify the following expressions.

a. $\sqrt{8}\sqrt{18}$

b. $\dfrac{\sqrt{54}}{\sqrt{24}}$

Solution.

a. $\sqrt{8}\sqrt{18} \overset{(1)}{=} \sqrt{8 \cdot 18} = \sqrt{144} = 12$

b. $\dfrac{\sqrt{54}}{\sqrt{24}} \overset{(2)}{=} \sqrt{\dfrac{54}{24}} = \sqrt{\dfrac{9}{4}} = \dfrac{3}{2}$ □

EXAMPLE 2. Simplify the expression

$$\frac{\sqrt{x^4y^{-6}z^3}}{\sqrt{x^2y^3z}}$$

assuming that x, y, and z are positive.

Solution.

$$\frac{\sqrt{x^4y^{-6}z^3}}{\sqrt{x^2y^3z}} \overset{(2)}{=} \sqrt{\frac{x^4y^{-6}z^3}{x^2y^3z}} = \sqrt{\frac{x^2z^2}{y^9}} = \sqrt{\frac{x^2z^2}{y^8}\frac{1}{y}}$$

$$= \sqrt{\left(\frac{xz}{y^4}\right)^2\frac{1}{y}} \overset{(1)}{=} \sqrt{\left(\frac{xz}{y^4}\right)^2}\sqrt{\frac{1}{y}} = \frac{xz}{y^4}\sqrt{\frac{1}{y}}$$

$$= \frac{xz}{y^4}\frac{1}{\sqrt{y}} = \frac{xz}{y^4\sqrt{y}} \quad \square$$

CAUTION: Despite the product and quotient rules for radicals given in (1) and (2), a corresponding sum rule fails. In fact, $\sqrt{a+b} \neq \sqrt{a} + \sqrt{b}$ (unless $a = 0$ or $b = 0$; see Exercise 70). To support this claim, we notice that

$$\sqrt{16+9} = \sqrt{25} = 5 \quad \text{but} \quad \sqrt{16} + \sqrt{9} = 4 + 3 = 7$$

so that

$$\sqrt{16+9} \neq \sqrt{16} + \sqrt{9}$$

Let us observe that the square root and the absolute value (defined in Section 0.2) are intimately related by the formula

$$\boxed{\sqrt{a^2} = |a| \quad \text{for any real number } a} \tag{3}$$

After all, if $a \geq 0$, $\sqrt{a^2} = a = |a|$. But if $a < 0$, then since $-a > 0$ and $(-a)^2 = a^2$, we conclude that

$$\sqrt{a^2} = -a = |a|$$

More generally, one can prove that for any real number a and any positive integer m,

$$\boxed{\sqrt{a^{2m}} = |a|^m}$$

(See Exercise 69.) Thus

$$\sqrt{\pi^{12}} = \pi^6 \quad \text{and} \quad \sqrt{(-11)^{10}} = 11^5$$

Also

$$\sqrt{a^6} = |a|^3$$

*n*th Roots

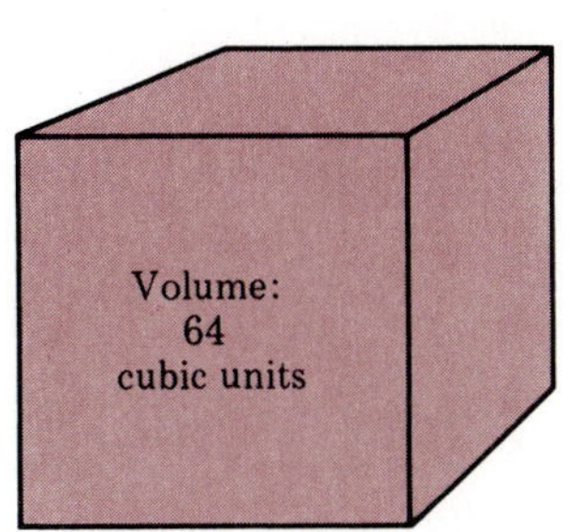

FIGURE 0.12

If the volume of a cube is 64 cubic units, then any side of the cube has length 4 (Figure 0.12), and we write $\sqrt[3]{64} = 4$. Similarly, for any real number a we define $\sqrt[3]{a}$ to be the number b such that $b^3 = a$, that is,

$$\sqrt[3]{a} = b \quad \text{if and only if} \quad b^3 = a$$

For example,

$$\sqrt[3]{-27} = -3 \quad \text{because} \quad (-3)^3 = -27$$

The number $\sqrt[3]{a}$ is called the ***cube root*** (or ***third root***) of a. Notice that unlike the square root, the cube root is defined for *all* real numbers.

More generally, for any integer $n \geq 2$ we define the ***nth root*** $\sqrt[n]{a}$ of a as follows:

$$\boxed{\sqrt[n]{a} = b \quad \text{if and only if} \quad b^n = a \quad \begin{cases} \text{for } a \geq 0 \text{ and } b \geq 0 \text{ if } n \text{ is even} \\ \text{for any real number } a \text{ if } n \text{ is odd} \end{cases}}$$

Thus $\sqrt[4]{81} = 3$ because $3^4 = 81$ and $3 \geq 0$, whereas $\sqrt[5]{-32} = -2$ since $(-2)^5 = -32$. If $n = 2$, then $\sqrt[n]{a}$ becomes $\sqrt[2]{a}$, which is normally written $\sqrt{a}$. The number n in $\sqrt[n]{a}$ is called the ***index*** of the root; the index of $\sqrt{a}$ is 2.

CAUTION: Notice carefully that when n is an even integer, $\sqrt[n]{a}$ is defined only for nonnegative values of a.

The same kinds of laws hold for general nth roots as for square roots.

Laws of nth Roots

Let a and b be real numbers and m and n positive integers. Each of the following formulas is valid for all values of a and b for which both sides of the equation are defined.

i. $\sqrt[n]{ab} = \sqrt[n]{a}\sqrt[n]{b}$ ii. $\sqrt[n]{\dfrac{a}{b}} = \dfrac{\sqrt[n]{a}}{\sqrt[n]{b}}$ iii. $\sqrt[m]{\sqrt[n]{a}} = \sqrt[mn]{a}$

iv. $(\sqrt[n]{a})^n = a$ v. $\sqrt[n]{a^n} = \begin{cases} |a| & \text{if } n \text{ is even} \\ a & \text{if } n \text{ is odd} \end{cases}$

Simplifying nth roots is similar to simplifying square roots, but with nth roots we use the laws listed above and factor out nth powers of numbers or variables where possible.

EXAMPLE 3. Simplify the following expressions.

a. $\sqrt[5]{2}\sqrt[5]{-16/243}$ b. $\sqrt{\sqrt[3]{729}}$ c. $\dfrac{\sqrt[4]{32x^8y^6}}{\sqrt[4]{x^2y^2}}$

Solution.

a. $$\sqrt[5]{2}\sqrt[5]{-16/243} \overset{(i)}{=} \sqrt[5]{-32/243} = -\sqrt[5]{32/243} = -\sqrt[5]{2^5/3^5} = -\sqrt[5]{(2/3)^5} \overset{(v)}{=} -\frac{2}{3}$$

b. $$\sqrt{\sqrt[3]{729}} \overset{(iii)}{=} \sqrt[6]{729} = \sqrt[6]{3^6} \overset{(v)}{=} 3$$

c. $$\frac{\sqrt[4]{32x^8y^6}}{\sqrt[4]{x^2y^2}} \overset{(ii)}{=} \sqrt[4]{\frac{32x^8y^6}{x^2y^2}} = \sqrt[4]{32x^6y^4} = \sqrt[4]{(2^4x^4y^4)(2x^2)}$$
$$\overset{(i)}{=} \sqrt[4]{2^4x^4y^4}\sqrt[4]{2x^2} \overset{(v)}{=} 2|xy|\sqrt[4]{2x^2}$$ □

Rational Exponents

We have discussed integral powers and roots of real numbers, that is, numbers of the form a^n and $\sqrt[n]{a}$ where n is an integer, but as yet we have not defined what we would mean by, say, $7^{2/5}$. However, in such studies as the efficiency of internal combustion engines, numbers like $7^{2/5}$ really do arise, so we will now define rational powers of numbers.

We begin by defining $a^{1/n}$ for any positive integer n. If the Laws of Integral Exponents are to remain valid for rational exponents, then we must have

$$(a^{1/n})^n = a^{(1/n)n} = a^1 = a = (\sqrt[n]{a})^n$$

which implies that $a^{1/n} = \sqrt[n]{a}$. Thus we make the following definition:

$$a^{1/n} = \sqrt[n]{a} \quad \begin{cases} \text{for } a \geq 0 \text{ if } n \text{ is even} \\ \text{for any } a \text{ if } n \text{ is odd} \end{cases}$$

EXAMPLE 4. Compute the following powers.

a. $4^{1/2}$ b. $27^{1/3}$ c. $1^{1/6}$ d. $\left(-\frac{1}{32}\right)^{1/5}$

Solution.

a. $4^{1/2} = \sqrt{4} = 2$

b. $27^{1/3} = \sqrt[3]{27} = 3$

c. $1^{1/6} = \sqrt[6]{1} = 1$

d. $\left(-\frac{1}{32}\right)^{1/5} = \sqrt[5]{-\frac{1}{32}} = -\frac{1}{2}$ □

Recall that any rational number r can be written as m/n, where m and n are integers with $n > 0$ and m/n in lowest terms. Thus to define a^r it suffices to define $a^{m/n}$, where m and n are integers and $n > 0$. If the Laws of Integral Exponents are to remain valid for rational exponents, then we must have

$$(a^{1/n})^m = a^{(1/n)m} = a^{m/n} \quad \text{and} \quad (a^m)^{1/n} = a^{m(1/n)} = a^{m/n}$$

In the special case in which $m = 2$, the laws of nth roots imply that

$$(a^2)^{1/n} = \sqrt[n]{a^2} = \sqrt[n]{a} \cdot \sqrt[n]{a} = (\sqrt[n]{a})^2 = (a^{1/n})^2$$

More generally, one can show that if 2 is replaced by an arbitrary integer, the formula

$$(a^m)^{1/n} = (a^{1/n})^m$$

is valid. This leads us to the following definition:

$$a^{m/n} = (a^{1/n})^m = (a^m)^{1/n} \tag{4}$$

CAUTION: If n is even and m is odd, then the number $a^{m/n}$ is defined only for $a \geq 0$ (so that the middle expression of (4) is meaningful). Similarly, if $m \leq 0$, then a must be different from 0 in order for $a^{m/n}$ to be defined.

Observe from (4) that the value of $a^{m/n}$ may be computed as either $(a^{1/n})^m$ or $(a^m)^{1/n}$. However, one of the expressions may be easier to compute than the other, as we will see in the following example.

EXAMPLE 5. Compute the following powers.

a. $4^{3/2}$ b. $(-8)^{5/3}$ c. $(\sqrt{32})^{-2/5}$

Solution.

a. $4^{3/2} = (4^{1/2})^3 = 2^3 = 8$

b. $(-8)^{5/3} = [(-8)^{1/3}]^5 = (-2)^5 = -32$

c. $(\sqrt{32})^{-2/5} = [(\sqrt{32})^{-2}]^{1/5} = \left[\left(\frac{1}{\sqrt{32}}\right)^2\right]^{1/5} = \left(\frac{1}{32}\right)^{1/5} = \frac{1}{2}$ □

Observe that we computed $4^{3/2}$ as $(4^{1/2})^3$ in (a). We could easily have computed $4^{3/2}$ alternatively as $(4^3)^{1/2} = (64)^{1/2} = 8$. However, it would have been more difficult to compute $(-8)^{5/3}$ in (b) as $[(-8)^5]^{1/3}$, because 8^5 is such a large number.

On a calculator, rational powers are computed using the key marked ^. Roots (except square roots, for which there is normally an alternative procedure) are computed by converting radicals to fractional powers. For example, in order to compute $\sqrt[5]{(1.39)^3}$, one would calculate $(1.39)^{(3/5)}$. In general,

$$\sqrt[m]{a^n} = a^{n/m}$$

The next example gives an application of rational exponents.

EXAMPLE 6. In a gasoline internal combustion engine with so-called combustion ratio b, the efficiency e of the engine is given by

$$e = 1 - \frac{1}{b^{2/5}}$$

Determine the value of the efficiency if $b = 7$.

Solution. We find by calculator that (to two significant digits)

$$e = 1 - \frac{1}{7^{2/5}} \approx 0.54$$

Thus the efficiency is approximately 54%. □

After studying logarithms in Chapter 4, we will be able to define a^r when r is *any real number*, irrational as well as rational. In the meantime we will assume that r is rational.

All the previous laws of exponents remain valid for rational exponents. They are listed below.

Laws of Rational Exponents

Let a and b be real numbers and r and s rational. Each of the following formulas is valid for all values of a and b for which all expressions in the formula are defined.

i. $a^r a^s = a^{r+s}$ ii. $(a^r)^s = a^{rs} = (a^s)^r$ iii. $(ab)^r = a^r b^r$

iv. $\left(\frac{a}{b}\right)^r = \frac{a^r}{b^r} = a^r b^{-r}$ v. $a^{-r} = \frac{1}{a^r}$ vi. $\frac{a^r}{a^s} = a^{r-s}$

EXAMPLE 7. Simplify the expression $(x^{1/3})^2(x^4y^6)^{1/3}$.

Solution.

$$\begin{aligned}(x^{1/3})^2(x^4y^6)^{1/3} &\overset{\text{(iii)}}{=} (x^{1/3})^2(x^4)^{1/3}(y^6)^{1/3}\\ &\overset{\text{(ii)}}{=} x^{2/3}x^{4/3}y^{6/3}\\ &\overset{\text{(i)}}{=} x^{6/3}y^2 = x^2y^2 \quad \square\end{aligned}$$

EXERCISES 0.4

In Exercises 1–10, simplify the expression.

1. $\sqrt{\frac{1}{25}}$
2. $\sqrt{1.69}$
3. $\sqrt{\frac{1}{4}\cdot\frac{1}{36}}$
4. $\sqrt{3 \times 10^{12}}$
5. $\sqrt{6}\sqrt{12}$
6. $\sqrt{32}\sqrt{72}$
7. $\sqrt{5 \times 10^5}\sqrt{20 \times 10^7}$
8. $\frac{\sqrt{63}}{\sqrt{21}}$
9. $\frac{\sqrt{75}}{\sqrt{147}}$
10. $\frac{\sqrt{6 \times 10^5}}{\sqrt{2 \times 10^{11}}}$

In Exercises 11–22, assume that all letters denote positive numbers. Simplify the expressions.

11. $\sqrt{24x^6y^{-4}}$
12. $\sqrt{3rs^{-3}t^2}\sqrt{27r^3s^5t^6}$
13. $\sqrt{8r/s^2}\sqrt{16s^2/r^4}$
14. $\frac{\sqrt{c^2d^6}}{\sqrt{4c^3d^{-4}}}$
15. $\frac{\sqrt{c^{-10}d^{-12}}}{\sqrt{c^{14}d^{-3}}}$
16. $\sqrt{x^2 - 2xy + y^2}$
17. $\sqrt{(4a - 7b)^4}$
18. $\sqrt{(9a^2 - 11b^3)^4}$
19. $\sqrt{\sqrt{16a^4b^8}}$
20. $\sqrt{\sqrt{9a^8/b^{10}}}$
21. $\sqrt{ab}\left(\frac{1}{\sqrt{b}} + \frac{1}{\sqrt{a}}\right)$
22. $\left(\sqrt{x} - \frac{1}{\sqrt{x}}\right)^2$

In Exercises 23–28, simplify the expression.

23. $\sqrt[3]{27}$
24. $\sqrt[3]{-1/64}$
25. $\sqrt[3]{0.000125}$
26. $\sqrt[4]{256}$
27. $\sqrt[4]{0.0016}$
28. $\sqrt[5]{243}$

In Exercises 29–34, assume that all letters denote positive numbers. Simplify the expressions.

29. $\sqrt[3]{a^3b^6}$
30. $\sqrt[4]{16a^4b^{12}}$
31. $\frac{1}{\sqrt[3]{-125x^3y^6z}}$
32. $\frac{\sqrt[4]{(x+y)^4}}{\sqrt[4]{81x^{12}y^8}}$
33. $\sqrt[4]{t\sqrt[3]{t^9}}$

34. $\sqrt[3]{a+b}\,\sqrt[3]{a^2-ab+b^2}$

In Exercises 35–38, simplify the given expression. Do not assume that *a*, *b*, *c*, *x*, and *y* are necessarily positive.

35. $\sqrt[4]{a^{12}b^8}$ **36.** $\sqrt[4]{\dfrac{a^4}{b^{16}c^{20}}}$

37. $\sqrt[4]{\dfrac{a^9b^7}{b^3a}}$ **38.** $\sqrt[6]{\dfrac{x^6+y^6}{(x+y)^{18}}}$

In Exercises 39–44, simplify the given expression.

39. $8^{5/3}$ **40.** $16^{3/2}$ **41.** $125^{-2/3}$

42. $(-64)^{-4/3}$ **43.** $\left(\dfrac{9}{25}\right)^{-3/2}$ **44.** $(10^{-5})^{2/5}$

In Exercises 45–58, assume that all letters denote positive numbers. Simplify the given expression.

45. $x^{1/3}x^{2/5}x^{4/15}$ **46.** $x^{-4/3}x^{7/5}x^{-13/9}$

47. $(36a^3b^4)^{3/2}$ **48.** $(x^3y^6z^9)^{2/3}$

49. $(x^3y^6z^8)^{2/3}$ **50.** $\left(\dfrac{8a^4}{27b^2}\right)^{2/3}$

51. $\left(\dfrac{16}{z^3}\right)^{-3/4}$ **52.** $\left(\dfrac{x^4y^7}{z^5}\right)^{-3/7}$

53. $(\sqrt{pq})^{2/3}$ **54.** $(\sqrt[3]{p^2+q^2})^{3/2}$

55. $\left(\dfrac{z^{1/3}\sqrt{x-y}}{2(x-y)}\right)^6$ **56.** $\sqrt[3]{\dfrac{b}{27a^3}}$

57. $\sqrt{b^3}\,\sqrt[3]{b^2}$

58. $(a^2+b^2)^{2/3}-\dfrac{a^2}{\sqrt[3]{a^2+b^2}}$

In Exercises 59–63, use a calculator to approximate the given expression.

59. $4^{2/5}$ **60.** $10^{-1/4}$ **61.** $\pi^{1/3}$

62. $(\sqrt{2})^{1/5}$ **63.** $(1.27)^{-4/9}$

In Exercises 64–65, use a calculator to see whether it gives the same value for each of the two numbers.

64. $\dfrac{\sqrt{(2.34)^4(1.79)^3}}{\sqrt{(5.21)^3(4.08)^2}}$ and $\sqrt{\dfrac{(2.34)^4(1.79)^3}{(5.21)^3(4.08)^2}}$

65. $\dfrac{\sqrt[3]{(3.29)^4(-1.136)^5}}{\sqrt[3]{671.209}}$ and $\sqrt[3]{\dfrac{(3.29)^4(-1.136)^5}{671.209}}$

In Exercises 66–68, write the statement as an equation.

66. The length *s* of a side of a cube is the cube root of the volume *V*.

67. The radius *r* of a sphere is equal to the square root of the quantity obtained by dividing the surface area *S* by 4π.

68. The radius *r* of a sphere is the cube root of the quantity obtained by dividing 3 times the volume *V* by 4π.

69. Show that for any real number *a* and any positive integer *m* we have $\sqrt{a^{2m}}=|a|^m$.

70. Show that if $a>0$ and $b>0$, then

$$\sqrt{a+b}\neq\sqrt{a}+\sqrt{b}$$

(*Hint:* Square both $\sqrt{a+b}$ and $(\sqrt{a}+\sqrt{b})$.)

71. a. Show that $\sqrt{a^2b^2}=|a||b|$ for any real numbers *a* and *b*.
b. Use part (*a*) and (3) to show that $|ab|=|a||b|$.

72. According to Einstein's Theory of Relativity, the mass of an object moving with velocity *v* is

$$\frac{m_0}{\sqrt{1-v^2/c^2}}$$

where *c* is the velocity of light and where m_0 is the "rest mass" of the object, that is, the mass when the velocity is 0. Find the mass of an object moving with velocity $c/3$ if its rest mass is 5 grams.

73. Find the mass of an electron whose rest mass is 9.11×10^{-28} grams, assuming that the electron is moving with velocity $c/3$.

74. In order for a satellite to go around the earth in a circular orbit of radius *r*, its velocity *v* in miles per hour must be given by

$$v=\sqrt{GM/r}$$

where *G* is the universal gravitational constant and *M* is the mass of the earth. Taking $GM=1.237\times10^{12}$, approximate the velocity a satellite must have in order to go around the earth in a circular orbit with radius 4096 miles (so the satellite is approximately 100 miles above the surface of the earth).

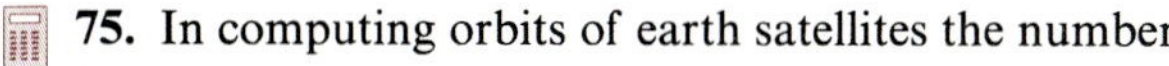

75. In computing orbits of earth satellites the number

$$\left(\frac{1.237 \times 10^{12}}{4\pi^2}\right)^{1/3}$$

appears. Use a calculator to approximate its value.

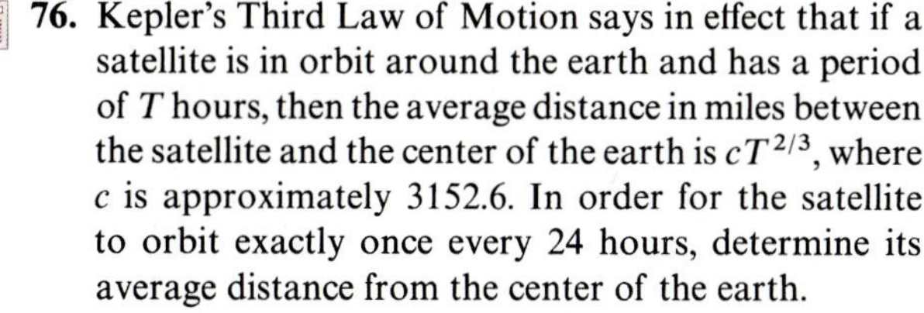

76. Kepler's Third Law of Motion says in effect that if a satellite is in orbit around the earth and has a period of T hours, then the average distance in miles between the satellite and the center of the earth is $cT^{2/3}$, where c is approximately 3152.6. In order for the satellite to orbit exactly once every 24 hours, determine its average distance from the center of the earth.

77. When a diatomic gas (such as oxygen) undergoes an adiabatic change (one in which no heat is lost from the gas), the pressure p and volume V of the gas must change in such a way that $pV^{1.4}$ remains constant. If the volume doubles, how must the pressure change?

78. The efficiency e of a diesel engine is given by

$$e = 1 - \frac{1}{1.4}\,\frac{\dfrac{1}{r^{1.4}} - \dfrac{1}{b^{1.4}}}{\dfrac{1}{r} - \dfrac{1}{b}}$$

where r and b denote the expansion and compression ratios, respectively. Reasonable values for r and b are $r = 5$ and $b = 15$. Approximate the value of e when $r = 5$ and $b = 15$.

0.5 POLYNOMIALS

We have already used letters such as a, b, c, x, y, and z to represent real numbers. In the remaining sections of this chapter we will focus on combinations of expressions involving such letters.

The basic building block of the expressions we will consider is the ***monomial*** ax^k, where a is a specified real number and k is a specified nonnegative integer, but x is unspecified. The number a is called the **coefficient** of the monomial, and x is called a **variable**. Examples of monomials are

$$16x^2, \qquad -\sqrt{2}x^6, \quad \text{and} \quad -\frac{3}{5}x^{58}$$

Since $x^1 = x$ and $x^0 = 1$ for $x \neq 0$, we normally write

$$ax \quad \text{for} \quad ax^1 \quad \text{and} \quad a \quad \text{for} \quad ax^0$$

When we add two monomials ax^k and bx^m, we obtain the ***binomial***

$$ax^k + bx^m$$

It follows that

$$x - 2, \qquad 2x^2 + x, \quad \text{and} \quad x^8 - 3x^3$$

are binomials.

More generally, the sum of a finite number of monomials is a ***polynomial*** (or ***polynomial expression***). The general form of a polynomial is

$$a_n x^n + a_{n-1} x^{n-1} + \cdots + a_1 x + a_0$$

where n is a nonnegative integer, x is a variable, and the numbers a_n, $a_{n-1}, \ldots, a_1, a_0$ are constants called the ***coefficients*** of the polynomial. Examples of polynomials are

$$5, \qquad x - \frac{4}{3}, \qquad x^2 + 6x + 9, \qquad x^3 - 27, \quad \text{and} \quad -\sqrt{7}x^6 - x^4$$

Notice that whereas the number n in x^n is an exponent, the number n in a_n simply indicates that a_n is the coefficient of x^n.

The monomials that make up a polynomial are called ***terms*** of the polynomial. Thus, the terms of $x^2 - 6x + 1$ are x^2, $-6x$, and 1. If $a_n \neq 0$, then the coefficient a_n of $a_n x^n$ is the ***leading coefficient*** of the polynomial and n is the ***degree*** of the polynomial. Consequently the degree of $x^3 - 27$ is 3 and the leading coefficient is 1; likewise, the degree of $-\sqrt{7}x^6 - x^4$ is 6 and the leading coefficient is $-\sqrt{7}$. Finally, the number a_0 is called the ***constant term*** of the polynomial. The constant term may be 0, as in the polynomial $2x^3 - x^2 + x$.

Polynomials are classified according to their degrees. The polynomial 0 has no degree attached to it. For the other polynomials we have:

Polynomial	*Degree*	*Form*	*Example*
constant	0	$a_0 \; (a_0 \neq 0)$	$-\frac{4}{7}$
linear	1	$a_1 x + a_0 \; (a_1 \neq 0)$	$2x - 8$
quadratic	2	$a_2 x^2 + a_1 x + a_0 \; (a_2 \neq 0)$	$4x^2 - 6x + \sqrt{3}$
cubic	3	$a_3 x^3 + a_2 x^2 + a_1 x + a_0 \; (a_3 \neq 0)$	$x^3 - \frac{5}{2}x^2 + \pi$
nth-degree	n	$a_n x^n + a_{n-1} x^{n-1} + \cdots + a_1 x + a_0 \quad (a_n \neq 0)$	$x^n + 1$

CAUTION: In a polynomial, the exponents of the powers of x must be nonnegative integers; rational powers and negative powers of x are not allowed. Thus $4x^{3/2} + 3x - 7$ is not a polynomial.

We say that two polynomials are ***equal*** if they have the same degree and if the corresponding powers of x have the same coefficients. In particular, the two polynomials

$$x^2 - 5x + 4 \quad \text{and} \quad ax^2 + bx + c$$

are equal if and only if $a = 1$, $b = -5$, and $c = 4$.

In a given polynomial, if we replace the variable x by a particular real number, the resulting expression is meaningful and represents a real number. Thus if we replace x by 2 in the polynomial $x^2 + 6x + 9$, then the polynomial becomes $2^2 + 6(2) + 9$, which simplifies to $4 + 12 + 9$, or 25. The process of replacing x by a particular real number is called ***substitution***. The number obtained when a real number c is substituted into a polynomial is called the ***value*** of the polynomial for $x = c$.

EXAMPLE 1. Find the values of $x^2 + 6x + 9$ for $x = 0$ $x = -\frac{4}{3}$, and $x = \sqrt{3}$.

Solution. We find in turn that

$$\text{for}\quad x = 0,\qquad x^2 + 6x + 9 = 0^2 + 6(0) + 9 = 0 + 0 + 9 = 9$$

$$\text{for}\quad x = -\frac{4}{3},\qquad x^2 + 6x + 9 = \left(-\frac{4}{3}\right)^2 + 6\left(-\frac{4}{3}\right) + 9$$

$$= \frac{16}{9} - 8 + 9 = \frac{25}{9}$$

$$\text{for}\quad x = \sqrt{3},\qquad x^2 + 6x + 9 = (\sqrt{3})^2 + 6(\sqrt{3}) + 9$$

$$= 3 + 6\sqrt{3} + 9 = 12 + 6\sqrt{3} \qquad \square$$

So far we have used the letter x for the variable of a polynomial. However, it is possible to use other letters as well. Thus $y^2 + 6y + 9$ is a polynomial, as is $t^2 + 6t + 9$. Occasionally we refer to a polynomial whose variable is y as a ***polynomial in y***. In this vein, $y^2 + 6y + 9$ is a polynomial in y. Next we observe that if we substitute a number such as $-\frac{4}{3}$ for y in $y^2 + 6y + 9$, we obtain

$$\left(-\frac{4}{3}\right)^2 + 6\left(-\frac{4}{3}\right) + 9$$

which has the value $\frac{25}{9}$. This is the same value we found in Example 1. The reason it is the same is that $x^2 + 6x + 9$ and $y^2 + 6y + 9$ have the same degree and the same coefficients for corresponding powers. The fact that different letters have been used for the variable in the two polynomials is immaterial to the value of the polynomial when a real number is substituted.

Polynomials appear with great frequency in mathematics and in the sciences, including discussions of the height of an object under the influence of gravity alone. For example, if a ball is dropped from a height of h_0 feet, then (until it hits the ground) its height h above ground after t seconds is given by

$$\boxed{h = -16t^2 + h_0} \qquad (1)$$

where h is in feet and t is in seconds.

EXAMPLE 2. Suppose a ball is dropped from a balcony 86 feet above the ground. How high will it be after 2 seconds?

Solution. Using (1) with $h_0 = 86$, we have

$$h = -16t^2 + 86$$

Substituting $t = 2$, we obtain $h = -16(2^2) + 86 = -64 + 86 = 22$, so the ball will be 22 feet above the ground at that instant. □

Addition and Subtraction of Polynomials

We can combine numbers by adding, subtracting, multiplying, and dividing them. We can do the same with polynomials because they represent numbers. All the rules (including the commutative, associative, and distributive laws) that apply to real numbers may be applied to polynomials as well. In particular, the sum of any two polynomials in x can be obtained by adding the coefficients of like powers of x, which frequently can be done most easily by lining up terms with like powers of x vertically.

EXAMPLE 3. Find the sum $(x^3 - 2x^2 + 4x - 1) + (-2x^5 -- 3x^3 + \sqrt{5}x^2 + 3)$.

Solution. By lining up terms with like powers of x vertically and using the laws mentioned above, we find that

$$\begin{array}{rrrrr} & x^3 & - \quad 2x^2 & + 4x & - 1 \\ -2x^5 & - 3x^3 & + \sqrt{5}\,x^2 & & + 3 \\ \hline -2x^5 & - 2x^3 & + (-2 + \sqrt{5})x^2 & + 4x & + 2 \end{array}$$

□

Adding polynomials can also be accomplished in a horizontal format. In doing so we combine all terms with like powers of x:

$$\begin{aligned} &(x^3 - 2x^2 + 4x - 1) + (-2x^5 - 3x^3 + \sqrt{5}x^2 + 3) \\ &\quad = -2x^5 + (x^3 - 3x^3) + (-2x^2 + \sqrt{5}x^2) + 4x + (-1 + 3) \\ &\quad = -2x^5 - 2x^3 + (-2 + \sqrt{5})x^2 + 4x + 2 \end{aligned}$$

Which format we use, vertical or horizontal, depends on personal preference. It is probably easier to keep coefficients of like powers of the variable straight with the vertical format, but it takes more space on the page of a book. Therefore, except in this section, we will generally adopt the horizontal format.

We subtract polynomials by subtracting terms with like powers.

EXAMPLE 4. Find the difference $(x^4 - 6x^3 - 2x + 1) - (x^3 - 2x^2 - 3x - 5)$.

Solution. Subtracting terms with like powers, we find that

$$\begin{array}{r} x^4 - 6x^3 \qquad\quad - 2x + 1 \\ x^3 - 2x^2 - 3x - 5 \\ \hline x^4 - 7x^3 + 2x^2 + \ x + 6 \end{array} \quad \square$$

Multiplication of Polynomials

To multiply two polynomials we use the law of exponents

$$x^m x^n = x^{m+n}$$

in conjunction with the distributive law. One elementary product is

$$\boxed{(ax + b)(cx + d) = acx^2 + (ad + bc)x + bd}$$

The next example involves a product of more complicated polynomials.

EXAMPLE 5. Find the product $(2x^2 - 3x - 4)(x^3 - 6x^2 + 1)$.

Solution. We multiply in a vertical fashion:

$$\begin{array}{rrrrrr} & & & 2x^2 & - 3x & - 4 \\ & & x^3 & - 6x^2 & + 1 & \\ \hline & & & 2x^2 & - 3x & - 4 \\ & - 12x^4 & + 18x^3 & + 24x^2 & & \\ 2x^5 & - 3x^4 & - 4x^3 & & & \\ \hline 2x^5 & - 15x^4 & + 14x^3 & + 26x^2 & - 3x & - 4 \end{array} \quad \square$$

Observe that the degree of the product in Example 5 is 5, which is the sum $2 + 3$ of the degrees 2 and 3 of the factors $2x^2 - 3x - 4$ and $x^3 - 6x^2 + 1$, respectively. In general, the degree of the product of two polynomials is the sum of the degrees of the two polynomials.

Polynomials in Two Variables

A polynomial in the two variables x and y is a sum of monomials (or terms) of the form cx^ky^m, where c is a constant, x and y are variables, and k and m are nonnegative integers. Examples are

$$2x + y^2, \quad x^2 + 2xy + y^2, \quad \text{and} \quad x^4 + 4x^3y^2 + 9xy^3 + y^6$$

One can define a polynomial in the three variables x, y, and z, or in more than three variables, in the same way. A polynomial in more than one variable is usually referred to as a ***polynomial in several variables***.

Adding and multiplying polynomials of several variables proceeds as with ordinary polynomials. Thus

$$(2x^2 - 6x^2y + 3xy^2) + (-x^2 + 3x^2y - y^3) = x^2 - 3x^2y + 3xy^2 - y^3$$

and

$$(x^2 - xy + y)(3x^2 - 4y) = 3x^4 - 3x^3y - x^2y + 4xy^2 - 4y^2$$

Next we list identities for a few frequently occurring products. The first three essentially appeared as (2), (3), and (4), respectively, in Section 0.1.

$$(x + y)^2 = x^2 + 2xy + y^2 \tag{2}$$

$$(x - y)^2 = x^2 - 2xy + y^2 \tag{3}$$

$$(x + y)(x - y) = x^2 - y^2 \tag{4}$$

$$(x + y)^3 = x^3 + 3x^2y + 3xy^2 + y^3 \tag{5}$$

$$(x - y)^3 = x^3 - 3x^2y + 3xy^2 - y^3 \tag{6}$$

In each of these formulas, x or y may be replaced by another letter, by a specific number, or even by a more complicated expression.

EXAMPLE 6. Find the following products.

a. $(x - 3z)^2$ b. $(w^2 + 2z^5)^3$

Solution.

a. By (3), with $3z$ substituted for y, we have

$$(x - 3z)^2 = x^2 - 2x(3z) + (3z)^2 = x^2 - 6xz + 9z^2$$

b. By (5), with w^2 substituted for x and $2z^5$ substituted for y,

$$\begin{aligned}(w^2 + 2z^5)^3 &= (w^2)^3 + 3(w^2)^2(2z^5) + 3(w^2)(2z^5)^2 + (2z^5)^3 \\ &= w^6 + 6w^4z^5 + 12w^2z^{10} + 8z^{15} \quad \square\end{aligned}$$

EXERCISES 0.5

In Exercises 1–2, find the value of the given polynomial for $x = 0$, $x = 2$, and $x = -1$.

1. $2x^2 - 5x + 3$

2. $x^3 - 3x^2 + 3x - 1$

In Exercises 3–16, perform the indicated operations and then simplify.

3. $(3x^2 - 2x - 1) + (4x^2 + 2x - 5)$

4. $(5x^4 + 3x^2 - 7) + (6x^3 + 5x^2 - 4x + 9)$

5. $(2x^3 + \frac{1}{2}x^2 + 4x) - (3x^3 - \frac{1}{2}x^2 + 2x - 4)$

6. $(\sqrt{3}x^3 - \sqrt{2}x^2) - (2x^3 - \sqrt{2}x^2 - x - 1)$

7. $(2x - 3)(4x - 5)$

8. $(\frac{1}{2}x + 3)(\frac{1}{3}x + 4)$

9. $(3x - 2)(-x + 5) + (2x - 1)(5x - 7)$

10. $(x + 3)^2 - (x - 3)^2$

11. $(2x - 1)^2 - 4(x - 2)^2$

12. $y^2(4y^3 - y^2) + (y^2 - 1)^2$

13. $(2 - s)(s - 4)^2$

14. $(x^2 + 2x + 4)(x - 2)$

15. $(2y^4 - 4y^2 + 8)(y^2 + 2)$

16. $(y^3 + 3y - 1)(2y^3 - y^2 - 2)$

In Exercises 17–32, perform the indicated operations and then simplify.

17. $(x^2 + 2xy + y^2) + (3x^2 - xy + y^2)$

18. $(x^2 + 2xy) - (xy + y^2)$

19. $(2x^2 - y^3)^2$

20. $(\frac{1}{2}x + \frac{1}{4}y)^2$

21. $(2x - 3y)^3$

22. $(p + 2q^2)^3$

23. $(x + h)^2 - x^2$

24. $(x + h)^3 - x^3$

25. $(x - y)(x^2 + xy + y^2)$

26. $(x + y)(x^2 - xy + y^2)$

27. $(x + y + z)(x + y - z)$

28. $(x + 2y - 3z)^2$

29. $(u^{1/2} + v^{1/2})^2$

30. $(u^{1/3} + v^{1/3})^3$

31. $\left(\frac{1}{r} - \frac{1}{s}\right)^2$

32. $\left(\frac{1}{r} + \frac{1}{2s}\right)^2$

33. Show that

$$x(x + y)(x + 2y)(x + 3y) = (x^2 + 3xy + y^2)^2 - y^4$$

34. Show that $x^3 - y^3 = (x - y)^3 + 3xy(x - y)$.

***35.** By letting $x = 3y - 1$, find numbers x and y such that

$$y(6 - y) = x^3 - x \neq 0$$

36. It turns out that for many integer values of x, the polynomial $4x^3 - 28x + 25$ is the square of an integer. Show that the value of this polynomial is the square of an integer for $x = 0, 1, -1, 2, -2, 3, -3, 4$, and 11.

37. An integer $p \geq 2$ is **prime** if the only positive integer divisors of p are 1 and p. In the eighteenth century, the mathematician Leonhard Euler observed that the value of the polynomial $x^2 - x + 41$ is prime for every integer up to and including 40. Show that the value of this polynomial is prime for $x = 40$.

38. Show that if x is replaced by $y - b/(3a)$, then the cubic polynomial $ax^3 + bx^2 + cx + d$ is transformed into a cubic polynomial of the form $ay^3 + ey + f$ (so the coefficient of y^2 has become 0).

***39.** Using Figure 0.13, give a geometric interpretation of (5) in case x and y are positive numbers.

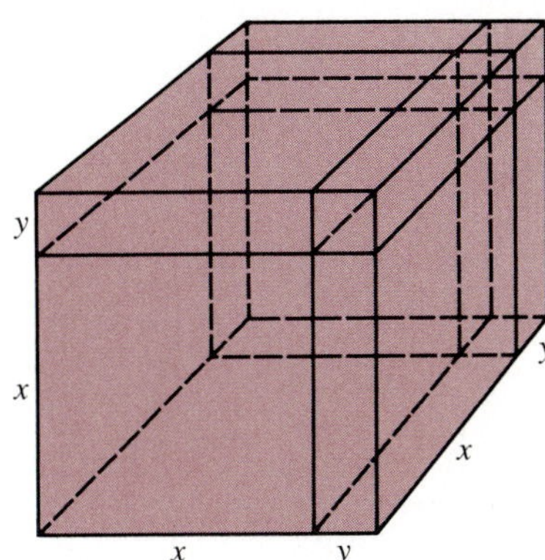

FIGURE 0.13

40. Suppose a rock is dropped from a bridge 70 feet above ground. Using (1), determine how high the rock will be after
a. 1 second
b. 2 seconds
c. 3 seconds (*Hint:* Can the height of the rock be negative?)

41. A fireworks rocket is launched in a vertical direction. Until it hits the ground again, the rocket's height h above the ground after t seconds is given by

$$h = -16t^2 + 96t$$

Compare the heights of the rocket after 1 second and after 5 seconds.

42. Two baseballs are thrown straight downward at the same instant, one from a window 50 feet above ground, and the other from a window 100 feet above ground. The first ball is thrown with a speed of 30 feet per second, so while it descends, its height t seconds later is given by $-16t^2 - 30t + 50$. The second ball is thrown with a speed of 45 feet per second, so while it descends, its height t seconds later is given by $-16t^2 - 45t + 100$. Find an expression for the distance between the two baseballs until the first hits the ground.

0.6 FACTORING POLYNOMIALS

In the preceding section we multiplied polynomials. Now we consider the opposite procedure: writing a polynomial as a product of other polynomials, called ***factors***. For example, since

$$4x^2 + 12x + 8 = 4(x + 2)(x + 1)$$

it follows that 4, $x + 2$, and $x + 1$ are factors of $4x^2 + 12x + 8$. Similarly,

$$x^4 - 6x^3 + 9x^2 = x^2(x - 3)^2$$

so that x^2 and $(x - 3)^2$ are factors of $x^4 - 6x^3 + 9x^2$. Of course, since $x^2 = x \cdot x$ and $(x - 3)^2 = (x - 3)(x - 3)$, we have

$$x^4 - 6x^3 + 9x^2 = x \cdot x(x - 3)(x - 3)$$

so that x and $x - 3$ are also factors of $x^4 - 6x^3 + 9x^2$. The process of rewriting a polynomial as the product of factors is called ***factoring***, and factoring is important in the analysis of properties of polynomials and quotients of polynomials. Our interest lies in finding factors of degree 1 or higher, which are called ***nontrivial factors***.

Factors of Certain Quadratic Polynomials

To determine which polynomials have nontrivial factors is a difficult problem, and actually finding them can range from easy to impossible (and is likely to be hard if the degree of the polynomial is large). Even when the degree of the polynomial is 2, finding factors can be involved; the general discussion in this case must wait until Section 1.2. Since at this point we wish to become familiar with factors of polynomials, we will now look at polynomials of degree 2 whose coefficients are integers.

Let us set out to factor $x^2 + 9x + 8$ so that

$$x^2 + 9x + 8 = (x + a)(x + b) \tag{1}$$

Since $(x + a)(x + b) = x^2 + (a + b)x + ab$, we can write (1) alternatively as

$$x^2 + 9x + 8 = x^2 + (a + b)x + ab$$

which implies that $a + b = 9$ and $ab = 8$. Since $ab = 8$ and we are assuming that a and b are integers, the possible choices for a and b as a pair are

$$\begin{array}{cc} 1 \text{ and } 8 & 2 \text{ and } 4 \\ -1 \text{ and } -8 & -2 \text{ and } -4 \end{array}$$

But $a + b = 9$, so our choice is restricted to 1 and 8. Thus $a = 1$ and $b = 8$ (or $a = 8$ and $b = 1$). Either possibility yields the factors $x + 1$ and $x + 8$, so that

$$x^2 + 9x + 8 = (x + 1)(x + 8)$$

In general, if we wish to factor a polynomial of the form $x^2 + dx + e$ in such a way that the factors have integer coefficients, then we must find integers a and b such that

$$x^2 + dx + e = (x + a)(x + b) \tag{2}$$

Since

$$(x + a)(x + b) = x^2 + (a + b)x + ab \tag{3}$$

we can write (2) alternatively as

$$x^2 + dx + e = x^2 + (a + b)x + ab$$

which implies that

$$a + b = d \quad \text{and} \quad ab = e \tag{4}$$

EXAMPLE 1. Factor the polynomial $x^2 - 7x + 12$.

Solution. If a and b are integers such that

$$x^2 - 7x + 12 = (x + a)(x + b)$$

then the equations in (4) become

$$a + b = -7 \quad \text{and} \quad ab = 12$$

Since $ab = 12$, the possibilities for a and b as a pair are

1 and 12	2 and 6	3 and 4
−1 and −12	−2 and −6	−3 and −4

Of these, only -3 and -4 satisfy $a + b = -7$. Thus $a = -3$ and $b = -4$ (or $a = -4$ and $b = -3$). We conclude that

$$x^2 - 7x + 12 = (x - 3)(x - 4) \quad \square$$

Two special forms of $x^2 + dx + e$ are

$$x^2 + 2ax + a^2 \quad (\text{where } d = 2a \text{ and } e = a^2)$$

and

$$x^2 - a^2 \quad (\text{where } d = 0 \text{ and } e = -a^2)$$

Using (2) and (4) of Section 0.5, with a instead of y, we can find factors for these special polynomials:

$$x^2 + 2ax + a^2 = (x + a)^2 \tag{5}$$

$$x^2 - a^2 = (x + a)(x - a) \tag{6}$$

EXAMPLE 2. Factor the following polynomials.

a. $x^2 + 6x + 9$ b. $x^2 - 16$ c. $49 - y^2$

Solution.

a. By (5), with $a = 3$, we have

$$x^2 + 6x + 9 = x^2 + 2(3x) + 3^2 = (x + 3)^2$$

b. By (6), with $a = 4$, we have

$$x^2 - 16 = x^2 - 4^2 = (x + 4)(x - 4)$$

c. By (6), with $x = 7$ and y substituted for a, we find that

$$49 - y^2 = 7^2 - y^2 = (7 + y)(7 - y) \quad \square$$

In spite of the results of Examples 1 and 2, there are polynomials of degree 2 that do not have any nontrivial factors with real coefficients. To prove this we consider the polynomial $x^2 + c^2$, with $c \neq 0$. If $x - a$ were a factor of $x^2 + c^2$, then we would have

$$x^2 + c^2 = (x - a)(\text{polynomial})$$

But the right side is 0 for $x = a$, whereas the left side is positive for every value of x (because $c \neq 0$). Consequently $x^2 + c^2$ has no non-trivial factors. Therefore none of the polynomials

$$x^2 + 1, \qquad x^2 + 3 \quad \text{and} \quad x^4 + 2$$

has a nontrivial factor. More generally, the following result is true:

> The binomial $x - a$ is a factor of a polynomial if and only if the value of the polynomial for $x = a$ is 0.

EXAMPLE 3. Show that $x + \frac{1}{2}$ is a factor of $3x^2 - \frac{13}{2}x - 4$.

Solution. First we observe that $x + \frac{1}{2} = x - (-\frac{1}{2})$. Then we find the value of $3x^2 - \frac{13}{2}x - 4$ for $x = -\frac{1}{2}$:

$$3\left(-\frac{1}{2}\right)^2 - \frac{13}{2}\left(-\frac{1}{2}\right) - 4 = \frac{3}{4} + \frac{13}{4} - 4 = 0$$

By the comment preceding the example, $x + \frac{1}{2}$ is a factor of $3x^2 - \frac{13}{2}x - 4$. □

A two-variable version of (3) is

$$x^2 + (a + b)xy + aby^2 = (x + ay)(x + by)$$

as you can verify by multiplying out the right side. Thus if we wish to factor $x^2 + dxy + ey^2$ so as to have

$$x^2 + dxy + ey^2 = (x + ay)(x + by)$$

we must once again have

$$a + b = d \quad \text{and} \quad ab = e$$

which appeared in (4).

EXAMPLE 4. Factor $x^2 - 2xy - 8y^2$.

Solution. If a and b are integers such that

$$x^2 - 2xy - 8y^2 = (x + ay)(x + by)$$

then $ab = -8$. It follows that the choices for a and b as a pair are

$$\begin{array}{cc} 1 \text{ and } -8 & 2 \text{ and } -4 \\ -1 \text{ and } 8 & -2 \text{ and } 4 \end{array}$$

Since $a + b = -2$, we conclude that $a = 2$ and $b = -4$ (or $a = -4$ and $b = 2$). Either possibility yields the factors $x + 2y$ and $x - 4y$, so that

$$x^2 - 2xy - 8y^2 = (x + 2y)(x - 4y) \quad □$$

If the leading coefficient of the polynomial is not 1, then similar techniques can be used.

EXAMPLE 5. Factor $3x^2 + 2x - 1$.

Solution. If the factors are to have integer coefficients, then the coefficients of x in the factors must be 3 and 1 or -3 and -1, and the constant terms must be 1 or -1. By trial and error we find that

$$3x^2 + 2x - 1 = (3x - 1)(x + 1) \quad □$$

Factoring General Polynomials

Although it is impossible to give instructions for factoring polynomials in general, we can give guidelines in certain cases. In one type of polynomial there is a nontrivial factor of every term. Such a factor is called a ***common factor***.

EXAMPLE 6. Factor the polynomial $5x^4 + 15x^2$.

Solution. Here $5x^2$ is a common factor, and we factor it out, obtaining

$$5x^4 + 15x^2 = 5x^2(x^2 + 3) \quad \square$$

In another type of polynomial there is a common factor not of every term but of each of several groups of terms together comprising the polynomial. Such common factors can then be factored out by grouping them together. The method is known as ***factoring by grouping***.

EXAMPLE 7. Factor the following polynomials by grouping.

a. $x^3 - 3x^2 + 2x - 6$ b. $2x^3 + 5x^2 - 6x - 15$

c. $ax^2 + bxy - axy - by^2$

Solution.

a. We observe that $x - 3$ is a factor of $x^3 - 3x^2$ and of $2x - 6$, and consequently

$$\begin{aligned} x^3 - 3x^2 + 2x - 6 &= (x^3 - 3x^2) + (2x - 6) \\ &= x^2(x - 3) + 2(x - 3) \\ &= (x^2 + 2)(x - 3) \end{aligned}$$

b. Since

$$2x^3 + 5x^2 = x^2(2x + 5) \quad \text{and} \quad -6x - 15 = -3(2x + 5)$$

it follows that $2x + 5$ is a common factor and that

$$\begin{aligned} 2x^3 + 5x^2 - 6x - 15 &= x^2(2x + 5) - 3(2x + 5) \\ &= (x^2 - 3)(2x + 5) \end{aligned}$$

c. We find that

$$ax^2 + bxy = x(ax + by) \quad \text{and} \quad -axy - by^2 = -y(ax + by)$$

Consequently

$$\begin{aligned} ax^2 + bxy - axy - by^2 &= x(ax + by) - y(ax + by) \\ &= (x - y)(ax + by) \quad \square \end{aligned}$$

Observe that in the solution of (b) we could further factor $x^2 - 3$ and obtain

$$x^2 - 3 = (x - \sqrt{3})(x + \sqrt{3})$$

but since we are interested only in factors with integer coefficients at this time, we left the answer as it was.

Finally, we factor the polynomials $x^n - a^n$ and $x^n + a^n$, which appear frequently in mathematics:

$$x^n - a^n = (x - a)(x^{n-1} + ax^{n-2} + a^2x^{n-3} + \cdots + a^{n-2}x + a^{n-1}) \text{ for } n \text{ positive} \quad (7)$$

$$x^n + a^n = (x + a)(x^{n-1} - ax^{n-2} + a^2x^{n-3} - \cdots - a^{n-2}x + a^{n-1}) \text{ for } n \text{ positive and odd} \quad (8)$$

For $n = 3$, these formulas become

$$x^3 - a^3 = (x - a)(x^2 + ax + a^2) \quad (9)$$

$$x^3 + a^3 = (x + a)(x^2 - ax + a^2) \quad (10)$$

EXAMPLE 8. Factor

a. $x^3 - 27$ b. $x^5 + 32$

Solution.

a. Using (9) with $a = 3$, we obtain

$$x^3 - 27 = x^3 - 3^3 = (x - 3)(x^2 + 3x + 9)$$

b. Using (8) with $n = 5$ and $a = 2$, we have

$$x^5 + 32 = x^5 + 2^5 = (x + 2)(x^4 - 2x^3 + 4x^2 - 8x + 16) \quad \square$$

CAUTION: Notice that (8) is valid only if n is an odd integer. If n is even, then $x + a$ is *not* a factor of $x^n + a^n$, although there may be nontrivial factors of $x^n + a^n$ (see Exercise 54).

EXERCISES 0.6

In Exercises 1–14, factor the given polynomial.

1. $x^2 + 8x + 12$
2. $x^2 - 2x - 3$
3. $x^2 - 7x + 6$
4. $t^2 + t - 12$
5. $2t^2 - 6t - 8$
6. $4y^2 - 1$
7. $21 - 10b + b^2$
8. $x^2 - 4x + 4$
9. $a^2 - 14a + 49$
10. $a^2 - 5a + \frac{25}{4}$
11. $x^2 + 11x + \frac{121}{4}$
12. $49z^2 - 36$
13. $16 - 9z^2$
14. $x^2 - 2\sqrt{2}x + 2$

In Exercises 15–20, factor the given polynomial.

15. $2x^2 + 7x + 3$
16. $5x^2 + 4x - 1$
17. $-x^2 + 5x - 6$
18. $7x^2 + 8x - 12$
19. $7x^2 + 17x - 12$
20. $40x^2 + 14x - 45$

In Exercises 21–28, factor the given polynomial.

21. $x^2 + 2xy + y^2$
22. $5x^2 + 10xy - 40y^2$
23. $x^2 - 4y^2$
24. $12x^2 + xy - y^2$
25. $x^4 - 2x^2y^2 + y^4$
26. $x^4 + 5x^2y^2 + 6y^4$

27. $5x^2 - 14xy - 3y^2$ **28.** $2x^2 - 5xy + 3y^2$

In Exercises 29–40, factor the given polynomial by finding a common factor or by grouping.

29. $x^7 - x^6$ **30.** $15x^4 + 3x^3$

31. $(3x + 5)x^2 + (3x + 5)x$

32. $x^2 - 1 + 3(x - 1)$

33. $(x^2 - 5)^2 - 8(x^2 - 5) + 16$

34. $6x^3 + 12x^2 + 24x + 48$

35. $x^3 + x^2 - x - 1$

36. $x^5 + 2x^4 - x^3 - 2x^2$

37. $(3x + 2y)^2 + (3x + 2y) - 12$

38. $4x^2 + 4xy + y^2 - 16$

39. $x^2 - 10x + 25 - 4y^2$

40. $x^2 - 12x + 36 - 16y^2$

In Exercises 41–48, factor the given polynomial.

41. $x^3 + 1$ **42.** $x^3 - 1$ **43.** $x^3 - 8$

44. $8x^3 - 27$ **45.** $8x^3 - 1$ **46.** $y^4 - 16$

47. $32x^5 - 1$ **48.** $t^5 + 1$

In Exercises 49–52, determine whether the given polynomial has the factor $x - a$ for the specified value of a.

49. $5x^5 + 4x^2 + 1; \quad a = -1$

50. $x^3 - 3x^2 - 6x + 9; \quad a = 3$

51. $x^6 - 7x^3 - 3x - 4; \quad a = 2$

52. $-x^4 - 3x^3 - 4x^2 - 5x - 2; \quad a = -2$

53. Show that

$$x^4 + 1 = (x^2 - \sqrt{2}x + 1)(x^2 + \sqrt{2}x + 1).$$

54. Show that

$$x^4 + a^4 = (x^2 - \sqrt{2}ax + a^2)(x^2 + \sqrt{2}ax + a^2).$$

55. Find the only value of x for which

$$\frac{x^3 - x^2 - x + 1}{x^3 - x^2 + x - 1} = 0$$

***56.** Suppose that $a \neq 0$ and that one factor of $ax^2 - bx + a$ is $3x - 2$. Find the ratio a/b.

57. There are exactly two real numbers a such that $x - a$ is a factor of $x^8 + x^6 + x^4 + x^2 - 340 = 0$. One of the numbers is 2. Find the other number.

***58.** Show that $x^2 + x + 1$ cannot be factored as $(x + c)(x + d)$, where c and d are real numbers, even if c and d are not required to be integers. (*Hint:* Show that if $x^2 + x + 1 = (x + c)(x + d)$, then $c + d = 1$ and $cd = 1$. Then show from these equations that $0 < c < 1$ and $0 < d < 1$. But then $cd < 1$, which contradicts the equation $cd = 1$.)

0.7 RATIONAL EXPRESSIONS

In Section 0.5 we added, subtracted, and multiplied polynomials, and in each case the result was a polynomial. When we divide one polynomial by another, the result in general is not a polynomial. Quotients of polynomials are ***rational expressions***. Some examples are

$$\frac{1}{x+1}, \quad \frac{x^2 - 5x + 4}{x - 3}, \quad \text{and} \quad \frac{xy}{3x^2 + 4y^2}$$

A rational expression bears the same relationship to a polynomial as a rational number does to an integer. As with rational numbers, the polynomial appearing on top of a rational expression is its ***numerator***, and the polynomial on the bottom is its ***denominator***.

We have already discussed substituting real numbers into polynomials. We can also substitute real numbers into rational expressions, but we must use caution, since division by 0 is undefined. For example, it is not possible to substitute the number -1 into

$$\frac{1}{x+1}$$

since the result would be $\frac{1}{0}$, which is meaningless. Similarly, since $x^3 + 3x^2 + 2x = x(x+1)(x+2)$, it is not possible to substitute 0, -1, or -2 into

$$\frac{-2x+1}{x^3+3x^2+2x}$$

In general we can substitute into a rational expression any number for which the denominator is not 0. For example, we can substitute 2 for x in the preceding rational expression and obtain

$$\frac{-2(2)+1}{2^3+3(2)^2+2(2)} = -\frac{3}{24} = -\frac{1}{8}$$

In the same way that we normally reduce a fraction to lowest terms, we can also reduce a rational expression to lowest terms by canceling factors common to numerator and denominator. In reducing a rational expression we first identify factors common to numerator and denominator and then use the law

$$\frac{ac}{bc} = \frac{a}{b} \tag{1}$$

to cancel the common factors.

EXAMPLE 1. Reduce the following rational expressions to lowest terms.

a. $\dfrac{x^2-4}{x-2}$ b. $\dfrac{x^4+3x^3+3x^2+x}{(x+1)(x^2-1)}$

Solution.

a. By (6) in Section 0.6,

$$x^2 - 4 = (x+2)(x-2)$$

Therefore we use (1) to cancel, obtaining

$$\frac{x^2-4}{x-2} = \frac{(x+2)(x-2)}{x-2} = x+2$$

b. We observe that

$$x^4 + 3x^3 + 3x^2 + x = x(x+1)^3$$

and

$$(x+1)(x^2-1) = (x+1)(x+1)(x-1) = (x+1)^2(x-1)$$

Consequently by (1),

$$\frac{x^4 + 3x^3 + 3x^2 + x}{(x+1)(x^2-1)} = \frac{x(x+1)^3}{(x+1)^2(x-1)} = \frac{x(x+1)}{x-1} \qquad \square$$

Combinations of Rational Expressions

All the rules for adding, subtracting, multiplying, and dividing rational numbers hold for rational expressions. Multiplying and dividing rational expressions are usually less complicated than adding and subtracting them, because we do not need to employ common denominators. For products and quotients of rational expressions we apply the formulas

$$\boxed{\frac{a}{b} \cdot \frac{c}{d} = \frac{ac}{bd}} \tag{2}$$

and

$$\boxed{\frac{a/b}{c/d} = \frac{a}{b} \div \frac{c}{d} = \frac{ad}{bc}} \tag{3}$$

respectively, and cancel all common factors to reduce the expression to lowest terms.

EXAMPLE 2. Find the quotient $\dfrac{x+1}{x-1} \div \dfrac{x^2+2x+1}{x^3-1}$ and reduce it to lowest terms.

Solution. First we notice that

$$x^2 + 2x + 1 = (x+1)^2 \quad \text{and} \quad x^3 - 1 = (x-1)(x^2+x+1)$$

Using (3), we find that

$$\begin{aligned}
\frac{x+1}{x-1} \div \frac{x^2+2x+1}{x^3-1} &\overset{(3)}{=} \frac{x+1}{x-1} \cdot \frac{x^3-1}{x^2+2x+1} \\
&\overset{(2)}{=} \frac{(x+1)(x^3-1)}{(x-1)(x^2+2x+1)} \\
&= \frac{(x+1)(x-1)(x^2+x+1)}{(x-1)(x+1)^2}
\end{aligned}$$

Finally, we cancel common factors in the last expression, obtaining

$$\frac{x+1}{x-1} \div \frac{x^2+2x+1}{x^3-1} = \frac{x^2+x+1}{x+1} \qquad \square$$

When we add or subtract rational expressions whose denominators have no nontrivial common factors, we first use the rule

$$\boxed{\frac{a}{b} + \frac{c}{d} = \frac{ad+bc}{bd}} \tag{4}$$

or

$$\boxed{\frac{a}{b} - \frac{c}{d} = \frac{ad-bc}{bd}} \tag{5}$$

and then simplify if possible.

EXAMPLE 3. Find the sum $\dfrac{x+1}{x-2} + \dfrac{x+2}{x+3}$, and simplify.

Solution. Using (4), we have

$$\begin{aligned}\frac{x+1}{x-2} + \frac{x+2}{x+3} &= \frac{(x+1)(x+3)+(x-2)(x+2)}{(x-2)(x+3)} \\ &= \frac{(x^2+4x+3)+(x^2-4)}{(x-2)(x+3)} \\ &= \frac{2x^2+4x-1}{(x-2)(x+3)} \qquad \square\end{aligned}$$

When adding rational expressions with common factors in the denominators, we may simplify the algebraic computations by using a ***least common denominator*** of the denominators in the original rational expressions—that is, a polynomial of lowest degree that is a multiple of the denominators of both rational expressions. (This is reminiscent of the procedure applied when we added fractions by utilizing the least common denominator of two numbers in Section 0.1.)

EXAMPLE 4. Express $\dfrac{1}{x^2+5x+4} - \dfrac{1}{x^2+8x+16}$ as a rational expression in lowest terms.

Solution. Since

$$x^2 + 5x + 4 = (x + 4)(x + 1)$$

and

$$x^2 + 8x + 16 = (x + 4)^2$$

a least common denominator is $(x + 4)^2(x + 1)$. Consequently

$$\begin{aligned}\frac{1}{x^2 + 5x + 4} - \frac{1}{x^2 + 8x + 16} &= \frac{1}{(x + 4)(x + 1)} - \frac{1}{(x + 4)^2}\\ &= \frac{x + 4}{(x + 4)^2(x + 1)} - \frac{x + 1}{(x + 4)^2(x + 1)}\\ &= \frac{(x + 4) - (x + 1)}{(x + 4)^2(x + 1)}\\ &= \frac{3}{(x + 4)^2(x + 1)} \quad \square\end{aligned}$$

Similar techniques apply if there is more than one variable.

EXAMPLE 5. Express $\dfrac{1}{x^2y} + \dfrac{1}{xy^2}$ as a rational expression in lowest terms.

Solution. The common denominator we use is x^2y^2. As a result,

$$\begin{aligned}\frac{1}{x^2y} + \frac{1}{xy^2} &= \frac{y}{x^2y^2} + \frac{x}{x^2y^2}\\ &= \frac{y + x}{x^2y^2} \quad \square\end{aligned}$$

Combinations of expressions that are not necessarily polynomials but are nevertheless obtained by adding, subtracting, multiplying, dividing, or taking roots of polynomials are called ***algebraic expressions***. The same general rules apply when combining algebraic expressions. In the following example, we will reduce the numerator and the denominator separately and then combine, noting that

$$\boxed{\frac{a/b}{c/d} = \frac{a}{b} \div \frac{c}{d} = \frac{a}{b} \cdot \frac{d}{c}} \qquad (6)$$

EXAMPLE 6. Simplify $\dfrac{1 - \dfrac{1}{x+1}}{1 + \dfrac{1}{x-1}}$.

Solution. First we rewrite the numerator:

$$1 - \frac{1}{x+1} = \frac{x+1}{x+1} - \frac{1}{x+1} = \frac{(x+1)-1}{x+1} = \frac{x}{x+1}$$

The denominator is altered similarly:

$$1 + \frac{1}{x-1} = \frac{x-1}{x-1} + \frac{1}{x-1} = \frac{(x-1)+1}{x-1} = \frac{x}{x-1}$$

Therefore by (6),

$$\frac{1 - \dfrac{1}{x+1}}{1 + \dfrac{1}{x-1}} = \frac{\dfrac{x}{x+1}}{\dfrac{x}{x-1}} = \frac{x}{x+1} \cdot \frac{x-1}{x} = \frac{x-1}{x+1} \qquad \square$$

Rationalizing the Denominator

If a fraction such as $1/\sqrt{a}$ contains one or more radicals, it is common to alter it so that the denominator contains no radicals. The procedure, called ***rationalizing the denominator***, involves multiplying the fraction by 1 written in a special way. For $1/\sqrt{a}$ we would multiply by 1 written as $\sqrt{a}/\sqrt{a}$, which yields

$$\frac{1}{\sqrt{a}} = \frac{1}{\sqrt{a}} \cdot \frac{\sqrt{a}}{\sqrt{a}} = \frac{\sqrt{a}}{\sqrt{a}\sqrt{a}} = \frac{\sqrt{a}}{a}$$

The denominator of the last fraction contains no radicals. For example,

$$\frac{3}{\sqrt{2}} = \frac{3}{\sqrt{2}} \cdot \frac{\sqrt{2}}{\sqrt{2}} = \frac{3\sqrt{2}}{2} = \frac{3}{2}\sqrt{2}$$

If the denominator of an algebraic expression contains radicals such as $\sqrt{x} + \sqrt{b}$, we would multiply by

$$\frac{\sqrt{x} - \sqrt{b}}{\sqrt{x} - \sqrt{b}}$$

EXAMPLE 7. Simplify the fraction

$$\frac{\sqrt{x}-\sqrt{b}}{\sqrt{x}+\sqrt{b}}$$

Solution. We multiply the fraction by $(\sqrt{x}-\sqrt{b})/(\sqrt{x}-\sqrt{b})$, which yields

$$\frac{\sqrt{x}-\sqrt{b}}{\sqrt{x}+\sqrt{b}} = \frac{\sqrt{x}-\sqrt{b}}{\sqrt{x}+\sqrt{b}} \cdot \frac{\sqrt{x}-\sqrt{b}}{\sqrt{x}-\sqrt{b}}$$

$$= \frac{x - 2\sqrt{x}\sqrt{b} + b}{x - b} = \frac{x - 2\sqrt{xb} + b}{x - b} \qquad \square$$

EXERCISES 0.7

In Exercises 1–8, reduce the rational expression to lowest terms.

1. $\dfrac{x+1}{x^2+5x+4}$
2. $\dfrac{x^2+3x-28}{x^2-3x-4}$
3. $\dfrac{9y^2-4}{3y^2-3y+\frac{2}{3}}$
4. $\dfrac{2b^2+b-3}{8b^2+2b-15}$
5. $\dfrac{s^4-13s^2+36}{s^2-7s+12}$
6. $\dfrac{s^4-5s^2-36}{s^2-7s+12}$
7. $\dfrac{x^2-y^2}{x^3-y^3}$
8. $\dfrac{x^2+y^2}{x^4-y^4}$

In Exercises 9–16, write as one rational expression in simplified form.

9. $\dfrac{x^2-1}{x^2+x-2} \cdot (x^2-4)$
10. $\dfrac{x^2-16}{(x^2+3x-28)^2} \, (x^2-49)$
11. $\dfrac{y^2}{y+3} \cdot \dfrac{y^2+y-6}{y}$
12. $\dfrac{y+3}{y-2} \cdot \dfrac{y^2-4}{y+3}$
13. $\dfrac{z^2-9}{z^2-4} \cdot \dfrac{z^2+6z+9}{z^2-4z+3}$
14. $\dfrac{a^2+5a-50}{a^2-1} \div \dfrac{a^2-7a+10}{a^2+6a+5}$
15. $\dfrac{x^2-2xy+y^2}{x^2+xy+y^2} \div \dfrac{x^2-y^2}{x^3-y^3}$
16. $\dfrac{x^2+2xy+y^2}{x^2+xy+y^2} \div \dfrac{x^2+y^2}{x^3-y^3}$

In Exercises 17–34, write as one rational expression in simplified form.

17. $\dfrac{1}{x} + \dfrac{2}{x-1}$
18. $\dfrac{x}{x-3} + \dfrac{4}{x+2}$
19. $\dfrac{5}{y^2-9} + \dfrac{3}{y+3}$
20. $\dfrac{y}{y^2+5y-24} + \dfrac{1}{3-y}$
21. $\dfrac{z+1}{z-1} + \dfrac{z-1}{z+1}$
22. $\dfrac{t^2+1}{t^2-1} + \dfrac{1}{1-t}$
23. $\dfrac{4}{t-1} + \dfrac{t+7}{t^2-4t+3}$
24. $\dfrac{v-1}{v+1} - \dfrac{1-7v}{v^2-2v-3}$
25. $\dfrac{v+5}{v-4} - \dfrac{12v+6}{v^2-2v-8}$
26. $\dfrac{a}{a-b} + \dfrac{b}{b-a}$

27. $1 - \dfrac{a^2}{a^2 + b^2}$

28. $\dfrac{1}{p - q} + \dfrac{1}{p + q} - \dfrac{1}{p^2 - q^2}$

29. $\dfrac{1}{p} + \dfrac{1}{q} + \dfrac{1}{r}$

30. $\dfrac{q}{p} + \dfrac{r}{q} + \dfrac{p}{r}$

31. $\dfrac{1}{h}\left(\dfrac{1}{x + h} - \dfrac{1}{x}\right)$

32. $\dfrac{1}{h}\left[\dfrac{1}{(x + h)^2} - \dfrac{1}{x^2}\right]$

33. $\left(\dfrac{1}{y} - \dfrac{1}{x}\right) \div \left(\dfrac{1}{y} + \dfrac{1}{x}\right)$

34. $\left(\dfrac{x}{y} + \dfrac{y}{x}\right) \div \dfrac{x^2 + y^2}{xy}$

In Exercises 35–42, simplify the expression.

35. $\dfrac{\dfrac{1}{x} + x}{\dfrac{2}{x} + 1}$

36. $\dfrac{\dfrac{1}{x} - 1}{x^2 + 2}$

37. $\dfrac{x^2 + \dfrac{4}{x}}{x + \dfrac{4}{x^2}}$

38. $\dfrac{\dfrac{x + 4}{x} - 1}{3 - \dfrac{4 - x}{x}}$

39. $\dfrac{\dfrac{x - y}{x}}{\dfrac{x^2 - y^2}{xy}}$

40. $\dfrac{\dfrac{x}{y} - \dfrac{y}{x}}{\dfrac{1}{x} + \dfrac{1}{y}}$

41. $\dfrac{\dfrac{x^2}{y} - \dfrac{y^2}{x}}{\dfrac{1}{x} + \dfrac{1}{y}}$

42. $\dfrac{1}{\sqrt{x}} - \dfrac{1}{\sqrt{y}}$

In Exercises 43–48, rationalize the denominator.

43. $\dfrac{1}{\sqrt{3}}$

44. $\dfrac{\sqrt{6}}{3 + \sqrt{6}}$

45. $\dfrac{1}{\sqrt{x} - 9}$

46. $\dfrac{1}{\sqrt{x} + 6}$

47. $\dfrac{1}{\sqrt{x} - \sqrt{y}}$

48. $\dfrac{\sqrt{x} - \sqrt{y}}{\sqrt{x} + \sqrt{y}}$

Occasionally one needs to "rationalize the numerator" of an expression. If the numerator has the form $\sqrt{a} - \sqrt{b}$, this can be accomplished by multiplying the expression by 1 in the form

$$\frac{\sqrt{a} + \sqrt{b}}{\sqrt{a} + \sqrt{b}}$$

in order to remove all radicals from the numerator.

In Exercises 49–52, rationalize the numerator.

49. $\dfrac{\sqrt{x + 3} - \sqrt{3}}{x}$

50. $\dfrac{1 - \sqrt{1 - x^2}}{x^2}$

51. $\dfrac{\sqrt{x} - \sqrt{y}}{\sqrt{x} + \sqrt{y}}$

52. $\dfrac{x - \sqrt{x^2 - a^2}}{x + \sqrt{x^2 - a^2}}$

53. If p denotes the object distance and q the image distance of a simple lens, then the focal length is

$$\frac{1}{\dfrac{1}{p} + \dfrac{1}{q}}$$

Simplify this expression.

54. If three resistors having resistances R_1, R_2, and R_3 are connected in parallel, the resistance of the combination is

$$\frac{1}{\dfrac{1}{R_1} + \dfrac{1}{R_2} + \dfrac{1}{R_3}}$$

Simplify this expression.

KEY TERMS

real number
- natural number
- integer
- rational number
- irrational number

factor

real line
- origin
- positive direction
- coordinate

distance

absolute value

base

exponent

scientific notation

square root

radical

cube root

*n*th root

polynomial
- monomial
- coefficient
- binomial
- leading coefficient
- degree
- constant term
- polynomial in several variables

rational expression
- numerator
- denominator
- rational expression in lowest terms
- least common denominator
- rationalizing the denominator

algebraic expression

KEY FORMULAS

$$(a + b)(c + d) = ac + bc + ad + bd$$

$$(a + b)^2 = a^2 + 2ab + b^2$$

$$(a - b)^2 = a^2 - 2ab + b^2$$

$$\frac{a}{b} + \frac{c}{d} = \frac{ad + bc}{bd}$$

$$\frac{a}{b} \cdot \frac{c}{d} = \frac{ac}{bd}$$

$$\frac{a}{b} \div \frac{c}{d} = \frac{ad}{bc}$$

$$x^2 - a^2 = (x + a)(x - a)$$

$$x^3 - a^3 = (x - a)(x^2 + ax + a^2)$$

$$x^3 + a^3 = (x + a)(x^2 - ax + a^2)$$

$$x^n - a^n = (x - a)(x^{n-1} + ax^{n-2} + a^2x^{n-3} + \cdots + a^{n-2}x + a^{n-1})$$

$$x^n + a^n = (x + a)(x^{n-1} - ax^{n-2} + a^2x^{n-3} - \cdots - a^{n-2}x + a^{n-1}) \ (n \text{ odd})$$

$$a^r a^s = a^{r+s}$$

$$(a^r)^s = a^{rs} = (a^s)^r$$

$$(ab)^r = a^r b^r$$

$$a^{-r} = \frac{1}{a^r}$$

$$|a| = \begin{cases} a \text{ if } a \geq 0 \\ -a \text{ if } a < 0 \end{cases}$$

REVIEW EXERCISES

In Exercises 1–8, calculate the value of the given expression.

1. $9^{-3/2}$

2. $\left(-\dfrac{1}{8}\right)^{5/3}$

3. $\dfrac{2^{-3}}{2^{-7}}$

4. $(\sqrt[3]{-3})^6$

5. $\sqrt{0.09}$

6. $3^4 \cdot 3^{-2}$

7. $\sqrt[3]{\dfrac{-8}{27}}$

8. $\sqrt[5]{\dfrac{1}{32}}$

In Exercises 9–10, write out the given statement using the symbols for inequalities.

9. $a - \sqrt{5}$ is nonnegative.

10. 0.2 is greater than $|3x + 1|$.

In Exercises 11–12, find the distance between a and b.

11. $a = 2.7, b = -1.6$

12. $a = -\frac{4}{5}, b = -\frac{2}{3}$

In Exercises 13–14, write the given number in scientific notation.

13. 159,000

14. 0.00314

In Exercises 15–24, simplify the given expression.

15. $\left(\dfrac{1}{2}a^{-2}\right)^3 a^4$

16. $(\sqrt{3} + \sqrt{2})^2$

17. $\sqrt{2} + \sqrt{50}$

18. $\dfrac{1}{2}(\sqrt{a+b} + \sqrt{a-b})^2$

19. $\dfrac{(15a^3)^2}{(15a^2)^3}$

20. $(\sqrt[3]{2x})^9$

21. $(x^5y^{-2}z^6)^{3/2}$

22. $\dfrac{(a+b)^2 - (a-b)^2}{(a+b)^2 + (a-b)^2}$

23. $\dfrac{a - \dfrac{a^2}{a+b}}{b - \dfrac{b^2}{a+b}}$

24. $\dfrac{\dfrac{x}{x+y} + \dfrac{y}{x-y}}{x^2 + y^2}$

In Exercises 25–28, approximate the value of the given expression.

25. $(.0143)^3(.00012)^5$

26. $\sqrt{\dfrac{2.03}{9.34 - 5.21}}$

27. $\sqrt[5]{(7.69 \times 10^{-2})^4}$

28. $\dfrac{(0.1)^{2/3}}{(0.2)^{2/3} - (0.3)^{2/3}}$

In Exercises 29–34, perform the indicated operation and simplify.

29. $(\frac{1}{2}a - \frac{2}{3})^2$

30. $(3x + 4)^2 - (3x - 5)^2$

31. $(x^{1/3} - x^{-1/3})^3$

32. $(y/x - 2x/y)^2$

33. $(2x^2 + 5y)^3$

34. $(x^2 + y^2 - 2z^2)^2$

In Exercises 35–36, assume that $a > 1$, and determine whether the expression is positive or negative.

35. $a - \sqrt{a^2 + 4}$

36. $a - \sqrt{a^2 - 1}$

In Exercises 37–52, factor the given polynomial.

37. $x^2 - 6x - 27$

38. $y^2 + 8y - 105$

39. $t^2 - 20t + 100$

40. $4u^2 + 4u + 1$

41. $12x^2 - 11x + 2$

42. $x^3 - 2x^2 + 2x - 4$

43. $(y - 1)^5 - 4(y - 1)^3$

44. $x^3 - x^2 - 90x$

45. $z^4 + z^2 - 2$

46. $z^6 - z^4$

47. $x^3 + x^2 + x + 1$

48. $x^3y + xy^3$

49. $x^2 + 2xy - 35y^2$

50. $9x^4 - 6x^2y^2 + y^4$

51. $16x^4 - 1$

52. $x^8 - 256$

In Exercises 53–56, find the value of the given expression at the given number.

53. $4x^3 - 2x^2 + 3x - 19$; 1

54. $\dfrac{3x - 5}{2x + 1}$; 2

55. $\dfrac{3x^2 + 1}{-4x^3 - 11}$; -3

56. $\dfrac{x^2}{x^6 - 2}$; $\sqrt{2}$

In Exercises 57–70, write as one rational expression in simplified form.

57. $\dfrac{x^2 - 6x + 5}{x^2 - 3x - 10}$

58. $\dfrac{x^2 + 2xy + y^2}{x^3 - x^2y - y^2x + y^3}$

59. $\dfrac{4x^2 - 1}{x^2 + x - 12} \cdot \dfrac{x^2 - 5x + 6}{6x^2 + x - 1}$

60. $\left(\dfrac{1}{x^2} - \dfrac{1}{x^3}\right)\left(\dfrac{1}{x} + \dfrac{1}{x^2}\right)$

61. $\dfrac{x^2 - 4}{x^2 - x - 6} \div \dfrac{x^2 + 2x - 8}{x^2 + x - 12}$

62. $\dfrac{3}{2x - 1} + \dfrac{2}{3x - 2}$

63. $\dfrac{x}{(x - 1)^2} - \dfrac{1}{x^2 - 1}$

64. $\dfrac{3x - 4}{x^2 + x} - \dfrac{5x - 2}{x^2 - x}$

65. $\dfrac{1 - 4x^{-2}}{1 - 3x^{-1} + 2x^{-2}}$

66. $1 - \dfrac{x}{x - y}$

67. $\dfrac{x}{x - y} - \dfrac{y}{x + y}$

68. $\dfrac{\dfrac{2}{x} + \dfrac{3}{y}}{\dfrac{1}{x} - \dfrac{2}{y}}$

69. $\dfrac{\dfrac{1}{x} - \dfrac{1}{x^2}}{\dfrac{1}{x} + \dfrac{1}{x^2}}$

70. $\dfrac{\sqrt{x}}{\sqrt{x} - \sqrt{y}} + \dfrac{\sqrt{y}}{\sqrt{x} + \sqrt{y}}$

In Exercises 71–72, simplify the given expression by rationalizing the denominator.

71. $\dfrac{1}{\sqrt{5} - 2}$

72. $\dfrac{\sqrt{2a} - \sqrt{8b}}{\sqrt{2a} + \sqrt{8b}}$

In Exercises 73–75, write the statement as an equation.

73. The surface area S of a rectangular box with height h and square base of side s is equal to the sum of twice the square of s and four times the product of s and h.

74. The surface area S of a cylindrical can (including top and bottom) of radius r and height h is the sum of 2π times the radius squared and 2π times the product of the radius and height.

75. The area A of an equilateral triangle is the product of $\sqrt{3}/4$ and the square of the length s of a side.

76. Use the formula in Exercise 74 to approximate the surface area of a cylindrical can (including top and bottom) with radius 1.25 and height 4.37. Give your answer in decimal form, rounding off to 4 significant digits.

77. Use the formula in Exercise 75 to approximate the area of an equilateral triangle with side length $\frac{1}{3}$. Give your answer in decimal form, rounding off to 4 significant digits.

78. Observe that

$$\frac{1}{2} = \frac{1}{3} + \frac{1}{2 \cdot 3} \quad \text{and} \quad \frac{1}{3} = \frac{1}{4} + \frac{1}{3 \cdot 4}$$

These equations are examples of a more general equation involving positive integers n and $n + 1$.

a. Write down such an equation.

b. Prove your equation is valid.

79. What are the possible values of $\dfrac{a - b}{|a - b|}$?

80. Prove that

$$\left(\frac{x - y}{2}\right)^2 = \left(\frac{x + y}{2}\right)^2 - xy$$

81. A positive integer is called a ***perfect number*** if it is the sum of its positive divisors less than itself. Show that the following are perfect numbers.

a. 6 b. 28 c. 496

82. Let a and b be real numbers with $b \neq -1$, and let

$$x = \frac{a^2}{1 + b^3}, \quad y = bx, \quad \text{and} \quad z = ax$$

Show that $x^3 + y^3 = z^2$.

83. Let u and v be positive numbers with $u > v$, and let

$$a = 2uv, \quad b = u^2 - v^2, \quad \text{and} \quad c = u^2 + v^2$$

The numbers $a, b,$ and c are known as a ***Pythagorean triple***.

a. Show that $c^2 = a^2 + b^2$. (This is the reason for the name of the triple; after all, it follows that there is a right triangle whose sides have lengths a, b and c.)

b. Find the Pythagorean triples arising from the choices

i. $u = 2, v = 1$ ii. $u = 3, v = 1$

iii. $u = 3, v = 2$

1 EQUATIONS AND INEQUALITIES

Consider the following two problems:

1. Suppose a swimmer dives from a platform 64 feet above a deep pool. If we disregard air resistance, how long would it take the swimmer to reach the pool?
2. A human fever is generally regarded to be an oral temperature exceeding 98.6 degrees Fahrenheit. What temperatures in degrees Celsius correspond to a fever?

The answers to these problems are not obvious. In this chapter we use equations and inequalities to solve such problems. (See Example 1 in Section 1.3 and Example 6 in Section 1.6.)

Equations arise by setting two given algebraic expressions equal to one another; inequalities arise by making one given algebraic expression less than another. As you see, equations and inequalities are close relatives of one another. They are also extremely important in mathematics. In fact, it could almost be said that higher mathematics revolves about the study of equations and inequalities. Moreover, there are many applications of equations and inequalities outside mathematics. These applications range from the solutions of simple, everyday problems to the formulation of the deepest principles of modern science. In this chapter we will illustrate some of the simpler applications of equations and inequalities.

1.1 LINEAR EQUATIONS

Before we study linear equations themselves, we need some terminology concerning equations in general. First of all, an ***equation*** is a statement that two expressions are equal. In Chapter 0 we encountered equations such as

$$x^2 + 6x + 9 = (x + 3)^2 \quad \text{and} \quad \frac{x^2 - 4}{x - 2} = x + 2 \tag{1}$$

and in the present chapter we will see equations such as

$$2x + 7 = 5x - 3 \quad \text{and} \quad 3x^2 - 4x + 1 = 0 \tag{2}$$

However, the role the variable x plays in (1) is different from its role in (2). Indeed, the first equation in (1) is valid for all values of x, and the second equation in (1) is valid for all values of x except 2 (for which the expression on the left side is undefined). In contrast, the equations in (2) are valid only for very special values of x. In the course of this chapter we will show that the equation $2x + 7 = 5x - 3$ is valid only for $x = \frac{10}{3}$ and that the equation $3x^2 - 4x + 1 = 0$ is valid only for $x = 1$ or $x = \frac{1}{3}$.

We call an equation an ***identity*** if it is valid for all values of the variable that make each expression in the equation meaningful. Thus the equations in (1) are identities. In contrast, an equation is called ***conditional*** if it is valid only for special values of the variable. Both equations in (2) are conditional.

Two conditional equations are said to be ***equivalent*** if they are valid for precisely the same values of the variable. For example, the following pairs of equations are equivalent:

$x^2 - 6x = 0$ and $t^2 - 6t = 0$ (Only the variable letters are different.)

$\frac{9}{5}x + 32 = 68$ and $\frac{9}{5}x = 36$ (The same constant is added to both sides of one equation to obtain the other equation.)

$3x = 17$ and $x = \frac{17}{3}$ (Both sides of one equation are multiplied by the same constant to obtain the other equation.)

The equations of main interest in this chapter will be conditional equations. Because conditional equations are valid only for special values of the variable, our goal will be to determine those special values of the variable, called ***solutions*** (or sometimes ***roots***) of the equation. In this terminology, 1 and $\frac{1}{3}$ are solutions of $3x^2 - 4x + 1 = 0$, because if we substitute 1 or $\frac{1}{3}$ for x in the equation, the resulting equation is valid. The process of finding all the solutions of a given equation is called ***solving the equation***. Generally speaking, when we set out to solve a given equation whose solutions are not obvious, we will try to obtain an equivalent equation that is more easily solved.

Linear Equations

The simplest kind of conditional equation is a ***linear equation***, that is, an equation that is equivalent to either

$$\boxed{ax = b \quad \text{or} \quad ax + b = 0}$$

where a and b are fixed real numbers and $a \neq 0$. The following are examples of linear equations:

$$x = 3, \qquad 5x = 2, \qquad \pi x - \frac{7}{2} = 0, \quad \text{and} \quad 2x + 7 = 5x - 3$$

The general procedure for solving a linear equation is to

i. Put all terms containing x (or the given variable) on one side of the equation.
ii. Put all other terms on the other side of the equation.
iii. Simplify the resulting equation to solve for x.

This procedure yields a unique solution of the given equation.

EXAMPLE 1. Solve the equation $2x + 7 = 5x - 3$ for x.

Solution. In order to have all terms containing x on the left side and all other terms on the right, we first subtract $5x$ from both sides:

$$\begin{aligned} 2x + 7 &= 5x - 3 \\ (2x + 7) - 5x &= -3 \\ -3x + 7 &= -3 \end{aligned}$$

Then we subtract 7 from both sides, and finally solve the resulting equation:

$$\begin{aligned} -3x &= -3 - 7 = -10 \\ x &= \frac{-10}{-3} = \frac{10}{3} \end{aligned}$$

$$\textit{Check:} \quad 2\left(\frac{10}{3}\right) + 7 = \frac{20}{3} + 7 = \frac{41}{3} \text{ and } 5\left(\frac{10}{3}\right) - 3 = \frac{50}{3} - 3 = \frac{41}{3}$$

Therefore the solution is $\frac{10}{3}$. □

CAUTION: It is always possible to make mistakes. For that reason it is a good idea to check the proposed solution by substituting it into the original equation, as we have done in Example 1 and will continue to do.

Linear equations arise in many applications. For example, when we convert the temperature F in degrees Fahrenheit to the corresponding temperature C in degrees Celsius, the temperatures F and C are related by the equation

$$\frac{9}{5}C + 32 = F \tag{3}$$

Suppose we wish to determine the number of degrees Celsius corresponding to 68 degrees Fahrenheit. This means that we wish to solve the linear equation

$$\frac{9}{5}C + 32 = 68 \tag{4}$$

for C. According to (i) and (ii), we would subtract 32 from each side of (3) to obtain

$$\frac{9}{5}C = 68 - 32 = 36$$

Dividing each side by $9/5$ (or, equivalently, multiplying by $5/9$), yields

$$\frac{5}{9}\left(\frac{9}{5}C\right) = \frac{5}{9}(36)$$

and thus

$$C = 20$$

We should check to see that our result is correct: $\frac{9}{5}(20) + 32 = 36 + 32 = 68$. Our conclusion is that 20 degrees Celsius corresponds to 68 degrees Fahrenheit.

The next example involves an equation that does not look at all like a linear equation. Yet by appropriate manipulation it turns out to be equivalent to a linear equation, which can be solved by the process outlined in (i)–(iii).

EXAMPLE 2. Solve $\dfrac{x+2}{x-1} = 2 - \dfrac{x-1}{x+3}$ for x.

Solution. First we multiply both sides by $(x-1)(x+3)$ in order to remove the terms $x-1$ and $x+3$ from the denominators. Then we manipulate the resulting equation in order to obtain the solution:

$$\frac{x+2}{x-1} = 2 - \frac{x-1}{x+3}$$

$$\left(\frac{x+2}{x-1}\right)(x-1)(x+3) = 2(x-1)(x+3) - \left(\frac{x-1}{x+3}\right)(x-1)(x+3)$$

$$(x+2)(x+3) = 2(x-1)(x+3) - (x-1)^2$$

$$x^2 + 5x + 6 = 2(x^2 + 2x - 3) - (x^2 - 2x + 1)$$

$$x^2 + 5x + 6 = x^2 + 6x - 7$$

$$-x = -13$$

$$x = 13$$

Check: $\dfrac{13+2}{13-1} = \dfrac{15}{12} = \dfrac{5}{4}$ and $2 - \dfrac{13-1}{13+3} = 2 - \dfrac{12}{16} = \dfrac{5}{4}$

Thus 13 is the solution. □

Equations Containing Absolute Values

Consider the equation $|x-3| = 2$. If the absolute value were not present, the equation would be the linear equation $x - 3 = 2$, which we know how to solve. Nevertheless, the equation $|x-3| = 2$ is not linear, so we need a new method to solve it. Since $|x-3| = x-3$ or $|x-3| = -(x-3)$, depending on whether $x - 3 \geq 0$ or $x - 3 < 0$, it follows that in order to solve the equation $|x-3| = 2$ we need to consider two cases: $|x-3| = x-3$ and $|x-3| = -(x-3)$. Each of these cases yields a linear equation, which we can solve by the method already discussed.

EXAMPLE 3. Solve the equation $|x-3| = 2$ for x.

Solution. As we suggested above, we consider the following two cases:

a. $$|x-3| = x-3$$

which yields $x - 3 = 2$, and thus $x = 5$.

b. $$|x-3| = -(x-3)$$

which yields $-(x-3) = 2$, or equivalently, $-x + 3 = 2$. Therefore $-x = -1$, so that $x = 1$.

Check: $|5-3| = |2| = 2$ and $|1-3| = |-2| = 2$

Thus 5 and 1 are the solutions of the given equation. □

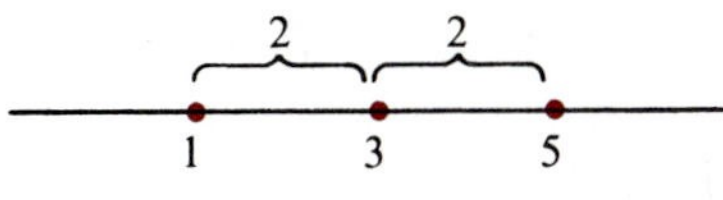

FIGURE 1.1

For a more geometric solution of Example 3, recall that $|a-b|$ is the distance between a and b. Therefore solving the equation $|x-3| = 2$ is equivalent to finding the values of x for which the distance between x and 3 equals 2. We conclude that $x = 5$ and $x = 1$ are the solutions (Figure 1.1).

Equations with Two Variables

Consider the equation

$$2x + 5 = 3y - 7 \tag{5}$$

This equation contains two variables, x and y, and thus is quite different from the other equations we have studied in this section. But suppose we assign a numerical value, say 6, to y. Then (5) becomes

$$2x + 5 = 3(6) - 7$$

which reduces to

$$2x + 5 = 11$$

itself a linear equation in x that can be solved by the methods of this section. In fact, no matter what numerical value we assign to y, we obtain a linear equation in x that can be solved. But rather than solve every such equation individually, we can treat y as an unspecified constant and solve for x. We use this idea as we solve the equation in (5).

EXAMPLE 4. Solve the equation

$$2x + 5 = 3y - 7$$

for x.

Solution. We treat y as a constant and solve the equation for x:

$$2x + 5 = 3y - 7$$
$$2x = (3y - 7) - 5 = 3y - 12$$
$$x = \frac{1}{2}(3y - 12) = \frac{3}{2}y - 6$$

Check: $2(\frac{3}{2}y - 6) + 5 = 3y - 12 + 5 = 3y - 7$

Thus $\frac{3}{2}y - 6$ is the solution of the given equation. □

Applications of Linear Equations

One of the reasons algebra is so important is that very often it can be used to help solve problems in such disciplines as physics, engineering, economics, and geometry. Such problems are frequently formulated verbally. For that reason they are sometimes called ***word problems***, but we prefer to call them ***applied problems***. Although you may think that some of the problems we will present are rather contrived, they are part of mathematical folklore, and serve to illustrate the fact that mathematics can indeed be applied in everyday life.

To use the mathematics we have developed in solving an applied problem, we will translate the problem into mathematical language and then solve the resulting mathematical problem, which in this section will always be a linear equation. Because of the extreme breadth of applied problems, there is no hope of prescribing a single detailed procedure that works for all of them. After discussing and solving various applied problems, we will give a synopsis of our general method of solving such problems. It should serve as a guideline to help you in solving other applied problems.

For the first application, suppose that one invests a sum of money P (called the ***principal***) in a savings account that draws simple interest at an annual rate r. This means that after t years the interest I earned by the account is given by the formula

$$\boxed{I = Prt} \tag{6}$$

Four letters appear in (6). If three of them are assigned specific numerical values, then we can solve for the fourth. Normally the interest rate is expressed as a percentage. Thus if the interest rate is 7%, then $r = 0.07$. As an example of the use of (6), suppose that \$2000 is invested in a savings account which draws simple interest at an annual rate of 7%. Then after 1 year the interest I earned is given by

$$I = (2000)(0.07)(1) = 140$$

and after 5 years the interest earned is given by

$$I = (2000)(0.07)(5) = 700$$

EXAMPLE 5. (Simple Interest Problem) Suppose that \$16,000 is deposited in a bank in the following way: \$10,000 into a U.S. Treasury note earning 9% simple interest annually, and \$6000 into a passbook account earning 5% simple interest annually. How long will it take to earn \$12,000 in interest?

Solution. Let

$$t = \text{the number of years required to earn \$12,000 in interest}$$

After t years the amount of interest in dollars earned from the Treasury note will be $(10{,}000)(0.09)(t)$, and the amount earned from the passbook account will be $(6000)(0.05)(t)$. Since the total amount of interest to be earned is \$12,000, we have the linear equation

$$(10{,}000)(0.09)(t) + (6000)(0.05)(t) = 12{,}000$$

This simplifies to

$$900t + 300t = 12{,}000$$
$$1200t = 12{,}000$$
$$t = 10$$

Check: After 10 years the Treasury note will have earned (10,000)(0.09)(10) dollars in interest, and the passbook account will have earned (6000)(0.05)(10) dollars in interest. Accordingly their combined earned interest will be given by

$$(10{,}000)(0.09)(10) + (6000)(0.05)(10) = 9000 + 3000 = 12{,}000$$

Therefore it will take 10 years to earn $12,000 in interest. □

In our next example we discuss a geometric problem to which algebraic techniques apply.

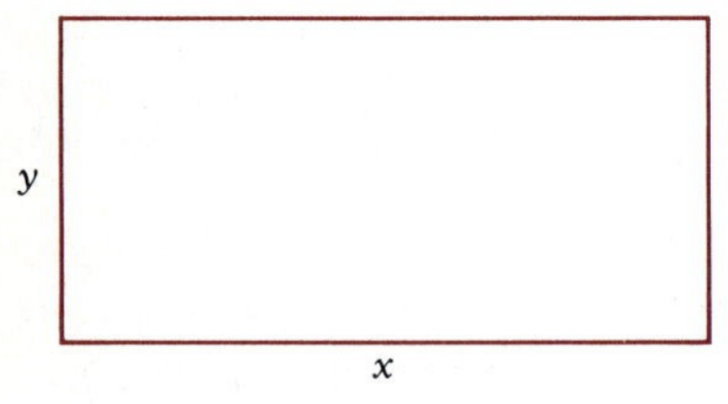

Length = 2 × Width
Perimeter = 270 feet

FIGURE 1.2

EXAMPLE 6. (Geometric Problem) A rectangular plot of land has a perimeter of 270 feet, and its length is twice its width (Figure 1.2). Find the dimensions of the plot.

Solution. Let

$$x = \text{the length in feet of the plot}$$
$$y = \text{the width in feet of the plot}$$

Since the perimeter $2x + 2y$ is 270 feet, we have the equation

$$2x + 2y = 270 \tag{7}$$

Since the length is twice the width,

$$x = 2y \tag{8}$$

When we substitute $2y$ for x in (7), we obtain the linear equation

$$2(2y) + 2y = 270$$

which reduces to

$$6y = 270$$
$$y = 45$$

Now it follows from (8) that $x = 2(45) = 90$.

$$\textit{Check:}\quad \text{perimeter} = 2x + 2y = 2(90) + 2(45) = 270$$
$$\text{length} = x = 90 = 2y = 2 \cdot \text{width}$$

Therefore the plot is 90 feet long and 45 feet wide. □

Suppose a car travels at the speed of 30 miles per hour. Then in one hour the car travels 30 miles, in two hours (30)(2) or 60 miles, in three hours (30)(3) or 90 miles, and so forth. In general the car travels $30t$ miles in t hours. Now suppose that the car (or any other object) travels at a constant speed, or rate, r. Then the distance d traveled during t units of time is given by

$$d = rt \tag{9}$$

For example, if a car moves at the rate r of 30 miles per hour, then the distance d in miles that the car travels during a time period of t hours is given by

$$d = 30t$$

so that, for instance, in $\frac{1}{2}$ hour the distance would be $d = (30)(\frac{1}{2}) = 15$ (miles).

EXAMPLE 7. (Velocity Problem) Two cars that are initially 10 miles apart travel toward each other, one at the rate of 20 miles per hour and the other at the rate of 30 miles per hour. How far must the faster car travel before the two cars meet?

Solution. Let t be the time (in hours) required for the two cars to meet. By (9), the slower car travels $20t$ miles and the faster car travels $30t$ miles (Figure 1.3). Since together they travel 10 miles until they meet, we have

$$20t + 30t = 10$$
$$50t = 10$$
$$t = \frac{1}{5}$$

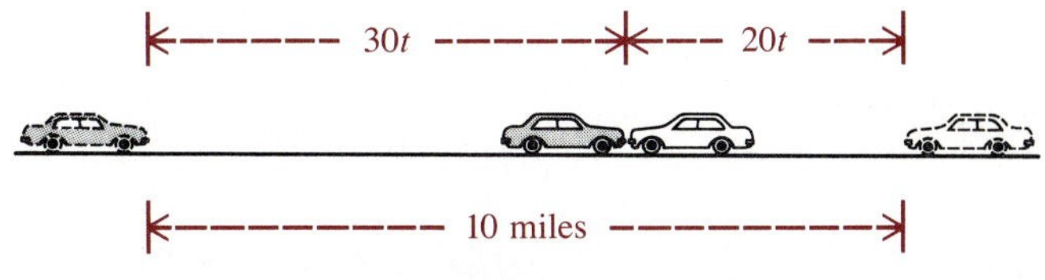

FIGURE 1.3

Because the rate $r = 30$ for the faster car, we know by (9) that the faster car travels $(30)(\frac{1}{5}) = 6$ miles before the two cars meet. □

CAUTION: In order for (9) to apply, the units in which r is measured must be compatible with the units in which d and t are measured. In the problems of this section, distance is measured in miles, time in hours, and the rate in miles per hour. Notice that if two of the three letters are given specific values, then we can solve for the third.

We turn now to a different type of rate problem.

EXAMPLE 8. (Work Problem) Suppose it takes an adult 2 hours to mow a lawn, and a child 3 hours to do the same job. How long would it take them to mow the lawn at the same time with two lawn mowers?

Solution. Let t be the length of time it takes the adult and the child to mow the lawn together. Then the part of the lawn that they can mow in one hour is $1/t$. But in one hour the adult can mow $\frac{1}{2}$ of the lawn and the child $\frac{1}{3}$ of the lawn. Therefore

$$\underbrace{\frac{1}{t}}_{\substack{\text{part mowed}\\ \text{together in one hour}}} = \underbrace{\frac{1}{2}}_{\substack{\text{part mowed by}\\ \text{adult in one hour}}} + \underbrace{\frac{1}{3}}_{\substack{\text{part mowed by}\\ \text{child in one hour}}}$$

Solving for t, we find that

$$\frac{1}{t} = \frac{5}{6}, \quad \text{or} \quad t = \frac{6}{5}$$

Check: The work done by the adult and the child in $\frac{6}{5}$ hours is given by

$$\frac{1}{2}\left(\frac{6}{5}\right) + \frac{1}{3}\left(\frac{6}{5}\right) = \frac{3}{5} + \frac{2}{5} = 1$$

Therefore it would take $\frac{6}{5}$ hours (that is, 1 hour and 12 minutes) for the adult and the child to mow the lawn together. □

Our final example is a so-called mixture problem.

EXAMPLE 9. (Mixture Problem) A radiator contains 6 quarts of fluid consisting of 40% antifreeze and 60% water (by volume). How much of the mixture should be drained off and replaced by pure antifreeze in order to obtain a mixture containing 60% antifreeze?

Solution. Since we wish to end up with a 6 quart mixture with 60% antifreeze, the amount of antifreeze at the end will have to be (0.6)6 quarts. Let us go through the process, step by step:

1. At the beginning we have 6 quarts of liquid in the radiator, with 40% antifreeze, so the initial amount of antifreeze is (0.4)6 quarts.
2. After we drain off x quarts of the liquid, there is $(0.4)x$ less antifreeze in the radiator.
3. Next we add x quarts of pure antifreeze to the radiator.

Figure 1.4 depicts the process, which yields the following equation for the amount of antifreeze in the radiator:

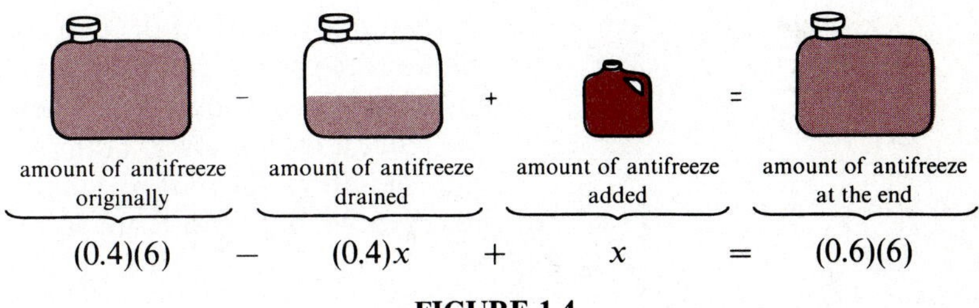

$$(0.4)(6) - (0.4)x + x = (0.6)(6)$$

FIGURE 1.4

We solve this linear equation as follows:

$$2.4 - 0.4x + x = 3.6$$

$$0.6x = 1.2$$

$$x = \frac{1.2}{0.6} = 2$$

Check: If 2 quarts are removed and replaced by antifreeze, then the amount of antifreeze afterwards will be

$$(0.4)(6 - 2) + 2 = (0.4)(4) + 2 = 3.6$$

quarts. Since there are 6 quarts in the radiator and $(3.6)/6 = 0.6$, the percentage of antifreeze in the radiator at the end will be 60%.

Therefore 2 quarts of the mixture should be drained off and replaced by antifreeze to make 60% of the mixture antifreeze. □

Solving Applied Problems

The following guidelines should help you solve other applied problems, including the exercises that follow.

1. After reading the problem carefully, choose a variable for the quantity to be determined. If necessary, choose auxiliary variables for other quantities appearing in the problem.
2. From the information given in the problem, write down any equations that must be satisfied by the variables. This step may require ingenuity. A picture is sometimes helpful.
3. Eliminate all the auxiliary variables from the equations. The goal is to obtain an equation that contains only the variable for the quantity to be determined.
4. Solve the equation obtained in step 3.
5. Check the solution to be sure that it really is a solution of the given applied problem.

Before going on to the exercises, you might find it helpful to reread the solutions of Examples 5–9 in this section in order to see how we followed the guidelines listed above. And when you attempt to solve applied problems, don't give up too quickly. Translating applied problems and the information in them into mathematical equations takes time, and the more you try and practice, the more successful you will become at solving them.

EXERCISES 1.1

In Exercises 1–6, solve the equation.

1. $x - 2 = 5$

2. $4y + 5 = 9$

3. $4z - 2 = 3 - 7z$

4. $2(1 + z) = 3z + 5$

5. $1.3t + 5.2 = 2.6 - 7.8t$

6. $\frac{1}{3}(1 - 2t) + \frac{1}{6}(3t + 6) = 0$

In Exercises 7–16, solve the equation.

7. $x^2 - 5x + 6 = x^2 + 3x + 2$

8. $x^2 - 6 = -(3x - x^2)$

9. $(x - 1)(x + 2) = (x + 3)(-4 + x)$

10. $(y + 2)^2 = y^2 - 4$

11. $\dfrac{1}{x} + 3 = \dfrac{2}{x} + 4$

12. $\dfrac{1}{1 - t} = \dfrac{1}{1 + t}$

13. $u^{1/3} - u^{-2/3} = 4u^{1/3} + 6u^{-2/3}$

14. $\dfrac{w - 2}{w + 3} = \dfrac{w + 1}{w - 1}$

15. $\dfrac{z - 2}{z - 1} + 3 = \dfrac{4z + 1}{z + 2}$

16. $\dfrac{1}{1 - t} + \dfrac{2}{1 - t^2} = \dfrac{3}{1 + t}$

In Exercises 17–22, solve the given equation.

17. $|x - 4| = 3$

18. $|x + 7| = 2$

19. $|-2x + 5| = 7$

20. $|\frac{2}{3}x + 4| = 2$

21. $|\frac{1}{2}x + 3| = |\frac{2}{3}x - 3|$

22. $|2x - 3| = x$

In Exercises 23–28, solve for x.

23. $2x = 6y - 1$

24. $3x - 2 = 4y$

25. $0.3x - 0.2y = 1.8x + 3.3y - 0.1$

26. $4x + 2y = 5$

27. $-\frac{1}{2}x + 1 = 3y + 5$

28. $\frac{1}{4}x - \frac{1}{2}y - 3 = 0$

In Exercises 29–30, solve for y.

29. $x = 2y + 7$

30. $\frac{1}{3}x - \frac{1}{2}y - 2 = 0$

31. Find a value of a such that -4 is a solution of the equation $2x + 3 - 4a = x + 7$.

32. Find a value of a such that 3 is a solution of the equation

$$\frac{1}{2-x} + \frac{a}{x+2} = \frac{1}{x^2-4}$$

33. If x denotes the length of an object in inches and y the length in centimeters, then x and y satisfy the equation

$$y = 2.54x$$

Approximate the number of inches in one meter (100 centimeters). Round your answer to four significant digits.

34. A student's grades from three examinations are 82, 64, and 91. In order to obtain an average of 80, what must the student's grade on the fourth test be?

35. In order to raise the average of the three tests in the preceding exercise by 3 points, what must the score on the fourth test be?

36. Karen's father is four times as old as Karen. In 6 years his age will be 10 years more than double Karen's age at that time. How old are they both now?

37. The sum of the ages of two children is 14. Two years ago the age of one child exceeded twice the age of the other by one year. How old are they now?

38. Find two numbers whose sum is 29 and whose difference is 5.

39. Find three consecutive odd integers whose sum is 147.

40. Receipts from 6500 tickets to a basketball game totaled \$14,800. If student tickets cost \$2 each and nonstudent tickets cost \$3 each, how many of each were sold?

41. A collection of 64 coins containing only nickels and dimes is worth \$4.50. How many of each are there?

42. A professor has a list of problems to assign to the students in a class. If 3 problems are assigned to each student, there will be 33 problems left over, but 15 more problems would be needed in order to be able to assign 5 problems to each person in the class. How many students are there in the class?

43. Suppose the cost of a digital clock, including a 4% sales tax, is \$26.52. Find the cost of the clock without the tax.

44. According to the provisions of a will, a sum of money is to be divided among four people. One person is to receive one half of the sum, a second person is to receive one third of the sum, a third is to receive one twelfth of the sum, and a fourth person is to receive the remainder. How large must the sum of money be in order that the fourth person receives \$1500?

45. A mutual fund pays an 11% dividend per year. If the dividend is \$467.50 after the first year, how much was originally invested?

46. Suppose \$3000 was invested in two savings accounts, one at 6% simple interest and the other at 8% simple interest. If the interest accumulated in one year was \$196, how much was deposited in each account?

47. Suppose that \$6000 was invested in two savings accounts, one at 7% simple interest and the other at 8% simple interest. How much was put into each account if both accounts yielded the same amount of interest for the first year.

48. The perimeter of a rectangle is 100 inches, and the width is two thirds of the length. Find the dimensions of the rectangle.

49. The perimeter of a rectangular driveway is 106 feet, and the difference between the length and width is 23 feet. Find the length and width of the driveway.

50. The perimeter of a rectangle is 160 meters. If a new rectangle is formed by doubling the length of one pair of sides and decreasing the length of the remaining pair of sides by 30 meters, the perimeter remains the same. What were the dimensions of the original rectangle?

51. The perimeter of a rectangular rose garden is 56 feet. Determine its dimensions if three such gardens placed side by side would form a square.

52. The perimeter of an isosceles triangle is 33, and one side is three fourths as long as the other two sides. Determine the length of the shortest side.

53. The perimeter of an isosceles trapezoid is 42 inches. The shortest two sides, which are the same length, are 7 inches shorter than the next longer side and 15 inches shorter than the longest side. Determine the length of the shortest sides.

54. Light of sufficiently high frequency can dislodge electrons from their associated atoms. This effect is known as the ***photoelectric effect***. The kinetic energy E of an electron so dislodged is given by

$$E = hv - \omega$$

where h is Planck's constant, v is the frequency of the light, and ω is the binding energy of the electron. Solve the equation for the frequency v.

55. The object distance p, image distance q, and focal length f of a simple lens satisfy the ***lens equation***

$$\frac{1}{p} + \frac{1}{q} = \frac{1}{f}$$

(Figure 1.5). Solve the equation for p.

56. A child tries to reach a dime that appears to be in two feet of water. However, the apparent depth of an object in a substance is the actual depth of the object divided by the index of refraction n of the liquid. If the index of refraction for water is 1.333, find the actual depth of the dime.

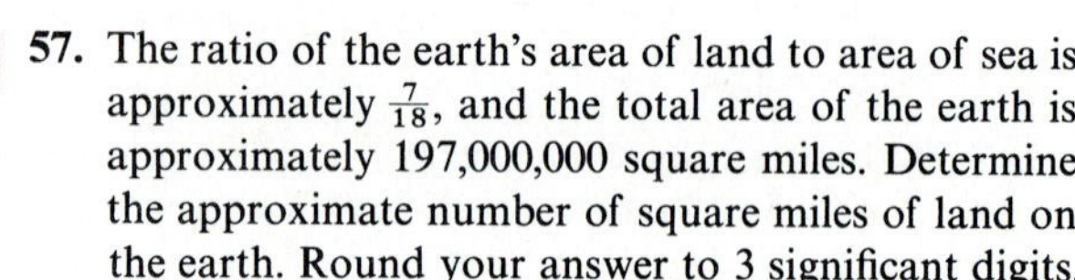

57. The ratio of the earth's area of land to area of sea is approximately $\frac{7}{18}$, and the total area of the earth is approximately 197,000,000 square miles. Determine the approximate number of square miles of land on the earth. Round your answer to 3 significant digits.

58. Two bicyclists are 4 miles apart and travel toward each other. One bicycles at the rate of 8 miles per hour, and the other bicycles at 12 miles per hour. After how long will they be exactly one mile apart?

59. A child hikes along a trail through a forest at 2.5 miles per hour. If a parent sets out after the child half an hour later at a pace of 4 miles per hour, after how long will the parent overtake the child?

60. A jet traveled from Philadelphia to San Francisco at an average velocity of 500 miles per hour, and because of the west-to-east tailwind it made the return trip at an average velocity of 600 miles per hour. The return trip took 48 minutes less than the trip west. What is the distance between Philadelphia and San Francisco?

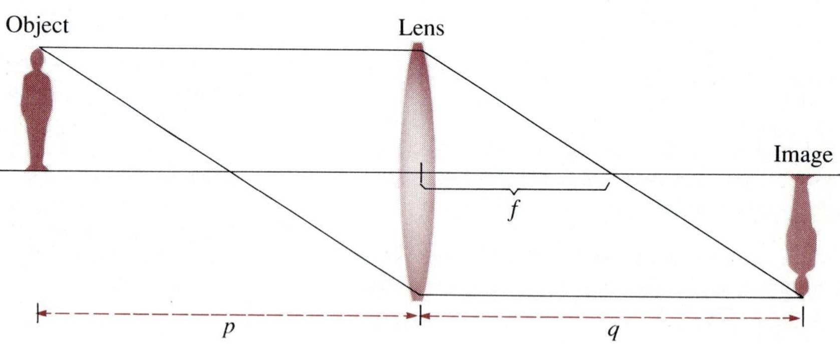

FIGURE 1.5

61. A farmer sets out to walk to town 13 miles away at a rate of 4 miles per hour. After a while the farmer is picked up and driven the rest of the way at an average speed of 40 miles per hour. If the total trip takes 1 hour, how far does the farmer walk?

62. Two runners in the 26-mile Boston marathon travel at rates of 6 and 8 miles per hour, respectively. Suppose that they begin at the same instant.
   a. How far back will the slower runner be when the faster one completes the course?
   b. How long will it take for the two runners to be 4.5 miles apart?

63. At noon the front end of a freight train moving at a constant speed enters a tunnel 275 meters long. Ten seconds later the back end of the train enters the tunnel. After another 11 seconds the back end of the train leaves the tunnel. Find the length and the speed of the train.

64. One picker can harvest a strawberry patch in 2 hours, and a second picker can harvest the patch in 2.5 hours. How long would it take for both pickers to harvest the patch together?

65. One gasoline truck can fill a storage tank in 20 minutes. Another can fill the same tank in 30 minutes. How long would it take to fill the tank using both trucks simultaneously?

66. Suppose it takes 4 hours for Pam to shovel the snow from the driveway and 4.5 hours for Sam. How long would it take them to do it together, with one shovel each?

67. Suppose Bram joins Pam and Sam, and Bram alone can shovel the driveway in 6 hours. How long would it take the three to shovel the driveway together, with three shovels?

68. Suppose it would take Mary and Jane 3 hours and 45 minutes to paint a porch together. If Mary could paint it alone in 6 hours, how long would it take Jane to paint it alone?

69. Suppose in Example 9 that 80% antifreeze is added rather than pure antifreeze. How much of the radiator mixture should be drained off and replaced by the new 80% antifreeze liquid to obtain a mixture containing 60% antifreeze?

70. A sample of 20 pounds of sea water has a 11.5% salt content. How much fresh water must be added to produce a mixture with a 6.5% salt content? Round your answer to the nearest tenth of a gallon.

71. How much water must be evaporated from 8 gallons of a saline solution with a 20% salt content in order for it to have a 25% salt content?

72. One alloy is 40% silver, and another alloy is 30% silver. How much of each should be used to produce 50 pounds of an alloy that is 36% silver?

73. Unleaded gasoline for cars has octane ratings (percent by volume of iso-octane in one gallon of gasoline). Normally, the octane ratings have been approximately 87 for regular gasoline and 92 for super, or ethyl, gasoline. Suppose a gas tank contains 6 gallons of regular with an octane rating of 87. To achieve a mixture with an octane rating of 90, how many gallons of super with an octane rating of 92 must be added?

74. A small car dealership has five times as many cars as trucks on the lot. By adding 15 more of each vehicle type, the company can bring the ratio of cars to trucks to the desired level of 2 to 1. How many cars and how many trucks are on the lot?

*75. Little is known about the personal life of the mathematician Diophantus of Alexandria, who is thought to have lived in the third century A.D. However, in an epitaph to Diophantus, it is stated that he lived one sixth of his life in childhood, one twelfth in his youth, and one seventh more as a bachelor. Five years after his marriage a son was born who died four years before Diophantus, at half of Diophantus's final age. Assuming that all this is true, determine the age of Diophantus when he died.

76. The ***harmonic mean*** of two numbers a and b with nonzero sum is

$$\frac{2ab}{a+b}$$

*a. Suppose a car travels from A to B with speed v_1 and returns with speed v_2. Show that the round trip could be made in the same length of time if the car traveled at a constant speed equal to the harmonic mean of v_1 and v_2.

b. If v_1 in part (a) is 40 miles per hour and v_2 is 60 miles per hour, find the constant speed that yields the same time for the round trip.

1.2 QUADRATIC EQUATIONS AND THE QUADRATIC FORMULA

A ***quadratic equation*** is an equation of the form

$$\boxed{ax^2 + bx + c = 0, \quad \text{with } a \neq 0} \tag{1}$$

or any equation that can be put into the form of (1) by shifting all nonzero terms to the left side of the equation. Thus

$$x^2 = 9, \qquad x^2 - 4x - 3 = 0, \quad \text{and} \quad 3t^2 = 2t - 1$$

are quadratic equations.

Quadratic equations occur in physical applications. For instance, suppose that a ball is thrown upward at 64 feet per second from a dormitory roof 48 feet above the ground. If the air resistance is negligible, the laws of motion imply that until it hits the ground, the ball's height above the ground x seconds after it is thrown is $-16x^2 + 64x + 48$ feet. Now when the ball hits the ground, its height is 0, so x satisfies

$$-16x^2 + 64x + 48 = 0$$

or equivalently,

$$x^2 - 4x - 3 = 0 \tag{2}$$

But this is a quadratic equation. What value(s) of x satisfy the equation in (2)? You might see if you can answer the question before reading on.

Our goal is to give a method of determining all real solutions of an arbitrary quadratic equation. In contrast to a linear equation, which always has one and only one real solution, a quadratic equation can have two, one, or no real solutions, depending on the constants occurring in the equation.

If we can find real numbers p and q such that $x - p$ and $x - q$ are factors of $ax^2 + bx + c$, that is,

$$ax^2 + bx + c = a(x - p)(x - q) \tag{3}$$

then p and q are evidently solutions of (1), because

$$ap^2 + bp + c = a\overbrace{(p - p)}^{0}(p - q) = 0$$

and

$$aq^2 + bq + c = a(q - p)\overbrace{(q - q)}^{0} = 0$$

Moreover, p and q are the only solutions of the given equations, because if

$$ax^2 + bx + c = a(x - p)(x - q) = 0$$

then since $a \neq 0$ by hypothesis, the Zero Property (see Section 0.1) implies that either $x - p = 0$ or $x - q = 0$, that is, $x = p$ or $x = q$.

For example, since

$$x^2 + 2x - 3 = (x - 1)(x + 3)$$

the solutions of the equation $x^2 + 2x - 3 = 0$ are 1 and -3.

Solving a quadratic equation by this method depends on our ability to factor a quadratic polynomial. Now we will derive a formula for the solutions of a quadratic equation that does not depend on factoring but nevertheless is easily applied.

The Quadratic Formula

In order to find the solutions of the equation

$$x^2 - 4x - 3 = 0 \tag{4}$$

in (2), we begin in what may at first seem a roundabout way by adding 3 to both sides, to eliminate the constant term on the left side of the equation. This yields

$$x^2 - 4x = 3$$

Then we add a suitable constant to both sides so that the left side is the square $(x + r)^2$ of a linear polynomial. In this case the constant to add is 4, because $x^2 - 4x + 4 = (x - 2)^2$. Adding 4 to both sides, we obtain

$$x^2 - 4x + 4 = 3 + 4$$

which is equivalent to

$$(x - 2)^2 = 7 \tag{5}$$

To solve for x is now simple. By taking square roots of both sides of (5), we find that

$$x - 2 = \sqrt{7} \quad \text{or} \quad x - 2 = -\sqrt{7}$$

which means that

$$x = 2 + \sqrt{7} \quad \text{or} \quad x = 2 - \sqrt{7}$$

Thus $2 + \sqrt{7}$ and $2 - \sqrt{7}$ are the solutions of the equation in (4).

If x represents time and $x^2 - 4x - 3$ represents the motion of the ball mentioned at the outset of this section, then the time at which the ball hits the ground is a nonnegative number that satisfies the equation $x^2 - 4x - 3 = 0$. Since $2 - \sqrt{7} < 0$ and thus cannot represent any time after the ball was thrown, it follows that the ball hits the ground $2 + \sqrt{7}$ seconds (approximately 4.6 seconds) after it was thrown.

The process leading from the given equation in (4) to the equation in (5) is called ***completing the square***, because the terms involving x^2 and x become a part of the complete square of the form $(x + r)^2$.

EXAMPLE 1. Find all real solutions of the quadratic equation

$$x^2 - 7x + 6 = 0$$

by first completing the square.

Solution. We follow the procedure detailed above:

$$x^2 - 7x = -6$$

$$x^2 - 7x + \frac{49}{4} = -6 + \frac{49}{4} = -\frac{24}{4} + \frac{49}{4} = \frac{25}{4}$$

$$\left(x - \frac{7}{2}\right)^2 = \frac{25}{4} = \left(\frac{5}{2}\right)^2$$

Taking square roots, we find that

$$x - \frac{7}{2} = \frac{5}{2} \quad \text{or} \quad x - \frac{7}{2} = -\frac{5}{2}$$

so that

$$x = \frac{7}{2} + \frac{5}{2} = \frac{12}{2} = 6 \quad \text{or} \quad x = \frac{7}{2} - \frac{5}{2} = \frac{2}{2} = 1$$

Therefore the proposed solutions are 6 and 1.

Check: For $x = 6$, $6^2 - 7(6) + 6 = 36 - 42 + 6 = 0$

For $x = 1$, $1^2 - 7(1) + 6 = 1 - 7 + 6 = 0$

Consequently 6 and 1 are the solutions of the given equation. □

Notice that the answer obtained in Example 1 by completing the square is the same as would be obtained by factoring using methods of Section 0.6. However, by completing the square we need not use trial and error to arrive at the solution.

To obtain all solutions of a quadratic equation

$$ax^2 + bx + c = 0, \quad \text{with} \quad a \neq 0 \tag{6}$$

we factor out a from the terms involving x^2 and x, and then complete the square as we did in Example 1:

$$ax^2 + bx + c = 0$$

$$ax^2 + bx = -c$$

$$a\left(x^2 + \frac{b}{a}x\right) = -c$$

$$a\left(x^2 + \frac{b}{a}x + \frac{b^2}{4a^2} - \frac{b^2}{4a^2}\right) = -c$$

$$a\left(x^2 + \frac{b}{a}x + \frac{b^2}{4a^2}\right) - \frac{b^2}{4a} = -c$$

$$a\left(x + \frac{b}{2a}\right)^2 = \frac{b^2}{4a} - c = \frac{b^2 - 4ac}{4a}$$

$$\left(x + \frac{b}{2a}\right)^2 = \frac{b^2 - 4ac}{4a^2} \tag{7}$$

If $b^2 - 4ac \geq 0$, we can find the real solutions of (7) by taking square roots of both sides of (7), which yields

$$x + \frac{b}{2a} = \frac{\sqrt{b^2 - 4ac}}{2a} \quad \text{or} \quad x + \frac{b}{2a} = -\frac{\sqrt{b^2 - 4ac}}{2a}$$

or equivalently,

$$x = -\frac{b}{2a} + \frac{\sqrt{b^2 - 4ac}}{2a} \quad \text{or} \quad x = -\frac{b}{2a} - \frac{\sqrt{b^2 - 4ac}}{2a}$$

Thus if $b^2 - 4ac \geq 0$, the real solutions of (7) and hence of (6) are

$$x = \frac{-b + \sqrt{b^2 - 4ac}}{2a} \quad \text{and} \quad x = \frac{-b - \sqrt{b^2 - 4ac}}{2a} \tag{8}$$

The formulas in (8) are often combined to give one formula:

The Quadratic Formula

$$x = \frac{-b \pm \sqrt{b^2 - 4ac}}{2a}$$

Because the square root has been defined only for nonnegative numbers, the quadratic formula yields real solutions of the quadratic equation $ax^2 + bx + c = 0$ only if $b^2 - 4ac$, called the ***discriminant*** of the quadratic equation, is nonnegative. In Section 1.4 we will discuss solutions that are not real numbers.

The number of real solutions of $ax^2 + bx + c = 0$ given by the quadratic formula depends on the value of the discriminant:

i. If $b^2 - 4ac > 0$, the quadratic formula provides two real solutions.
ii. If $b^2 - 4ac = 0$, the quadratic formula provides one real solution.
iii. If $b^2 - 4ac < 0$, the quadratic formula provides no real solutions.

In practice, we determine the real solutions of $ax^2 + bx + c = 0$ by applying the quadratic formula in the following way. We first substitute the values of a, b, and c into the formula

$$x = \frac{-b \pm \sqrt{b^2 - 4ac}}{2a} \tag{9}$$

regardless of the value of $b^2 - 4ac$, and then we simplify $b^2 - 4ac$, noting whether it is positive, zero, or negative. If $b^2 - 4ac \geq 0$, we further simplify the right side of (9) when possible to find the value(s) of x; if $b^2 - 4ac < 0$, we conclude that no real solutions exist.

The power of the quadratic formula is that it enables us by numerical computation to tell how many real solutions a given quadratic equation has and to compute the values of any real solutions that exist. The following examples illustrate this point.

EXAMPLE 2. Find all real solutions of the following equations.

a. $3x^2 - 4x + 1 = 0$ b. $6x^2 + \sqrt{2}x - 2 = 0$

Solution.

a. By the quadratic formula, with $a = 3$, $b = -4$, and $c = 1$, any real solutions are given by

$$x = \frac{-(-4) \pm \sqrt{(-4)^2 - 4(3)(1)}}{2(3)} = \frac{4 \pm \sqrt{4}}{6} = \frac{4 \pm 2}{6}$$

Since
$$\frac{4+2}{6} = 1 \quad \text{and} \quad \frac{4-2}{6} = \frac{1}{3}$$

there are two proposed solutions of the given equation: 1 and $\frac{1}{3}$.

Check: For $x = 1$, $3(1)^2 - 4(1) + 1 = 3 - 4 + 1 = 0$

For $x = \frac{1}{3}$, $3\left(\frac{1}{3}\right)^2 - 4\left(\frac{1}{3}\right) + 1 = \frac{1}{3} - \frac{4}{3} + 1 = 0$

Consequently 1 and $\frac{1}{3}$ are the solutions of the given equation.

b. By the quadratic formula, with $a = 6$, $b = \sqrt{2}$ and $c = -2$, any real solutions are given by

$$x = \frac{-\sqrt{2} \pm \sqrt{(\sqrt{2})^2 - 4(6)(-2)}}{2(6)}$$
$$= \frac{-\sqrt{2} \pm \sqrt{50}}{12} = \frac{-\sqrt{2} \pm \sqrt{25 \cdot 2}}{12}$$
$$= \frac{-\sqrt{2} \pm 5\sqrt{2}}{12}$$

Since

$$\frac{-\sqrt{2} + 5\sqrt{2}}{12} = \frac{4\sqrt{2}}{12} = \frac{1}{3}\sqrt{2}$$

and
$$\frac{-\sqrt{2} - 5\sqrt{2}}{12} = \frac{-6\sqrt{2}}{12} = -\frac{1}{2}\sqrt{2}$$

the two proposed solutions of the given equation are $\frac{1}{3}\sqrt{2}$ and $-\frac{1}{2}\sqrt{2}$.

Check: For $x = \frac{1}{3}\sqrt{2}$, $6\left(\frac{1}{3}\sqrt{2}\right)^2 + \sqrt{2}\left(\frac{1}{3}\sqrt{2}\right) - 2$

$$= 6\left(\frac{2}{9}\right) + \frac{2}{3} - 2 = 0$$

$$\text{For } x = -\frac{1}{2}\sqrt{2}, 6\left(-\frac{1}{2}\sqrt{2}\right)^2 + \sqrt{2}\left(-\frac{1}{2}\sqrt{2}\right) - 2$$
$$= 6\left(\frac{1}{2}\right) - 1 - 2 = 0$$

Consequently $\frac{1}{3}\sqrt{2}$ and $-\frac{1}{2}\sqrt{2}$ are the solutions of the given equation. □

EXAMPLE 3. Find all real solutions of the following equations.
a. $16^2 + 24x + 9 = 0$ b. $3t^2 = 2t - 1$

Solution.
a. By the quadratic formula, any real solutions are given by

$$x = \frac{-24 \pm \sqrt{(24)^2 - 4(16)(9)}}{2(16)} = \frac{-24 \pm \sqrt{0}}{32} = -\frac{3}{4}$$

Thus there is one proposed solution of the given equation: $-\frac{3}{4}$.

$$\textit{Check: } 16\left(-\frac{3}{4}\right)^2 + 24\left(-\frac{3}{4}\right) + 9$$
$$= 16\left(\frac{9}{16}\right) - 18 + 9 = 0$$

Consequently $-\frac{3}{4}$ is the solution of the equation.

b. If we subtract $2t - 1$ from both sides of the given equation, we obtain the equivalent equation

$$3t^2 - 2t + 1 = 0$$

whose solutions, if any, are precisely the solutions of the given equation. But by the quadratic formula, any real solutions are given by

$$t = \frac{-(-2) \pm \sqrt{(-2)^2 - 4(3)(1)}}{2(3)} = \frac{2 \pm \sqrt{-8}}{6}$$

Since $-8 < 0$, there are no real solutions. □

One consequence of the quadratic formula is the fact that any quadratic equation

$$ax^2 + bx + c = 0$$

for which a and c have opposite signs has two real solutions. The reason is that in this case, $-4ac > 0$, so that the discriminant $b^2 - 4ac > 0$. In contrast, if a and c have the same sign, then the equation can have two, one, or no real solutions. These possibilities occur in Examples 2 and 3.

Occasionally we encounter an equation that is not quadratic but can be transformed into an equivalent quadratic equation by means of an algebraic manipulation.

EXAMPLE 4. Find all real solutions of $1 + \frac{1}{x^2} = \frac{3}{x}$.

Solution. We multiply both sides of the equation by x^2 and then solve the resulting quadratic equation:

$$x^2 + 1 = 3x$$

$$x^2 - 3x + 1 = 0$$

By the quadratic formula,

$$x = \frac{-(-3) \pm \sqrt{(-3)^2 - 4(1)(1)}}{2(1)} = \frac{3 \pm \sqrt{5}}{2}$$

Therefore the proposed solutions are $(3 + \sqrt{5})/2$ and $(3 - \sqrt{5})/2$. We could check here that these are actually the solutions of the given equation, but the calculations are tedious. □

In conclusion, we emphasize the close connection between nontrivial factors of the quadratic polynomial $ax^2 + bx + c$ and solutions of the quadratic equation $ax^2 + bx + c = 0$. Indeed, if

$$ax^2 + bx + c = a(x - p)(x - q)$$

then p and q are the solutions of the equation $ax^2 + bx + c = 0$. If $b^2 - 4ac \geq 0$, then conversely, the solutions p and q of the equation $ax^2 + bx + c = 0$ that are given by

$$p = \frac{-b + \sqrt{b^2 - 4ac}}{2a} \quad \text{and} \quad q = \frac{-b - \sqrt{b^2 - 4ac}}{2a}$$

provide the factorization

$$ax^2 + bx + c = a(x - p)(x - q)$$

of the quadratic polynomial $ax^2 + bx + c$.

EXERCISES 1.2

In Exercises 1–16, solve the given equation by rearranging (if necessary) and then factoring.

1. $x^2 + 4x + 4 = 0$
2. $x^2 - x - 2 = 0$
3. $x^2 - 50x + 625 = 0$
4. $y^2 - 2y = 3$
5. $9y^2 - 12y + 4 = 0$
6. $y^2 + 3y - 4 = 0$
7. $y^2 - 2\sqrt{2}y = -2$
8. $2y^2 + 18 = -12y$
9. $4x^2 + 2x + \frac{1}{4} = 0$
10. $9x^2 + \frac{1}{9} = 2x$
11. $5x + 4 = -x^2$
12. $6x - 9x^2 = 1$
13. $x^2 + 3x + 5 = 15$
14. $2t^2 + 42 = 20t$
15. $2 + \dfrac{1}{x^2} = \dfrac{3}{x}$
16. $25x + \dfrac{4}{x} = 20$

In Exercises 17–22, solve the given equation by completing the square.

17. $x^2 - 2x - 6 = 0$
18. $y^2 = 4y + 12$
19. $y^2 - 10y = 2$
20. $3t^2 - 6t = 12$
21. $2x^2 + 4x - 3 = 0$
22. $4x - x^2 = -3$

In Exercises 23–46, find all real solutions (if any) of the given equation by using the quadratic formula.

23. $x^2 - 3x + 1 = 0$
24. $x^2 + 3x + 1 = 0$
25. $x^2 + x + 1 = 0$
26. $x^2 + x - 4 = 0$
27. $x^2 - 2x = 4$
28. $x^2 = 4\sqrt{3}x - 12$
29. $3 + 2x = 2x^2$
30. $2x^2 + 3x - 5 = 0$
31. $1 - 6w + 3w^2 = 0$
32. $6w^2 + 5w = 1$
33. $6x^2 + 5x = -1$
34. $3x^2 - 4\sqrt{3}x + 4 = 0$
35. $3x^2 - 4\sqrt{3}x - 4 = 0$
36. $5y^2 + 7y + 5 = 0$
37. $4y^2 + 4\sqrt{2}y = 2$
38. $4y^2 + 4\sqrt{2}y + 2 = 0$
39. $t^2 - t + 3 = 0$
40. $t + \dfrac{1}{t} = 6$
41. $\dfrac{5}{t^2} + \dfrac{3}{t} - 1 = 0$
42. $\dfrac{3}{t^2} - \dfrac{5}{t} = 2$
43. $\dfrac{x-1}{2x-3} = \dfrac{x+2}{x-2}$
44. $\dfrac{x}{x^2-1} = 4$
45. $(x - 1)(x - 2) = 3(x + 1) + 4$
46. $(x + 3)(x + 7) = 3(x - 2) + 5$
47. Find the values of b for which there is exactly one solution of the equation $x^2 + bx + 9 = 0$.
48. Find all values of b such that $x^2 + bx + 5 = 0$ has exactly one real solution.
49. Find the value of b such that the real solutions of $x^2 + bx - 8 = 0$ are negatives of each other.
50. Show that no matter what the value of b is, there are two distinct real solutions of $x^2 + bx - 9 = 0$.
51. Find the values of b for which the real solutions of $x^2 + bx - 8 = 0$ differ by 6.
52. Show that if x_1 and x_2 are real solutions of the equation $x^2 + bx + c = 0$, then $x_1 + x_2 = -b$ and $x_1x_2 = c$.

In Exercises 53–56, use a calculator and the quadratic formula to approximate the solutions of the given equation.

53. $3x^2 - 17x - 23 = 0$
54. $4x^2 + 100x + 3 = 0$
55. $3.14x^2 - 1.3x - 4.59 = 0$
56. $x^2 - 3\sqrt{2}x + \sqrt{5} = 0$
57. Suppose the equation $6x^2 + 18x + c = 0$ has exactly one root. Determine the root and the value of c.
58. Suppose the equation $ax^2 + bx + c = 0$ has two distinct nonzero roots r and s. Find an expression for the sum
$$\frac{1}{r} + \frac{1}{s}$$
in terms of a, b, and c.
59. Suppose the equation $x^2 - 2x + c = 0$ has two roots, one of which is twice the other. Determine the value of c.

60. The ancient Greeks thought that the rectangles with the most pleasing shapes are those for which the ratio of the width to the length is the same as the ratio of the length to the sum of the length and the width, that is,

$$\frac{w}{l} = \frac{l}{l + w} = \frac{1}{1 + \dfrac{w}{l}}$$

where l is the length and w is the width of the rectangle. If we let $x = w/l$, the above equation becomes

$$x = \frac{1}{1 + x} \tag{10}$$

The ratio x is called the ***golden ratio*** (or ***golden section***) because of the interpretation the ancient Greeks associated with it. Determine the golden ratio by solving (10).

61. The ratio of the width to the length of the rectangle in Figure 1.6 is the golden ratio, and the length of the rectangle is 4 centimeters. Approximate the width to 3 significant digits.

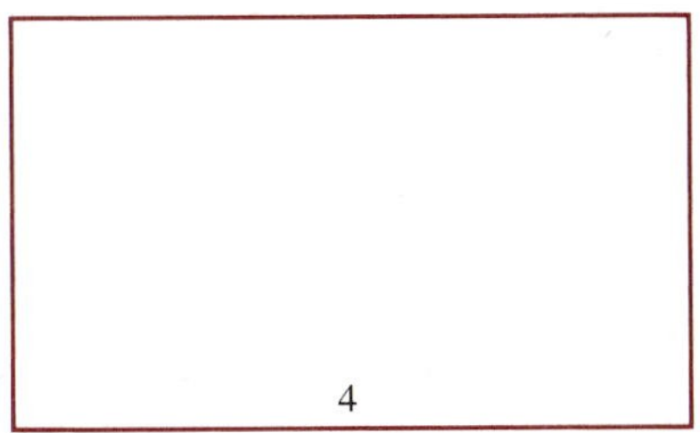

Rectangle whose sides are in the golden ratio.

FIGURE 1.6

1.3 APPLICATIONS OF QUADRATIC EQUATIONS

In Section 1.1 we solved applied problems that led to *linear* equations when translated into mathematical language. However, not all applied problems can be solved by means of linear equations. In this section we will consider applied problems that lead to *quadratic* equations.

The general method of solving applied problems used in Section 1.1 still applies. However, in order to facilitate working the examples, we will not write out our checks of the proposed solutions of the equations. Nevertheless, we will need to see which of any proposed solutions qualify as solutions of the given applied problems.

Vertical Motion

Suppose that an object is thrown or propelled or dropped, and from that moment on moves vertically under the sole influence of gravity. For all practical purposes, this would be the case, for example, if a person jumps from a diving platform or throws a rock straight up.

Let us suppose that time is measured in seconds and that at some convenient moment (such as when the object begins moving under the sole influence of gravity) the time is 0. Suppose also that the height is measured in feet, with h_0 the ***initial height*** above ground level at the instant $t = 0$, and finally, assume that the velocity is measured in feet per second, with v_0 denoting the velocity at time $t = 0$. Then v_0 is the ***initial velocity*** of the object. We take v_0 to be positive if the object is moving upward at time $t = 0$ and negative if the object is moving downward at that time. For example, if a ball is hurled upward at 40 feet per second, then $v_0 = 40$, whereas if the ball is thrown downward at 40 feet per second, then $v_0 = -40$. If the ball is dropped (from rest), then $v_0 = 0$.

The laws of motion imply that the height h of such an object above ground level at time $t \geq 0$ is given by

$$h = -16t^2 + v_0 t + h_0 \tag{1}$$

It follows that if the constants h_0 and v_0 are known and if h is given, then (1) is a quadratic equation in the variable t. Of course, (1) is valid only while the object is under the sole influence of gravity (for instance, before it hits the ground). By using (1) we can solve many problems concerning vertical motion.

EXAMPLE 1. Suppose a swimmer dives from a platform 64 feet above a deep pool. How long will the swimmer fall before touching the water?

Olympic champion diver Greg Louganis executing a high dive.

Solution. We take $t = 0$ at the instant the swimmer begins to descend, the initial velocity v_0 to be 0, and the initial height h_0 to be 64. Since the height h will be 0 at the instant the swimmer first touches the water, we must find the time at which $h = 0$. To accomplish this we use (1) with $h = 0$, $v_0 = 0$, and $h_0 = 64$:

$$0 = -16t^2 + (0)t + 64$$

which reduces to

$$16t^2 = 64$$

or

$$t^2 = 4$$

Therefore $t = 2$ or $t = -2$. Consequently 2 and -2 are proposed solutions of the problem. Since $t = 0$ when the swimmer starts the descent, the proposed solution -2, which corresponds to a time *before* the dive, does not apply. Thus the swimmer touches the water after 2 seconds. □

CAUTION: In Example 1 we derived the mathematical equation $t^2 = 4$ from the given physical conditions. After we found the solutions 2 and -2 of the equation, we rejected the solution -2. It was not rejected because it was not a solution of the equation, but because it did not satisfy the physical condition that $t \geq 0$. Therefore in solving an applied problem it is critical that we scrutinize solutions arising from the mathematics in order to see which if any are actually solutions of the given applied problem, that is, which solutions actually satisfy the conditions relating to the given applied problem.

EXAMPLE 2. A rocket loaded with fireworks is to be shot vertically upward from the ground with an initial velocity of 160 feet per second. The fireworks are to be detonated at a height of 384 feet while they are still on the rise. How long after takeoff should detonation occur?

Solution. Let $t = 0$ at takeoff. Our problem is to find the time at which the height h of the rocket is 384 feet and the rocket is rising. We are assuming that the rocket is shot upward from the ground level, which means that $h_0 = 0$. By assumption the initial velocity is given by $v_0 = 160$, so that by (1) we need to solve the equation

$$384 = -16t^2 + 160t + 0$$

or

$$16t^2 - 160t + 384 = 0 \tag{2}$$

To solve (2) we have the following equivalent equations:

$$16(t^2 - 10t + 24) = 0$$

$$t^2 - 10t + 24 = 0$$

$$(t - 4)(t - 6) = 0$$

Therefore $t = 4$ or $t = 6$. Thus the rocket is at a height of 384 feet at 4 seconds and again at 6 seconds. Since the rocket is ascending at the former time and descending at the latter time, detonation should occur 4 seconds after the rocket is launched. □

Geometric Applications

One of the most important results known from antiquity is the Pythagorean Theorem, named after the outstanding Greek mathematician Pythagoras,

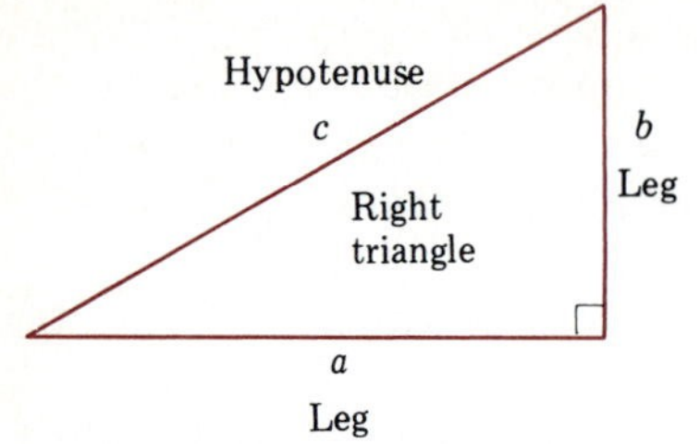

The Pythagorean Theorem: $a^2 + b^2 = c^2$

FIGURE 1.7

who lived around 550 B.C. Recall that a right triangle is a triangle two of whose sides are perpendicular to each other. Assume that the lengths of these two sides, called the ***legs*** of the triangle, are a and b (Figure 1.7) and the length of the third side, called the ***hypotenuse*** of the triangle, is c. Then the lengths of the three sides are related by the following formula:

Pythagorean Theorem

$$a^2 + b^2 = c^2$$

The Pythagorean Theorem is basic to arguments involving right triangles.

EXAMPLE 3. Suppose that the lengths of the legs of a right triangle differ by one inch, and the hypotenuse is one inch longer than the longer leg. Determine the lengths of all the sides of the triangle.

Solution. Let

$$x = \text{the length in inches of the shorter leg}$$

Then the other leg is $x + 1$ inches long, and the hypotenuse is $x + 2$ inches long. By the Pythagorean Theorem,

$$x^2 + (x + 1)^2 = (x + 2)^2$$

We solve this equation by using the following equivalent equations:

$$\begin{aligned} x^2 + (x^2 + 2x + 1) &= x^2 + 4x + 4 \\ 2x^2 + 2x + 1 &= x^2 + 4x + 4 \\ x^2 - 2x - 3 &= 0 \\ (x - 3)(x + 1) &= 0 \end{aligned}$$

Therefore $x = 3$ or $x = -1$. Since all lengths must be nonnegative, it follows that the shorter leg is 3 inches long. Consequently the other leg is 4 inches long, and the hypotenuse is 5 inches long. □

Miscellaneous Applications

EXAMPLE 4. A rectangular painting has an area of 200 square inches. A frame 1 inch wide is added to each side of the painting, making the total area of painting and frame 261 square inches (Figure 1.8). What were the dimensions of the painting before the addition of the frame?

Solution. Let

$$x = \text{the width of the painting before framing}$$

$$y = \text{the height of the painting before framing}$$

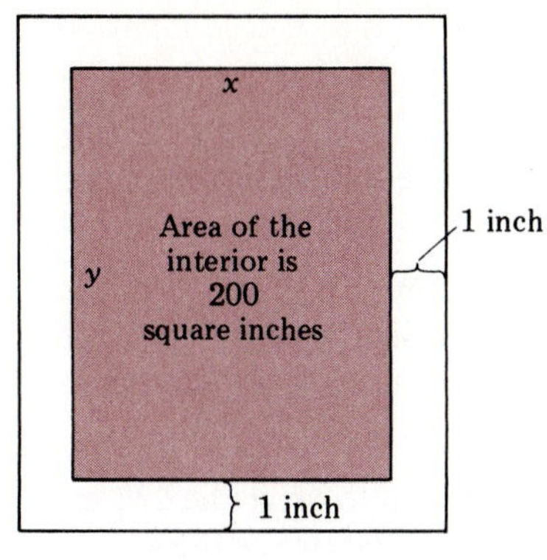

FIGURE 1.8

Then
$$xy = 200$$

so that

$$y = \frac{200}{x} \tag{3}$$

After the frame has been added, the width becomes $x + 2$ and the height becomes $y + 2$. Since the area of the painting plus the frame is by assumption 261 square inches, this means that

$$(x + 2)(y + 2) = 261 \tag{4}$$

Equation (4) has two variables, x and y. Since they are related by (3), we substitute for y in (4) to obtain

$$(x + 2)\left(\frac{200}{x} + 2\right) = 261$$

which is an equation in the single variable x, and hence an equation we know how to treat. Multiplying out and then multiplying both sides by x, we obtain the following equivalent equations:

$$200 + \frac{400}{x} + 2x + 4 = 261$$

$$2x - 57 + \frac{400}{x} = 0$$

$$2x^2 - 57x + 400 = 0$$

By the quadratic formula,

$$x = \frac{-(-57) \pm \sqrt{(-57)^2 - 4(2)(400)}}{2(2)} = \frac{57 \pm \sqrt{49}}{4} = \frac{57 \pm 7}{4}$$

Since

$$\frac{57 + 7}{4} = 16 \quad \text{and} \quad \frac{57 - 7}{4} = \frac{25}{2}$$

it follows that $x = 16$ or $x = \frac{25}{2}$. Now by (3),

$$\text{if } x = 16, \quad \text{then} \quad y = \frac{200}{16} = \frac{25}{2}$$

$$\text{if } x = \frac{25}{2}, \quad \text{then} \quad y = \frac{200}{25/2} = 16$$

Consequently the dimensions of the painting before framing were 16 inches by $\frac{25}{2}$ inches. □

EXAMPLE 5. Pat drives the 432 miles between Boston and Washington, D.C., in one hour less than Dean and at an average speed of 6 miles per hour faster than Dean. How fast does each drive?

Solution. Let t_P be the time in hours it takes Pat to drive the 432 miles, and let t_D be the corresponding time for Dean. Since Pat drives the 432 miles in one hour less than Dean,

$$t_P = t_D - 1 \tag{5}$$

Now we apply the formula $d = rt$ relating distance, rate, and time that appeared in (9) of Section 1.1. If r is the rate at which Dean drives, then by assumption

$$432 = rt_D$$

so that

$$t_D = \frac{432}{r} \tag{6}$$

Since Pat drives 6 miles per hour faster than Dean, Pat's rate is $r + 6$. Therefore

$$432 = (r + 6)t_P \tag{7}$$

Substituting (5) into (7), and then (6) into the resulting equation, we obtain

$$432 = (r + 6)(t_D - 1) = (r + 6)\left(\frac{432}{r} - 1\right)$$

Multiplying each side by r, we find that

$$432r = (r + 6)\left(\frac{432}{r} - 1\right)r$$

$$432r = (r + 6)(432 - r)$$

$$432r = 432r + 2592 - r^2 - 6r$$

$$r^2 + 6r - 2592 = 0$$

By the quadratic formula,

$$r = \frac{-6 \pm \sqrt{6^2 - 4(1)(-2592)}}{2(1)} = \frac{-6 \pm \sqrt{36 + 10{,}368}}{2}$$
$$= \frac{-6 \pm \sqrt{10{,}404}}{2} = \frac{-6 \pm \sqrt{4 \cdot 9 \cdot 289}}{2} = \frac{-6 \pm 102}{2}$$

Since

$$\frac{-6 + 102}{2} = \frac{96}{2} = 48 \quad \text{and} \quad \frac{-6 - 102}{2} = -\frac{108}{2} = -54$$

it follows that $r = 48$ or $r = -54$. Because speed must be nonnegative, we conclude that Dean travels at a rate of 48 miles per hour. Pat therefore travels at the rate of $48 + 6 = 54$ miles per hour. □

EXERCISES 1.3

1. An acrobat drops from a platform 180 feet above a deep pool. How long will it take for the acrobat to reach the pool?

2. A skyscraper window cleaner loses a pail 256 feet above the ground. How many seconds later does it pass by a window at the 112-foot level?

3. A rock is thrown down from a bridge 96 feet above the water. If the initial velocity is -16 feet per second, how long does it take the rock to hit the water?

4. A ball is dropped from a balcony 48 feet above the ground.
   a. How long will it take for the ball to hit the ground?
   b. How much sooner would the ball's flight end if it landed on an awning 7 feet above the ground?
   c. Approximate the answer in (b), rounding your answer to the nearest hundredth of a second.

5. A paintbrush is thrown straight up with a velocity of 48 feet per second toward a painter 32 feet higher up. How long a wait does the painter have before catching it?

6. A ball is thrown downward from a bridge 192 feet above a river. After 1 second the ball has traveled 80 feet. After how many more seconds will it hit the water?

7. The legs of a right triangle have lengths 5 and 6 inches. What is the length of the hypotenuse?

8. The length of the hypotenuse of a certain right triangle is 1 inch longer than one leg and is 8 inches longer than the other leg. Find the lengths of all three sides of the triangle.

9. The hypotenuse of a right triangle is 10 meters long, and the length of one leg exceeds the length of the other by one meter. How long is each leg?

10. A baseball diamond is a square 90 feet on a side (see Figure 1.9). What is the distance from home plate to second base?

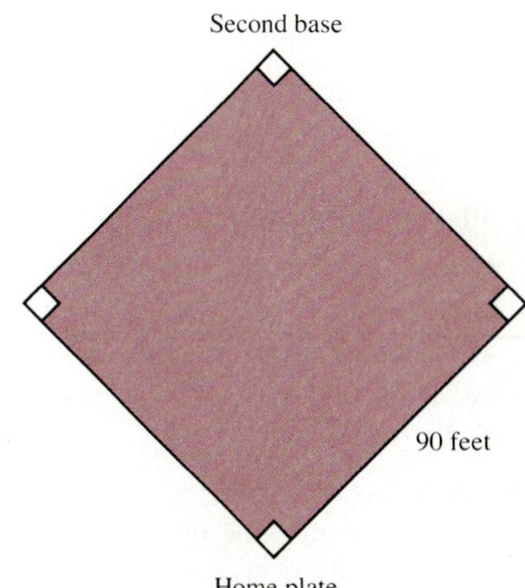

FIGURE 1.9

11. A plot of land is in the shape of a right triangle whose hypotenuse is 1300 feet long and one of whose legs is 1200 feet long. How many feet of fencing are required to enclose the plot?

12. A ladder 13 feet long is placed so that its base is 5 feet from a wall and its top is 12 feet above the ground (see Figure 1.10). If the ladder slips so that the base rests 7 feet from the wall, how far does the top of the ladder fall?

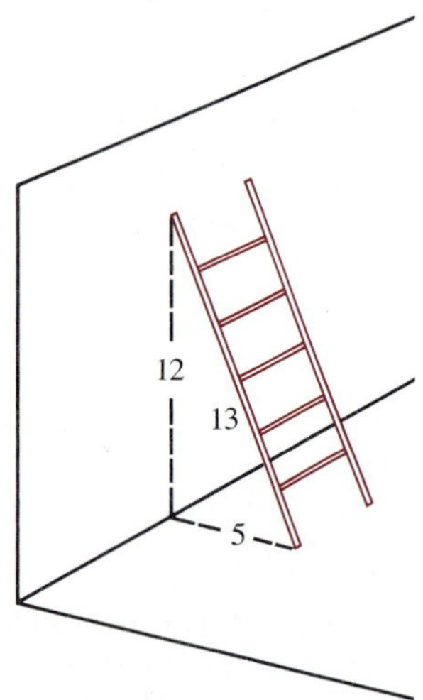

FIGURE 1.10

13. A car travels 8 miles per hour faster than a truck and travels 160 miles in one hour less than the truck. How fast is the car traveling?

14. Two jets leave St. Louis at the same time, one traveling north at 520 miles per hour and the other traveling west at 450 miles per hour. How far apart are they after 2 hours?

15. Two airplanes begin 1000 miles apart and fly along lines that intersect at right angles. One plane flies an average of 100 miles per hour faster than the other. If the planes meet after 2 hours, how fast do the airplanes fly?

16. A jet leaves Chicago at noon and travels south at 600 miles per hour. One hour later a second jet leaves Chicago and travels east at 520 miles per hour. How far apart are the jets at 2:30 P.M.?

17. A pilot wishes to make a round trip between Los Angeles and San Francisco in 5 hours. The distance between the airports is 420 miles, and it is anticipated that there will be a head wind of 30 miles per hour as the plane flies north and a tail wind of 40 miles per hour when the plane returns. At what constant air speed must the plane be flown to achieve the goal?

18. A baker must make a delivery by truck to a store 12.5 miles from the bakery. By increasing the normal speed of the truck by 5 miles per hour, the baker could reduce the delivery time by 5 minutes. How fast does the baker normally drive?

19. The average speed of a commuter on a 20-mile trip into downtown Chicago is 16 miles per hour slower at rush hour than at midday, and the trip takes 20 minutes longer. What are the two rates?

20. Find two consecutive odd integers whose product is 195.

21. Find two consecutive integers whose product is 1056.

22. One positive number is 3 more than twice a second positive number and the sum of their squares is 194. What are the numbers?

23. Find the two points on the y axis that are a distance of 6 units from the point (4, 0) on the x axis.

24. A rectangular garden has an area of 1750 square feet and is 15 feet longer than it is wide. Find its dimensions.

25. The perimeter of a rectangle is 32 inches, and the area is 63 square inches. What are the dimensions of the rectangle?

26. A rectangular corral adjoins a barn. The corral has an area of 2352 square feet and is enclosed by a fence 140 feet long. If the barn forms one side of the corral, find the possible dimensions of the corral.

27. A rectangular lawn is 80 feet long and 60 feet wide. How wide a strip must be mowed around the lawn for half of the lawn to be cut?

28. A rectangular cloth measures 20 inches by 24 inches. We wish to embroider a strip of equal width on each side of the cloth in such a way that the cloth with embroidery will be rectangular with an area of 672 square inches. How wide a strip must we embroider?

29. A wire 44 inches long is cut into two pieces, each of which is bent into the form of a square. If the sum of the areas of the squares is 65 square inches, how long were the pieces of wire?

30. A square sheet of metal has sides of length 8 inches. A square piece is cut from each of the corners and the edges are folded up to form a pan (Figure 1.11). If the area of the base of the pan is equal to the sum of the areas of the sides of the pan, what is the length of the sides of the squares that were cut out?

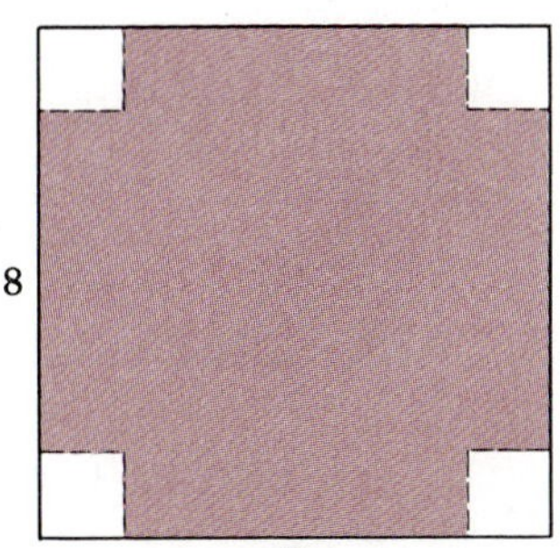

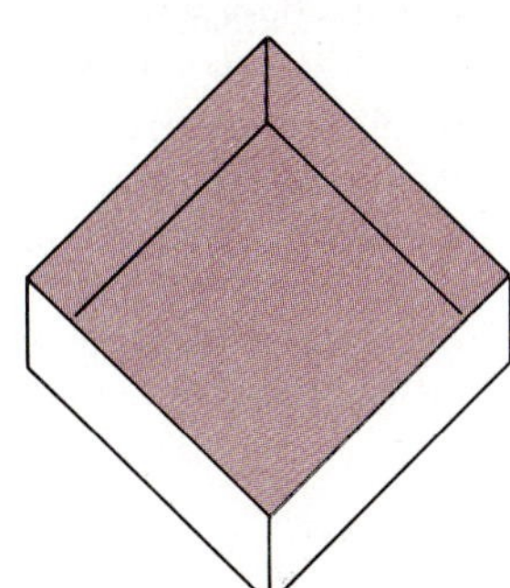

FIGURE 1.11

31. The volume of a cylindrical tin can is 48π cubic inches. If the can is 3 inches tall, find the radius of the base. (*Hint:* The volume is given by $V = \pi r^2 h$, where V, r, and h denote the volume, radius of the base, and height, respectively.)

32. A circular pool has a perimeter of 125 meters. A path of constant width is to surround the pool. The total area of the pool and path is to be 2000 square meters.

a. Find the width of the path.

b. Approximate the width of the path to the nearest hundredth of a meter.

33. On a certain rainy day the manager of a store estimates that if the price of umbrellas is set at p dollars, then $10(10 - p)$ umbrellas will be sold. Suppose the store acquires umbrellas at a cost of \$3 each.

a. What are the prices at which the store can sell umbrellas and neither make nor lose money on the sale (that is, at which the profit is 0)? How many umbrellas will be sold at those prices?

b. What is the minimum price at which the store can sell umbrellas and make a profit of exactly \$100? How many umbrellas will be sold at the price?

34. One can calculate the distance from ground level to water level in a well by dropping a stone from the ground and measuring the time t_0 elapsed until the splash is heard at ground level. Now $t_0 = t_1 + t_2$, where t_1 is the time it takes for the stone to hit the water and t_2 is the time it takes the echo to return to ground level. If the distance is s, then $s = 16t_1^2$, so that

$$t_1 = \sqrt{\frac{s}{16}} = \frac{1}{4}\sqrt{s}$$

With the assumption that sound travels at 1100 feet per second, the distance is also given by $s = 1100t_2$, which means that

$$t_2 = \frac{s}{1100}$$

Consequently

$$t_0 = t_1 + t_2 = \frac{1}{4}\sqrt{s} + \frac{s}{1100}$$

Approximate the distance if the time between drop and echo is 5 seconds. Round your answer to the nearest hundredth of a foot.

35. When an 18-foot tall stalk of bamboo is broken, the top portion of the stalk bends over and touches the ground 6 feet from the base of the stalk (Figure 1.12). How far from the ground was the bamboo broken? (This problem appeared in Chinese algebra book in 1261.)

36. If a cannon is fired at an angle of 30° with respect to the ground, then until the ball hits the ground, the height y of the cannon ball after t seconds is given by $y = \frac{1}{2}v_0 t - 16t^2$, where v_0 is the initial speed of the

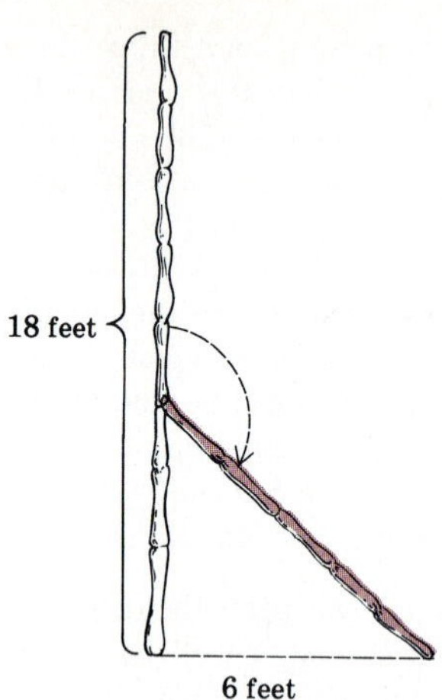

FIGURE 1.12

ball. Determine how long the cannon ball is in the air before hitting the ground.

*37. A geometric proof of the Pythagorean Theorem is suggested by Figure 1.13. Let the hypotenuse have length c and the legs lengths a and b with $a \leq b$. Assuming that a square of side c can be disected into four copies of the right triangle and a square of side $b - a$, and using the fact that the area of the triangle is $\frac{1}{2}ab$, prove that $c^2 = a^2 + b^2$.

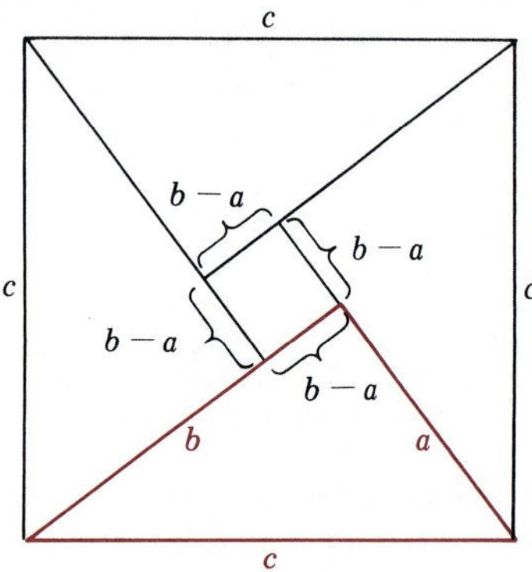

FIGURE 1.13

38. In 1876, James Garfield, who later became President, published a proof of the Pythagorean Theorem in the New England Journal of Education. His proof entailed letting ABC be a right triangle and placing a congruent triangle CDE, as in Figure 1.14, so that ACD is a straight line. He then showed that $a^2 + b^2 = c^2$ by equating the area of the trapezoid $ADEB$ with the sum of the areas of triangles ABC, CDE, and BCE. Recreate his proof, filling in the details. (*Hint:* Observe that angle BCE is a right angle.)

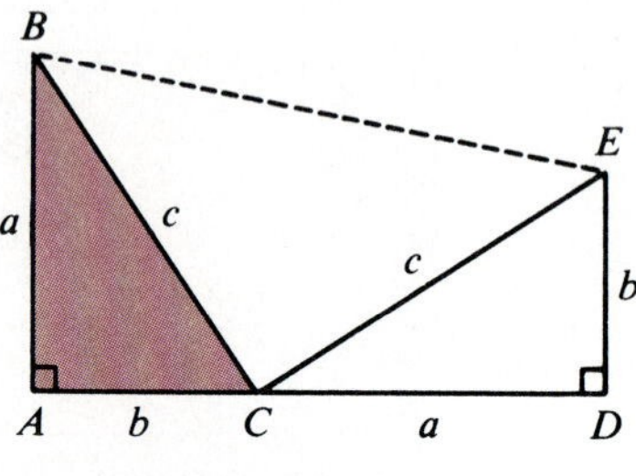

FIGURE 1.14

*39. A proof of the Pythagorean Theoren that uses similar triangles is suggested by Figure 1.15. Let the given right triangle have sides of length a, b, and c, with c the length of the hypotenuse. Using the fact that triangle ABC is similar to each of triangles ADC and BCD, prove that $c^2 = a^2 + b^2$. (*Hint:* Find expressions for a^2 and b^2 using similar triangles, and then add them, using the fact that $s + t = c$.)

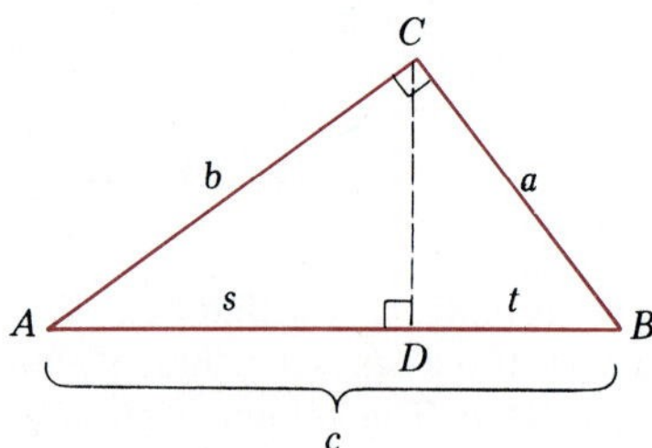

FIGURE 1.15

1.4 COMPLEX NUMBERS

In Section 1.2 we studied the real solutions of quadratic equations of the form

$$ax^2 + bx + c = 0, \quad \text{with } a \neq 0 \tag{1}$$

We found that if $b^2 - 4ac \geq 0$, then the real solutions of (1) are given by the quadratic formula

$$x = \frac{-b \pm \sqrt{b^2 - 4ac}}{2a} \tag{2}$$

However, if $b^2 - 4ac < 0$, then $\sqrt{b^2 - 4ac}$, and hence the formula in (2), are meaningless to us at this point because we have not defined square roots of negative numbers.

In fact, if $b^2 - 4ac < 0$, then there are no real solutions of (1). Consider, for example, the equation

$$x^2 + 1 = 0$$

for which $b^2 - 4ac = -4 < 0$. Since $x^2 + 1 \geq 1 > 0$ for all real numbers x, there are no real solutions of $x^2 + 1 = 0$. Therefore if we wish to have solutions for $x^2 + 1 = 0$, as well as for other quadratic equations of the form (1), it will be necessary to introduce new (non-real) numbers.

In this section we will introduce complex numbers, and show how they can be used to define square roots of negative numbers. It will follow that the formula in (2) will then be meaningful, even if $b^2 - 4ac < 0$. Moreover, it turns out that (2) will provide solutions for any real values of a, b, and c.

We begin our study of complex numbers by introducing a new number i with the property that

$$\boxed{i^2 = -1}$$

Notice that i is *not* a real number, since the square of any real number is always nonnegative. Nevertheless, i is a solution of $x^2 + 1 = 0$, since $i^2 + 1 = (-1) + 1 = 0$. In order to have solutions for the other quadratic equations for which $b^2 - 4ac < 0$, such as

$$x^2 + 9 = 0 \quad \text{and} \quad x^2 + x + 1 = 0 \tag{3}$$

we introduce numbers of the form $a + bi$, where a and b are any real numbers, and call $a + bi$ a ***complex number***.* Later we will show that every

* In some other fields, especially electrical engineering, j is used in place of i, since i refers to electric current.

quadratic equation (in particular, either of those in (3)) has a solution among the complex numbers. We will also show that i may be identified with a pair of real numbers.

The real number a is called the ***real part*** of $a + bi$, and the real number b is called the ***imaginary part*** of $a + bi$. Thus the real part of $-2 + 3i$ is -2 and the imaginary part is 3. Two complex numbers $a + bi$ and $c + di$ are said to be ***equal*** if $a = c$ and $b = d$, that is, if their real parts are equal and their imaginary parts are equal. If a is a real number, then $a + 0i$ is identified with the real number a, and we write a in place of $a + 0i$; in particular, $0 + 0i$ is written 0. Similarly, we write bi instead of $0 + bi$, and we call bi a ***pure imaginary number***, or just an imaginary number. Thus $\frac{1}{4}i$ and $-i$ are pure imaginary numbers. Sometimes, to avoid confusion, we write $a + bi$ as $a + ib$.

Addition and Multiplication of Complex Numbers

The value of complex numbers to mathematicians and scientists is greatly enhanced by the fact that it is possible to define an addition and multiplication for them that retain the usual laws of arithmetic (such as commutativity and associativity). For the ***sum*** of two complex numbers $a + bi$ and $c + di$ we have

$$\boxed{(a + bi) + (c + di) = (a + c) + (b + d)i} \qquad (4)$$

In other words, we add two complex numbers by adding their real parts and their imaginary parts separately and joining the results as indicated.

EXAMPLE 1. Write the following in the form $a + bi$.

a. $(4 + 2i) + (-3 + 5i)$ b. $(3 - 6i) + (-1 + 6i)$

Solution.

a. By (4),

$$(4 + 2i) + (-3 + 5i) = (4 - 3) + (2 + 5)i = 1 + 7i$$

b. By (4),

$$(3 - 6i) + (-1 + 6i) = (3 - 1) + (-6 + 6)i = 2 + 0i = 2 \quad \square$$

From (4) it follows that

$$(a + bi) + (-a - bi) = (a - a) + (b - b)i = 0 + 0i = 0$$

so $-a - bi$ is the additive inverse of $a + bi$ and is denoted $-(a + bi)$. Thus

$$\boxed{-(a + bi) = -a - bi}$$

For the ***difference*** of two complex numbers $a + bi$ and $c + di$ we have

$$(a + bi) - (c + di) = (a - c) + (b - d)i \tag{5}$$

EXAMPLE 2. Write the difference $(5 + 3i) - (-2 + 4i)$ in the form $a + bi$.

Solution. By (5),

$$(5 + 3i) - (-2 + 4i) = [5 - (-2)] + (3 - 4)i = 7 - i \quad \square$$

To introduce the product of two complex numbers $a + bi$ and $c + di$, we multiply as if the numbers were polynomials in i and use the fact that $i^2 = -1$:

$$\begin{aligned}(a + bi)(c + di) &= ac + adi + bci + bdi^2 \\ &= ac + (ad + bc)i - bd\end{aligned}$$

By rearranging the terms on the right side of the last line, we obtain the ***product*** of $a + bi$ and $c + di$:

$$(a + bi)(c + di) = (ac - bd) + (ad + bc)i \tag{6}$$

It is not necessary to memorize formula (6), because to find the product $(a + bi)(c + di)$ we can simply multiply the individual numbers in the parentheses, as we did above, and use the fact that $i^2 = -1$.

EXAMPLE 3. Write the product of $4 - 5i$ and $-2 - \mathrm{i}$ in the form $a + bi$.

Solution. Multiplying out, we find that

$$\begin{aligned}(4 - 5i)(-2 - i) &= (4)(-2) + (4)(-i) + (-5i)(-2) + (-5\mathrm{i})(-i) \\ &= -8 - 4i + 10i + 5i^2 \\ &= -8 - 4i + 10i - 5 \\ &= -13 + 6i \quad \square\end{aligned}$$

Among products of complex numbers are powers of i. We list some of these powers below.

$$\begin{aligned}
i^1 &= i & i^5 &= i^4 i = (1)i = i \\
i^2 &= -1 & i^6 &= i^4 i^2 = (1)i^2 = -1 \\
i^3 &= i^2 i = (-1)i = -i & i^7 &= i^4 i^3 = (1)i^3 = -i \\
i^4 &= i^2 i^2 = (-1)(-1) = 1 & i^8 &= i^4 i^4 = (1)(1) = 1
\end{aligned}$$

As this list suggests, the powers of i repeat every 4 powers, and because $i^4 = 1$, we can compute any integral power of i. For example,

$$i^{15} = i^4 \cdot i^4 \cdot i^4 \cdot i^3 = (1)(1)(1)(-i) = -i$$

Although we will not do so, it is possible to prove that all the algebraic properties (such as commutativity and associativity) of real numbers remain true for addition and multiplication of complex numbers. Thus all the algebraic formulas we derived for real numbers hold for complex numbers. For example,

$$(v + w)^3 = v^3 + 3v^2w + 3vw^2 + w^3 \tag{7}$$

for any complex numbers v and w.

EXAMPLE 4. Let $z = 2 - 3i$. Write z^3 in the form $a + bi$.

Solution. We use (7) with $v = 2$ and $w = -3i$:

$$\begin{aligned} z^3 &= (2 - 3i)^3 \\ &= 2^3 + 3(2^2)(-3i) + 3(2)(-3i)^2 + (-3i)^3 \\ &= 8 - 36i + 54i^2 - 27i^3 \\ &= 8 - 36i - 54 + 27i \\ &= -46 - 9i \quad \square \end{aligned}$$

Square Roots of Negative Numbers

Since $i^2 = -1$, we may think of i as the square root of -1. But observe that

$$(-i)^2 = i^2 = -1$$

so that we could just as well think of $-i$ as the square root of -1. Similarly, if c is any positive real number, then

$$(\sqrt{c}i)^2 = -c \quad \text{and} \quad (-\sqrt{c}i)^2 = -c \tag{8}$$

We call $\sqrt{c}i$ the ***principal square root*** of $-c$ and write

$$\boxed{\sqrt{-c} = \sqrt{c}i} \tag{9}$$

Thus

$$\sqrt{-1} = i, \qquad \sqrt{-4} = \sqrt{4}i = 2i, \quad \text{and} \quad \sqrt{-27} = \sqrt{27}i = 3\sqrt{3}i$$

It follows from (9) that negative numbers as well as nonnegative numbers have square roots when complex numbers are admitted.

CAUTION: If c and d are positive real numbers, then

$$\sqrt{cd} = \sqrt{c}\sqrt{d}$$

However, this equation is *not* valid if c and d are negative numbers. Indeed,

$$\sqrt{(-4)(-9)} = \sqrt{36} = 6$$

whereas by (9),

$$\sqrt{-4}\sqrt{-9} = (\sqrt{4}i)(\sqrt{9}i) = (2i)(3i) = 6i^2 = -6$$

Thus

$$\sqrt{(-4)(-9)} \neq \sqrt{(-4)}\sqrt{(-9)}$$

Solutions of a Quadratic Equation with Real Coefficients

Complex numbers were introduced to afford solutions of polynomial equations. We begin with equations of the form

$$x^2 + c = 0 \quad \text{with} \quad c > 0$$

If we let $x = \sqrt{c}i$ or $x = -\sqrt{c}i$, then by (8),

$$x^2 = -c, \qquad \text{or equivalently,} \quad x^2 + c = 0$$

Consequently $\sqrt{c}i$ and $-\sqrt{c}i$ are solutions of $x^2 + c = 0$ if $c > 0$, and in fact they are the only solutions of the equation.

EXAMPLE 5. Find the solutions of $x^2 + 9 = 0$.

Solution. By our preceding comments, the solutions of $x^2 + 9 = 0$ are $\sqrt{9}i$ and $-\sqrt{9}i$, that is, $3i$ and $-3i$. □

Now we turn to solutions of the quadratic equation

$$ax^2 + bx + c = 0$$

where a, b, and c are real numbers and $a \neq 0$. The solutions can be obtained by the quadratic formula

$$x = \frac{-b \pm \sqrt{b^2 - 4ac}}{2a} \tag{10}$$

If $b^2 - 4ac \geq 0$, this is just the quadratic formula appearing in Section 1.2, and the solutions are real numbers. However, if $b^2 - 4ac < 0$, then $\sqrt{b^2 - 4ac}$ is the square root of a negative number, and therefore we can use (9) to identify the solutions. We illustrate the procedure in Examples 6 and 7.

EXAMPLE 6. Find the solutions of $x^2 + x + 1 = 0$.

Solution. The solutions are given by (10):

$$x = \frac{-1 \pm \sqrt{(1)^2 - 4(1)(1)}}{2} = \frac{-1 \pm \sqrt{-3}}{2}$$

By (9) with $c = 3$, we have $\sqrt{-3} = \sqrt{3}i$, so that

$$x = \frac{-1 \pm \sqrt{3}i}{2} = -\frac{1}{2} \pm \frac{\sqrt{3}}{2}i$$

Consequently there are two solutions, $-\frac{1}{2} + \frac{1}{2}\sqrt{3}i$ and $-\frac{1}{2} - \frac{1}{2}\sqrt{3}i$. □

EXAMPLE 7. Find the solutions of $3x^2 - 4x + 6 = 0$.

Solution. By (10) and (9), the solutions are given by

$$\begin{aligned} x &= \frac{-(-4) \pm \sqrt{(-4)^2 - 4(3)(6)}}{2(3)} \\ &= \frac{4 \pm \sqrt{-56}}{6} = \frac{4 \pm \sqrt{56}i}{6} = \frac{4 \pm \sqrt{4 \cdot 14}i}{6} \\ &= \frac{4 \pm 2\sqrt{14}i}{6} = \frac{2}{3} \pm \frac{1}{3}\sqrt{14}i \end{aligned}$$

Therefore there are two solutions, $\frac{2}{3} + \frac{1}{3}\sqrt{14}i$ and $\frac{2}{3} - \frac{1}{3}\sqrt{14}i$. □

As before, you should check proposed solutions to validate them. For the solutions obtained in Example 6, the check would be as follows:

$$\left(-\frac{1}{2} + \frac{\sqrt{3}}{2}i\right)^2 + \left(-\frac{1}{2} + \frac{\sqrt{3}}{2}i\right) + 1 = \left(\frac{1}{4} - \frac{1}{2}\sqrt{3}i - \frac{3}{4}\right) + \left(-\frac{1}{2} + \frac{\sqrt{3}}{2}i\right) + 1 = 0$$

$$\left(-\frac{1}{2} - \frac{\sqrt{3}}{2}i\right)^2 + \left(-\frac{1}{2} - \frac{\sqrt{3}}{2}i\right) + 1 = \left(\frac{1}{4} + \frac{1}{2}\sqrt{3}i - \frac{3}{4}\right) + \left(-\frac{1}{2} - \frac{\sqrt{3}}{2}i\right) + 1 = 0$$

Most calculators will not compute the square root of a negative number. Thus if we desire to approximate say, $\sqrt{-14}$, we would compute $\sqrt{14}$ and then affix i:

$$\sqrt{-14} = \sqrt{14}i \approx 3.741657387i$$

Conjugates

Closely related to any complex number $a + bi$ is its ***conjugate*** $a - bi$, often denoted by $\overline{a + bi}$. For example,

$$\overline{2 + 7i} = 2 - 7i \quad \text{and} \quad \overline{4 - \tfrac{1}{3}i} = 4 + \tfrac{1}{3}i$$

From the definition of the conjugate we see that a real number is equal to its conjugate. Thus

$$\overline{-5} = \overline{-5 + 0i} = -5 - 0i = -5$$

Moreover, the conjugate of a pure imaginary number is its negative. For example,

$$\overline{6i} = -6i$$

Observe that $a + bi = a - bi$ if and only if $b = 0$, which happens if and only if $a + bi$ is real. Thus a + bi is equal to its conjugate if and only if $a + bi$ is a real number.

Since

$$\overline{-\frac{1}{2} + \frac{\sqrt{3}}{2}i} = -\frac{1}{2} - \frac{\sqrt{3}}{2}i$$

it follows from the solution of Example 6 that the two solutions of the equation $x^2 + x + 1 = 0$ are conjugates of one another. Similarly, the solutions of $3x^2 - 4x + 6 = 0$ in Example 7 are also conjugates of one another. More generally, if z is a solution of $ax^2 + bx + c = 0$, where a, b, and c are real numbers, then it can be shown that the conjugate $\bar{z}$ is also a solution. Consequently, if we know that z is one non-real solution of $ax^2 + bx + c = 0$, then we know that $\bar{z}$ is the other (non-real) solution. (Of course, if the solution z is a real number, then $\bar{z} = z$ and hence yields no new solution.)

EXERCISES 1.4

In Exercises 1–18, perform the indicated operation. Express your answer in the form $a + bi$.

1. $(5 + 6i) + (-3 + 2i)$

2. $(-1 + 3i) + (9 - 8i)$

3. $5 + (-3 + 2i)$

4. $3i + (-2 - 5i)$

5. $\left(\frac{1}{2} + \frac{1}{3}i\right) + \left(\frac{3}{2} - \frac{1}{6}i\right)$

6. $\left(\frac{4}{3} - \frac{1}{5}i\right) + \left(-\frac{2}{3} + \frac{1}{5}i\right)$

7. $(3\sqrt{2} - \sqrt{5}i) + (4\sqrt{2} - 5\sqrt{5}i)$

8. $(-\sqrt{2} + \sqrt{3}i) + \left(\frac{1}{\sqrt{2}} - \frac{1}{\sqrt{3}}i\right)$

9. $(25 + 13i) - (15 + 7i)$

10. $\left(\frac{3}{2} - \frac{1}{2}i\right) - \left(\frac{1}{6} - \frac{3}{2}i\right)$

11. $(2 - 3i)(1 + 3i)$

12. $(1 - 2i)(3 - 2i)$

13. $(-4 - 3i)\left(\frac{1}{2} - i\right)$

14. $\left(2 - \frac{1}{4}i\right)(4 - 2i)$

15. $\left(\frac{1}{3} + \frac{1}{2}i\right)\left(\frac{1}{3} - \frac{1}{2}i\right)$

16. $\left(\frac{1}{3} + \frac{1}{2}i\right)\left(\frac{1}{3} + \frac{1}{2}i\right)$

17. $\left(\sqrt{2} + \frac{1}{\sqrt{2}}i\right)(2 + \sqrt{2}i)$

18. $\left(\sqrt{2} + \frac{1}{\sqrt{2}}i\right)\left(\sqrt{2} - \frac{1}{\sqrt{2}}i\right)$

In Exercises 19–26, evaluate the given expression. Express your answer in the form $a + bi$.

19. $\sqrt{-4}$ **20.** $\sqrt{-9}$

21. $\sqrt{-8}$ **22.** $\sqrt{-27}$

23. $\sqrt{i^2}$ **24.** $\sqrt{i^4}$

25. $\dfrac{-4 + \sqrt{-48}}{2}$ **26.** $\dfrac{3 - \sqrt{-36}}{2}$

In Exercises 27–42, find the solutions of the given equation. Express the solutions in the form $a + bi$.

27. $x^2 + 4 = 0$ **28.** $x^2 + 25 = 0$

29. $x^2 + 12 = 0$ **30.** $x^2 + 28 = 0$

31. $x^2 - 2x + 5 = 0$ **32.** $x^2 + 2x + 10 = 0$

33. $2x^2 - 2x + 1 = 0$ **34.** $5x^2 + 8x + 5 = 0$

35. $x^2 - x + 1 = 0$ **36.** $x^2 + 2x + 3 = 0$

37. $x^2 - 2x + 3 = 0$ **38.** $x^2 + x + 2 = 0$

39. $x^2 - 3x + 5 = 0$ **40.** $2x^2 - 7x + 10 = 0$

41. $-3x^2 + 2x - 2 = 0$ **42.** $-\frac{1}{2}x^2 - 3x - 5 = 0$

In Exercises 43–44, find the exact solutions of the given equation. Then approximate the solutions, rounding both the real and imaginary parts to the nearest thousandth.

43. $x^2 + 2x + \pi = 0$ **44.** $\sqrt{2}x^2 + x + 2 = 0$

In Exercises 45–52, find the conjugate of the given complex number. Express your answer in the form $a + bi$.

45. $2 + 3i$ **46.** $4 - 5i$ **47.** $-\frac{1}{2} + \frac{1}{3}i$

48. $-\frac{2}{3} - \frac{4}{5}i$ **49.** $9i$ **50.** 13

51. $\overline{2 - 3i}$ **52.** $\overline{4i + 1}$

53. Show that $1 + i$ is a solution of the equation $x^2 - 2ix - 2 = 0$, but that the conjugate of $1 + i$ is not a solution of the equation. (Thus if a polynomial does not have real coefficients, then the conjugate of a root of the polynomial need not be a root of the polynomial.)

1.5 OTHER TYPES OF EQUATIONS

In the first four sections of this chapter we discussed linear and quadratic equations and found formulas for their solutions. Rather than moving on to third-degree equations, whose solutions are generally much more difficult to obtain, we will devote this section to several general types of equations that can be modified and then solved by methods we have already described. To simplify matters, we will only seek real solutions of the equations.

Quadratic-Type Equations

Each of the equations

$$(3 + x)^2 - 5(3 + x) + 4 = 0,$$

$$x^4 - 7x^2 + 10 = 0,$$

and

$$x - 4\sqrt{x} + 4 = 0$$

has the feature that if we substitute u for a suitable expression in x, the resulting equation in u is a quadratic equation. As a result, such an equation is called an ***equation of quadratic type***. It is solved by first solving the associated quadratic equation in u and then solving for x in terms of u.

EXAMPLE 1. Find all solutions of $(3 + x)^2 - 5(3 + x) + 4 = 0$.

Solution. If we substitute u for $3 + x$, the equation becomes

$$u^2 - 5u + 4 = 0$$

which can be factored to yield

$$(u-4)(u-1)=0$$

Therefore $u=4$ or $u=1$. Since $u=3+x$, we find that $x=u-3$. Consequently if $u=4$ then $x=4-3=1$, and if $u=1$ then $x=1-3=-2$. It follows that 1 and -2 are the proposed solutions of the given equation.

Check: For $x=1, (3+1)^2-5(3+1)+4=16-20+4=0$

For $x=-2, (3+(-2))^2-5(3+(-2))+4$

$=1-5+4=0$

Thus 1 and -2 are the solutions. □

EXAMPLE 2. Find all solutions of $x^4-7x^2+10=0$.

Solution. This time we substitute u for x^2, which in particular implies that $u \geq 0$. The given equation becomes

$$u^2-7u+10=0, \quad \text{with} \quad u \geq 0$$

which yields

$$(u-5)(u-2)=0, \quad \text{with} \quad u \geq 0$$

The solutions of the associated equation in u are 5 and 2, both of which are positive. Since $u=x^2$, it follows that $x=\sqrt{u}$ or $x=-\sqrt{u}$, so that the proposed solutions of the given equation are $\sqrt{5}$, $-\sqrt{5}$, $\sqrt{2}$, and $-\sqrt{2}$.

Check: For $x=\sqrt{5}, (\sqrt{5})^4-7(\sqrt{5})^2+10=25-35+10=0$

For $x=-\sqrt{5}, (-\sqrt{5})^4-7(-\sqrt{5})^2+10=25-35+10=0$

For $x=\sqrt{2}, (\sqrt{2})^4-7(\sqrt{2})^2+10=4-14+10=0$

For $x=-\sqrt{2}, (-\sqrt{2})^4-7(-\sqrt{2})^2+10=4-14+10=0$

Thus $\sqrt{5}$, $-\sqrt{5}$, $\sqrt{2}$, and $-\sqrt{2}$ are the solutions of the given equation. □

If the equation given in Example 2 had been

$$x^4+7x^2+10=0 \tag{1}$$

then we would have substituted u for x^2 as before, and the associated equation in u would have been

$$u^2+7u+10=0, \quad \text{with} \quad u \geq 0$$

so that

$$(u+5)(u+2)=0, \quad \text{with} \quad u \geq 0 \tag{2}$$

But both of the solutions -5 and -2 of the equation in (2) are negative, so there are no nonnegative solutions that satisfy (2). Consequently there are no real solutions of the equation in (1).

EXAMPLE 3. Find all solutions of $x - 4\sqrt{x} + 4 = 0$.

Solution. If we substitute u for $\sqrt{x}$ and note that $u \geq 0$, then we find that the given equation is transformed into

$$u^2 - 4u + 4 = 0, \quad \text{with} \quad u \geq 0$$

which yields

$$(u - 2)^2 = 0, \quad \text{with} \quad u \geq 0$$

Therefore $u = 2$. Since $u = \sqrt{x}$, we have $x = u^2$, so that if $u = 2$, then $x = 2^2 = 4$. Consequently the proposed solution of the given equation is 4.

Check: For $x = 4, 4 - 4\sqrt{4} + 4 = 4 - 8 + 4 = 0$

Thus 4 is the only solution of the given equation. □

Equations Involving Radicals

As illustrated in Example 3, an equation involving radicals occasionally arises. Often the solutions are most easily ascertained if we first eliminate the radicals by substitution or by squaring both sides of the equation, and then solve the new equation. We must be careful to check all solutions of the new equation to determine which, if any, yield solutions of the original equation.

EXAMPLE 4. Find all solutions of $\sqrt{3x + 4} = 5$.

Solution. Let us square both sides of the equation and then solve for x:

$$\sqrt{3x + 4} = 5$$
$$3x + 4 = 25$$
$$3x = 21$$
$$x = \frac{21}{3} = 7$$

Check: For $x = 7, \sqrt{3(7) + 4} = \sqrt{25} = 5$

Therefore 7 is the only solution of the given equation. □

EXAMPLE 5. Find all solutions of the equation $\sqrt{4 - 3x} = x$.

Solution. Following the same procedure of squaring both sides of the equation, we obtain

$$\sqrt{4 - 3x} = x$$
$$4 - 3x = x^2$$

$$x^2 + 3x - 4 = 0$$

$$(x - 1)(x + 4) = 0$$

Therefore the proposed solutions are 1 and -4.

Check: For $x = 1$, $\sqrt{4 - 3(1)} = 1$

For $x = -4$, $\sqrt{4 - 3(-4)} = \sqrt{16} = 4 \neq -4$, so -4 is not a solution.

Consequently the only solution of the given equation is 1. □

In working the preceding example, we altered the original equation and found two solutions of a new equation, but one of these turned out not to be a real solution of the original equation! What went wrong? The answer is that we squared both sides of the original equation, and as frequently happens, squaring both sides of the equation did *not* lead to an equivalent equation. For a simple example in which squaring leads to a nonequivalent equation, observe that

$$x = 2$$

is *not* equivalent to the equation obtained when both sides are squared:

$$x^2 = 4$$

Indeed, $x = 2$ has one solution, 2, whereas $x^2 = 4$ has two solutions, 2 and -2.

All solutions of a given equation are retained when we square both sides, but other candidates, which actually are not solutions, may be introduced by squaring. A solution of an altered equation that does not satisfy the original equation is called an ***extraneous solution***. Thus -4 is an extraneous solution of the equation $\sqrt{4 - 3x} = x$ of Example 5. As we have seen, extraneous solutions can be introduced when we square both sides of an equation. The same is true if we raise both sides of an equation to any *even* power (see Exercise 65).

CAUTION: Because of the possibility of introducing extraneous solutions when we raise both sides of a given equation to an even power, it is doubly important to check proposed solutions in these cases.

In the next example we will cube each side of an equation. Yet no extra candidates for solutions will be introduced. More generally, when both sides of an equation are raised to an *odd* power, no extraneous real solutions are introduced.

EXAMPLE 6. Find all solutions of $\sqrt[3]{x^2 - 8} = 2$.

Solution. We cube both sides of the equation and then simplify:

$$\begin{aligned} \sqrt[3]{x^2 - 8} &= 2 \\ x^2 - 8 &= 8 \\ x^2 &= 16 \end{aligned}$$

Therefore the proposed solutions are 4 and -4.

Check: For $x = 4$, $\sqrt[3]{4^2 - 8} = \sqrt[3]{8} = 2$
For $x = -4$, $\sqrt[3]{(-4)^2 - 8} = \sqrt[3]{8} = 2$

Consequently 4 and -4 are the solutions of the given equation. □

In the following example we will square both sides of an equation twice, because squaring once does not clear the equation of radicals.

EXAMPLE 7. Find the solutions of $\sqrt{x + 2} - \sqrt{3x - 5} = 1$.

Solution. First we alter the equation so that one radical appears on each side of the equation. Then we square both sides and simplify:

$$\begin{aligned} \sqrt{x + 2} - \sqrt{3x - 5} &= 1 \\ \sqrt{x + 2} &= 1 + \sqrt{3x - 5} \\ x + 2 &= 1 + 2\sqrt{3x - 5} + (3x - 5) \\ -2x + 6 &= 2\sqrt{3x - 5} \\ -x + 3 &= \sqrt{3x - 5} \end{aligned}$$

Squaring again, we obtain

$$\begin{aligned} x^2 - 6x + 9 &= 3x - 5 \\ x^2 - 9x + 14 &= 0 \\ (x - 2)(x - 7) &= 0 \end{aligned}$$

Thus the proposed solutions are 2 and 7.

Check: For $x = 2$, $\sqrt{2 + 2} - \sqrt{3(2) - 5} = \sqrt{4} - \sqrt{1} = 1$

For $x = 7$, $\sqrt{7 + 2} - \sqrt{3(7) - 5} = \sqrt{9} - \sqrt{16} = -1 \neq 1$

Therefore 2 is a genuine solution of the given equation, whereas 7 is an extraneous solution. We conclude that the given equation has one real solution, 2. □

Equations Involving Integral Exponents

Some equations involving integral exponents can be solved for a variable by taking roots. The next two examples illustrate this feature.

EXAMPLE 8. Let T be the time in hours required for a planet to orbit its sun once, and let a be the average distance in miles between the planet and its sun. Kepler's Third Law states that $T^2 = ca^3$, where c is a nonzero constant. Solve the equation for a.

Solution. From the equation $T^2 = ca^3$ we have

$$a^3 = \frac{T^2}{c}$$

Therefore by taking cube roots of both sides, we obtain

$$a = \left(\frac{T^2}{c}\right)^{1/3} = \frac{T^{2/3}}{c^{1/3}} \quad \square$$

EXAMPLE 9. Solve the following equations for x.

a. $(x^2 - 1)^3 = 27$ b. $81x^4 = 16$

Solution.

a. We take cube roots and then solve for x:

$$x^2 - 1 = \sqrt[3]{27} = 3$$
$$x^2 = 3 + 1 = 4$$
$$x = 2 \quad \text{or} \quad -2$$

Check: For $x = 2$, $(2^2 - 1)^3 = 3^3 = 27$

For $x = -2$, $[(-2)^2 - 1]^3 = 3^3 = 27$

Consequently the solutions are 2 and -2.

b. Here we divide both sides by 81 and then take fourth roots:

$$x^4 = \frac{16}{81}$$
$$|x| = \sqrt[4]{\frac{16}{81}} = \sqrt[4]{\frac{2^4}{3^4}} = \frac{2}{3}$$

Therefore the proposed solutions are $\frac{2}{3}$ and $-\frac{2}{3}$.

Check: For $x = \frac{2}{3}$, $81(\frac{2}{3})^4 = 81(\frac{16}{81}) = 16$

For $x = -\frac{2}{3}$, $81(-\frac{2}{3})^4 = 81(\frac{16}{81}) = 16$

Consequently the solutions are $\frac{2}{3}$ and $-\frac{2}{3}$. $\square$

Other Equations

In this final part of the section we will analyze equations that do not fall into any of the earlier categories, but whose solutions can be found by means of factorization and the Zero Property (which appeared in Section 0.1).

EXAMPLE 10. Find all solutions of $x^3 - 6x^2 + 8x = 0$.

Solution. We can factor out an x from the equation, which yields

$$x(x^2 - 6x + 8) = 0$$

and then factor $x^2 - 6x + 8$ to obtain

$$x(x - 2)(x - 4) = 0$$

Thus by the Zero Property the proposed solutions are 0, 2, and 4.

Check: For $x = 0$, $0^3 - 6(0)^2 + 8(0) = 0$

For $x = 2$, $2^3 - 6(2)^2 + 8(2) = 8 - 24 + 16 = 0$

For $x = 4$, $4^3 - 6(4)^2 + 8(4) = 64 - 96 + 32 = 0$

Therefore 0, 2, and 4 are the solutions of the given equation. □

EXAMPLE 11. Find all solutions of $(x + 2)(2x^2 - 5x - 7) = 0$.

Solution. By the Zero Property the solutions of this equation consist of all solutions of $x + 2 = 0$ and all those of $2x^2 - 5x - 7 = 0$. Now for $x + 2 = 0$ we have the solution -2. For $2x^2 - 5x - 7 = 0$ we use the quadratic formula (or factor directly, if you wish):

$$x = \frac{-(-5) \pm \sqrt{(-5)^2 - 4(2)(-7)}}{2(2)} = \frac{5 \pm \sqrt{81}}{4} = \frac{5 \pm 9}{4}$$

Thus $x = \frac{14}{4} = \frac{7}{2}$ or $x = -\frac{4}{4} = -1$, so the proposed solutions of the given equation are $-2, \frac{7}{2}$, and -1.

Check: For $x = -2$, $(-2 + 2)[2(-2)^2 - 5(-2) - 7]$

$= (0)(8 + 10 - 7) = 0$

For $x = \frac{7}{2}$, $\left(\frac{7}{2} + 2\right)\left[2\left(\frac{7}{2}\right)^2 - 5\left(\frac{7}{2}\right) - 7\right]$

$= \left(\frac{7}{2} + 2\right)\left(\frac{49}{2} - \frac{35}{2} - 7\right) = \left(\frac{7}{2} + 2\right)(0) = 0$

$$\text{For } x = -1, (-1 + 2)[2(-1)^2 - 5(-1) - 7]$$
$$= (-1 + 2)(2 + 5 - 7) = 1(0) = 0$$

Therefore -2, $\frac{7}{2}$, and -1 are the solutions of the given equation. □

EXAMPLE 12. Find all solutions of $x^{3/2} = 2x^{1/2}$.

Solution. We alter the equation so that it becomes

$$x^{3/2} - 2x^{1/2} = 0$$

which is equivalent to

$$x^{1/2}(x - 2) = 0$$

Thus the proposed solutions are 0 and 2.

Check: For $x = 0$, $0^{3/2} = 0$ and $(2)(0)^{1/2} = 0$

For $x = 2$, $2^{3/2} = 2^{1+1/2} = 2 \cdot 2^{1/2}$

Consequently 0 and 2 are the solutions of the given equation. □

EXERCISES 1.5

In Exercises 1–16, find all real solutions (if any) of the given equation.

1. $(x + 2)^2 + 11(x + 2) + 18 = 0$
2. $(x^2 - 3)^2 - 5(x^2 - 3) - 14 = 0$
3. $(x - 2)^2 + x - 32 = 0$
4. $(x^2 + x)^2 - 5(x^2 + x) - 6 = 0$
5. $x^4 - 6x^2 + 8 = 0$
6. $x^4 - 8x^2 + 15 = 0$
7. $x^4 - 8x^2 - 9 = 0$
8. $x^4 + 5x^2 + 6 = 0$
9. $x^6 - 27x^3 - 28 = 0$
10. $x^8 - 4x^4 - 12 = 0$
11. $x - \sqrt{x} - 12 = 0$
12. $x + 7\sqrt{x} + 10 = 0$
13. $x^{1/2} + 3x^{1/4} - 18 = 0$
14. $x^{1/2} - 2x^{1/4} + 1 = 0$
15. $x^{2/3} - 3x^{1/3} + 2 = 0$
16. $x^{4/3} - 5x^{2/3} + 6 = 0$

In Exercises 17–32, find all real solutions of the given equation.

17. $\sqrt{2x + 4} = 6$
18. $\sqrt{1 - x} = 4$
19. $\sqrt{4 + x^2} = 3$
20. $\sqrt{x^2 + 2x} = 2\sqrt{2}$
21. $\sqrt{x^2 - 5x} = \sqrt{14}$
22. $\sqrt{2x + 3} = x$
23. $\sqrt{6x - 1} = 3x$
24. $\sqrt[3]{x^2 + 2} = 3$
25. $\sqrt[3]{3x^2 - 1} = 2$
26. $\sqrt{3 - x} = \sqrt{5 - x^2}$
27. $\sqrt{4x - 5} = \sqrt{x^2 - 2x}$
28. $\sqrt{2x - 1} = 3 + \sqrt{x - 5}$
29. $\sqrt{5 - x} + 1 = \sqrt{7 + 2x}$
30. $\sqrt{x + 1} + \sqrt{x - 1} = \sqrt{2x + 1}$
31. $\sqrt{3 - 2\sqrt{x}} = \sqrt{x}$

32. $\sqrt{10 + 3\sqrt{x}} = \sqrt{x} + 2$

In Exercises 33–46, find all real solutions of the given equation.

33. $8x^3 = 27$ **34.** $6x^3 = -16$

35. $27x^3 - 10 = 0$ **36.** $16x^4 - 0.0081 = 0$

37. $x^n = 2$ **38.** $n^n = 3^n x^n$

39. $x^{2n-1} + 1 = 0$ **40.** $(x^2 + 1)^2 = \frac{9}{4}$

41. $(x^2 - 1)^2 = 4$ **42.** $(x^2 - 9)^2 = 4$

43. $(x^3 - 27)^3 = -64$ **44.** $(16x^2 - 9)^4 = 0$

45. $x^9 + x^4 = 0$ **46.** $\left(1 + \frac{1}{x}\right)^{100} = 2$

In Exercises 47–52, find all real solutions of the given equation.

47. $x^5 + x^3 - 2x = 0$

48. $x^3 - 16x^2 + 48x = 0$

49. $(x^2 - 4)(x^2 - 6x + 8) = 0$

50. $(x^2 - 6)(x^2 + x - 2) = 0$

51. $(x^2 - 9)(x^2 + 16) = 0$

52. $(x^3 + 8)(4x^4 - 2x^2 + \frac{1}{4}) = 0$

In Exercises 53–56, determine all real solutions of the given equation by finding common factors.

53. $x^3 - x^2 + x - 1 = 0$

54. $x^3 - x^2 - x + 1 = 0$

55. $x^4 - 3x^3 - 4x^2 + 12x = 0$

56. $x^4 + 5x^3 - 9x^2 - 45x = 0$

In Exercises 57–64, determine all solutions (real as well as complex) of the given equation.

57. $x^4 - 1 = 0$ **58.** $81x^4 = 16$

59. $x^4 + 2x^2 + 1 = 0$ **60.** $x^4 + 5x^2 + 4 = 0$

61. $x^4 - 3x^2 - 4 = 0$ **62.** $x^6 - x^4 - 2x^2 = 0$

63. $x^3 - 1 = 0$ **64.** $x^3 - 8 = 0$

65. a. Determine the extraneous solution of the equation

$$\sqrt[4]{2x^2 - 1} = x$$

that we introduce if we solve it by raising both sides to the fourth power.

b. Let m be a positive integer. Determine the extraneous solution of the equation

$$\sqrt[4m]{2x^{2m} - 1} = x$$

that we introduce if we solve it by raising both sides to the $(4m)$th power.

66. The volume V of a sphere of radius r is given by $V = \frac{4}{3}\pi r^3$. Solve the equation for r.

67. The following equation occurs in the study of electricity:

$$E^2 = \frac{Q^2}{(1 + a^2)^3}$$

Solve the equation for a.

68. The equation

$$y = \frac{100\,kx^n}{1 + kx^n}$$

appears in the study of the saturation of hemoglobin with oxygen. Solve the equation for x.

69. The equation

$$V = \frac{\pi p r^4}{8\eta l}$$

expresses the rate of volume flow V (volume per unit time) of a fluid through a cylindrical tube in terms of the radius r, the length l, the pressure difference p at the ends of the tube, and the viscosity η of the fluid (Figure 1.16). Solve the equation for r.

70. The efficiency e of a gasoline internal combustion engine is given by

$$e = 1 - \frac{1}{b^{2/5}}$$

where b denotes the combustion ratio (that is, the ratio of the maximum to the minimum volume) in

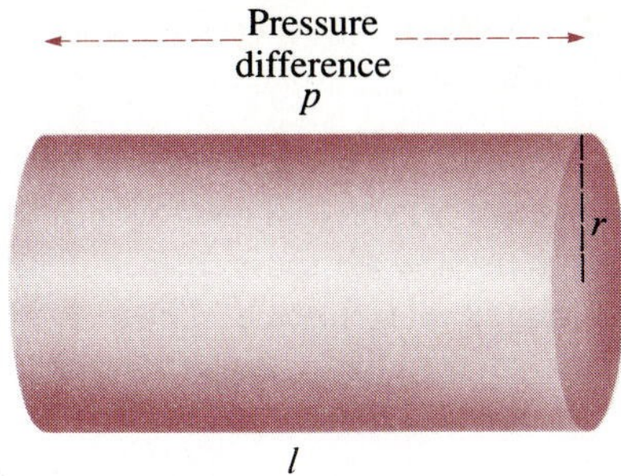

FIGURE 1.16

the cylinders. Assuming that the efficiency is 52%, approximate the combustion ratio b, and round your answer to the nearest hundredth.

71. About 150 B.C., the ancient Greek astronomer Hipparchus developed a scale for measuring the brightness of stars. The scale gave what is called the ***apparent magnitude*** of the stars visible to the naked eye, from first magnitude to sixth magnitude. The brightest 20 stars were assigned the first magnitude, the next fainter group the second magnitude, and so on until the very faintest stars visible to the eye were assigned the sixth magnitude. It turns out that, on the average, a first magnitude star is approximately 100 times as bright as a sixth-magnitude star, and moreover, there is a number c such that the average brightness of the stars of any given magnitude is c times the average brightness of stars of the next fainter magnitude.

a. Determine the number c.

b. Let d be the ratio of the average brightness of a second-magnitude star to a fifth-magnitude star. Solve for d in terms of c.

c. Approximate the numbers c and d, and round your answers to the nearest thousandth.

1.6 INEQUALITIES

So far in this chapter we have studied equations and their solutions. Although equations are fundamental to mathematics, so are inequalities. In the remainder of this chapter we will study inequalities and their solutions. Since inequalities of complex numbers are not defined, it follows that whenever we consider inequalities of numbers, we will assume that the numbers are real.

Intervals

The four basic inequalities involving real numbers a and b are

$$a < b, \qquad a \le b, \qquad a > b, \quad \text{and} \quad a \ge b$$

From Section 0.2 we know that

$$a < x < b \quad \text{means} \quad a < x \quad \text{and} \quad x < b \quad \text{simultaneously}$$

and

$$a > x \ge b \quad \text{means} \quad a > x \quad \text{and} \quad x \ge b \quad \text{simultaneously}$$

Other compound inequalities, which are defined similarly, include combinations of $<$ and $\le$ and combinations of $>$ and $\ge$.

CAUTION: We *never* use either $<$ or $\le$ together with $>$ or $\ge$ in the same compound inequality. Thus we never write an expression like $2 < 5 \ge 3$.

In all compound inequalities the inequality signs must open in the same direction ($<$ and $\leq$, for example).

With the basic inequalities and the basic compound inequalities we can describe nine categories of special sets of real numbers called ***intervals***. They are listed below. In our list we use the symbols ∞ (read "infinity") and $-\infty$ (read "minus infinity" or "negative infinity"). These two symbols *do not* represent real numbers but merely help us represent certain kinds of intervals.

Type of Interval	*Notation*	*Description*
Open interval	(a, b)	all x such that $a < x < b$
	(a, ∞)	all x such that $a < x$
	$(-\infty, a)$	all x such that $x < a$
	$(-\infty, \infty)$	all real numbers
Closed interval	$[a, b]$	all x such that $a \leq x \leq b$
	$[a, \infty)$	all x such that $a \leq x$
	$(-\infty, a]$	all x such that $x \leq a$
Half-open interval	$(a, b]$	all x such that $a < x \leq b$
	$[a, b)$	all x such that $a \leq x < b$

In the list, the symbols [and] indicate that the corresponding end-point is included in the set, whereas the symbols (and) indicate that the corresponding endpoint (if there is one) is not included in the set. The various kinds of intervals are described graphically on the real line in Figure 1.17.

The four intervals (a, b), $[a, b]$, $(a, b]$, and $[a, b)$ are ***bounded intervals***, and the remaining ones, each of which involves ∞ or $-\infty$, are ***unbounded intervals***.

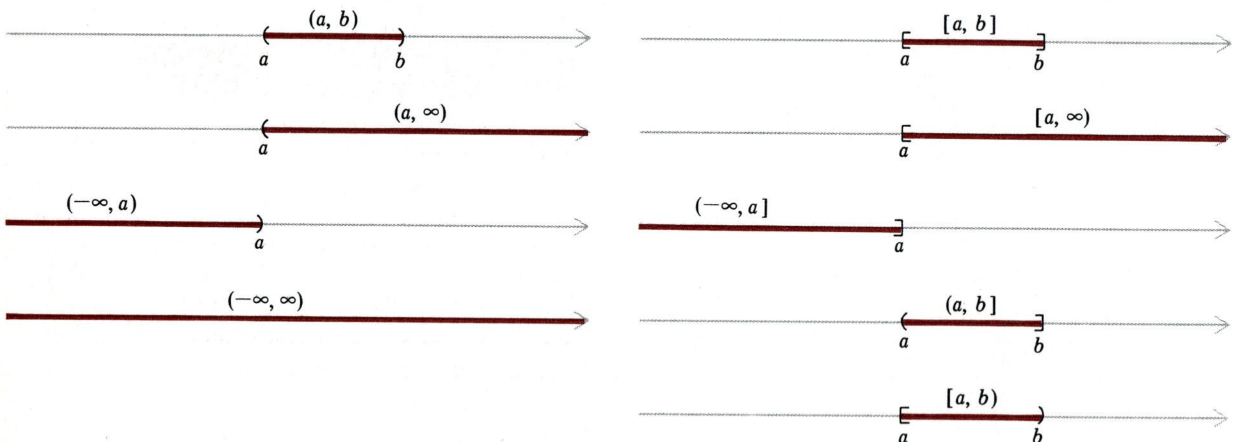

FIGURE 1.17

EXAMPLE 1. Determine whether each of the following intervals is open, closed, or half-open, and whether it is bounded or unbounded. Then locate the interval on the real line.

a. $(0, 3]$ b. $(-\infty, -2)$ c. $[-4, 1]$
d. $[-1, \infty)$ e. $(-\frac{3}{2}, -\frac{1}{3})$

Solution.

a. $(0, 3]$ is half-open and bounded.
b. $(-\infty, -2)$ is open and unbounded.
c. $[-4, 1]$ is closed and bounded.
d. $[-1, \infty)$ is closed and unbounded.
e. $(-\frac{3}{2}, -\frac{1}{3})$ is open and bounded.

The five intervals are shown in Figure 1.18. □

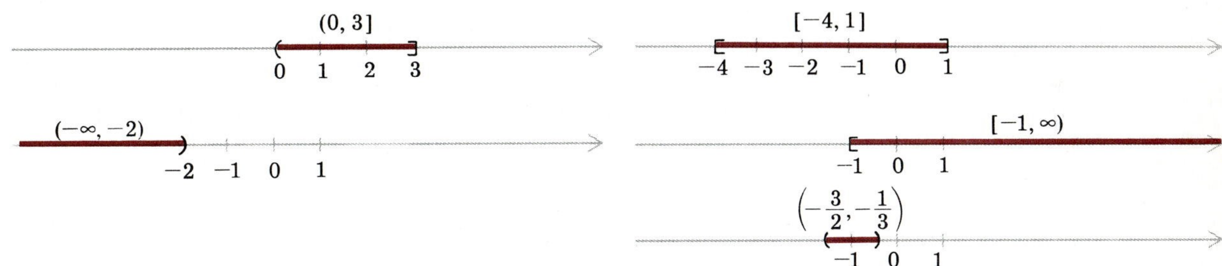

FIGURE 1.18

Basic Laws of Inequalities

In Section 0.2 we presented the law of trichotomy, which says that for any two given real numbers a and b, either $a < b$ or $a = b$ or $a > b$. We are ready now to present four more laws that will form the basis of mathematical computation involving inequalities. Throughout we will assume that a, b, c, and d are real numbers.

Laws of Inequalities

If $a < b$ and $b < c$, then $a < c$.
If $a < b$, then $a + c < b + c$.
If $a < b$ and $c > 0$, then $ac < bc$.
If $a < b$ and $c < 0$, then $ac > bc$.

Interchanging $<$ and $>$, we obtain the following versions.

If $a > b$ and $b > c$, then $a > c$.
If $a > b$, then $a + c > b + c$.
If $a > b$ and $c > 0$, then $ac > bc$.
If $a > b$ and $c < 0$, then $ac < bc$.

The laws listed above remain valid if $<$ is replaced by $\leq$ and if $>$ is replaced by $\geq$.

CAUTION: Note carefully that when both sides of an inequality are multiplied by a *negative* number, the sense of the inequality must be reversed (from $<$ to $>$, from $\leq$ to $\geq$, from $>$ to $<$, or from $\geq$ to $\leq$). Thus if $3 > -\frac{1}{2}x$, then $(-2)3 < (-2)(-\frac{1}{2}x)$, or equivalently, $-6 < x$.

Solutions of Linear Inequalities

Consider the inequality

$$5x + 2 < -3$$

with x a variable as usual. If $x = -2$, then x satisfies $5x + 2 < -3$, since

$$5(-2) + 2 = -10 + 2 = -8 < -3$$

so we say that -2 is a solution of the inequality. By contrast, if $x = 1$, then x does not satisfy $5x + 2 < -3$, because

$$5(1) + 2 = 7 > -3$$

so we say that 1 is not a solution of $5x + 2 < -3$. More generally, the ***solutions*** of a given inequality are the values of the variable that make the inequality valid. We will show below that the solutions of $5x + 2 < -3$ consist of all x such that $x < -1$, that is, they comprise the interval $(-\infty, -1)$. For simplicity, in this chapter we will only consider inequalities whose solutions consist of either an interval or a finite collection of intervals.

When we determine the solutions of a given inequality, we say that we ***solve the inequality***. This often involves finding a series of inequalities that are equivalent to the original inequality, by which we mean that they all have the same set of solutions as the original one has.

In this section we will solve ***linear inequalities***, that is, inequalities that are equivalent to an inequality having one of the forms

$$ax < b, \qquad ax \leq b, \qquad ax > b, \qquad \text{or} \qquad ax \geq b$$

We solve a linear inequality in a way analogous to the way we solved linear equations:

i. Put all terms containing x on one side of the inequality.
ii. Put all other terms on the other side of the inequality.
iii. Simplify the resulting inequality to solve for x.

EXAMPLE 2. Solve the inequality $5x + 2 < -3$.

Solution. The only term containing x is on the left side, so we need only subtract 2 from each side and then simplify:

$$5x + 2 < -3$$
$$5x < -5$$
$$\frac{1}{5}(5x) < \frac{1}{5}(-5)$$
$$x < -1$$

Since $5x + 2 < -3$ is equivalent to $x < -1$, it follows that x is a solution of the given equation if and only if $x < -1$. In other words, the solutions form the interval $(-\infty, -1)$. □

EXAMPLE 3. Solve the inequality $3x - 2 \geq 8 + 5x$.

Solution. Using (i)–(iii), we obtain the following equivalent inequalities:

$$3x - 2 \geq 8 + 5x$$
$$-2x - 2 \geq 8$$
$$-2x \geq 10$$
$$\left(-\frac{1}{2}\right)(-2x) \leq \left(-\frac{1}{2}\right)(10)$$
$$x \leq -5$$

In other words, the solutions of the original inequality consist of all $x \leq -5$, which means that they form the interval $(-\infty, -5]$. □

Observe that we reversed the sense of the inequality when we multiplied by the negative number $-\frac{1}{2}$ in the solution of Example 3.

Solutions of Composite Inequalities

Recall that the composite inequality

$$-2 < \frac{5 - x}{3} \leq 6x - 1 \qquad (1)$$

is a shorthand way of writing the pair of inequalities

$$-2 < \frac{5 - x}{3} \quad \text{and} \quad \frac{5 - x}{3} \leq 6x - 1 \qquad (2)$$

Therefore x is a solution of the composite inequality in (1) if and only if it is a solution of both inequalities in (2) simultaneously.

EXAMPLE 4. Solve the inequality $-2 < \dfrac{5-x}{3} \leq 6x - 1$.

Solution. We solve the given composite inequality by working separately on the two inequalities in (2):

$$\begin{array}{ll} -2 < \dfrac{5-x}{3} & \dfrac{5-x}{3} \leq 6x - 1 \\ (-2)(3) < \dfrac{5-x}{3}(3) & \dfrac{5-x}{3}(3) \leq (6x-1)(3) \\ -6 < 5 - x & 5 - x \leq 18x - 3 \\ x - 6 < 5 & 5 \leq 19x - 3 \\ x < 11 & 8 \leq 19x \\ & \dfrac{8}{19} \leq x \end{array}$$

Consequently the solutions consist of all values of x that satisfy $x < 11$ and $\frac{8}{19} \leq x$ simultaneously, that is, $\frac{8}{19} \leq x < 11$. Thus the solutions form the interval $[\frac{8}{19}, 11)$. □

Sometimes we can solve a composite inequality by performing the same operations on all members of the inequality. Of course, the basic laws of inequalities must be carefully observed. The procedure is illustrated in the next example.

EXAMPLE 5. Solve the composite inequality $-3 \leq \dfrac{2x+3}{-4} < 7$.

Solution. We have the following equivalent inequalities:

$$-3 \leq \frac{2x+3}{-4} < 7$$

$$(-3)(-4) \geq \left(\frac{2x+3}{-4}\right)(-4) > (7)(-4)$$

$$12 \geq 2x + 3 > -28$$

$$9 \geq 2x > -31$$

$$\frac{9}{2} \geq x > -\frac{31}{2}$$

Thus the solutions of the given inequality form the interval $(-\frac{31}{2}, \frac{9}{2})$. □

Our final example is an applied problem whose solution involves inequalities.

EXAMPLE 6. By common agreement a fever is any oral temperature greater than 98.6 degrees Fahrenheit. What temperatures in degrees Celsius correspond to a fever?

Solution. Recall from (3) in Section 1.1 that if F and C represent degrees Fahrenheit and Celsius respectively, then F and C are related by the formula

$$F = \frac{9}{5}C + 32$$

Since a fever corresponds to any Fahrenheit temperature F such that $F > 98.6$, the corresponding Celsius temperature must satisfy

$$\frac{9}{5}C + 32 > 98.6 \tag{3}$$

We need to determine the values of C for which (3) is valid, and this we do with the following equivalent inequalities:

$$\frac{9}{5}C + 32 > 98.6$$

$$\frac{9}{5}C > 66.6$$

$$C > 37$$

Therefore a fever is any temperature greater than 37 degrees Celsius. □

In the next section we will continue solving inequalities.

EXERCISES 1.6

In Exercises 1–8, identify the intervals as open, closed or half-open, and as bounded or unbounded.

1. $(-1, 1)$ **2.** $(-7, -6]$ **3.** $(-\infty, 4]$

4. $[6, 6.1)$ **5.** $(6, 6.01]$ **6.** $(-1, \infty)$

7. $[7, 7]$ **8.** $[0, \infty)$

In Exercises 9–14, write the inequality in interval form.

9. $-4 < x \leq 3$ **10.** $5 \leq x \leq 7$

11. $-1.1 < x < -0.9$ **12.** $x > -8$

13. $-1 \leq x < 1$ **14.** $x \leq -2$

In Exercises 15–24, solve for x and then express the solutions as an interval.

15. $-12x \geq -3$

16. $-7x - 2 \geq 0$

17. $4 - 3x > -1 - x$

18. $\frac{1}{2} + 2x \leq \frac{4}{3} - 5x$

19. $12 - 2x < 4(x - 6)$

20. $\frac{1}{3}(2x - 3) > 3(x + \frac{1}{3})$

21. $\dfrac{1-x}{2} \geq \dfrac{2+x}{-3}$

22. $\dfrac{5-2x}{7} \leq \dfrac{3x+4}{2}$

23. $x^2 \geq (x + 3)^2$

24. $(x - 1)^2 < (x + 2)^2$

In Exercises 25–32, solve for x and then express the solutions as an interval.

25. $3 > x + 5 > 0$

26. $-2 \leq x - 1 \leq 4$

27. $0 \leq 5(x + 3) < 10$

28. $0 < \frac{1}{2}(2x + 4) < \frac{1}{3}$

29. $-0.01 < x - 2 < 0.01$

30. $-10^{-4} < x - 2 < 10^{-4}$

31. $0 < x - a < d$

32. $-d < x - a < d$

33. Let a, b, c, and d be real numbers with b and d positive.

a. Prove that

$$\frac{a}{b} < \frac{c}{d} \quad \text{if and only if} \quad ad < bc$$

(*Hint:* Multiply both sides of the first inequality by bd.)

b. Use (a) to show that

$$\frac{22}{59} < \frac{3}{8}$$

34. Which temperatures in degrees Celsius correspond to the temperatures larger than 32 and smaller than 212 degrees Fahrenheit?

35. A snack bar manager estimates that if the price of hot dogs is set at x cents, where $20 \leq x \leq 200$, then $1000 - 5x$ hot dogs will be sold daily. If the manager must sell at least 400 hot dogs each day at a cost of at least 20 cents, what are the possible prices for hot dogs?

36. A farmer is willing to sell s bushels of corn if the price of a bushel is $2 + (s/100{,}000)$ dollars. If the government sets a ceiling of \$4.50 on the price of a bushel of corn, what are the possible numbers of bushels the farmer would sell?

37. A store manager figures that $10(10 - p)$ umbrellas can be sold at p dollars per umbrella. If at least 45 umbrellas are to be sold, what are the possible prices that can be charged for each one?

38. A certain car holds 21 gallons of gasoline and gets 22 miles per gallon. If the car runs out of gasoline after having traveled at least 330 miles during a day, what are the possible amounts of gasoline that were in the tank at the start of the day?

39. An ice cream manager estimates that at the price of x dollars per gallon, $50{,}000–10{,}000x$ gallons of ice cream will be sold every month. If the manager wishes to sell at least 30,000 gallons per month, what are the possible prices per gallon that the manager can charge?

40. A student pays \$25 per week for a room. Another room is available for \$22 per week. If it would cost \$15 to move to the second room, for what lengths of stay would it pay for the student to make the transfer to the new room?

1.7 MORE ON INEQUALITIES

This section begins where the preceding section ended. The method of solving the inequalities appearing in this section will rely heavily on the following rules:

$$ac > 0 \quad \text{if} \quad a > 0 \quad \text{and} \quad c > 0, \quad \text{or if} \quad a < 0 \quad \text{and} \quad c < 0 \tag{1}$$

$$ac < 0 \quad \text{if} \quad a < 0 \quad \text{and} \quad c > 0, \quad \text{or if} \quad a > 0 \quad \text{and} \quad c < 0 \tag{2}$$

These rules, which appeared in a slightly different form in Section 0.2, are consequences of the inequality laws of Section 1.6.

We begin by solving ***quadratic inequalities***, that is, inequalities that are equivalent to one of the following:

$$\begin{aligned} &ax^2 + bx + c > 0, \qquad ax^2 + bx + c \geq 0, \\ &ax^2 + bx + c < 0, \quad \text{or} \quad ax^2 + bx + c \leq 0 \end{aligned} \tag{3}$$

We can use the rules in (1) and (2) to solve any of the quadratic inequalities in (3), once we factor the quadratic polynomial into linear factors. The procedure is illustrated in the next example.

EXAMPLE 1. Solve the inequality $x^2 + x - 12 < 0$.

Solution. Since $x^2 + x - 12 = (x + 4)(x - 3)$, the given inequality is equivalent to

$$(x + 4)(x - 3) < 0$$

To solve this inequality we first find the intervals on which the components $x + 4$ and $x - 3$ are positive and those on which they are negative. Then we use this information along with (1) and (2) to determine the intervals on which the product $(x + 4)(x - 3)$ is positive and those on which it is negative. It is convenient to display the results in a diagram such as the one shown in Figure 1.19.

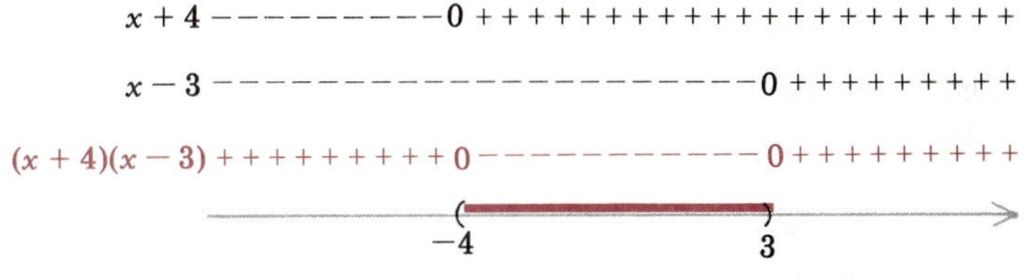

FIGURE 1.19

From the diagram we see that $(x + 4)(x - 3) < 0$ for x in $(-4, 3)$. The endpoints of $(-4, 3)$ are not included because $(x + 4)(x - 3) = 0$ for $x = -4$ and for $x = 3$. Thus the solutions of the given inequality form the interval $(-4, 3)$. □

In the solution of the preceding example, once we had factored the given inequality $x^2 + x - 12 < 0$ and had obtained

$$(x + 4)(x - 3) < 0$$

we could have used an alternative method. The method is based on the property that the polynomial has the same sign for all x between successive zeros. Since $x^2 + x - 12 = (x + 4)(x - 3)$ is zero when $x = -4$ and $x = 3$, it follows that $x^2 + x - 12$ has the same sign for all x in $(-\infty, -4)$, for all x in $(-4, 3)$, and for all x in $(3, \infty)$. Therefore we only need to compute the value of the polynomial at one "test" point in each of the intervals. Thus we prepare the following table:

Interval	Test point c	Sign of $x^2 + x - 12$ at c
$(-\infty, -4)$	-5	$+$
$(-4, 3)$	0	$-$
$(3, \infty)$	4	$+$

The fact that $x^2 + x - 12 < 0$ at the test point $c = 0$, along with our remarks preceding the table, implies that $x^2 + x - 12 < 0$ for all x in the interval $(-4, 3)$ and that $x^2 + x - 12 > 0$ for all x in $(-\infty, -4)$ or $(3, \infty)$. Thus we obtain the same set of solutions, namely the numbers in the interval $(-4, 3)$, as before. (Had we tried other test points, such as -2 and $\frac{1}{2}$, we would have arrived at the same conclusion.)

The rules in (1) and (2) concern products. Since any quotient a/c can be written as the product $a(1/c)$, and since $1/c$ has the same sign as c, we have the following rules for quotients:

$$\frac{a}{c} > 0 \quad \text{if} \quad a > 0 \quad \text{and} \quad c > 0, \quad \text{or if} \quad a < 0 \quad \text{and} \quad c < 0 \tag{4}$$

$$\frac{a}{c} < 0 \quad \text{if} \quad a < 0 \quad \text{and} \quad c > 0, \quad \text{or if} \quad a > 0 \quad \text{and} \quad c < 0 \tag{5}$$

EXAMPLE 2. Solve the inequality $\dfrac{2x - 1}{5x + 3} > 0$.

Solution. Notice that $2x - 1 = 0$ for $x = \frac{1}{2}$ and that $5x + 3 = 0$ for $x = -\frac{3}{5}$. Using this information along with (4), we prepare the diagram shown in Figure 1.20. The endpoint $-\frac{3}{5}$ is not a solution, since $\dfrac{2x - 1}{5x + 3}$ is not

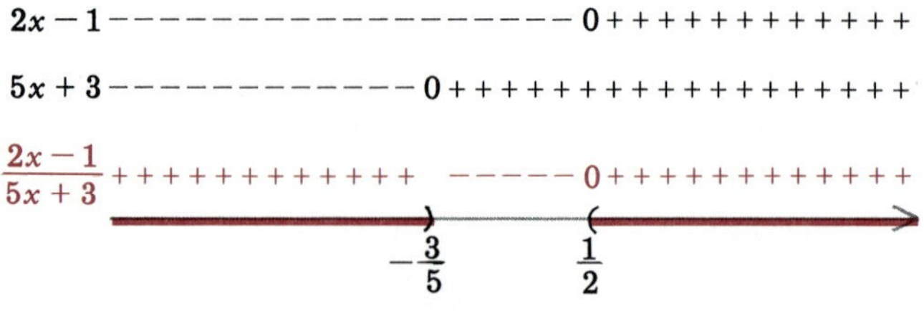

FIGURE 1.20

defined for $x = -\frac{3}{5}$. The endpoint $\frac{1}{2}$ is not a solution because

$$\frac{2\left(\frac{1}{2}\right) - 1}{5\left(\frac{1}{2}\right) + 3} = \frac{1 - 1}{\frac{5}{2} + 3} = 0$$

From these observations and the diagram we conclude that the solutions of the given inequality form the intervals $(-\infty, -\frac{3}{5})$ and $(\frac{1}{2}, \infty)$. □

EXAMPLE 3. Solve the inequality $\dfrac{2x - 1}{5x + 3} \geq 0$.

Solution. From Example 2 we know that $(2x - 1)/(5x + 3) > 0$ if and only if x is in $(-\infty, -\frac{3}{5})$ or in $(\frac{1}{2}, \infty)$. Since

$$\frac{2x - 1}{5x + 3} = 0$$

if and only if $2x - 1 = 0$, that is, if and only if $x = \frac{1}{2}$, we conclude that the set of solutions of the given inequality consists of $\frac{1}{2}$ along with the intervals $(-\infty, -\frac{3}{5})$ and $(\frac{1}{2}, \infty)$. Thus the solutions of the given inequality form the intervals $(-\infty, -\frac{3}{5})$ and $[\frac{1}{2}, \infty)$. □

EXAMPLE 4. Solve the inequality $\dfrac{6}{x - 2} \leq 2$.

Solution. We subtract 2 from both sides so that the right side is 0, and then simplify:

$$\frac{6}{x - 2} \leq 2$$

$$\frac{6}{x - 2} - 2 \leq 0$$

$$\frac{6}{x - 2} - 2\frac{x - 2}{x - 2} \leq 0$$

$$\frac{6 - 2(x - 2)}{x - 2} \leq 0$$

$$\frac{10 - 2x}{x - 2} \leq 0$$

$$\frac{2(5 - x)}{x - 2} \leq 0$$

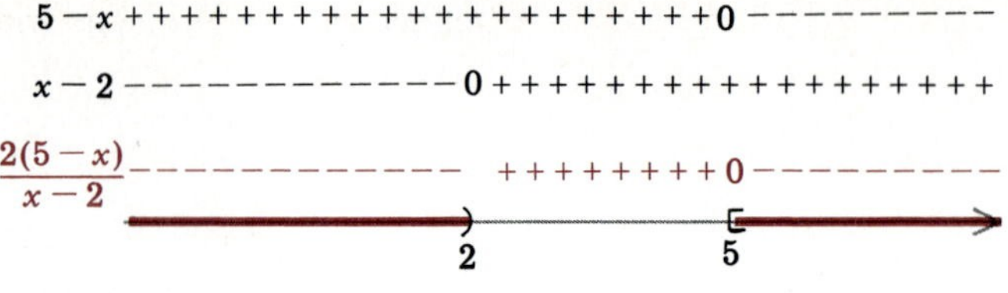

FIGURE 1.21

Now we prepare the diagram shown in Figure 1.21. From the diagram we see that

$$\frac{2(5-x)}{x-2} \leq 0$$

for x in $(-\infty, 2)$ and $[5, \infty)$. Thus the solutions of the given inequality form the intervals $(-\infty, 2)$ and $[5, \infty)$. □

Another way of solving inequalities such as the one in Example 4 involves eliminating the denominator. Since $x - 2$ can be either positive or negative, we multiply by $(x - 2)^2$, which is always nonnegative, and then rearrange:

$$\frac{6}{x-2} \leq 2$$

$$\frac{6}{x-2}(x-2)^2 \leq 2(x-2)^2$$

$$6(x-2) \leq 2(x^2 - 4x + 4)$$

$$6x - 12 \leq 2x^2 - 8x + 8$$

$$0 \leq 2x^2 - 14x + 20$$

$$0 \leq x^2 - 7x + 10$$

$$0 \leq (x-5)(x-2)$$

From a diagram analogous to Figure 1.21 we obtain the same solution as before (noting that 2 is not a solution of the given inequality).

So far all of our examples have contained two factors. However, the same general method works for any number of factors. In the next example, there are three factors.

EXAMPLE 5. Solve the inequality $(x + 3)(x + 1)(x - 2) < 0$.

Solution. We prepare the diagram shown in Figure 1.22. From it we see that the solutions of the given inequality form the intervals $(-\infty, -3)$ and $(-1, 2)$. □

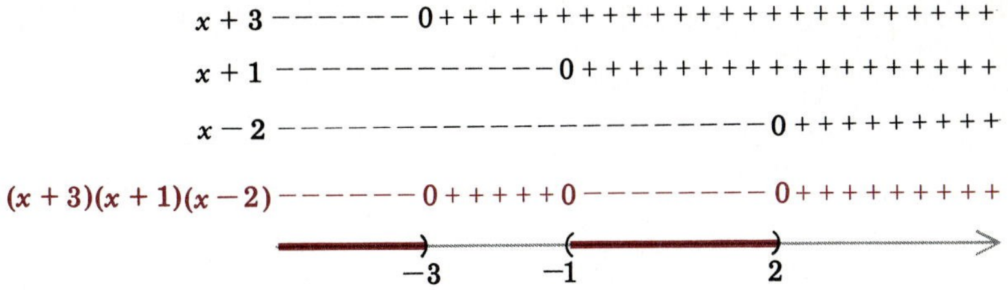

```
x + 3      ------0+++++++++++++++++++++++++
x + 1      ------------0+++++++++++++++++++
x − 2      ------------------0+++++++++
(x + 3)(x + 1)(x − 2) ------0+++++0--------0+++++++++
                 −3     −1        2
```

FIGURE 1.22

Before turning to an application of inequalities, we mention a further rule of inequalities that is occasionally of use:

$$\text{if} \quad a > 0 \quad \text{and} \quad a < b, \quad \text{then} \quad \frac{1}{a} > \frac{1}{b} \tag{6}$$

Thus if $1/(2x) > 7$ then since $7 > 0$, it follows from (6) that $2x < \frac{1}{7}$, so that $x < \frac{1}{14}$.

In Sections 1.1 and 1.3 we solved applied problems by means of equations. However, some applied problems can be solved by means of inequalities, as our next example illustrates.

EXAMPLE 6. A rock is thrown vertically upward from a height of 6 feet above ground with an initial velocity of 96 feet per second. During what time interval will the rock be more than 134 feet above ground?

Solution. By (1) in Section 1.3, with $v_0 = 96$ and $h_0 = 6$, the height h of the rock at time t is given by

$$h = -16t^2 + 96t + 6$$

We wish to find the values of t for which $h > 134$, which means that we must solve the inequality

$$-16t^2 + 96t + 6 > 134$$

Proceeding as in earlier solutions of inequalities we find that

$$\begin{aligned} -16t^2 + 96t - 128 &> 0 \\ -16(t^2 - 6t + 8) &> 0 \\ t^2 - 6t + 8 &< 0 \\ (t - 2)(t - 4) &< 0 \end{aligned}$$

We prepare the diagram shown in Figure 1.23. From the diagram we see that solutions of $(t - 2)(t - 4) < 0$ form the interval $(2, 4)$. Thus the rock will

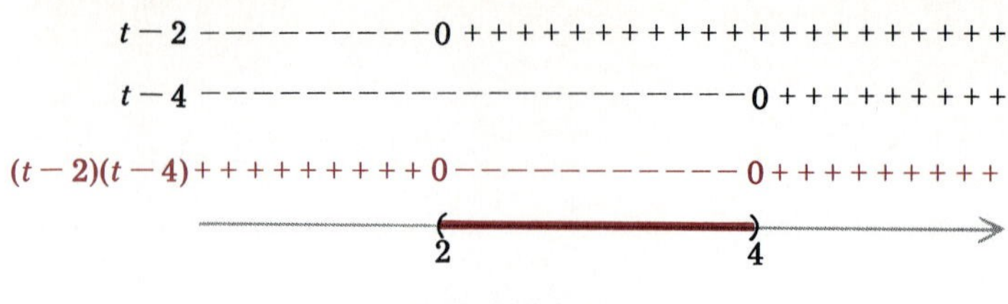

FIGURE 1.23

be more than 134 feet high during the time interval between 2 seconds and 4 seconds. □

Inequalities Involving Absolute Values

In addition to the kinds of inequalities we have already encountered, inequalities involving absolute values appear from time to time in advanced mathematics, primarily because of the relation of absolute value to distance between real numbers. In this section we will discuss and solve inequalities involving absolute values of real numbers.

Recall from Section 0.2 that

$$|x| = \begin{cases} x & \text{if} \quad x \geq 0 \\ -x & \text{if} \quad x < 0 \end{cases} = \text{the distance between } x \text{ and } 0 \tag{7}$$

It follows from (7) that if c is any positive number, then

$$|x| < c \text{ means that} \begin{cases} x < c & \text{if} \quad x \geq 0 \\ -x < c & \text{if} \quad x < 0 \quad (\text{so } x > -c \text{ if } x < 0) \end{cases} \tag{8}$$

Combining the two parts of the right side of (8), we conclude that

$$\boxed{|x| < c \quad \text{if and only if} \quad -c < x < c} \tag{9}$$

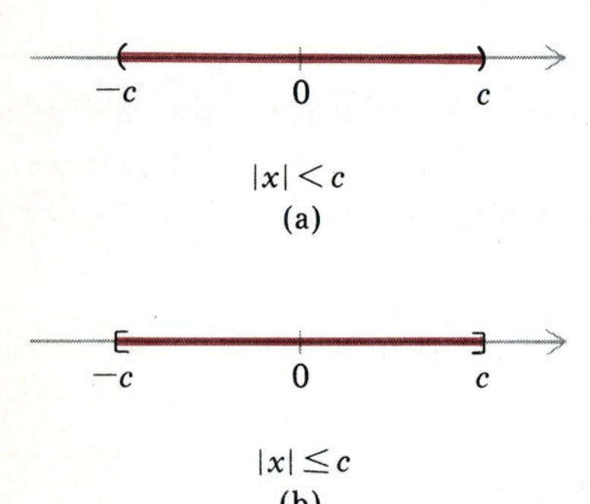

FIGURE 1.24

The inequality $|x| < c$ says simply that the distance between x and 0 is less than the positive number c (Figure 1.24a). For example, from (9) it follows that $|x| < 3$ means that $-3 < x < 3$, or equivalently, that the distance between x and 0 is less than 3.

By similar reasoning we find that for any nonnegative number c,

$$|x| \leq c \quad \text{if and only if} \quad -c \leq x \leq c \tag{10}$$

The inequality $|x| \leq c$ means that the distance between x and 0 is less than or equal to c (Figure 1.24b).

EXAMPLE 7. Solve the following inequalities.

a. $|x| < \frac{1}{3}$ b. $|x| \leq 2$

Solution.

a. By (9), the inequality $|x| < \frac{1}{3}$ is equivalent to $-\frac{1}{3} < x < \frac{1}{3}$, so the solutions form the open interval $(-\frac{1}{3}, \frac{1}{3})$.

b. By (10), the inequality $|x| \le 2$ is equivalent to $-2 \le x \le 2$, so the solutions form the closed interval $[-2, 2]$. □

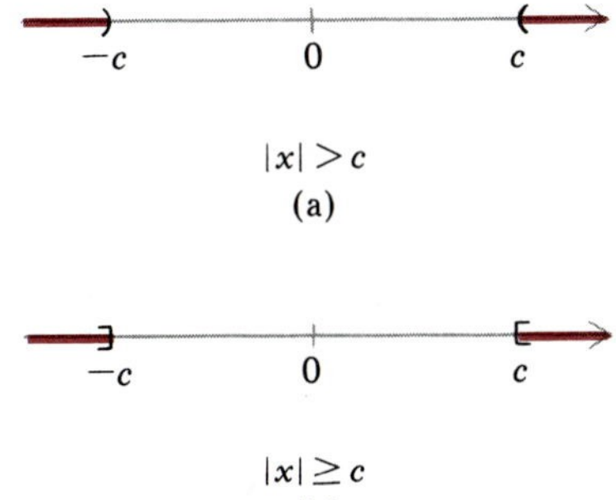

FIGURE 1.25

Now we turn to the inequalities $|x| > c$ and $|x| \ge c$. Since $|x| > c$ means that $|x| \le c$ is false, (10) implies that for any nonnegative number c,

$$|x| > c \quad \text{if and only if} \quad x > c \text{ or } x < -c \tag{11}$$

Similarly, (9) implies that for any nonnegative number c,

$$|x| \ge c \quad \text{if and only if} \quad x \ge c \text{ or } x \le -c \tag{12}$$

The inequality $|x| > c$ means that the distance between x and 0 is greater than c (Figure 1.25a), and the inequality $|x| \ge c$ means that the distance between x and 0 is greater than or equal to c (Figure 1.25b).

EXAMPLE 8. Solve the following inequalities.

a. $|x| > 4$ b. $|x| \ge \frac{1}{2}$

Solution.

a. By (11), the inequality $|x| > 4$ is equivalent to the statement that $x > 4$ or $x < -4$, so the solutions form the two open intervals $(-\infty, -4)$ and $(4, \infty)$.

b. By (12), the inequality $|x| \ge \frac{1}{2}$ is equivalent to the statement that $x \ge \frac{1}{2}$ or $x \le -\frac{1}{2}$, so the solutions form the two closed intervals $(-\infty, -\frac{1}{2}]$ and $[\frac{1}{2}, \infty)$. □

Recall that $|x - a|$ is the distance between the numbers x and a. Thus geometrically the inequality $|x - a| < c$ means that the distance between x and a is less than c. There are analogous geometric interpretations of the inequalities

$$|x - a| \le c, \qquad |x - a| > c, \quad \text{and} \quad |x - a| \ge c$$

EXAMPLE 9. Solve the following inequalities.

a. $|x - 7| < 2$ b. $|x - 7| \le 2$

Solution by the algebraic method.

a. By (9), the inequality $|x - 7| < 2$ is equivalent to

$$-2 < x - 7 < 2$$

Adding 7 to all three expressions, we obtain

$$5 < x < 9$$

Thus the solutions form the open interval $(5, 9)$.

b. By (10), the inequality $|x - 7| \leq 2$ is equivalent to

$$-2 \leq x - 7 \leq 2$$

Adding 7 to each expressions yields

$$5 \leq x \leq 9$$

Thus the solutions form the closed interval $[5, 9]$.

Solution by the geometric method.

a. Notice that x satisfies the inequality $|x - 7| < 2$ if and only if the distance between x and 7 is less than 2 units. On the real line we locate 7 and mark off 2 units to either side (Figure 1.26a). The numbers 5 and 9 are not solutions, since they both lie exactly 2 units from 7, but all numbers between 5 and 9 are solutions. Thus the solutions form the open interval $(5, 9)$.

b. Notice that x satisfies the inequality $|x - 7| \leq 2$ if the distance between x and 7 is less than or equal to 2. Thus 5 and 9 are solutions of the inequality, as are all numbers in between. Consequently the solutions comprise the closed interval $[5, 9]$ (Figure 1.26b). □

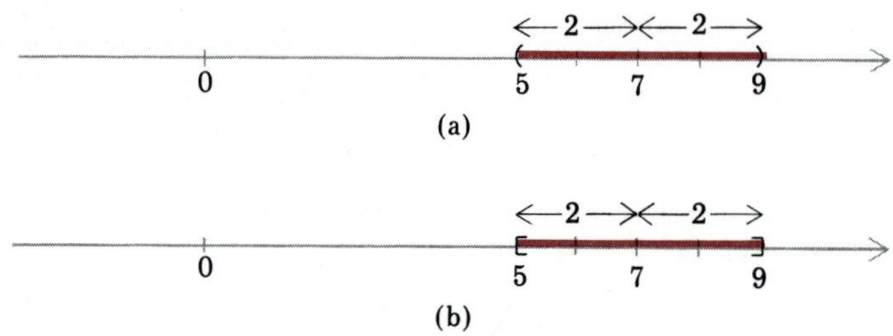

FIGURE 1.26

Other inequalities involving absolute values can be solved by similar methods.

EXAMPLE 10. Solve the inequality $|3x - 5| < \frac{1}{2}$.

Solution by the algebraic method. By (9), the inequality $|3x - 5| < \frac{1}{2}$ is equivalent to

$$-\frac{1}{2} < 3x - 5 < \frac{1}{2}$$

We can either treat this compound inequality as the pair of inequalities

$$-\frac{1}{2} < 3x - 5 \quad \text{and} \quad 3x - 5 < \frac{1}{2}$$

and solve the pair for x separately, or we can perform our alterations on the compound inequality itself. We will do the latter:

$$-\frac{1}{2} < 3x - 5 < \frac{1}{2}$$

$$-\frac{1}{2} + 5 < (3x - 5) + 5 < \frac{1}{2} + 5$$

$$\frac{9}{2} < 3x < \frac{11}{2}$$

$$\frac{9}{2}\left(\frac{1}{3}\right) < 3x\left(\frac{1}{3}\right) < \frac{11}{2}\left(\frac{1}{3}\right)$$

$$\frac{3}{2} < x < \frac{11}{6}$$

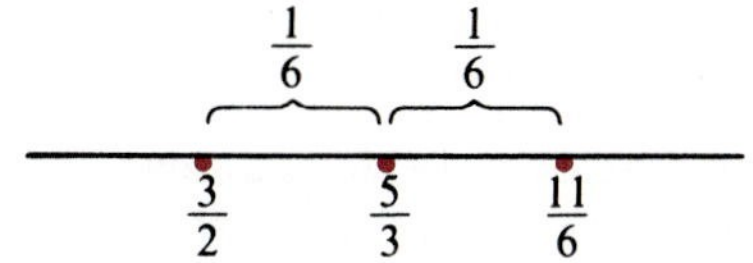

FIGURE 1.27

Therefore the solutions comprise the open interval $(\frac{3}{2}, \frac{11}{6})$.

Solution by the geometric method. $|3x - 5| < \frac{1}{2}$ is equivalent to $3|x - \frac{5}{3}| < \frac{1}{2}$, which is equivalent to $|x - \frac{5}{3}| < \frac{1}{6}$. Thus we seek the values of x whose distance from $\frac{5}{3}$ is less than $\frac{1}{6}$. Using Figure 1.27, we conclude that the solutions form the open interval $(\frac{3}{2}, \frac{11}{6})$. □

EXERCISES 1.7

In Exercises 1–28, solve the given inequality and express the solutions in terms of intervals.

1. $(x - 1)(x - 2) \geq 0$

2. $(x + 5)(x - 1) < 0$

3. $(x + 1)(x + 3) \leq 0$

4. $(x - \sqrt{2})(x + \sqrt{3}) \leq 0$

5. $(x - 3)^2 > 0$

6. $(x + 2)^2 \geq 0$

7. $x^2 < 4$

8. $x^2 \geq 9$

9. $(x - 2)^2 \leq 1$

10. $(x^2 - 3)^2 \leq 9$

11. $x^2 - 2x - 15 > 0$

12. $x^2 + 5x \leq 14$

13. $x^2 \leq -9x$

14. $x^2 > \sqrt{2}x$

15. $\frac{x - 4}{x + 5} > 0$

16. $\frac{2 - x}{3 + x} < 0$

17. $\frac{2x - 3}{3x + 6} \leq 0$

18. $\frac{2x - 1}{x} \geq 5$

19. $\frac{x}{x - 1} > \frac{x}{x + 1}$

20. $\frac{2x}{x - 3} \leq \frac{2x}{x - 6}$

21. $(x - 1)(x - 2)(x - 3) > 0$

22. $x(x + 3)(x + 5) \leq 0$

23. $(x - 7)(x^2 + 4) > 0$

24. $(x^2 - 1)(x^2 - 9) \leq 0$

25. $\dfrac{(2-x)(3+x)^2}{x-\frac{1}{2}} \geq 0$

26. $\dfrac{x+1}{x^2-x} < 0$

27. $\dfrac{x+1}{x^2-x} \geq -1$

28. $\dfrac{2x+3}{6x^2+1} \leq 3$

In Exercises 29–32, solve the given inequality and express the solutions in terms of intervals. You may need to use the quadratic formula to factor the left side.

29. $x^2 + x < 1$

30. $x^2 - 3x + 1 \geq 0$

31. $2x^2 + 4x + 1 < 0$

32. $3x^2 - 2x - 1 \leq 0$

In Exercises 33–38, solve the given inequality and locate the solutions on the real line.

33. $|x| < 4$

34. $|x| < 1.5$

35. $|x| \leq \frac{1}{5}$

36. $|x| \leq 3.2$

37. $|x| \geq \frac{9}{2}$

38. $|-x| > 3$

In Exercises 39–58, solve the given inequality.

39. $|x - 5| < 3$

40. $|x - 10| < 2$

41. $|x + 3| \leq 3$

42. $|x + 1| \leq 4$

43. $|7 - x| > 1$

44. $|\frac{1}{2} - x| > \frac{1}{4}$

45. $|x - 2| \geq \frac{1}{3}$

46. $|4 + x| \geq 6$

47. $|2x - 1| < 3$

48. $|3x - 2| < \frac{1}{2}$

49. $|3x - 2| > 0$

50. $|3x - 2| \geq 0$

51. $|x^2 - 1| < 1$

52. $|x^2 - 5| < 4$

53. $|x^2 - 9| < 27$

54. $|x^3 - 13| < 14$

55. $1 \leq |x - 5| < 2$

56. $0 < |x + 3| \leq 4$

57. $1 \leq |6x + 4| \leq 3$

58. $\frac{1}{2} < |\frac{3}{2} - 5x| < \frac{3}{4}$

59. Find all values of b for which the equation $x^2 + bx + 5 = 0$ has

a. no real solution

b. two real solutions

60. Find all values of b for which the equation $x^2 - 2bx + 6 = 0$ has

a. no real solution

b. two real solutions

61. Find the values of a such that 2 is a solution of the inequality

$$\frac{x-a}{x+a} \leq 3$$

62. Show that if $a < b$, then $a < (a+b)/2 < b$. (The number $(a+b)/2$ is called the ***arithmetic mean*** of a and b.)

63. Let $0 < a < b$.

a. Show that $(\sqrt{b} - \sqrt{a})^2 > 0$.

b. Using part (a), prove that $a < \sqrt{ab} < (a+b)/2$. (The number $\sqrt{ab}$ is called the ***geometric mean*** of a and b.)

64. a. Show that $x^2 > x$ for $x > 1$.

b. Show that $x^2 < x$ for $0 < x < 1$.

65. a. If $x^2 \geq 36$, is it necessarily true that $x \geq 6$? Explain.

b. If $x^3 \geq 64$, is it necessarily true that $x \geq 4$? Explain.

66. For what values of x is $1/x < x$?

67. A ball is thrown vertically upward from a height of 100 feet with an initial velocity of 80 feet per second. During what time interval is the height of the ball at least 4 feet?

68. A ball is dropped from a window 100 feet above ground. Between what times will the ball be between 84 and 36 feet above ground?

69. If the ball in Exercise 68 were thrown upward at 48 feet per second, when would the ball be between 36 and 100 feet above ground?

70. Suppose a bicycle shop can sell x bicycle replacement seats per month if the price is set at $12 - 0.2x$ dollars per seat. If the purchase price from the wholesaler is \$4 per seat and the shop manager desires to make a profit of at least \$75 per month, what are the possible numbers of seats that can be sold monthly, and at what prices?

71. Assume, as in Exercise 70, that the price per seat is $12 - 0.2x$ and the purchase price from the wholesaler is \$4 per seat. Show that the manager cannot make a profit of more than \$80.

72. A baker estimates that he can bake up to 1800 loaves of rye bread during a week and that he can sell x loaves weekly if he charges $100 - (1/20)x$ cents per loaf. If it costs 25 cents per loaf to produce rye bread, what are the possible numbers of loaves that will net a profit for the baker?

73. A rectangular sheet of metal is 16 inches long and 8 inches wide. A pan is to be made from the sheet by cutting out four square pieces, one from each corner of the sheet (Figure 1.28). If the area of the base is to be at least 48 square inches, what are the possible heights of the sides created by folding up the edges?

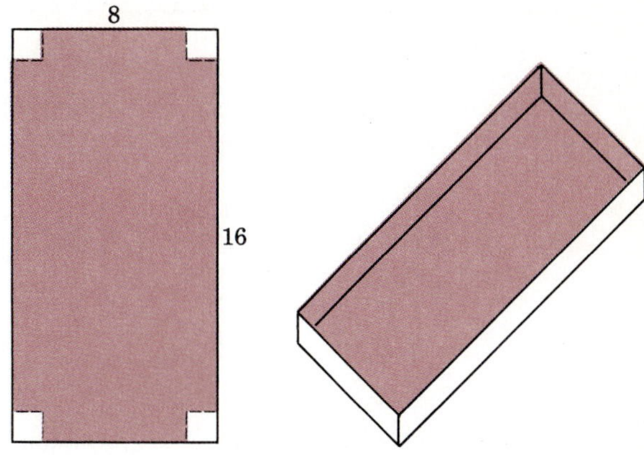

FIGURE 1.28

KEY TERMS

equation
- identity
- conditional equation
- equivalent equations
- linear equation
- quadratic equation

interval
- open interval
- closed interval
- half-open interval
- bounded interval
- unbounded interval

solution (root) of an equation
extraneous solution

completing the square
complex number
- real part
- imaginary part

principal square root
conjugate
solution of an inequality

KEY FORMULAS

$$x = \frac{-b \pm \sqrt{b^2 - 4ac}}{2a}$$ quadratic formula

$$a^2 + b^2 = c^2$$ Pythagorean Theorem

$$i^2 = -1$$

$$\sqrt{-c} = \sqrt{c}i$$

KEY LAWS

If $a < b$ and $c > 0$, then $ac < bc$.

If $a < b$ and $c < 0$, then $ac > bc$.

$ac > 0$ if $a > 0$ and $c > 0$, or if $a < 0$ and $c < 0$.

$ac < 0$ if $a < 0$ and $c > 0$, or if $a > 0$ and $c < 0$.

REVIEW EXERCISES

In Exercises 1–34, find all real solutions (if any) of the given equation.

1. $\frac{1}{2}x - 7 = 6$

2. $-3t + 2 = -4$

3. $\frac{1}{3}(x - 3) + 4(-x + 2) = -4$

4. $\dfrac{3}{x+3} = \dfrac{-9}{x-1}$

5. $\dfrac{3x+8}{8x-3} = -2$

6. $\dfrac{4x-5}{3-7x} = 2$

7. $(3y + 5)(-2y + 1) = 0$

8. $2\sqrt{y} = \dfrac{1}{\sqrt{y}}$

9. $|3x - 5| = 2$

10. $|\frac{1}{2}x + \frac{1}{3}| = \frac{2}{5}$

11. $5x^2 - 15 = 0$

12. $s^2 - s - 380 = 0$

13. $3y^2 + y + 7 = 0$

14. $4s^2 = 13s + 3$

15. $2s^2 - s - 2 = 0$

16. $4x^2 - 20x + 25 = 0$

17. $\dfrac{5}{x-2} = x + 2$

18. $3x - \dfrac{2}{x} = 6$

19. $2x(x + 4) = (x - 1)(x + 3)$

20. $2x(x + 1) = (x - 1)(x + 3)$

21. $x^4 - x^2 = 2$

22. $x^{5/2} - 2x^{3/2} + \sqrt{x} = 0$

23. $(2x - 1)^2 - 2x + 1 = 0$

24. $\sqrt{2x-1} = 1 - \sqrt{4x-1}$

25. $\sqrt{2-3x} = 1 + \sqrt{1-2x}$

26. $t^3 - 2t^2 = t$

27. $t^3 + 6t^2 = 9t$

28. $(3x - 5)(2x^2 + 13x - 7) = 0$

29. $2x^4 - x^3 - 2x^2 + x = 0$

30. $\dfrac{4x^2-3}{-17x+7} = 0$

31. $(6 - x^2)^2 - 4 = 0$

32. $(\frac{1}{8} - x^3)^3 = -729$

33. $(x^2 - 16)^4 = 256$

34. $2^{n-1} = x^n$

In Exercises 35–38, perform the indicated operation. Express your answer in the form $a + bi$.

35. $(2 - i) + (-6 + 7i)$

36. $(-3 - 4i) - (4 - 3i)$

37. $(3 + i)(3 - i)$

38. $\left(\frac{1}{2}+\frac{3}{2}i\right)\left(-\frac{1}{2}+\frac{5}{2}i\right)$

In Exercises 39–40, evaluate the given expression. Express your answer in the form $a + bi$.

39. $\sqrt{-12}$

40. $\dfrac{2-\sqrt{-24}}{4}$

In Exercises 41–44, find the (complex) solutions of the given equation. Express your solutions in the form $a + bi$.

41. $x^2 + 27 = 0$

42. $x^2 + 2x + 4 = 0$

43. $3x^2 + 3x + 1 = 0$

44. $4x^2 + 6x + 3 = 0$

In Exercises 45–46, find the conjugate of the given complex number. Express your answer in the form $a + bi$.

45. $-4 - 3i$

46. $6i - 1$

In Exercises 47–64, solve the given inequality and express the solutions in terms of intervals.

47. $4x - 7 \le 3$

48. $-2x + 3 > 6$

49. $-2x + 3 > -6$

50. $-7 \le -3x + 2 < 0$

51. $y^2 \le (y + 1)^2$

52. $(2y + 3)(y - \sqrt{3}) < 0$

53. $x^2 - 9x + 20 > 0$

54. $x^2 - 4x + 2 \le 0$

55. $3t^2 + 7t < -2$

56. $t(2t - 1) \ge (t + 1)^2 + 3$

57. $\dfrac{2x+1}{-5x+2} \geq 0$

58. $\dfrac{5x-2}{2x+3} < -1$

59. $\dfrac{1+8x}{x-3} \geq 2$

60. $(2y+1)(y+2)(y-5) > 0$

61. $|x| \geq \sqrt{2}$

62. $|x-3| < 0.5$

63. $|-2x+5| \leq 13$

64. $4 < |-3x-8| \leq 6$

65. Find the values of a such that 0 is a solution of the equation

$$\frac{1}{2x-1} + \frac{3a}{x+1} = 4a$$

66. Find the values of a such that 1 is a solution of the equation $a^2x - 2ax = 3$.

67. For what values of a is -1 a solution of the following inequality?

$$\frac{2x-a}{3x+a} < -2$$

68. The average of five numbers is 3. When the smallest number is deleted, the average of the remaining numbers is 4. Find the smallest of the five numbers.

69. If the sales tax is 5% and amounts to \$3.18 on a pair of sunglasses, what is the total cost of the glasses (item cost plus tax)?

70. The hottest and coldest outdoor temperatures ever recorded on the surface of the earth were 136.4 and -126.9 degrees Fahrenheit, respectively. What are the corresponding temperatures in degrees Celsius? Round your answer off to the nearest tenth of a degree. (*Hint:* Use (3) of Section 1.1.)

71. Suppose that a circle of radius 9 is tangent to a circle of smaller radius, and that there is a line segment of length 12 that is tangent to both circles (Figure 1.29). Find the radius r of the smaller circle. (*Hint:* Draw a line through the center of the smaller circle that is parallel to the common tangent.)

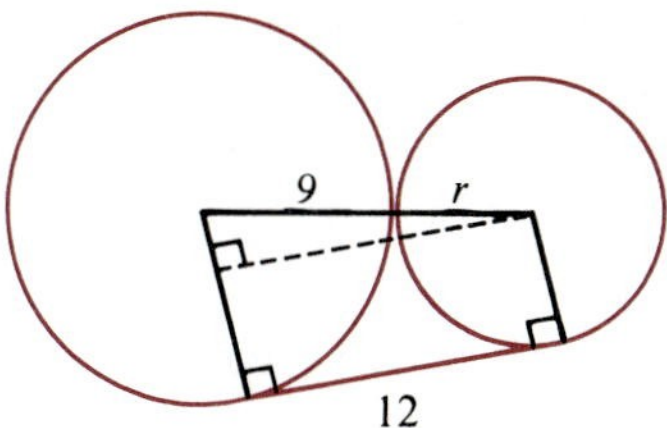

FIGURE 1.29

72. Two boats traveling at the same speed depart from the same port at the same time. One travels north for a while and then heads straight for a second port. The other travels 10 miles south and then 20 miles west and arrives at the second port at exactly the same time as the first boat. How far north did the first boat travel before heading for the second port?

73. A Coast Guard boat is 5 miles north of Bermuda, and a boat suspected of carrying smugglers is 25 miles west of Bermuda and moving directly toward Bermuda. If both boats travel at the same speed, how far will the Coast Guard boat have to travel in order to intercept the other boat? (*Hint:* See Figure 1.30.)

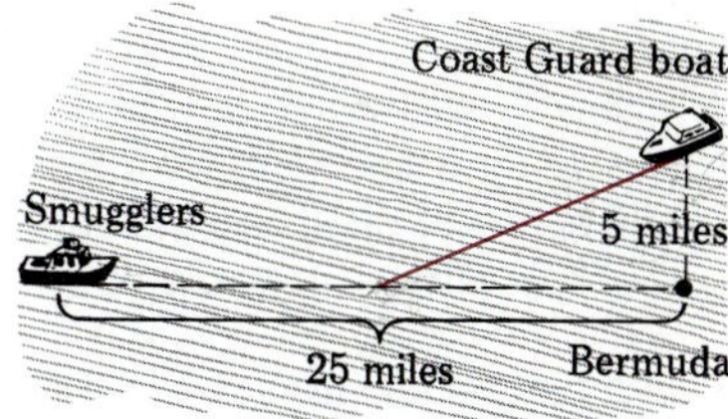

FIGURE 1.30

74. It takes two hours longer for an airplane flying at a constant air speed to fly 750 miles against a 50 mph headwind than it does to make the return trip at the same air speed with a 50 mph tailwind. Determine the air speed of the airplane.

75. If 3 people all mowing at the same rate can mow 4 acres in 8 hours, how long will it take 4 people all mowing at the same rate to mow 3 acres?

76. Bill and Jane can do a job in 20 minutes. If Bill worked at twice his normal rate, they could do the job in 15 minutes. How long would it take Jane to do the job alone?

77. A ball is thrown vertically upward from a height of 12 feet with an initial velocity of 48 feet per second. During what time interval is the ball at least 44 feet high?

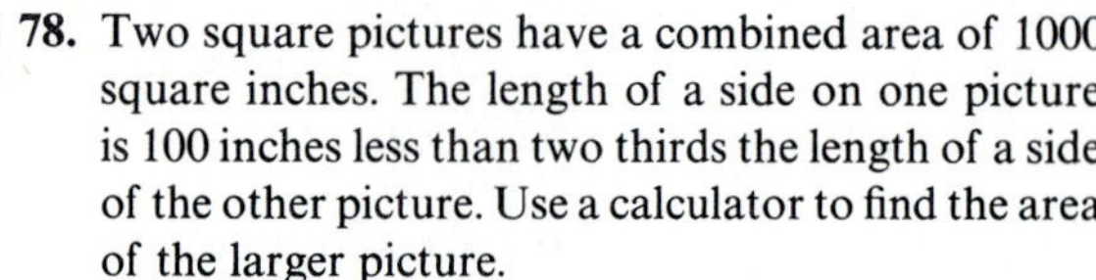

78. Two square pictures have a combined area of 1000 square inches. The length of a side on one picture is 100 inches less than two thirds the length of a side of the other picture. Use a calculator to find the area of the larger picture.

79. A group of students plans to buy a refrigerator costing \$180 and to divide the cost equally. If they can convince two more students to join them (with all paying equal amounts), the cost to each of the original students will decrease by \$15. How many students are in the original group?

80. A plane has a cruising speed of 420 miles per hour, and the wind velocity averages 60 miles per hour. Assuming that the flight is with the wind in one direction and against it in the other, how far can the plane travel in a round trip of 7 hours?

81. A printer charges for producing any desired number of copies of a book in the following way. The printing charge depends only on the number of copies printed, and the charge for typesetting is fixed, independent of the number of books to be printed. If the printer is willing to make 20,000 copies of a certain book for \$26,000 and to make 30,000 copies of the same book for \$34,000, find the cost of typesetting the book.

***82.** A circular pond has a radius of 8 feet, and there is a reed in the middle of the pond that protrudes 4 feet above the water. When the top of the reed is pulled to the side (without bending), the tip just reaches the edge of the pond. How tall is the reed?

83. A walkway cuts diagonally across a 40-foot square yard (see Figure 1.31). If the area of the walkway is 304 square feet, find the width of the walkway.

***84.** A person delivers telephone books to three apartment buildings. The first building requires one more than half of all the telephone books. The second building requires one more than half of all the remaining telephone books. The third building requires one more than half of all the telephone books still remaining. If there are five telephone books left over, how many were there to begin with?

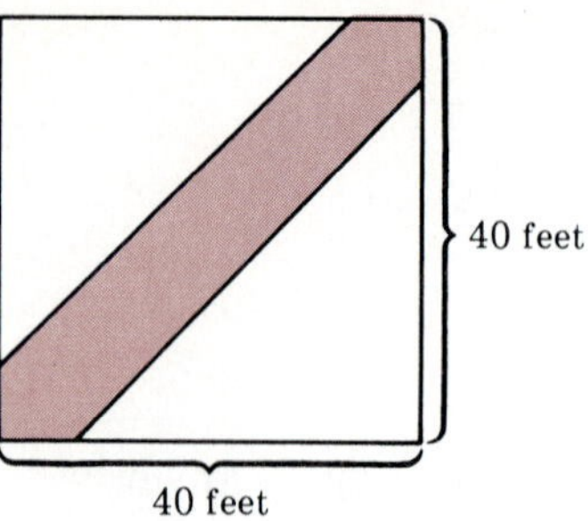

FIGURE 1.31

85. A tank contained 4 gallons of a mixture of antifreeze and water. After 1 gallon of the mixture was drained off and replaced by 1 gallon of pure antifreeze, the new mixture consisted of 60% antifreeze and 40% water. What percentage of the original mixture was antifreeze?

86. The volume V of a cone of radius r and height h is given by $V = \frac{1}{3}\pi r^2 h$. Solve the equation for r.

87. If p denotes the object distance, q the image distance, and f the focal length of a simple lens, then $f = \dfrac{pq}{p+q}$.

a. If $p = 5$ and $f = 3$, determine q.

b. Suppose p, q, and f are given, with $q = 4$. Assume that if p were tripled, then f would be doubled. Find the value of p.

88. Let $0 < a < b$. Then the harmonic mean h of a and b is given by the equation

$$\frac{1}{h} = \frac{1}{2}\left(\frac{1}{a} + \frac{1}{b}\right)$$

Solve the equation for b.

89. Recall from Section 1.1 that if P dollars are deposited into an account earning simple interest at an annual rate r, then the amount of interest after n years is Prn, and therefore the amount A_n in the account after n years is given by

$$A_n = P + Prn$$

Solve the equation for r.

90. If, in Exercise 89, the interest were compounded

annually, then the formula for A_n would be

$$A_n = P(1 + r)^n$$

Solve the equation for r.

91. The heat Q that results when x moles of sulfuric acid are mixed with y moles of water is given by

$$Q = \frac{17{,}860xy}{1.798x + y}$$

Solve the equation for x.

92. Under certain conditions the percentage efficiency E of an internal combustion engine is given by

$$E = 100\left(1 - \frac{v}{V}\right)^{0.4}$$

where V and v are, respectively, the maximum and minimum volumes of air in each cylinder.
a. Solve the equation for v.
b. Solve the equation for V.

93. A coin dropped from a bridge 15 feet above the water lands 2 feet above the water on a seat of a boat. Approximately how long does the coin fall? Round your answer to the nearest hundredth of a second.

***94.** A trapezoid with area 45 square meters has parallel sides of lengths 6 meters and 9 meters, respectively (Figure 13.2a). Find the area of the large triangle formed when the two nonparallel sides are extended until they meet (1.32b).

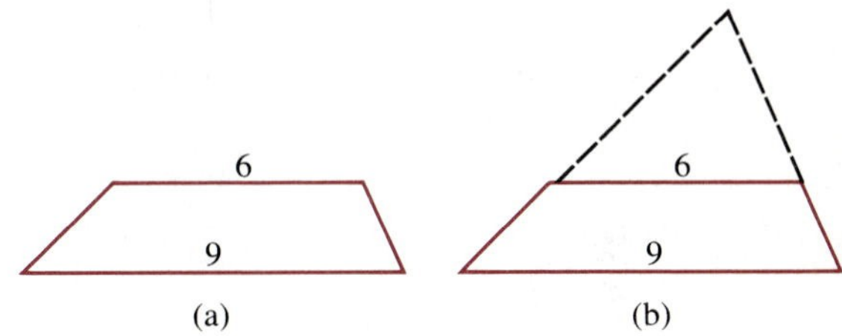

FIGURE 1.32

2 FUNCTIONS AND THEIR GRAPHS

It is common to associate various pairs of quantities with one another. For example, one can associate the volume of a spherical balloon with the corresponding radius. Thus by knowing the radius, one can compute the volume of the balloon. Similarly, the weather service at an airport such as O'Hare International Airport charts the temperature during the day. To determine what the temperature was at the airport yesterday noon, we would only need to locate yesterday noon on the chart and then read off the temperature at that instant. The dependence of one quantity on another, such as volume on radius or temperature on the time of day, is described mathematically by a function, which is the main topic of this chapter.

One of the most important ways of describing a function involves a mathematical picture called a graph. So before we define and study functions, we will discuss notions related to graphs.

2.1 CARTESIAN COORDINATES FOR THE PLANE

In Section 0.2 we associated the points on a line with real numbers. Now we will associate the points in a plane with pairs of real numbers. This association will aid us in describing functions.

We begin by drawing two perpendicular lines in the plane, a horizontal one called the ***x axis*** and a vertical one called the ***y axis*** (Figure 2.1). These are the ***coordinate axes***, and they cross at the ***origin***, denoted by 0. Next we set up a coordinate system for the x axis so that points on it to the right of the origin correspond to positive numbers and points on it to the left of the origin correspond to negative numbers. Similarly we set up a coordinate

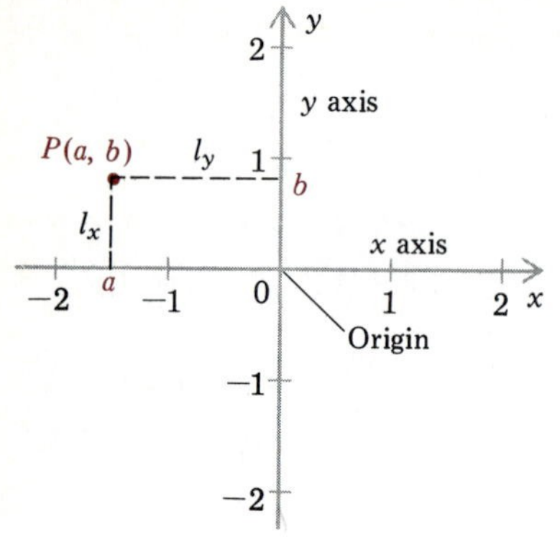

FIGURE 2.1

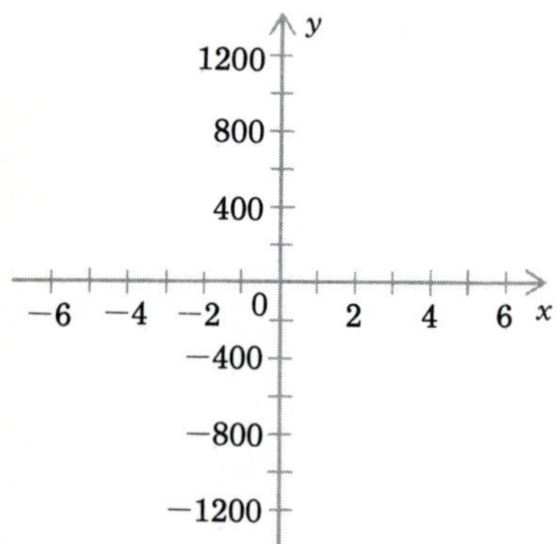

A coordinate system with different scales on the axes

FIGURE 2.2

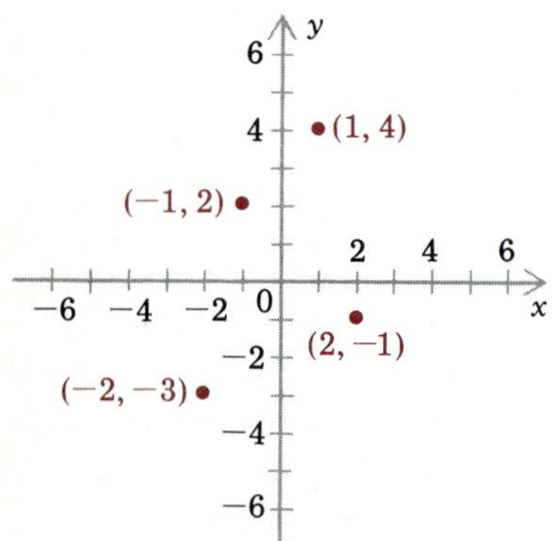

FIGURE 2.3

system for the y axis so that points on it above the origin correspond to positive numbers and points on it below the origin correspond to negative numbers. The positive directions on the coordinate axes are indicated by arrows (Figure 2.1).

To associate points in the plane with pairs of real numbers, we consider an arbitrary point P in the plane. Through P we draw the line l_x perpendicular to the x axis and the line l_y perpendicular to the y axis. Then l_x crosses the x axis at a point that corresponds to a real number a, called the ***x coordinate*** (or ***abscissa***) of P. Similarly, l_y crosses the y axis at a point corresponding to a real number b, called the ***y coordinate*** (or ***ordinate***) of P (Figure 2.1). Thus P determines the ***ordered pair*** of numbers a and b, written (a, b) and called an ordered pair because the x coordinate, a, precedes the y coordinate, b, in the pair.

Conversely, if (a, b) is a given ordered pair of real numbers, then the vertical line l_x that crosses the x axis at a and the horizontal line l_y that crosses the y axis at b meet at a single point P, and the x and y coordinates of P are a and b, respectively. Thus we have a correspondence between points in the plane and ordered pairs of numbers. If (a, b) corresponds to P, then we refer to a and b as the ***coordinates*** of P; we will occasionally write $P(a, b)$ for the point P whose coordinates are (a, b).

Such an association of the points in a plane with ordered pairs of real numbers is called a ***Cartesian*** (or ***rectangular***) ***coordinate system***, and a plane endowed with a Cartesian coordinate system is called a ***Cartesian plane***. The name "Cartesian" honors René Descartes (1596–1650), the French mathematician who became famous for applying algebraic techniques to the discipline of geometry. We will refer to Cartesian planes as "planes" and will frequently refer to points and ordered pairs of real numbers interchangeably. For example, we might refer to the ordered pair (3, 2) as a point.

The spacing between successive integers on an axis of the plane is called the ***scale*** of the axis. We can adjust the scales of the two coordinate axes to suit our needs; as Figure 2.2 illustrates, the scales of the two axes may be very different from one another.

When we draw a point in the plane corresponding to a given ordered pair of numbers, we say that we ***plot***, or ***sketch***, the point. In our first few examples, we will plot single points and also sets of points in the plane.

EXAMPLE 1. Determine the x and y coordinates of the following points, and then plot the points.

a. $(1, 4)$ b. $(-1, 2)$ c. $(2, -1)$ d. $(-2, -3)$

Solution.

a. The x coordinate is 1; the y coordinate is 4.
b. The x coordinate is -1; the y coordinate is 2.
c. The x coordinate is 2; the y coordinate is -1.
d. The x coordinate is -2; the y coordinate is -3.

The points are plotted in Figure 2.3 □

Turning to sets of points in the plane, we first observe that there are infinitely many points (x, y) such that $x = 0$ (that is, whose x coordinate is 0), and they constitute the y axis. Likewise, the points (x, y) such that $y = 0$ (that is, whose y coordinate is 0) constitute the x axis.

EXAMPLE 2. Sketch the set of points (x, y) in the plane satisfying

a. $x = 2$ b. $y = -1$

Solution.

a. If $x = 2$, then the x coordinate of (x, y) is 2, so the point lies on the vertical line 2 units to the right of the y axis (Figure 2.4a).

b. If $y = -1$, then the y coordinate of (x, y) is -1, so the point lies on the horizontal line 1 unit below the x axis (Figure 2.4b). □

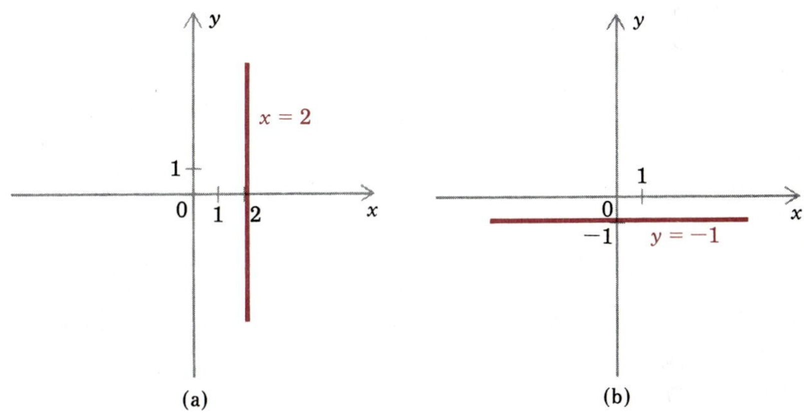

FIGURE 2.4

In general, if a is any real number, then the set of all points (x, y) satisfying the equation $x = a$ is a vertical line, and the set of all points (x, y) satisfying the equation $y = a$ is a horizontal line.

EXAMPLE 3. Sketch the set of points (x, y) in the plane satisfying $(x - 2)(y + 1) = 0$.

Solution. Observe that $(x - 2)(y + 1) = 0$ if and only if either $x - 2 = 0$ or $y + 1 = 0$, that is, if and only if $x = 2$ or $y = -1$. But the points (x, y) that satisfy $x = 2$ and those that satisfy $y = -1$ are sketched in Figures 2.4a and b. Thus (x, y) satisfies the given equation if and only if it lies on either of the two lines shown in Figure 2.5. □

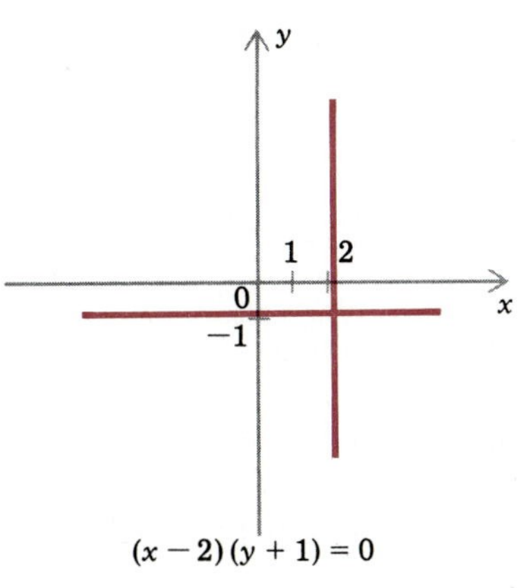

$(x - 2)(y + 1) = 0$

FIGURE 2.5

We can describe more extensive regions of the plane by means of inequalities involving coordinates of the points.

EXAMPLE 4. Sketch the set of points (x, y) in the plane satisfying

a. $x > 2$ b. $y > -1$

Solution.

a. If (x, y) satisfies $x > 2$, then the x coordinate of (x, y) is greater than 2, so that (x, y) lies to the right of the vertical line $x = 2$. Thus the points (x, y) satisfying $x > 2$ are the points to the right of, but not on, the line $x = 2$ (Figure 2.6a).

b. If $y > -1$, then the y coordinate of (x, y) is greater than -1, so that (x, y) lies above the horizontal line $y = -1$. Thus the points (x, y) satisfying $y > -1$ are the points above, but not on, the line $y = -1$ (Figure 2.6b). □

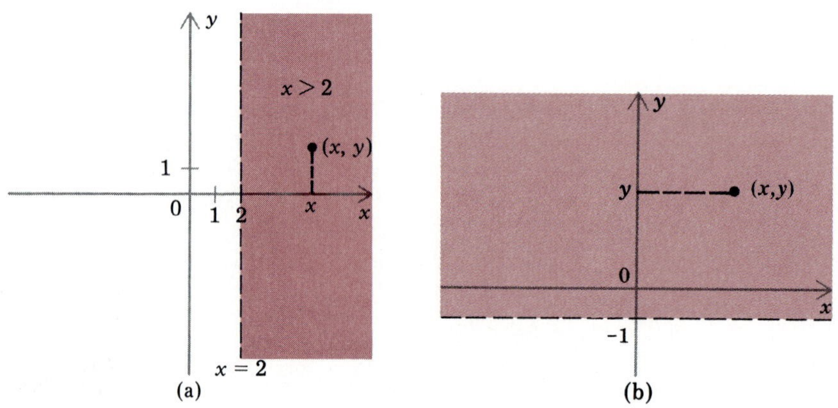

FIGURE 2.6

Had we desired the points (x, y) satisfying $x \geq 2$ in part (a) of Example 4, we would have found them to be the points either to the right of or on the line $x = 2$.

The points (x, y) satisfying $x > 0$ are the points to the right of the y axis (Figure 2.7a), and the points satisfying $y > 0$ are the points above the x axis (Figure 2.7b). Thus the points in the plane satisfying *both* $x > 0$ and $y > 0$ are the points in the upper righthand fourth of the plane, determined by the coordinate axes and called the ***first quadrant***, or ***quadrant I***. The other three quadrants are also identified by the positivity or negativity of the coordinates; all four quadrants are identified in Figure 2.8.

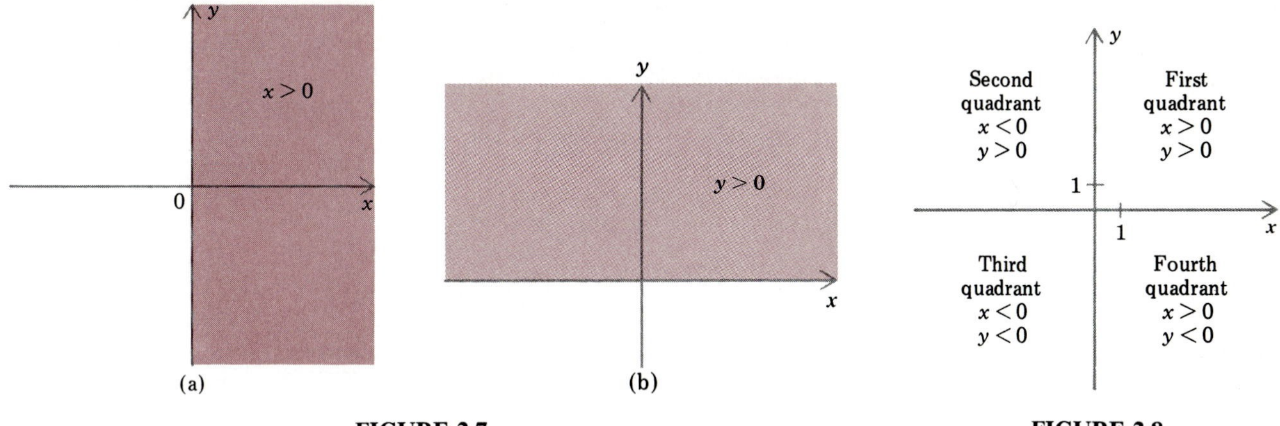

FIGURE 2.7

FIGURE 2.8

EXAMPLE 5. Sketch the set of points (x, y) in the plane satisfying $(x - 2)(y + 1) > 0$.

Solution. Notice that $(x - 2)(y + 1) > 0$ if and only if either $x - 2 > 0$ and $y + 1 > 0$, or $x - 2 < 0$ and $y + 1 < 0$. This is equivalent to $x > 2$ and $y > -1$, or $x < 2$ and $y < -1$. From Figure 2.6a and b we deduce that the points (x, y) satisfying $x > 2$ and $y > -1$ are those in region I of Figure 2.9. In a similar way we find that the points (x, y) satisfying $x < 2$ and $y < -1$ are those in region II of Figure 2.9. Combining the information obtained so far, we find that the points satisfying the given inequality are those shaded in Figure 2.9. □

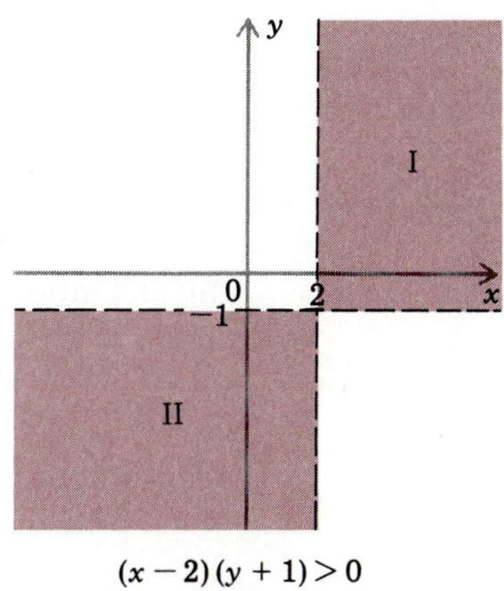

$(x - 2)(y + 1) > 0$

FIGURE 2.9

The Distance Between Two Points

If two points are on the same horizontal or vertical line in the plane, then the distance between them is defined to be the distance between their x coordinates or their y coordinates, respectively. Thus the distance between $P(x_1, y_1)$ and $Q(x_2, y_1)$ is $|x_2 - x_1|$, and the distance between $P(x_1, y_1)$ and $R(x_1, y_2)$ is $|y_2 - y_1|$ (Figure 2.10).

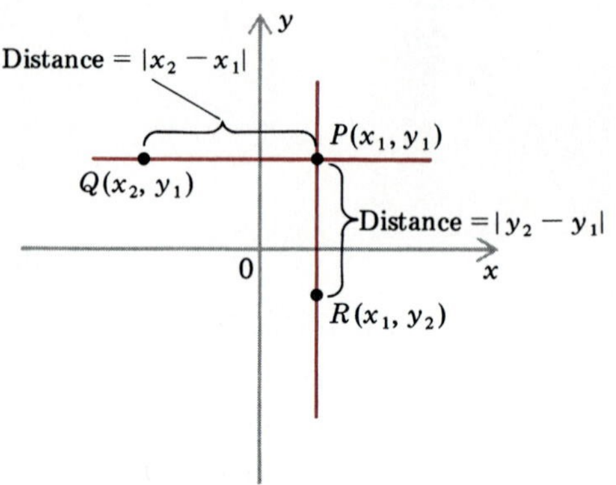

FIGURE 2.10

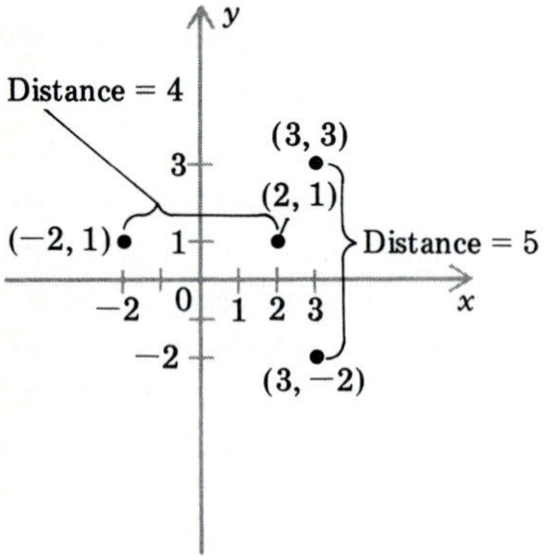

FIGURE 2.11

$$d(P, Q) = \sqrt{(x_2 - x_1)^2 + (y_2 - y_1)^2}$$

FIGURE 2.12

EXAMPLE 6. Find the distances between the following pairs of points, illustrated in Figure 2.11.

a. $(-2, 1)$ and $(2, 1)$ b. $(3, 3)$ and $(3, -2)$

Solution.

a. The distance between $(-2, 1)$ and $(2, 1)$ is $|2 - (-2)| = |4| = 4$.

b. The distance between $(3, 3)$ and $(3, -2)$ is $|-2 - 3| = |-5| = 5$. □

We define the ***distance*** $d(P, Q)$ between any two points $P(x_1, y_1)$ and $Q(x_2, y_2)$ in the plane by the following formula:

Distance Formula

$$d(P, Q) = \sqrt{(x_2 - x_1)^2 + (y_2 - y_1)^2}$$

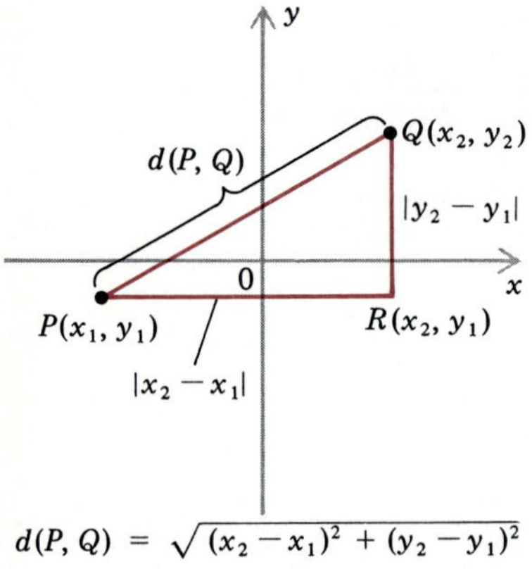

This definition is based on the Pythagorean Theorem, which relates the lengths of the sides of a right triangle (see Figure 2.12).

EXAMPLE 7. Find the distance between $(2, 6)$ and $(4, -1)$.

Solution. By the distance formula, with $P = (2, 6)$ and $Q = (4, -1)$, we have

$$d(P, Q) = \sqrt{(4 - 2)^2 + (-1 - 6)^2}$$
$$= \sqrt{2^2 + (-7)^2} = \sqrt{53} \quad □$$

The Midpoint of a Line Segment

Let $P(x_1, y_1)$ and $Q(x_2, y_2)$ be two distinct points in the plane, so that P and Q determine a line segment (Figure 2.13). Then the coordinates (x, y) of the midpoint M of that line segment can be determined from x_1, y_1, x_2, and y_2. We will not derive the formula for the coordinates but will simply state it:

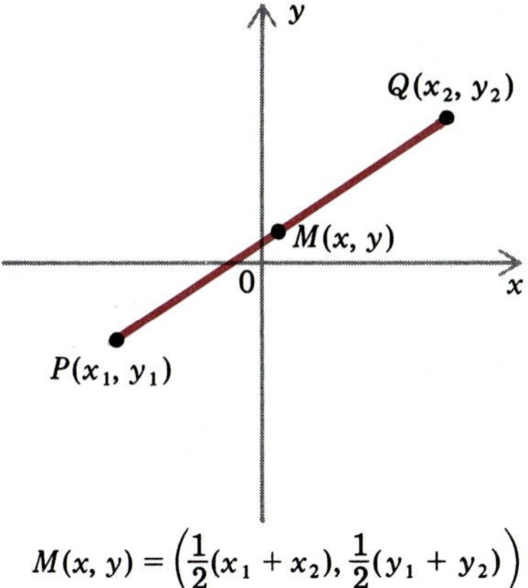

$$M(x, y) = \left(\frac{1}{2}(x_1 + x_2), \frac{1}{2}(y_1 + y_2)\right)$$

FIGURE 2.13

Midpoint Formula

$$M(x, y) = \left(\frac{1}{2}(x_1 + x_2), \frac{1}{2}(y_1 + y_2)\right)$$

Thus the x coordinate of the midpoint is the average of the x coordinates of the two points P and Q, and the y coordinate of the midpoint is the average of the y coordinates of the two points.

EXAMPLE 8. Find the coordinates of the midpoint of the line segment joining (3, 4) and (5, −2) (Figure 2.14).

Solution. By the midpoint formula the coordinates are given by

$$\left(\frac{1}{2}(3 + 5), \frac{1}{2}[4 + (-2)]\right)$$

which reduces to (4, 1). □

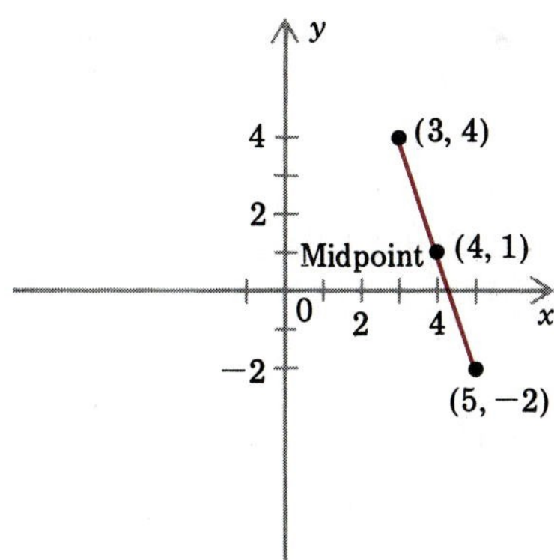

FIGURE 2.14

EXERCISES 2.1

1. Set up a coordinate system, and plot the points $(-2, 0)$, $(1, 2)$, $(-3, 1)$, $(-1.5, -2.5)$, $(0, -\sqrt{2})$, $(1, -2)$.

2. Set up a coordinate system, and plot the points $(100, 200)$, $(-150, 100)$, $(-200, 200)$, $(300, -150)$, $(400, 0)$, $(0, -300)$.

In Exercises 3–22, set up a coordinate system and sketch the set of points (x, y) in the plane satisfying the given equation or inequality.

3. $y = 5$ **4.** $x = -2$

5. $x > 1$ **6.** $y < -1$

7. $xy = 0$ **8.** $(x - 3)(y + 2) = 0$

9. $(\frac{1}{2}x + 1)(2 - y) = 0$ **10.** $x^2 = 1$

11. $y^2 = 4$ **12.** $x^2 + x = 0$

13. $y^3 - y = 0$ **14.** $y^2 - 8y + 16 = 0$

15. $x^2 - 3x - 4 = 0$ **16.** $x^2 + 5x - 14 = 0$

17. $y^2 = 4y + 12$ **18.** $x^2 \geq 1$

19. $x^2 < 1$ **20.** $xy > 0$

21. $xy < 0$ **22.** $x^2 + y^2 = 0$

In Exercises 23–34, find the distance between the two given points.

23. $(-1, 2)$ and $(-1, 4)$ **24.** $(10, 12)$ and $(5, 12)$

25. $(5, 12)$ and $(0, 0)$ **26.** $(3, 3)$ and $(5, 1)$

27. $(4, -3)$ and $(-2, -5)$ **28.** $(-1, -3)$ and $(-7, 4)$

29. $(4, -3)$ and $(-1, 9)$ **30.** $(\frac{1}{4}, \frac{1}{4})$ and $(-\frac{1}{4}, -\frac{1}{4})$

31. $(3, \sqrt{2})$ and $(-\sqrt{2}, 4)$

32. $(0.1, -0.1)$ and $(0.4, 0.3)$

33. $(0, 0)$ and $(3a, 4a)$, where $a \geq 0$

34. $(0, 0)$ and $(3a, 4a)$, where $a < 0$

In Exercises 35–42, find the coordinates of the midpoint of the line segment joining the two given points.

35. $(4, 2)$ and $(8, 10)$ **36.** $(-1, 3)$ and $(5, -1)$

37. $(0, 4)$ and $(0, 10)$ **38.** $(-2, 4)$ and $(3, 10)$

39. $(3.2, 1.4)$ and $(-1.8, 4.3)$

40. $(-2.2, -0.8)$ and $(3.6, 2.5)$

41. $(-a, a)$ and (a, a) **42.** $(1, 1)$ and $(\frac{2}{3}a, \frac{1}{4}a)$

43. Determine the points on the x axis that are 2 units from the point $(4, 1)$.

44. Determine the points on the y axis that are 3 units from the point $(2, 0)$.

45. Let $P(a, b)$ and $Q(c, d)$ be distinct points. Find a formula for the coordinates of the point two thirds of the distance from P to Q.

46. A ***rhombus*** is a polygon whose four sides are equal in length. Show that the points $(0, 0)$, $(3, 0)$, $(2, \sqrt{5})$, and $(5, \sqrt{5})$ are the vertices of a rhombus.

47. A triangle is ***isosceles*** if two of its sides have equal length. Show that the points $(0, 4)$, $(3, -1)$, and $(3 + 3\sqrt{2}, 3)$ are the vertices of an isosceles triangle.

Exercises 48–50 illustrate how algebra can be used to prove geometric facts.

48. Show that the line segments joining the midpoints of the sides of an equilateral triangle form an equilateral triangle. (*Hint:* Let the three vertices of the given triangle be $(0, 0)$, $(a, 0)$, and $(a/2, \sqrt{3}\,a/2)$.)

49. Show that the diagonals of a square intersect at their midpoints. (*Hint:* Set up a coordinate system as in Figure 2.15. Then compute the coordinates of the midpoint M_1 of the diagonal joining P_1 and P_3, and the midpoint M_2 of the diagonal joining P_2 and P_4. Finally, show that $M_1 = M_2$.)

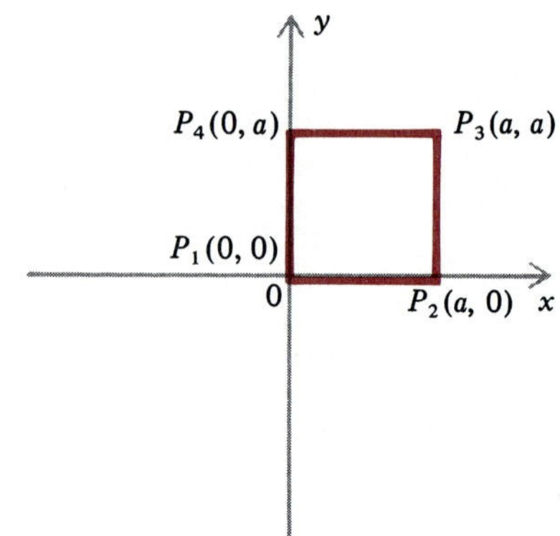

FIGURE 2.15

50. The following instructions were found at the intersection of Walnut and Broad Streets in Philadelphia: "To find the treasure chest, go west 2 kilometers, then south 1 kilometer, then east 3 kilometers, then south 2 kilometers, and, finally, go half the distance in the direction of the starting point at the intersection of Walnut and Broad." What is the (straight-line) distance between the intersection and the treasure chest?

51. The infield in a baseball field is square, 90 feet on a side. What is the distance between home plate and second base?

2.2 LINES IN THE PLANE

In this section we analyze straight lines in the plane. For convenience we will separate the collection of straight lines in the plane into two categories: vertical and nonvertical lines.

Consider the vertical line drawn in Figure 2.16. Notice that the x coordinate of each point on the line is -1. Indeed, a point (x, y) is on the line if and only if $x = -1$. As a result, we say that the equation $x = -1$ is an equation of the given vertical line.

More generally, if c is any fixed number, then we call the equation $x = c$ an equation of the vertical line whose x intercept is c.

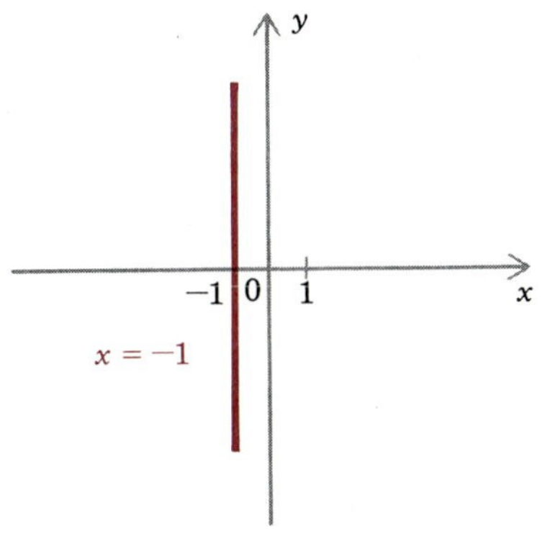

FIGURE 2.16

EXAMPLE 1. Find equations for the vertical lines l_1, l_2, and l_3 appearing in Figure 2.17.

Solution. We have the following:

$$l_1\!: x = -4 \qquad l_2\!: x = \frac{5}{3}, \quad \text{and} \quad l_3\!: x = \pi \qquad \square$$

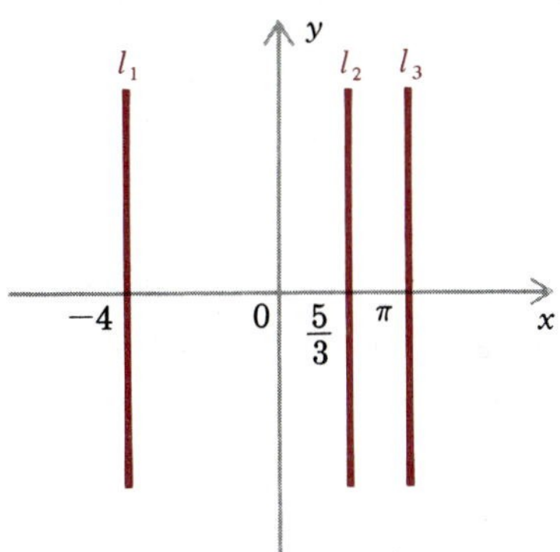

FIGURE 2.17

This is really all there is to say about vertical lines. Henceforth we will restrict our discussion to nonvertical lines.

Slope of a Nonvertical Line

Let l be any nonvertical line in the plane, and let $P_1(x_1, y_1)$ and $P_2(x_2, y_2)$ be any two distinct points on l. Since l is not vertical, $x_1 \neq x_2$. It turns out that the ratio

$$\frac{y_2 - y_1}{x_2 - x_1}$$

depends only on the line l and *not* on the particular points P_1 and P_2. To see this, observe that if $P_3(x_3, y_3)$ and $P_4(x_4, y_4)$ are any two points on l with P_3 distinct from P_4, then triangles P_1QP_2 and P_3RP_4 are similar

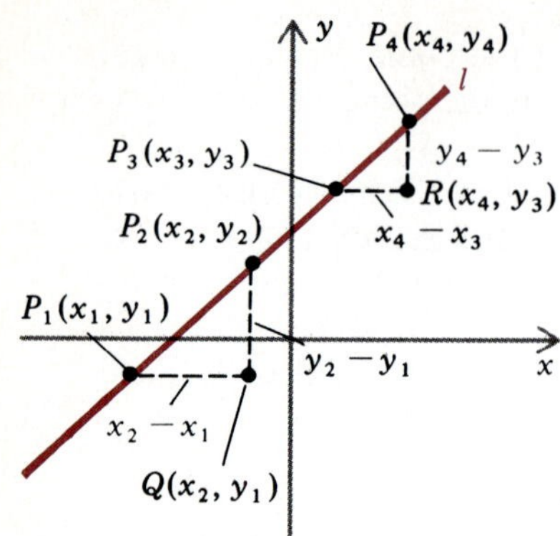

FIGURE 2.18

(Figure 2.18), so that

$$\frac{y_2 - y_1}{x_2 - x_1} = \frac{y_4 - y_3}{x_4 - x_3}$$

This common ratio is called the ***slope*** of l and is usually denoted by m. Thus if $P_1(x_1, y_1)$ and $P_2(x_2, y_2)$ are any two distinct points on l, then the slope of l is given by the following equation:

Slope of a Line

$$m = \frac{y_2 - y_1}{x_2 - x_1} \quad (1)$$

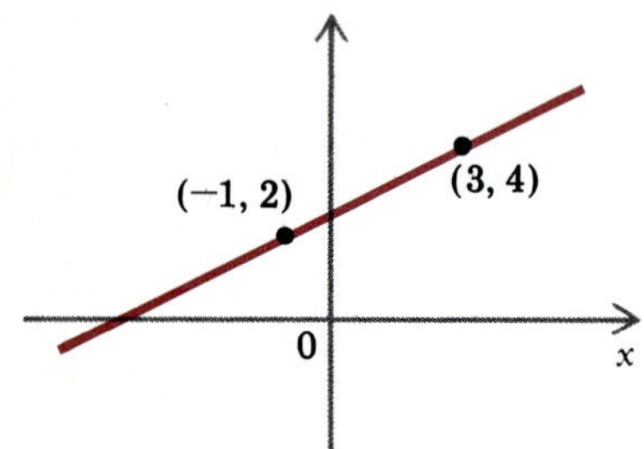

FIGURE 2.19

EXAMPLE 2. Find the slope of the line l that contains the points $(-1, 2)$ and $(3, 4)$ (Figure 2.19).

Solution. If we let $(-1, 2)$ be $P_1(x_1, y_1)$ and $(3, 4)$ be $P_2(x_2, y_2)$, then by (1) the slope m of l is given by

$$m = \frac{4 - 2}{3 - (-1)} = \frac{2}{4} = \frac{1}{2} \quad \square$$

CAUTION: We have not defined the slope of a vertical line, nor will we do so. If one tries to apply (1) to a vertical line $x = c$, one obtains the meaningless expression

$$\frac{y_2 - y_1}{c - c}$$

Next we will use the notion of slope to lead us to two types of equations for nonvertical lines—the point–slope equation and the slope–intercept equation.

Point–Slope Equation of a Line

Suppose that we are given the slope m and a point $P_1(x_1, y_1)$ on a nonvertical line l. We would like to find an equation that is satisfied by any point $P(x, y)$ on l. If we apply (1) with $P_2(x_2, y_2)$ replaced by $P(x, y)$, we find that

$$m = \frac{y - y_1}{x - x_1}$$

which is equivalent to the following equation:

Point-Slope Equation of a Line

$$y - y_1 = m(x - x_1) \quad (2)$$

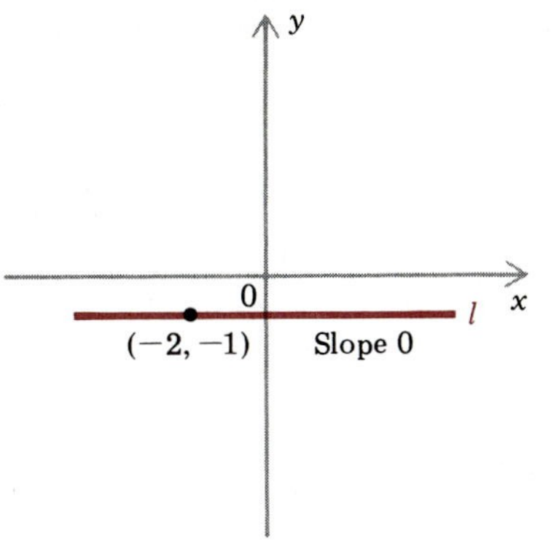

FIGURE 2.20

A point–slope equation involves the slope m and a single point $P_1(x_1, y_1)$ on the line. A point $P(x, y)$ is on the line if and only if x and y satisfy (2).

EXAMPLE 3. Find a point–slope equation of the line l that has slope 0 and passes through the point $(-2, -1)$. Then sketch the line.

Solution. Using (2) with $m = 0$, $x_1 = -2$, and $y_1 = -1$, we obtain the point–slope equation

$$y - (-1) = 0(x - (-2))$$

To sketch l we first simplify the equation:

$$y + 1 = 0$$
$$y = -1$$

Consequently l is horizontal and crosses the y axis at the point $(0, -1)$ (Figure 2.20). □

In Example 3 we found that the given line, which has slope 0, is horizontal. By the same method we deduce that any line having slope 0 and passing through, say (x_1, y_1), has an equation of the form

$$y - y_1 = 0(x - x_1)$$

which simplifies to

$$y - y_1 = 0$$

and then to

$$y = y_1 \tag{3}$$

It follows from (3) that any line with slope 0 is horizontal.

EXAMPLE 4. Find a point–slope equation of the line that passes through the point $(3, 2)$ and has the given slope. Then sketch the line.

a. $\frac{1}{2}$ b. 3 c. -1

Solution. In each case we use (2):

a. We have $m = \frac{1}{2}$, $x_1 = 3$, and $y_1 = 2$, so that a point–slope equation of the line is

$$y - 2 = \frac{1}{2}(x - 3)$$

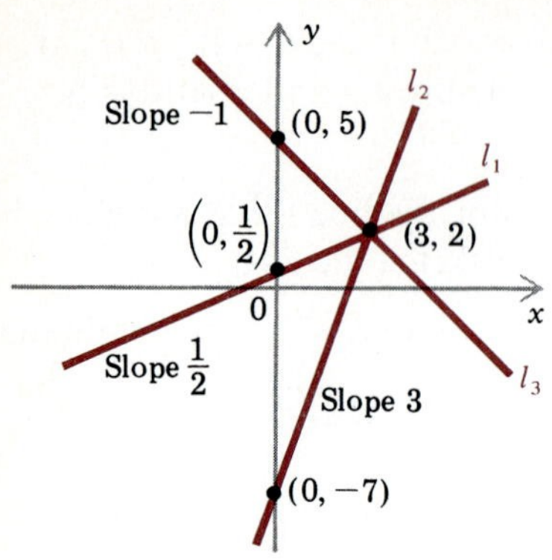

FIGURE 2.21

which simplifies to

$$y - 2 = \frac{1}{2}x - \frac{3}{2}$$

or

$$y = \frac{1}{2}x + \frac{1}{2}$$

In order to sketch the graph, we find a second point on the line by setting $x = 0$ and noting that $y = \frac{1}{2}$. Therefore $(0, \frac{1}{2})$ is a second point, and the line, l_1, can be drawn (Figure 2.21).

b. This time $m = 3$, $x_1 = 3$, and $y_1 = 2$, so that

$$y - 2 = 3(x - 3)$$

which simplifies to

$$y - 2 = 3x - 9$$

or

$$y = 3x - 7$$

For the graph we again find a second point on the line by letting $x = 0$ and noting that $y = -7$. Therefore $(0, -7)$ is a second point, and the line, l_2, can be drawn (Figure 2.21).

c. Now we have $m = -1$, $x_1 = 3$, and $y_1 = 2$, so that

$$y - 2 = -1(x - 3)$$

which simplifies to

$$y - 2 = -x + 3$$

or

$$y = -x + 5$$

As before, we let $x = 0$ and find that $y = 5$, so that $(0, 5)$ is a second point on the line. The line, l_3, is drawn in Figure 2.21. □

Let us tie together the results of Examples 3 and 4 and the accompanying Figures 2.20 and 2.21. First, if $m = 0$, then the line is horizontal. Next, if $m > 0$, then the line moves upward from left to right. Finally, if $m < 0$, then the line moves downward from left to right. Moreover, the larger the absolute value of the slope, the steeper the line is.

If we know the slope m and a point (x_1, y_1) on a given line l, then a point–slope equation of l is $y - y_1 = m(x - x_1)$. Thus the point on l whose x coordinate is $x_1 + 1$ satisfies

$$y - y_1 = m(x_1 + 1 - x_1) = m$$

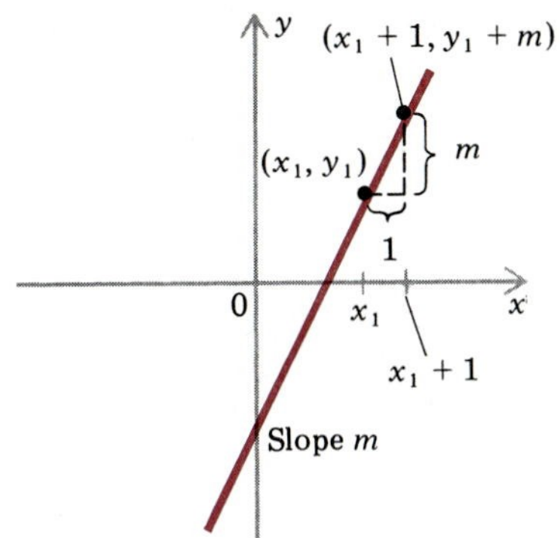

FIGURE 2.22

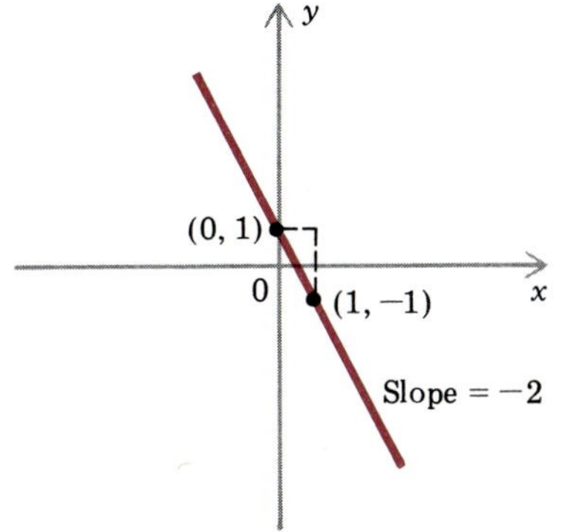

FIGURE 2.23

so that

$$y = y_1 + m$$

Therefore an increase of 1 unit in the value of x (from x_1 to $x_1 + 1$) causes a change of m units in the value of y (from y_1 to $y_1 + m$), and the point $(x_1 + 1, y_1 + m)$ also lies on l (Figure 2.22). We will use these ideas to help sketch the line discussed in the next example.

EXAMPLE 5. Sketch the line that contains the point (0, 1) and has slope -2.

Solution. We can sketch the line once we have two points on it. We are given one point on the line, (0, 1). To find a second point we use the fact that the line has slope -2. Thus if we start at (0, 1) and *increase* the value of x by 1, then the value of y must *decrease* by 2. This yields the point $(0 + 1, 1 - 2)$, that is, $(1, -1)$, on the line. Now we can sketch the line (Figure 2.23). □

In order to find an equation of the line l that contains two given distinct points, we first use (1) to determine the slope of l, and then use a point–slope equation.

EXAMPLE 6. Find an equation of the line l that contains the points $(2, -2)$ and $(3, -1)$. Then show that (5, 1) is on l.

Solution. By (1) the slope m of l is given by

$$m = \frac{-1 - (-2)}{3 - 2} = \frac{1}{1} = 1$$

Thus if we apply the point–slope equation of the line with $m = 1$ and $P(x_1, y_1) = (2, -2)$, we obtain

$$y - (-2) = 1(x - 2)$$

which is equivalent to

$$y + 2 = x - 2$$

To show that (5, 1) is on l we let $x = 5$ and $y = 1$ and see that the equation is valid:

$$1 + 2 = 5 - 2 \quad \square$$

Slope–Intercept Equation of a Line

A nonvertical line l must cross the y axis at some point. The y coordinate of that point is called the ***y intercept*** of l and is usually denoted by b (Figure 2.24). Thus $(0, b)$ is the point of intersection of the y axis and l. Now suppose that the nonvertical line l has slope m and y intercept b. If we let $P_1(x_1, y_1) = (0, b)$ and apply the formula for the point–slope equation of a

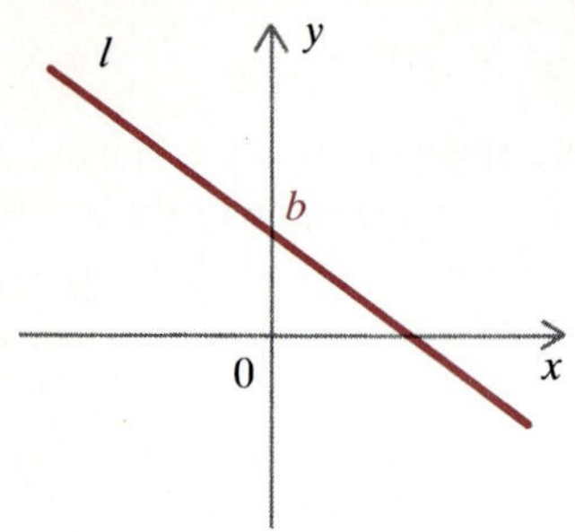

FIGURE 2.24

line, we find that $P(x, y)$ is on the line l provided that

$$y - b = m(x - 0)$$

which simplifies to give us the following equation:

Slope–Intercept Equation of a Line

$$y = mx + b \tag{4}$$

The slope–intercept equation of a line involves the slope m and the y intercept b of the line. As before, (x, y) is on the line if and only if (4) is satisfied.

EXAMPLE 7. Find the slope–intercept equation of the line that has slope -2 and y intercept $\frac{1}{2}$. Then sketch the line.

Solution. By (4) with $m = -2$ and $b = \frac{1}{2}$, the slope–intercept equation is

$$y = -2x + \frac{1}{2} \tag{5}$$

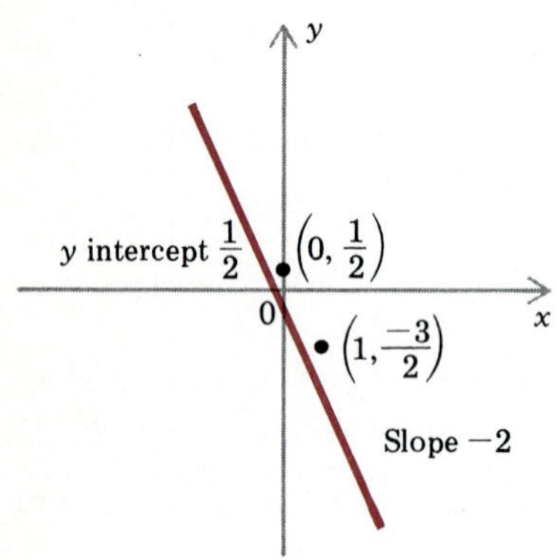

FIGURE 2.25

Since the y intercept is $\frac{1}{2}$, we know that $(0, \frac{1}{2})$ is one point on the line. Taking $x = 1$ in equation (5), we find that

$$y = -2(1) + \frac{1}{2} = -\frac{3}{2}$$

Thus $(1, -\frac{3}{2})$ is a second point on the line, which is sketched in Figure 2.25. □

We have presented two types of equations for a nonvertical line: point–slope and slope–intercept. Other types of equations appear in Exercises 45, 48, and 50. However, all equations for a given line are equivalent, and information that may be explicit in one equation of the line may be obtained from any other. For example, the y intercept b of a line is explicitly written in the slope–intercept equation $y = mx + b$, but it does not appear explicitly in a point–slope equation $y - y_1 = m(x - x_1)$ unless $x_1 = 0$ or $m = 0$. Nevertheless we can determine the y intercept from *any* equation of the line by setting $x = 0$ and solving for y.

EXAMPLE 8. Find the y intercept of the line with equation $y - 5 = -4(x + 2)$.

Solution. Setting $x = 0$ and solving for y, we obtain

$$y - 5 = -4(0 + 2) = -8$$
$$y = -8 + 5 = -3$$

Thus the y intercept is -3. □

Parallel and Perpendicular Lines

Among the geometric properties of lines that are easily studied by means of slopes of lines, the two most important are the properties of parallelism and perpendicularity:

> Let l_1 and l_2 be nonvertical lines with slopes m_1 and m_2.
>
> i. The lines l_1 and l_2 are parallel if and only if $m_1 = m_2$.
> ii. The lines l_1 and l_2 are perpendicular if and only if $m_1 m_2 = -1$.

We will not prove the results just stated, although they depend only on the definition of slope and basic geometric facts.

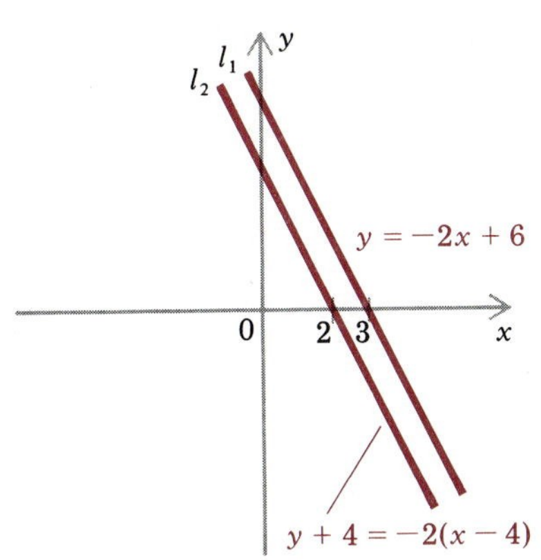

Parallel lines

FIGURE 2.26

EXAMPLE 9. Show that the lines

$$l_1: y = -2x + 6 \quad \text{and} \quad l_2: y + 4 = -2(x - 4)$$

are parallel.

Solution. The equation of l_1 is the slope–intercept equation of l_1; from it we see that the slope of l_1 is -2. The equation of l_2 is a point–slope equation of l_2; from it we see that the slope of l_2 is also -2. Since the two slopes are equal, the lines are parallel by (i) (Figure 2.26). □

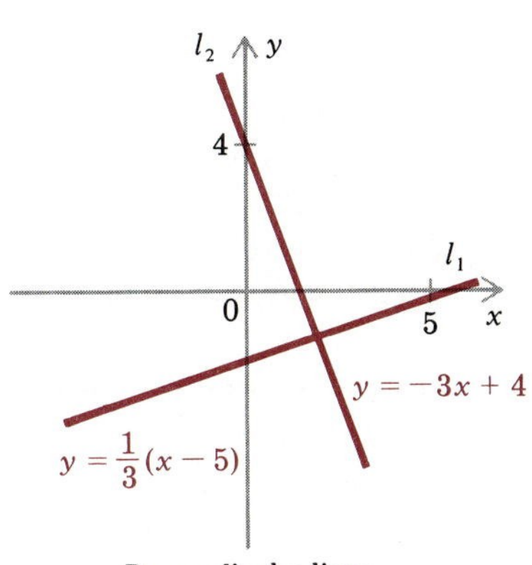

Perpendicular lines

FIGURE 2.27

EXAMPLE 10. Show that the lines

$$l_1: y = \frac{1}{3}(x - 5) \quad \text{and} \quad l_2: y = -3x + 4$$

are perpendicular.

Solution. The equation of l_1 is a point–slope equation; from it we see that l_1 has slope $m_1 = \frac{1}{3}$. The equation of l_2 is the slope–intercept equation of l_2; from it we see that l_2 has slope $m_2 = -3$. Since

$$m_1 m_2 = \frac{1}{3}(-3) = -1$$

it follows from (ii) that l_1 and l_2 are perpendicular (Figure 2.27). □

EXAMPLE 11. Find an equation of the line l that contains the point $(\frac{1}{2}, \frac{2}{3})$ and is parallel to the line with equation $y = -4x + 7$.

Solution. The slope of the line with equation $y = -4x + 7$ is -4. It follows from (i) that l also has slope -4. Since we are given a point on l and we now know the slope of l, we use the point–slope equation in (2) with $m = -4$, $x_1 = \frac{1}{2}$, and $y_1 = \frac{2}{3}$ to obtain the equation

$$y - \frac{2}{3} = -4\left(x - \frac{1}{2}\right)$$

for l. □

EXERCISES 2.2

In Exercises 1–4, find the slope of the line that contains the points P_1 and P_2.

1. $P_1 = (4, 3)$, $P_2 = (6, 9)$
2. $P_1 = (-5, -4)$, $P_2 = (-9, -7)$
3. $P_1 = (\frac{1}{4}, \frac{1}{6})$, $P_2 = (\frac{3}{4}, \frac{1}{2})$
4. $P_1 = (1.57, 2.86)$, $P_2 = (1.84, 2.56)$

In Exercises 5–6, approximate the slope of the line that contains the points P_1 and P_2. Round your answer to four significant digits.

5. $P_1 = (1.023, -2.179)$, $P_2 = (3.625, 4.008)$
6. $P_1 = (-0.0045, -0.1342)$, $P_2 = (0.9762, 0.1234)$

In Exercises 7–12, find a point–slope equation of the line that has slope m and passes through the point P. Then sketch the line.

7. $m = 0$, $P = (3, -4)$
8. $m = 2$, $P = (-2, -1)$
9. $m = \frac{2}{3}$, $P = (3, 0)$
10. $m = -4$, $P = (-\frac{1}{2}, 2)$
11. $m = -1$, $P = (0, \frac{3}{2})$
12. $m = \sqrt{2}$, $P = (\sqrt{2}, 3)$

In Exercises 13–18, find an equation of the line that contains the points P_1 and P_2.

13. $P_1 = (0, 1)$, $P_2 = (2, 5)$
14. $P_1 = (0, -1)$, $P_2 = (3, 8)$
15. $P_1 = (-1, -2)$, $P_2 = (-2, 2)$
16. $P_1 = (2, 3)$, $P_2 = (0, 0)$
17. $P_1 = (2, -1)$, $P_2 = (2, 6)$
18. $P_1 = (4, -3)$, $P_2 = (2, 4)$

In Exercises 19–24, find the slope–intercept equation of the line that has slope m and y intercept b. Then sketch the line.

19. $m = 0$, $b = -2$
20. $m = 3$, $b = 4$
21. $m = -\frac{4}{5}$, $b = -1$
22. $m = 1$, $b = \pi$
23. $m = -\sqrt{3}$, $b = \frac{1}{2}$
24. $m = 0.25$, $b = 0.4$

In Exercises 25–32, find the slope m and the y intercept b (if any) of the line with the given equation. Then sketch the line.

25. $y = 3$
26. $x = -4$
27. $y = 3x - \frac{1}{2}$
28. $3y = x - 4$
29. $2x + 3y = 6$
30. $\frac{1}{2}y - 2x = 2$
31. $3.6x - 1.2y = 0$
32. $y - 7 = 3(x - 2)$

In Exercises 33–44, determine whether the two lines having the given equations are parallel, perpendicular, or neither.

33. $x = 0$; $y = 0$
34. $y = 3$; $x = -4$
35. $y = x$; $y = -x$
36. $y = 2x$; $y = -\frac{1}{2}x$
37. $y = 2x$; $y = -2x$
38. $2x - 3y = 4$; $y = -\frac{3}{2}x$
39. $\frac{1}{2}y = x$; $y = 2(x - 4)$

40. $x = -y;\ y = 4x + 2$

41. $3x - 5y = 4;\ 2x - 6y = 3$

42. $4x + 6y = 5;\ 6x + 9y = 1$

43. $y = 0.01(x - 1);\ y + 100x = 50$

44. $y - 5 = \frac{1}{2}(x + 3);\ x - 2y = 6$

45. Let $P_1(x_1, y_1)$ and $P_2(x_2, y_2)$ be two distinct points not on the same vertical line, and let the line l pass through P_1 and P_2. Then $P(x, y)$ is on l provided that

$$(x_2 - x_1)(y - y_1) = (y_2 - y_1)(x - x_1) \qquad (6)$$

This equation is called a ***two-point equation*** of l.
a. Show that P_1 and P_2 satisfy (6).
b. Use (6) to obtain a point–slope equation of l.

46. Find a two-point equation of the line that passes through each of the following pairs of points. Then sketch the line.
a. $(4, 2)$ and $(3, -3)$
b. $(-1, -2)$ and $(-5, 0)$
c. $(\frac{1}{2}, \frac{3}{2})$ and $(-\frac{5}{2}, \frac{3}{2})$

47. Let l be a nonhorizontal line. The x coordinate of the point at which l crosses the x axis is the ***x intercept*** of l, and is usually denoted by a (Figure 2.28). Thus $(a, 0)$ is on l. The x intercept is determined by setting $y = 0$ in an equation of l and then solving for x. Determine the x intercept of the lines having the following equations.
a. $y = 6x - 2$ b. $y - 2 = \frac{1}{2}(x + 1)$ c. $x = -\frac{3}{4}$

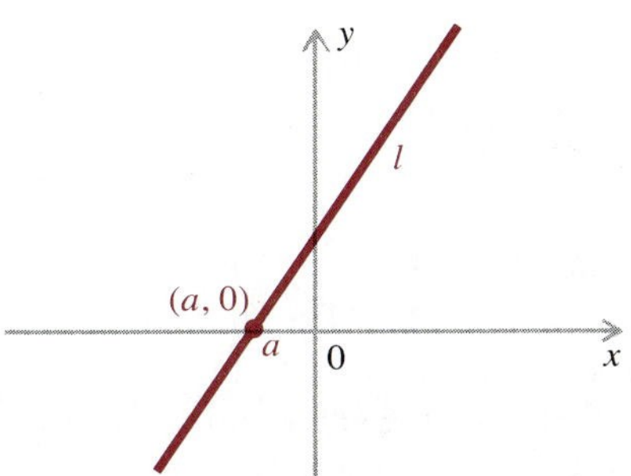

FIGURE 2.28

48. Let l be a line that is neither horizontal nor vertical, and let its x and y intercepts be a and b, respectively. Then (x, y) is on l provided that

$$\frac{x}{a} + \frac{y}{b} = 1$$

This equation is called the ***two-intercept equation*** of l. Show that the slope of l is $-b/a$ by writing the equation in the equivalent slope–intercept form.

49. Find the two-intercept equation of the line that has x intercept a and y intercept b. Then sketch the line.
a. $a = -1, b = 2$ b. $a = 3, b = \frac{1}{2}$
c. $a = 0.7, b = 0.3$ d. $a = -2, b = -3$

50. *Any* line l can be described by an equation of the form

$$Ax + By = C$$

where either $A \neq 0$ or $B \neq 0$. This equation is called a ***general linear equation***.
a. Determine the x intercept a (if any) and the y intercept b (if any) of l.
b. Determine the slope (if any) of l.

51. Determine which of the following lines contain the point $(-2, 4)$.
a. $y = 2x + 8$ b. $y = 2x - 8$
c. $y + 2 = -2(x - 1)$ d. $y - 2 = 2(x + 1)$
e. $3x + 4y = 22$ f. $3x + 4y = 10$

52. Determine which of the following lines contain the point $(-3, -1)$.
a. $y = 2x - 5$ b. $y = 2x + 5$
c. $y + 1 = \frac{1}{2}(x + 3)$ d. $y - 1 = \frac{1}{2}(x - 3)$
e. $-2y + y = -5$ f. $2x - y = -5$

53. Suppose that a line l is to pass through $(2, 3)$. What slope must it have in order to pass through $(5, -1)$ as well?

54. Show that a line whose equation is $Ax + By = 0$ passes through the origin. If $B \neq 0$, what is the slope of the line?

55. Find an equation of the line that passes through the point $(-1, -4)$ and has slope -2.

56. Find an equation of the line that passes through the point $(3, 0)$ and has y intercept $-\frac{1}{3}$.

57. Find an equation of the line that has x and y intercepts -1 and 4, respectively.

58. Line l_1 passes through the points $(-2, 1)$ and $(4, -1)$, and line l_2 passes through the points $(5, -3)$ and $(-7, -6)$. Which line has the greater slope?

59. Find an equation of the line that passes through the point $(-2, 5)$ and is parallel to the line with equation $4x + 3y = 1$.

60. Find an equation of the line that passes through the point $(3, -3)$ and is perpendicular to the line with equation $y = 6x - 2$.

61. Find the point of intersection of the lines having equations $2x + y = 0$ and $x - y = -3$. (*Hint:* First solve for y in terms of x in the first equation, and then substitute for y in the second equation.)

62. Find the point of intersection of the lines having equations $2x - 4y = 3$ and $x - y = 1$.

63. Show that the lines having the following equations determine a rectangle: $y = 2x + 1$, $y = 2x + 5$, $y = -\frac{1}{2}x + 1$, $y = -\frac{1}{2}x - 3$.

64. Show that the figure determined by the four points $(2, 1)$, $(3, 3)$ $(2, 5)$, and $(1, 3)$ is a parallelogram by proving that the opposite sides are parallel.

65. Show that the figure determined by the four points $(3, 2)$, $(4, 3)$, $(3, 4)$, and $(2, 3)$ is a square.

66. Show that the figure determined by the three points $(1, 1)$, $(5, 8)$ and $(-1, 5)$ is a right triangle.

2.3 FUNCTIONS

Frequently temperature in degrees Fahrenheit is associated with the corresponding temperature in degrees Celsius, and the area of a circle is associated with the radius of the circle. In mathematics such associations are called functions.

DEFINITION 2.1 A ***function*** consists of a domain and a rule. The ***domain*** is a collection of real numbers. The ***rule*** assigns to each number in the domain one and only one number. The total collection of numbers that a function assigns to the numbers in its domain is called the ***range*** of the function.

Functions are normally denoted by the letters f or g and occasionally by other letters such as h, F, G, or H (in the same part of the alphabet). The value assigned by a function f to a member x of its domain is written $f(x)$ and is read "f of x" or "the value of f at x." If f assigns $\sqrt{2}$ to the number -1, we would write $f(-1) = \sqrt{2}$, whereas if f assigns $\frac{2}{3}$ to 0.4, then we would write $f(0.4) = \frac{2}{3}$. A function f is like a machine that applies the rule of f to each number x in the domain of f and thereby produces the number $f(x)$ in the range of f (Figure 2.29).

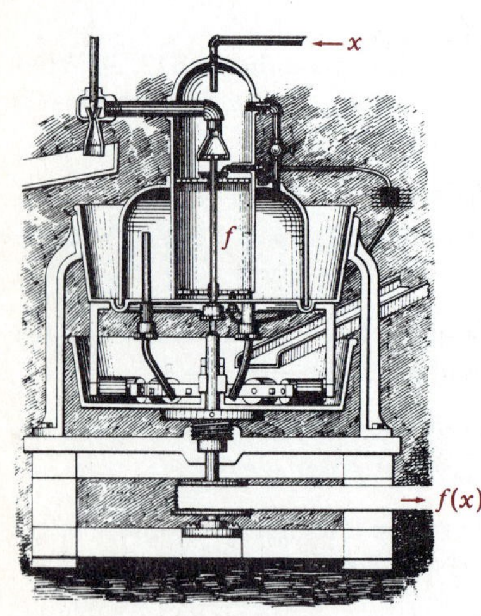

FIGURE 2.29

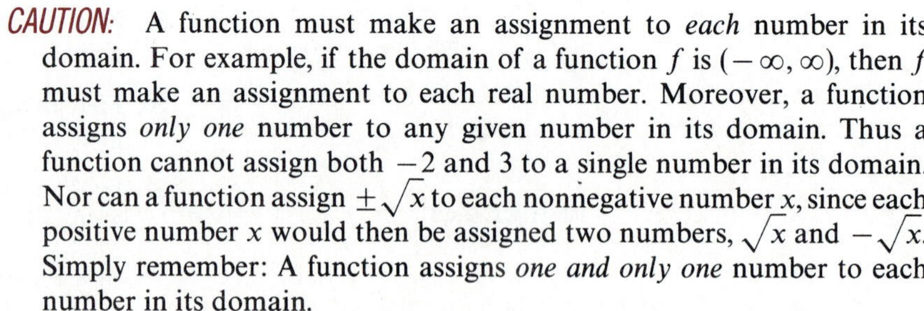

CAUTION: A function must make an assignment to *each* number in its domain. For example, if the domain of a function f is $(-\infty, \infty)$, then f must make an assignment to each real number. Moreover, a function assigns *only one* number to any given number in its domain. Thus a function cannot assign both -2 and 3 to a single number in its domain. Nor can a function assign $\pm\sqrt{x}$ to each nonnegative number x, since each positive number x would then be assigned two numbers, $\sqrt{x}$ and $-\sqrt{x}$. Simply remember: A function assigns *one and only one* number to each number in its domain.

Examples of Functions

Our first example of a function f is obtaned by assigning the number 4 to each real number. Thus we write

$$f(x) = 4 \quad \text{for all real } x \tag{1}$$

For the particular values $-\pi$, 0, and $\sqrt{3}$ of x we have

$$f(-\pi) = 4, \qquad f(0) = 4, \quad \text{and} \quad f(\sqrt{3}) = 4$$

The range of f consists of the single number 4, which is assigned to each number.

More generally, if a function f assigns the same number c to each real number x, then we write

$$\boxed{f(x) = c \quad \text{for all real } x}$$

and call f a ***constant function***. Its range consists of the single number c. The function described in (1) is thus a constant function, with $c = 4$.

Next, let us recall from Section 2.2 that the slope–intercept equation of a nonvertical line is

$$y = mx + b$$

Such an equation leads us to define a function f by

$$f(x) = mx + b \quad \text{for all real } x$$

Because of the relationship between f and the line $y = mx + b$, f is called a ***linear function***. Thus if

$$f(x) = 2x - 3 \quad \text{for all real } x \quad \text{and} \quad g(x) = -x + \frac{1}{2} \quad \text{for all real } x$$

then f and g are linear functions.

Now let f be the function that assigns $x^2 + 3x - 1$ to real number x. Then we write

$$f(x) = x^2 + 3x - 1 \quad \text{for all } x \tag{2}$$

EXAMPLE 1. For the function given in (2), determine the value of

a. $f(-4)$ b. $f(x + h)$ c. $\dfrac{f(x+h) - f(x)}{h}$

Solution.

a. We let $x = -4$ in (2):

$$f(-4) = (-4)^2 + 3(-4) - 1 = 16 - 12 - 1 = 3$$

b. We substitute $x + h$ for x in (2):

$$\begin{aligned} f(x+h) &= (x+h)^2 + 3(x+h) - 1 \\ &= x^2 + 2xh + h^2 + 3x + 3h - 1 \end{aligned}$$

c. We use the result of (b), and simplify:

$$\begin{aligned} \frac{f(x+h) - f(x)}{h} &= \frac{(x^2 + 2xh + h^2 + 3x + 3h - 1) - (x^2 + 3x - 1)}{h} \\ &= \frac{2xh + h^2 + 3h}{h} = 2x + h + 3 \quad \square \end{aligned}$$

Next let f be the function that assigns $\sqrt{x}$ to each nonnegative real number x. Then f is given by

$$\boxed{f(x) = \sqrt{x} \quad \text{for all} \quad x \geq 0} \tag{3}$$

This is the ***square root function***. Its domain consists of all nonnegative numbers, as does its range.

EXAMPLE 2. Let $f(x) = \sqrt{x}$. Determine the value of

a. $f(2)$ b. $f(\frac{1}{9})$ c. $f(x^2)$

Solution.

a. Letting $x = 2$, we have $f(2) = \sqrt{2}$.

b. Letting $x = \frac{1}{9}$, we have

$$f\left(\frac{1}{9}\right) = \sqrt{\frac{1}{9}} = \frac{1}{3}$$

c. Substituting x^2 for x, we have

$$f(x^2) = \sqrt{x^2} = |x| \quad \square$$

We remark that although $\sqrt{x}$ is defined for nonnegative values of x, by contrast $\sqrt{x^2}$ is defined for all numbers x. Thus from part (c) of Example 2 we find that

$$\sqrt{(-3)^2} = |-3| = 3$$

For another example of a function, let g be the function that assigns $(x-1)/(x+3)$ to each real number x except -3. Then

$$g(x)=\frac{x-1}{x+3} \quad \text{for} \quad x \neq -3 \tag{4}$$

EXAMPLE 3. For the function g given in (4), determine the value of

a. $g(0)$ b. $g\left(\frac{1}{x}\right)$

Solution.

a. Here we let $x=0$ in (4) and obtain

$$g(0)=\frac{0-1}{0+3}=-\frac{1}{3}$$

b. We substitute $1/x$ for x in (4):

$$g\left(\frac{1}{x}\right)=\frac{\frac{1}{x}-1}{\frac{1}{x}+3}=\frac{1-x}{1+3x} \quad \square$$

Notice that -3 is not in the domain of the function g defined by (4), so $g(-3)$ is undefined. If we attempted to compute $g(-3)$ by means of the formula in (4), we would obtain

$$g(-3) \stackrel{?}{=} \frac{-3-1}{-3+3} \stackrel{?}{=} \frac{-4}{0}$$

but $-4/0$ is undefined.

We say that two functions are ***equal*** if they have the same domains and if to any given number in their common domains their rules associate the same number. Thus the functions g and h given by

$$g(x)=\frac{x-1}{x+3} \quad \text{for} \quad x \neq -3 \quad \text{and} \quad h(t)=\frac{t-1}{t+3} \quad \text{for} \quad t \neq -3$$

are equal, since their domains both consist of all real numbers different from -3 and their rules associate the same number to any given number in their common domain. As these examples illustrate, the specific letter that is used for the variable in describing a function is irrelevant.

Two functions are *not equal* if either their domains or their rules are distinct. Consequently if

$$g(x) = \frac{x-1}{x+3} \quad \text{for} \quad x \neq -3$$

and

$$G(x) = \frac{x-1}{x+3} \quad \text{for} \quad x \geq 0 \tag{5}$$

then g and G are *not* equal, because the domain of g contains all negative numbers except -3, whereas the domain of G contains no negative numbers.

The rule of a function can be given in two or more parts. For example, consider the function f whose rule is given by

$$f(x) = \begin{cases} x^5 - 3 & \text{for } x < 2 \\ \sqrt{x} & \text{for } x \geq 4 \end{cases} \tag{6}$$

The domain of f consists of the intervals $(-\infty, 2)$ and $[4, \infty)$. The rule by which a number is assigned to a number x in the domain of f depends on whether x is in $(-\infty, 2)$ or in $[4, \infty)$.

EXAMPLE 4. For the function f given by (6), find the value of

a. $f(-1)$ b. $f(16)$

Solution.

a. Since $-1 < 2$, we let $x = -1$ in the top expression on the right of (6):

$$f(-1) = (-1)^5 - 3 = -4$$

b. Since $16 > 4$, we let $x = 16$ in the bottom expression on the right of (6):

$$f(16) = \sqrt{16} = 4 \quad \square$$

An Alternative Way of Describing a Function

Suppose f is the function that associates the area of a circle with its radius x and is defined by

$$f(x) = \pi x^2 \quad \text{for} \quad x \geq 0$$

Then a second way of describing f is obtained by writing y for $f(x)$:

$$y = \pi x^2 \quad \text{for} \quad x \geq 0 \tag{7}$$

In (7) each nonnegative value of x determines a particular value of y, and in this way y "depends" on x, so y is called a ***dependent variable***. In contrast, x is an ***independent variable***.

Functions that describe physical relationships are frequently presented in variable notation; moreover, the letters used for variables usually relate to the physical aspects of the quantities. Since (7) represents the area corresponding to any given radius, the function is frequently given in variable notation by

$$A = \pi r^2 \quad \text{for} \quad r \geq 0 \tag{8}$$

where A stands for area and r for radius. In a like manner, we could describe the function that assigns the temperature C in degrees Celsius to any given temperature F in degrees Fahrenheit by

$$C = \frac{5}{9}(F - 32) \quad \text{for} \quad F \geq -459.67 \tag{9}$$

Notice that the domain of this function consists of all numbers greater than or equal to -459.67, which represents "absolute zero," theoretically the lowest Fahrenheit temperature possible. The variables in (8) and (9) were chosen to remind us of the physical quantities they represent.

Most functions we will encounter express $f(x)$ (or y) in terms of x. When giving such a formula, we normally specify the numbers in the domain directly after the formula, as in (1)–(9). However, when the domain is to consist of all numbers for which the formula is meaningful, we normally omit mention of the domain. Thus we may write

$$f(x) = x^2 + 3x - 1 \quad \text{and} \quad g(x) = \frac{x - 1}{x + 3} \tag{10}$$

respectively, for the functions presented in (2) and (4), since each of their domains consists of all numbers for which the expression on the right side of the equation makes sense. Nevertheless we cannot omit mention of the domain of G in

$$G(x) = \frac{x - 1}{x + 3} \quad \text{for} \quad x \geq 0$$

(which appeared in (5)) because $(x - 1)/(x + 3)$ makes sense for negative as well as nonnegative numbers. If the domain is understood and the rule of the function is simple, we sometimes designate the function by an expression only. Thus we could refer to the functions in (10) as $x^2 - 3x + 1$ and $(x - 1)/(x + 3)$, respectively.

When we do not specify the domain outright, we still need to be aware of the domain. Sometimes it is important to determine precisely those real numbers that are members of the domain.

EXAMPLE 5. Let

$$f(x) = \frac{x+2}{x^2-9}$$

Determine the domain of f.

Solution. A number x is in the domain of f if and only if it does not make the denominator 0, that is, if and only if $x^2 - 9 \neq 0$. Since $x^2 - 9 = 0$ means that $x^2 = 9$, and thus $x = 3$ or $x = -3$, it follows that the domain consists of all real numbers except 3 and -3. □

EXAMPLE 6. Find the domain of each of the given functions.

a. $f(x) = \sqrt{x^2 - 16}$ b. $g(x) = \dfrac{1}{\sqrt{x^2 - 16}}$

Solution.

a. Since the square root is defined only for nonnegative numbers, $\sqrt{x^2 - 16}$ is defined only for $x^2 - 16 \geq 0$, or equivalently, $x^2 \geq 16$. Thus the domain of f consists of all numbers in the intervals $(-\infty, -4]$ and $[4, \infty)$.

b. Since

$$g(x) = \frac{1}{\sqrt{x^2 - 16}} = \frac{1}{f(x)}$$

the domain of g consists of all numbers in the domain of f *except* those for which $f(x) = 0$, that is, $x^2 - 16 = 0$, or $x^2 = 16$. Therefore the domain of g consists of all numbers in the intervals $(-\infty, -4)$ and $(4, \infty)$. □

Example 6 illustrates two facts that help in determining domains of functions defined by formulas:

1. Square roots of negative numbers are undefined.
2. Division by zero is undefined.

An Alternative Definition of Function

An alternative definition of function that is in common use employs the notion of ordered pair introduced in Section 2.1. To set the stage for the alternative definition, let f be a function. For any number x in the domain of f, we combine the numbers x and $f(x)$ to form the ordered pair $(x, f(x))$. Notice that because f is a function, there is precisely one pair whose first entry is x, namely, $(x, f(x))$. The collection of all such pairs $(x, f(x))$ for x in the domain of f identifies the function f.

Conversely, any set of ordered pairs (x, y) of numbers with the property that the first entries in any two distinct pairs are different defines a unique function f: The domain of f is the collection of first entries in the pairs, and for each such pair (x, y), $f(x)$ is the number y. In this framework the collection of ordered pairs (x, x^2) defines the function given by the formula $f(x) = x^2$, and the collection of ordered pairs $(x, \sqrt{x})$ for $x \geq 0$ defines the square root function.

These comments lead to the following alternative definition of function.

DEFINITION 2.2 A ***function*** is a nonempty collection of ordered pairs of real numbers no two of which have the same first entry.

Although we have two definitions of function, we prefer to rely only on the first, the one appearing as Definition 2.1.

EXERCISES 2.3

In Exercises 1–12, find the indicated values of the given function.

1. $f(x) = -14$; $f(-2)$, $f(\sqrt{3})$, $f(-x)$

2. $f(x) = x^2 - 3$; $f(5)$, $f(-4)$, $f(x-2)$

3. $f(t) = t^2 - 3t + 4$; $f(0)$, $f(-1)$, $f(t^2)$

4. $f(t) = \dfrac{t+4}{t-1}$; $f(2)$, $f\left(\dfrac{1}{3}\right)$, $f\left(\dfrac{1}{t}\right)$

5. $f(u) = \dfrac{4u^2 - 2u}{u^3 - 0.09}$; $f(10^{-1})$, $f(.3)$

6. $g(x) = \dfrac{\dfrac{1}{x} - 1}{\dfrac{1}{x} + 2}$; $g(1)$, $g(-2)$, $g\left(\dfrac{1}{x}\right)$

7. $g(x) = \sqrt{x+5}$; $g(-1)$, $g(2)$, $g(x^2 - 5)$

8. $F(z) = \sqrt[3]{z-4}$; $F(12)$, $F(-4)$, $F(z+a)$

9. $F(z) = 2z^{1/3} - z^{3/2}$; $F(3)$, $F(30)$

10. $F(z) = \sqrt{\dfrac{2z^2 - 1}{5z^2 - 1}}$; $F(-1)$, $F(1)$, $F(\sqrt{z})$

11. $f(x) = \begin{Bmatrix} -x & \text{for } x < 1 \\ 0 & \text{for } x > 3 \end{Bmatrix}$; $f(-5)$, $f(\pi)$

12. $f(x) = \begin{Bmatrix} \dfrac{x-1}{x+2} & \text{for } x \neq -2 \\ 1 & \text{for } x = -2 \end{Bmatrix}$; $f(2)$, $f(-2)$

In Exercises 13–20, find the values of y at the given values of x.

13. $y = -2x + 3$; $x = -4$, $x = 0$

14. $y = 3x^2 - 2x + 7$; $x = -2$, $x = -\frac{1}{3}$

15. $y = \dfrac{3x-4}{2x^2 + 3x - 5}$; $x = 0$, $x = \frac{1}{2}$

16. $y = \dfrac{x^2 - 1}{x^2 + 1}$; $x = 1$, $x = -3$

17. $y = \dfrac{1}{x} - \dfrac{1}{x^2}$; $x = -1$, $x = \frac{1}{3}$

18. $y = \sqrt{\dfrac{x}{x-5}}$; $x = 9$, $x = 6$, $x = -\frac{10}{9}$

19. $y = \dfrac{\sqrt[3]{x}}{\sqrt[3]{x} - 1}$; $x = -8$, $x = 54$

20. $y = 3|x|\sqrt{1-x}$; $x = 1, -2$

In Exercises 21–42, find the domain of the given function.

21. $f(x) = 4x^6 + \sqrt{3}x^3 - \dfrac{1}{3}x^2 + 1$

22. $f(x) = \dfrac{1}{x-7}$

23. $g(x) = \dfrac{3}{4x} - \dfrac{4}{2x-5}$

24. $g(x) = \dfrac{x+1}{x^2-4}$

25. $g(x) = \dfrac{x+2}{x^2+4}$

26. $g(x) = \dfrac{x-3}{x^2-9}$

27. $G(z) = \dfrac{1}{(z-1)(z+6)}$

28. $G(z) = \dfrac{1}{z^2+2z+1}$

29. $G(z) = \dfrac{1}{z^4+z^2+3}$

30. $f(x) = \dfrac{1}{\sqrt{x}}$

31. $g(x) = \dfrac{1}{\sqrt{-x}}$

32. $G(x) = \dfrac{1}{\sqrt{x-5}}$

33. $y = \sqrt{x^2-4}$

34. $y = \dfrac{1}{\sqrt{x^2-4}}$

35. $y = \sqrt{\dfrac{x}{x-1}}$

36. $y = \sqrt{x(x-3)}$

37. $y = \sqrt[3]{x-1}$

38. $y = \sqrt{\sqrt{x}-1}$

39. $y = \sqrt{x - \dfrac{1}{\sqrt{x}}}$

40. $y = \sqrt{\dfrac{x^2-1}{x^2+1}}$

41. $f(x) = \dfrac{|x|}{x}$

42. $f(x) = \begin{cases} 2x-3 & \text{for } x \le 0 \\ x + \dfrac{5}{x} & \text{for } x > 1 \end{cases}$

In Exercises 43–46, let x and h be real numbers. Find $\dfrac{f(x+h)-f(x)}{h}$ and simplify.

43. $f(x) = 2x^2 + 1$

44. $f(x) = x^3$

45. $f(x) = \dfrac{1}{x}$

46. $f(x) = -\dfrac{1}{x^2}$

47. Determine which of the following functions have a range that contains -2.

a. $f(x) = \sqrt{x+1}$ b. $f(x) = \sqrt{1-x}$

c. $f(x) = \dfrac{x+2}{x^2-4}$ *d. $f(x) = \dfrac{x-2}{x^2+4}$

48. Determine which of the following collections of pairs of real numbers define a function according to Definition 2.2.

a. all (x, y) with $x^2 + y^2 = 1$ and $y \ge 0$
b. all (x, y) with $x^2 + y^2 = 1$ and $x \ge 0$
c. all (x, y) with $x = y^3$
d. all (x, y) with $x = \sqrt{y}$

49. Let $f(x) = (x-2)^2 + 3(x-2) - 1$ and $g(x) = x^2 - 3x - 3$. Determine whether or not $f = g$.

50. Let

$$f(x) = \frac{|x|}{x} \quad \text{and} \quad g(x) = \begin{cases} -1 & \text{for } x < 0 \\ 1 & \text{for } x > 0 \end{cases}$$

Determine whether or not $f = g$.

51. Let

$$f(x) = \frac{1}{x - \sqrt{x^2-1}} \quad \text{and} \quad g(x) = x + \sqrt{x^2-1}$$

Show that $f = g$.

52. Let $f(x) = \dfrac{2x-3}{x-2}$. Show that $f(f(x)) = x$.

53. Let $f(x) = \dfrac{1}{x+1}$. Show that $f\left(\dfrac{1}{x}\right) = xf(x)$.

54. Let $f(x) = \dfrac{1+x}{1-x}$. Show that $f\left(\dfrac{1}{x}\right) = -f(x)$.

In Exercises 55–57, approximate the values of the given function.

55. $f(x) = 2.3x^3 + \pi x - 1.7$; $f(3.5)$, $f(-0.2)$, $f(13.9)$

56. $g(t) = 6.54t^2 - \dfrac{0.55}{t}$; $g(1.21)$, $g(6.01)$, $g(4.97)$, $g(0.3)$

57. $h(z) = \sqrt{z^2+3}$; $h(2.3)$, $h(\frac{4}{3})$, $h(\sqrt{2}+1)$

58. Let f be a function that satisfies the equation $xf(x) + f(4-x) = x^2 - x + 5$.

a. Find $f(2)$. *b. Find $f(1)$ and $f(3)$.

59. Define a function that expresses the circumference of a circle as a function of

a. the radius b. the diameter

60. Define a function that expresses the area of a square as a function of

a. the length of a side of the square
b. the length of a diagonal

61. One inch is 2.54 centimeters. Define a function that converts

a. from inches to centimeters
b. from centimeters to inches

62. One leg of a right triangle is twice as long as the other.

a. Define a function that expresses the area of the triangle in terms of the length of the smaller leg.

b. Define a function that expresses the length of the hypotenuse in terms of the smaller leg.

63. A person 5 feet tall casts a shadow created by a lamp 20 feet high. Define a function that expresses the length L of the shadow in terms of the distance x between the base of the lamp and the person's feet (Figure 2.30).

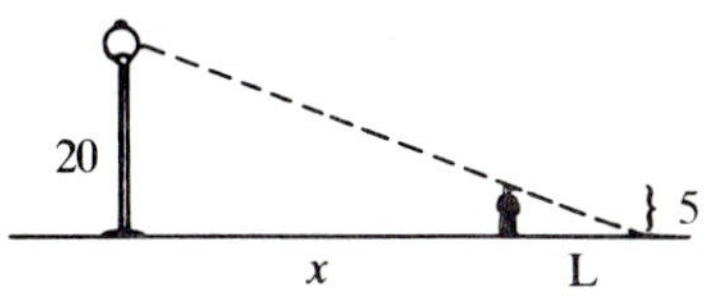

FIGURE 2.30

64. The Kelvin temperature scale can be obtained from the Celsius scale by adding 273.15 to temperatures given in Celsius.

a. Write a formula for Kelvin temperatures K in terms of Celsius temperatures C.

b. Write a formula for Kelvin temperatures K in terms of Fahrenheit temperatures F.

c. Convert 100° Celsius to degrees Kelvin.

d. Convert 68°F to degrees Kelvin.

65. Within limits, the length of a copper rod is essentially a linear function of the temperature of the rod. Suppose that at 10°C the length is 15 centimeters and at 40°C the length is 15.04 centimeters.

a. Find a formula for the length L in centimeters of the rod in terms of the temperature in degrees Celsius.

b. At what temperature Celsius would the length of the rod be 15.1 centimeters?

66. Suppose a pill is proposed to have the shape of a cylinder of radius $\frac{1}{8}$ inch and length l inches, capped on each end by a hemisphere of radius $\frac{1}{8}$ inch (Figure 2.31). Find a formula for the volume V in terms of the height l.

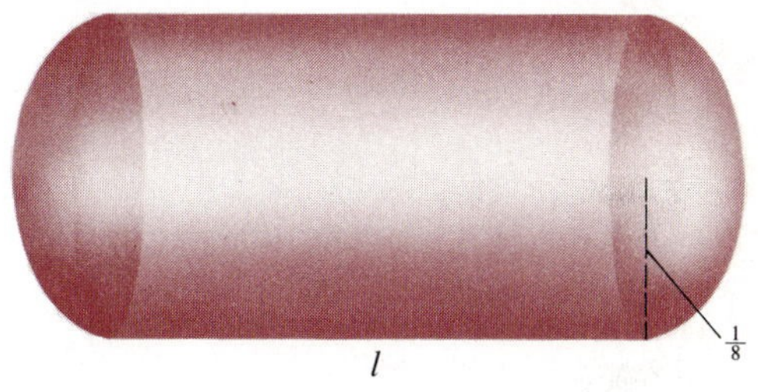

FIGURE 2.31

2.4 FUNCTIONS AS MODELS

Experimental data collected by scientists usually consist of long lists of numbers. For example, if a ball is dropped into a deep well and the distance traveled by the ball is ascertained each tenth of a second, then the following data might result:

Time elapsed (in seconds):	.1	.2	.3	.4	.5	.6
Depth (in feet):	.16	.64	1.44	2.56	4.00	5.76

Aside from the fact that measurements are at best only approximations to reality, it is unwieldy to handle long lists of numbers. Moreover, it is hard to extrapolate precisely most values not listed in a table. For instance, how far will the ball have traveled in 1.43 seconds or in $\sqrt{2}$ seconds?

For these reasons scientists usually construct "models" for the phenomena they investigate. This means that they formulate physical laws, formulas, or functions that express the dependence of certain quantities on other quantities. One such formula gives the distance an object dropped from rest will fall in t seconds if influenced only by gravity:

$$f(t) = 16t^2 \quad \text{for} \quad t \geq 0 \tag{1}$$

Here t is measured in seconds and $f(t)$ in feet. With such a formula we can confidently compute the values mentioned in the preceding paragraph.

EXAMPLE 1. Using the formula in (1), determine the distance traveled by the ball during

a. 1.43 seconds b. $\sqrt{2}$ seconds

Solution.

a. By (1) with $t = 1.43$, the distance traveled in 1.43 seconds is given by

$$f(1.43) = 16(1.43)^2 = 32.7184 \text{ (feet)}$$

b. By (1) with $t = \sqrt{2}$, the distance traveled in $\sqrt{2}$ seconds is given by

$$f(\sqrt{2}) = 16(\sqrt{2})^2 = 16(2) = 32 \text{ (feet)} \quad \square$$

A second model arises from Sir Isaac Newton's famous Second Law of Motion, which states that if an object has a mass of m kilograms, then the force F (in newtons) required to accelerate it at a rate of a (meters per second squared) is the product of the mass and the acceleration. This is expressed by the formula

$$F = ma \tag{2}$$

Under the assumption that the mass is constant, the force F is a function of the acceleration.

EXAMPLE 2. Use (2) to determine the force required to accelerate a moped with a mass of 100 kilograms at a rate of 3 meters per second squared.

Solution. By (2), with $m = 100$ and $a = 3$, we have

$$F = (100)(3) = 300 \text{ (newtons)} \quad \square$$

Variation and Proportionality

Certain types of models have occurred so frequently in the sciences that they have prompted some special terminology. For example, if c is a fixed nonzero number and if x and y are related by the formula

$$y = cx \tag{3}$$

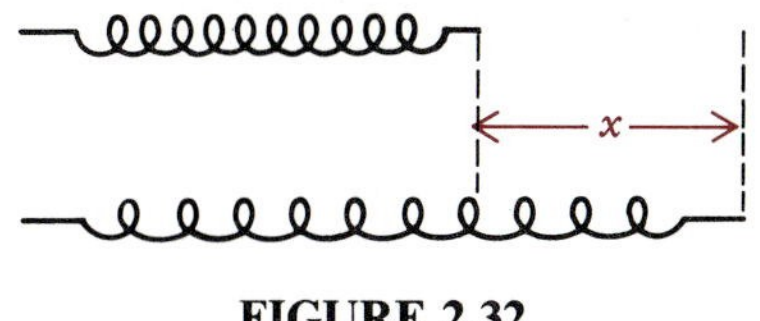

FIGURE 2.32

for all x in a given collection of real numbers, then y is said to be ***proportional*** to (or to be ***directly proportional*** to or to ***vary directly*** as) x. In this case c is called the ***constant of proportionality***; sometimes the letter k is used in place of c. Thus length measured in centimeters is proportional to length measured in inches, because if x denotes the length of an object in inches and y the length in centimeters, then x and y satisfy the equation

$$y = 2.54x \quad \text{for} \quad x \geq 0$$

Likewise, Hooke's Law states that within certain bounds the force y needed in order to keep a spring stretched x units beyond its natural length is given by the formula

$$y = kx$$

where k is the so-called ***spring constant*** of the spring. (See Figure 2.32.) In this case the force is proportional to the distance the spring is stretched.

EXAMPLE 3. Ohm's Law states that the current i flowing in an electrical circuit is proportional to the voltage V applied to the circuit. Express Ohm's Law by means of a formula.

Solution. Using (3) with V and i replacing x and y, respectively, we may write

$$i = cV \tag{4}$$

where c is a constant. □

In passing, we mention that the voltage V in Example 3 can be either positive or negative (corresponding to the two directions the current can flow through the circuit). The constant c is the ***conductivity*** of the wire, and its reciprocal, R, is the ***resistance*** of the wire. Thus (4) could also be written as

$$i = \frac{V}{R}$$

EXAMPLE 4. The volume V of a spherical balloon is proportional to the cube of the radius r. Express this fact by means of a formula involving V and r.

Solution. We use (3), with V and r^3 substituted for y and x, respectively. This leads us to the formula

$$V = cr^3$$

If appropriate units are used, then $c = \frac{4}{3}\pi$, so that

$$V = \tfrac{4}{3}\pi r^3 \quad \square$$

EXAMPLE 5. One of Albert Einstein's most important contributions to physics was his observation that mass and energy are equivalent. According to Einstein's theory, when mass is converted to energy, the amount E of energy released is proportional to the amount m of mass converted. Express this result by means of a formula.

Solution. Using (3) and realizing that mass must be nonnegative, we have

$$E = km \quad \text{for} \quad m \geq 0 \tag{5}$$

where we have used k for the constant of proportionality. Actually, if appropriate units are used, then $k = c^2$, where c is the speed of light in a vacuum. Thus (5) becomes

$$E = mc^2 \quad \text{for} \quad m \geq 0$$

perhaps the most famous physical formula of all time. □

If c is a nonzero constant and x and y satisfy

$$y = \frac{c}{x} \tag{6}$$

for all x in a given set of real numbers, then y is said to be ***inversely proportional*** to x. Notice that if $c > 0$, then an increase in x corresponds to a decrease in y. Again, c is the ***constant of proportionality***.

EXAMPLE 6. Boyle's Law states that if the temperature of a gas is constant, then the pressure p of the gas is inversely proportional to the volume V of the gas. Express Boyle's Law by means of a formula.

Solution. Using formula (6) and realizing that the volume of gas must be positive, we may express Boyle's Law as

$$p = \frac{c}{V} \quad \text{for} \quad V > 0$$

where, as usual, c is a constant. □

Joint Proportionality

Frequently a physical quantity depends on several other quantities, rather than just one. The terms "proportional" and "inversely proportional" may also be employed in such cases. For example, if c is a nonzero constant and if x, y, and z are related by the formula

$$z = cxy$$

then we say that z is ***jointly proportional*** (or more simply, ***proportional***) to x and y. If

$$z = \frac{cx}{y}$$

then we say that z is ***proportional*** to x and ***inversely proportional*** to y.

EXAMPLE 7. The volume V of a cylinder is jointly proportional to the square of the radius r of the base and to the height h of the cylinder. Express this fact by means of a formula involving r and h.

Solution. We have

$$V = cr^2h$$

In fact, the constant c of proportionality is π, which yields the formula

$$V = \pi r^2h$$

for the volume of the cylinder. □

The terminology of joint proportionality or inverse proportionality extends to any number of variables. In the next example we have a quantity that is jointly proportional to two other quantities and inversely proportional to a third.

EXAMPLE 8. Newton's Law of Gravitation states that the force of attraction F between two objects is proportional to each of their masses m_1 and m_2 and is inversely proportional to the square of the distance r between them. Express this law by means of a formula.

Solution. Using G as the constant of proportionality, we have

$$F = \frac{Gm_1m_2}{r^2}$$

The constant G is known as the ***universal gravitational constant***. □

EXERCISES 2.4

1. The perimeter p of a square is proportional to the length s of a side. Express this fact by means of a formula. (The constant is 4.)

2. The circumference C of a circle is proportional to the diameter d. Express this fact by means of a formula. (The constant is π.)

3. The surface area S of a cube is proportional to the square of the side length s. Express this fact by means of a formula. (The constant is 6.)

4. The water pressure p at a point is proportional to the depth h of the point below the surface of the water. Express this fact by means of a formula.

5. Planck's quantum theory of light states (in part) that a light ray may be considered as a particle called a photon. The energy E of the photon is proportional to the frequency v of the light ray. Express this result by means of a formula. (The constant of proportionality, known as ***Planck's constant***, is denoted by h.)

6. Joule's Law states that the rate P at which heat is produced in a wire is proportional to the square of the current i flowing in the wire. Express Joule's Law by means of a formula.

7. In the study of motion it is sometimes assumed that the air resistance R on an object is proportional to the square of the velocity v of the object. Express this assumption by means of a formula.

8. Use the result of Example 4 to solve this problem.
   a. Would three spherical lead balls of radius 2 inches produce enough lead for a single spherical lead ball of radius 3 inches? Explain your answer.
   b. Would the three spherical balls of part (a) produce enough lead for one cube of side length 4.7 inches? Explain your answer.

9. The area A of an equilateral triangle is proportional to the square of the length s of a side (Figure 2.33).
   a. Express this fact by means of a formula. (The constant is $\sqrt{3}/4$.)
   b. If the area of an equilateral triangle is $9\sqrt{3}$ square centimeters, determine the length of a side.
   c. Approximate the length of a side if the area is 6.1 square meters. Round your answer to 2 significant digits.

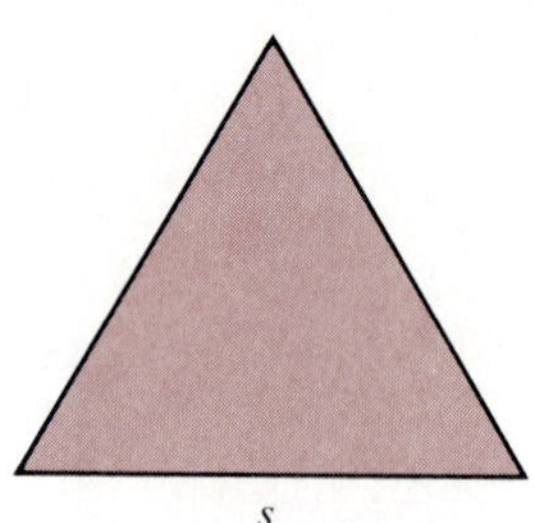

FIGURE 2.33

10. The volume V of a cube is proportional to the cube of the length s of a side.
    a. Express this fact by means of a formula. (The constant is 1.)
    b. By how much is the volume increased when the length of a side is increased from 3 inches to 4 inches?

11. The length l_f in feet of an object is proportional to the length l_m in meters.
    a. Express this fact by means of a formula. (The constant is $\frac{1250}{381}$.)
    b. Express the fact that the length l_y in yards of an object is proportional to the length l_m in meters by means of a formula. Using (a), determine the constant of proportionality.

12. Distance D_k in kilometers is proportional to distance D_m in miles.
    a. Express this fact by means of a formula. (The constant is exactly 1.609344.)
    b. Express by means of a formula the fact that the distance in miles is proportional to distance in kilometers. Approximate the value of the constant of proportionality.
    c. Use (a), with the numerical value for the constant, to determine the approximate number of kilometers in 93×10^6 miles (which is often taken as the average distance between the earth and sun). Round your answer to 2 significant digits.

13. For each gram of hydrogen that is converted to helium in a star, approximately 0.0277 gram is converted to energy. Use the result of Example 5 to compute the amount of energy released for each gram of hydrogen converted to helium. (*Hint:* The speed

of light is approximately 3×10^{10} centimeters per second. Your answer will be in gram centimeters square per second squared, commonly called "ergs.")

14. According to Hubble's Law, the velocity v with which a galaxy is receding from any other galaxy is proportional to the distance d between the galaxies. (Hubble's Law is sometimes viewed as evidence for the "big bang" theory, which holds that the universe originated with a gigantic explosion.)
   a. Express Hubble's Law by means of a formula.
   b. If a galaxy whose distance from the earth is 10 million light years (that is, 9.46×10^{19} kilometers) is receding from the earth with a velocity of 170 kilometers per second, find the approximate value of the constant of proportionality.
   c. How fast would a galaxy recede if it were located 15 million light years from earth?

15. The kinetic energy K of an object of given mass m is proportional to the square of the speed v of the object.
   a. Express this fact by means of a formula. (The constant of proportionality in this case is equal to one half the mass of the object, if appropriate units are used.)
   b. Suppose two bundles of newspapers, one weighing 10 kilograms and the other 30 kilograms, hit a wall with velocities v_1 and v_2 meters per second, respectively. If both bundles possess the same kinetic energy, what is the relationship between their velocities?

16. To a very good approximation, the time (or period) T it takes a pendulum to make one complete swing is proportional to the square root of the length l of the pendulum (Figure 2.34).

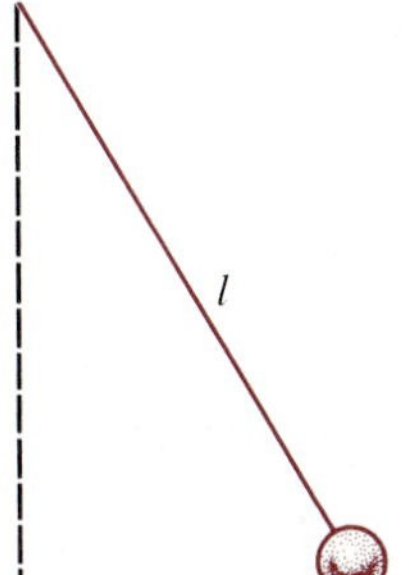

FIGURE 2.34

   a. Express this fact by means of a formula.
   b. To double the period, how must the length of the pendulum be changed?

17. The wavelength λ of a light ray is inversely proportional to its frequency ν (the number of waves per second). Express this fact by means of a formula. (The constant of proportionality in this case is c, the velocity of light.)

18. Kepler's Third Law of Planetary Motion states that the square of the period T of a planet (the time required for the planet to make one revolution about the sun) is proportional to the cube of the average distance a from the planet to the sun.
   a. Express Kepler's Third Law by means of a formula.
   b. If the period is 365 days and the average distance is approximately 93×10^6 miles, approximate the constant c.
   c. Approximate the constant c in (b) if the period is given in seconds and the average distance in kilometers (see Exercise 12).

19. Coulomb's Law states that the electric force F exerted by one charged particle on another is proportional to the charges q_1 and q_2 on the two particles and is inversely proportional to the square of the distance r between the two particles. Express Coulomb's Law by means of a formula.

20. The ***escape velocity*** on a planet is the velocity that a spaceship must possess when leaving the surface of the planet in order to escape the gravitational field of the planet without firing any of the ship's rockets. The ***escape energy*** of a spaceship on a planet is the amount of energy required to impart escape velocity to the spaceship.
   a. The escape energy E of a spaceship is proportional to the square of the escape velocity v_e and to the mass m of the spaceship. Express this fact by means of a formula.
   b. The escape velocity on a star is defined as for a planet. The escape velocity v_e on a star that is contracting is inversely proportional to the square root of the radius r of the star. Express this fact by means of a formula. (If the radius of the star becomes so small that the escape velocity exceeds the velocity of light, not even light can escape from the star, and the star becomes a ***black hole***.)

2.5 THE GRAPH OF A FUNCTION

When a function is described by means of a formula, the formula can tell us much about the function. But pictures are also associated with functions and can themselves give much information about the function. A pictorial representation of a function is called a graph.

DEFINITION 2.3 Let f be a function. Then the set of all points $(x, f(x))$ such that x is in the domain of f is called the ***graph*** of f (Figure 2.35), and we say that such a point $(x, f(x))$ is ***on the graph of f***.

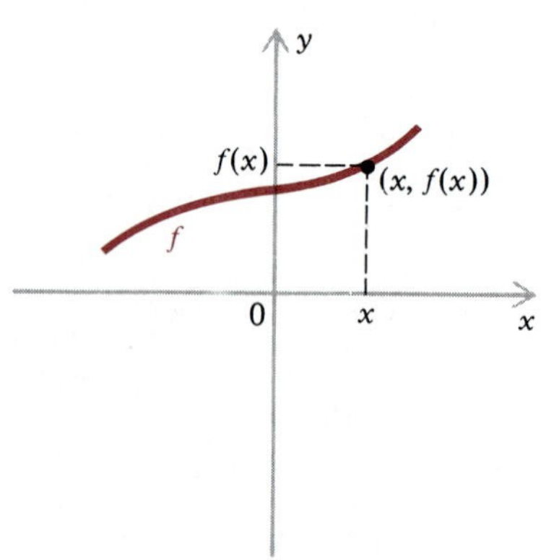

The graph of a function f

FIGURE 2.35

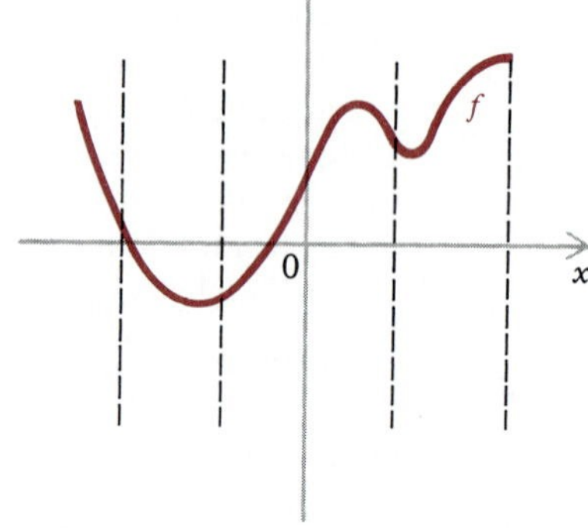

Vertical lines intersect the graph of a function f at most once

FIGURE 2.36

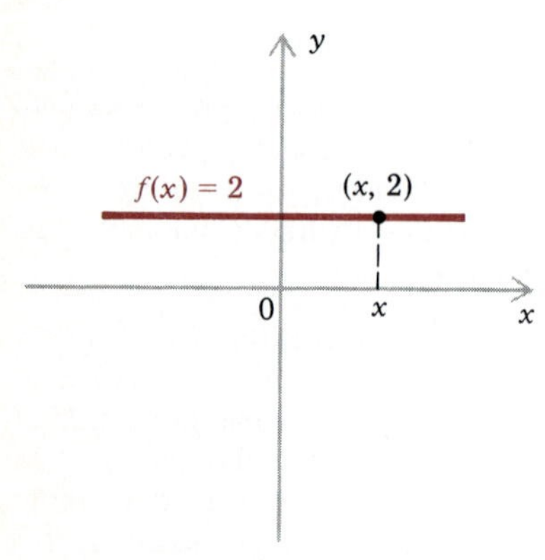

FIGURE 2.37

Since no two distinct points on the graph of a function can have the same x coordinate (see Definition 2.2), and since a vertical line is characterized by all its points having the same x coordinate, it follows that no vertical line can intersect the graph of a function more than once (Figure 2.36). You might look quickly at the graphs that follow and see that this is true.

EXAMPLE 1. Let $f(x) = 2$. Sketch the graph of f.

Solution. Notice that if x is any real number, then $(x, f(x))$ is the point $(x, 2)$, which means that the y coordinate of each point on the graph of f is 2. Thus the graph is the horizontal line drawn in Figure 2.37. □

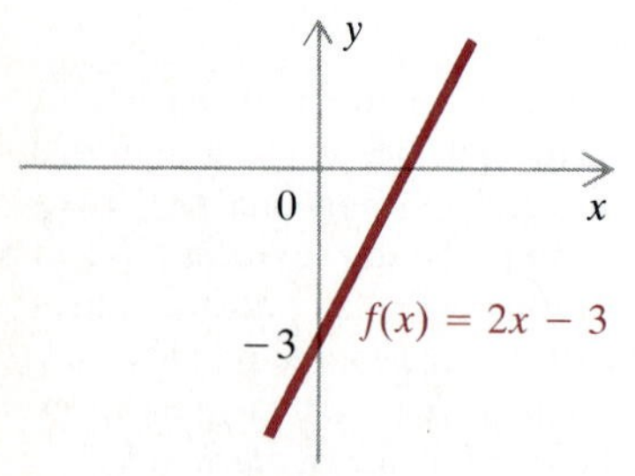

FIGURE 2.38

If f is any constant function, say $f(x) = c$, then as in Example 1, the graph of f is a horizontal line that crosses the y axis at the point whose y coordinate is c.

EXAMPLE 2. Let $f(x) = 2x - 3$. Sketch the graph of f.

Solution. By definition, (x, y) is on the graph of f if and only if $y = 2x - 3$. But $y = 2x - 3$ is the slope-intercept equation of the line with slope 2 and intercept -3. Thus the graph is the line appearing in Figure 2.38. □

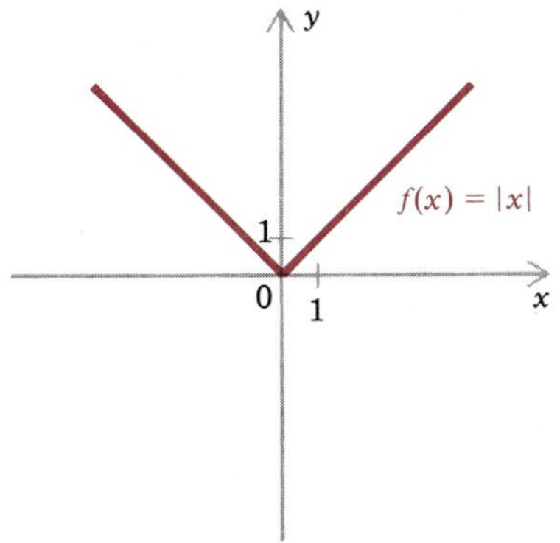

The graph of the absolute value function

FIGURE 2.39

More generally, if $f(x) = mx + b$, then the graph of f is the line whose slope is m and whose y intercept is b.

EXAMPLE 3. Sketch the graph of the ***absolute value function***:

$$f(x) = |x|$$

Solution. Since $f(x) = x$ for $x \geq 0$, it follows that the part of the graph of f to the right of the y axis coincides with the line $y = x$. Analogously, since $f(x) = -x$ for $x < 0$, the part of the graph of f to the left of the y axis coincides with the line $y = -x$. Combining this information, we obtain the graph of f (Figure 2.39). □

EXAMPLE 4. Let $f(x) = \sqrt{x}$. Sketch the graph of f.

Solution. To get an idea of what the graph looks like, let us plot a few of its points with the help of the following table:

x	0	$\frac{1}{4}$	1	4	9	16
$f(x)$	0	$\frac{1}{2}$	1	2	3	4

Connecting the corresponding points with a smooth curve, we obtain the graph of the square root function (Figure 2.40). □

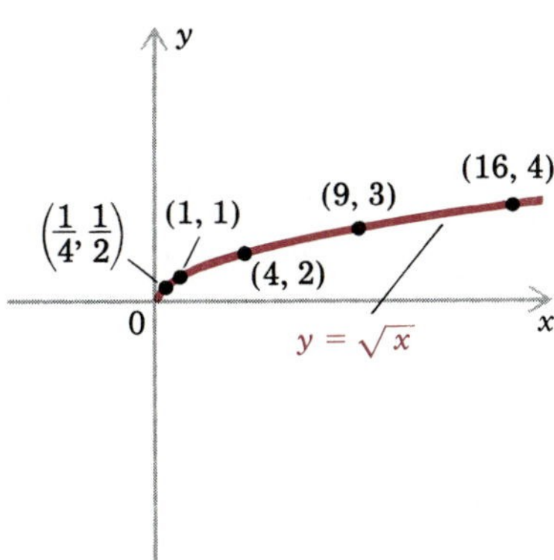

The graph of the square root function

FIGURE 2.40

EXAMPLE 5. Let $f(x) = x^2$. Sketch the graph of f.

Solution. As in the solution of Example 4, we first make a table of values for f:

x	-3	-2	-1	0	1	2	3
$f(x)$	9	4	1	0	1	4	9

Notice that as x grows, the value $f(x)$ seems to grow ever faster. We obtain the graph sketched in Figure 2.41. □

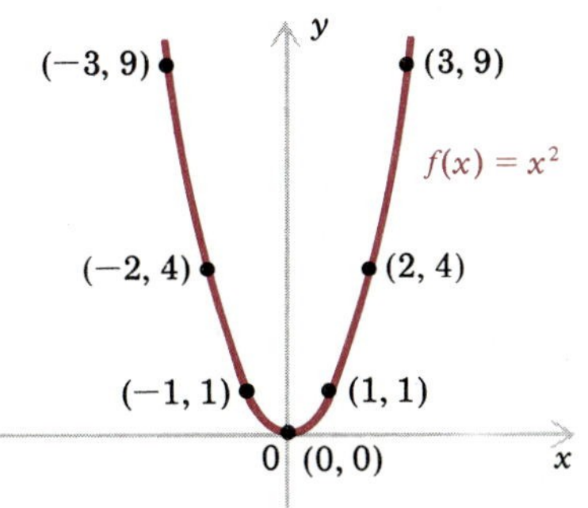

FIGURE 2.41

EXAMPLE 6. Let $g(x) = x^2 - 4$ and $h(x) = x^2 + 1$. Sketch the graphs of g and h.

Solution. Let $f(x) = x^2$, as in Example 5. We will compare the graphs of g and h with that of f. Notice that for each value of x, $g(x)$ is exactly 4 units less than the corresponding value for $f(x)$. Therefore we conclude that the graph of g has the same shape as the graph of f in Example 5, but is shifted down 4 units, as indicated in Figure 2.42. Similarly, the graph of h has the same shape as the graph of f in Example 5, but is shifted up 1 unit. Therefore the graph of h is as indicated in Figure 2.43. □

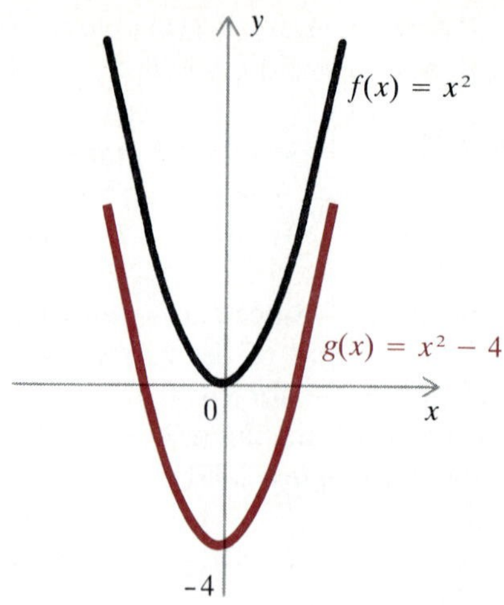

FIGURE 2.42

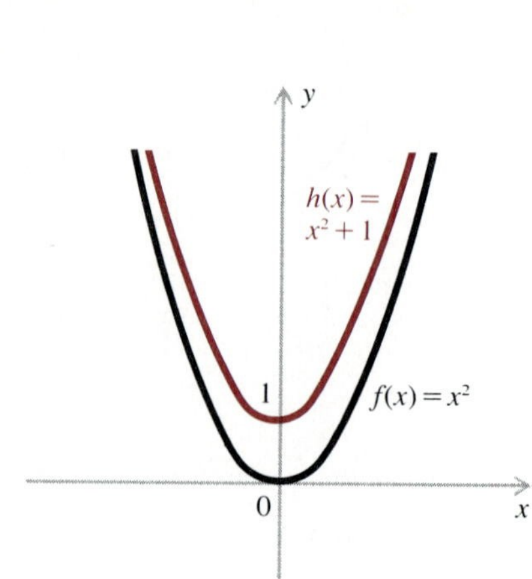

FIGURE 2.43

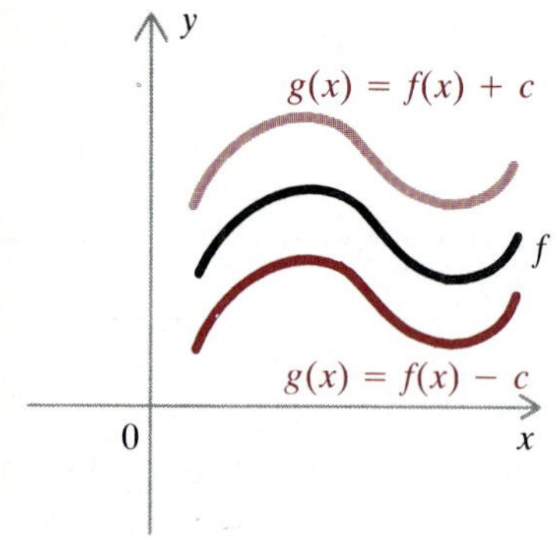

Vertical shifts of the graph of f

FIGURE 2.44

The graphs of g and h appearing in Example 6 represent the graph of f shifted vertically, and as such are called ***vertical shifts*** of the graph of f. In general, vertical shifts can be obtained as follows (see Figure 2.44):

If $g(x) = f(x) + c$ and $c > 0$	then the graph of g is the graph of f shifted up c units.
If $g(x) = f(x) - c$ and $c > 0$	then the graph of g is the graph of f shifted down c units.

Next we turn to horizontal shifts of graphs.

EXAMPLE 7. Let $g(x) = (x - 1)^2$ and $h(x) = (x + 2)^2$. Sketch the graphs of g and h.

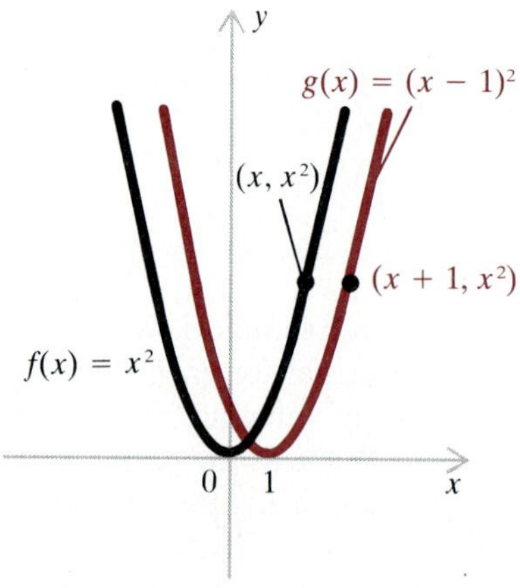

FIGURE 2.45

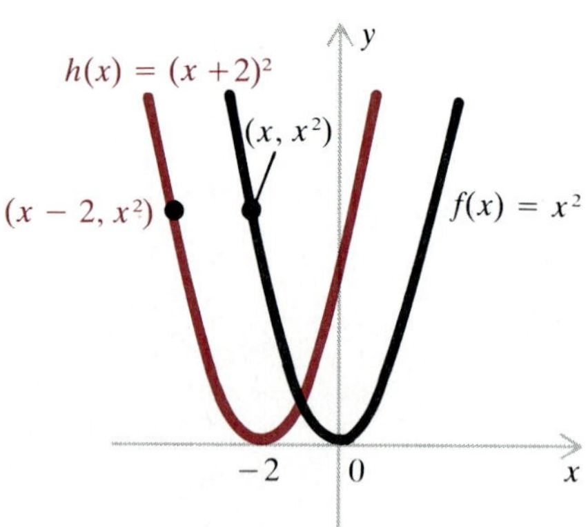

FIGURE 2.46

Solution. Once again let $f(x) = x^2$. As in Example 6, we will compare the graphs of g and h with that of f. Because $g(x + 1) = [(x + 1) - 1]^2 = x^2$, it follows that $(x + 1, x^2)$ is on the graph of g, for each real number x. However, $(x + 1, x^2)$ lies 1 unit to the right of (x, x^2), which is on the graph of f. We conclude that the graph of g is 1 unit to the right of the graph of f (Figure 2.45). Similarly, $h(x - 2) = [(x - 2) + 2]^2 = x^2$, so that $(x - 2, x^2)$ is on the graph of h, and lies 2 units to the left of (x, x^2), which is on the graph of f. We find that the graph of h is 2 units to the left of the graph of f (Figure 2.46). □

The graphs of g and h appearing in Example 7 represent the graph of f shifted horizontally, and as such are called ***horizontal shifts*** of the graph of f. Notice that $g(x) = f(x - 1)$, and to obtain the graph of g we shift the graph of f to the right 1 unit. Similarly, $h(x) = f(x + 2)$, and to obtain the graph of h we shift the graph of f to the left 2 units. In general, horizontal shifts can be obtained as follows (see Figure 2.47):

If $g(x) = f(x - c)$ and $c > 0$	then the graph of g is the graph of f shifted to the right c units.
If $g(x) = f(x + c)$ and $c > 0$	then the graph of g is the graph of f shifted to the left c units.

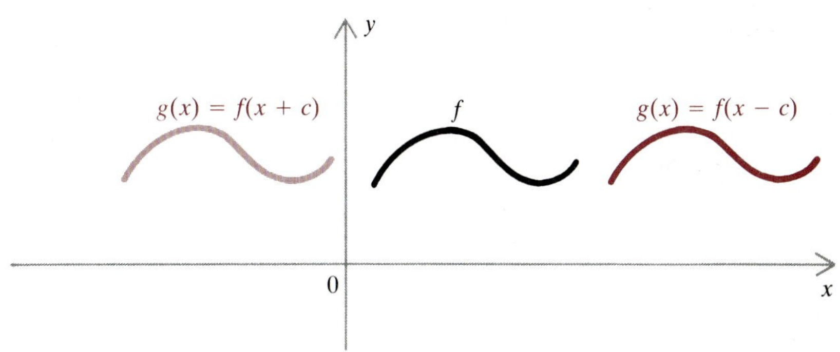

Horizontal shifts of the graph of f

FIGURE 2.47

Having discussed vertical and horizontal shifts of graphs, we now turn to a reflection of a graph.

EXAMPLE 8. Let $g(x) = -x^2$. Sketch the graph of g.

Solution. Let $f(x) = x^2$, as before. Then

$$g(x) = -x^2 = -f(x)$$

Notice that $(x, -x^2)$ is on the graph of g, for each real number x. However, $(x, -x^2)$ and (x, x^2) are symmetric with respect to the x axis. Therefore the

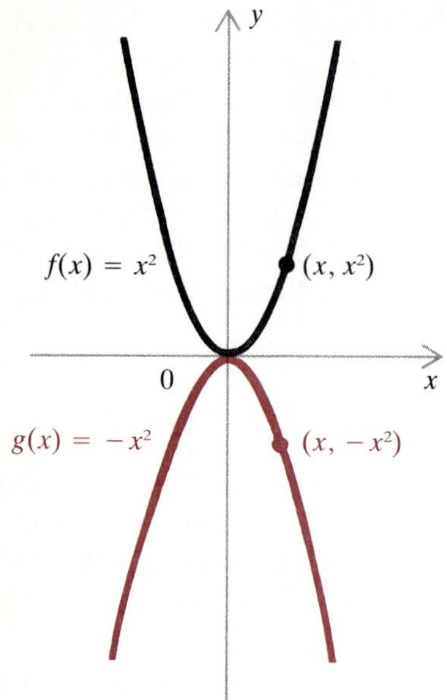

FIGURE 2.48

graph of g can be obtained by reflecting the graph of f through the x axis (Figure 2.48). □

More generally, if f is any function and $g(x) = -f(x)$, then the graphs of f and g are symmetric with respect to the x axis, and we say that the graph of g is the ***reflection*** of the graph of f through the x axis (Figure 2.49).

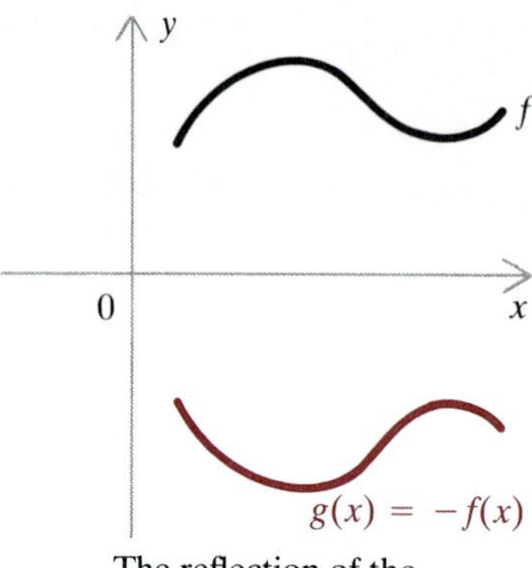

The reflection of the graph of f through the x axis

FIGURE 2.49

EXAMPLE 9. Let

$$g(x) = \begin{cases} 2 - x & \text{for} \quad x < 0 \\ x^2 & \text{for} \quad x \geq 0 \end{cases}$$

Sketch the graph of g.

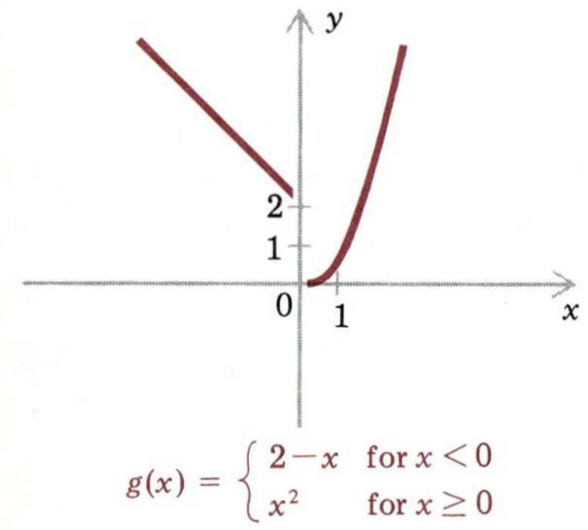

FIGURE 2.50

Solution. Since the formula for g is in two parts, one for x in $(-\infty, 0)$ and the other for x in $[0, \infty)$, we will determine the corresponding parts of the graph one at a time. First, $g(x) = 2 - x$ for $x < 0$, so the part of the graph of g to the left of the y axis is a portion of the line with slope -1 and y intercept 2. Second, $g(x) = x^2$ for $x \geq 0$, so that the part of the graph of g to the right of the y axis coincides with the corresponding part of the graph of f in Example 5 (where $f(x) = x^2$). Combining this information, we obtain the graph of g (Figure 2.50). □

Symmetry

Symmetry is a quality of art and plays a central role in architecture. One of the most famous examples of symmetry in architecture is the Taj Mahal. The notion of symmetry also relates to graphs of functions in mathematics.

We say that the graph of a function is ***symmetric with respect to the y axis*** if the portion of the graph to the right of the y axis is mirrored to the left of the y axis. In mathematical terms this means that if (x, y) is on the graph, so is $(-x, y)$ (Figure 2.51). The graphs of the functions in

The Taj Mahal, Agra, India

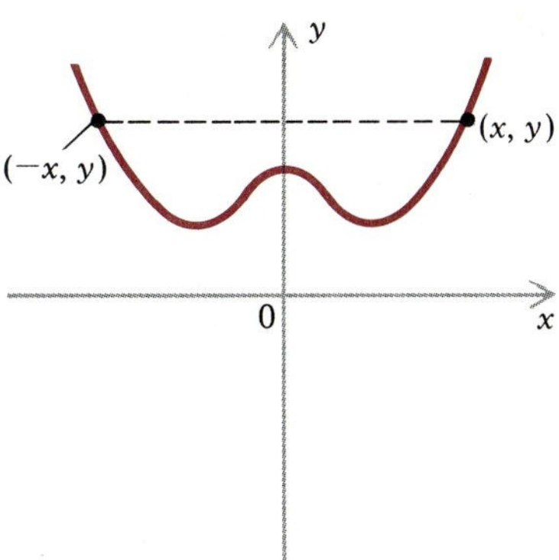

Symmetry with respect to the y axis

FIGURE 2.51

Figures 2.39, 2.41–2.43, and 2.48 have this property, as you can check. In each case the reason is that

$$f(-x) = f(x) \quad \text{for all } x \text{ in the domain of } f \tag{1}$$

A function that satisfies (1) is called an ***even function***. Thus, $|x|$, x^2, $x^2 - 4$, $x^2 + 1$, and $-x^2$ are all even functions.

Suppose that n is an even integer, so that $n = 2m$ for some integer m. Then

$$(-x)^n = (-x)^{2m} = [(-x)^2]^m = (x^2)^m = x^{2m} = x^n$$

This means that any function that can be described by a formula containing only even powers of x is an even function, and its graph is symmetric with respect to the y axis.

EXAMPLE 10. For each of the following functions, determine whether or not f is even.

a. $f(x) = -x^4 + 2x^2$ b. $f(x) = x^2 - x$

Solution.

a. Because $(-x)^4 = x^4$ and $(-x)^2 = x^2$, we have

$$f(-x) = -(-x)^4 + 2(-x)^2 = -x^4 + 2x^2 = f(x)$$

so that by (1), f is an even function.

b. Observe that $f(x) = x^2 - x$ and $f(-x) = (-x)^2 - (-x) = x^2 + x$. Since

$$x^2 - x = x^2 + x \quad \text{only if} \quad x = 0$$

it follows that $f(-x) \neq f(x)$ for $x \neq 0$. Thus f is not an even function. □

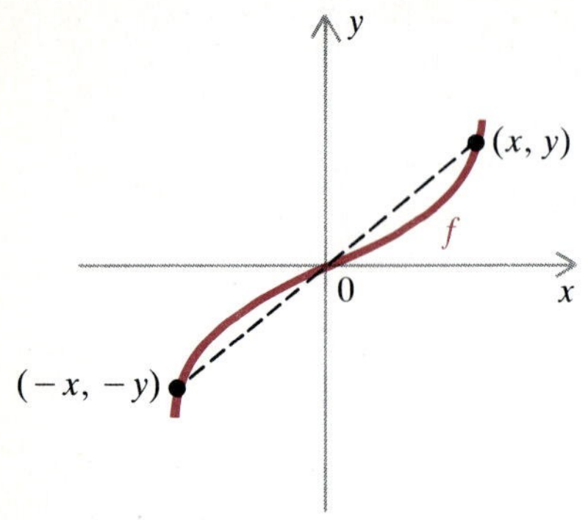

Symmetry with respect to the origin

FIGURE 2.52

If

$$f(-x) = -f(x) \quad \text{for all } x \text{ in the domain of } f$$

then f is called an ***odd function***. For example, the function x is an odd function. The graph of an odd function is ***symmetric with respect to the origin***, which means that if (x, y) is on the graph, then so is $(-x, -y)$ (Figure 2.52).

EXAMPLE 11. Let $f(x) = x^3$. Show that f is an odd function, and then sketch its graph.

Solution. First of all,

$$(-x)^3 = -x^3$$

so that $f(-x) = -f(x)$, and thus f is an odd function. Next, we assemble a table of values for f:

x	-3	-2	-1	0	1	2	3
$f(x)$	-27	-8	-1	0	1	8	27

Plotting the corresponding points and connecting them smoothly, we obtain the graph in Figure 2.53. □

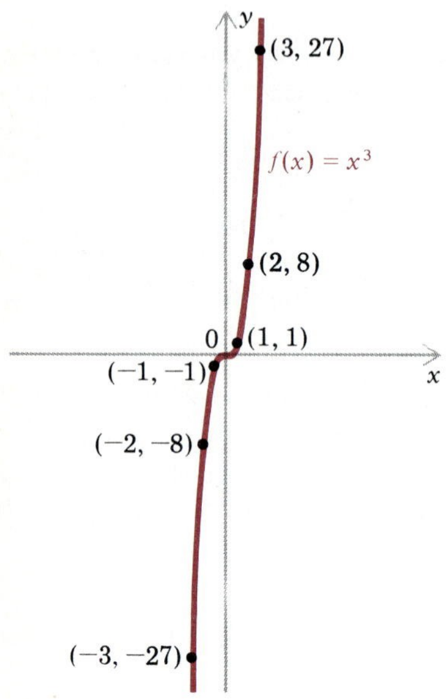

FIGURE 2.53

We can see more generally that functions such as

$$f(x) = 2x^3 - \frac{1}{x}$$

which have only odd powers of x and no constant term, are odd functions. In particular, if $f(x) = x^n$, where n is an odd integer, then f is an odd function.

Intercepts

Our custom so far has been to plot several points before trying to sketch the graph of a function. This raises the question of which points we should plot. Since we always begin a figure with the coordinates axes, it is reasonable to plot the points of the graph that lie on either of the coordinate axes.

A ***y intercept*** of the graph of a function f is the y coordinate of any point where the graph of f meets the y axis (Figure 2.54). Since $x = 0$ for any point (x, y) on the y axis, it follows that the graph of a function has a y intercept if and only if 0 is in the domain of f, in which case $f(0)$ is the one and only y intercept. Analogously, an ***x intercept*** of the graph of a function is the x coordinate of any point where the graph meets the x axis (Fig-

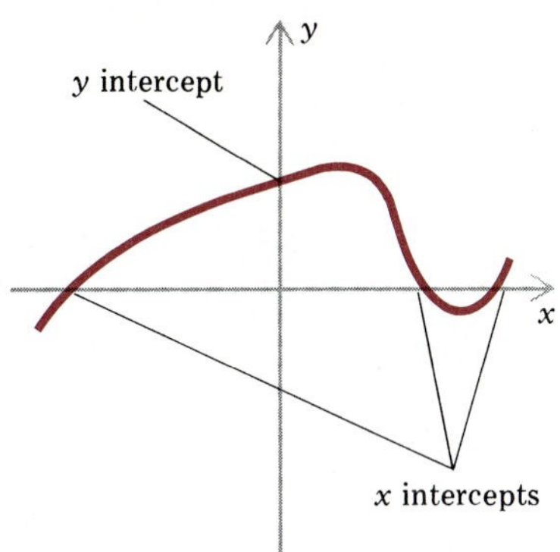

FIGURE 2.54

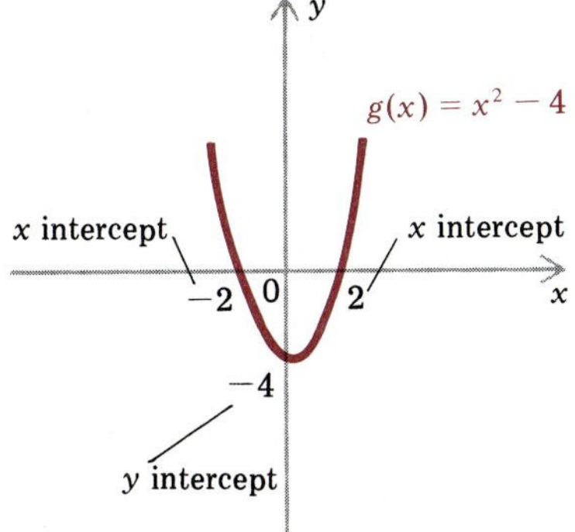

FIGURE 2.55

ure 2.54). Since $y = 0$ for any point (x, y) on the x axis, it follows that the x intercepts of the graph of f are the values of x (if any) for which $f(x) = 0$. In brief,

> i. $f(0)$ is the y intercept (if 0 is in the domain of f).
> ii. x is an x intercept if $f(x) = 0$.

EXAMPLE 12. Let $g(x) = x^2 - 4$. Find the y and x intercepts of the graph of g.

Solution. Since $g(0) = -4$, we know the y intercept is -4. To find the x intercepts we must find those values of x for which $g(x) = 0$, that is, for which $x^2 - 4 = 0$. But these are the values of x for which $x^2 = 4$, which yields $x = 2$ and $x = -2$. Consequently these are two x intercepts, 2 and -2; Figure 2.55 supports this claim. □

If we consider the function given by $h(x) = x^2 + 4$, we reach very different conclusions about intercepts. On the one hand, $h(0) = 4$, so the y intercept of h is 4. On the other hand, $h(x) = 0$ if and only if $x^2 + 4 = 0$. But since there are no real solutions of $x^2 + 4 = 0$, the graph of h has no x intercepts (see Figure 2.56). Thus we see that there may or may not be x intercepts for the graph of a given function. Similarly, there may or may not be a y intercept, depending on whether or not 0 is in the domain.

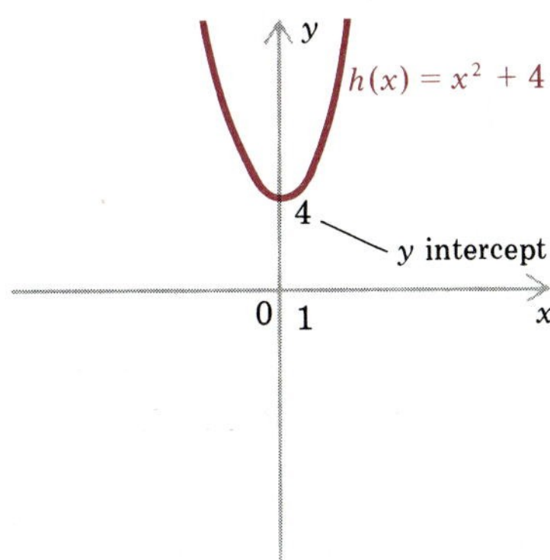

FIGURE 2.56

EXAMPLE 13. Let $f(x) = x^2 + 2x - 3$. Find the x and y intercepts of the graph of f.

Solution. Since $f(0) = 0^2 + 0 - 3 = -3$, the y intercept is -3. To find the x intercepts we solve the equation $f(x) = 0$ for x. This leads to the equation

$$x^2 + 2x - 3 = 0$$

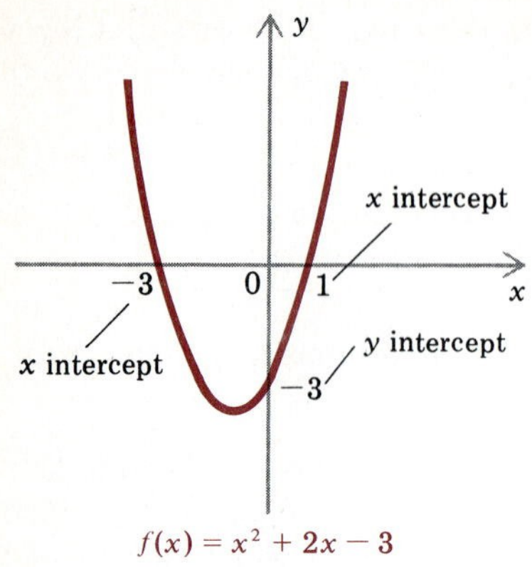

FIGURE 2.57

which, after factorization, becomes

$$(x + 3)(x - 1) = 0$$

Thus we see that $f(x) = 0$ if $x = -3$ or $x = 1$. This means that the x intercepts are -3 and 1. The graph of f, which we will learn how to sketch in Section 3.1, is shown in Figure 2.57. □

EXAMPLE 14. Let $f(x) = 1/x$. Show that the graph of f has no x or y intercepts.

Solution Since division by 0 is undefined, 0 is not in the domain of f, so there is no y intercept. Next, notice that $1/x$ is never 0, so that there are no values of x for which $f(x) = 0$. This means that there are no x intercepts. The graph of f, which we will be able to sketch later (see Section 3.4), is shown in Figure 2.58 and supports the claim that no intercepts exist. □

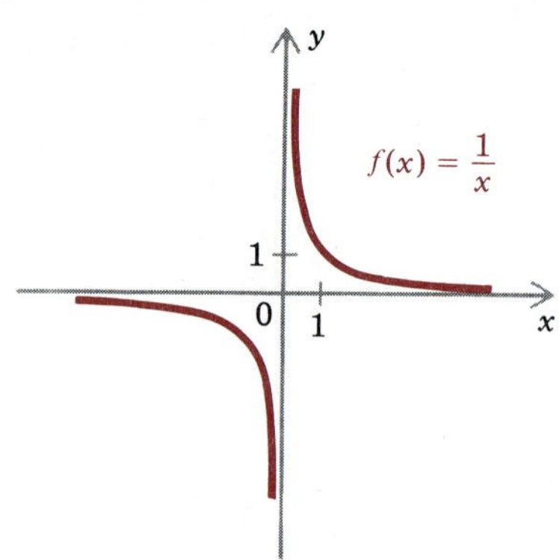

A graph with no intercepts

FIGURE 2.58

EXERCISES 2.5

In Exercises 1–24, sketch the graph of the given function and indicate any intercepts.

1. $f(x) = x + 3$
2. $f(x) = -2x + 5$
3. $f(x) = x^2 + 1$
4. $f(x) = x^2 - 3$
5. $f(x) = (x + 1)^2$
6. $f(x) = (x - 3)^2$
7. $g(x) = |x| - 2$
8. $g(x) = |x| + 4$
9. $g(x) = |x - 2|$
10. $g(x) = |x + 4|$
11. $h(x) = \sqrt{x} - 1$
12. $h(x) = \sqrt{x} + 3$
13. $h(x) = \sqrt{x - 1}$
14. $h(x) = \sqrt{x + 2}$
15. $f(x) = \dfrac{1}{x} - 3$
16. $f(x) = \dfrac{1}{1 + x}$
17. $f(x) = -2x^2$
18. $f(x) = (x + 2)^2 - 3$
19. $f(x) = x^2 + 4x + 2$ (*Hint:* Complete the square.)
20. $f(x) = x^2 - 2x + 3$ (*Hint:* Complete the square.)
21. $h(x) = \begin{cases} -x & \text{for } x \le 0 \\ -2x & \text{for } x > 0 \end{cases}$
22. $h(x) = \begin{cases} x^2 & \text{for } x \le 0 \\ x & \text{for } x > 1 \end{cases}$

23. $h(x) = \begin{cases} \dfrac{1}{x} & \text{for } x < -1 \\ 3x & \text{for } x \geq -1 \end{cases}$

24. $h(x) = \begin{cases} |x| & \text{for } x < 1 \\ x^2 & \text{for } x \geq 1 \end{cases}$

In Exercises 25–30, determine all intercepts of the graph of the given function.

25. $f(x) = x^2 - 4x$

26. $f(x) = x^2 + 4x$

27. $f(x) = x^3 + 6x^2 + 9x$

28. $f(x) = \dfrac{x+1}{x+2}$

29. $g(x) = \dfrac{x^2 - 1}{x^2 + 2x - 3}$

30. $g(x) = ||x| - 2|$

In Exercises 31–38, sketch the graph and indicate whether the graph is symmetric with respect to the y axis, the origin, or neither.

31. $f(x) = 5$

32. $f(x) = 2x$

33. $f(x) = 2x - 1$

34. $g(x) = -|x|$

35. $g(x) = |x| - 4$

36. $g(x) = \dfrac{|x|}{x}$

37. $h(x) = x^2 - \dfrac{1}{2}$

38. $h(x) = \dfrac{1}{x} + 2$

In Exercises 39–48, determine which functions are even, which are odd, and which are neither.

39. $f(x) = x^2 + x$

40. $f(x) = \dfrac{x^2}{x^4 + 2}$

41. $f(x) = x^3 + 4x^5$

42. $f(x) = x - \dfrac{1}{x}$

43. $g(x) = x^2 - \dfrac{1}{x}$

44. $g(x) = \sqrt{x^2 - 1}$

45. $g(x) = \dfrac{x+3}{x-5}$

46. $h(x) = \dfrac{x^2+3}{x-5}$

47. $h(x) = \dfrac{x^2+1}{x^2-1}$

48. $h(x) = \sqrt{x^2 + 3}$

49. Determine which of the graphs in Figure 2.59 are graphs of functions.

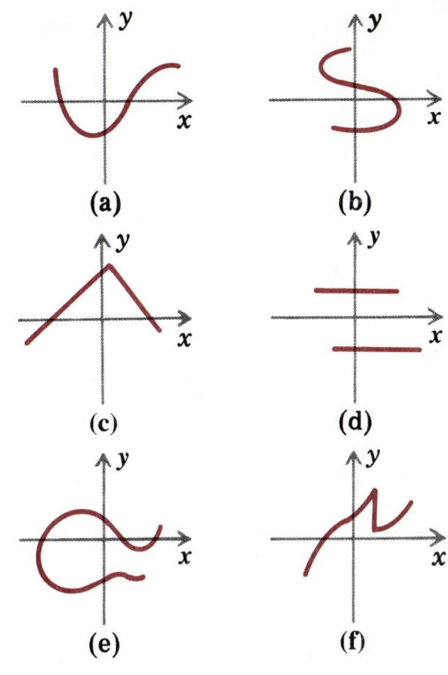

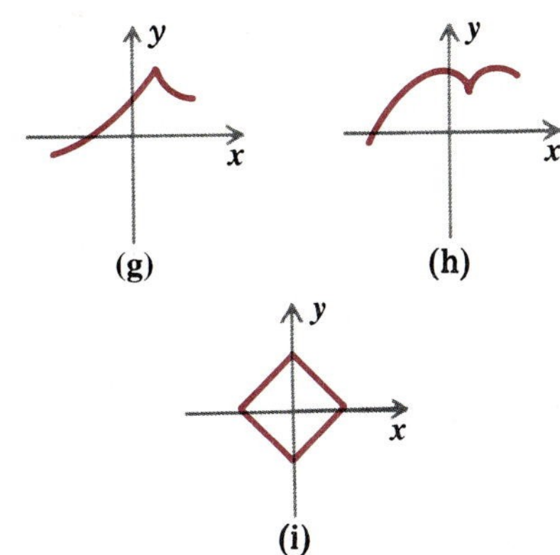

FIGURE 2.59

50. Are there any nonzero functions that are both even and odd? Give reasons for your answer.

51. For any real number x, let $[x]$ denote the largest integer less than or equal to x. Thus $[2] = 2$, $[-\frac{7}{5}] = -2$, and $[\pi] = 3$. Let

$$f(x) = [x]$$

Then f is called the ***greatest integer function***, or the ***staircase function***. Sketch the graph of f.

2.6 COMPOSITION OF FUNCTIONS

When we compute the value $h(x)$ of a function h we frequently perform the computation in more than one step. For example, if

$$h(x) = \sqrt{x + 5} \tag{1}$$

then we obtain $h(x)$ for $x \geq -5$ by first computing $x + 5$ and then taking the square root of $x + 5$. Thus

$$h(4) = \sqrt{4 + 5} = \sqrt{9} = 3$$

If we let

$$f(x) = x + 5 \quad \text{and} \quad g(x) = \sqrt{x}$$

then for $x \geq 5$ we have

$$\sqrt{x + 5} = \sqrt{f(x)} = g(f(x))$$

so that

$$h(x) = g(f(x))$$

Similarly, if

$$h(x) = \frac{1}{x - 4} \tag{2}$$

then to compute $h(x)$ for any x such that $x - 4 \neq 0$, that is, for $x \neq 4$, we first compute $x - 4$ and then take its reciprocal. Therefore, if

$$f(x) = x - 4 \quad \text{and} \quad g(x) = \frac{1}{x}$$

then for $x \neq 4$ we have

$$\frac{1}{x - 4} = \frac{1}{f(x)} = g(f(x))$$

so once again

$$h(x) = g(f(x))$$

In general, if f and g are two functions, then we define the ***composite function*** $g \circ f$ to be the function whose rule is given by

$$\boxed{(g \circ f)(x) = g(f(x))} \tag{3}$$

and whose domain consists of all the numbers x in the domain of f such that $f(x)$ is in the domain of g. (This is just the set of all numbers x for which the right side of (3) is meaningful.) Thus each of the functions h defined in (1) and (2) may be expressed as the composite of two functions f and g.

EXAMPLE 1. Let $h(x) = \dfrac{1}{(x+3)^2}$. Express h as the composite $g \circ f$ of two functions f and g.

Solution. If we let

$$f(x) = x + 3 \quad \text{and} \quad g(x) = \frac{1}{x^2}$$

then for $x \neq -3$ we have

$$h(x) = \frac{1}{(x+3)^2} = \frac{1}{(f(x))^2} = g(f(x))$$

Thus $h = g \circ f$. □

In the solution of Example 1, if we had let

$$f(x) = (x+3)^2 \quad \text{and} \quad g(x) = \frac{1}{x}$$

then it would still have been true that $h = g \circ f$. In general, there are many ways of writing a given function as the composite of two other functions.

EXAMPLE 2. Let $f(x) = \sqrt{x}$ and $g(x) = x^2 - 1$. Find the domain and rule of each of the following composite functions.

a. $g \circ f$ b. $f \circ g$

Solution.

a. The domain of $g \circ f$ consists of all numbers x in the domain of f such that $f(x)$ is in the domain of g. But since the domain of g consists of all numbers, it follows that $f(x)$ is in the domain of g for any number x in the domain of f. Therefore the domain of $g \circ f$ is the domain of f, which is the set of all nonnegative numbers. The rule of $g \circ f$ is given by

$$(g \circ f)(x) = g(f(x)) = g(\sqrt{x}) = (\sqrt{x})^2 - 1 = x - 1 \tag{4}$$

b. The domain of $f \circ g$ consists of all numbers x in the domain of g such that $g(x)$ is in the domain of f, which consists of all nonnegative numbers. Thus the domain of $f \circ g$ consists of all numbers x such that $x^2 - 1 \geq 0$, which is equivalent to $x^2 \geq 1$. But $x^2 \geq 1$ for x in

$(-\infty, -1]$ or $[1, \infty)$. We conclude that the domain of $f \circ g$ consists of the intervals $(-\infty, -1]$ and $[1, \infty)$. The rule of $f \circ g$ is given by

$$(f \circ g)(x) = f(g(x)) = f(x^2 - 1) = \sqrt{x^2 - 1} \quad \square$$

CAUTION: Notice that in Example 2 the functions $g \circ f$ and $f \circ g$ are not the same. This is usually the case. Notice also that the right side of (4) is meaningful for all real numbers x, whereas the domain of $g \circ f$ consists only of the nonnegative numbers. Therefore we must be careful when specifying the domain of a composite function.

EXERCISES 2.6

In Exercises 1–6, express h as the composite $g \circ f$ of two functions f and g (neither of which is equal to h).

1. $h(x) = 2(x - 1)$

2. $h(x) = -7\sqrt{x}$

3. $h(x) = \dfrac{3}{x^4}$

4. $h(x) = \dfrac{-5}{x + 4}$

5. $h(t) = \dfrac{1}{2\sqrt{t}}$

6. $h(t) = (t + 1)^{1/3}$

In Exercises 7–20, find the domain and rule of $g \circ f$.

7. $f(x) = x - 1;\ g(x) = 2x^2 + x + 1$

8. $f(x) = 2x - 1;\ g(x) = x^2$

9. $f(x) = \dfrac{1}{2x};\ g(x) = \sqrt{x}$

10. $f(x) = \sqrt{x};\ g(x) = \dfrac{1}{2x}$

11. $f(x) = x - 1;\ g(x) = \sqrt{x} + 1$

12. $f(x) = \sqrt{x};\ g(x) = \dfrac{1}{2x - 4}$

13. $f(x) = \dfrac{1}{2x};\ g(x) = \dfrac{1}{x^2 - 1}$

14. $f(x) = \dfrac{x^2 - 1}{x^2 + 1};\ g(x) = \dfrac{1}{x}$

15. $f(x) = \dfrac{1}{x};\ g(x) = \dfrac{1}{x}$

16. $f(x) = \dfrac{x - 1}{x + 1};\ g(x) = \dfrac{x + 1}{x - 1}$

17. $f(x) = \dfrac{x - 1}{x + 1};\ g(x) = \dfrac{x + 3}{x - 2}$

18. $f(x) = \sqrt{x^2 + 1};\ g(x) = \sqrt{x^2 - 1}$

19. $f(x) = \sqrt{x^2 + 1};\ g(x) = \sqrt{x^2 - 4}$

20. $f(x) = \dfrac{2x}{x - 1};\ g(x) = \sqrt{2x - 4}$

21. Let $f(x) = x$. Show that $g \circ f = g$ and $f \circ g = g$ for any function g.

22. Let $f(x) = \dfrac{x}{x - 1}$. Show that $(f \circ f)(x) = x$ for all $x \neq 1$.

23. Let $f(x) = \dfrac{1}{1 - x}$. Show that $(f \circ f \circ f)(x) = x$ for all x except 0 and 1. (By definition, $(f \circ f \circ f)(x) = [f \circ (f \circ f)](x)$.)

24. Suppose the domain of f is $[0, 1)$ and $g(x) = f(x + 10)$. Find the domain of g.

25. Suppose the domain of f is $[0, 1)$ and $g(x) = f(10 - x)$. Find the domain of g.

26. Suppose that domain of f is $(-\pi/2, \pi/2)$ and $g(x) = f(x - \pi/6)$. Find the domain of g.

27. Let $f(x) = mx + b$, where m and b are constants. Let $g(x) = f(x + 1) - f(x)$. Show that g is a constant function, and determine the constant.

28. Let $f(x) = ax^2 + bx + c$, where a, b, and c are constants. Show that if

$$g(x) = f(x + 1) - f(x) \text{ and } G(x) = g(x + 1) - g(x)$$

then G is a constant function that is independent of b and c.

29. The volume $V(r)$ of a spherical balloon of radius r is given by

$$V(r) = \frac{4}{3}\pi r^3$$

If the radius of the balloon is increasing with time t according to the formula

$$r(t) = \frac{3}{2}t^2 \quad \text{for} \quad t \geq 0$$

find a formula for the volume of the balloon at any time $t \geq 0$.

30. Suppose a car is traveling 40 miles per hour, so the distance D_m in miles it travels in t hours is given by

$$D_m(t) = 40t$$

If m represents miles and k kilometers, then the conversion from miles to kilometers is given by

$$k = g(m) = 1.609344m$$

Find a formula for the distance D_k in kilometers as a function of time.

2.7 INVERSES OF FUNCTIONS

We can convert from degrees Celsius to degrees Fahrenheit by means of the formula

$$F = \frac{9}{5}C + 32 \tag{1}$$

By solving this equation for C we obtain a formula for converting from degrees Fahrenheit to degrees Celsius:

$$C = \frac{5}{9}(F - 32) \tag{2}$$

In order to explore further the relationship between (1) and (2), we view (1) and (2) as defining the functions f and g given by

$$f(x) = \frac{9}{5}x + 32 \quad \text{and} \quad g(x) = \frac{5}{9}(x - 32) \tag{3}$$

Notice that if $y = f(x)$, then

$$y = \frac{9}{5}x + 32$$

so that

$$g(y) = g\left(\frac{9}{5}x + 32\right) = \frac{5}{9}\left[\left(\frac{9}{5}x + 32\right) - 32\right] = \frac{5}{9}\left(\frac{9}{5}x\right) = x$$

Thus if $y = f(x)$ then $x = g(y)$. A similar calculation shows that the converse is true: if $x = g(y)$, then $y = f(x)$. We express this relationship between f and g by saying that g is an inverse of f.

DEFINITION 2.4 Let f be a function. Then f ***has an inverse*** if there is a function g such that

i. the domain of g is the range of f
ii. for all x in the domain of f and all y in the range of f,

$$f(x) = y \quad \text{if and only if} \quad g(y) = x \tag{4}$$

Under these conditions g is an ***inverse*** of f.

Notice that in Definition 2.4, g is a *function*. Its domain is specified by (i) and its rule by (ii). From this it follows that g is unique. To emphasize its connection with f, we write f^{-1} for g and call f^{-1} ***the inverse of f***. With g replaced by f^{-1}, (4) becomes

$$f(x) = y \quad \text{if and only if} \quad f^{-1}(y) = x \tag{5}$$

for all x in the domain of f and all y in the range of f. We also find that if f has an inverse, then f^{-1} has f as inverse, so f and f^{-1} are inverses of each other. Thus we may rewrite (5) with f and f^{-1} interchanged, obtaining

$$f^{-1}(x) = y \quad \text{if and only if} \quad f(y) = x \tag{6}$$

CAUTION: Not every function has an inverse. At the end of this section we will show that, among other functions, x^2 does not have an inverse.

From (5) we obtain two important formulas relating a function f and its inverse f^{-1}:

$$f^{-1}(f(x)) = x \text{ for all } x \text{ in the domain of } f \tag{7}$$

$$f(f^{-1}(y)) = y \text{ for all } y \text{ in the range of } f \tag{8}$$

EXAMPLE 1. Let $f(x) = 5x^3$, and assume that we already know that f has an inverse and that

$$f^{-1}(y) = \sqrt[3]{\frac{y}{5}}$$

Verify (7) and (8) from the formulas given for f and f^{-1}.

Solution. For (7) we have

$$f^{-1}(f(x)) = f^{-1}(5x^3) = \sqrt[3]{\frac{5x^3}{5}} = \sqrt[3]{x^3} = x$$

and for (8) we have

$$f(f^{-1}(y)) = f\left(\sqrt[3]{\frac{y}{5}}\right) = 5\left(\sqrt[3]{\frac{y}{5}}\right)^3 = 5\left(\frac{y}{5}\right) = y \quad \square$$

Because we usually write formulas for functions in terms of x (rather than y), we will also usually write the formula for f^{-1} in terms of x. Thus in Example 1 the formula for the inverse f^{-1} would normally be written as

$$f^{-1}(x) = \sqrt[3]{\frac{x}{5}}$$

Finding Formulas for Inverses

In Example 1 we assumed the formula for f^{-1} on faith. Now we will describe a procedure that sometimes allows us to determine a formula for the inverse of a function f, provided that f is given by a simple formula. The individual steps in the procedure are:

> Step 1. Write $y = f(x)$.
> Step 2. Solve the equation in Step 1 for x in terms of y.
> Step 3. In the formula for x arising from Step 2, replace x by $f^{-1}(y)$.
> Step 4. Replace each y in the result of Step 3 by x.

Let us see how the method works for the temperature conversion function f appearing in (3).

EXAMPLE 2. Let $f(x) = \frac{9}{5}x + 32$. Find a formula for f^{-1}.

Solution. Following the steps listed above, we obtain

Step 1:
$$y = \frac{9}{5}x + 32$$

Step 2:
$$\begin{cases} 5y = 9x + 160 \\ 9x = 5y - 160 = 5(y - 32) \\ x = \frac{5}{9}(y - 32) \end{cases}$$

Step 3:
$$f^{-1}(y) = \frac{5}{9}(y - 32)$$

Step 4:
$$f^{-1}(x) = \frac{5}{9}(x - 32) \quad \square$$

Notice that f^{-1} is the function g appearing in (3), so g is shown to be the inverse of f in a second way.

EXAMPLE 3. Let $f(x) = 2 - x^3$. Find a formula for f^{-1}.

Solution. Again we use the steps given above:

Step 1: $y = 2 - x^3$

Step 2: $\begin{cases} x^3 = 2 - y \\ x = \sqrt[3]{2 - y} \end{cases}$

Step 3: $f^{-1}(y) = \sqrt[3]{2 - y}$

Step 4: $f^{-1}(x) = \sqrt[3]{2 - x}$ □

Because mistakes are easy to make, it is advisable to check either that $f(f^{-1}(x)) = x$ or that $f^{-1}(f(x)) = x$ after obtaining a formula for f^{-1}. Using this advice for the function in Example 3, we have

$$\begin{aligned} f(f^{-1}(x)) &= 2 - (f^{-1}(x))^3 \\ &= 2 - (\sqrt[3]{2 - x})^3 \\ &= 2 - (2 - x) = x \end{aligned}$$

Graphs of Inverses

To find a method of graphing the inverse of a function f, we begin by observing that if (1, 3) is on the graph of f, then $f(1) = 3$; therefore by the definition of f^{-1}, it follows that $f^{-1}(3) = 1$, which in turn means that (3, 1) is on the graph of f^{-1}. More generally, for any real numbers a and b, if (a, b) is on the graph of f, then (b, a) is on the graph of f^{-1}, and vice versa. In particular, $(x, f(x))$ is on the graph of f for any x in the domain of f, so that by the preceding comments $(f(x), x)$ is on the graph of f^{-1}. Notice that $(x, f(x))$ and $(f(x), x)$ are opposite one another with respect to the line $y = x$ (Figure 2.60). For this reason we say that $(x, f(x))$ and $(f(x), x)$ are ***symmetric with respect to the line y = x***, and when we pass from one of these points to the other, we say that we ***reflect*** the point through the line $y = x$. Therefore we can obtain the graph of f^{-1} by reflecting all the points on the graph of f through the line $y = x$, or as we usually say, by reflecting the graph of f through the line $y = x$ (Figure 2.60). The graphs of f and f^{-1} are said to be symmetric with respect to the line $y = x$.

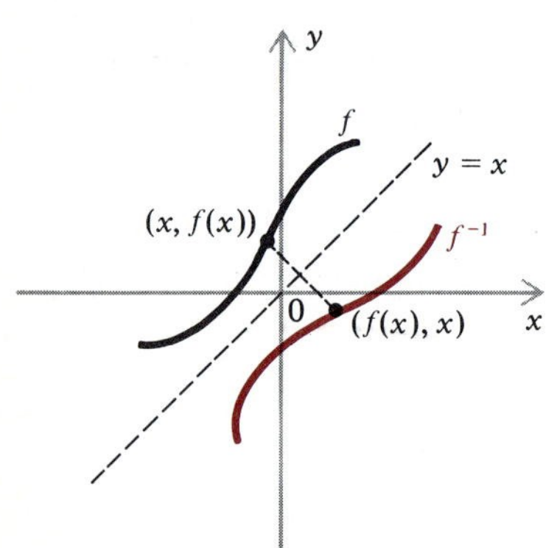

The graphs of f and f^{-1} are symmetric with respect to the line $y = x$.

FIGURE 2.60

EXAMPLE 4. Let $f(x) = -2x + 3$. First sketch the graph of f, and then obtain the graph of f^{-1} by reflecting the graph of f through the line $y = x$.

Solution. The graph of f is the line with slope -2 and y intercept 3. It is sketched in Figure 2.61. Next we draw the line $y = x$, and finally we sketch the graph of f^{-1} by reflecting through the line $y = x$ (Figure 2.61). □

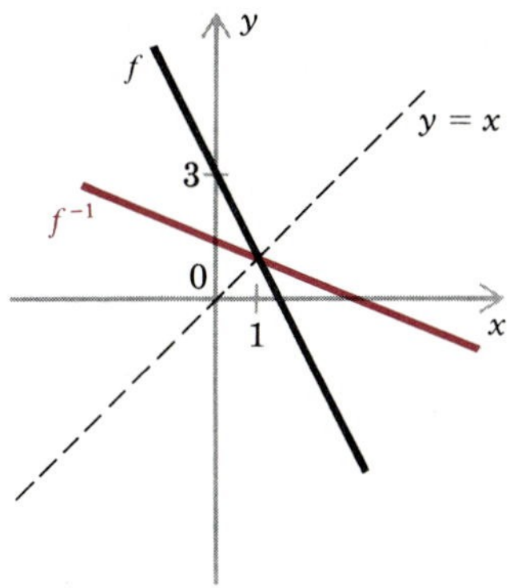

FIGURE 2.61

There is a geometric method for telling which functions f have inverses: f has an inverse if and only if no horizontal line intersects the graph of f more than once. As you can easily see from Figure 2.62, this implies that the functions x^2 and $|x|$ do not have inverses. In algebraic terms, a function has an inverse if and only if the following condition holds:

For any x and z in the domain of f,

$$\text{if } x \neq z, \quad \text{then} \quad f(x) \neq f(z) \tag{9}$$

A function with the property in (9) is said to be ***one-to-one*** because each member of the range is identified with exactly one member of the domain. Thus if $f(x) = x^3$, then f is one-to-one because

$$\text{if } x \neq z, \quad \text{then} \quad f(x) = x^3 \neq z^3 = f(z)$$

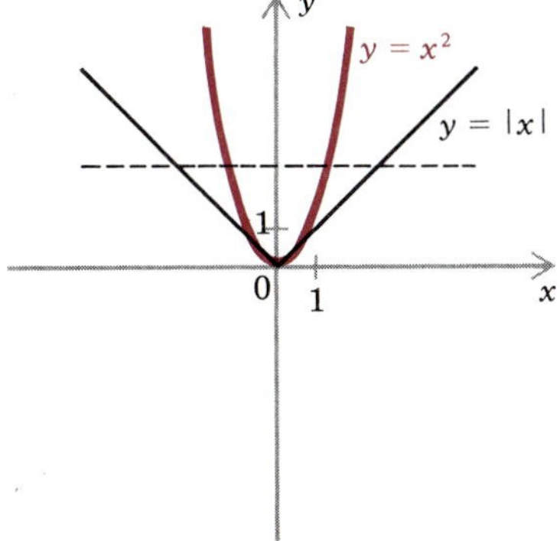

The functions x^2 and $|x|$ do not have inverses

FIGURE 2.62

By contrast, if $f(x) = x^2$, then $g(x) = g(-x)$ for *all* x, so that g is *not* one-to-one.

It would be tedious to have to examine each pair x and z in the domain of f in order to see if (9) held. However, if we can show that either

$$f \text{ is } \textbf{increasing}, \text{ that is, if } x < z \text{ then } f(x) < f(z) \tag{10}$$

or

$$f \text{ is } \textbf{decreasing}, \text{ that is, if } x < z \text{ then } f(x) > f(z) \tag{11}$$

then (9) is automatically satisfied, so that f has an inverse. We observe that geometrically (10) means that the graph of f rises from left to right (Figure 2.63a), whereas (11) means that the graph of f falls from left to right (Figure 2.63b).

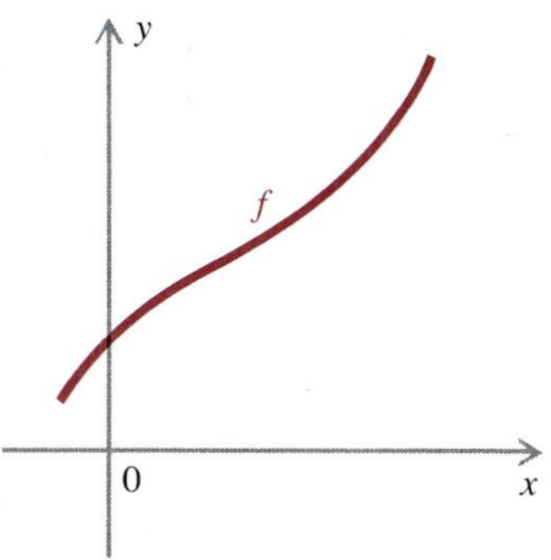

The graph of an increasing function rises from left to right.
(a)

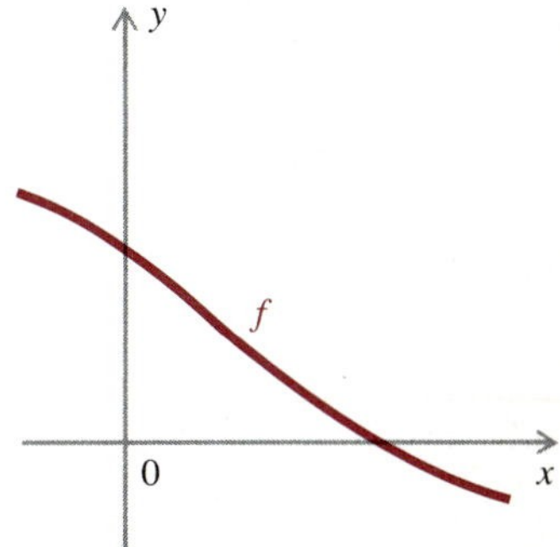

The graph of a decreasing function falls from left to right.
(b)

FIGURE 2.63

We can use (10) and (11) to show that a given function f has an inverse even if we may be unable to find a formula for the inverse function.

EXAMPLE 5. Let $f(x) = x^3 + x^5$. Show that f has an inverse function.

Solution. By the discussion above, it suffices to show that f is increasing. To that end, suppose that $x < z$. Then $x^3 < z^3$ and $x^5 < z^5$, so that

$$f(x) = x^3 + x^5 < z^3 + z^5 = f(z)$$

Therefore (10) is satisfied, which means that f is increasing, and hence has an inverse. □

We observe that we have no way of deriving an equation for the inverse of f in Example 5. Yet we are still able to show that f *has* an inverse.

EXERCISES 2.7

In Exercises 1–22, find a formula for f^{-1}.

1. $f(x) = 2x$

2. $f(x) = 2x + 1$

3. $f(x) = -\frac{1}{2}(x - 3)$

4. $f(x) = \pi x - \sqrt{2}$

5. $f(x) = x^3$

6. $f(x) = x^5$

7. $f(x) = 3x^3 - 5$

8. $f(x) = \pi x^2$ for $x \geq 0$

9. $f(x) = -x^5 + 1$

10. $f(x) = \frac{1}{x}$

11. $f(x) = \sqrt{x}$

12. $f(x) = \sqrt{x} + 5$

13. $f(x) = \sqrt{x - 3}$

14. $f(x) = 7 + \sqrt{2x - 1}$

15. $f(x) = \frac{1}{\sqrt{x}}$

16. $f(x) = \sqrt[3]{x + 5}$

17. $f(x) = \frac{1}{\sqrt[3]{4 - 2x}}$

18. $f(x) = \frac{-4}{x^3}$

19. $f(x) = \frac{x - 1}{x + 1}$

20. $f(x) = \frac{-x + 3}{2x - 5}$

21. $f(x) = \frac{2x^3 - 1}{x^3 + 3}$

22. $f(x) = \frac{x^5}{-x^5 + 3}$

In Exercises 23–32, sketch the graphs of f and f^{-1} in the same coordinate system.

23. $f(x) = x + 1$

24. $f(x) = 3x$

25. $f(x) = 3(x + 1)$

26. $f(x) = 3x + 1$

27. $f(x) = -2x + 4$

28. $f(x) = \sqrt{x}$

29. $f(x) = \frac{1}{x}$

30. $f(x) = -\frac{1}{x}$

31. $f(x) = \sqrt{x} + 1$

32. $f(x) = \sqrt{x} - 1$

In Exercises 33–40, determine whether f is increasing, decreasing, or neither.

33. $f(x) = 3x - 5$

34. $f(x) = -4x + 7$

35. $f(x) = x^3 - x^5$

36. $f(x) = x^2$

37. $f(x) = \sqrt{x}$

38. $f(x) = \sqrt{x} + x^2$

39. $f(x) = \frac{1}{x}$

40. $f(x) = \frac{1}{x}$ for $x < 0$

41. Show that each of the following functions is equal to its own inverse:

a. $f(x) = x$ b. $f(x) = \frac{1}{x}$ c. $f(x) = -x$

d. $f(x) = \sqrt{1 - x^2}$ for $0 \leq x \leq 1$

42. Does a constant function have an inverse? Explain your answer.

43. The area A of an equilateral triangle of side length s is given by $A = \frac{\sqrt{3}}{4}s^2$.

a. Find a formula for the side length in terms of the area.
b. Determine the value of s for which $A = 16$.

44. The surface area S of a cylinder with height 4 and radius r, and including both top and bottom is given by

$$S = 2\pi r^2 + 8\pi r$$

Find a formula for the radius r in terms of the surface area S.

2.8 ITERATES OF FUNCTIONS

In the preceding section we discussed the application of composites to the notion of inverse functions, which will play an important role in Chapters 4 and 5. Now we apply the composite in an entirely different way.

Let us consider the number 2. On our calculator we take the square root $\sqrt{2}$, its square root $\sqrt{\sqrt{2}}$, and continue taking square roots. What appears on the calculator screen are, successively, the numbers

$$1.414213562, 1.189207115, 1.090507733, 1.044273782, 1.021897149, \ldots \quad (1)$$

Notice that with successive square roots the displayed number decreases and is at least 1. After 31 applications of the square root the number displayed on our calculator is so close to 1 that the calculator gives the answer as 1. The same kind of behavior occurs for successive square roots of any number that is larger than 1.

If we let $f(x) = \sqrt{x}$ for $x \geq 0$, then the numbers listed in (1) are 2, $f(2)$, $f(f(2))$, $f(f(f(2)))$, $f(f(f(f(2))))$, $f(f(f(f(f(2)))))$, ... These numbers are called the iterates of 2 for the function f. More generally, if f is any function whose range is contained in its domain (as happens with $\sqrt{x}$), then the numbers

$$a, f(a), f(f(a)), f(f(f(a))), f(f(f(f(a)))), \ldots$$

are called the ***iterates of a for f***. For convenience we write

$$f^{[2]}(a) = f(f(a))$$

$$f^{[3]}(a) = f(f(f(a)))$$

In general, for any positive integer n, the ***nth iterate of a for f*** is given by

$$f^{[n]}(a) = \underbrace{f(f(f(\cdots(f}_{n \text{ of these}}(a)\underbrace{)\cdots)))}_{n \text{ of these}}$$

Observe that for any integer $n \geq 2$,

$$f^{[n]}(a) = f(f^{[n-1]}(a))$$

so we obtain $f^{[n]}(a)$ by finding the value of f at $f^{[n-1]}(a)$.

EXAMPLE 1. Describe the behavior of the iterates of 0 for f and g where

a. $f(x) = x^2 + 1$ b. $g(x) = x^2 + \frac{1}{4}$

Solution.

a. The initial few iterates of 0 for $x^2 + 1$ are

$$\begin{aligned} &0 \\ f(0) &= 0^2 + 1 = 1 \\ f^{[2]}(0) &= f(f(0)) = f(1) = 1^2 + 1 = 2 \\ f^{[3]}(0) &= f(f^{[2]}(0)) = f(2) = 2^2 + 1 = 5 \\ f^{[4]}(0) &= f(f^{[3]}(0)) = f(5) = 5^2 + 1 = 26 \\ f^{[5]}(0) &= f(f^{[4]}(0)) = f(26) = 26^2 + 1 = 677 \end{aligned}$$

Since the square of a large number is an even larger number, it is apparent that successive iterates increase without bound.

b. The initial few iterates of 0 for $x^2 + \frac{1}{4}$ are

$$\begin{aligned} &0 \\ g(0) &= \frac{1}{4} \\ g^{[2]}(0) &= g(g(0)) = g\left(\frac{1}{4}\right) = .3125 \\ g^{[3]}(0) &= g(g^{[2]}(0)) = g(.3125) = .34736525 \end{aligned}$$

Continuing the process repeatedly, we obtain

$$\begin{aligned} g^{[30]}(0) &= .4716857877 \\ g^{[100]}(0) &= .4906042201 \end{aligned}$$

The iterates grow, but very slowly. It turns out that they approach .5. However, numerous iterations would be required before the iterates are so close to .5 that the calculator would just display .5. □

Finding iterates of a function can be tedious unless one utilizes a computer program or special features of some calculators. Nevertheless, iterates can behave very interestingly, as the exercises suggest.

Finding Iterates Graphically

We can analyze the iterates of a number a for a function f by means of a graph. First we draw the graph of f and the line $y = x$ on the same coordinate

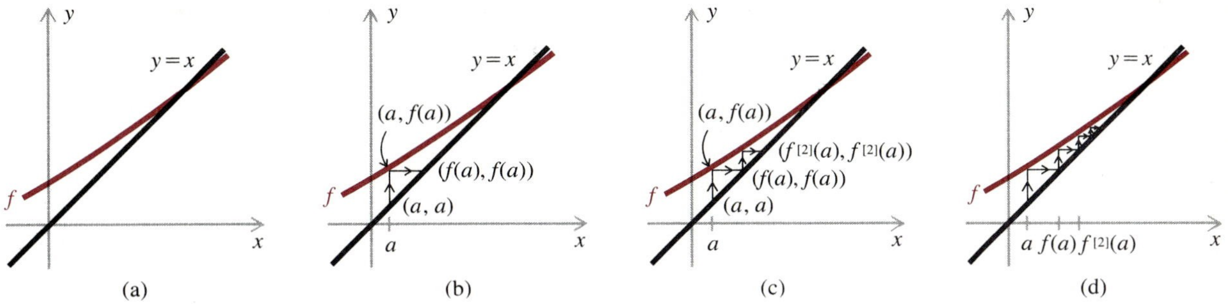

FIGURE 2.64

plane (Figure 2.64a). Next we draw a vertical line from the point (a, a) on the line $y = x$ to the point $(a, f(a))$ on the graph of f, and then a horizontal line to the point $(f(a), f(a))$ (Figure 2.64b). The effect of the vertical and horizontal lines is to travel from (a, a) to $(f(a), f(a))$. Repeating the process yields the point $(f^{[2]}(a), f^{[2]}(a))$ (Figure 2.64c). Continuing in this way, we can generate $f(a)$, $f^{[2]}(a)$, $f^{[3]}(a)$, and so forth, which are the numbers on the x-axis lying directly above or below the respective points on the line $y = x$ (Figure 2.64d).

The process just outlined can be continued indefinitely to obtain as many iterates as we like. Analysis of iterates by means of graphs such as those in Figure 2.64 is called ***graphical analysis*** of the iterates. In many instances sketching a few iterates of a given number for a function f can yield a recognizable pattern for the iterates in general. To substantiate this claim, we return to the functions appearing in Example 1 and submit them to graphical analysis.

EXAMPLE 2. Use graphical analysis to discover the behavior of the iterates of 0 for f and g, where

a. $f(x) = x^2 + 1$ b. $g(x) = x^2 + \dfrac{1}{4}$

Solution.

a. The graph of f appears in Figure 2.65a. Since

$$x < 1 \le x^2 + 1 \text{ if } |x| < 1, \quad \text{and} \quad x \le x^2 < x^2 + 1 \text{ if } |x| \ge 1$$

it follows that

$$x < x^2 + 1 = f(x) \quad \text{for all } x$$

Therefore the graph of f lies above the line $y = x$. Figure 2.65b shows that successive iterates of 0 grow without bound.

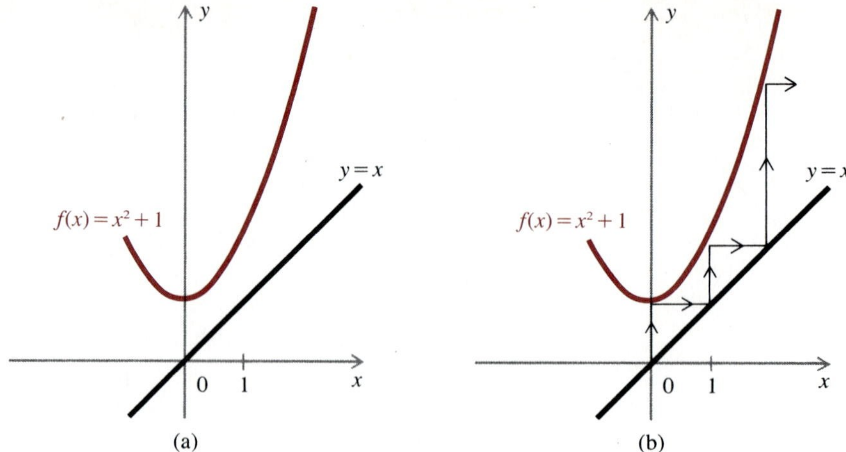

FIGURE 2.65

b. The graph of g appears in Figure 2.66a. For any real number x,

$$x^2 - x + \frac{1}{4} = \left(x - \frac{1}{2}\right)^2 \geq 0 \tag{2}$$

so that

$$g(x) = x^2 + \frac{1}{4} \geq x$$

Also, $g(x) = x$ if and only if $x^2 + \frac{1}{4} = x$, or equivalently, $x^2 - x + \frac{1}{4} = 0$, which by (2) means that $x = \frac{1}{2}$. Therefore, the graph of g lies above the line

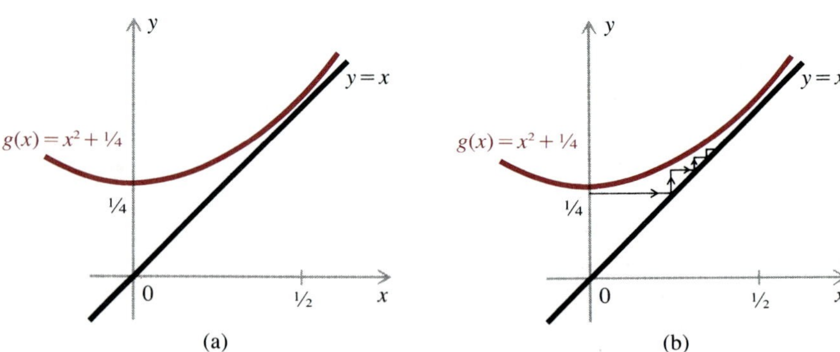

FIGURE 2.66

$y = x$, and touches the line at the point $(\frac{1}{2}, \frac{1}{2})$ (Figure 2.66a). Now Figure 2.66b confirms that the iterates of 0 approach $\frac{1}{2}$, as part (b) of Example 1 indicates. □

Notice in part (b) of Example 2 that the iterates of 0 under g approach $\frac{1}{2}$, which is the number that satisfies the equation $g(x) = x$, and which corresponds to the point of intersection of the graph of g and the line $y = x$ (Figure 2.66a). This turns out to be true in general: If the iterates of a number under a function f approach a number p, then p satisfies the equation $f(p) = p$. Such a number p is called a ***fixed point*** of f.

Iterates of even simple functions such as $f_c(x) = x^2 + c$, where c is a constant, have attracted much attention in the past decade. In fact, the iterates of such functions are the point of departure for a new subject called chaos.

EXERCISES 2.8

In Exercises 1–4, analyze the behavior of the iterates of a for the given function f by calculating iterates.

1. $f(x) = \sqrt{x}$; $a = \frac{1}{2}$; $a = 5$; $a = 30$; $a = 1000$

2. $f(x) = x^2$; $a = .5$; $a = 1$; $a = -1$; $a = 2$

3. $f(x) = x^2 + \frac{1}{8}$; $a = 0$; $a = .5$

4. $f(x) = x^2 + \frac{1}{2}$; $a = 0$; $a = .5$; $a = -.5$

In Exercises 5–8, analyze the behavior of the iterates of a for the given function by graphical analysis.

5. $f(x) = \sqrt{x}$; $a = .2$; $a = 3$

6. $f(x) = x^2$; $a = \frac{1}{3}$; $a = -\frac{1}{2}$

7. $f(x) = x^2 + \frac{1}{2}$; $a = 0$; $a = 1$

8. $f(x) = x - x^2$; $a = .3$; $a = 1$ (*Hint:* See Figure 2.67).

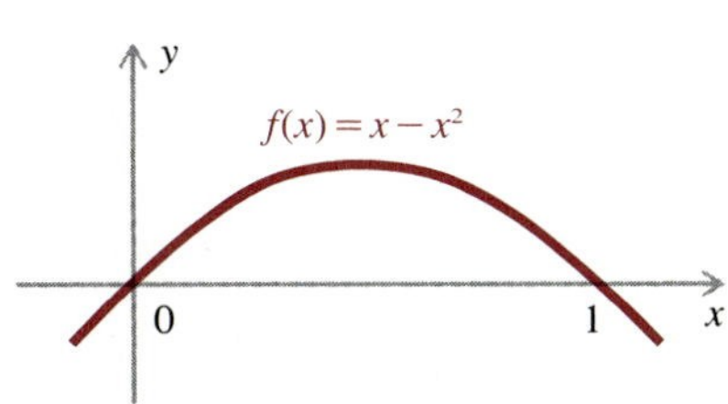

FIGURE 2.67

In Exercises 9–10, find the fixed points of f.

9. $f(x) = 2x - 2x^2$

10. $f(x) = x^2 + \frac{1}{8}$

In Exercises 11–15, calculate at least n iterates of the given number, and discover the behavior of the iterates.

11. $f(x) = 2x - 2x^2$; $n = 12$; $a = .5$; $a = .4$

12. $f(x) = 3.2x - 3.2x^2$; $n = 20$; $a = .5$; $a = .8$

13. $f(x) = 3.5x - 3.5x^2$; $n = 30$; $a = .5$; $a = .2$

14. $f(x) = 3.85x - 3.85x^2$; $n = 100$; $a = .5$; $a = .73$

15. $f(x) = 4x - 4x^2$; $n = 300$; $a = .5$; $a = .3$; $a = .8$

16. Let $f(x) = \dfrac{1}{x}$. Without using a calculator, find the pattern of the iterates of any nonzero number a.

17. Let $f(x) = \dfrac{1}{1 - x}$. Without using a calculator, find the pattern of the iterates of any number a different from 0 and 1.

KEY TERMS

coordinate axes
- x axis
- y axis

coordinates
- x coordinate (abscissa)
- y coordinate (ordinate)

origin
ordered pair
Cartesian coordinate system
quadrant
distance between points
slope
point–slope equation
slope–intercept equation
linear equation
function
- domain
- rule
- range
- even
- odd
- composite
- inverse

variable
- dependent variable
- independent variable

proportion
- directly proportional
- inversely proportional
- jointly proportional
- constant of proportionality

graph of a function
symmetry
- with respect to the y axis
- with respect to the origin

intercept
- x intercept
- y intercept

nth iterate
fixed point

KEY FORMULAS

$$d(P, Q) = \sqrt{(x_2 - x_1)^2 + (y_2 - y_1)^2}$$

$$M(x, y) = (\tfrac{1}{2}(x_1 + x_2), \tfrac{1}{2}(y_1 + y_2))$$

$$m = \frac{y_2 - y_1}{x_2 - x_1}$$

$$f^{-1}(f(x)) = x \quad \text{and} \quad f(f^{-1}(y)) = y$$

REVIEW EXERCISES

1. Determine the distance between each of the following pairs of points.
   a. $(-1, -3)$ and $(-4, 3)$
   b. $(\sqrt{2}, \frac{1}{2}\sqrt{2})$ and $(3\sqrt{2}, -2\sqrt{2})$

2. For each of the following functions, determine which of the numbers -2, 2, -4, and 4 are in the domain.

   a. $f(x) = \dfrac{x+2}{x-2}$ b. $f(x) = \dfrac{x+2}{x^2-4}$

In Exercises 3–10, determine the domain of the function.

3. $f(x) = \dfrac{1}{x+3}$

4. $f(x) = \dfrac{x-2}{x+5} - \dfrac{x+1}{x-2}$

5. $g(x) = \dfrac{1}{x^2 + 4x - 7}$

6. $g(x) = \sqrt{x + \pi}$

7. $h(x) = \sqrt{6x - 4}$

8. $h(x) = |x - 3|$

9. $h(x) = \sqrt{|x| - 5}$

10. $f(x) = \begin{cases} x^2 & \text{for } x < -2 \\ 3 & \text{for } -2 < x \leq 1 \\ \sqrt{x-4} & \text{for } 5 < x \end{cases}$

11. Show that if $a \neq 0$, then the domain and range of $\sqrt{x+a}$ are distinct.

12. Let $f(x) = \dfrac{3x-4}{2x+5}$. Find

a. $f(0)$ b. $f(-1)$ c. $f\left(\dfrac{3}{4}\right)$ d. $f\left(-\dfrac{2}{x}\right)$

13. Let $f(x) = x^2 + 3ax - 5$, and suppose that $f(-2) = 1$. Determine a.

14. Let $f(t) = 1 - t^2$. Find $\dfrac{f(b) - f(a)}{b - a}$.

15. Let $f(x) = \dfrac{16}{x^2 - \sqrt{x^4 - 16}}$.

a. Find the domain of f.
b. Let $g(x) = x^2 + \sqrt{x^4 - 16}$. Show that $f = g$.

16. Let

$$f(x) = \sqrt{x+1} - \sqrt{x-2}$$

and

$$g(x) = \frac{1}{\sqrt{x+1} + \sqrt{x-2}}$$

Determine whether or not $f = g$.

In Exercises 17–26, sketch the graph of the function, noting any intercepts and symmetry.

17. $f(x) = 2x - 5$

18. $f(x) = -\frac{1}{2}x - \frac{3}{2}$

19. $f(x) = x^2 - 6$

20. $f(x) = 2(x+2)^2$

21. $f(x) = ||x| - 2|$

22. $f(x) = \dfrac{3}{x}$

23. $g(x) = \sqrt{-x}$

24. $g(x) = \sqrt{x+1}$

25. $g(x) = \begin{cases} x + 4 & \text{for } x \leq -1 \\ -4 - x^2 & \text{for } x \geq 1 \end{cases}$

26. $g(x) = \begin{cases} -x & \text{for } x < -1 \\ 2 & \text{for } x = -1 \\ 3x & \text{for } x \geq 0 \end{cases}$

In Exercises 27–32, find $f \circ g$ and $g \circ f$ for the given functions f and g.

27. $f(x) = \dfrac{25}{x^2}$ and $g(x) = 5x$

28. $f(x) = \dfrac{1}{x+2}$ and $g(x) = x - 2$.

29. $f(x) = x^3$ and $g(x) = \sqrt{2x+3}$

30. $f(x) = \dfrac{x+4}{x-3}$ and $g(x) = x^2 + 3$

31. $f(x) = |x|$ and $g(x) = \sqrt{x^2 - 4}$

32. $f(x) = \dfrac{2-3x}{4+x} = g(x)$

33. Let $f(x) = \dfrac{x+1}{x-1}$. Show that $(f \circ f)(x) = x$ for all $x \neq 1$, and thus that f is its own inverse.

34. Let $f(x) = 1/x$ and $g(x) = \dfrac{x^2 - 1}{x^2 + 1}$. Show that $(g \circ f)(x) = -g(x)$.

35. Let $f(x) = \sqrt{x^2 - 3}$ and $g(x) = \sqrt{x^2 + 9}$. Determine whether or not $f \circ g = g \circ f$. Explain your answer.

36. Let $f(x) = \sqrt{2 + x}$.

a. Use a calculator to approximate the following numbers.

i. $f(\sqrt{2})$ ii. $(f \circ f)(\sqrt{2})$ iii. $(f \circ f \circ f)(\sqrt{2})$

b. Can you guess the value that

$$\underbrace{(f \circ f \circ \cdots \circ f)}_{n \text{ of these}}(\sqrt{2})$$

approaches as n increases without bound?

In Exercises 37–40, find a formula for the inverse of the given function.

37. $f(x) = \dfrac{1}{x+3}$

38. $f(x) = \dfrac{5}{x^{1/3} + 4}$

39. $f(x) = \dfrac{4x - 1}{3x + 2}$ **40.** $f(x) = \sqrt{x - 2}$

In Exercises 41–42, analyze the behavior of the iterates of a for the given function f by calculating iterates.

41. $f(x) = \sqrt[3]{x}$; $a = \frac{1}{2}$; $a = 2$; $a = -3$

42. $f(x) = x^2 + \frac{5}{36}$; $a = 0$; $a = 1$

In Exercises 43–44, calculate at least n iterates of the given number, and discover the behavior of the iterates.

43. $f(x) = 3.3x - 3.3x^2$; $n = 20$; $a = .5$; $a = .2$

44. $f(x) = 3.9x - 3.9x^2$; $n = 100$; $a = .5$; $a = .9$

45. Determine the distance between the points $(-\sqrt{a}, -\sqrt{b})$ and $(\sqrt{a}, \sqrt{b})$.

46. Find the points on the line $y = 2x + 1$ that are 1 unit from the point (0, 3).

47. Find the points on the line $x - 2y = 3$ that are 5 units from the point (7, 2).

48. Find an equation of all points (x, y) that are 4 units from the point $(-2, -3)$.

49. Find an equation of the circle centered at $(\frac{1}{2}, -\frac{1}{4})$ and passing through $(0, -\frac{1}{4} + \frac{3}{4}\sqrt{7})$.

50. A line l has x intercept -2 and y intercept -4.
a. Write the slope-intercept equation of l.
b. Write a point-slope equation of l.

51. Find an equation of the line that is parallel to the line $x + 2y = 3$ and passes through the point $(-1, -3)$.

52. Find an equation of the line that is perpendicular to the line $3x - 2y = 6$ and passes through the point (4, 6).

53. Find an equation of the line consisting of all points (x, y) that are equidistant from the points (3, 1) and $(-1, 4)$.

54. Determine whether or not the points $(-1, -1)$, (1, 3), and (57, 115) lie on a straight line.

55. The converse of the Pythagorean Theorem states that if the lengths a, b, and c of the sides of a triangle satisfy the equation $c^2 = a^2 + b^2$, then the triangle is a right triangle. Use this result to determine whether or not the three points (2, 1), $(1, -2)$, and $(-2, 3)$ form the vertices of a right triangle.

56. Let M be the midpoint of the line segment joining the points $(-2, 1)$ and $(-1, 3)$. Find the distance between M and the origin.

57. Prove that the midpoint of the hypotenuse of a right triangle is equidistant from all three vertices. (*Hint:* Let the two legs lie on the coordinate axes.)

58. Prove that the line joining the midpoints of two sides of a triangle is parallel to the third side (Figure 2.68). (*Hint:* Set up the coordinate system so that the vertices of the triangle are $P_1(0, 0)$, $P_2(a, 0)$, and $P_3(b, c)$.)

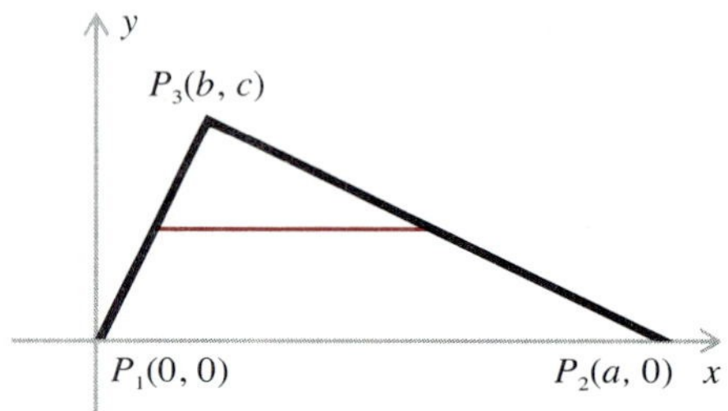

FIGURE 2.68

59. Prove that the points $(-2, 2)$, $(-1, -1)$, (1, 3), and (2, 0) are the vertices of a square.

60. A cylindrical can has a height of 5 inches, and a top and bottom. In terms of the radius, the surface area S of the can is given by

$$S = 2\pi(r^2 + 5r) \quad \text{for} \quad r \geq 0$$

Write a formula for the radius in terms of the surface area. (*Hint:* $r^2 + 5r = (r + \frac{5}{2})^2 - \frac{25}{4}$.)

61. The illumination L from a lamp is directly proportional to the intensity I of the light bulb and is inversely proportional to the square of the distance D from the bulb.
a. Express this fact by means of a formula.
b. If one moves from 12 feet to 4 feet away from a light source, how is the illumination of the person affected?

62. The safe load L of a horizontal beam supported at both ends is jointly proportional to the width w and the square of the depth d and is inversely proportional to the length l of the beam. (Figure 2.69).

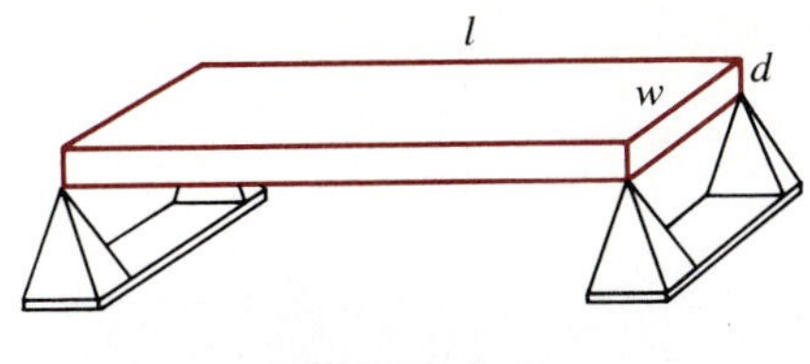

FIGURE 2.69

a. Express this fact by means of a formula.
b. If the safe load on such a beam 2 inches wide, 8 inches deep, and 10 feet long is 1000 pounds, find the safe load on a beam 4 inches wide, 10 inches deep, and 12 feet long.

3 POLYNOMIAL AND RATIONAL FUNCTIONS

Any polynomial expression defines a function whose domain consists of all real numbers. For example, the polynomial expression $2x^4 - x^3 + 3x^2 + \sqrt{2}$ defines the function f whose rule is given by

$$f(x) = 2x^4 - x^3 + 3x^2 + \sqrt{2}$$

In general, if n is a nonnegative integer and $a_n, a_{n-1}, a_{n-2}, \ldots, a_1$, and a_0 are constants with $a_n \neq 0$, then a function f is defined by the formula

$$f(x) = a_n x^n + a_{n-1}x^{n-1} + a_{n-2}x^{n-2} + \cdots + a_1 x + a_0$$

Such a function is called a ***polynomial function***. This chapter will be mainly devoted to the graphs of polynomial functions and quotients of polynomial functions, which are called ***rational functions***.

Special polynomial and rational functions describe many physical quantities and properties. Examples are the shape of a suspension bridge or a mirror in a telescope, the gravitational force exerted on a satellite by the earth, and the trajectory of a golf ball. These types of functions will also help us solve problems such as the following:

> A rancher has 2 miles of fencing and wishes to fence in a rectangular grazing field with an area as large as possible. What should the dimensions of the field be? (See Example 3 of Section 3.2.)

3.1 QUADRATIC FUNCTIONS

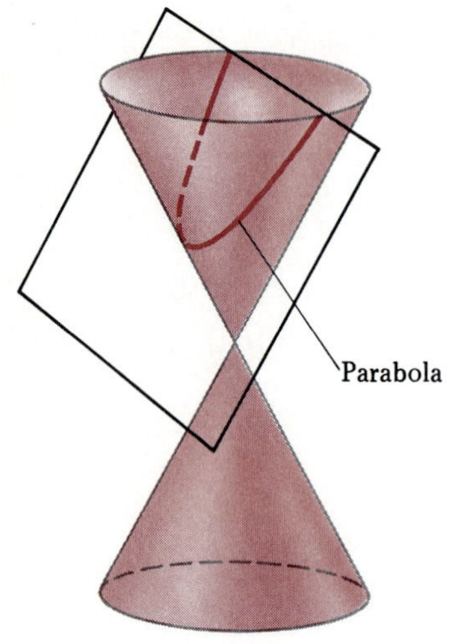

FIGURE 3.1

Because they have the lowest degree, constant functions and linear functions are the simplest polynomial functions. Nonzero constant functions have degree 0, and linear functions have degree 1. The next simplest kind of polynomial functions are those of degree 2, called ***quadratic functions***. A quadratic function is usually given in the form

$$f(x) = ax^2 + bx + c$$

where a, b, and c are constants with $a \neq 0$. The graph of a quadratic function is called a ***parabola***. It has been known since antiquity that this type of curve results when a cone is sliced by a plane (Figure 3.1). The emphasis in this section will be on sketching parabolas.

Let f and g be the quadratic functions defined by

$$f(x) = x^2 \quad \text{and} \quad g(x) = -x^2$$

The graphs of these functions were sketched in Section 2.5, and are reproduced in Figure 3.2a, b. Notice that the graph of x^2 opens upward, whereas the graph of $-x^2$ opens downward.

Next, if

$$f(x) = ax^2 \quad \text{with} \quad a \neq 0$$

then the graph of f has the same general shape as that of x^2 or $-x^2$, depending on whether $a > 0$ or $a < 0$. However, the graph is flatter if $|a| < 1$ and is more pointed if $|a| > 1$ (Figure 3.3a, b). In each case the graph is symmetric with respect to the y axis, which is referred to as the ***axis*** of the parabola. The point (0, 0), which is the lowest point on the graph of f if $a > 0$ and is the highest point on the graph if $a < 0$, is called the ***vertex*** of the parabola.

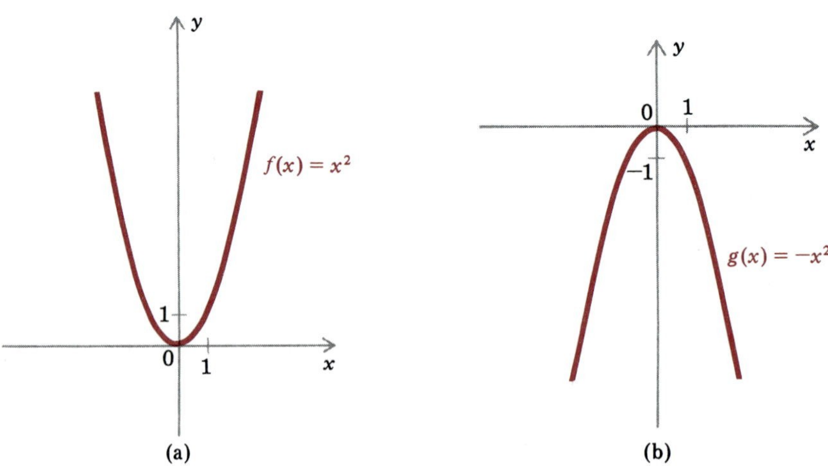

FIGURE 3.2

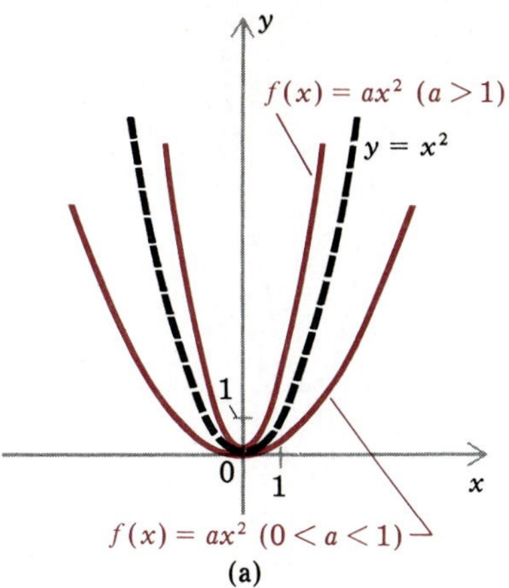

(a)

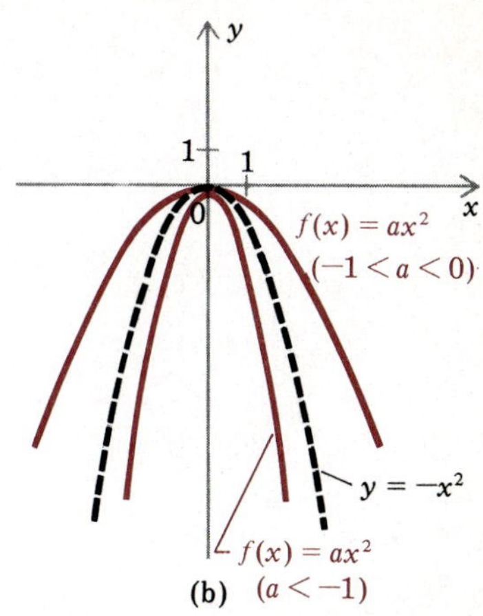

(b)

FIGURE 3.3

EXAMPLE 1. Sketch the graph of each of the following functions.

a. $f(x) = 2x^2$ b. $g(x) = -\dfrac{1}{3}x^2$

Solution.

a. The y coordinate of each point on the graph of f is twice the y coordinate of the corresponding point on the graph of x^2 (see Figure 3.2a). This leads us to the sketch in Figure 3.4a, with a few points plotted for assistance.

b. The y coordinate of each point on the graph of g is $\frac{1}{3}$ the y coordinate of the corresponding point on the graph of $-x^2$ (see Figure 3.2b). This leads us to the sketch in Figure 3.4b, again with a few points plotted for assistance. □

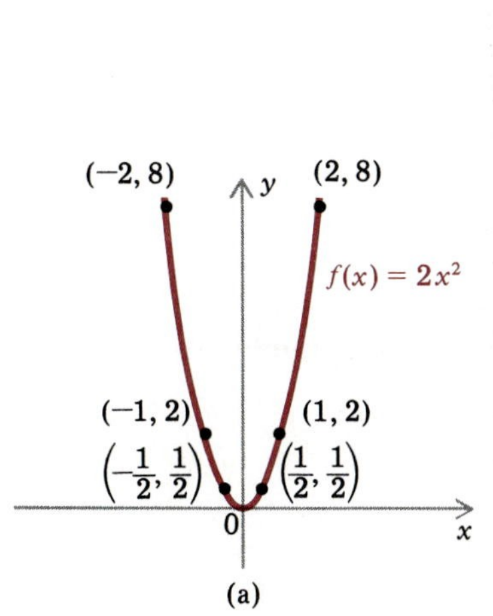

(a)

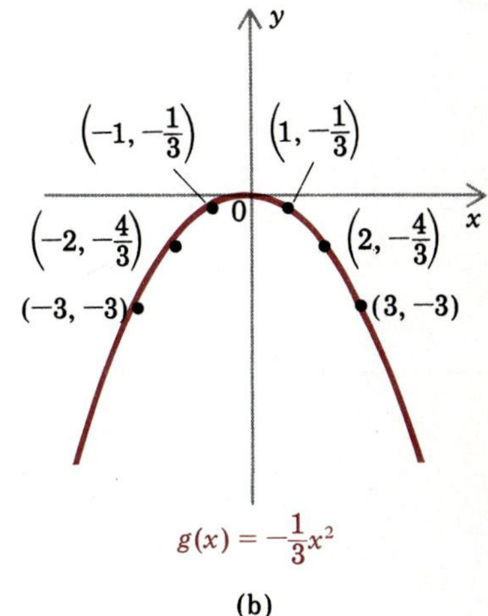

(b)

FIGURE 3.4

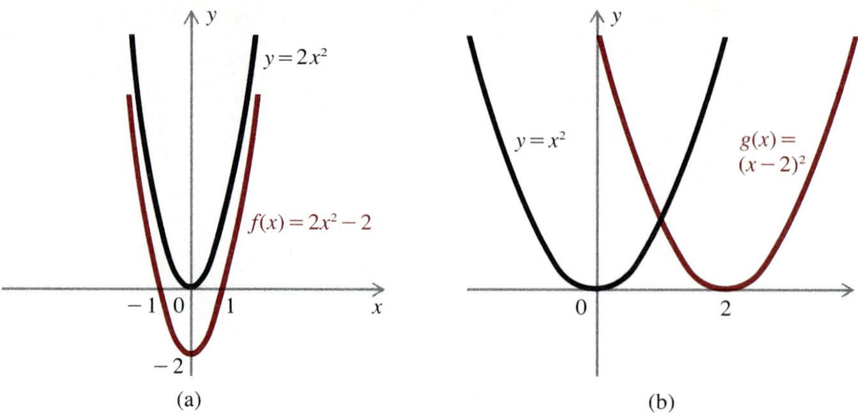

FIGURE 3.5

Recall from Section 2.5 that if we know the graph of a function f, then the graph of $y = f(x) + c$ is obtained by shifting the graph of f vertically (c units upward if $c > 0$ and $|c|$ units downward if $c < 0$). Similarly, if $g(x) = f(x - c)$, then the graph of g is obtained by shifting the graph of f horizontally (c units to the right if $c > 0$ and $|c|$ units to the left if $c < 0$). Thus if

$$f(x) = 2x^2 - 2 \quad \text{and} \quad g(x) = (x - 2)^2$$

then the graphs of f and g are as in Figures 3.5a and b.

The Graph of a General Quadratic Function

Suppose that we have an arbitrary quadratic function, given by

$$f(x) = ax^2 + bx + c \quad \text{with} \quad a \neq 0 \tag{1}$$

In order to sketch the graph of f, we complete the square in $ax^2 + bx + c$, as we did in Section 1.2, and obtain

$$f(x) = a\left(x + \frac{b}{2a}\right)^2 + c - \frac{b^2}{4a} \tag{2}$$

At first glance it seems no easier to draw the graph of f from (2) than from (1). However, let us simplify the right side of (2) by letting

$$h = -\frac{b}{2a} \quad \text{and} \quad k = c - \frac{b^2}{4a} \tag{3}$$

Then (2) becomes

$$f(x) = a(x - h)^2 + k$$

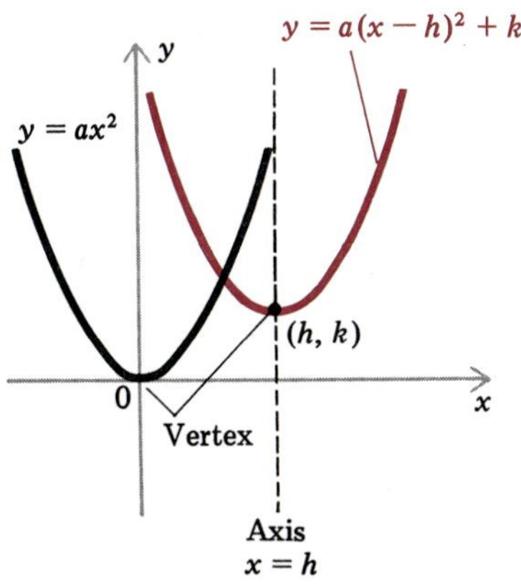

FIGURE 3.6

The graph of $a(x-h)^2 + k$ is obtained from the graph of ax^2 by a vertical shift of $|k|$ units (upward if $k > 0$ and downward if $k < 0$) and a horizontal shift of $|h|$ units (to the right if $h > 0$ and to the left if $h < 0$) (Figure 3.6). For example, if

$$f(x) = -\frac{1}{3}(x+4)^2 + 2$$

then the graph of f can be obtained by shifting the graph of $-\frac{1}{3}x^2$ upward 2 units and to the left 4 units.

When we shift from the graph of ax^2 to the graph of $a(x-h)^2 + k$, the vertex (0, 0) of ax^2 is shifted to the point (h, k) and the axis $x = 0$ is shifted to the line $x = h$ (Figure 3.6). Thus (h, k) is called the ***vertex*** of the graph of $a(x-h)^2 + k$, and the line $x = h$ is called the ***axis*** of the graph. The vertex is the lowest point on the graph if $a > 0$ and is the highest point if $a < 0$.

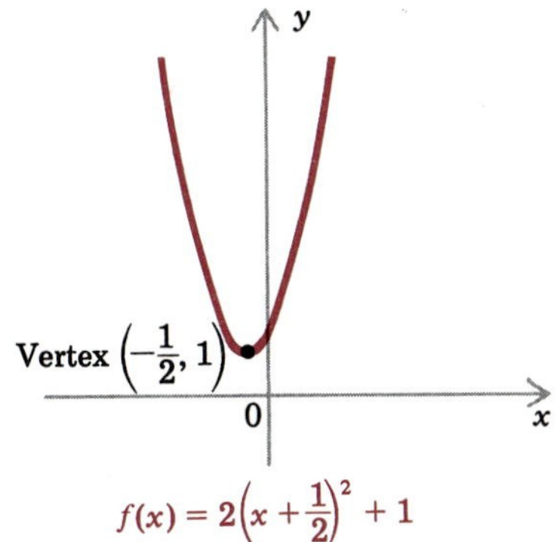

FIGURE 3.7

EXAMPLE 2. Let $f(x) = 2(x + \frac{1}{2})^2 + 1$. Sketch the graph of f, and locate the vertex of the parabola.

Solution. The graph of f can be obtained by shifting the graph of $2x^2$ upward 1 unit and to the left $\frac{1}{2}$ unit (Figure 3.7). The vertex is $(-\frac{1}{2}, 1)$. □

Let us return to the formula

$$f(x) = ax^2 + bx + c$$

and use the results of the preceding discussion to outline a procedure for sketching the graph of f. As is usually the case, it helps to plot a few points, so our outline includes finding the intercepts.

i. Find the y intercept by calculating $f(0)$. Plot the corresponding point.
ii. Find the x intercepts (if any) by solving $f(x) = 0$ for x. Plot the corresponding points (if any).
iii. Complete the square of the right side to express $f(x)$ in the form

$$f(x) = a(x-h)^2 + k \tag{4}$$

iv. From the new equation notice that the graph of f is a parabola whose vertex is (h, k) and whose axis is $x = h$. Plot the vertex, (h, k).
v. Obtain the graph of f by shifting the graph of ax^2 vertically $|k|$ units (upward if $k > 0$ and downward if $k < 0$) and horizontally $|h|$ units (to the right if $h > 0$ and to the left if $h < 0$). The graph will contain the points already plotted and will open upward if $a > 0$ and downward if $a < 0$.

From (3) it follows that the vertex and axis of the parabola are given (in terms of the original coefficients a, b, and c) by

$$\text{Vertex:} \quad \left(-\frac{b}{2a}, c - \frac{b^2}{4a}\right)$$

$$\text{Axis:} \quad x = -\frac{b}{2a}$$

But the vertex and axis are usually simpler to find after we have completed the square. As a result, when we set out to sketch the graph of a quadratic function, we usually first complete the square, and then sketch the graph by reading off the information we need from the expression for $f(x)$.

EXAMPLE 3. Let $f(x) = x^2 + 2x - 3$. Sketch the graph of f, and locate the vertex of the parabola.

Solution. Since $f(0) = -3$, the y intercept is -3. To find the x intercepts we solve the equation $f(x) = 0$ for x:

$$x^2 + 2x - 3 = 0$$

$$(x + 3)(x - 1) = 0$$

Consequently

$$x = -3 \quad \text{or} \quad x = 1$$

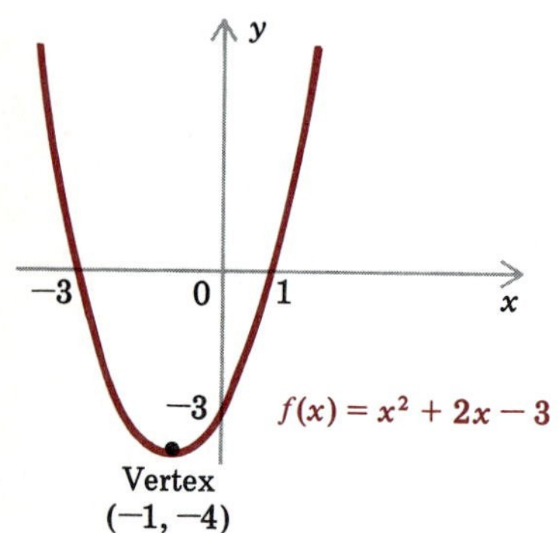

FIGURE 3.8

Therefore the x intercepts are -3 and 1. Now we complete the square:

$$\begin{aligned} f(x) &= (x^2 + 2x) - 3 \\ &= (x^2 + 2x + 1 - 1) - 3 \\ &= (x^2 + 2x + 1) - 4 \\ &= (x + 1)^2 - 4 \end{aligned}$$

In the notation of (4) we have $a = 1$, $h = -1$, and $k = -4$, so the vertex is $(-1, -4)$. Thus we obtain the graph of f by shifting the graph of x^2 downward 4 units and to the left 1 unit (Figure 3.8). □

EXAMPLE 4. Sketch the graph of

$$y = -3x^2 + 12x - 8 \tag{5}$$

and locate the vertex of the parabola.

Solution. Since $y = -8$ for $x = 0$, the y intercept is -8. The x intercepts are obtained by setting $y = 0$ and solving for x by the quadratic formula:

$$-3x^2 + 12x - 8 = 0$$

$$x = \frac{-12 \pm \sqrt{(12)^2 - 4(-3)(-8)}}{2(-3)} = \frac{-12 \pm 4\sqrt{3}}{-6} = 2 \pm \frac{2}{3}\sqrt{3}$$

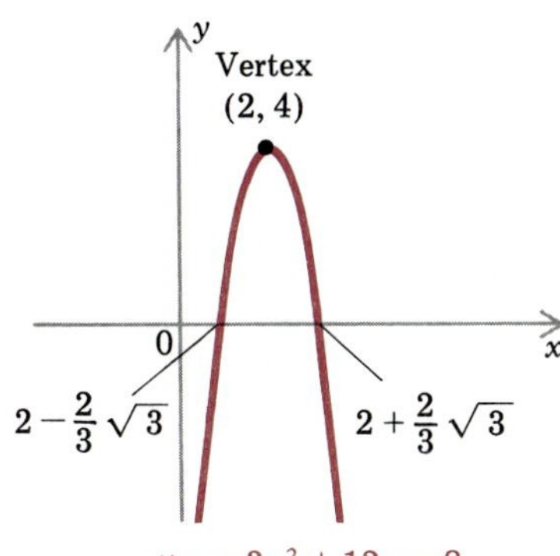

FIGURE 3.9

Therefore the x intercepts are $2 + \frac{2}{3}\sqrt{3}$ and $2 - \frac{2}{3}\sqrt{3}$. Completing the square in (5), we find that

$$\begin{aligned} y &= -3x^2 + 12x - 8 \\ &= -3(x^2 - 4x) - 8 \\ &= -3(x^2 - 4x + 4) + 12 - 8 \\ &= -3(x - 2)^2 + 4 \end{aligned}$$

Using the notation of (4), we have $a = -3$, $h = 2$, and $k = 4$, so the vertex is (2, 4). Thus we obtain the graph of the given equation by shifting the graph of $-3x^2$ upward 4 units and to the right 2 units (Figure 3.9). □

In Section 3.2 we will continue our study of quadratic functions. However, the main point of interest will be the largest, or smallest, value assumed by such a function.

EXERCISES 3.1

In Exercises 1–20, sketch the graph of the given function. In each case, determine the vertex and all intercepts of the parabola.

1. $f(x) = \frac{3}{4}x^2$
2. $f(x) = -3x^2$
3. $f(x) = x^2 + 1$
4. $f(x) = x^2 - \frac{9}{4}$
5. $f(x) = -2x^2 + 8$
6. $f(x) = 1 - 4x^2$
7. $f(x) = -\frac{3}{2}(x - 1)^2$
8. $f(x) = \frac{1}{2}(x + 3)^2$
9. $f(x) = \frac{3}{2}(x + \frac{1}{3})^2 - \frac{1}{3}$
10. $f(x) = 2(x - \frac{1}{2})^2 + \frac{3}{2}$
11. $g(x) = x^2 + 2x$
12. $g(x) = x^2 - 2x$
13. $g(x) = x^2 - 2x + 2$
14. $f(x) = x^2 + 2x + 2$
15. $f(x) = -x^2 + 6x - 5$
16. $f(x) = 2x^2 + 4x + 5$
17. $f(x) = 1 - 2x - 3x^2$
18. $y = \frac{1}{2}x^2 + x + 1$
19. $y = (x + 1)(2x - 1)$
20. $y = (3x + 1)(x - 1)$

In Exercises 21–24, find an equation of the sketched parabola. (*Hint:* Use (4).)

21. The parabola in Figure 3.10a.
22. The parabola in Figure 3.10b.
23. .The parabola in Figure 3.10c.
24. The parabola in Figure 3.10d.

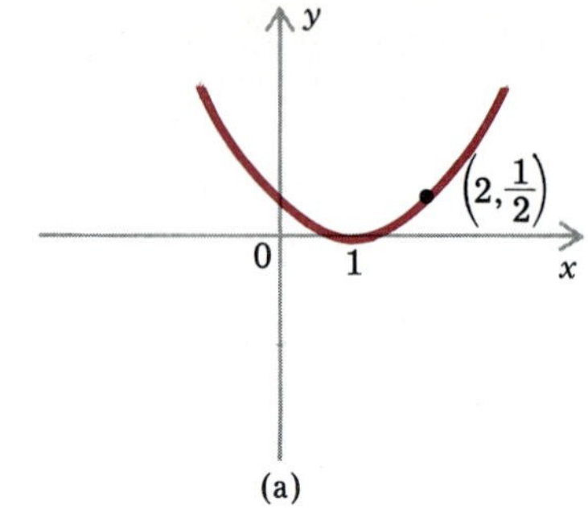

(a)

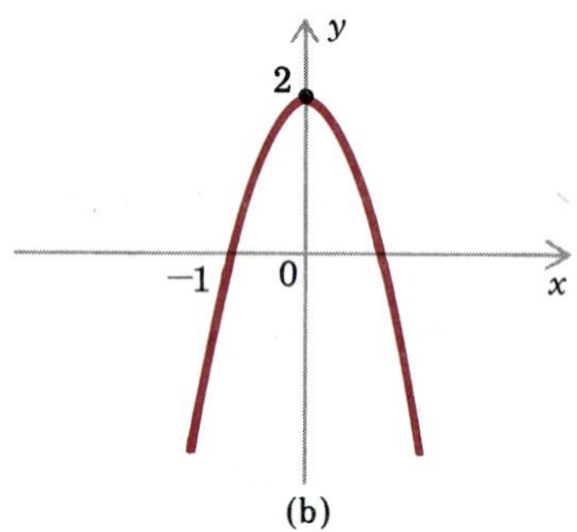

(b)

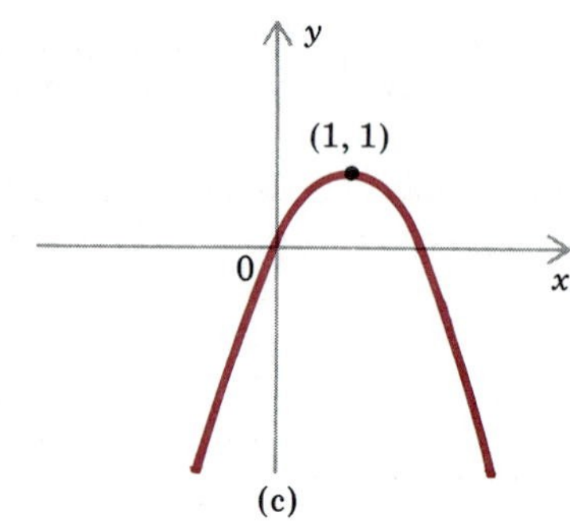

(c)

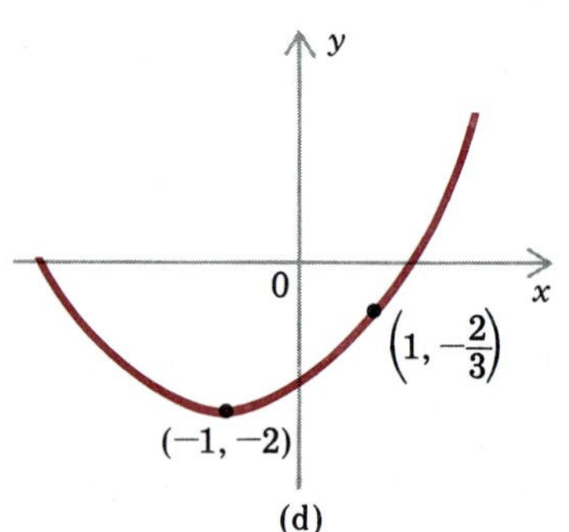

(d)

FIGURE 3.10

25. Let $f(x) = -2x^2 + 3$. Find an equation of the function g whose graph is obtained from the graph of f by
    a. shifting upward 1 unit
    b. shifting downward 3 units
    c. shifting to the right $\frac{1}{2}$ unit
    d. shifting to the left 4 units

26. Let $g(x) = -x^2 + 2x - 5$. Find an equation of the function f whose graph is obtained from the graph of g by
    a. shifting upward $\frac{3}{2}$ units and to the left 4 units.
    b. shifting downward 2 units and to the right $\frac{1}{2}$ unit

27. For what value of a does the graph of the equation $y = ax^2 + 6x + 3$ pass through the point $(-1, 0)$?

28. For what value of b does the graph of the equation $y = -\frac{1}{2}x^2 + bx + \frac{1}{3}$ pass through the point $(-2, -1)$?

29. For what value of c does the graph of the equation $y = -x^2 + x + c$ pass through the point $(2, 6)$?

30. In each part find the value of c for which the graph of f touches but does not cross the x axis.
    a. $f(x) = \frac{1}{2}x^2 - 2x + c$
    b. $f(x) = x^2 - 3x + c$
    c. $f(x) = 3x^2 + 10x + c$

31. Let $f(x) = ax^2 + c$. Prove that the graph of f has
    a. two x intercepts if a and c have opposite signs
    b. no x intercept if a and c have the same sign

32. Let $f(x) = 2x^2 - 8x + 3$. Determine the points on the graph of f that lie on the horizontal line 18 units above the vertex.

33. A parabolic arch is 16 feet tall at its highest point and spans 36 feet at its base. Find its height 9 feet from the plane on which its axis is located (Figure 3.11).

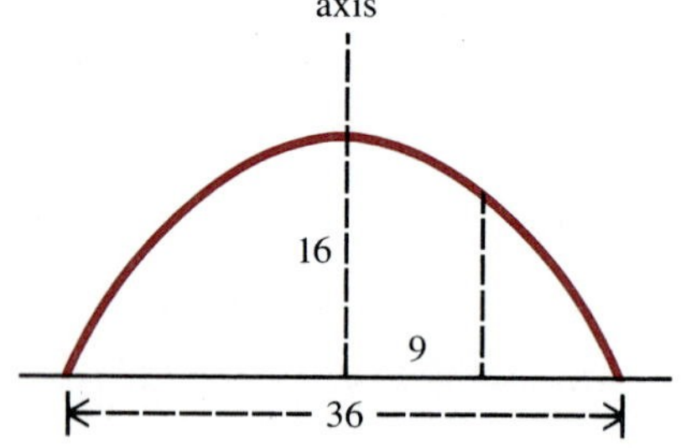

FIGURE 3.11

3.2 MAXIMUM AND MINIMUM VALUES OF QUADRATIC FUNCTIONS

From Section 3.1 we know that the graph of any quadratic function f is a parabola. If

$$f(x) = ax^2 + bx + c \quad \text{with} \quad a \neq 0 \tag{1}$$

then we also know that the vertex of the parabola is

$$\left(-\frac{b}{2a}, c - \frac{b^2}{4a}\right) \tag{2}$$

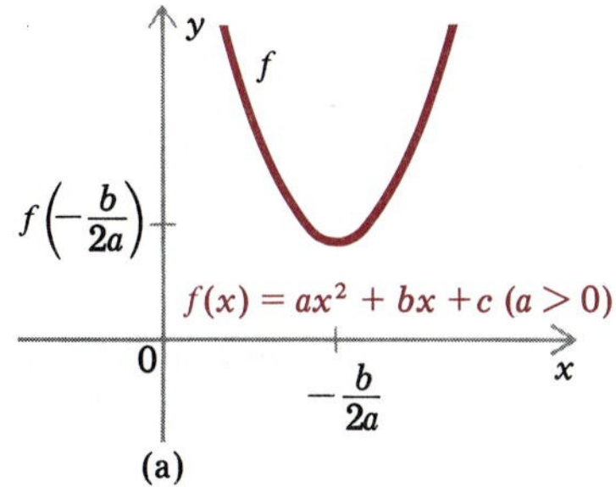

(a)

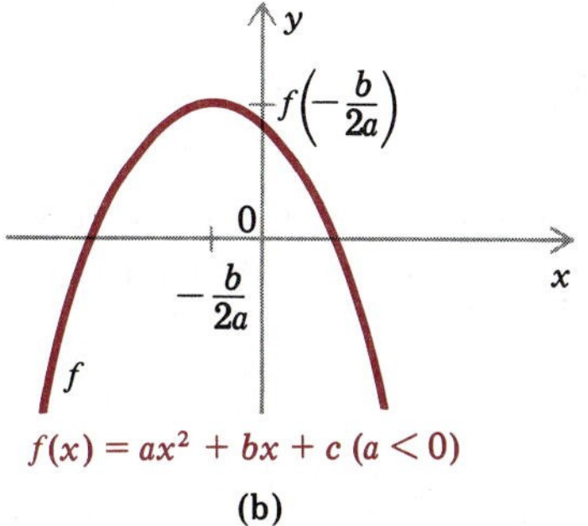

(b)

FIGURE 3.12

The vertex is the lowest point on the graph if $a > 0$ and is the highest point if $a < 0$. It follows that

$$\boxed{\text{if } a > 0, \quad \text{then} \quad f\left(-\frac{b}{2a}\right) \leq f(x) \quad \text{for all } x}$$

so $f(-b/2a)$ is called the ***minimum value*** of f (Figure 3.12a). Analogously,

$$\boxed{\text{if } a < 0, \quad \text{then} \quad f\left(-\frac{b}{2a}\right) \geq f(x) \quad \text{for all } x}$$

and $f(-b/2a)$ is the ***maximum value*** of f (Figure 3.12b). Thus every quadratic function has either a minimum or a maximum value, depending on the sign of the leading coefficient (a in (1)).

To compute the minimum or maximum value of such a function f, we can calculate $f(-b/2a)$, either by substituting $-b/2a$ for x in (1) or by calculating $c - b^2/4a$ (see (2)).

EXAMPLE 1. Let $f(x) = 2x^2 - 3x + 1$. Verify that f assumes a minimum value, and compute the minimum value.

Solution. In this example $f(x)$ has the form of (1) with $a = 2$, $b = -3$, and $c = 1$. Since $a > 0$, it follows that $f(-b/2a)$ is the minimum value. Now

$$-\frac{b}{2a} = -\frac{(-3)}{2(2)} = \frac{3}{4}$$

Consequently to find the minimum value of $f(x)$, we calculate $f(\frac{3}{4})$:

$$f\left(\frac{3}{4}\right) = 2\left(\frac{3}{4}\right)^2 - 3\left(\frac{3}{4}\right) + 1 = \frac{9}{8} - \frac{9}{4} + 1 = -\frac{1}{8}$$

Therefore the minimum value of f is $-\frac{1}{8}$. □

EXAMPLE 2. Let $f(x) = -x^2 + \sqrt{2}x + \frac{1}{2}$. Verify that f assumes a maximum value, and compute the maximum value.

Solution. Here $f(x)$ has the form of (1) with $a = -1$, $b = \sqrt{2}$, and $c = \frac{1}{2}$. Since $a < 0$, $f(-b/2a) = c - b^2/4a$ is the maximum value of f. Next we calculate $c - b^2/4a$:

$$c - \frac{b^2}{4a} = \frac{1}{2} - \frac{(\sqrt{2})^2}{4(-1)} = \frac{1}{2} - \left(-\frac{1}{2}\right) = 1$$

Consequently the maximum value of f is 1. □

Applications

In applications it is frequently desirable to maximize or minimize a variable quantity, such as area, profit, or cost. If the quantity to be maximized or minimized can be expressed as a quadratic function of another quantity, the discussion in this section provides us with a method for finding the maximum or minimum value.

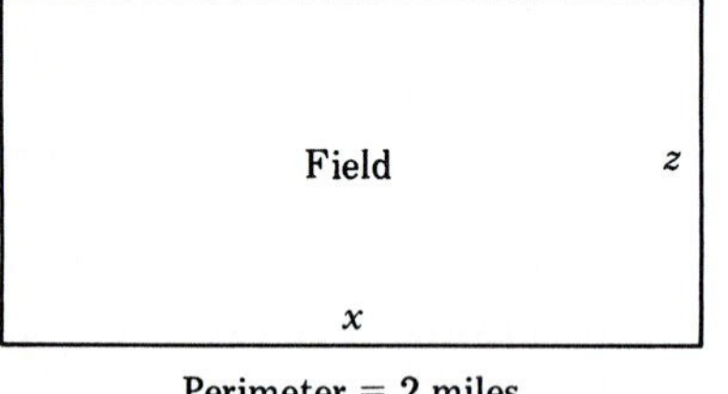

Perimeter = 2 miles

FIGURE 3.13

EXAMPLE 3. A rancher has 2 miles of fencing and wishes to fence in a rectangular grazing field with an area as large as possible. What should the dimensions of the field be?

Solution. Since the rancher has 2 miles of fencing, the perimeter of the rectangle will be 2 miles. Let x and z be the dimensions in miles of any rectangle with a perimeter of 2 miles (Figure 3.13). Then

$$2x + 2z = 2$$

so that

$$x + z = 1$$

or equivalently,

$$z = 1 - x$$

Since $x \geq 0$ and $z \geq 0$, it follows that $0 \leq x \leq 1$. Next we rewrite the area xz of the rectangle solely in terms of x:

$$xz = x(1 - x) = x - x^2 = -x^2 + x$$

If we let

$$f(x) = -x^2 + x \quad \text{for} \quad 0 \leq x \leq 1$$

then $f(x)$ represents the area of a field one of whose sides has length x. Thus we wish to maximize $f(x)$. Now $f(x)$ has the form of (1) with $a = -1$, $b = 1$, and $c = 0$. Since $a < 0$, we know that there is a maximum value and that the maximum value occurs for

$$x = -\frac{b}{2a} = -\frac{1}{2(-1)} = \frac{1}{2}$$

But if $x = \frac{1}{2}$, then

$$z = 1 - \frac{1}{2} = \frac{1}{2}$$

Thus the maximum area results if the rectangle is a square with sides of length $\frac{1}{2}$ mile. □

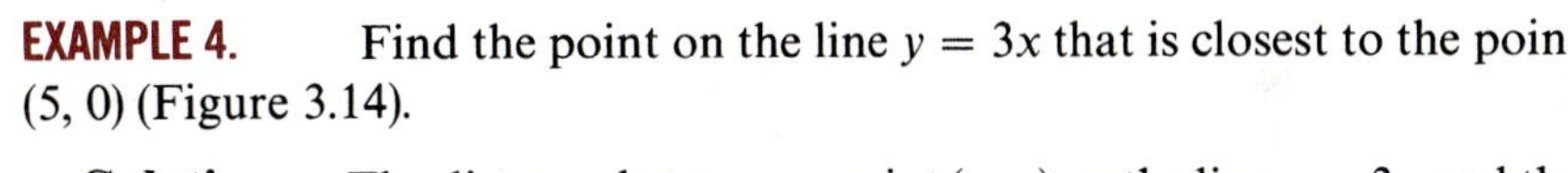

EXAMPLE 4. Find the point on the line $y = 3x$ that is closest to the point $(5, 0)$ (Figure 3.14).

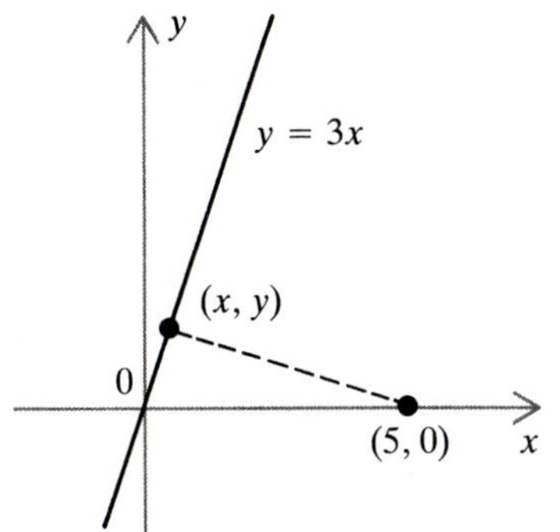

FIGURE 3.14

Solution. The distance between a point (x, y) on the line $y = 3x$ and the point $(5, 0)$ is given by

$$\sqrt{(x-5)^2 + (3x-0)^2}$$

so we let

$$\begin{aligned} f(x) &= \sqrt{(x-5)^2 + (3x-0)^2} \\ &= \sqrt{(x^2 - 10x + 25) + 9x^2} \\ &= \sqrt{10x^2 - 10x + 25} \end{aligned}$$

We desire to find the minimum value of $f(x)$. However, f is not a quadratic function. But if we let

$$g(x) = 10x^2 - 10x + 25$$

then since $g(x) = (f(x))^2$, the minimum value of f occurs for the same number x as the minimum value of g. Now g is a quadratic function, and $g(x)$ has the form of (1) with $a = 10$, $b = -10$, and $c = 25$. Since $a > 0$, there is a minimum value, which occurs for

$$x = -\frac{b}{2a} = -\frac{-10}{2(10)} = \frac{1}{2}$$

Therefore $y = 3(\frac{1}{2}) = \frac{3}{2}$, so the closest point is $(\frac{1}{2}, \frac{3}{2})$. □

Notice that in the preceding example, we did not need to find the minimum value of f, but only the number x at which the minimum value occurs.

EXAMPLE 5. On a rainy day the manager of a store estimates that 40 umbrellas can be sold at \$9 each, and that for each nickel decrease in price, one more umbrella can be sold (which corresponds to 20 more umbrellas sold for each dollar decrease in price). Assuming that this is true and that umbrellas cost the store \$3 each, determine the selling price that maximizes the profit. How many umbrellas will be sold at that price?

Solution. We will express all monetary quantities in dollars. Let x be the price of each umbrella and z the number of umbrellas sold. Then the number z of umbrellas sold at price x equals 40 (the number that would be sold at \$9) plus 20 times the amount (in dollars) that the price has been decreased from \$9. Thus

$$z = 40 + 20(\text{decrease from } 9) = 40 + 20(9 - x)$$

or equivalently,

$$z = 220 - 20x$$

Now the revenue R resulting from the sale of the z umbrellas at the price x is given by

$$R = xz = x(220 - 20x) = 220x - 20x^2$$

Since each umbrella costs \$3, the cost C of the z umbrellas sold is given by

$$C = 3z = 3(220 - 20x) = 660 - 60x$$

As a result, the profit P is given by

$$\begin{aligned} P = R - C &= (220x - 20x^2) - (660 - 60x) \\ &= -20x^2 + 280x - 660 \end{aligned}$$

Thus if we let

$$f(x) = -20x^2 + 280x - 660$$

then we wish to find the maximum value of f. Since $f(x)$ has the form of (1) with $a = -20$, $b = 280$, and $c = -660$, it follows that the maximum value of f occurs for

$$x = -\frac{b}{2a} = -\frac{280}{2(-20)} = 7$$

Consequently each umbrella should be priced at \$7 in order to maximize profit. If $x = 7$, then

$$z = 220 - 20(7) = 80$$

so 80 umbrellas will be sold at the optimal price of \$7. □

Problems concerning the motion of an object moving vertically under the sole influence of gravity involve quadratic functions. Recall from (1) in Section 1.3 that the height at time t of an object moving solely under the influence of gravity is given by

$$h(t) = -16t^2 + v_0 t + h_0 \tag{3}$$

where v_0 is the initial velocity and h_0 is the initial height. Since h is a quadratic function, if we know the initial velocity and initial height of an object, we can determine the maximum height.

EXAMPLE 6. A tape measure is thrown vertically upward from an initial height of 6 feet with an initial velocity of 32 feet per second toward a third floor balcony 21 feet above ground. Does the tape measure reach the balcony?

Solution. Taking $v_0 = 32$ and $h_0 = 6$ in (3), we have

$$h(t) = -16t^2 + 32t + 6$$

which has the form of (1) with $a = -16$, $b = 32$, and $c = 6$, and with t and h replacing x and f, respectively. Since $a < 0$, we know that h assumes a maximum value for

$$t = -\frac{b}{2a} = -\frac{32}{2(-16)} = 1$$

Thus the tape measure attains its maximum height after 1 second. The fact that

$$\begin{aligned} h(1) &= -16(1)^2 + 32(1) + 6 \\ &= -16 + 32 + 6 = 22 \end{aligned}$$

implies that the maximum height attained by the tape measure is 22 feet. Therefore the tape measure does reach the balcony. □

In solving maximum and minimum problems, the following approach is frequently effective:

i. After reading the problem carefully, choose a letter for the quantity to be maximized or minimized. Also choose auxiliary variables for the other quantities appearing in the problem.
ii. Express the quantity to be maximized or minimized in terms of the auxiliary variables.
iii. Choose one auxiliary variable x to serve as master variable, and use the information given in the problem to express all other auxiliary variables in terms of x.
iv. Use the results of steps (ii) and (iii) to express the quantity to be maximized or minimized in terms of x alone.
v. Use the theory of this section to find the desired maximum or minimum value.

EXERCISES 3.2

In Exercises 1–10, determine whether the given quadratic function assumes a maximum value or a minimum value. Then find the maximum or the minimum value.

1. $f(x) = 3x^2 + 5$

2. $f(x) = x^2 + x - \frac{1}{2}$

3. $f(x) = 20x^2 - 40x$

4. $f(x) = -x^2 + 2x + 3$

5. $g(x) = x - \frac{1}{4}x^2$

6. $g(x) = -5x^2 - x + 1$

7. $g(t) = 16t^2 + 64t + 36$

8. $g(t) = -\frac{1}{4}t^2 + t - 13$

9. $y = t^2 + 3t$

10. $y = \frac{1}{2}x^2 - 3x + \frac{3}{2}$

11. A rancher has 2 miles of fencing and wishes to fence in a rectangular grazing field with an area as large as possible by erecting a fence on only 3 sides, a river forming the other side (Figure 3.15). What should the dimensions of the rectangle be?

12. A farmer plans to fence in two adjacent rectangular grazing fields of the same dimensions, one for cows and the other for horses, with a common fence separating the two fields. If the farmer has 900 meters of fencing, how large can the area of each field be?

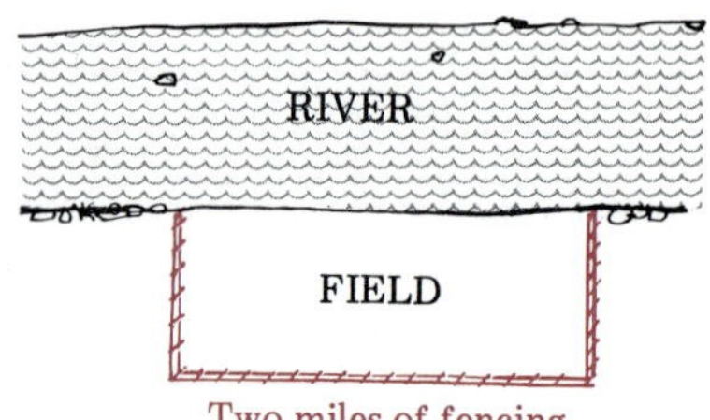

FIGURE 3.15

13. A piece of wire 1 meter long is to be cut into two pieces, each of which is then to be bent into the shape of a square (Figure 3.16). How should the wire be cut

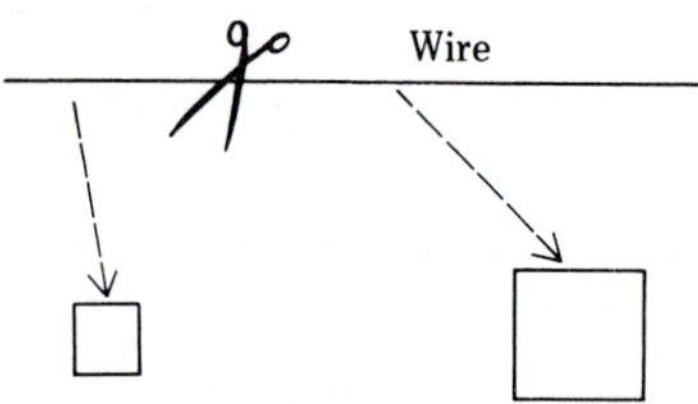

FIGURE 3.16

if the sum of the areas of the two squares is to be minimized?

14. Find the dimensions of the largest rectangle that can be inscribed in the triangle sketched in Figure 3.17.

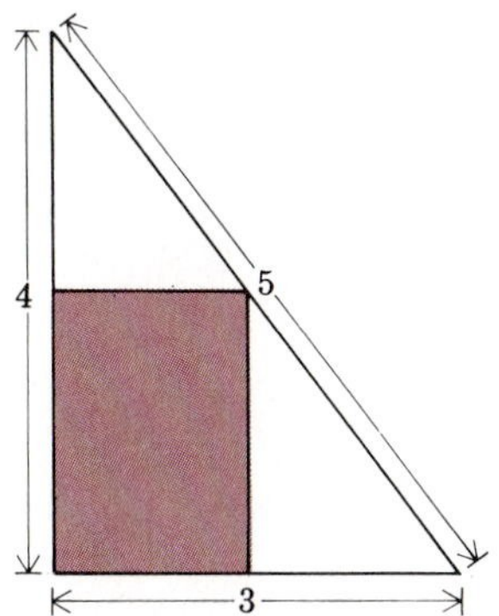

FIGURE 3.17

15. A rectangle is to have a perimeter of 12 inches. Determine the lengths of the sides that will yield the largest area.

16. The base of a given isosceles triangle is 12, and the height is 8. Find the dimensions of the rectangle with maximum area that can be inscribed in the triangle if the base of the rectangle is to lie along the base of the triangle (Figure 3.18).

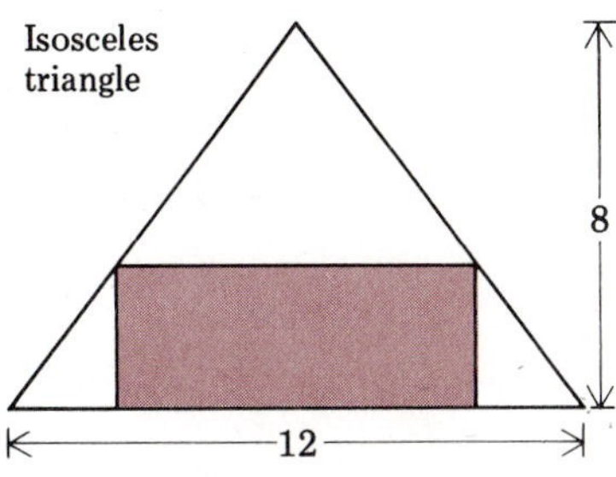

FIGURE 3.18

17. Find the dimensions of the rectangle whose perimeter is 16 and whose diagonal is as small as possible.

18. Find two numbers whose sum is 24 and whose product is as large as possible.

19. Find two numbers whose sum is -16 and whose product is as large as possible.

20. Find two numbers whose difference is 6 and whose product is as small as possible.

21. Find two numbers whose product is as large as possible, under the condition that the sum of one of the numbers and 3 times the other is 48.

22. Find two numbers whose sum is 36 and the sum of whose squares is as small as possible. Then determine the sum of the two squares.

23. Find the point on the line $2x + y = 2$ that is closest to the point $(0, -3)$.

24. Find the point on the line $y = 10 - 3x$ that is closest to the origin.

25. Find the points on the parabola $y = x^2$ that are closest to the point $(0, 1)$. (*Hint:* The square of the distance is a quadratic function of x^2. First find the value of x^2 for which the distance is minimal.)

26. The Hot-Cake Company can sell 100 pounds of pancake mix a day at \$1.80 a pound, and it believes that sales will decrease $\frac{1}{5}$ pound for each penny increase in price per pound. For what price would the revenue be maximized? What will the corresponding revenue be?

27. The Fuzzy-Wuzzy Fabric Shop can sell 36 yards of denim per day if it charges \$2 per yard, and it figures that each increase of \$.25 per yard will reduce sales of denim by 3 yards. Determine the price per yard that will maximize the daily revenue from the sale of denim.

28. A construction company estimates that it can build and sell x houses in a certain development if it sets the price of each home at $100{,}000 - 1000x$ dollars. If each house costs \$60,000 to construct, how many houses must the company sell to maximize its profit? How much must the company charge for each house in order to maximize its profit?

29. Suppose a baseball is thrown vertically upward from a height of 64 feet with an initial velocity of 48 feet per second. Find the maximum height attained by the baseball and the length of time it takes the ball to attain that height.

3.3 GRAPHS OF HIGHER-DEGREE POLYNOMIAL FUNCTIONS

In Section 3.1 we discussed the graph of a quadratic function. The graph is a parabola, and with a minimum of calculation we can determine its vertex, intercepts, and general shape. Now we turn to higher-degree polynomial functions. The higher the degree, the less accurate our graph is apt to be. Nevertheless, there are techniques that allow us to sketch the general shape of many such functions.

First we will analyze the graphs of polynomials of the form ax^n, and then we will study the graphs of other higher-degree polynomial functions.

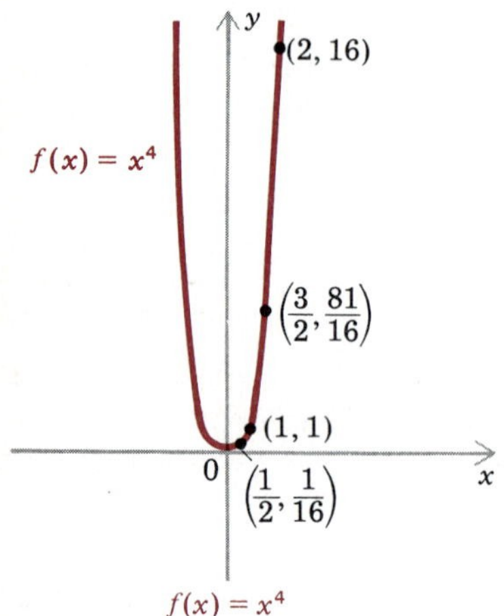

FIGURE 3.19

EXAMPLE 1. Let $f(x) = x^4$. Sketch the graph of f.

Solution. Since

$$f(-x) = (-x)^4 = (-1)^4x^4 = x^4 = f(x)$$

the graph of f is symmetric with respect to the y axis. We plot a few points on the graph, using the following table:

x	0	$\frac{1}{2}$	1	$\frac{3}{2}$
$f(x)$	0	$\frac{1}{16}$	1	$\frac{81}{16}$

Next we draw a smooth curve through the points we have plotted, and finally we use the symmetry of the graph to obtain the sketch in Figure 3.19. □

The shapes of the graphs of x^2 and x^4 appear generally similar (Figure 3.20), and in fact the shapes of the graphs of x^2 and x^n are generally similar for any *positive even* integer n. More particularly, they are symmetric with respect to the y axis (since $(-x)^n = x^n$ if n is even), have their minimum value of 0 at 0, and lie above the x axis (because $x^n \geq 0$ if n is even).

A similar pattern emerges with respect to the graph of x^3 (see Figure 3.21) and the graph of x^n for any odd integer $n \geq 3$. Here the graphs are

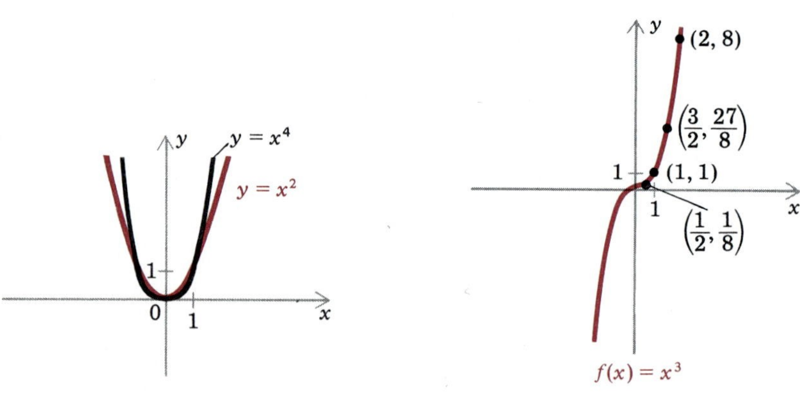

FIGURE 3.20

FIGURE 3.21

symmetric with respect to the origin (since $(-x)^n = -x^n$ if n is odd) and have neither maximum nor minimum values.

If a is a nonzero constant, the graphs of ax^n and x^n are related to each other in the same way as the graphs of ax^2 and x^2 are. The following example illustrates this fact.

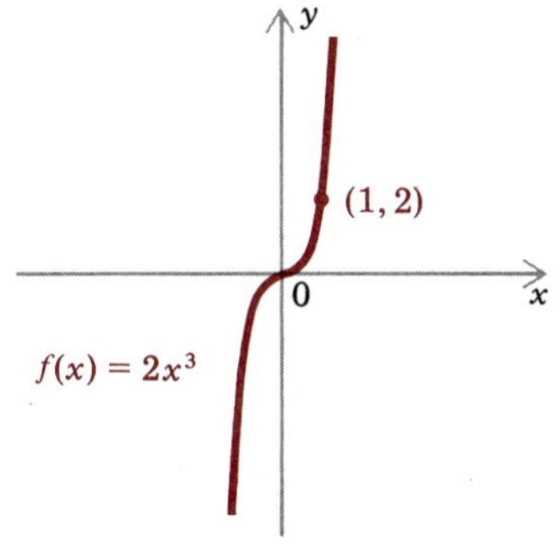

FIGURE 3.22

EXAMPLE 2. Sketch the graphs of the following functions.

a. $f(x) = 2x^3$ b. $g(x) = -\frac{1}{3}x^4$

Solution.

a. Since we already have the graph of x^3 in Figure 3.21, to obtain the graph of f we need only multiply the y coordinate of each point on the graph of x^3 by 2, which has the effect of "pulling" the graph of x^3 away from the x axis by a factor of 2 (Figure 3.22).
b. The graph of x^4 is shown in Figure 3.19. To obtain the graph of g, just multiply the y coordinate of each point on the graph of x^4 by $-\frac{1}{3}$, which has the effect of reflecting the graph of x^4 through the x axis and "pulling" it toward the x axis by a factor of 3 (Figure 3.23). □

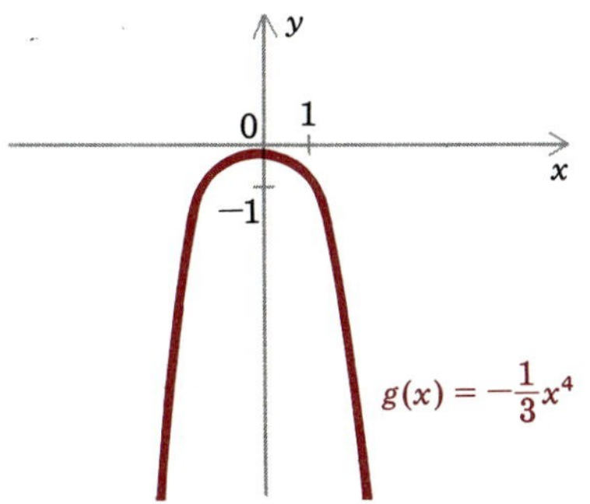

FIGURE 3.23

Sometimes the graph of a polynomial can be obtained by a vertical and/or horizontal shift of the graph of a polynomial of the form ax^n.

EXAMPLE 3. Let $f(x) = (x-2)^3 + 1$. Sketch the graph of f.

Solution. The graph of f can be obtained by shifting the graph of x^3 upward 1 unit and to the right 2 units (Figure 3.24). □

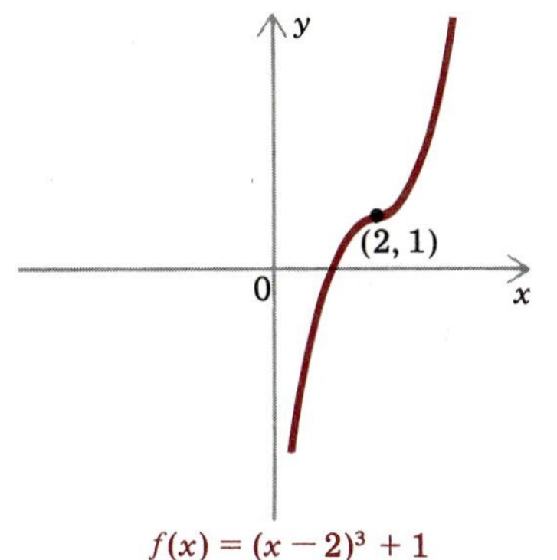

FIGURE 3.24

Now we turn to graphs of other polynomial functions of degree greater than 2. We could plot as many points on the graph of such a function as we like by calculating the values of the function for various numbers in the domain, but this would be tedious and, moreover, would not indicate any possible relationship between the graphs of different polynomial functions. In the discussion that follows we will concentrate instead on such features as how a graph looks far away from the y axis and where it is above and where below the x axis.

Let

$$f(x) = a_n x^n + a_{n-1}x^{n-1} + a_{n-2}x^{n-2} + \cdots + a_1 x + a_0 \quad \text{with} \quad a_n \neq 0 \quad (1)$$

Far away from the y axis the graph of f has the general shape of the graph of the first term, $a_n x^n$. The reason is that when x is large, $a_n x^n$ dominates $a_{n-1}x^{n-1} + a_{n-2}x^{n-2} + \cdots + a_1 x + a_0$. Thus, far away from the y axis the graphs of $x^3 - x$ and x^3 have the same general shape, and likewise, far away from the y axis the graphs of $2x^4 - 8x^3 + 8x^2$ and $2x^4$ have the same general shape.

To sketch the portions of the graph of the function in (1) nearer the y axis, we are helped enormously if we can factor $f(x)$ as

$$f(x) = a_n(x - c_1)(x - c_2) \cdots (x - c_n) \tag{2}$$

The numbers $c_1, c_2, \ldots, c_n$ appearing in (2) are the x intercepts of f, because $f(x) = 0$ if and only if

$$a_n(x - c_1)(x - c_2) \cdots (x - c_n) = 0$$

which happens if and only if any one of the factors on the left is 0. In other words, $f(x) = 0$ if and only if x is one of the numbers $c_1, c_2, \ldots, c_n$.

Some polynomial functions are easily factored by the methods of Section 0.6. To illustrate, we let

$$f(x) = x^3 - x$$

Factoring, we find that

$$x^3 - x = x(x^2 - 1) = x(x - 1)(x + 1)$$

If we write x as $x - 0$, then

$$f(x) = (x - 0)(x - 1)(x + 1)$$

It follows that the x intercepts of f are 0, 1, and -1, so the graph of f intersects the x axis at $(0, 0)$, $(1, 0)$, and $(-1, 0)$.

Knowing the x intercepts allows us to determine the intervals on which $f(x)$ is positive and those on which $f(x)$ is negative. We see how this is done in the following example.

EXAMPLE 4. Let $f(x) = x^3 - x$. Sketch the graph of f.

Solution. First of all, the y intercept is 0, since $f(0) = 0$. Next,

$$f(x) = x(x - 1)(x + 1)$$

so the x intercepts are 0, 1, and -1 (as we said above). Now the x intercepts divide the x axis into the subintervals $(-\infty, -1)$, $(-1, 0)$, $(0, 1)$, and $(1, \infty)$. To determine whether $f(x)$ is positive or negative on each of these subintervals, we study the factors x, $x - 1$, and $x + 1$ individually on each subinterval. Figure 3.25 gives the results.

In constructing Figure 3.25 we used the fact that $x - c < 0$ if $x < c$, and $x - c > 0$ if $x > c$. We also used the fact that $f(x) > 0$ if an even number of

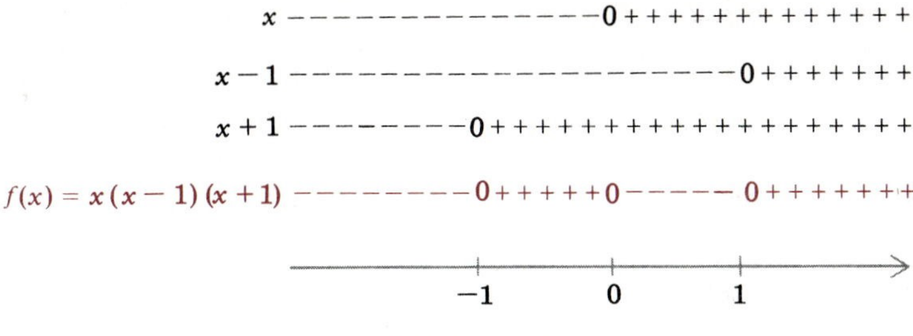

FIGURE 3.25

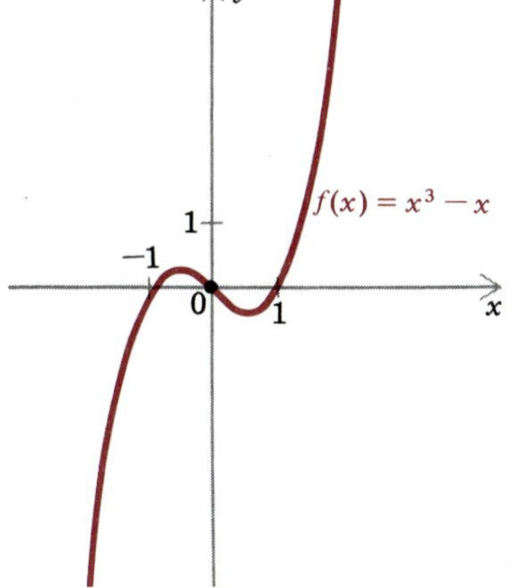

FIGURE 3.26

the factors of $f(x)$ are negative and the rest are positive, and $f(x) < 0$ if an odd number of the factors are negative and the rest are positive. At this point we know the x intercepts and the intervals on which $f(x)$ is positive or negative. Moreover, because

$$f(-x) = (-x)^3 - (-x) = -x^3 + x = -(x^3 - x) = -f(x)$$

the graph is symmetric with respect to the origin. Finally, far away from the y axis the graph has the general shape of the graph of x^3 (see Figure 3.21). Consequently we can draw a smooth curve for the graph of f (Figure 3.26). □

The fact that the "hilltop" to the left of the y axis in Figure 3.26 occurs for $x = -\sqrt{3}/3$, and the "valley bottom" to the right of the y axis occurs for $x = \sqrt{3}/3$, can be proved by methods of calculus. Even without this specific information we were able to give a reasonably accurate sketch of the graph of f.

Notice that the graph of f in Example 4 *crosses* the x axis at the points corresponding to the x intercepts. Example 5 shows that the graph of a function may touch, but not cross, the x axis at a point corresponding to an x intercept.

EXAMPLE 5. Let $g(x) = x^3 + 2x^2 + x$. Sketch the graph of g.

Solution. Since $g(0) = 0$, the y intercept is 0. Next we factor $g(x)$:

$$g(x) = x^3 + 2x^2 + x = x(x^2 + 2x + 1) = x(x + 1)^2$$

Therefore the x intercepts are 0 and -1, and we prepare the chart shown in Figure 3.27, which gives the signs of x and $(x + 1)^2$, as well as the resulting signs of $g(x)$, on the intervals $(-\infty, -1)$, $(-1, 0)$, and $(0, \infty)$:

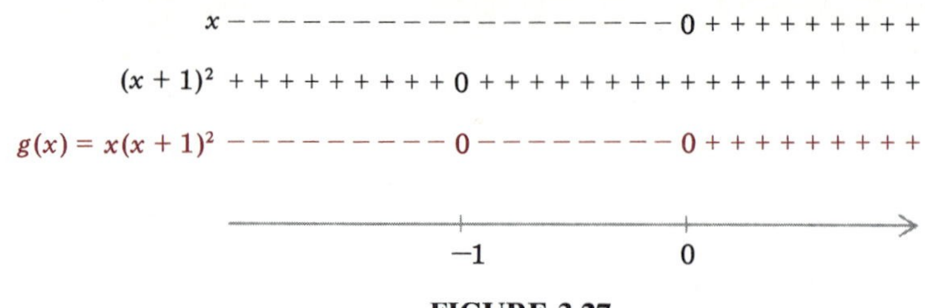

FIGURE 3.27

Using the information in Figure 3.27, along with the fact that far away from the y axis the graph of g has the same general shape as the graph of x^3, we sketch the graph of g in Figure 3.28. □

Through these examples we have developed a procedure that facilitates drawing a rough sketch of the graph of any polynomial function f that we can factor:

i. Find the y intercept, $f(0)$.
ii. Factor $f(x)$ to obtain a formula in the form of (2).
iii. Obtain the x intercepts from the factorization in (ii), and note the intervals into which the x intercepts divide the real line.
iv. Prepare a chart that tells on which of the intervals in (iii) the value $f(x)$ is positive and on which negative.
v. Determine the general shape of the graph far away from the y axis (the same as the shape of the graph of $a_n x^n$).
vi. Using the information obtained in (i)–(v), along with any symmetry of the graph, sketch the graph of f.

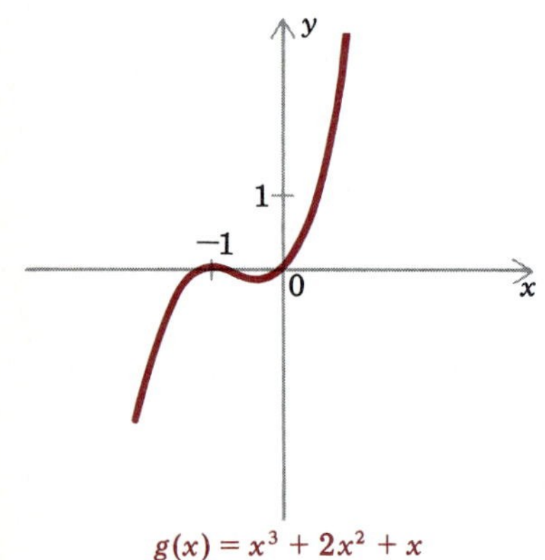

$g(x) = x^3 + 2x^2 + x$

FIGURE 3.28

Notice in Figure 3.28 that the graph of g *touches*, but does not *cross*, the x axis at -1. The reason is that in the factorization of g, $x + 1$ appears *twice*, that is, $(x + 1)^2$ appears in the factorization. Because $(x + 1)^2 \geq 0$ for all x, $g(x)$ does not change sign at -1. In general, if c is an x intercept of a polynomial function f and if $x - c$ appears an even number of times in the factorization of f, then $f(x)$ does not change sign at c and the graph of f merely touches the x axis at c, whereas if $x - c$ appears an odd number of times, then $f(x)$ changes sign at c and the graph of f crosses the x axis at c. For example, if

$$f(x) = 2x^4 - 8x^3 + 8x^2 \tag{3}$$

then

$$f(x) = 2x^2(x^2 - 4x + 4) = 2x^2(x - 2)^2 \tag{4}$$

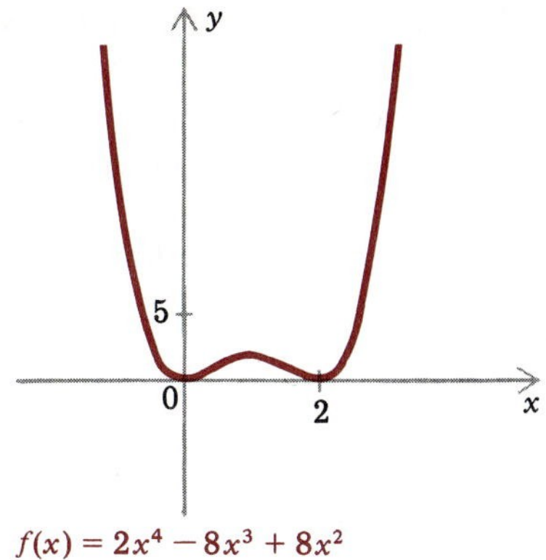

$f(x) = 2x^4 - 8x^3 + 8x^2$

FIGURE 3.29

Therefore the x intercepts are 0 and 2. Since both x and $x - 2$ appear twice in the factorization (4) of f, it follows that $f(x) \geq 0$ for all x, and hence the graph of f does not cross the x axis. The graph of f is shown in Figure 3.29.

Graphs of Nonfactored Polynomials

The polynomials appearing in Examples 4 and 5 and in (3) are easy to factor. However, it may not be so easy to find a factor $x - c$ of a given polynomial $f(x) = a_n x^n + a_{n-1}x^{n-1} + \cdots + a_1 x + a_0$. Nevertheless, if all the coefficients of $f(x)$ are integers, and if c is also an integer, then by a result to be proved in Section 10.5, $x - c$ can be a factor of $f(x)$ only if c divides the constant term a_0 of $f(x)$. Thus if $f(x) = x^3 + x - 2$ and if c is an integer, then $x - c$ can be a factor of $f(x)$ only if c divides -2, that is, $c = 1, -1, 2$ or -2. Checking which, if any, of these values of c might actually yield a factor can be facilitated by the Factor Theorem, which we mentioned in Section 0.6 and will prove in Section 10.2:

THEOREM 3.1
The Factor Theorem

Let $f(x)$ be any polynomial. Then $x - c$ is a factor of $f(x)$ if and only if $f(c) = 0$.

In the examples and exercises that follow, $f(x)$ will have integer coefficients, and we will assume that there are factors of the form $x - c$, where c is an integer.

EXAMPLE 6. Let $f(x) = x^3 + x - 2$. Find a factor of $f(x)$.

Solution. We look for a factor of the form $x - c$, where c is an integer. By the comments preceding the example, the only possible values of c are 1, -1, 2, and -2. Now

$$f(1) = 1^3 + 1 - 2 = 0 \qquad f(-1) = (-1)^3 + (-1) - 2 = -4$$

$$f(2) = 2^3 + 2 - 2 = 8 \qquad f(-2) = (-2)^3 + (-2) - 2 = -12$$

Since $f(1) = 0$, the Factor Theorem implies that $x - 1$ is a factor of $f(x)$. □

Having discovered one factor of a polynomial function $f(x)$, we can write $f(x)$ in the form

$$f(x) = (x - c)(\text{polynomial function})$$

by performing long division. For example, since we showed in Example 6 that $x - 1$ is a factor of $x^3 + x - 2$, we can divide as follows:

$$\begin{array}{rl}
 & x^2 + x + 2 \\
x - 1 & \overline{)\, x^3 \qquad\;\; + x - 2} \\
 & \underline{x^3 - x^2} \\
 & \qquad x^2 + x \\
 & \qquad \underline{x^2 - x} \\
 & \qquad\qquad 2x - 2 \\
 & \qquad\qquad \underline{2x - 2} \\
 & \qquad\qquad\qquad 0
\end{array}$$

Therefore $f(x) = (x - 1)(x^2 + x + 2)$.

Notice that the polynomial $x^2 + x + 2$ cannot be factored as the product of two first degree polynomials with real coefficients because the quadratic formula yields real solutions only if $(1)^2 - 4(1)(2) \geq 0$; however, $(1)^2 - 4(1)(2) < 0$. It follows that when we factor $x^3 + x - 2$ we obtain $(x - 1)(x^2 + x - 2)$ as our final factorization.

In general, suppose we wish to factor a polynomial $f(x)$ as far as possible. If we find a first factor $x - c$ of $f(x)$, then $f(x)$ can be written in the form

$$f(x) = (x - c)g(x), \text{ where } g(x) \text{ is a polynomial with real coefficients}$$

Frequently it is possible to use the Factor Theorem again to find a factor $x - d$ of $g(x)$, so that

$$g(x) = (x - d)h(x), \text{ where } h(x) \text{ is a polynomial with real coefficients}$$

and thus

$$f(x) = (x - c)g(x) = (x - c)(x - d)h(x)$$

This process may be repeated until no further factorization is discovered. Sometimes it is even possible to continue this process until $f(x)$ is factored "completely," in the sense that $f(x)$ is expressed as the product of first degree polynomials. For instance, in Example 4 we factored $x^3 - x$ completely: $x^3 - x = x(x - 1)(x + 1)$.

EXAMPLE 7. Let $f(x) = 3x^3 + 7x^2 - 4$. Factor $f(x)$ completely, and then sketch the graph of f.

Solution. Any integer roots of $f(x)$ must be among 1, -1, 2, -2, 4, and -4. We find that

$$\begin{aligned}
f(1) &= 3(1)^3 + 7(1)^2 - 4 = 6 \neq 0 \\
f(-1) &= 3(-1)^3 + 7(-1)^2 - 4 = 0 \\
f(2) &= 3(2)^3 + 7(2)^2 - 4 = 48 \neq 0 \\
f(-2) &= 3(-2)^3 + 7(-2)^2 - 4 = 0 \\
f(4) &= 3(4)^3 + 7(4)^2 - 4 = 300 \neq 0 \\
f(-4) &= 3(-4)^3 + 7(-4)^2 - 4 = -84 \neq 0
\end{aligned}$$

Since $f(-1) = 0$ and $f(-2) = 0$, it follows that $x + 1$ and $x + 2$ are factors of $f(x)$. Next we divide $3x^3 + 7x^2 - 4$ by $x + 1$:

$$\begin{array}{r|l}
 & 3x^2 + 4x - 4 \\
\hline
x + 1 & 3x^3 + 7x^2 \qquad - 4 \\
 & 3x^3 + 3x^2 \\
\hline
 & \qquad 4x^2 \\
 & \qquad 4x^2 + 4x \\
\hline
 & \qquad\quad - 4x - 4 \\
 & \qquad\quad - 4x - 4 \\
\hline
 & \qquad\qquad 0
\end{array}$$

Thus

$$f(x) = (x + 1)(3x^2 + 4x - 4)$$

Now $x + 2$ is a factor of $f(x)$ and hence is a factor of $3x^2 + 4x - 4$, so we divide $3x^2 + 4x - 4$ by $x + 2$:

$$\begin{array}{r|l}
 & 3x - 2 \\
\hline
x + 2 & 3x^2 + 4x - 4 \\
 & 3x^2 + 6x \\
\hline
 & \quad -2x - 4 \\
 & \quad -2x - 4 \\
\hline
 & \qquad 0
\end{array}$$

Finally, we have the complete factorization:

$$f(x) = (x + 1)(x + 2)(3x - 2) = 3(x + 1)(x + 2)\left(x - \frac{2}{3}\right)$$

Consequently the x intercepts are -1, -2, and $\frac{2}{3}$. Next we prepare the chart in Figure 3.30 that tells where $x + 1$, $x + 2$, and $3x - 2$ are positive and where they are negative:

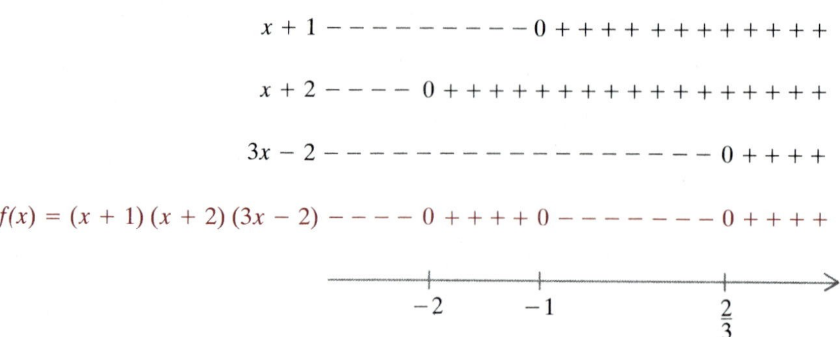

FIGURE 3.30

Using this information, we can sketch the graph of f (Figure 3.31). □

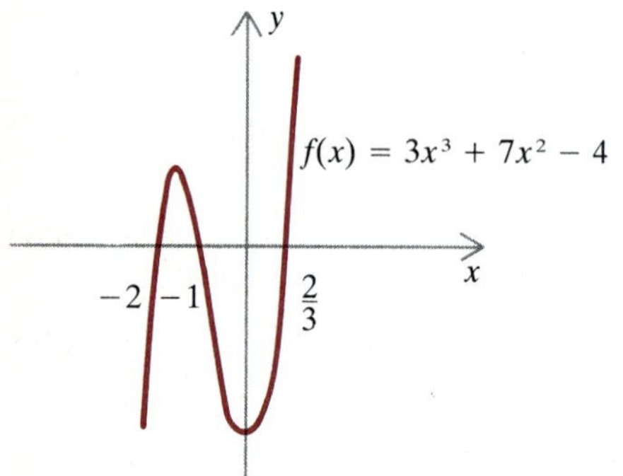

FIGURE 3.31

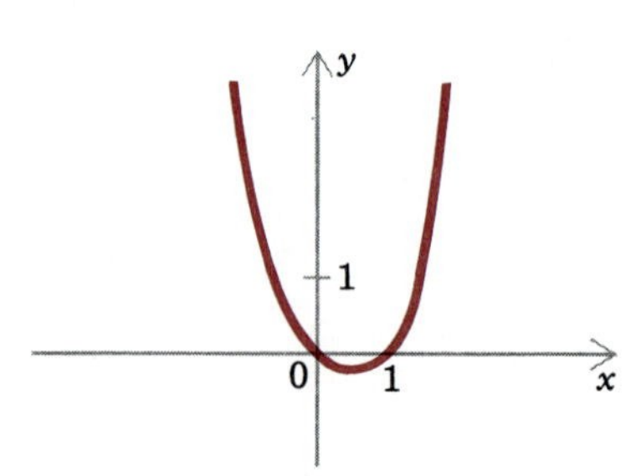

A plausible graph of $x^4 - x^3$

FIGURE 3.32

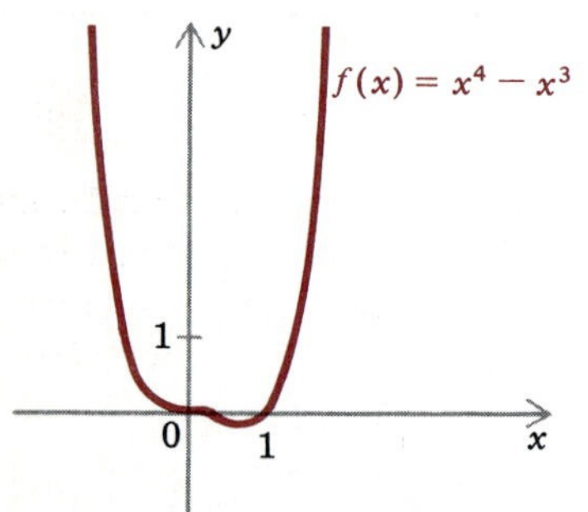

FIGURE 3.33

CAUTION: The methods of this section do not give precise, detailed information about the local shape of a graph. For example, if

$$f(x) = x^4 - x^3$$

then

$$f(x) = x^3(x - 1)$$

so by using the procedure outlined above we would probably draw the graph appearing in Figure 3.32. A more precise sketch of the graph appears in Figure 3.33. Such refinements in the graph can be obtained by methods of calculus, or by calculating a large number of values of f.

EXERCISES 3.3

In Exercises 1–24, sketch the graph of the given function, and indicate all intercepts.

1. $f(x) = x^3 + 1$

2. $f(x) = 1 - x^4$

3. $f(x) = \frac{1}{4} + 2x^3$

4. $f(x) = 4 - \frac{1}{4}x^4$

5. $f(x) = -2(x + 1)^3$

6. $g(x) = \frac{1}{8}(x - 2)^4 - 2$

7. $g(x) = x^5 + 1$

8. $g(x) = -\frac{1}{81}(x - 3)^6$

9. $g(x) = (x + 1)(x - 1)(x - 2)$

10. $f(x) = x(x + 1)^3$

11. $f(x) = -x^2(x + 1)(x + 2)$

12. $f(x) = 2x^2(x - 4)$

13. $f(x) = x^2(x - 1)^2$

14. $f(x) = x^3(x - 1)^2$

15. $g(x) = (x^2 - 1)(x + 1)^2$

16. $g(x) = (x^2 - 3x + 2)(x^2 - 4)$

17. $g(x) = x^3 - x^2 - 2x$

18. $g(x) = x^3 - 4x^2 + 4x$

19. $f(x) = x^4 - x^2$

20. $f(x) = -x^5 - x^4 + 6x^3$

21. $f(x) = x^5 - x^4$

22. $y = x^5 - x^3$

23. $y = x^4 - 5x^2 + 4$

24. $y = -x^4 + 2x^3 + x^2 - 2x$

In Exercises 25–30, verify that the linear polynomial is a factor of the other polynomial.

25. $x^3 + 2x + 3; x + 1$

26. $x^3 - 2x^2 - 9; x - 3$

27. $x^3 - 9x^2 + 14x + 24; x - 4$

28. $x^3 - 6x^2 + 4x + 8; x - 2$

29. $x^4 + x^3 - 6x^2 - 4x + 8; x + 2$

30. $2x^3 + 5x^2 + 8x + 3; x + \frac{1}{2}$

In Exercises 31–36, factor $f(x)$ completely, and then sketch the graph of f.

31. $f(x) = x^3 + 4x^2 + x - 6$

32. $f(x) = 2x^3 + 5x^2 + x - 2$

33. $f(x) = 3x^3 - 4x^2 - 5x + 2$

34. $f(x) = x^3 + 5x^2 + 3x - 9$

35. $f(x) = 2x^3 + 3x^2 - 5x - 6$

36. $f(x) = -4x^3 + 8x^2 - 5x + 1$

37. Find the number c such that the graph of the equation $y = x^4 - x^2 + 3x + c$ passes through (1, 5).

38. Find the number c such that the graph of the equation $y = 3x^3 - x^2 + 2x + c$ has y intercept 7.

39. Let $f(x) = ax^3 - 2x^2 + x - 6$. If -1 is an x intercept of the graph of f, find the value of a.

40. Let $f(x) = x^3 + bx^2 + 3x + 9$. If 3 is an x intercept of the graph of f, find the value of b.

3.4 GRAPHS OF RATIONAL FUNCTIONS

Now that we have studied graphs of polynomial functions at some length, we turn to graphs of quotients of polynomials. As we mentioned in the introduction to Chapter 3, the quotient of two polynomials is called a rational function. In other words, if P and Q are polynomials, then the function f defined by

$$f(x) = \frac{P(x)}{Q(x)}$$

is a rational function. The domain of f consists of all real numbers x such that $Q(x) \neq 0$ (since we cannot divide by 0). Ironically, those numbers that are *not* in the domain of a rational function f frequently play an essential role in sketching the graph of f. Let us illustrate this with the graph of $1/x$.

EXAMPLE 1. Let $f(x) = 1/x$. Sketch the graph of f.

Solution. The domain of f consists of all real numbers x except 0. There are no x or y intercepts, as we showed in Example 14 of Section 2.5. Moreover,

$$f(-x) = \frac{1}{-x} = -\frac{1}{x} = -f(x)$$

and thus the graph of f is symmetric with respect to the origin. Next we construct tables of values of $f(x)$ for small and for large positive values of x:

x	1	$\frac{1}{10}$	$\frac{1}{100}$	$\frac{1}{1000}$
$f(x)$	1	10	100	1000

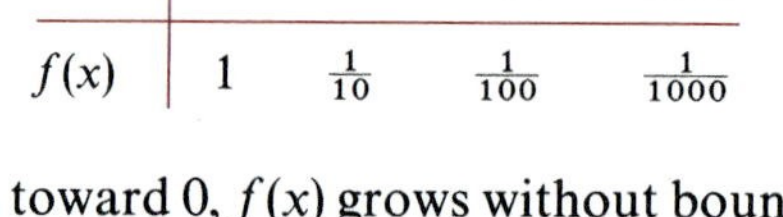

	1	10	100	1000
$f(x)$	1	$\frac{1}{10}$	$\frac{1}{100}$	$\frac{1}{1000}$

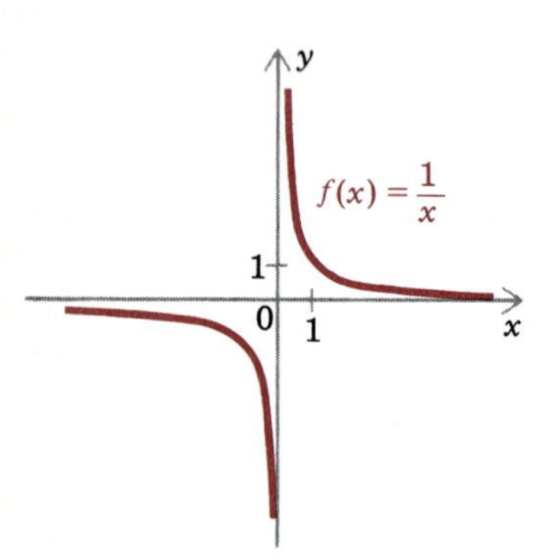

FIGURE 3.34

The first table suggest that as x shrinks toward 0, $f(x)$ grows without bound; accordingly, the part of the graph of f corresponding to small positive values of x is far from the x axis. From the second table we conclude that as x grows without bound, $f(x)$ shrinks toward 0; thus the part of the graph of f corresponding to large positive values of x is close to the x axis. Combining this information and then using the symmetry with respect to the origin, we obtain the graph of f shown in Figure 3.34. □

Because $1/x$ becomes arbitrarily large if x is positive and approaches 0 (that is, the point $(x, 1/x)$ on the graph gets arbitrarily far from the x axis as x approaches 0 from the right), the vertical line $x = 0$ is called a vertical asymptote of the graph. In general, we say that a vertical line $x = a$ is a ***vertical asymptote*** of the graph of the function g if either (or both) of the following conditions hold:

1. As x approaches a from the left or from the right (or both), $g(x)$ is positive and becomes arbitrarily large (Figure 3.35a, b).
2. As x approaches a from the left or from the right (or both), $g(x)$ is negative and becomes arbitrarily large in absolute value (Figure 3.35c, d).

The graph in Figure 3.35e satisfies both condition 1 and condition 2.

It can be proved that if P and Q are polynomial functions such that $P(a) \neq 0$ and $Q(a) = 0$, then the line $x = a$ is a vertical asymptote of the graph of the rational function P/Q. Thus both the rational functions

$$\frac{x}{(x-2)^3} \quad \text{and} \quad \frac{x^3 + 2x - 1}{(x-2)(x-4)}$$

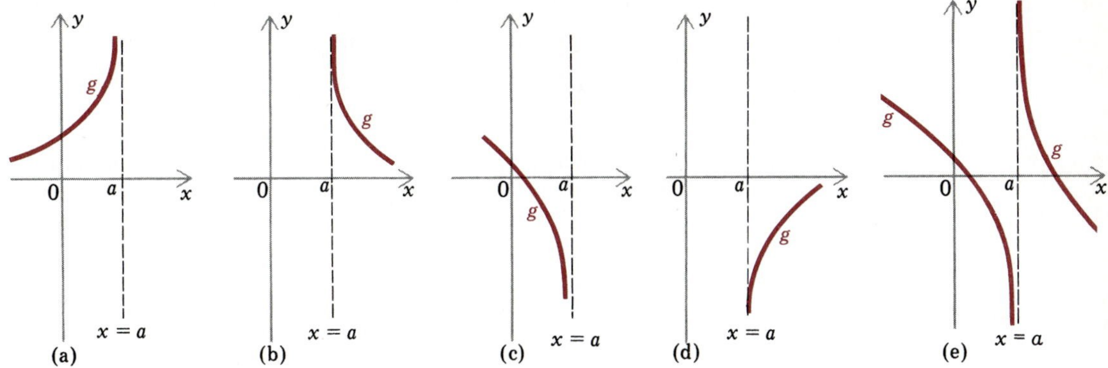

The line $x = a$ is a vertical asymptote of the graph of g.

FIGURE 3.35

have the same vertical asymptote $x = 2$, and the second function also has the vertical asymptote $x = 4$.

Recall that $1/x$ approaches 0 if x is positive and becomes arbitrarily large (so that the part of the graph far away from the y axis is close to the line $y = 0$). We say that the horizontal line $y = 0$ is a horizontal asymptote of the graph. In general, we say that a horizontal line $y = c$ is ***horizontal asymptote*** of the graph of a function g if either (or both) of the following conditions hold:

3. If x is positive and becomes arbitrarily large, $g(x)$ gets close to c (Figure 3.36).

4. If x negative and becomes arbitrarily large in absolute value, $g(x)$ gets close to c (Figure 3.37).

If $g(x) = x^2/(x^2 + 1)$ and $c = 1$, then g satisfies both condition 3 and condition 4, and hence the line $y = 1$ is a horizontal asymptote by either condition (Figure 3.38).

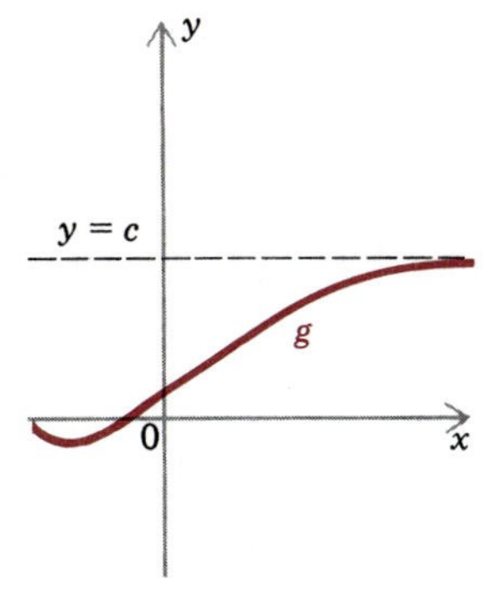

FIGURE 3.36

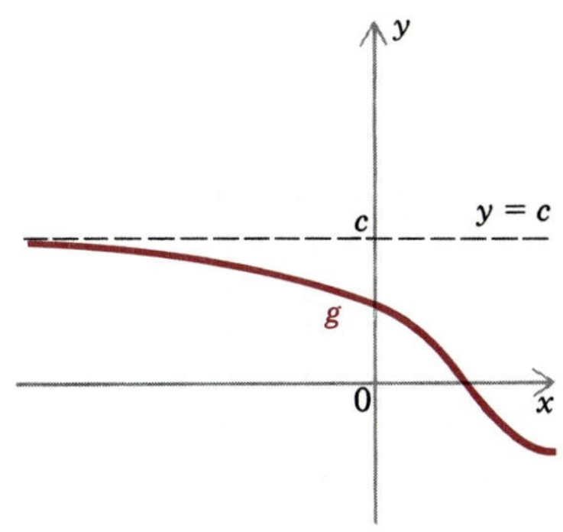

FIGURE 3.37

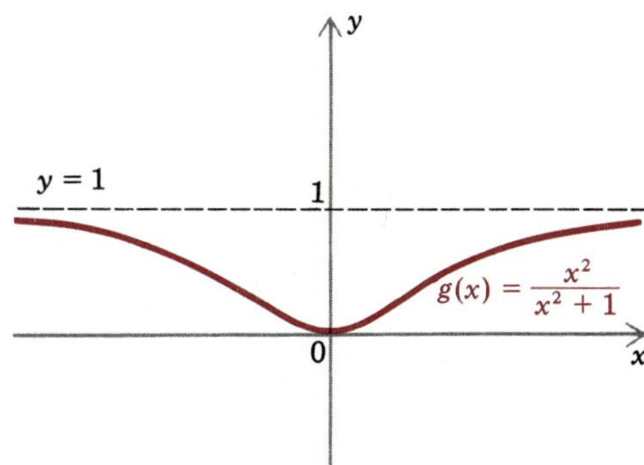

FIGURE 3.38

Asymptotes play a major role in graphing most rational functions that are not polynomials, as we will presently see.

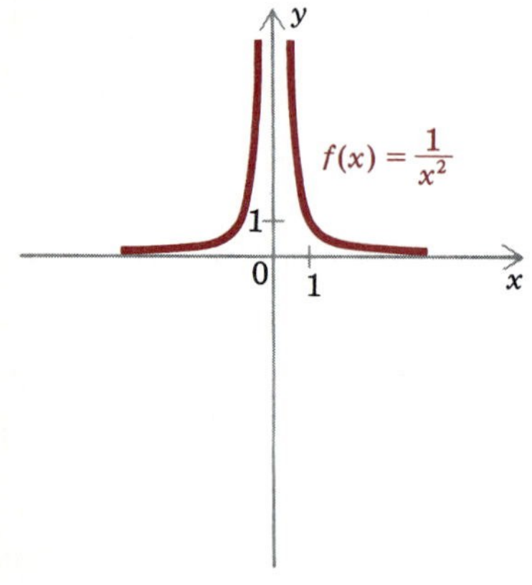

FIGURE 3.39

EXAMPLE 2. Let $f(x) = 1/x^2$. Sketch the graph of f.

Solution. The domain consists of all $x \neq 0$, and $1/x^2$ is never 0. Consequently there are no x or y intercepts. Next, observe that

$$f(-x) = \frac{1}{(-x)^2} = \frac{1}{x^2} = f(x)$$

so the graph of f is symmetric with respect to the y axis. Notice that if x is positive and gets close to 0, so does x^2, and consequently $1/x^2$ becomes very large. Therefore the vertical line $x = 0$ is a vertical asymptote of the graph. Similarly, if x is positive and becomes arbitrarily large, so does x^2, and thus $1/x^2$ approaches 0. Consequently the horizontal line $y = 0$ is a horizontal asymptote of the graph. Combining this information and using the symmetry of the graph with respect to the y axis, we draw the graph of f in Figure 3.39. □

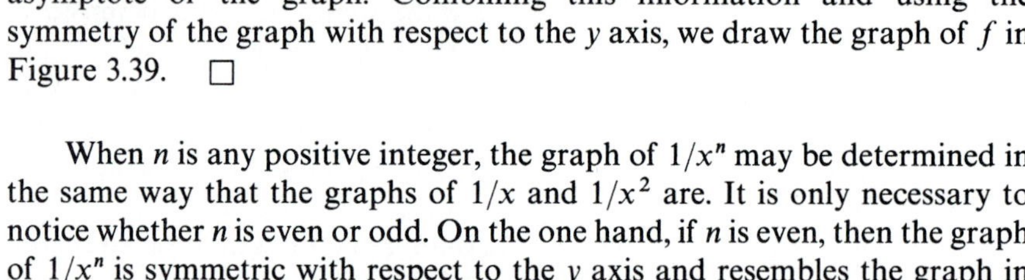

When n is any positive integer, the graph of $1/x^n$ may be determined in the same way that the graphs of $1/x$ and $1/x^2$ are. It is only necessary to notice whether n is even or odd. On the one hand, if n is even, then the graph of $1/x^n$ is symmetric with respect to the y axis and resembles the graph in Figure 3.40a. On the other hand, if n is odd, then the graph of $1/x^n$ is symmetric with respect to the origin and resembles the graph in Figure 3.40b.

Finally, if a is a nonzero constant, then the graph of $1/(x - a)^n$ has the same appearance as the graph of $1/x^n$ does, except that the vertical asymptote is $x = a$ (instead of $x = 0$), and the graph is shifted horizontally so as to be centered about the line $x = a$.

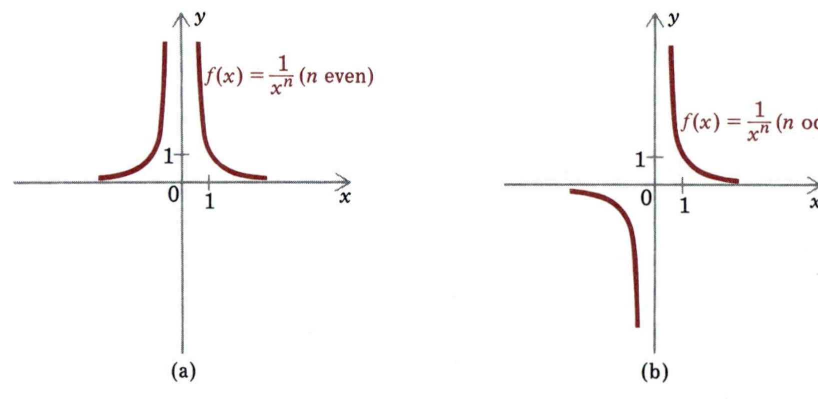

FIGURE 3.40

EXAMPLE 3. Sketch the graphs of the following functions, and identify their vertical asymptotes.

a. $f(x) = \dfrac{1}{(x-2)^3}$ b. $g(x) = \dfrac{1}{(x+4)^4}$

Solution.

a. The graph of f has the vertical asymptote $x = 2$, and the graph is as sketched in Figure 3.41, with y intercept $-\frac{1}{8}$, since

$$f(0) = \frac{1}{(-2)^3} = -\frac{1}{8}$$

b. Since $x + 4 = x - (-4)$, the vertical asymptote is $x = -4$. The graph is as sketched in Figure 3.42. The y intercept is $\frac{1}{256}$, since

$$g(0) = \frac{1}{4^4} = \frac{1}{256} \quad \square$$

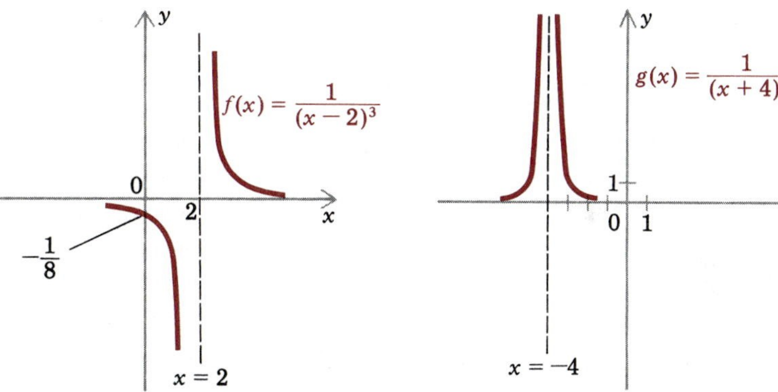

FIGURE 3.41

FIGURE 3.42

Graphs of Factored Rational Functions

If we can factor both numerator and denominator of a rational function, we can sometimes obtain a reasonably accurate sketch of the graph with the techniques already at hand. These include finding the domain and any intercepts, locating where the function is positive and where negative, and determining any asymptotes and symmetry of the graph.

EXAMPLE 4. Let $f(x) = \dfrac{1}{(x-2)(x+3)}$. Sketch the graph of f.

Solution. The domain consists of all numbers except 2 and -3. The y intercept is $-\frac{1}{6}$, since

$$f(0) = \frac{1}{(-2)(3)} = -\frac{1}{6}$$

The numerator of $f(x)$ cannot be 0, so that $f(x) \neq 0$ for all x in the domain. Therefore there are no x intercepts. The vertical asymptotes are determined from the values at which the denominator is 0, which are $x = 2$ and $x = -3$. Next, we prepare the chart in Figure 3.43 to determine the sign of $f(x)$ on the three intervals $(-\infty, -3)$, $(-3, 2)$, and $(2, \infty)$, into which -3 and 2 divide the x axis.

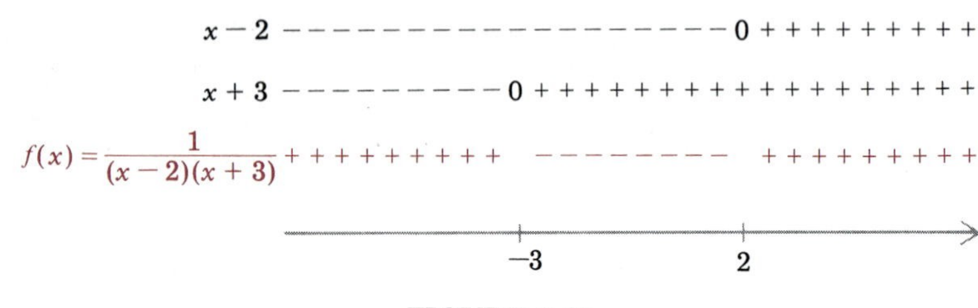

FIGURE 3.43

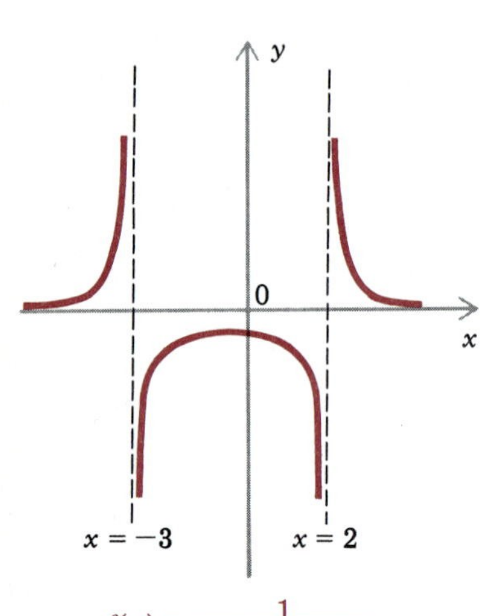

$f(x) = \dfrac{1}{(x-2)(x+3)}$

FIGURE 3.44

As our final preparation for sketching the graph, we observe that $f(x)$ approaches 0 if x is positive and becomes large, or if x is negative and becomes large in absolute value. This yields the horizontal asymptote $y = 0$. Because there is no symmetry, we are ready to use the information we have assembled and sketch the graph of f in Figure 3.44. □

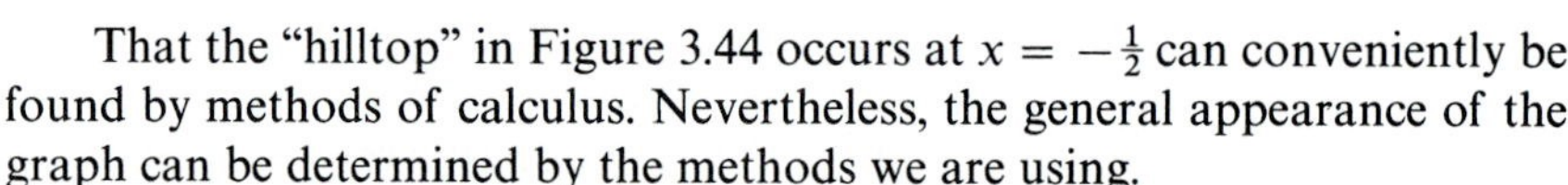

That the "hilltop" in Figure 3.44 occurs at $x = -\frac{1}{2}$ can conveniently be found by methods of calculus. Nevertheless, the general appearance of the graph can be determined by the methods we are using.

Now we will see when an arbitrary rational function P/Q has horizontal asymptotes. Let

$$P(x) = a_n x^n + a_{n-1}x^{n-1} + \cdots + a_1 x + a_0 \quad \text{with } a_n \neq 0$$

and

$$Q(x) = b_m x^m + b_{m-1}x^{m-1} + \cdots + b_1 x + b_0 \quad \text{with } b_m \neq 0$$

so that

$$\frac{P(x)}{Q(x)} = \frac{a_n x^n + a_{n-1}x^{n-1} + \cdots + a_1 x + a_0}{b_m x^m + b_{m-1}x^{m-1} + \cdots + b_1 x + b_0}$$

Then

i. If $n < m$, then the line $y = 0$ is a horizontal asymptote.
ii. If $n = m$, then the line $y = a_n/b_m$ is a horizontal asymptote.
iii. If $n > m$, then there is no horizontal asymptote. Instead, far away from the y axis the graph of P/Q has the general appearance of the graph of

$$\frac{a_n}{b_m}x^{n-m}$$

whose graph was described in Section 3.3.

In the next example, a different kind of asymptote will appear.

EXAMPLE 5. Let $f(x) = \dfrac{(1-x^2)(x+2)}{x^2+2x}$. Sketch the graph of f.

Solution. The denominator $x^2 + 2x$ can be rewritten as $x(x + 2)$, which implies that the domain of f consists of all numbers except 0 and -2. Since 0 is not in the domain, there is no y intercept. Now we factor the numerator and denominator:

$$f(x) = \frac{(1-x^2)(x+2)}{x^2+2x} = \frac{(1-x)(1+x)(x+2)}{x(x+2)}$$

Canceling the factor $x + 2$ in the numerator and denominator yields

$$f(x) = \frac{(1-x)(1+x)}{x} \quad \text{for} \quad x \neq 0, -2 \tag{1}$$

From (1) we see that -1 and 1 are x intercepts. Since the denominator is 0 for $x = 0$ but the numerator is not 0 for $x = 0$, the line $x = 0$ is a vertical asymptote. Using (1), we assemble the chart in Figure 3.45 that shows the

		-1		0		1	
$1-x$	+ + + + + +	+	+ + + + + + +	+	+ + + + + + +	0	− − − − − −
$1+x$	− − − − − −	0	+ + + + + + +	+	+ + + + + + +	+	+ + + + + +
x	− − − − − −	−	− − − − − − −	0	+ + + + + + +	+	+ + + + + +
$f(x) = \dfrac{(1-x)(1+x)}{x}$	+ + + + + +	0	− − − − − −		+ + + + + + +	0	− − − − − −

FIGURE 3.45

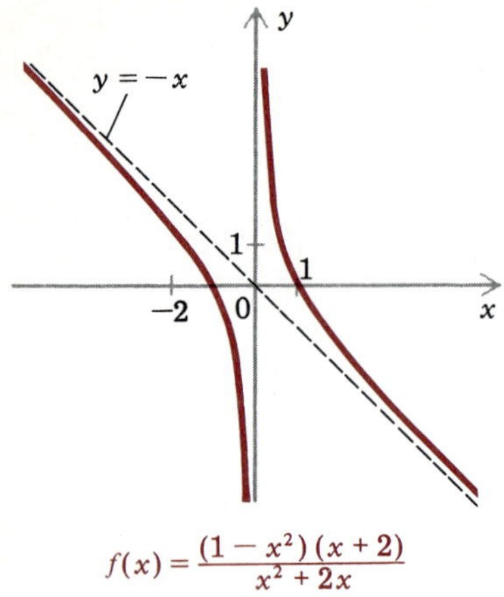

FIGURE 3.46

sign of $f(x)$ on each of the intervals $(-\infty, -1)$, $(-1, 0)$, $(0, 1)$ and $(1, \infty)$ determined by the vertical asymptote and the x intercepts.

By (1),

$$f(x) = \frac{(1-x)(1+x)}{x} = \frac{1-x^2}{x} = -x + \frac{1}{x} \tag{2}$$

so that far from the y axis, the graph of f resembles the graph of $-x$, and consequently there is no horizontal asymptote. Finally, from (2) we find that

$$f(-x) = \frac{1-(-x)^2}{-x} = -\frac{1-x^2}{x} = -f(x)$$

so that the graph of f is symmetric with respect to the origin. Now we are ready to draw the graph of f (Figure 3.46). □

The line $y = -x$ in Figure 3.46 is called the ***oblique asymptote*** of the graph of f, because far away from the y axis the graph of f approaches the line $y = -x$, which is neither horizontal nor vertical. The graph of any rational function for which the degree of the numerator is one more than the degree of the denominator has an oblique asymptote.

In our final example of this section we will sketch the graph of a rational function f that has the factor $x^2 + 1$ that itself cannot be factored. The polynomial $x^2 + 1$ is positive for every value of x, and the hence does not create any intercepts or vertical asymptotes in the graph and, in particular, does not affect the sign of $f(x)$.

EXAMPLE 6. Let $f(x) = \dfrac{x^2+1}{4x^2-1}$. Sketch the graph of f.

Solution. The y intercept is -1, because $f(0) = -1$. There are no x intercepts, because $x^2 + 1 > 0$ for all x in the domain. Since

$$f(x) = \frac{x^2+1}{4x^2-1} = \frac{x^2+1}{4(x^2-\frac{1}{4})} = \frac{x^2+1}{4(x+\frac{1}{2})(x-\frac{1}{2})}$$

it follows that the lines $x = -\frac{1}{2}$ and $x = \frac{1}{2}$ are vertical asymptotes (and that $-\frac{1}{2}$ and $\frac{1}{2}$ are not in the domain of f). The sign of $f(x)$ on each of the intervals $(-\infty, -\frac{1}{2})$, $(-\frac{1}{2}, \frac{1}{2})$, and $(\frac{1}{2}, \infty)$ determined by the vertical asymptotes is given in the chart in Figure 3.47.

We also find that f satisfies possibility (ii) above, so that since $\frac{1}{4}x^{2-2} = \frac{1}{4}$, it follows that $y = \frac{1}{4}$ is a horizontal asymptote. Finally, $f(-x) = f(x)$, as you can easily verify, so the graph of f is symmetric with respect to the y axis. With all the assembled information we can draw the graph of f (Figure 3.48). □

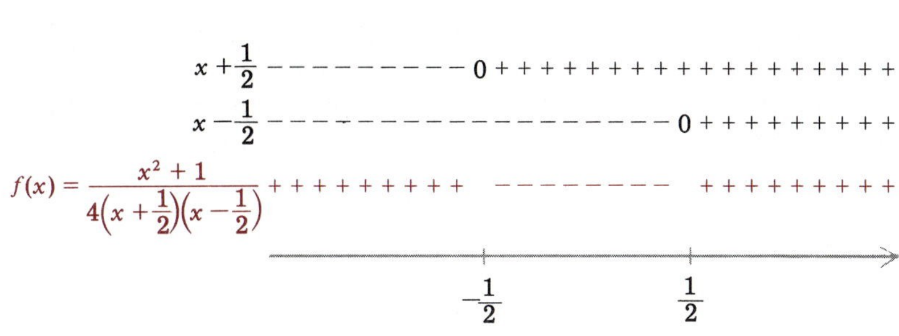

FIGURE 3.47

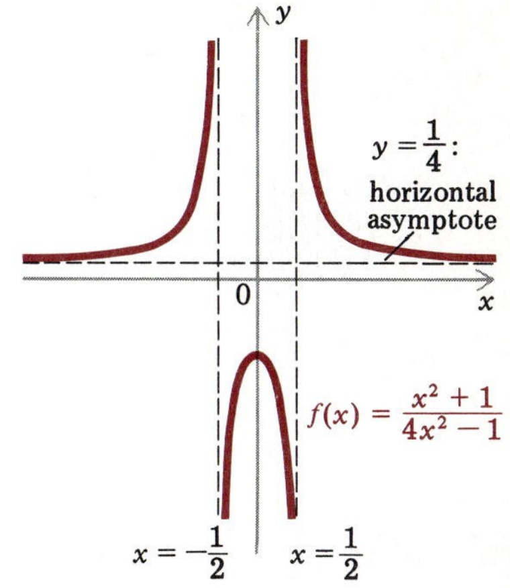

FIGURE 3.48

The following 8 steps, which we have used in the preceding examples, can help in graphing a rational function $f = P/Q$, where P and Q are polynomial functions.

1. If it has not been done already, factor the numerator $P(x)$ and the denominator $Q(x)$ into linear polynomials of the form $x - a$, quadratic polynomials that cannot be factored (such as $x^2 + 1$), and constants.
2. Determine the domain of f, which consists of all x for which $Q(x) \neq 0$. (Notice that a number a is *not* in the domain of f if and only if $x - a$ is a factor of $Q(x)$.)
3. Determine the intercepts (if any). The y intercept is $f(0)$ if 0 is in the domain of f. The x intercepts are the numbers a in the domain of f such that $x - a$ is a factor of the numerator.
4. Cancel any factors common to both numerator and denominator.
5. Determine the vertical asymptotes (if any). The line $x = a$ is a vertical asymptote if a factor of $x - a$ remains in the denominator after all cancellation has been completed in step 4.
6. Determine the sign of $f(x)$ on each of the intervals into which the x intercepts and the vertical asymptotes divide the x axis.
7. Determine the general appearance of the graph far away from the y axis as follows:
   Let

$$f(x) = \frac{a_n x^n + a_{n-1} x^{n-1} + \cdots + a_1 x + a_0}{b_m x^m + b_{m-1} x^{m-1} + \cdots + b_1 x + b_0}$$

   a. If $n < m$, then the line $y = 0$ is a horizontal asymptote.
   b. If $n = m$, then the line $y = a_n/b_n$ is a horizontal asymptote.

c. If $n > m$, then far away from the y axis the graph of f resembles the graph of

$$y = \frac{a_n}{b_m} x^{n-m}$$

In case $n - m = 1$, the latter graph is a straight line.

8. Use the information in steps 1–7, along with any symmetry of the graph, to sketch the graph of f.

EXERCISES 3.4

In Exercises 1–24, sketch the graph of the given function. Indicate all intercepts and asymptotes, as well as the behavior for large values of the variable.

1. $f(x) = -\dfrac{1}{x}$

2. $f(x) = \dfrac{1}{x - 1}$

3. $f(x) = \dfrac{4}{2 - x}$

4. $f(x) = 2 + \dfrac{1}{x^2}$

5. $f(x) = -\dfrac{1}{(x + 1)^2}$

6. $f(x) = \dfrac{1}{x^3} - 1$

7. $g(x) = -\dfrac{1}{4(x + \frac{1}{2})^4}$

8. $g(x) = \dfrac{3}{x^2 - 1}$

9. $g(x) = \dfrac{x}{x - 2}$

10. $g(x) = \dfrac{x - 1}{x + 1}$

11. $g(x) = \dfrac{-x}{x^2 - 4}$

12. $f(x) = -\dfrac{3}{x^2 + x}$

13. $f(x) = \dfrac{2}{(x - 1)(x + 2)}$

14. $f(x) = \dfrac{x(x + 2)}{(x - 1)(x^2 - 4)}$

15. $f(x) = \dfrac{x}{x^2 + 4x + 3}$

16. $f(x) = \dfrac{2x - 1}{x^2 - x - 2}$

17. $f(x) = \dfrac{x^2}{2x^2 - 3x - 2}$

18. $y = \dfrac{4x^2 + 4x + 1}{x}$

19. $y = \dfrac{4x^2 + 4x + 1}{x^2}$

20. $f(t) = \dfrac{t^2 + 1}{t + 1}$

21. $g(t) = \dfrac{-3t - 1}{t^2 + 4}$

22. $g(t) = \dfrac{t^4 - 1}{t^2 - 1}$

23. $f(x) = \dfrac{3x^2 + 6x + 3}{x^3 + 4x^2 + x + 4}$

24. $f(x) = \dfrac{x + \frac{1}{2}}{4x^3 + 4x^2 + x}$

3.5 GRAPHS OF EQUATIONS

Thus far in Chapter 3 we have concentrated on rational functions and their graphs. The importance of functions and their graphs cannot be overemphasized; yet equations involving x and y that do not necessarily represent functions play a role in mathematics, and so do their graphs; it is therefore well to study some of the more common ones. The ***graph of an equation*** in x and y is the collection of all ordered pairs (x, y) of real numbers that satisfy the equation.

We begin with circles. A ***circle*** of radius r centered at the point (a, b) in the plane is by definition the collection of all points (x, y) whose distance from (a, b) is r. From the distance formula (in Section 2.1), the distance between (a, b) and (x, y) is

$$\sqrt{(x-a)^2+(y-b)^2}$$

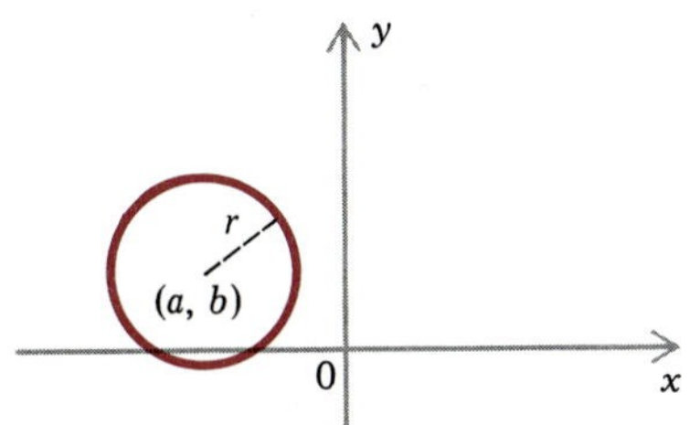

FIGURE 3.49

Therefore (x, y) is on the circle if and only if

$$\sqrt{(x-a)^2+(y-b)^2}=r$$

or, when both sides are squared,

$$\boxed{(x-a)^2+(y-b)^2=r^2} \tag{1}$$

In summary, the graph of (1) is the circle of radius r at (a, b) (Figure 3.49). If the center of the circle is (0, 0), then (1) becomes

$$\boxed{x^2+y^2=r^2} \tag{2}$$

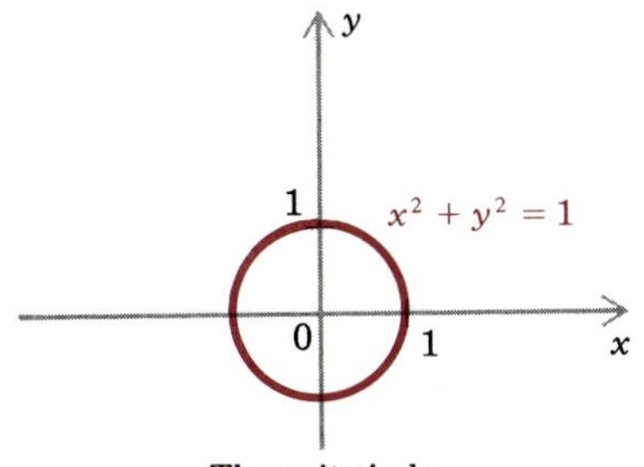

The unit circle

FIGURE 3.50

EXAMPLE 1. Find an equation of the circle of radius 1 centered at (0, 0). Then sketch the circle.

Solution. Using (2) with $r = 1$, we obtain the equation

$$x^2+y^2=1$$

The circle is sketched in Figure 3.50. □

The circle $x^2 + y^2 = 1$ is usually called the ***unit circle***; it will play a prominent role in Chapter 5.

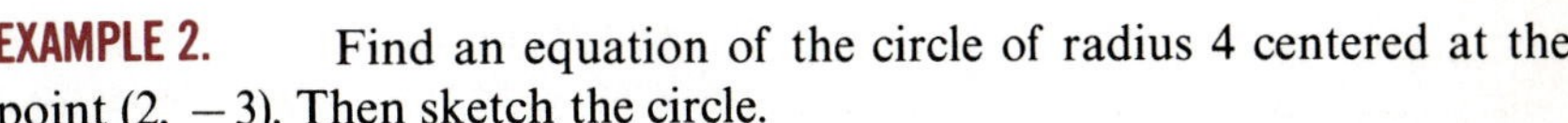

EXAMPLE 2. Find an equation of the circle of radius 4 centered at the point (2, −3). Then sketch the circle.

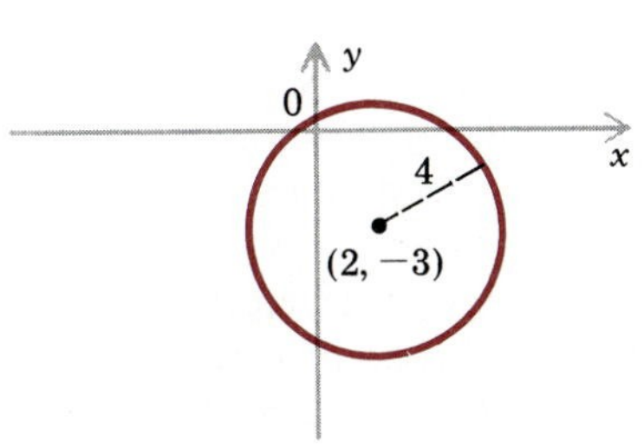

FIGURE 3.51

Solution. By (1) an equation is given by

$$(x-2)^2+(y-(-3))^2=4^2$$

which simplifies to

$$(x-2)^2+(y+3)^2=16$$

The circle is sketched in Figure 3.51. □

EXAMPLE 3. Find an equation of the circle whose center is (−1, −2) and which passes through the point (3, 1).

Solution. The radius r of the circle is the distance between the center $(-1, -2)$ and the point $(3, 1)$ on the circle. By the distance formula,

$$r = \sqrt{(3-(-1))^2 + (1-(-2))^2} = \sqrt{16+9} = \sqrt{25} = 5$$

Thus by (1) the equation we seek is

$$(x-(-1))^2 + (y-(-2))^2 = 5^2$$

which simplifies to

$$(x+1)^2 + (y+2)^2 = 25$$

The circle is sketched in Figure 3.52. □

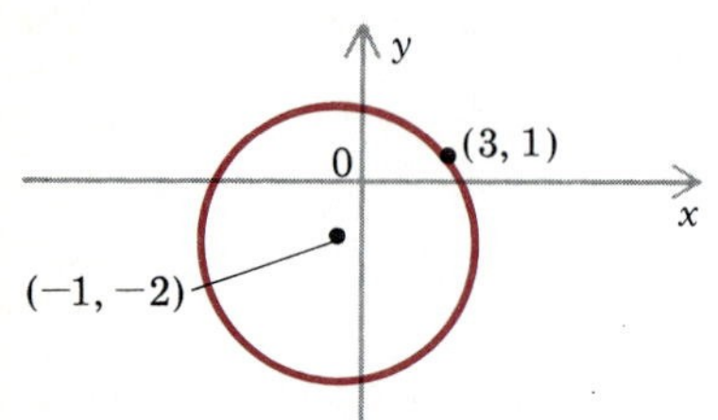

FIGURE 3.52

Symmetry in Graphs

In Section 2.5 we discussed symmetry for graphs of functions with respect to the y axis and the origin. Now we discuss various types of symmetry for graphs of equations. We begin by sketching the graph of a simple equation.

EXAMPLE 4. Sketch the graph of $y^2 = x$.

Solution. First we make a table:

x	0	1	4	9
y	0	1 and -1	2 and -2	3 and -3

Then we plot the corresponding points (x, y) on the graph and connect the points with a smooth curve (Figure 3.53). □

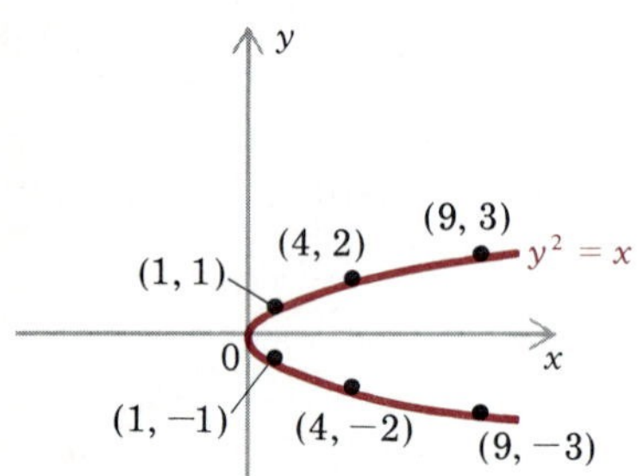

FIGURE 3.53

The graph of $y^2 = x$ in Figure 3.53 has the feature that it is symmetric with respect to the x axis, that is, the part above the x axis is mirrored below the x axis, and vice versa. Indeed, if the point (x, y) is on the graph, then the point $(x, -y)$, which lies opposite to (x, y) across the x axis, also lies on the graph (Figure 3.53). More generally, we say that the graph of an equation is ***symmetric with respect to the x axis*** if $(x, -y)$ is on the graph whenever (x, y) is (Figure 3.54).

The definition of symmetry with respect to the y axis given in Section 2.5 for graphs of functions applies equally well to graphs of equations. The graph of an equation is ***symmetric with respect to the y axis*** if $(-x, y)$ is on the graph whenever (x, y) is (Figure 3.55). For example, the circle $x^2 + (y-1)^2 = 2$ is symmetric with respect to the y axis (Figure 3.56).

Similarly, the definition of symmetry with respect to the origin given in Section 2.5 for graphs of functions transfers to graphs of equations: The graph of an equation is ***symmetric with respect to the origin*** if $(-x, -y)$ is on the graph whenever (x, y) is on the graph (Figure 3.57).

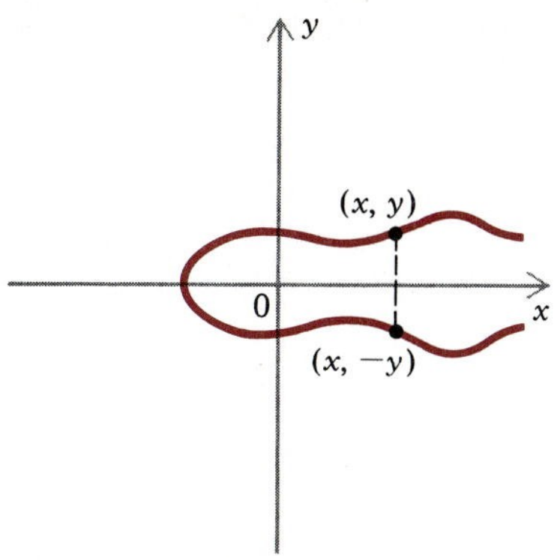

Symmetry with respect to the x axis

FIGURE 3.54

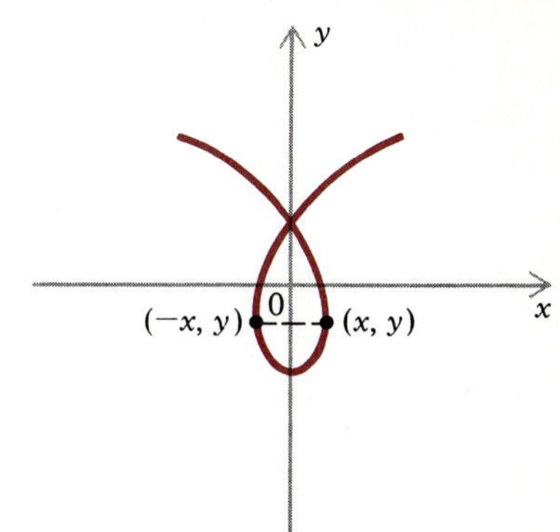

Symmetry with respect to the y axis

FIGURE 3.55

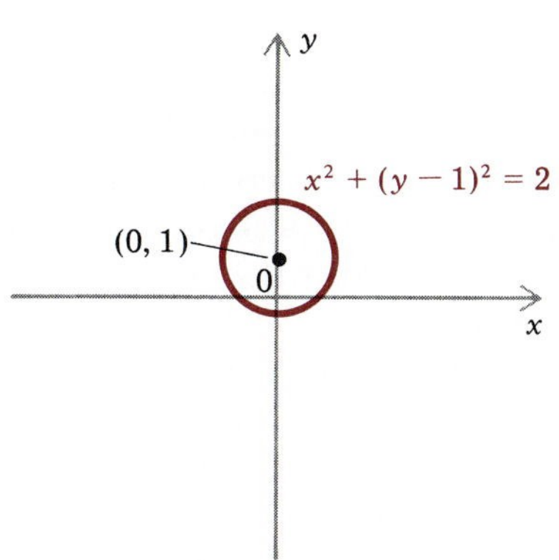

A circle that is symmetric with respect to the y axis.

FIGURE 3.56

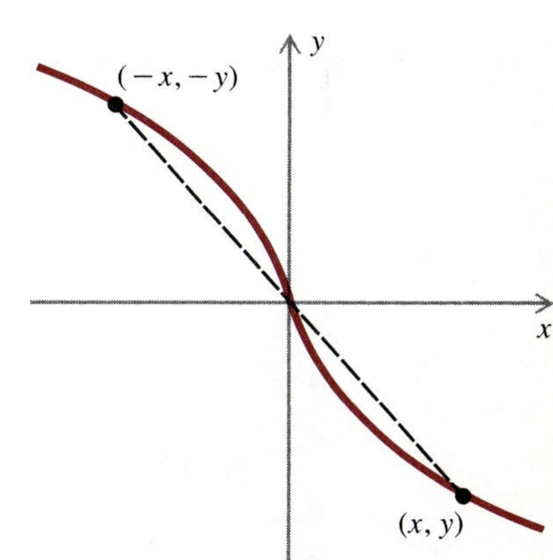

Symmetry with respect to the origin.

FIGURE 3.57

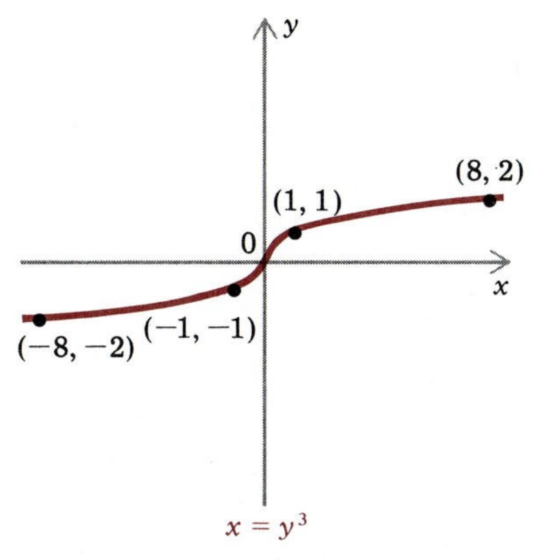

FIGURE 3.58

EXAMPLE 5. Sketch the graph of $x = y^3$.

Solution. With the table

x	-8	-1	0	1	8
y	-2	-1	0	1	2

we are able to plot the corresponding points and connect them with a smooth curve (Figure 3.58). We notice that the graph is symmetric with respect to the origin, since if $x = y^3$, then $-x = -y^3 = (-y)^3$. □

The graph of an equation need not be symmetric with respect to either of the axes or the origin (Figure 3.59). In contrast, the graph of an equation

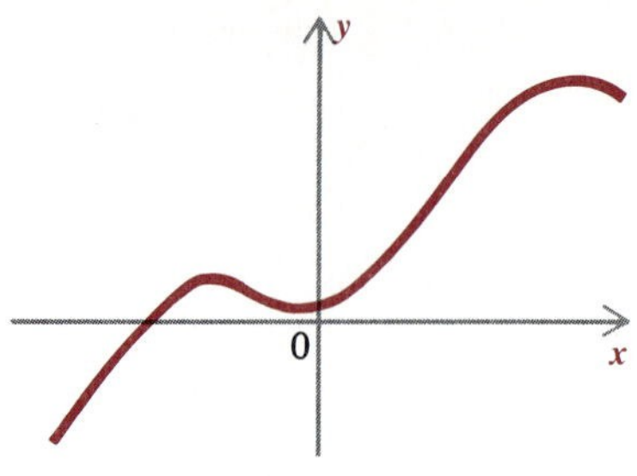

FIGURE 3.59

can be symmetric with respect to *both* axes as well as to the origin (see Figure 3.50). Finally, we mention that symmetry can facilitate sketching of graphs, because a portion of the graph then determines the remainder of the graph.

Intercepts

In keeping with our definition in Section 2.5 of the y intercept for a function, we define a ***y intercept*** of the graph of an equation to be the y coordinate of a point at which the graph meets the y axis (Figure 3.60). Since any point on the y axis has x coordinate 0, the y intercepts (if any) can be found by setting $x = 0$ in the equation and then solving for y. Similarly, an ***x intercept*** of the graph of an equation is the x coordinate of a point at which the graph meets the x axis (Figure 3.60), and the x intercepts (if any) can be found by setting $y = 0$ in the equation and then solving for x.

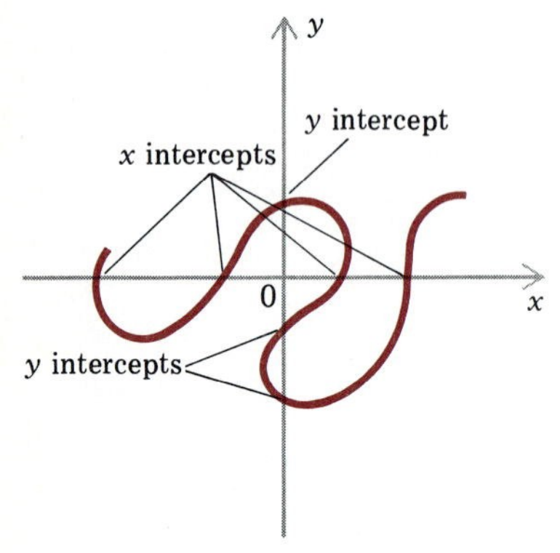

FIGURE 3.60

EXAMPLE 6. Find the intercepts of the graph of the equation $4x^2 + y^2 = 4$.

Solution. To find the y intercepts we set $x = 0$ and solve y:

$$4(0)^2 + y^2 = 4$$
$$y^2 = 4$$
$$y = 2 \quad \text{or} \quad y = -2$$

To find the x intercepts we set $y = 0$ and solve for x:

$$4x^2 + (0)^2 = 4$$
$$x^2 = 1$$
$$x = 1 \quad \text{or} \quad x = -1$$

Thus the y intercepts are 2 and -2, and the x intercepts are 1 and -1. □

EXAMPLE 7. Find the intercepts of the graph of the equation $x^2 - y^2 = 1$.

Solution. To find the y intercepts (if any) we set $x = 0$ and obtain

$$0^2 - y^2 = 1$$
$$y^2 = -1$$

Since this equation has no real solutions, there are no y intercepts. (Thus the graph of $x^2 - y^2 = 1$ does not meet the y axis.) To find the x intercepts we set $y = 0$ and solve for x. We find that

$$x^2 - (0)^2 = 1$$
$$x^2 = 1$$
$$x = 1 \quad \text{or} \quad x = -1$$

Thus the x intercepts are 1 and -1. □

EXAMPLE 8. Show that the graph of $y^2 - 5 = x + \dfrac{10}{x}$ has no intercepts.

Solution. There are no y intercepts because $10/x$ is not defined for $x = 0$, and hence the given equation is meaningless for $x = 0$. To search for x intercepts, we set $y = 0$ and attempt to solve the equation for x:

$$(0)^2 - 5 = x + \frac{10}{x}$$
$$-5x = x^2 + 10$$
$$x^2 + 5x + 10 = 0 \tag{3}$$

If we attempt to use the quadratic formula, we obtain

$$x = \frac{-5 \pm \sqrt{5^2 - 4(1)(10)}}{2(1)} = \frac{-5 \pm \sqrt{-15}}{2}$$

Since $-15 < 0$, (3) has no real solutions, and consequently the given equation has no x intercepts. □

EXERCISES 3.5

In Exercises 1–8, find an equation of the given circle and then sketch the circle.

1. The circle with radius 3 and center (0, 0).
2. The circle with radius 5 and center (3, 6).
3. The circle with radius 2 and center $(-1, 4)$.
4. The circle with radius $\frac{1}{2}$ and center $(-1, -2)$.

5. The circle with center (0, 0) that passes through (4, −1).

6. The circle with center (5, 12) that passes through (0, 0).

7. The circle with center (−2, 3) that passes through (1, −1).

8. The circle with center (−3, −1) that passes through (0, 4).

In Exercises 9–20, sketch the graph of the given equation and label all intercepts. Determine whether the graph possesses symmetry with respect to either axis or the origin.

9. $y^2 - 6y + 9 = 0$

10. $6x^2 + x - 2 = 0$

11. $x = |y|$

12. $|y| = |x|$

13. $y^2 = x - 1$

14. $x^2 = y^4$

15. $x = -y^3$

16. $x^3 = y^2$

17. $x^2 + y^2 = 4$

18. $(x - 1)^2 + (y + 2)^2 = 4$

19. $x = \sqrt{y}$

20. $x = \begin{cases} \sqrt{-y} & \text{for } y < 0 \\ \sqrt{y} & \text{for } y \geq 0 \end{cases}$

In Exercises 21–32, determine the intercepts (if any) of the graph of the given equation.

21. $x^2 + 4y^2 = 1$

22. $2y^2 - 3x^2 = 1$

23. $x - y^2 = 3$

24. $\frac{1}{x} + \frac{1}{y} = 1$

25. $y^2 = \sqrt{x^2 - 1}$

26. $y^2 - 4 = x + \frac{4}{x}$

27. $y^2 = |x - 5|$

28. $y + \frac{6}{y} = x^2 - 1$

29. $y - \frac{6}{y} = x^2 + 1$

30. $\frac{2}{x} - \frac{4}{x^2} = \pi xy - 5y^2 - 1$

31. $|x - 2| = |y + 1|$

32. $|x - 2| + y^2 = 4$

3.6 THE PARABOLA

Conic sections have been of interest to mathematicians since the third century B.C., when the Greek geometer Apollonius wrote eight treatises about conic sections and became known as the "Great Geometer." A ***conic section*** is formed when a double right circular cone is sliced by a plane. The intersection can be a ***parabola***, an ***ellipse*** (a special case of which is a circle), or a ***hyperbola***, and in very special cases it can be a point, a line, or two intersecting lines (called ***degenerate*** conic sections). Figure 3.61 depicts the various kinds of conic sections.

A conic section is a curve in a plane that cuts a double cone. If we think of the plane as the Cartesian plane, then the conic section can be described as the graph of an equation in x and y. In this and the next two sections we will associate the three main kinds of conic sections—parabola, ellipse, and hyperbola—with special types of equations, and we will sketch the graphs of such equations.

The graph of an equation of the form $y = ax^2 + bx + c$, with $a \neq 0$, is a parabola, as defined in Section 3.1, and has a graph as in Figure 3.62a, b. If the graph is turned on its side, then it remains a parabola and its equation has

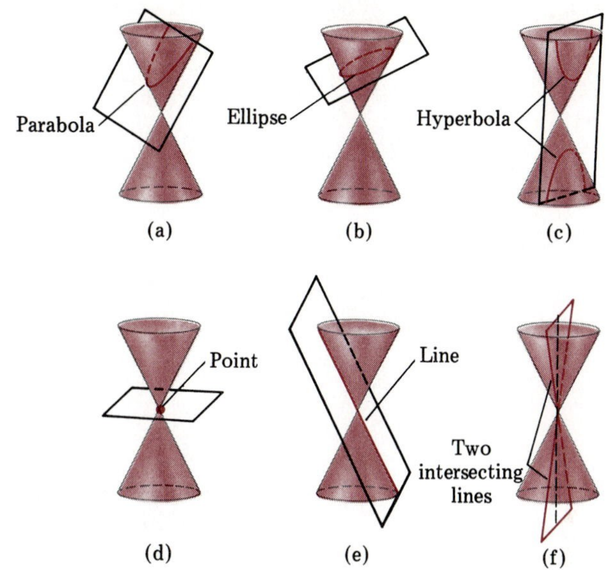

FIGURE 3.61

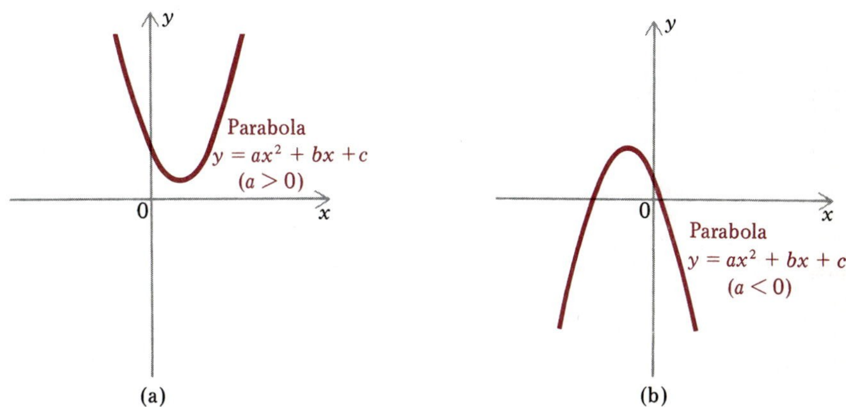

FIGURE 3.62

the form

$$x = ay^2 + by + c \quad \text{with} \quad a \neq 0 \tag{1}$$

We sketched the graph of one such equation, $x = y^2$, in Figure 3.53. In general, to sketch the graph of (1) we complete the square on the right side of (1) to obtain an equation of the form

$$x = a(y - k)^2 + h, \quad \text{or equivalently,} \quad x - h = a(y - k)^2 \tag{2}$$

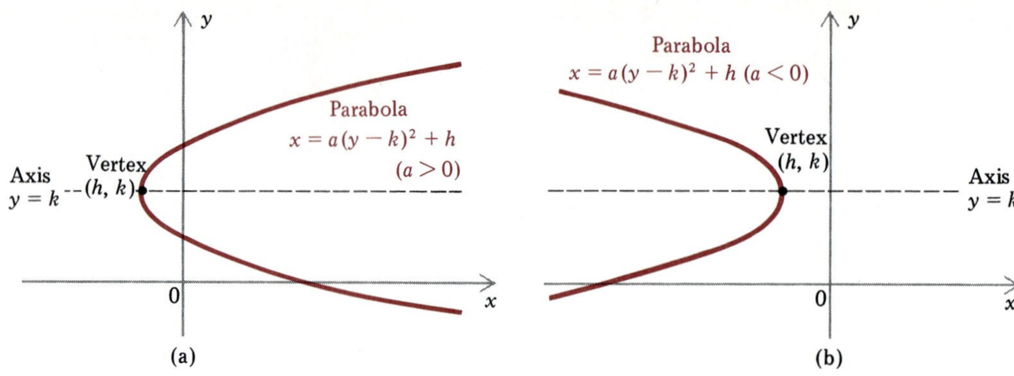

FIGURE 3.63

Then the graph opens to the right of the point (h, k) if $a > 0$ and to the left of (h, k) if $a < 0$ (Figure 3.63a, b). The point (h, k) is the ***vertex*** of the parabola. Moreover, the graph is symmetric with respect to the line $y = k$, called the ***axis*** of the parabola.

EXAMPLE 1. Sketch the parabola having equation

$$x = 2y^2 - 4y + 5$$

and note the intercepts, vertex, and axis of the parabola.

Solution. The x intercept is 5 because if $y = 0$, then $x = 5$. To determine any y intercepts, we set $x = 0$ in the given equation and solve for y:

$$2y^2 - 4y + 5 = 0 \tag{3}$$

$$y = \frac{-(-4) \pm \sqrt{(-4)^2 - 4(2)(5)}}{2(2)} = \frac{4 \pm \sqrt{-24}}{4}$$

Since $-24 < 0$, there is no real number y satisfying equation (3), and hence there are no y intercepts. To help locate the vertex of the parabola, we complete the square:

$$\begin{aligned} x &= 2y^2 - 4y + 5 \\ &= 2(y^2 - 2y) + 5 \\ &= 2(y^2 - 2y + 1) - 2 + 5 \\ &= 2(y - 1)^2 + 3 \end{aligned}$$

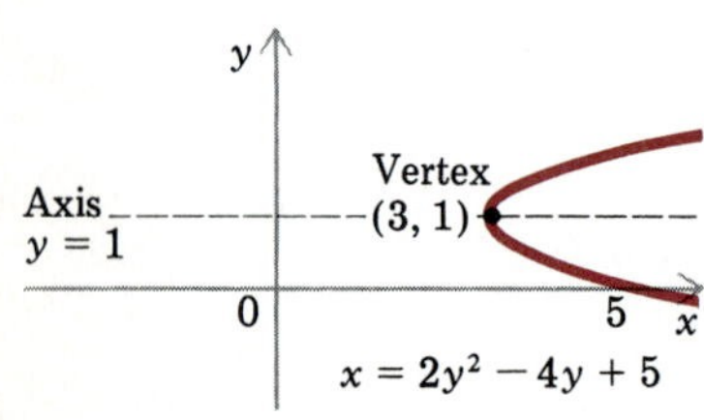

FIGURE 3.64

In the notation of (2) we have $a = 2$, $h = 3$, and $k = 1$. Therefore the vertex is the point $(3, 1)$, the axis is the line $y = 1$, and the parabola opens to the right because $a = 2 > 0$. The parabola is sketched in Figure 3.64. □

EXAMPLE 2. Sketch the parabola having equation

$$x = -y^2 - 4y - 3$$

and note the intercepts, vertex, and axis of the parabola.

Solution. The x intercept is -3 because if $y = 0$, then $x = -3$. To find any y intercepts, we set $x = 0$ in the given equation and solve for y:

$$\begin{aligned} -y^2 - 4y - 3 &= 0 \\ y^2 + 4y + 3 &= 0 \\ (y + 3)(y + 1) &= 0 \\ y = -3 \quad \text{or} \quad y &= -1 \end{aligned}$$

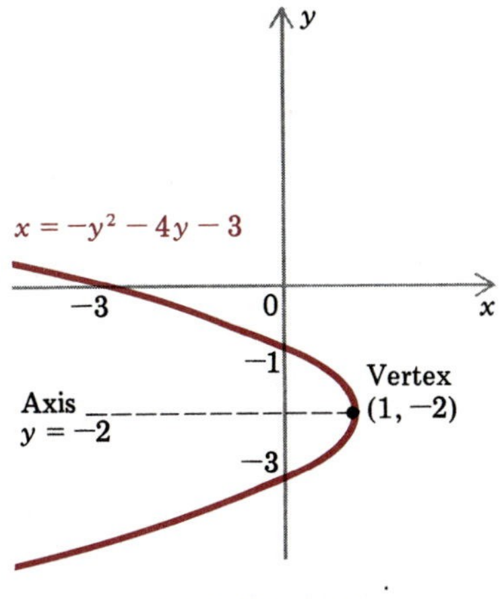

FIGURE 3.65

As a result, the y intercepts are -3 and -1. Next we complete the square on the right side of the given equation:

$$\begin{aligned} x &= -y^2 - 4y - 3 \\ &= -(y^2 + 4y) - 3 \\ &= -(y^2 + 4y + 4) + 4 - 3 \\ &= -(y + 2)^2 + 1 \end{aligned}$$

This time when we use the notation of (2), we find that $a = -1$, $h = 1$, and $k = -2$. Therefore the vertex is $(1, -2)$, and the axis of the parabola is the line $y = -2$. Since $a = -1 < 0$, the parabola opens to the left. The parabola is sketched in Figure 3.65. □

The graph of any equation of the form

$$Ax^2 + By^2 + Cx + Dy = E$$
$$\text{with } A = 0, B \neq 0, C \neq 0 \quad \text{or} \quad A \neq 0, B = 0, D \neq 0 \tag{4}$$

is a parabola. By using the technique of completing the square, one can rewrite (4) as either

$$\boxed{x = a(y - k)^2 + h} \tag{5}$$

or

$$\boxed{y = a(x - h)^2 + k} \tag{6}$$

In each case the vertex of the parabola is (h, k). For (5) the axis is the horizontal line $y = k$, and for (6) the axis is the vertical line $x = h$. In either case the parabola is symmetric with respect to its axis.

EXAMPLE 3. Sketch the parabola having equation

$$2x^2 - 12x - 3y = -6$$

noting the vertex and axis of the parabola.

Solution. The given equation has the form of (4) with $A = 2$, $B = 0$, $C = -12$, $D = -3$, and $E = -6$. Since $B = 0$, we can put the given equation into the form of (6) by completing the square:

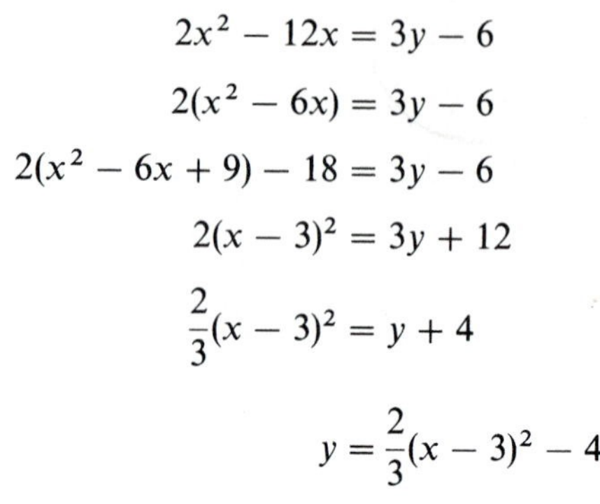

$$\begin{aligned} 2x^2 - 12x &= 3y - 6 \\ 2(x^2 - 6x) &= 3y - 6 \\ 2(x^2 - 6x + 9) - 18 &= 3y - 6 \\ 2(x - 3)^2 &= 3y + 12 \\ \frac{2}{3}(x - 3)^2 &= y + 4 \end{aligned}$$

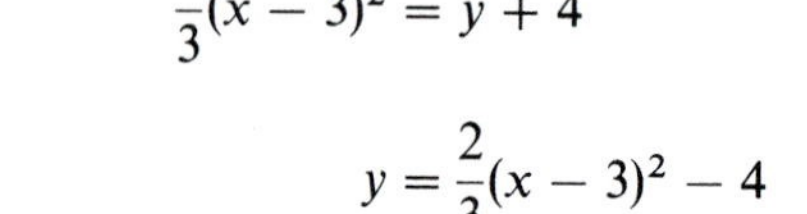

$$y = \frac{2}{3}(x - 3)^2 - 4$$

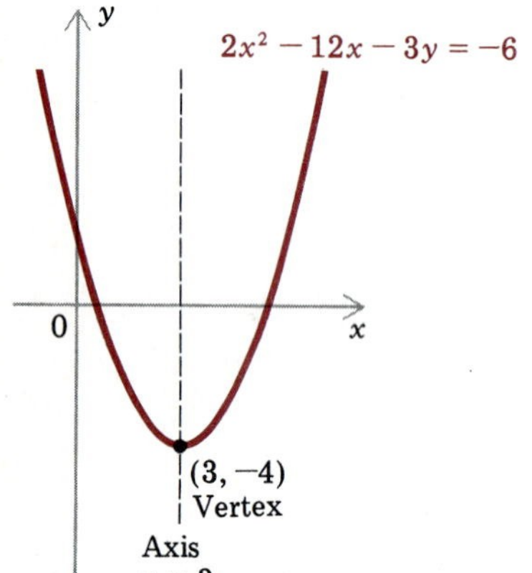

FIGURE 3.66

Thus the vertex is $(3, -4)$, and the axis is the vertical line $x = 3$. The graph is sketched in Figure 3.66. □

There is an alternative, geometric way of defining the parabola that is geometric in nature. Let l be a fixed line in the plane and F a fixed point not on l. The set of all points P in the plane equidistant from l and F is a parabola (Figure 3.67). The line l is called the ***directrix*** and the point F the ***focus*** of the parabola.

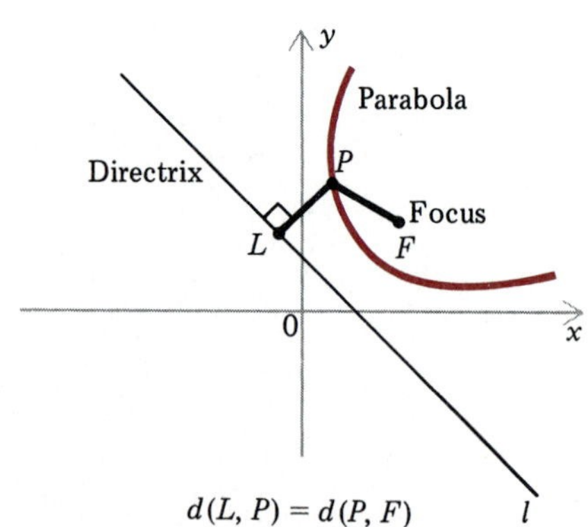

FIGURE 3.67

There are many theoretical and practical applications of parabolas. We list several of them here.

FIGURE 3.68a

1. When an object moves under the sole influence of the earth's gravity and eventually strikes the earth, its path is normally parabolic (or linear if the motion is vertical). Thus the path of a golfball, baseball, or football in flight would be parabolic if there were no air resistance.
2. The shapes of certain mirrors can be obtained by revolving a parabola about its axis. Such mirrors, called ***parabolic mirrors***, are found in automobile headlights and reflecting telescopes. In a headlight all light rays emitted from a light source at the focus are reflected from the mirror along lines parallel to the axis of the parabola. In a reflecting telescope such as the one at Mt. Palomar and various radio telescopes, all incoming rays from a star, which are essentially parallel to the axis, are reflected by the mirror and then pass through the focus, where the eyepiece of the telescope should be located (see Figure 3.68a).
3. A suspended cable tends to hang in the shape of a parabola if the weight of the cable is small compared to the weight it supports and if the weight is uniformly distributed along the cable. Bridges that are supported by cables in this fashion are called ***suspension bridges*** (see Figure 3.68b). All

FIGURE 3.68b

the larger bridges in the world, and most of the more famous ones, are suspension bridges; examples are the Golden Gate Bridge, the Brooklyn Bridge, and the Delaware Memorial Bridge.

EXERCISES 3.6

In Exercises 1–14, sketch the parabola, noting the vertex and the axis of the parabola.

1. $x = 3y^2$

2. $x + \frac{1}{2}y^2 = 0$

3. $x + 5 = -\frac{1}{3}(y - 1)^2$

4. $x - 1 = 4(y + 2)^2$

5. $y^2 = x - y$

6. $x = y^2 + 2y - 4$

7. $x = 2y^2 - 2y + 1$

8. $5x - 19 = -y(y + 2)$

9. $y + 4x^2 - 24x + 34 = 0$

10. $y + x^2 - 8x + 22 = 0$

11. $2x - y^2 + 6y - 7 = 0$

12. $3x + 2y^2 + 4y + 8 = 0$

13. $x^2 + 4x = 3y + 5$

14. $(x - y)^2 + (x + y)^2 = 2x^2 + 4y - 2x$

In Exercises 15–20, find an equation of the parabola having the given properties.

15. Vertex (0, 0); axis: y axis; (2, 5) on the parabola

16. Vertex (0, 0); axis: x axis; (2, 5) on the parabola

17. Vertex $(-2, 5)$; axis: $x = -2$; (0, 7) on the parabola

18. Vertex $(1, -4)$; axis: $y = -4$; $(-1, -1)$ on the parabola

19. Vertex (3, 4); (2, 5) and (4, 5) on the parabola

20. Vertex $(-1, 1)$; (3, 0) and $(0, \frac{1}{2})$ on the parabola

In Exercises 21–24, sketch the graph of the given equation. In each case the graph is a degenerate conic section.

21. $x^2 - 9y^2 + 2x + 18y = 8$

22. $4x^2 + 3y^2 - 24x - 18y + 63 = 0$

23. $36x^2 + 36x + 11 = -18y^2 + 12y$

24. $3x^2 + 6\sqrt{3}x = 2y^2 - 16y + 23$

25. For what value of c will the parabola with equation $cx = (y + 2)^2 - 6$ pass through the point $(-2, 4)$?

***26.** For what value of c will the axis of the parabola with equation $2x - 5 = (1/c)y^2 + 4y + 9$ be the line $y = 3$?

27. Determine all points on the parabola with equation $3x - 4y^2 + 2y = 7$ that have x coordinate 3.

28. Determine all points on the parabola with equation $x^2 - 2x - y = -\frac{1}{2}$ that have y coordinate 3.

29. In a suspension bridge the horizontal distance between the supports is the ***span*** of the bridge, and the vertical distance between the points where the cable is attached to the supports and the lowest point of the cable is the ***sag*** (Figure 3.69). The span of the George Washington Bridge is 3500 feet, and its sag is 316 feet. Find an equation of the parabola that represents the cable.

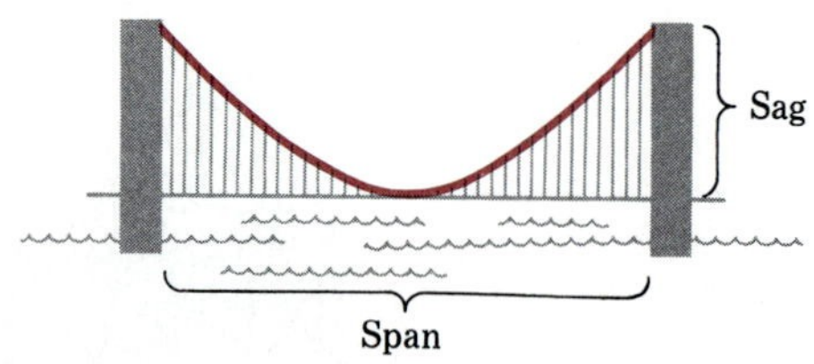

Suspension bridge

FIGURE 3.69

3.7 THE ELLIPSE

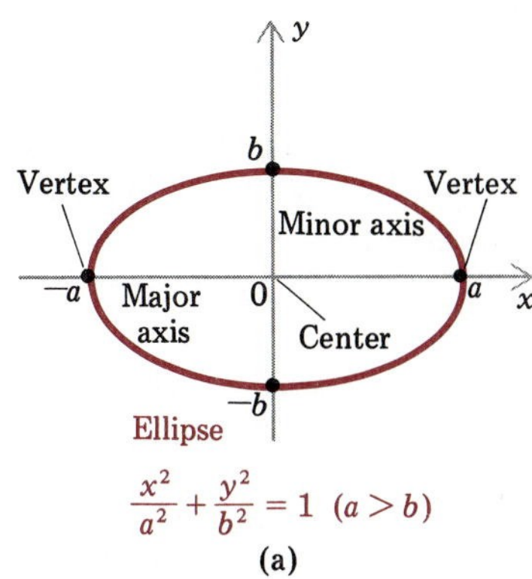

(a)

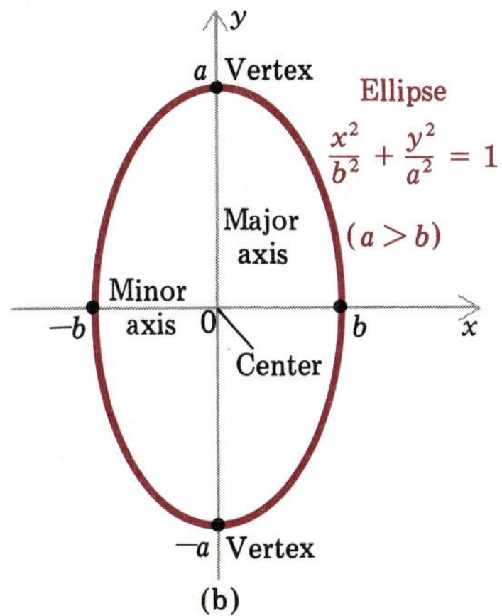

(b)

FIGURE 3.70

The graph of an equation of the form

$$Ax^2 + By^2 = E \quad \text{with } A, B, \text{ and } E \text{ of the same sign} \tag{1}$$

is an ellipse. Dividing each side of the equation in (1) by E allows us to rewrite the equation in the standard form

$$\frac{x^2}{a^2} + \frac{y^2}{b^2} = 1 \quad \text{or} \quad \frac{x^2}{b^2} + \frac{y^2}{a^2} = 1 \quad \text{where} \quad a \geq b > 0 \tag{2}$$

Ellipses given in standard form are symmetric with respect to both axes as well as the origin, which is called the ***center*** of the ellipse (Figure 3.70a, b). Setting $y = 0$ in the first equation of (2), and $x = 0$ in the second equation, yields $-a$ and a for two intercepts of the ellipse. The corresponding points of the ellipse are called ***vertices*** of the ellipse, and the line segment joining the vertices is the ***major axis*** of the ellipse. The length of the major axis is $2a$ (Figure 3.70a, b). The remaining two intercepts of the ellipse are obtained by setting $x = 0$ in the first equation of (2) and $y = 0$ in the second, which results in the intercepts $-b$ and b. Although the corresponding points have no special name, the line segment joining those points is called the ***minor axis*** of the ellipse. The minor axis has length $2b$ (Figure 3.70a, b). Since $a \geq b$ by assumption in (2),

the length of the major axis $\geq$ the length of the minor axis

EXAMPLE 1. Sketch the ellipse having equation

$$\frac{x^2}{9} + \frac{y^2}{4} = 1$$

and note the major and minor axes.

Solution. Since the coefficient in the denominator of x^2 is larger than that of y^2, we compare the given equation with the first equation of (2). This yields $a^2 = 9$, so that $a = \sqrt{9} = 3$, and it also yields $b^2 = 4$, so that $b = \sqrt{4} = 2$. It follows that the x intercepts are -3 and 3 and the y intercepts are -2 and 2. The ellipse is sketched in Figure 3.71. □

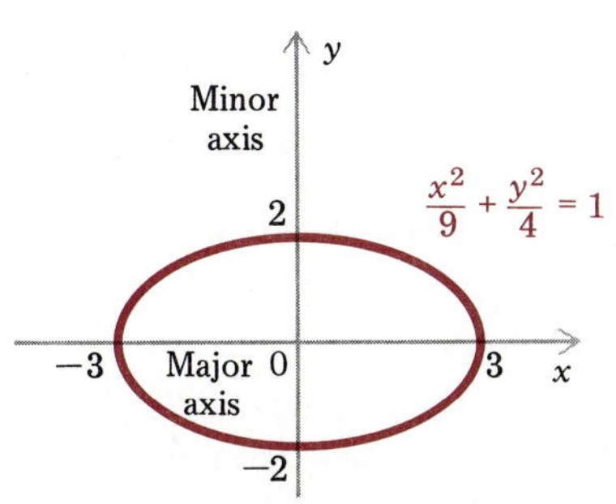

FIGURE 3.71

EXAMPLE 2. Sketch the ellipse having equation

$$16x^2 + 2y^2 = 32$$

and note the major and minor axes.

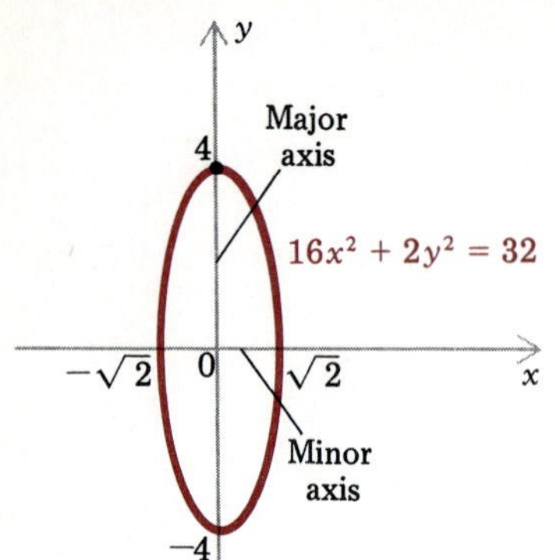

FIGURE 3.72

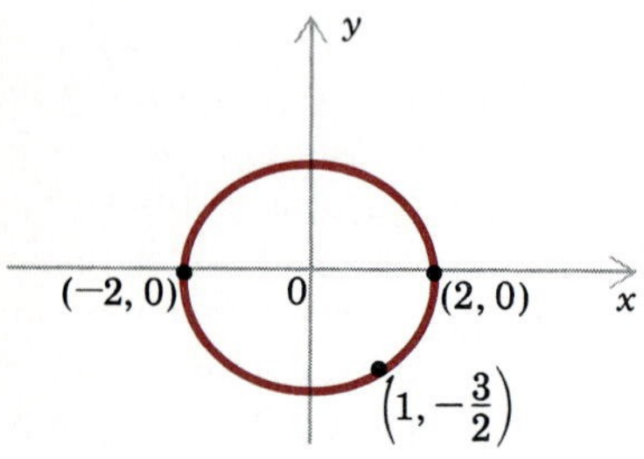

FIGURE 3.73

Solution. In order to use (2), we first divide each side by 32 to obtain

$$\frac{x^2}{2} + \frac{y^2}{16} = 1$$

Since the coefficient in the denominator of y^2 is larger than that of x^2, we compare our new equation with the second equation of (2). We find that $a = \sqrt{16} = 4$ and $b = \sqrt{2}$. Therefore the x intercepts are $-\sqrt{2}$ and $\sqrt{2}$, and the y intercepts are -4 and 4. The ellipse is sketched in Figure 3.72. □

EXAMPLE 3. Find an equation in standard form for the ellipse whose vertices are $(-2, 0)$ and $(2, 0)$ and which passes through the point $(1, -\frac{3}{2})$ (Figure 3.73).

Solution. Since the vertices lie on the x axis, the equation we seek has the form

$$\frac{x^2}{a^2} + \frac{y^2}{b^2} = 1$$

The value of a is 2, because the vertices are $(-2, 0)$ and $(2, 0)$. We will be done once we find the numerical value of b^2. To that end we notice that the point $(1, -\frac{3}{2})$ is on the ellipse, so that

$$\frac{1^2}{4} + \frac{(-3/2)^2}{b^2} = 1$$

Solving for b^2, we find that

$$\frac{(-3/2)^2}{b^2} = 1 - \frac{1}{4} = \frac{3}{4}$$

and consequently

$$b^2 = \left(-\frac{3}{2}\right)^2 \cdot \frac{4}{3} = \frac{9}{4} \cdot \frac{4}{3} = 3$$

Therefore the equation we seek is

$$\frac{x^2}{4} + \frac{y^2}{3} = 1 \quad \square$$

The ellipses in Figures 3.71 and 3.73 are wider than they are tall, whereas the ellipse in Figure 3.72 is taller than it is wide. One can tell immediately from an equation of an ellipse in standard form whether it is wider or taller: if the number under x^2 is larger than the number under y^2, the ellipse is wider; if the number under y^2 is larger than the number under x^2, the ellipse is taller. In the event that the numbers under x^2 and y^2 are the same, the ellipse is a circle, because in that case the equation becomes

$$\frac{x^2}{a^2} + \frac{y^2}{a^2} = 1$$

or equivalently,

$$x^2 + y^2 = a^2$$

Thus a circle is indeed a special case of an ellipse.

Except in very special cases that will not concern us here, the graph of an equation of the form

$$Ax^2 + By^2 + Cx + Dy = E \quad \text{with } A, B, \text{ and } E \text{ of the same sign} \tag{3}$$

is an ellipse whose center can be any point in the plane, depending on the values of the constants. By completing squares, one can rearrange (3) so that it has the form

$$\frac{(x-h)^2}{a^2} + \frac{(y-k)^2}{b^2} = 1 \quad \text{or} \quad \frac{(x-h)^2}{b^2} + \frac{(y-k)^2}{a^2} = 1 \quad \text{where } a \geq b > 0 \tag{4}$$

The ***center*** of the ellipse is the point (h, k), and the ellipse is symmetric with respect to the two lines $x = h$ and $y = k$. For an ellipse defined by the first equation in (4), the ***vertices*** are $(h - a, k)$ and $(h + a, k)$ (Figure 3.74a), and the

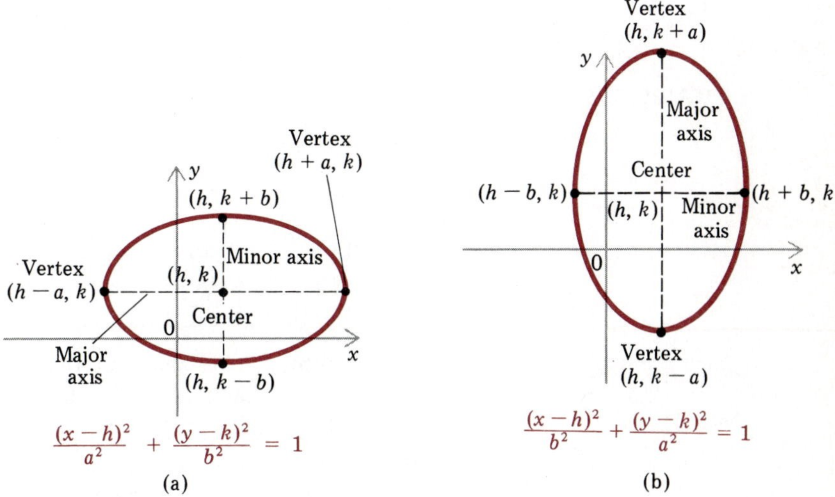

FIGURE 3.74

line segment joining the vertices is the ***major axis*** of the ellipse, whereas the line segment joining the points $(h, k - b)$ and $(h, k + b)$ is the ***minor axis*** of the ellipse. Analogously, for an ellipse defined by the second equation in (4), the vertices are $(h, k - a)$ and $(h, k + a)$ (Figure 3.74b), with the line joining them the ***major axis***, and the line joining the points $(h - b, k)$ and $(h + b, k)$ the ***minor axis*** of the ellipse. Again the length of the major axis is $2a$, and the length of the minor axis is $2b$.

EXAMPLE 4. Sketch the ellipse having equation

$$9x^2 + 54x + y^2 - 4y = -76$$

and note the center and vertices, as well as the major and minor axes.

Solution. We complete squares and then rearrange so that the right side is 1:

$$(9x^2 + 54x) + (y^2 - 4y) = -76$$
$$9(x^2 + 6x) + (y^2 - 4y) = -76$$
$$9(x^2 + 6x + 9) - 81 + (y^2 - 4y + 4) - 4 = -76$$
$$9(x + 3)^2 + (y - 2)^2 = -76 + 81 + 4 = 9$$
$$\frac{(x + 3)^2}{1} + \frac{(y - 2)^2}{9} = 1$$

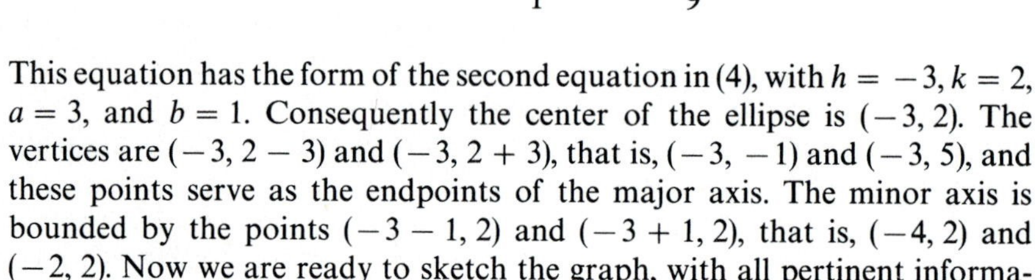

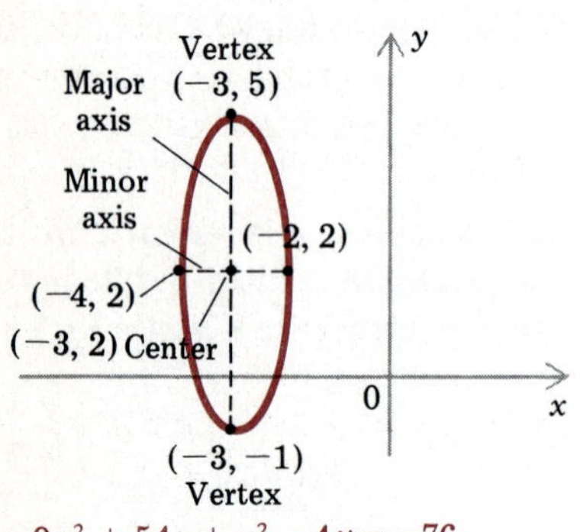

FIGURE 3.75

This equation has the form of the second equation in (4), with $h = -3, k = 2$, $a = 3$, and $b = 1$. Consequently the center of the ellipse is $(-3, 2)$. The vertices are $(-3, 2 - 3)$ and $(-3, 2 + 3)$, that is, $(-3, -1)$ and $(-3, 5)$, and these points serve as the endpoints of the major axis. The minor axis is bounded by the points $(-3 - 1, 2)$ and $(-3 + 1, 2)$, that is, $(-4, 2)$ and $(-2, 2)$. Now we are ready to sketch the graph, with all pertinent information on it (Figure 3.75). □

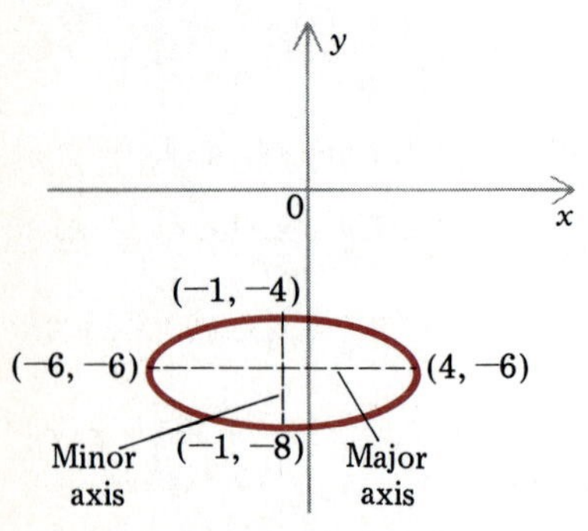

FIGURE 3.76

EXAMPLE 5. Find an equation of the ellipse whose major axis extends from $(-6, -6)$ to $(4, -6)$ and whose minor axis extends from $(-1, -8)$ to $(-1, -4)$ (Figure 3.76).

Solution. The center of the ellipse is the midpoint of the major (or minor) axis, which is $(-1, -6)$. Since the major axis is horizontal, the equation we seek has the form

$$\frac{(x - (-1))^2}{a^2} + \frac{(y - (-6))^2}{b^2} = 1$$

or equivalently,

$$\frac{(x+1)^2}{a^2} + \frac{(y+6)^2}{b^2} = 1$$

The distance between $(-6, -6)$ and $(4, -6)$ is $4 - (-6) = 10$, and this is the length of the major axis, which is also $2a$. Therefore $2a = 10$, so that $a = 5$. Similarly, the distance between $(-1, -8)$ and $(-1, -4)$ is $-4 - (-8) = 4$, and this is the length of the minor axis, which is $2b$. Therefore $2b = 4$, so that $b = 2$. Consequently the equation we seek is

$$\frac{(x+1)^2}{25} + \frac{(y+6)^2}{4} = 1 \quad \square$$

For a geometric description of an ellipse, we let F_1 and F_2 be points in the plane and k a number greater than the distance between F_1 and F_2. The set of all points P in the plane such that

$$d(F_1, P) + d(F_2, P) = k$$

is an ellipse (Figure 3.77a). The points F_1 and F_2 are called the ***foci*** of the ellipse. In the event that F_1 and F_2 are the same point, the ellipse is a circle (Figure 3.77b).

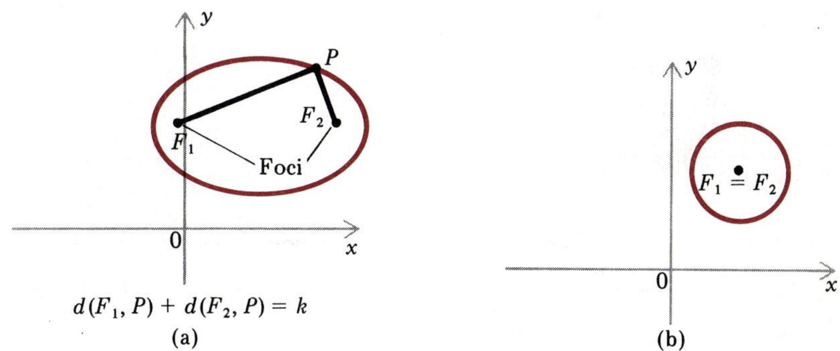

FIGURE 3.77

We end our discussion of ellipses with a few applications:

1. According to Kepler's laws of planetary motion, a planet moves in an elliptical orbit with the sun at one of the foci. Some comets, such as Halley's comet, also move in elliptical orbits about the sun. Halley's comet returns to the earth's vicinity approximately every 76 years and last appeared in the sky during the spring of 1986.

2. An electron in an atom normally moves in an approximately elliptical orbit.
3. The dome of a whispering gallery has the shape of an ellipse that has been revolved about its major axis. Sound emanating from one focus bounces from anywhere on the dome and is then directed to the second focus. It is said that John C. Calhoun used this phenomenon in the Statuary Hall of the Capitol in Washington, D.C., in order to eavesdrop on his adversaries.
4. Ellipses are aesthetically pleasing and are used in the design of buildings and the construction of formal gardens.

EXERCISES 3.7

In Exercises 1–14, sketch the ellipse, noting the center, the major and minor axes, and the vertices.

1. $\frac{x^2}{16} + \frac{y^2}{4} = 1$
2. $x^2 + \frac{y^2}{4} = 1$
3. $\frac{x^2}{3} + \frac{y^2}{3} = 1$
4. $4x^2 + y^2 = 1$
5. $9x^2 + 4y^2 = 36$
6. $2x^2 + y^2 = 2$
7. $\frac{(x-4)^2}{9} + \frac{(y+3)^2}{16} = 1$
8. $\frac{(x+4)^2}{2} + \frac{(y-1)^2}{25} = 1$
9. $(x+2)^2 + 4(y+1)^2 = 16$
10. $4x^2 + y^2 + 6y + 5 = 0$
11. $4x^2 - 8x + y^2 + 4y = 8$
12. $9x^2 + 18x + 4y^2 + 16y = 11$
13. $25x^2 + 16y^2 + 150x - 96y = 31$
14. $4x^2 + 9y^2 - 16x + 54y + 96 = 0$

In Exercises 15–24, find an equation of the ellipse having the given properties.

15. Center: (0, 0); x intercept: 2; y intercept: -3
16. Center: (0, 0); x intercept: -4; y intercept: 1
17. Center: (0, 0); vertex: (0, 4); $(\sqrt{3}, 2)$ on the ellipse
18. Center: $(-1, 2)$; vertex: (3, 2); (0, 4) on the ellipse
19. Center: (3, 2); (3, 7) and $(-4, 2)$ on the ellipse
20. Center: $(-1, -3)$; $(-1, 5)$ and $(3, -3)$ on the ellipse
21. Vertices: $(-2, -3)$ and $(-2, -7)$; length of minor axis: 2
22. Vertices: $(6, -1)$ and $(-10, -1)$; length of minor axis: $\frac{1}{4}$
23. Center: $(-\frac{3}{2}, \frac{1}{2})$; length of major axis: 5; length of minor axis: 1; minor axis horizontal
24. Major axis from (1, 1) to (1, 10); minor axis from $(0, \frac{11}{2})$ to $(2, \frac{11}{2})$
25. Find equations for the two distinct ellipses each of which has center (0, 4), major axis of length 3, and minor axis of length 2.
26. Find equations for the two distinct ellipses each of which has center $(-3, -2)$, major axis of length $\frac{3}{2}$, and minor axis of length $\frac{1}{2}$.
27. Show that if the major and minor axes of an ellipse have the same length, the ellipse is a circle.
28. For what value of c does the ellipse with equation $9(x-1)^2 + 4(y+2)^2 = 36c$ pass through the origin?
29. For what values of h does the ellipse with equation $(x-h)^2/9 + (y+5)^2/4 = 1$ pass through the point $(4, -4)$?
30. Let a and b be positive numbers. The area of the ellipse having an equation in the form of either of the equations in (2) is πab.
    a. Show that the above result implies that the area of the circle with radius r and center at the origin is πr^2.
    b. Find the area of the ellipse in Example 1.

31. a. Let a and b be positive numbers, and let u and v be numbers such that $u^2 + v^2 \neq 0$. Show that the point

$$\left(\frac{au}{\sqrt{u^2+v^2}}, \frac{bv}{\sqrt{u^2+v^2}}\right)$$

lies on the ellipse having equation

$$\frac{x^2}{a^2} + \frac{y^2}{b^2} = 1$$

b. Taking $u = 4$ and $v = -3$ in part (a), find the corresponding point on the ellipse having equation

$$\frac{x^2}{36} + \frac{y^2}{16} = 1$$

32. The ***eccentricity*** e of an ellipse with equation in the form of (4) is given by

$$e = \sqrt{1 - \frac{b^2}{a^2}}$$

a. Show that $0 \leq e < 1$.

b. Show that $e = 0$ if and only if the ellipse is a circle.

c. Find a formula for b in terms of a and e.

33. Find the eccentricities of the following ellipses.

a. $\dfrac{x^2}{4} + \dfrac{y^2}{16} = 1$ b. $\dfrac{x^2}{16} + \dfrac{y^2}{4} = 1$

c. $\dfrac{x^2}{4} + \dfrac{y^2}{9} = 1$ d. $x^2 + y^2 = 1$

34. A planetary orbit is elliptical and is frequently identified by the length $2a$ of the major axis and by the eccentricity e of the orbit. For the earth's orbit around the sun, the numbers $2a$ and e are approximately 299,000,000 (kilometers) and 0.17, respectively.

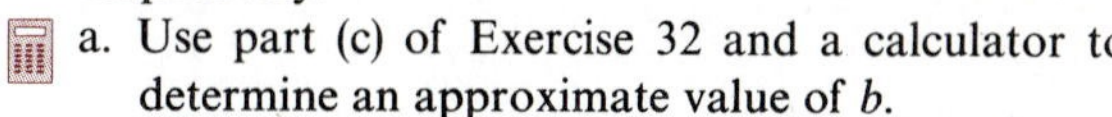

a. Use part (c) of Exercise 32 and a calculator to determine an approximate value of b.

b. Using the given information, find an equation in the form

$$\frac{x^2}{a^2} + \frac{y^2}{b^2} = 1$$

that approximately represents the earth's orbit around the sun.

3.8 THE HYPERBOLA

The last of the major kinds of conic sections is the hyperbola. The graph of an equation of the form

$$Ax^2 + By^2 = E \quad \text{with } A \text{ and } B \text{ of different signs and } E \neq 0$$

is a hyperbola, and by arranging the constants suitably we can write it in one of the two standard forms

$$\frac{x^2}{a^2} - \frac{y^2}{b^2} = 1 \quad \text{or} \quad \frac{y^2}{a^2} - \frac{x^2}{b^2} = 1 \quad \text{with} \quad a > 0 \quad \text{and} \quad b > 0$$

The corresponding hyperbolas are sketched in Figure 3.78a, b. They are symmetric with respect to both axes and the origin, which is called the ***center*** of the hyperbolas.

Notice that the hyperbola given by

$$\frac{x^2}{a^2} - \frac{y^2}{b^2} = 1 \tag{1}$$

has no y intercepts but has x intercepts a and $-a$. The points $(a, 0)$ and $(-a, 0)$ are called the ***vertices*** of the hyperbola, and the line segment determined by the vertices is called the ***transverse axis*** of the hyperbola. The hyperbola also has two oblique asymptotes, as we now show. Let us first consider the part of the graph in the first quadrant, corresponding to $x > 0$ and $y > 0$. Solving the equation in (1) for y (with $y > 0$), we obtain

$$\frac{y^2}{b^2} = \frac{x^2}{a^2} - 1$$

$$y^2 = b^2\left(\frac{x^2}{a^2} - 1\right)$$

$$y = b\sqrt{\frac{x^2}{a^2} - 1} = b\sqrt{\frac{x^2}{a^2}\left(1 - \frac{a^2}{x^2}\right)} = \frac{bx}{a}\sqrt{1 - \frac{a^2}{x^2}}$$

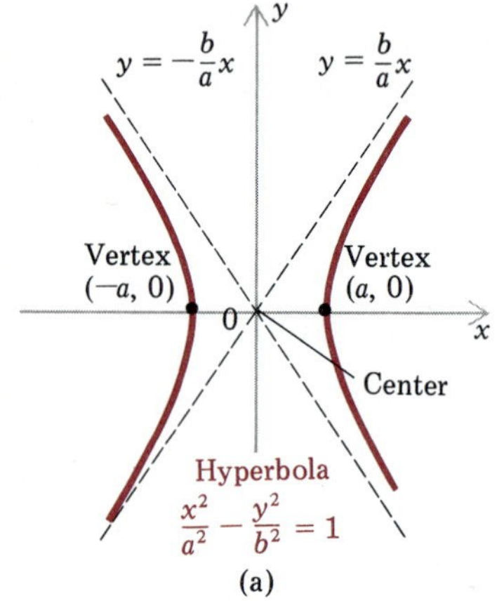

(a)

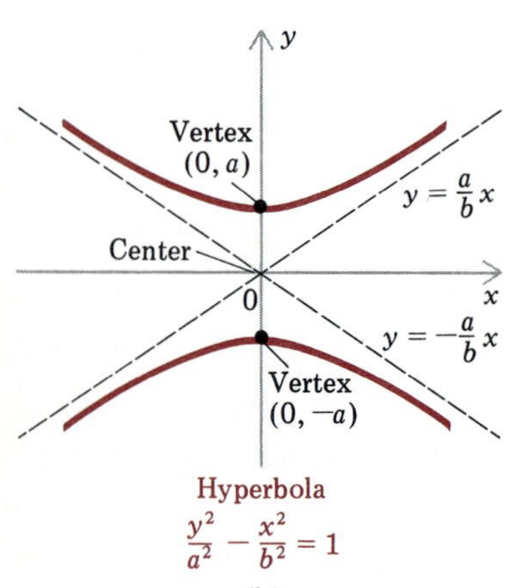

(b)

FIGURE 3.78

If x is positive and becomes arbitrarily large, a^2/x^2 gets close to 0, so $1 - (a^2/x^2)$ and hence $\sqrt{1 - (a^2/x^2)}$ get close to 1. Thus for large positive values of x, y is close to bx/a. This means that the line

$$y = \frac{b}{a}x$$

is an oblique asymptote of the part of the graph in the first quadrant.

Now let us consider the portions of the graph in the other quadrants. By the symmetry of the hyperbola with respect to the origin, the same line is an oblique asymptote of the portion of the hyperbola in the third quadrant (Figure 3.78a). It now follows from the symmetry of the graph with respect to the axes that the line

$$y = -\frac{b}{a}x$$

is an oblique asymptote of the part of the graph in the second and fourth quadrants, as is also indicated in Figure 3.78a.

The hyperbola given by

$$\boxed{\frac{y^2}{a^2} - \frac{x^2}{b^2} = 1} \qquad (2)$$

has no x intercepts but has the two y intercepts a and $-a$. In this case the points $(0, a)$ and $(0, -a)$ are called the ***vertices*** of the hyperbola and the line segment determined by the vertices is called the ***transverse axis***. In addition, the hyperbola has the two oblique asymptotes

$$y = \frac{a}{b}x \quad \text{and} \quad y = -\frac{a}{b}x$$

as indicated in Figure 3.78b.

One way to associate equations (1) and (2) with the corresponding graphs is this: If x comes first (as in (1)), then the graph crosses the x axis; if y comes first (as in (2)), then the graph crosses the y axis. It is also useful to note that, whereas the vertices help in graphing points closest to the center of the hyperbola, the asymptotes help in graphing points on the hyperbola far from the center.

EXAMPLE 1. Sketch the hyperbola having equation

$$\frac{x^2}{4} - y^2 = 1$$

and note the vertices and asymptotes.

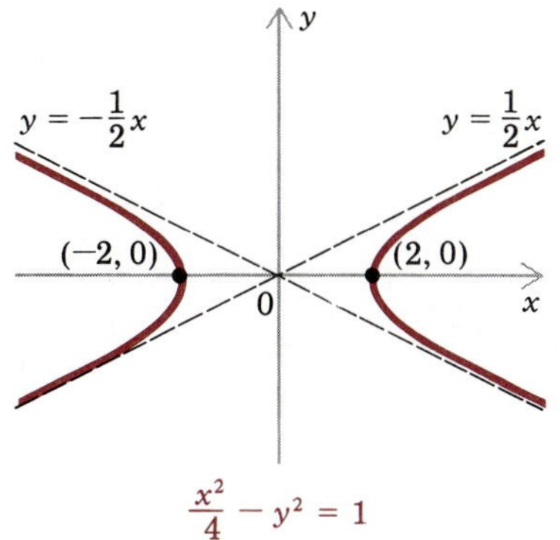

FIGURE 3.79

Solution. Since the given equation is equivalent to

$$\frac{x^2}{4} - \frac{y^2}{1} = 1$$

it has the form of equation (1) with $a = 2$ and $b = 1$. Therefore the vertices are $(-2, 0)$ and $(2, 0)$, and the asymptotes are

$$y = \frac{1}{2}x \quad \text{and} \quad y = -\frac{1}{2}x$$

The hyperbola is sketched in Figure 3.79. □

EXAMPLE 2. Sketch the hyperbola having equation

$$16y^2 - 9x^2 = 144$$

and note the vertices and asymptotes.

Solution. Dividing both sides of the equation by 144, we obtain

$$\frac{y^2}{9} - \frac{x^2}{16} = 1$$

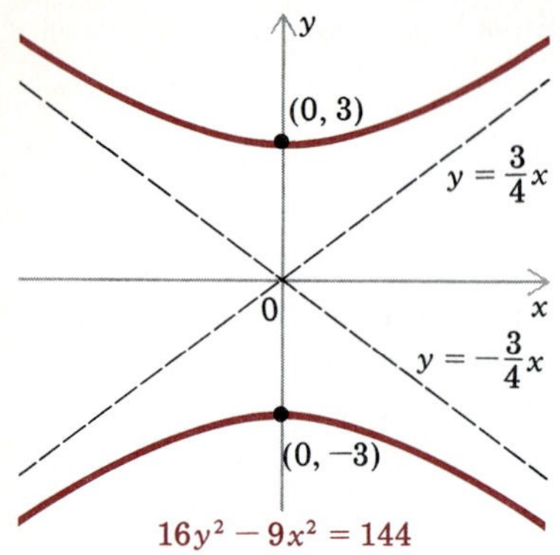

FIGURE 3.80

This equation has the form of (2) with $a = 3$ and $b = 4$. The vertices are $(0, -3)$ and $(0, 3)$, and the asymptotes are

$$y = \frac{3}{4}x \quad \text{and} \quad y = -\frac{3}{4}x$$

The hyperbola is sketched in Figure 3.80. □

Except in very special cases, the graph of any equation of the form

$$Ax^2 + By^2 + Cx + Dy = E \quad \text{with } A \text{ and } B \text{ of different signs} \tag{3}$$

is a hyperbola whose center can be any point in the plane, depending on the values of the constants. After completing squares and rearranging the constants, we can rewrite (3) either as

$$\frac{(x-h)^2}{a^2} - \frac{(y-k)^2}{b^2} = 1 \tag{4}$$

whose graph has vertices $(h - a, k)$ and $(h + a, k)$ and asymptotes

$$y - k = \frac{b}{a}(x - h) \quad \text{and} \quad y - k = -\frac{b}{a}(x - h)$$

(Figure 3.81a), or as

$$\frac{(y-k)^2}{a^2} - \frac{(x-h)^2}{b^2} = 1 \tag{5}$$

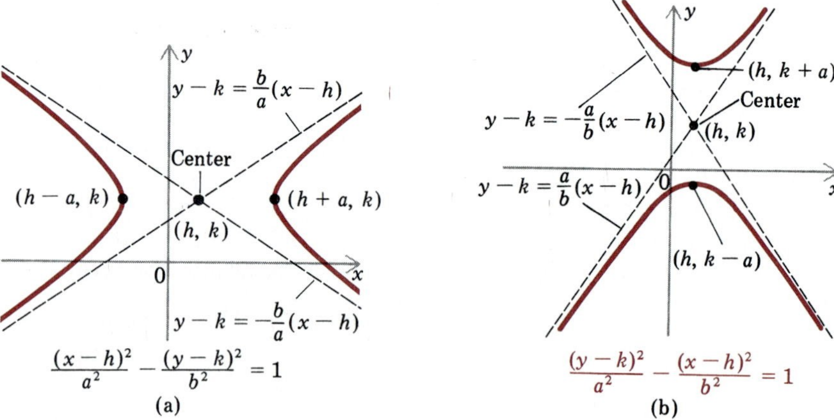

FIGURE 3.81

whose graph has vertices $(h, k - a)$ and $(h, k + a)$ and asymptotes

$$y - k = \frac{a}{b}(x - h) \quad \text{and} \quad y - k = -\frac{a}{b}(x - h)$$

(Figure 3.81b). In either case, the point (h, k) is the ***center*** of the hyperbola, and the hyperbola is symmetric with respect to the lines $x = h$ and $y = k$.

EXAMPLE 3. Sketch the hyperbola having equation

$$9x^2 - 25y^2 - 18x - 50y = -209$$

and note the center, vertices, and the asymptotes.

Solution. In order to put the equation into the form of (4) or (5), we complete squares:

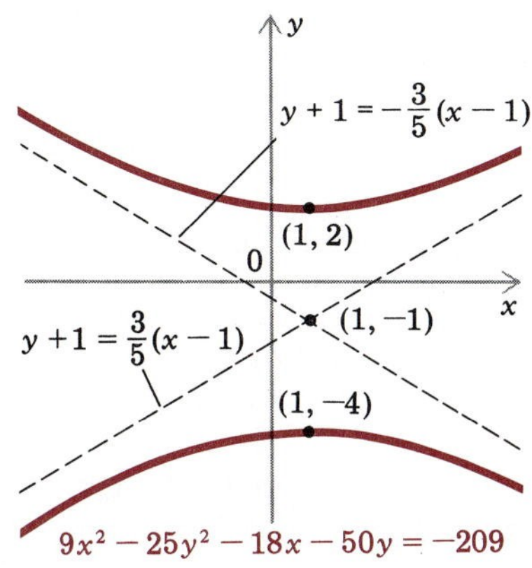

FIGURE 3.82

$$(9x^2 - 18x) - (25y^2 + 50y) = -209$$
$$9(x^2 - 2x) - 25(y^2 + 2y) = -209$$
$$9(x^2 - 2x + 1) - 9 - 25(y^2 + 2y + 1) + 25 = -209$$
$$9(x - 1)^2 - 25(y + 1)^2 = -209 + 9 - 25 = -225$$

Now we divide both sides by -225 and obtain

$$\frac{(y + 1)^2}{9} - \frac{(x - 1)^2}{25} = 1 \tag{6}$$

Equation (6) has the form of (5) with $h = 1$, $k = -1$, $a = 3$, and $b = 5$. As a result, the center is $(1, -1)$ and the vertices are $(1, -1 - 3) = (1, -4)$ and $(1, -1 + 3) = (1, 2)$. The asymptotes are

$$y + 1 = \frac{3}{5}(x - 1) \quad \text{and} \quad y + 1 = -\frac{3}{5}(x - 1)$$

The hyperbola is sketched in Figure 3.82. □

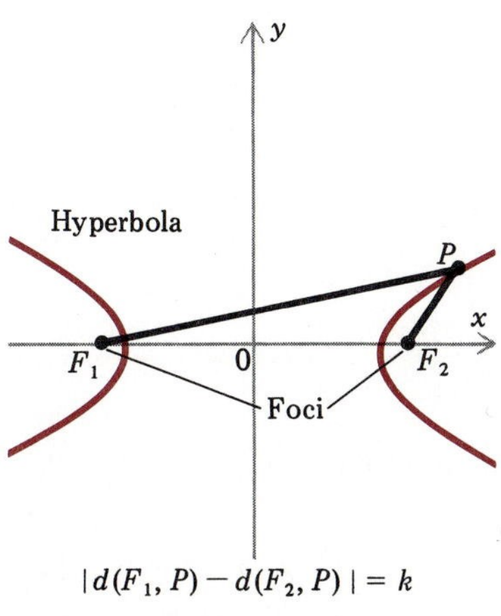

FIGURE 3.83

Like the parabola and ellipse, the hyperbola has a geometric description. Let F_1 and F_2 be two distinct points in the plane, and let k be a positive number less than the distance between F_1 and F_2. Then the set of all points P in the plane such that

$$|d(F_1, P) - d(F_2, P)| = k$$

is a hyperbola (Figure 3.83). The points F_1 and F_2 are called the ***foci*** of the hyperbola.

We mention a few applications of hyperbolas.

1. Comets that do not move in elliptical orbits around the sun almost always move in hyperbolic orbits. (In theory they can also move in parabolic orbits.) Such comets would not repeatedly orbit the earth as Halley's Comet does.
2. Boyle's Law, relating the pressure p and the volume V of a perfect gas at constant temperature, states that $pV = c$, where c is a constant. The graph of p as a function of V is a hyperbola.
3. Hyperbolas can be used to locate the source of a sound heard at three different locations.

Summary

As we mentioned at the outset of Section 3.6, the conic sections are generated by cutting a double cone by various planes (see Figure 3.61). Complementing this relationship between the different conic sections is the fact that our equations for the conic sections can be described in a unified way. More precisely, the graph of any equation of the form

$$Ax^2 + By^2 + Cx + Dy = E$$

is a conic section. The non-degenerate ones can be described as follows:

1. If $A \neq 0$ and $B = 0$, or if $A = 0$ and $B \neq 0$, then the graph is a parabola.
2. If A and B have the same sign, then the graph is an ellipse.
3. If A and B have different signs, then the graph is a hyperbola.

Our short introduction to conic sections gives you some of the basic information about them and their associated equations. Most books on calculus have a more extensive treatment of conic sections.

EXERCISES 3.8

In Exercises 1–14, sketch the hyperbola, noting the center, vertices, and asymptotes.

1. $x^2 - y^2 = 1$

2. $y^2 - x^2 = 4$

3. $\dfrac{x^2}{9} - \dfrac{y^2}{4} = 1$

4. $\dfrac{y^2}{25} - \dfrac{x^2}{16} = 1$

5. $25y^2 - 4x^2 = 100$

6. $4x^2 - y^2 = 4$

7. $\dfrac{(x+2)^2}{16} - \dfrac{(y+1)^2}{25} = 1$

8. $\dfrac{(y-1)^2}{1/4} - x^2 = 1$

9. $4y^2 - x^2 + 8y = 0$

10. $16x^2 - y^2 - 32x - 6y = 57$

11. $9x^2 - 16y^2 - 54x + 225 = 0$

12. $x^2 - 4y^2 - 4x - 8y = 4$

13. $9x^2 - 4y^2 - 18x - 24y = 63$

14. $9x^2 - 4y^2 - 18x - 24y = 26$

In Exercises 15–20, find an equation of the hyperbola with the given properties.

15. Center: (0, 0); x intercept: 2; asymptote: $y = -3x$

16. Center: (0, 0); y intercept: $-\frac{1}{2}$; asymptote: $y = \frac{3}{2}x$

17. Center: (0, 0); vertex: (0, 4); $(-4, -3\sqrt{3})$ on the hyperbola

18. Center: $(-1, 2)$; vertex: $(-1, -3)$; $(-3, -4)$ on the hyperbola

19. Center: $(2, -3)$; $(0, -3)$ and $(6, -3 + 2\sqrt{3})$ on the hyperbola

20. Center: $(7, -6)$; $(7, -1)$ on the hyperbola; asymptote: $y + 6 = \frac{5}{7}(x - 7)$

21. For what values of c does the hyperbola having equation $x^2 + 2x - c^2y^2 + y = 2$ pass through the point (1, 2)?

22. For what values of h does the hyperbola having equation

$$\frac{(x-h)^2}{16} - \frac{(y+1)^2}{36} = 1$$

pass through the point $(\sqrt{10}, 1)$?

23. a. Let a and b be positive numbers, and let u and v be numbers such that $u^2 - v^2 > 0$. Show that the point

$$\left(\frac{au}{\sqrt{u^2 - v^2}}, \frac{bv}{\sqrt{u^2 - v^2}}\right)$$

lies on the hyperbola having equation

$$\frac{x^2}{a^2} - \frac{y^2}{b^2} = 1$$

b. Taking $u = 13$ and $v = -5$ in part (a), find the corresponding point on the hyperbola having equation

$$\frac{x^2}{4} - \frac{y^2}{9} = 1$$

24. A hyperbola having an equation of the form

$$\frac{x^2}{a^2} - \frac{y^2}{b^2} = 1 \quad \text{or} \quad \frac{y^2}{b^2} - \frac{x^2}{a^2} = 1 \qquad (7)$$

with $a = b$ is called an ***equilateral hyperbola.*** Show that a hyperbola satisfying an equation of the form in (7) is equilateral if and only if its asymptotes are perpendicular to one another.

25. The ***eccentricity*** e of a hyperbola with an equation in the form of (4) or (5) is given by

$$e = \sqrt{1 + \frac{b^2}{a^2}}$$

Find the eccentricities of the following hyperbolas:

a. $\frac{x^2}{4} - \frac{y^2}{16} = 1$ b. $\frac{x^2}{16} - \frac{y^2}{4} = 1$

c. $\frac{x^2}{4} - \frac{y^2}{9} = 1$ d. $y^2 - x^2 = 1$

In Exercises 26–37, determine the type of conic section represented by the given equation. Determine the vertex of each parabola and the center of each ellipse or hyperbola.

26. $2x^2 - 4x - 3y + 3 = 0$

27. $x^2 + 3y^2 = 3$

28. $9x^2 - 4y^2 - 54x + 32y - 53 = 0$

29. $12y^2 - 12y + 8x + 15 = 0$

30. $x^2 + 6x - y + 11 = 0$

31. $9x^2 + 16y^2 + 72x - 96y + 287 = 0$

32. $9x^2 = 4y^2 - 8y + 3$

33. $x^2 + 6x + 9y^2 - 27 = 0$

34. $3y^2 + 6y - 2x = 3$

35. $25x^2 - 2y^2 = 100x + 8y - 142$

36. $9x^2 + 9y^2 + 6x - 24y = 28$

37. $9x^2 - 4y^2 - 54x + 32y - 19 = 0$

3.9 APPROXIMATING ZEROS OF FUNCTIONS

In Section 2.5 we defined an x intercept of the graph of a function f to be the x coordinate of any point where the graph of f meets the x axis (Figure 3.84). Thus c is an x intercept of f if $f(c) = 0$. Such numbers are also called *zeros* or *roots* of f.

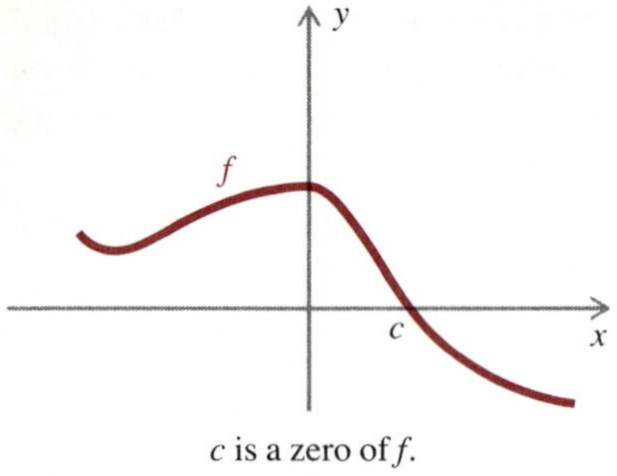

c is a zero of f.

FIGURE 3.84

A given function f may or may not have any zeros. For example, if $f(x) = x^2 + 1$, then f has no zeros because $f(x) = x^2 + 1 > 0$ for each real number x. Moreover, even if the graph of f has zeros, finding them may not be so easy. Indeed, suppose that

$$f(x) = x^5 + x + 1 \tag{1}$$

Does f have a zero? If so, how can we find a zero? In this section we will present two methods for approximating zeros of polynomials. The first is the bisection method, and can, in theory, be utilized without a calculator. The second method involves use of the zoom feature of a graphics calculator.

Bisection Method

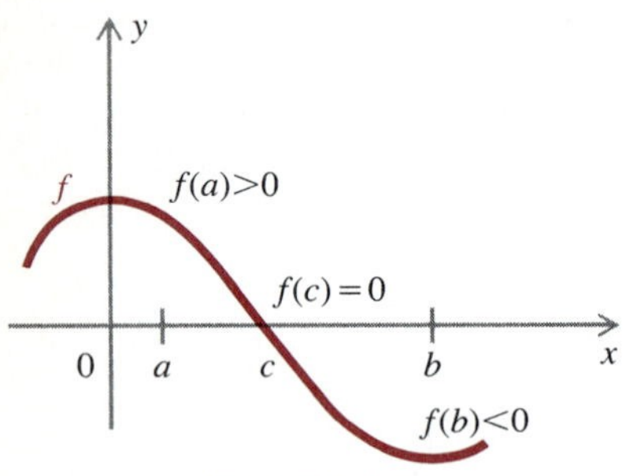

FIGURE 3.85

One of the really important theorems appearing in mathematics is the Intermediate Value Theorem, which tells us that if f is a polynomial function and if a and b are real numbers such that $f(a)$ and $f(b)$ have opposite signs (so that either $f(a) < 0$ and $f(b) > 0$ or else $f(a) > 0$ and $f(b) < 0$), then there is a number c between a and b such that $f(c) = 0$ (Figure 3.85). In other words, if a polynomial function assumes a positive value and a negative value, then it must have a zero somewhere in between. For example, let us reconsider the polynomial function in (1). Notice that $f(-1) = -1$ and $f(1) = 3$. Since $f(-1) < 0$ and $f(1) > 0$, it follows from the Intermediate Value Theorem that there is a number c between -1 and 1 such that $f(c) = 0$. Thus we now know that f has a zero. Unfortunately, as is usually the case, we cannot find the numerical value of c exactly. However, we can approximate the value of c by means of the bisection method, which we consider next.

The bisection method is based on the ancient Roman proverb "Divide and Conquer." We will first describe the procedure for the function in (1):

$$f(x) = x^5 + x + 1$$

Since $f(-1) < 0$ and $f(1) > 0$, the Intermediate Value Theorem assures us that there is a zero c of f in the interval $[-1, 1]$. Let x_1 denote the midpoint, 0, of the interval $[-1, 1]$. Then $f(x_1) = f(0) > 0$, so that $x_1 \neq c$. But whether $x_1 < c$ or $x_1 > c$, we know that $|x_1 - c| < 1$ (Figure 3.86a). Since $f(x_1) = f(0) > 0$ and $f(-1) < 0$, it follows by the Intermediate Value Theorem that there is a zero c of f in the interval $[-1, 0]$. Let x_2 denote the midpoint, $-\frac{1}{2}$, of the interval $[-1, 0]$. Then $|x_2 - c| < \frac{1}{2}$ (Figure 3.86b). The process can be continued until we obtain a number that is as close to c as we desire. We can assemble a table that gives the information obtained at each step of the process:

Interval	Length	Midpoint	Sign of f(midpoint)
$[-1, 1]$	2	0	+
$[-1, 0]$	1	$-\frac{1}{2}$	+

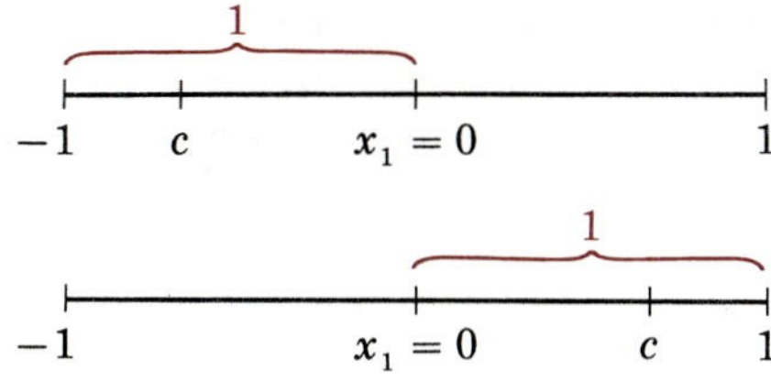

$|x_1 - c| < \frac{1}{2}$ whether $x_1 > c$ or $x_1 < c$

(a)

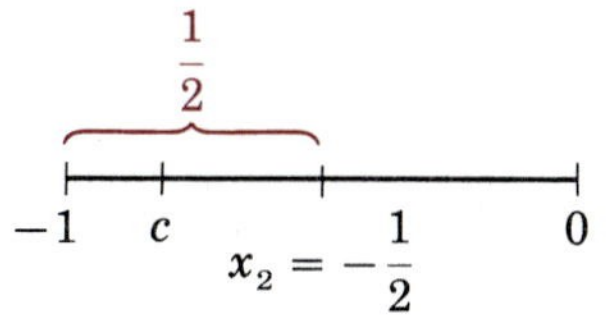

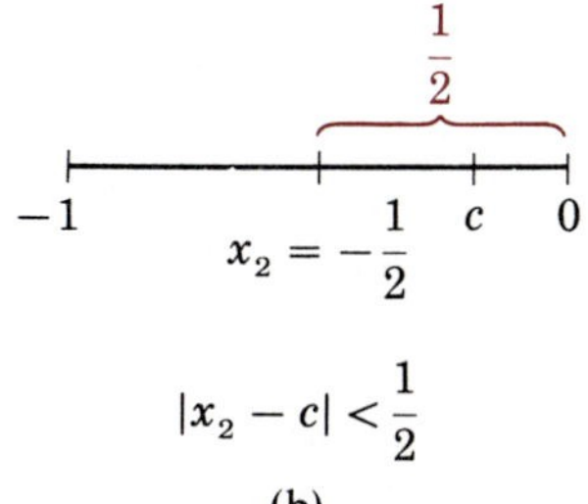

$|x_2 - c| < \frac{1}{2}$

(b)

FIGURE 3.86

When does the process terminate? If we seek a point that is guaranteed to be within, say, .01 of the zero c of f, then we stop when the length of the interval in the table is less than .02, since the midpoint of the interval will be less than .01 from each number in the subinterval (and the Intermediate Value Theorem assures that there is at least one zero of f in the interval). In general, if we wish to approximate a zero of f with error less than a given number e, then we stop when the length of the interval is less than $2e$, and take the approximate zero to be the midpoint of that final interval.

Notice that at each stage the interval I under scrutiny is half as large as the preceding interval (that is the basis of "Divide and Conquer"). For the next interval we choose the left or the right half of I, whichever has endpoints a and b such that $f(a)$ and $f(b)$ have opposite signs. And the process continues with this half of I as the interval under scrutiny.

EXAMPLE 1. Let $f(x) = x^3 + 2x + 2$. Use the bisection method to find a number in $[-1, 0]$ that is within .1 of a zero c of f.

Solution. Since $f(-1) = -1$ and $f(0) = 2$, the Intermediate Value Theorem guarantees a zero of f in the interval $[-1, 0]$. We assemble the following table:

Interval	Length	Midpoint	Sign of f(midpoint)
$[-1, 0]$	1	$-\frac{1}{2}$	+
$[-1, -\frac{1}{2}]$	.5	$-\frac{3}{4}$	+
$[-1, -\frac{3}{4}]$	.25	$-\frac{7}{8}$	−
$[-\frac{7}{8}, -\frac{3}{4}]$	.125	$-\frac{13}{16}$	

Since the length of the interval $[-\frac{7}{8}, -\frac{3}{4}]$ is .125, its midpoint, $-\frac{13}{16}$, is within $\frac{1}{2}(.125)$ and hence within .1 of a zero c of f that lies in the interval $[-1, 0]$. Figure 3.87 displays the successive midpoints x_1, x_2, x_3, and x_4. □

$\left[-1, -\frac{1}{2}\right]$ $\left[-\frac{7}{8}, -\frac{3}{4}\right]$

-1 $-\frac{7}{8}$ $-\frac{3}{4}$ $-\frac{1}{2}$ 0

$\left[-1, -\frac{3}{4}\right]$

FIGURE 3.87

That the bisection method leads to an arbitrarily close approximation to the zero of f in a given interval is assured because at each step the preceding interval is cut in half, and the Intermediate Value Theorem guarantees that a zero lies in one half or the other. If the midpoint of the interval happens to be the zero, so much the better. The calculations required for the successful implementation of the bisection method can obviously be facilitated by use of a calculator or computer.

Approximating a Zero by Graphics Calculator

Closely related to the bisection method is another method that can be implemented on a graphics calculator such as the Texas Instruments TI-81. This second method relies on graphing the function and then reading off the coordinates of the cursor as it moves along the graph of the function. On the TI-81 the method proceeds as follows:

Suppose we wish to approximate a zero c of f by a number x such that x and c are closer to each other than a distance e, that is, $|x - c| < e$. Then

1. Graph the function f on the TI-81. You should then be able to locate approximately where the graph touches the x axis. If there is more than one zero, you can choose the zero c you prefer to approximate.
2. First use the TRACE and then the $\langle$ and $\rangle$ keys to move the cursor along the graph until you locate adjacent points (x_1, y_1) and (x_2, y_2) near $(c, 0)$ with y_1 and y_2 having opposite signs. (By the Intermediate Value Theorem, a zero of f will lie between x_1 and x_2.)
3. If $|x_2 - x_1| < 2e$, then take the midpoint $\frac{1}{2}(x_1 + x_2)$ as the desired approximation x of c that is within a distance e from c. Otherwise, place the cursor on (x_1, y_1) or (x_2, y_2), whichever has the smaller absolute value for its y coordinate. Then zoom in and repeat Step 2 until $|x_2 - x_1| < 2e$.

We illustrate the method in the following example.

EXAMPLE 2. Let $f(x) = x^4 + 3x - 1$. Find a number in $[0, 1]$ that is guaranteed to be within .001 of a zero of f.

Solution. Using the standard zoom to graph f, we find that there are apparently two zeros, one in the interval $[-2, -1]$ and the other in the interval $[0, 1]$. We are to approximate the zero in $[0, 1]$ with error at most .001. Thus we have $e = .001$ and therefore must repeat step 2 until $|x_2 - x_1| < .002$. We assemble the following table as we zoom in and approximate the zero by using $\langle$ and $\rangle$ buttons:

Zoom	x_1	y_1	x_2	y_2	$\lvert x_2 - x_1\rvert$
1	.31578947	−.0426869	.52631579	.65568097	≈.21
2	.32631579	−.0097142	.34736842	.05666526	≈.021
3	.32736842	−.0064093	.32947368	2.0479E−4	≈.0021
4	.32936842	−1.261E−4	.32957895	5.3564E−4	≈.0002

With the fourth zoom we find that $|x_2 - x_1| < .002$. Therefore the midpoint x between x_1 and x_2, which is given by

$$x = \frac{1}{2}(.32936842 + .32957895) = .329473685$$

is such that $|x - c| < .001$, and hence x is the desired approximation of c. □

It is not necessary to record the values of y_1 and y_2, as we did in Example 2. Moreover, the solution of Example 2 yields a number x that is very close to c. How close is $f(x)$ to 0? On our TI-81 we find that

$$f(x) = 2.04789076\text{E}-4 = .000204789076$$

Thus we have found a number x that is close to a zero of f, and in addition we find that $f(x)$ is close to 0.

Approximating Solutions of Inequalities by Graphics Calculator

In the same manner as we found approximate zeros of functions, we can use a graphics calculator to find approximate solutions of inequalities that are more complex than those discussed in Chapter 1. We illustrate the procedure with an example.

EXAMPLE 3. Approximate to within .01 the numbers x for which $x^3 < 3x + 1$.

Solution. One way of proceeding is to let $f(x) = x^3 - 3x - 1$, and approximate the values of x for which $f(x) < 0$. From the graph of f we see that f has three zeros, b, c, and d, with $b < c < d$ (Figure 3.88). We see also that $f(x) < 0$ if and only if x lies in $(-\infty, b)$ or in (c, d). When we zoom in on the zero b, as in the solution of Example 2, we obtain

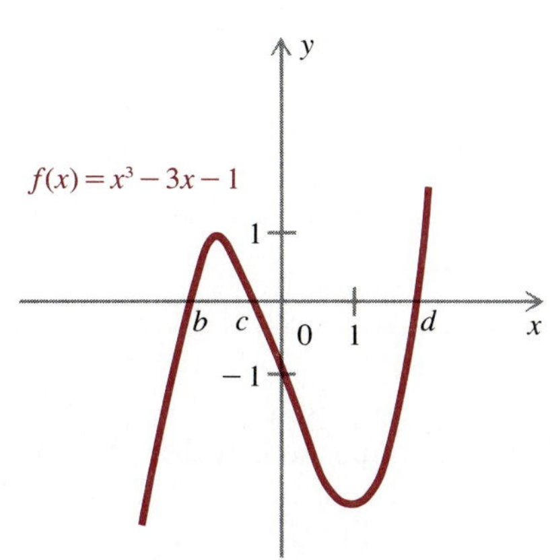

FIGURE 3.88

Zoom	x_1	y_1	x_2	y_2	$\lvert x_2 - x_1\rvert$
1	−1.578947	−.1995918	−1.368421	.54279049	.210526
2	−1.552632	−.0849756	−1.5	.125	.052632
3	−1.532895	−.0032601	−1.519737	.04922621	.013158

Taking the midpoint of the interval joining the final values of x_1 and x_2 yields the following value for b:

$$b \approx \frac{1}{2}(-1.532895 - 1.519737) = -1.526316$$

In the same way we find that

$$c \approx \frac{1}{2}(-.3355263 - .3486842) = -.34210525$$

and

$$d \approx \frac{1}{2}(1.875 + 1.8881579) = 1.88157895$$

Consequently the numbers x for which $x^3 < 3x + 1$ are approximately the numbers in either $(-\infty, -1.526316)$ or $(-.34210525, 1.88157895)$. □

We could alternatively solve Example 3 by first letting $f(x) = x^3$ and $g(x) = 3x + 1$, then graphing f and g on the graphics calculator, and finally zooming in on the intersections of the two graphs.

EXERCISES 3.9

In Exercises 1–4, use the bisection method to find a number in the given interval that is within .1 of the zero of f.

1. $f(x) = x^3 + x + 1; [-1, 0]$
2. $f(x) = x^3 - x + 1; [-2, -1]$
3. $f(x) = x^3 + x^2 + 1; [-2, -1]$
4. $f(x) = x^3 + 2x - 2; [0, 1]$

In Exercises 5–6, f has two zeros. Use the bisection method to find a number that is within .1 of each zero of f.

5. $f(x) = x^4 - 2x + 1$
6. $f(x) = -x^4 + x^3 + 3$

In Exercises 7–8, explain why the bisection method does not yield a zero of f.

7. $f(x) = x^4 - 2x^2 + 1$
8. $f(x) = x^4 + x^2 + .1$

In Exercises 9–14, use a graphics calculator to find a number that is within .001 of a zero of f. If f has more than one zero, approximate the one with smallest absolute value.

9. $f(x) = x^3 + x + 1$
10. $f(x) = x^3 - x + 1$
11. $f(x) = x^4 + x^3 + x^2 + x - 1$
12. $f(x) = x^4 - 3x^3 + 1$
13. $f(x) = x^5 - 2x^2 + 1$
14. $f(x) = x^6 - 2x^5 + 2x + 0.9$

In Exercises 15–20, use a graphics calculator to find the intervals for which the given inequalities hold. Approximate to within .01 accuracy.

15. $x^3 \le x^2 + 1$
16. $-x^3 + x^2 + 4x < 3$
17. $x^4 + x^3 \ge x - .3$
18. $x^4 + x^3 \ge x - .5$
19. $2x^3 + 3x - 1 < x^3 + x^2 - x - 2$
20. $x^4 - 2x^2 \le x^3 - x$
21. Find the values of c for which $x^4 + x^3 - x^2 - cx + 1 > 0$ for all real x. Approximate the values of c to within .01.

KEY TERMS

polynomial function
quadratic function
minimum value of a function
maximum value of a function
rational function
- vertical asymptote
- horizontal asymptote
- oblique asymptote

graph of an equation
- x intercept
- y intercept

symmetry
- with respect to the x axis
- with respect to the y axis
- with respect to the origin

bisection method
conic section
parabola
- axis
- vertex
- directrix
- focus

ellipse
- center
- vertex
- major axis
- minor axis
- focus

hyperbola
- center
- vertex
- transverse axis
- asymptote
- focus

REVIEW EXERCISES

In Exercises 1–4, sketch the graph of the given function, and identify the vertex, axis, and intercepts.

1. $f(x) = \frac{1}{4}x^2 - 4$

2. $f(x) = x^2 + 8x + 15$

3. $f(x) = -x^2 - 4x - 7$

4. $f(x) = x^2 - 3x + 2$

In Exercises 5–16, sketch the graph of the given function, and identify all intercepts and asymptotes.

5. $f(x) = 2 - x^3$

6. $f(x) = 4x^3 - x$

7. $f(x) = x^4 + 2$

8. $g(x) = -\dfrac{1}{(x-3)^2}$

9. $f(x) = -x^2(x-2)$

10. $f(x) = (x+1)(x-2)(x-3)$

11. $f(x) = x^4 - 10x^2 + 9$

12. $g(x) = \dfrac{2x}{x+3}$

13. $g(x) = \dfrac{x+1}{x-1}$

14. $y = \dfrac{(x-1)(x-3)}{(x+2)(x-4)}$

15. $y = \dfrac{x^2 + 3x + 2}{x^3 + 5x^2 + 4x}$

16. $y = \dfrac{x^2}{4x^2 - 1}$

In Exercises 17–18, factor $f(x)$ completely, and then sketch the graph of f.

17. $f(x) = 4x^3 - 16x^2 + 21x - 9$

18. $f(x) = 4x^3 + 8x^2 + x - 3$

In Exercises 19–20, use the bisection method to find a number in the given interval that is within .1 of the zero of f.

19. $f(x) = x^3 + 2x + 1$; $[-1, 0]$

20. $f(x) = x^5 - x^3 - 1$; $[0, 2]$

In Exercises 21–22, use a graphics calculator to find a number that is within .001 of a zero of f. If f has more than one zero, approximate the one with smallest absolute value.

21. $f(x) = x^4 - x^3 - 1$

22. $f(x) = x^5 - x^3 + 1$

In Exercises 23–30, sketch the graph of the equation, noting any intercepts and symmetry.

23. $(x + \sqrt{2})(y - \sqrt{3}) = 0$

24. $x^2 = y^2$

25. $|x - 4| = |y + 1|$

26. $x^2 + y^2 = 16$

27. $(x - 1)^2 = 4 - (y + 3)^2$

28. $7 + y^2 = x$

29. $y^2 = x + 3$

30. $x^2 + 4x - 12 = 0$

In Exercises 31–36, sketch the conic section, indicating all pertinent information (center, axes, vertices, asymptotes).

31. $x^2 + y^2 + 2x - 4y + 2 = 0$

32. $6x + y^2 + 6y = 0$

33. $x^2 - y^2 - x - y = 4$

34. $4x^2 + 25y^2 - 8x + 100y + 4 = 0$

35. $12x^2 + 72x + 72 = 9y^2 + 72y$

36. $18y^2 - 36y - 2x + 10 = 0$

37. Let $f(x) = 3x^2 + bx - 4$. One x intercept of the graph of f is 4. Determine the other x intercept.

38. Let $f(x) = 3x^2 + 2x - c$. Determine the value of c for which
a. the graph contains the point $(-1, 4)$.
b. the graph has exactly one x intercept.
c. the x intercepts are 6 units apart.

39. Find the values of h for which the point $(-5, 1)$ lies on the ellipse having equation

$$\frac{(x-h)^2}{25} + \frac{(y+5)^2}{100} = 1$$

40. Show that if u is any real number, then the point

$$\left(\frac{u^2-1}{u^2+1}, \frac{2u}{u^2+1}\right)$$

lies on the circle $x^2 + y^2 = 1$.

41. The Student Union plans to organize a group trip to Mexico during spring vacation. The cost per person is set at \$400 for 300 people, and it decreases \$1 for each additional person up to 100 additional people. What is the cost per person that brings the maximum revenue, and how many people must participate at that cost?

42. Find two numbers whose product is as large as possible if the sum of twice one number and 3 times the other is 36.

43. The height at time t of a cannon ball fired at a 30° angle with respect to the ground is given by

$$h = -16t^2 + \frac{1}{2}v_0 t$$

where v_0 is the initial, or muzzle, velocity. If the muzzle velocity is 1024 feet per second, determine the maximum height attained by the cannon ball.

44. If the muzzle velocity of the cannon in Exercise 43 is doubled, how is the maximum height affected?

45. A wire 2 feet long is to be cut into 2 pieces, one of which will be bent into the shape of a square and the other into the shape of a circle. Determine the lengths into which the wire must be cut if the sum of the areas of the square and circle is to be minimum.

46. What is the largest possible area of a rectangular tablecloth whose perimeter is 24 feet?

47. Let $f(x) = x^3 + x^2$ for $0 \le x \le 1$. Determine the dimensions of the rectangle of maximum area that has base along the x axis and can be inscribed in the shaded region in Figure 3.89. (*Hint:* The area can be expressed as a quadratic function of x^2. Find the value of x^2 that yields the maximum area.)

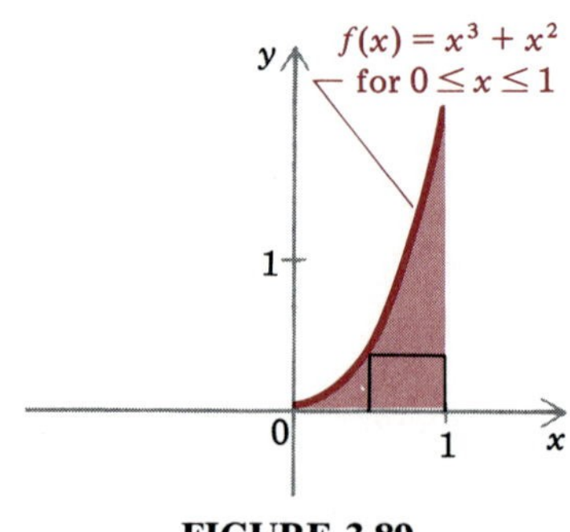

FIGURE 3.89

CUMULATIVE REVIEW I: CHAPTERS 0–3

In Exercises 1–4, simplify the given expression.

1. $\sqrt[3]{-\dfrac{81}{64}}$

2. $\left(\dfrac{x - x^{-1}}{x - x^{-2}}\right)^{-1}$

3. $\dfrac{(x\sqrt[3]{x}y^{2/5})^{15/2}}{x^{1/4}y^2}$

4. $(3x - y)^3 - (x + 2y)^3$

In Exercises 5–6, factor the given polynomial.

5. $(x - 3)^2(2x - 3)^4 + 12(x - 3)^3(2x - 3)^2$

6. $8x^3 - 12x^2 - 6x + 9$

In Exercises 7–10, write as one rational expression in simplified form.

7. $\dfrac{\dfrac{4}{x^2} - 2x}{3x + \dfrac{1}{x}}$

8. $\dfrac{6x - 3}{2x^3 - x^2 - 2x + 1}$

9. $\dfrac{x + 4}{x^2 - 4} - \dfrac{x - 4}{(x + 2)^2}$

10. $\dfrac{\sqrt{a + 3} + \sqrt{a}}{\sqrt{a + 3} - \sqrt{a}}$

In Exercises 11–16, find all solutions of the given equation.

11. $|3 - 4x| = 2$

12. $x^2 - \frac{3}{2}x - 2 = 0$

13. $\dfrac{2x^2 - 3}{4x - 1} = -2$

14. $\sqrt{x + 2} = 1 + \sqrt{1 - 3x}$

15. $3x^2 + 9 = 0$

16. $-x^2 + 2x - 2 = 0$

In Exercises 17–18, solve the given inequality and express the solution in terms of intervals.

17. $(4x - 5)(x^2 - 1) \le 0$

18. $\dfrac{x + 4}{2x - 1} \ge -2$

19. Let l be the line with equation $2x - 3y = -1$.
   a. Shown that the point $(-1, -1/3)$ is on l.
   b. Find an equation for the line l_1 that passes through the point $(-1, -1/3)$ and is perpendicular to l.

20. Consider the points $P = (\frac{1}{2}, -1)$ and $Q = (-\frac{3}{4}, -2)$.
   a. Find the distance between P and Q.
   b. Determine the slope of the line l that contains P and Q.
   c. Find an equation of l.
   d. Find an equation of the line l_1 that is parallel to l and passes through the point (2, 3).

21. Let $f(t) = \sqrt{4 - t^2}$.
   a. Determine the domain of f.
   b. Calculate any values of t for which $f(t) = t$.
   c. Calculate the values of t for which $f(f(t)) = t$.

22. Let $f(x) = \sqrt{x^2 - 16}$ and $g(x) = x^3 + 4$. Find the domain and rule of
   a. $g \circ f$
   b. $f \circ g$

23. Let $f(x) = (3 - 2x)/(1 + x)$. Find a formula for f^{-1}.

24. The equation

$$I = I_0\left(\frac{n - 1}{n + 1}\right)^2$$

occurs in the theory of the polarization of light. Solve the equation for n.

In Exercises 25–28, sketch the graph of the given function, indicating any intercepts, as well as vertices and asymptotes when applicable.

25. $f(x) = x^2 - 4x + 7$

26. $f(x) = x^3 - 4x^2 - 4x + 16$

27. $f(x) = x^2(x + 1)^3$

28. $f(x) = \dfrac{x - 2}{x^2 - 1}$

In Exercises 29–32, sketch the conic section, indicating all pertinent information (center, axes, vertices, asymptotes).

29. $\dfrac{(x - 2)^2}{9} + \dfrac{(y + 1)^2}{9} = 1$

30. $y^2 + 4y - 2x + 6 = 0$

31. $y^2 - 2x^2 - 2y - 4x = 17$

32. $x^2 + 4y^2 - 4y = 0$

33. Find an equation of the ellipse whose major axis is bounded by the points $(-1, 1)$ and $(-1, -5)$, and whose minor axis is bounded by the points $(-3, -2)$ and $(1, -2)$.

34. Find an equation of the hyperbola whose center is $(2, 4)$, one of whose vertices is $(2, 7)$, and one of whose asymptotes is the line $y - 4 = \sqrt{3}(x - 2)$.

35. Find a simple relationship that exists between a and b if

$$\sqrt{a^2 - ab} = \sqrt{3ab - 4b^2}$$

36. a. Let t and u be numbers greater than -1. Show that if $t < u$, then

$$\frac{t}{1+t} < \frac{u}{1+u}$$

b. Using part (a), show that for any two numbers x and y,

$$\frac{|x+y|}{1+|x+y|} \leq \frac{|x|}{1+|x|} + \frac{|y|}{1+|y|}$$

37. a. Show that $\dfrac{2}{x} = \dfrac{2}{x+1} + \dfrac{2}{x(x+1)}$.

b. The reciprocal $1/n$ of an integer n is called a ***unit fraction***. To simplify calculations, the ancient Egyptians expressed fractions as sums of unit fractions. Using (a) with x replaced by n, show that if n is an odd integer greater than 1, then $2/n$ is the sum of two distinct unit fractions. (*Hint:* If n is odd, then $n + 1$ is divisible by 2.)

c. Write $2/13$ as the sum of two unit fractions.

38. Bill Rodgers, the winner of the 1978 Boston Marathon, ran the 26.21875 mile distance in 2 hours, 10 minutes, and 13 seconds.

a. Determine his average speed.

b. Determine the average length of time it took him to run each mile.

39. It has been determined experimentally that the concentration C of a pesticide on orange leaves at any time t (one day or longer after spraying) is inversely proportional to the $3/2$ power of t. Express this by means of a formula.

40. The purchase price P of a given length of wire is proportional to the cross-sectional area x of the wire. The cost L due to power loss while the wire is in use is inversely proportional to x. Express the total cost C, which is the sum of P and L, in terms of x. (There will be two constants of proportionality, but they need not be equal.)

41. An open box is to be made from a rectangular piece of cardboard 24 inches long and 16 inches wide by cutting a square piece x inches on a side from each of the four corners and then bending up the sides. Write a formula for the volume V of the resulting box.

42. The rate R of emission of radiant energy from the surface of a body is directly proportional to the 4th power of the temperature T.

a. Express this fact by means of a formula.

b. The constant of proportionality varies according to the nature of the surface of the body. If R is measured in watts per square meter and the temperature in degrees Kelvin, then for copper the constant is approximately 1.70097×10^{-8}. Suppose the temperature on the surface of a copper wire is 300° Kelvin. Use a calculator to determine the rate R of emission of radiant energy. Write your answer in scientific notation.

43. French dressing consists of approximately 72% olive oil, 24% vinegar (or lemon juice), and the remainder salt and pepper. Suppose we have 1 cup of French dressing with 24% vinegar. How much vinegar must be added in order to yield a 25% level of vinegar?

44. Determine how many gallons of a 70% saline solution must be added to 8 gallons of a 20% saline solution in order to obtain a 30% saline solution.

45. A tank can be filled by two pipes together in one hour less than would be required by the larger pipe, and in 4 hours less then would be required by the smaller pipe. How long would it take the smaller pipe to fill the tank? (*Hint:* Let x be the number of hours required by the smaller pipe to fill the tank and y the number for the larger pipe. Express the numbers of hours required for the two pipes to fill the tank together in terms of x and y.)

46. A garden is 4 yards long and 3 yards wide. There is a crosswalk of uniform width, as in Figure I.1. If the area of the crosswalk is $3\frac{1}{4}$ square yards, how wide is the crosswalk?

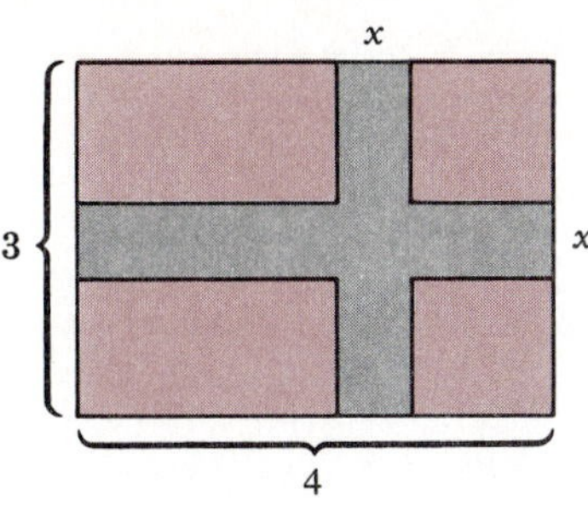

FIGURE I.1

47. When Old Faithful geyser in Yellowstone National Park first erupts, it can send water up some 150 feet.
   a. A water drop attains a maximum height of 150 feet above ground. Assuming no air resistance, determine the velocity of the drop when it leaves the ground.
   b. Using the information of part (a), determine the length of time the drop is in the air.

48. A university plans to rope off a rectangular field for a football game. It has 2000 feet of rope, and it plans to leave a 40-foot entrance free of rope. Find the dimensions of the field with the greatest possible area.

49. The center of gravity of a broad jumper during a given jump traces a curve given by

$$y = -\frac{5}{324}(x^2 - 16x - 260)$$

for all $x \geq 0$ such that $y \geq 0$. Here x and y are measured in feet.
   a. Determine the nonnegative values of x for which $y \geq 0$.
   b. Determine the maximum height of the center of gravity during the jump.

50. Find the value of x in the interval $[0, 2]$ for which the area of the triangle in Figure I.2 is as large as possible.

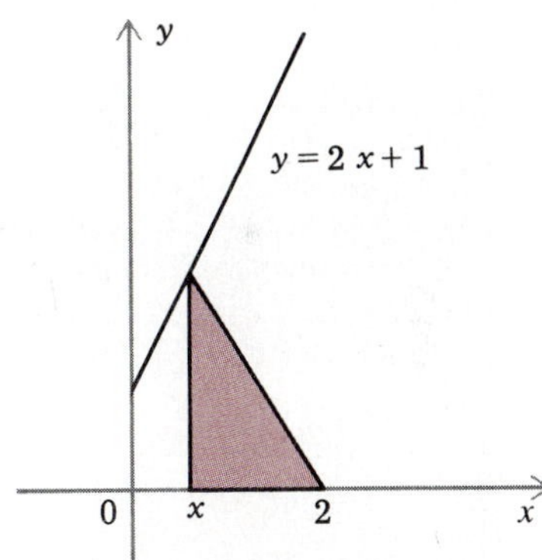

FIGURE I.2

4 EXPONENTIAL AND LOGARITHMIC FUNCTIONS

This chapter is devoted to two kinds of functions that cannot be formed by any of the elementary algebraic operations (addition, subtraction, multiplication, division, and roots). These are the exponential and logarithmic functions. They not only are extremely important in advanced mathematics but also occur with great frequency in the study of such varied topics as radioactivity, noise level, and atmospheric pressure.

For a sample of the applications of exponential and logarithmic functions, consider the following problems:

1. A mummy, known as Whiskey Lil, was discovered in 1955 in Chimney Cave, Lake Winnemucca, Nevada. By carbon dating it was learned that approximately 0.739 of the original C^{14} per gram was still present in 1955. When did Whiskey Lil die?
2. The great San Francisco earthquake of 1906 measured 8.3 on the Richter scale. In 1989 San Francisco suffered a less intense earthquake that measured 7.1 on the Richter scale. How many times more intense was the 1906 earthquake than the 1989 earthquake?

Exponential functions will help us solve the first problem (see Example 6 in Section 4.5), whereas logarithmic functions give us the necessary tools to solve the second problem (see Example 4 in Section 4.4).

Numerical answers to questions such as those mentioned above involve exponential or logarithmic functions, and generally they can only be given approximately in decimal form. For our computations we will use a calculator and give answers that are accurate to ten places.

4.1 EXPONENTIAL FUNCTIONS

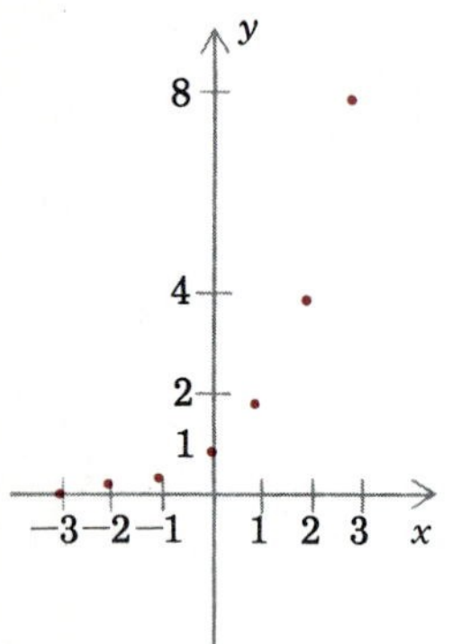

$y = 2^r$ for selected integer values of r
(a)

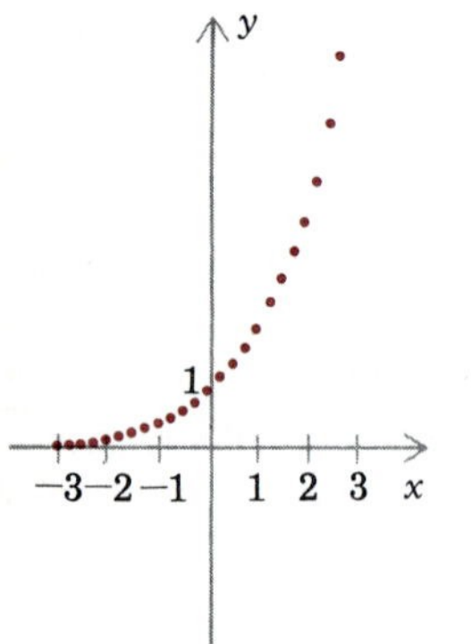

$y = 2^r$ for many rational values of r
(b)

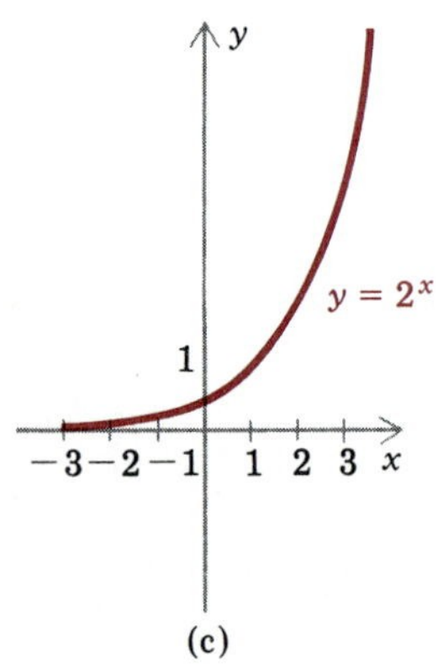

(c)

FIGURE 4.1

In Section 0.4 we defined the expression a^r for any $a > 0$ and any *rational* number r. However, until now we have not defined expressions such as $2^{\sqrt{3}}$, where the exponent is irrational. In this section we will give a meaning to a^x, in which x can be any number, irrational or rational.

To see how a^x can be defined for an arbitrary irrational number x, let us select a specific number $a > 0$, say $a = 2$. By plotting points from the table

r	-3	-2	-1	0	1	2	3
2^r	$\frac{1}{8}$	$\frac{1}{4}$	$\frac{1}{2}$	1	2	4	8

we obtain Figure 4.1a. The more points $(r, 2^r)$ we plot for other rational values of r, the more the points seem to fall on a smooth curve (Figure 4.1b). In fact, there is exactly one smooth curve that contains all points of the form $(r, 2^r)$, where r is rational (Figure 4.1c). As a result, for any number x we define 2^x to be the y coordinate of the point that lies on the curve directly above x on the x axis. From Figure 4.1c we see that the curve contains the point $(0, 1)$ and slopes upward to the right, which means that

$$\text{if } x < z, \quad \text{then} \quad 2^x < 2^z$$

Notice that if r is close to x, then 2^r is close to 2^x, and moreover, the closer r is to x, the closer 2^r is to 2^x. Thus we can think of 2^x as the "limiting value" of 2^r, where r is rational and r gets closer and closer to x. For example, if $x = \sqrt{3}$, we could let r have the values

$$1, \quad 1.7, \quad 1.73, \quad 1.732, \quad 1.73205, \quad 1.7320508, \ldots$$

which are obtained by taking more and more of the decimal expansion of $\sqrt{3}$, so that the successive values of r approach $\sqrt{3}$. The corresponding values of 2^r are

$$2^1, \quad 2^{1.7}, \quad 2^{1.73}, \quad 2^{1.732}, \quad 2^{1.73205}, \quad 2^{1.7320508}, \ldots$$

By calculator we obtain the following table for the numbers 2^r:

r	1	1.7	1.73	1.732	1.73205	1.7320508
2^r	2	3.249009585	3.317278183	3.321880096	3.321995226	3.321997068

It appears from the table that the decimal expansion of $2^{\sqrt{3}}$ begins 3.32199 . . .

The situation is analogous if a is any fixed number with $a > 1$. For rational values of r the points (r, a^r) lie on a unique smooth curve, and for any number x we define a^x to be the y coordinate of the point lying on that

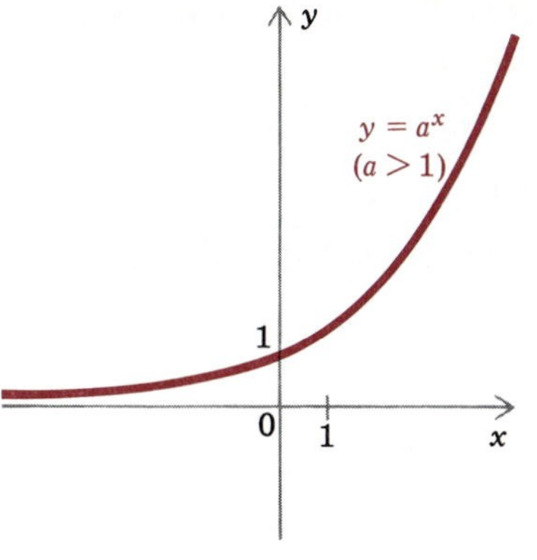

FIGURE 4.2

curve directly above the number x on the x axis (Figure 4.2). Notice also that the point (0, 1) is on the curve, and the curve slopes upward to the right, which means that

$$\text{for} \quad a > 1: \quad \text{if } x < z, \quad \text{then} \quad a^x < a^z \tag{1}$$

Moreover, if x is negative with $-x$ very large, then a^x is close to 0, whereas if x is positive and very large, then a^x is very large.

Now if $0 < a < 1$, then the method of defining a^x is the same. The main difference is that the smooth curve connecting the points (r, a^r) for rational values of r slopes downward to the right, which means that

$$\text{for} \quad 0 < a < 1: \quad \text{if } x < z, \quad \text{then} \quad a^x > a^z \tag{2}$$

(Figure 4.3). For example, $(\frac{1}{3})^2 > (\frac{1}{3})^5$. Moreover, if x is positive and very large, then a^x is close to 0, whereas if x is negative and $-x$ is very large, then a^x is very large.

Finally, if $a = 1$, then $a^r = 1^r = 1$ for any rational value of r, so the points (r, a^r) all lie on the horizontal line $y = 1$ (Figure 4.4). By defining 1^x to be the y coordinate of the point on the line $y = 1$ above x on the x axis, we find that $1^x = 1$ for any real number x.

Led by our discussion above, we define a^x for any fixed $a > 0$ and any number x as follows:

> $a^x =$ the y coordinate of the point that lies above the point x on the x axis and lies on the smooth curve that contains all points of the form (r, a^r), where r is a rational number.

No matter what the values of a and x are (so long as $a > 0$), a^x is the "limiting value" of a^r, where r is rational and r gets closer and closer to x.

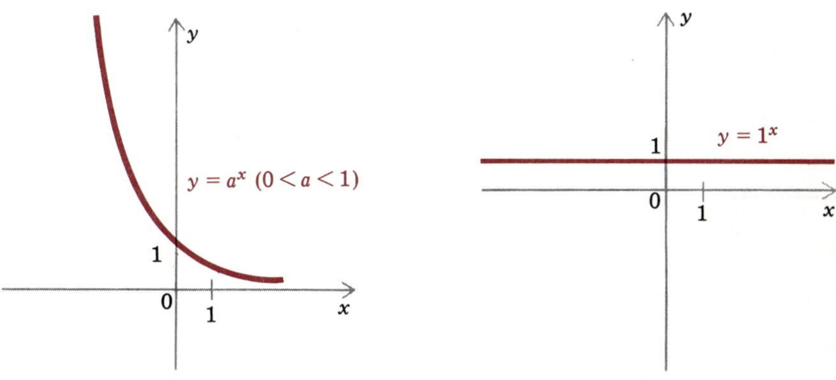

FIGURE 4.3

FIGURE 4.4

The "definition" we have just presented is intuitive, not a rigorous mathematical definition, which must to be left to more advanced texts. Nevertheless, we will abide by our "definition," which is consistent with the definition of a^r given earlier for rational values of r.

In the expression a^x the number a is called the ***base*** and x the ***exponent*** or ***power***. By holding the base fixed and allowing the exponent to vary, we obtain a function.

DEFINITION 4.1 Let $a > 0$. The function f defined by

$$f(x) = a^x$$

is called the ***exponential function with base a***.

For any $a > 0$ with $a \neq 1$, the following are properties of the function a^x and its graph:

1. $a^x > 0$ for all x, and the range of a^x consists of *all* positive numbers.
2. The y intercept is 1, and there is no x intercept.
3. The x axis is a horizontal asymptote of the graph of a^x.
4. If $a > 1$, then a^x is an increasing function (by (1)).
   If $0 < a < 1$, then a^x is a decreasing function (by (2)).

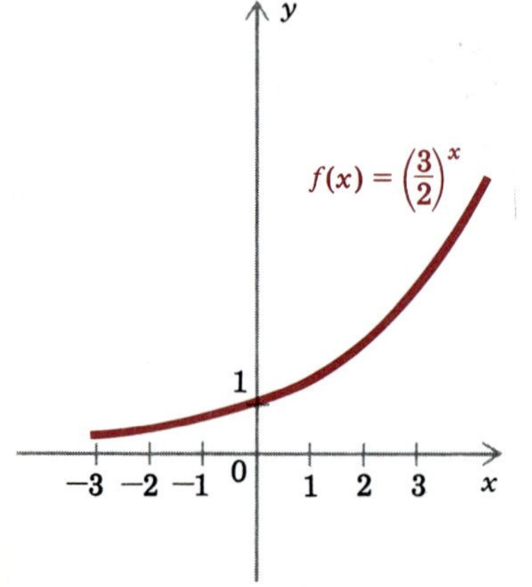

FIGURE 4.5

As we did earlier when $a = 2$, we can sketch the graph of the exponential function a^x for any specific positive value of a by computing a few values of a^x for specially chosen rational values of x and then connecting the corresponding points on the graph with a smooth curve. Of course, the more values we compute, the more accurate our sketch is likely to be.

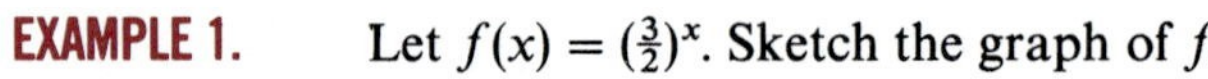

EXAMPLE 1. Let $f(x) = (\frac{3}{2})^x$. Sketch the graph of f.

Solution. Using the values in the table

x	-3	-2	-1	0	1	2	3
$f(x)$	$\frac{8}{27}$	$\frac{4}{9}$	$\frac{2}{3}$	1	$\frac{3}{2}$	$\frac{9}{4}$	$\frac{27}{8}$

we plot the corresponding points. Then we sketch a smooth curve through the points (Figure 4.5). □

EXAMPLE 2. Let $g(x) = (\frac{1}{3})^x$. Sketch the graph of g.

Solution. We make a short table of values:

x	-2	-1	0	1	2
$g(x)$	9	3	1	$\frac{1}{3}$	$\frac{1}{9}$

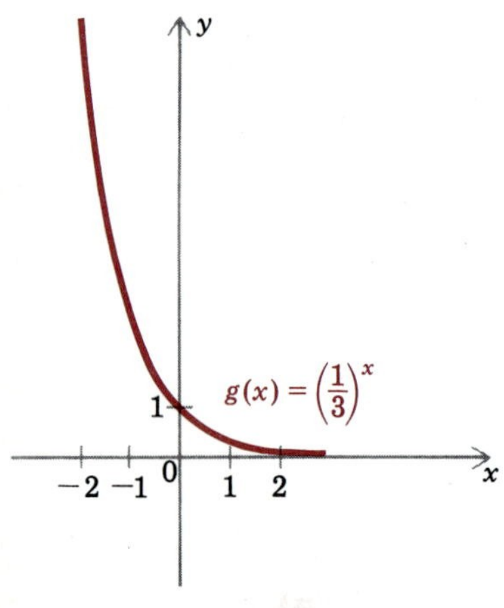

FIGURE 4.6

Then we plot the corresponding points and connect them with a smooth curve (Figure 4.6). □

The various laws of exponents that apply to a^x for rational values of x remain valid for arbitrary real values of x. Although we will not prove the laws, we list them below:

Laws of Exponents

Let a and b be fixed positive numbers. For any numbers x and y we have

i. $a^x a^y = a^{x+y}$ ii. $(a^x)^y = a^{xy}$

iii. $(ab)^x = a^x b^x$ iv. $1^x = 1$

v. $\left(\dfrac{a}{b}\right)^x = \dfrac{a^x}{b^x}$ vi. $\dfrac{a^x}{a^y} = a^{x-y}$

vii. $a^{-x} = \dfrac{1}{a^x} = \left(\dfrac{1}{a}\right)^x$ viii. $a^0 = 1$

EXAMPLE 3. Simplify $(4^{\sqrt{6}} \cdot 16^{\sqrt{6}})/(4^{3\sqrt{6}-1})$.

Solution. By the Laws of Exponents,

$$\frac{4^{\sqrt{6}} \cdot 16^{\sqrt{6}}}{4^{3\sqrt{6}-1}} = \frac{4^{\sqrt{6}}(4^2)^{\sqrt{6}}}{4^{3\sqrt{6}-1}} = \frac{4^{\sqrt{6}} \cdot 4^{2\sqrt{6}}}{4^{3\sqrt{6}-1}} = \frac{4^{\sqrt{6}+2\sqrt{6}}}{4^{3\sqrt{6}-1}}$$

$$= \frac{4^{3\sqrt{6}}}{4^{3\sqrt{6}} \cdot 4^{-1}} = \frac{1}{4^{-1}} = 4 \quad \square$$

From Law (vii) of Exponents,

$$3^{-x} = \left(\frac{1}{3}\right)^x$$

so that the graph of 3^{-x} is the same as the graph of $(\frac{1}{3})^x$, which appears in Figure 4.6. We will use this fact in the solution of the following example.

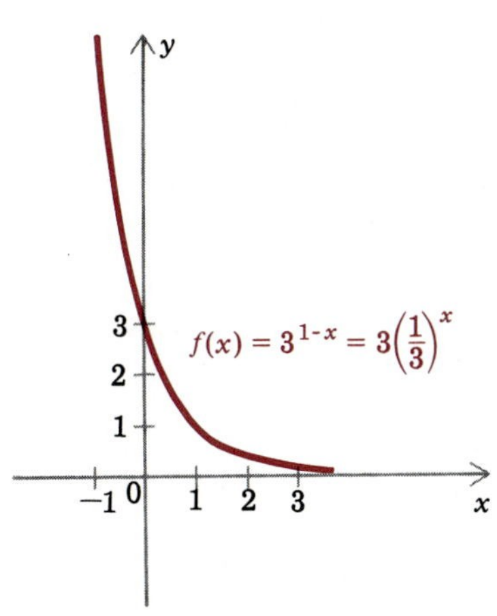

FIGURE 4.7

EXAMPLE 4. Let $f(x) = 3^{1-x}$. Sketch the graph of f.

Solution. By the Laws of Exponents and our remark above,

$$f(x) = 3^{1-x} = 3^1 3^{-x} = 3\left(\frac{1}{3}\right)^x$$

It follows that the graph of f can be obtained by multiplying the y coordinate of each point on the graph of $(\frac{1}{3})^x$ (shown in Figure 4.6) by 3. The resulting graph of f appears in Figure 4.7. $\square$

Another way to solve Example 4 is to observe that

$$3^{1-x} = \left(\frac{1}{3}\right)^{-(1-x)} = \left(\frac{1}{3}\right)^{x-1}$$

Thus the graph of 3^{1-x} is the same as the graph of $(1/3)^{x-1}$, which can be obtained by shifting the graph of $(1/3)^x$ one unit to the right. A comparison of Figures 4.6 and 4.7 bears this out.

The Exponential Function e^x

Of all the possible bases a for an exponential function, one outshines the others in importance. It is the one which was first called e by the Swiss mathematician Leonhard Euler (1707–1783). This number e is an irrational number whose decimal expansion begins

$$e = 2.71828182845904523536\ldots$$

One mathematical way of obtaining e is as the limit of numbers of the form $(1 + 1/x)^x$ for positive values of x that become arbitrarily large. For a few integral values of x we have

$$\left(1 + \frac{1}{10}\right)^{10} \approx 2.593742460 \quad \left(1 + \frac{1}{100}\right)^{100} \approx 2.704813829$$

$$\left(1 + \frac{1}{1000}\right)^{1000} \approx 2.716923932 \quad \left(1 + \frac{1}{10{,}000}\right)^{10{,}000} \approx 2.718145927$$

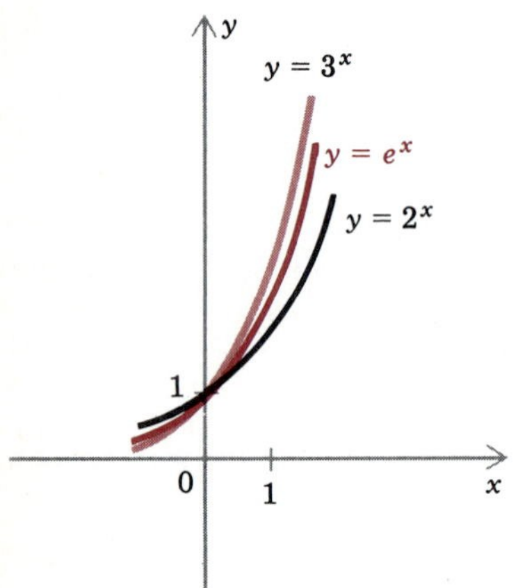

FIGURE 4.8

The function e^x is called the ***natural exponential function***, or simply the ***exponential function***. It is remarkable how often it arises both inside and outside of mathematics (see Sections 4.5 and 4.6 for a few applications). Since $2 < e < 3$, the graph of e^x has the same general shape as the graphs of 2^x and 3^x and lies between them (Figure 4.8).

Since e^x occurs so frequently in applications, it is useful to be able to approximate its value for various values of x. The method of obtaining e^x on a calculator varies, depending on the instrument. But the basic idea is the same. To evaluate $e^{0.256}$ on some calculators we would press the decimal key, then the 2, 5, and 6 keys and finally the e^x key. On others we would first press the e^x key and then the decimal, 2, 5, and 6 keys. Displayed on the calculator would be the approximate value 1.291752728, so that

$$e^{0.256} \approx 1.291752728$$

Next we will sketch the graph of a function related to the exponential function and basic to the study of probability.

EXAMPLE 5. Let $f(x) = e^{-(x^2)}$. Sketch the graph of f.

Solution. Because

$$f(-x) = e^{-(-x)^2} = e^{-(x^2)} = f(x)$$

f is even, and thus its graph is symmetric with respect to the y axis. Next we use a calculator to make a table of approximate values for several positive values of x:

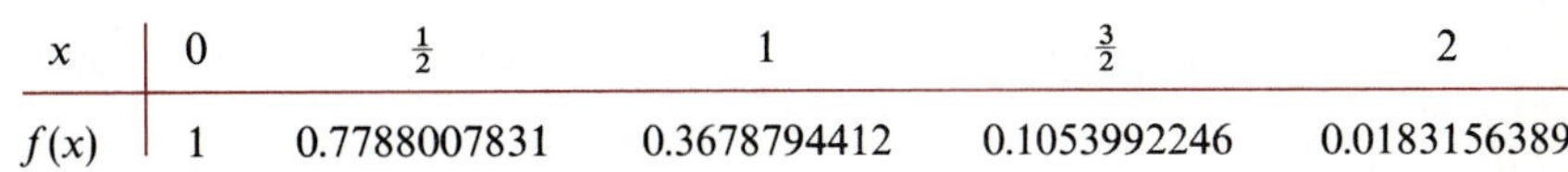

x	0	$\frac{1}{2}$	1	$\frac{3}{2}$	2
$f(x)$	1	0.7788007831	0.3678794412	0.1053992246	0.0183156389

After plotting the corresponding points, we connect them with a smooth curve to obtain the graph of f in Figure 4.9. □

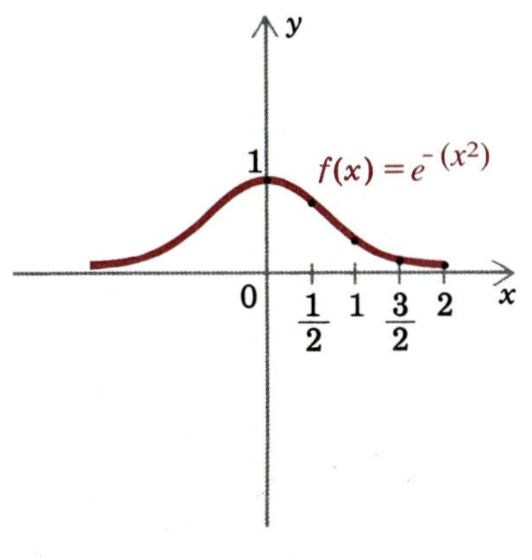

FIGURE 4.9

Exponential Equations

Let $a > 0$ with $a \neq 1$. From (1) and (2) it follows that if $x \neq z$, then $a^x \neq a^z$. This means that for $a > 0$ with $a \neq 1$,

$$\boxed{\text{if } a^x = a^z, \quad \text{then} \quad x = z} \tag{3}$$

Another way of expressing (3) is to say that such a function a^x is one-to-one. As a result, the function a^x has an inverse, a fact that will be of fundamental importance in the next section.

We can use (3) to solve certain equations involving exponential functions.

EXAMPLE 6. Solve the equation $9^x = 3^{x+\sqrt{2}}$ for x.

Solution. In order to use (3), we need the same base on both sides of the equation. Since $9 = 3^2$, the Laws of Exponents imply that

$$9^x = (3^2)^x = 3^{2x}$$

Thus the given equation is equivalent to

$$3^{2x} = 3^{x+\sqrt{2}}$$

From (3) we find that

$$2x = x + \sqrt{2}, \quad \text{so that} \quad x = \sqrt{2}$$

Therefore the solution of the given equation is $\sqrt{2}$. □

EXERCISES 4.1

In Exercises 1–14, sketch the graph of the given function.

1. $f(x) = 4^x$

2. $f(x) = \left(\frac{1}{2}\right)^x$

3. $f(x) = -(3^x)$

4. $f(x) = -3(2^x)$

5. $f(x) = 1^{x+2}$

6. $f(t) = 1 + 2^t$

7. $g(t) = 3 - 2^{-t}$

8. $g(x) = e^{-x}$

9. $g(x) = 2^{-2x}$

10. $h(x) = 3^{x+1}$

11. $h(x) = \sqrt{2}(2^{x+1/2})$

12. $g(x) = (0.5)^{2x-1}$

13. $g(x) = \left(\frac{1}{2}\right)^{|x|}$

14. $g(x) = \left(\frac{1}{2}\right)^{-|x|}$

In Exercises 15–22, simplify the given expression.

15. $\dfrac{\pi^{\sqrt{5}}\pi^{3-\sqrt{5}}}{\pi}$

16. $\dfrac{12^{2e}12^{3-3e}}{12^{-e} \cdot 144}$

17. $\dfrac{5^{\sqrt{6}}25^{3\sqrt{3}}}{125^{\sqrt{12}}}$

18. $(\sqrt{2^{\sqrt{2}}})^2$

19. $(2^{\sqrt{2}})^{-\sqrt{2}}$

20. $\left(\dfrac{1}{7^{\sqrt{5}}}\right)^{2\sqrt{5}}$

21. $(\sqrt[3]{9})^{\pi}(\sqrt[3]{3})^{\pi}$

22. $\dfrac{(\sqrt{6})^{\sqrt{3}}(\sqrt{5})^{\sqrt{3}}}{(\sqrt{10})^{\sqrt{3}}}$

In Exercises 23–32, find the base a if the graph of $y = a^x$ contains the given point.

23. $(3, 8)$

24. $(3, 27)$

25. $(1, 7)$

26. $(2, \frac{1}{4})$

27. $(2, \frac{1}{9})$

28. $(\frac{1}{2}, \frac{1}{2})$

29. $(-\frac{1}{2}, 2)$

30. $(3, -\frac{1}{64})$

31. $(\frac{1}{3}, \frac{1}{4})$

32. $(2, a)$

In Exercises 33–38, use a calculator to approximate the given number.

33. $e^{0.34}$

34. $e^{0.94}$

35. $e^{-1.2}$

36. e^{-3}

37. $\sqrt{e}$

38. $\sqrt[4]{e}$

In Exercises 39–42, use a calculator to determine the smallest integer x that satisfies the given inequality.

39. $2^{\sqrt{2x}} > 25$

40. $(1.1)^{\sqrt{2x}} > 3$

41. $x^{\sqrt{3}} > 200$

42. $x^{\sqrt{2/3}} > 13.3$

In Exercises 43–48, solve the given equation.

43. $2^{3x} = 2^{5x^2/4}$

44. $(5^x)^{x+1} = 5^{2x+12}$

45. $(\sqrt{3})^{x-1} = 3^{x+\sqrt{2}}$

46. $(6^{1/6})^{x^2} = (\sqrt{6})^{x\sqrt{5}}$

47. $(5^x)^{x+1} = (\sqrt{5})^{2x+10}$

48. $4^{\sqrt{x+1}} = 2^{3x-2}$

49. The ***hyperbolic sine*** function, whose value at x is denoted by $\sinh x$, and the ***hyperbolic cosine*** function, whose value at x is denoted by $\cosh x$, are given by

$$\sinh x = \frac{e^x - e^{-x}}{2}$$

$$\cosh x = \frac{e^x + e^{-x}}{2}$$

These functions are important in engineering. Show that for every number x,

$$(\cosh x)^2 - (\sinh x)^2 = 1$$

This is the fundamental equation relating the hyperbolic sine and cosine functions.

50. Use a calculator to determine which is larger, 3^π or π^3.

51. Use a calculator to determine which is larger, $(e^e)^e$ or $e^{(e^e)}$.

52. a. Find the value of e (which is e^1) on your calculator.
b. Find a positive integer n such that the value your calculator gives for $\left(1 + \frac{1}{n}\right)^n$ is the same as it gives for e.

53. In calculus it is shown that the sums of the form

$$1 + x + \frac{x^2}{2!} + \frac{x^3}{3!} + \cdots + \frac{x^n}{n!} \quad \text{where } n \text{ is a positive integer}$$

approach e^x as n increases without bound.
a. Using a calculator, show that if $x = 1$ and $n = 10$, then the sum agrees with $e^1 = e$ to eight places.
b. Using a calculator, determine to how many places $e^{1.5}$ and the sum agree if $x = 1.5$ and $n = 10$.

54. Let $a > 0$. Use Law (i) of Exponents twice to show that

$$a^{x+y+z} = a^x a^y a^z$$

55. Use (1) of this section to prove (2). (*Hint:* If $0 < a < 1$, then $1/a > 1$.)

56. Let $f(x) = e^{-ae^{-bx}}$, where a and b are positive constants. The graph of f, which is referred to as the ***Gompertz growth curve***, is used by actuaries in the preparation of life expectancy tables. Determine whether f is increasing or decreasing.

57. The air pressure $p(x)$ at a height of x feet above sea level is given approximately by

$$p(x) = e^{-(4.101 \times 10^{-4})x}$$

where $p(x)$ is given in "atmospheres." Determine the air pressure (rounded to 3 significant digits) at a height of

a. 29,028 feet (summit of Mt. Everest)
b. 14,110 feet (summit of Pikes Peak)
c. 1,250 feet (top of Empire State Building)
d. −282 feet (bottom of Death Valley)

58. One test that measures the thyroid condition in a human uses a small dose D of radioactive iodine I^{131} injected into the blood stream. The amount A remaining in the blood after t days is related to D by the formula

$$A(t) = De^{-0.0086t}$$

Suppose a dose is injected into the bloodstream at noon. Determine the percent in the bloodstream

a. at 1 P.M.
b. one half day later
c. one day later

59. Suppose the charge in coulombs on a capacitor changes with time according to the equation

$$C(t) = \frac{1}{500}\left(\frac{1}{50}\right)^{t/4}$$

Use a calculator to find the charge when $t = 1$.

4.2 LOGARITHMIC FUNCTIONS

Let $a > 0$ with $a \neq 1$. According to property 4 following Definition 4.1 in the preceding section, the exponential function a^x is increasing if $a > 1$ and decreasing if $0 < a < 1$. In either case it follows that the function a^x has an inverse.

DEFINITION 4.2 Let $a > 0$ with $a \neq 1$. The inverse of the function a^x is the ***logarithmic function to the base a***, and its value at any positive number x is denoted by $\log_a x$.

By (6) of Section 2.7,

$$\boxed{\log_a x = y \quad \text{if and only if} \quad a^y = x} \tag{1}$$

CAUTION: Since the domain of the function $\log_a x$ is identical to the range of the function a^x, which consists of the positive numbers, it follows that we cannot take the logarithm of 0 or any negative number.

The following examples illustrate the use of (1) in passing back and forth between equations involving exponents and equations involving logarithms.

EXAMPLE 1. Convert the following equations involving exponents to equations involving logarithms.

a. $10^2 = 100$ b. $81^{1/4} = 3$ c. $16 = 3^c$

Solution. In each case we use (1) to transform a statement of the form $a^x = y$ to one of the form $\log_a y = x$:

a. From $10^2 = 100$ we obtain $\log_{10} 100 = 2$.
b. From $81^{1/4} = 3$ we obtain $\log_{81} 3 = \frac{1}{4}$.
c. From $16 = 3^c$ we obtain $\log_3 16 = c$. □

EXAMPLE 2. Convert the following equations involving logarithms to equations involving exponents.

a. $\log_3 9 = 2$ b. $\log_2 32 = 5$ c. $\log_c 81 = 7$

Solution. In each part we use (1) to transform a statement of the form $\log_a y = x$ to one of the form $a^x = y$.

a. From $\log_3 9 = 2$ we obtain $3^2 = 9$.
b. From $\log_2 32 = 5$ we obtain $2^5 = 32$.
c. From $\log_c 81 = 7$ we obtain $c^7 = 81$. □

We can sometimes recognize the numerical value of a logarithm more easily by converting to exponents, as we illustrate now.

EXAMPLE 3. Evaluate the following expressions.

a. $\log_2 8$ b. $\log_3 3$ c. $\log_4 256$

Solution.

a. If $\log_2 8 = x$, then by (1), $2^x = 8$. But $2^3 = 8$, so that $x = 3$. Thus $\log_2 8 = 3$.
b. If $\log_3 3 = x$, then by (1), $3^x = 3$. But $3^1 = 3$, so that $x = 1$. Thus $\log_3 3 = 1$.
c. If $\log_4 256 = x$, then by (1), $4^x = 256$. But $4^4 = 256$, so that $x = 4$. Thus $\log_4 256 = 4$. □

EXAMPLE 4. Evaluate the following expressions.

a. $\log_{10} 1$ b. $\log_{25} \frac{1}{5}$ c. $\log_4 8$

Solution.

a. If $\log_{10} 1 = x$, then by (1), $10^x = 1$. But $10^0 = 1$, so that $x = 0$. Therefore $\log_{10} 1 = 0$.
b. If $\log_{25} \frac{1}{5} = x$, then by (1), $25^x = \frac{1}{5}$. But $25^{-1/2} = \frac{1}{5}$, so that $x = -\frac{1}{2}$. As a result, $\log_{25} \frac{1}{5} = -\frac{1}{2}$.
c. If $\log_4 8 = x$, then $4^x = 8$. To determine x, we need to write 8 as a power of 4. Since

$$8 = 4 \cdot 2 = 4^1 4^{1/2} = 4^{3/2}$$

it follows that $4^x = 8$ if $4^x = 4^{3/2}$, which means that $x = \frac{3}{2}$. Consequently $\log_4 8 = \frac{3}{2}$. □

Next we solve equations involving logarithms.

EXAMPLE 5. Solve the following equations for x.

a. $\log_{10} x = 0$ b. $\log_e x = 1$
c. $\log_x \pi = -1$ d. $\log_x 81 = 4$

Solution.

a. By (1) the equation $\log_{10} x = 0$ is equivalent to $10^0 = x$, so that $x = 10^0 = 1$. Thus the solution is 1.
b. By (1) the equation $\log_e x = 1$ is equivalent to $e^1 = x$, so that $x = e^1 = e$. Thus the solution is e.
c. By (1) the equation $\log_x \pi = -1$ is equivalent to $x^{-1} = \pi$, or $1/x = \pi$. Thus $x = 1/\pi$, so the solution is $1/\pi$.
d. By (1) the equation $\log_x 81 = 4$ is equivalent to $x^4 = 81$ with $x > 0$. But the only positive value of x for which $x^4 = 81$ is 3. Therefore 3 is the solution. □

By parts (a) and (b) of Example 5, $\log_{10} 1 = 0$ and $\log_e e = 1$, respectively. These are special cases of the following general formulas that hold for any $a > 0$ with $a \neq 1$:

$$\log_a 1 = 0 \tag{2}$$

$$\log_a a = 1 \tag{3}$$

EXAMPLE 6. Solve the following equations for x.

a. $\log_4 \dfrac{x}{5} = 3$ b. $\log_{10} (x - 6) = 2$

Solution.

a. By (1), the equation $\log_4 (x/5) = 3$ is equivalent to $4^3 = x/5$, so that

$$x = 5(4^3) = 5(64) = 320$$

Therefore the solution is 320.

b. By (1), the equation $\log_{10} (x - 6) = 2$ is equivalent to $10^2 = x - 6$, so that

$$x = 10^2 + 6 = 106$$

Therefore the solution is 106. □

The basic formulas

$$f(f^{-1}(x)) = x \quad \text{and} \quad f^{-1}(f(x)) = x$$

relating a function and its inverse, yield the formulas

$$a^{\log_a x} = x \quad \text{for } x > 0 \tag{4}$$

$$\log_a (a^x) = x \quad \text{for all } x \tag{5}$$

For example,

$$10^{\log_{10}\sqrt{7}} = \sqrt{7} \quad \text{and} \quad \log_e (e^{5.1}) = 5.1$$

Now suppose that x and z are positive numbers and

$$\log_a x = \log_a z$$

Then by (4),

$$x = a^{\log_a x} = a^{\log_a z} = z$$

Thus we have shown that if $\log_a x = \log_a z$, then $x = z$. The converse is obviously true, so we obtain the important result

$$\log_a x = \log_a z \quad \text{if and only if} \quad x = z \tag{6}$$

We can use (6) to solve certain equations involving logarithms.

EXAMPLE 7. Solve the equation $\log_2 (x^2 - 7) = \log_2 6x$.

Solution. By (6) with $a = 2$, the given equation is equivalent to the following equations:

$$x^2 - 7 = 6x$$

$$x^2 - 6x - 7 = 0$$

$$(x - 7)(x + 1) = 0$$

$$x = 7 \quad \text{or} \quad x = -1$$

Check: $\log_2 (7^2 - 7) = \log_2 42$ and $\log_2 (6 \cdot 7) = \log_2 42$
$\log_2 [(-1)^2 - 7]$ is undefined, since $(-1)^2 - 7 = -6$ is negative.

Therefore -1 is an extraneous solution, and consequently the solution of the given equation is 7. □

Graphs of Logarithmic Functions

Since every logarithmic function is the inverse of an exponential function, we can obtain the graph of a logarithmic function by reflecting the graph of the corresponding exponential function through the line $y = x$.

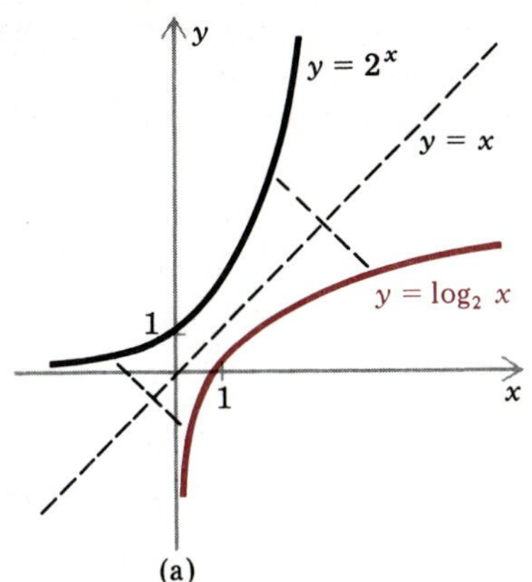

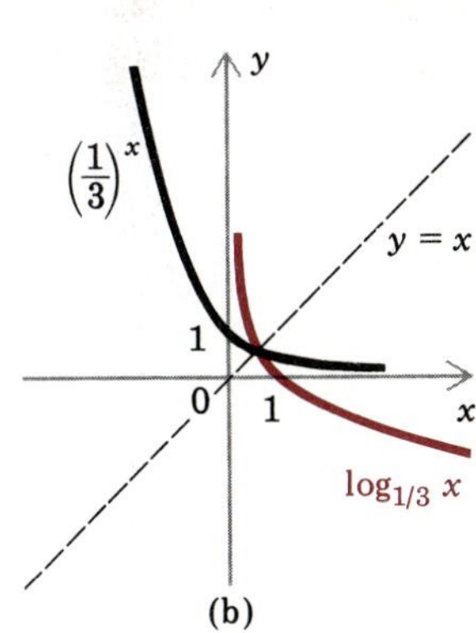

FIGURE 4.10

EXAMPLE 8. Sketch the graphs of the following functions:

a. $\log_2 x$ b. $\log_{1/3} x$

Solution. We sketched the graphs of 2^x and $(\frac{1}{3})^x$ in Figures 4.1c and 4.6, respectively. By reflecting these graphs through the line $y = x$, we obtain the graphs of $\log_2 x$ and $\log_{1/3} x$, respectively, in Figure 4.10a, b. □

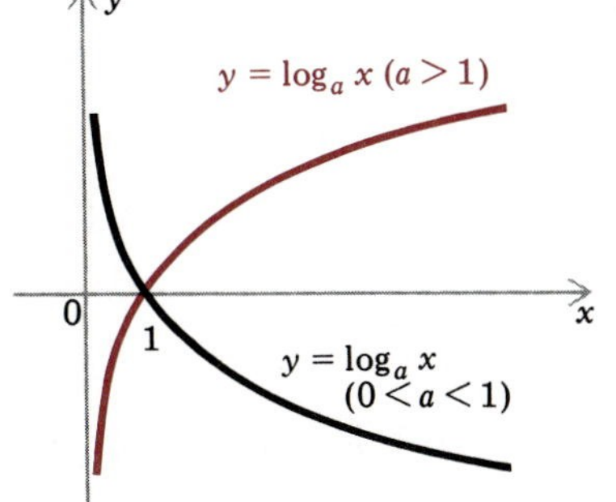

FIGURE 4.11

If $a > 0$ with $a \neq 1$, then the graph of the function $\log_a x$ is as in Figure 4.11, depending on whether $a > 1$ or $0 < a < 1$. The following features of the function $\log_a x$ and its graph are inherited from the corresponding features of the function a^x and its graph.

1. The range of $\log_a x$ consists of all numbers.
2. The x intercept is 1, and there is no y intercept.
3. The y axis is a vertical asymptote of the graph.
4. If $a > 1$, then $\log_a$ is an increasing function. If $0 < a < 1$, then $\log_a$ is a decreasing function.

We also observe that far to the right of the y axis the graph of the function $\log_a x$ becomes very flat, although the graph does not have a horizontal asymptote.

Special Bases

The base of a logarithmic function can be any positive number except 1, which means that $\sqrt{2}$ and $1/\pi$ are perfectly reasonable bases. However, from the viewpoint of mathematical theory and practical applications, the base is almost always greater than 1, and moreover, there are two bases that are far more widely used than any other: e and 10. In calculus and in other branches of mathematics, the logarithm to the base e plays a special role, and is called the ***natural logarithm function***. Its value at x is normally denoted by $\ln x$, so we would write

$$\ln x \quad \text{instead of} \quad \log_e x$$

(The expression "ln" stands for the Latin phrase "logarithmus naturalis.") In contrast, in numerical calculation the logarithm to the base 10 has historically reigned supreme, and it is called the ***common logarithmic function***. Its value at x is frequently denoted by $\log x$, so that we would write

$$\log x \quad \text{instead of} \quad \log_{10} x$$

EXERCISES 4.2

In Exercises 1–6, convert the exponential equation to a logarithmic equation.

1. $10^1 = 10$ **2.** $2^4 = 16$

3. $(\frac{1}{2})^{-5} = 32$ **4.** $343^{1/3} = 7$

5. $x^{-4} = 16$ **6.** $2^x = 9$

In Exercises 7–12, convert the logarithmic equation to an exponential equation.

7. $\log_{10} 1000 = 3$ **8.** $\log_{12} 1 = 0$

9. $\log_4 32 = 2.5$ **10.** $\log_{1.5} x = 3$

11. $\ln x = 2$ **12.** $\ln e^5 = x$

In Exercises 13–42, evaluate the given logarithm.

13. $\log_{10} 1000$ **14.** $\log_2 4$

15. $\log_2 16$ **16.** $\log_2 \frac{1}{4}$

17. $\log_3 \frac{1}{81}$ **18.** $\log_\pi \pi^{3/2}$

19. $\log_\pi 1$ **20.** $\log_\pi (1/\pi)$

21. $\log_\pi \pi^{3/2}$ **22.** $\log_{10} (10^e)$

23. $\ln (e^{-1})$ **24.** $\log_{10} \sqrt{10}$

25. $\log_8 2$ **26.** $\log_{16} 2$

27. $\log_{10} 0.01$ **28.** $\log_{10} 0.0001$

29. $\ln (1/e)$ **30.** $\log_{1/2} \frac{1}{8}$

31. $\log_{1/3} 81$ **32.** $\log_4 8$

33. $\log_4 32$ **34.** $\log_{27} 9$

35. $\log_{27} 81$ **36.** $\log_9 \frac{1}{27}$

37. $\log_{\sqrt{2}} 8$ **38.** $\log_{\sqrt{2}} \frac{1}{32}$

39. $\log_{2\sqrt{3}} 144$ **40.** $\log_2 (2^3 + 2^3)$

41. $\log_4 (8^3 + 8^3)$ **42.** $\log_{\sqrt{30}} (5^2 + 5)$

In Exercises 43–64, solve the given equation for x.

43. $\ln x = 0$ **44.** $\log_{10} x = 1$

45. $\log_2 x = 4$ **46.** $\log_\pi x = -1$

47. $\log_8 x = \frac{1}{3}$ **48.** $\log_{1/8} x = \frac{1}{3}$

49. $\log_4 x = -2$ **50.** $\log_{\sqrt{2}} x = 2$

51. $\log_{\sqrt{2}} x = 0$ **52.** $\log_{1.5} x = -3$

53. $\log_2 (x/3) = 2$ **54.** $\log_2 \sqrt{x} = 3$

55. $\log_4 (x - 3) = 3$ **56.** $\log_2 (4x - 7) = 4$

57. $\ln (x^2 - 8) = \ln 2x$

58. $\log_{10} \dfrac{x-1}{x-2} = \log_{10} \dfrac{x+1}{x+2}$

59. $\log_{10} \dfrac{x^2+4}{x} = \log_{10} \dfrac{x^2+4x+4}{x+1}$

60. $\log_3 |x - 1| = 2$ **61.** $\log_{10} |x^2 - 21| = 2$

62. $2^{\log_3 27} = x$ **63.** $3^{\log_2 x} = 27$

64. $4^x = (\frac{1}{4})^x$

In Exercises 65–70, solve the given equation for x.

65. $\log_x 16 = 2$ **66.** $\log_x 64 = 2$

67. $\log_x 125 = 3$ **68.** $\log_{x-1} 81 = 4$

69. $\log_{x^2} 16 = 2$ **70.** $\log_{x^2-2x} 64 = 2$

In Exercises 71–78, find the base a of the logarithmic function whose graph contains the given point.

71. $(4, 2)$ **72.** $(64, 3)$ **73.** $(3, \frac{1}{2})$

74. $(3, -\frac{1}{2})$ **75.** $(\frac{1}{27}, 3)$ **76.** $(\frac{1}{27}, -3)$

77. $(\frac{1}{2}, \frac{1}{2})$ **78.** $(\frac{1}{2}, -\frac{1}{2})$

In Exercises 79–86, sketch the graph of the given function.

79. $f(x) = 3 \log_2 x$ **80.** $f(x) = -\log_2 x$

81. $f(x) = \log_2 2x$ **82.** $f(x) = \log_2 |x|$

83. $f(x) = \log_4 (2 - x)$ **84.** $g(x) = \log_4 (x - 3)$

85. $f(x) = |\log_{1/3} x|$ **86.** $f(x) = -\log_{1/4} 3x$

87. Is $3 \log_{10} \frac{1}{6} > 2 \log_{10} \frac{1}{6}$? Explain your answer.

88. Show that if $\log_a b = 0$, then $b = 1$.

89. Suppose that you enter any positive number into a calculator. Describe what happens to the number displayed if you alternately press the ln and e^x keys.

4.3 LAWS OF LOGARITHMS AND CHANGE OF BASE

Just as there are laws of exponents, there are also laws of logarithms. The most important ones are presented below. All of them follow from the Laws of Exponents given in Section 4.1 and from the definition of logarithms.

Laws of Logarithms

Let a be a fixed positive number with $a \neq 1$. For any positive numbers x and y,

i. $\log_a (xy) = \log_a x + \log_a y$

ii. $\log_a \frac{1}{x} = -\log_a x$

iii. $\log_a \frac{x}{y} = \log_a x - \log_a y$

iv. $\log_a (x^c) = c \log_a x$ for any number c

v. $\log_a 1 = 0$

vi. $\log_a a = 1$

To prove (i), let $r = \log_a x$ and $s = \log_a y$. It follows from the definition of logarithm that $a^r = x$ and $a^s = y$. Therefore $xy = a^r a^s$. By the Laws of Exponents (Section 4.1), $a^r a^s = a^{r+s}$, so that $xy = a^{r+s}$. By another application of the definition of logarithm, it follows that $\log_a (xy) = r + s$. Thus

$$\log_a (xy) = r + s = \log_a x + \log_a y$$

which is (i). Hints for proving (ii)–(iv) are given in Exercises 50–52. Finally, we observe that (v) and (vi) are restatements, respectively, of (2) and (3) of Section 4.2.

The formula in (i) is frequently referred to as *the* Law of Logarithms. It contains the important feature of logarithms that the logarithm of a *product* of two numbers is the *sum* of the logarithms of the two numbers. In a similar vein, (iii) says that the logarithm of a *quotient* of two numbers is the *difference* of the logarithms of the two numbers.

CAUTION: There is no law of logarithms that simplifies either $\log_a (x + y)$ or $(\log_a x)/(\log_a y)$. In particular, $\log_a (x + y)$ is *almost never* equal to $\log_a x + \log_a y$, and $(\log_a x)/(\log_a y)$ is *almost never* equal to $\log_a x - \log_a y$.

To illustrate (i) and (iv), we observe that

$$\log_{10}(5x) = \log_{10} 5 + \log_{10} x$$

and

$$\log_2(3^{40}) = 40 \log_2 3$$

The Law of Logarithms (i) can be generalized to more than two variables. For three variables it reads as follows:

$$\boxed{\log_a xyz = \log_a x + \log_a y + \log_a z} \tag{1}$$

By combining (i) and (iii) we can obtain

$$\log_a \frac{xy}{z} = \log_a x + \log_a y - \log_a z \tag{2}$$

EXAMPLE 1. Express $\log_2 \frac{9}{2} + \log_2 \frac{8}{5} - \log_2 \frac{9}{8}$ as a single logarithm, and simplify your answer.

Solution. By (2),

$$\begin{aligned}\log_2 \frac{9}{2} + \log_2 \frac{8}{5} - \log_2 \frac{9}{8} &= \log_2 \left(\frac{\frac{9}{2} \cdot \frac{8}{5}}{\frac{9}{8}}\right) \\ &= \log_2 \left(\frac{9}{2} \cdot \frac{8}{5} \cdot \frac{8}{9}\right) \\ &= \log_2 \frac{32}{5} \quad \square\end{aligned}$$

It is possible to simplify the answer to Example 4 further, although the simplified version is not a single logarithm:

$$\log_2 \frac{32}{5} = \log_2 32 - \log_2 5 = \log_2 (2^5) - \log_2 5 = 5 - \log_2 5$$

Solving Equations Involving Logarithms

In the next two examples we will return to solving equations involving logarithms.

EXAMPLE 2. Solve the equation $3 \log_{10} x = \log_{10} 8$ for x.

Solution. We use (iv) above, and then (6) of Section 4.2:

$$\begin{aligned} 3 \log_{10} x &= \log_{10} 8 \\ \log_{10} (x^3) &= \log_{10} 8 \\ x^3 &= 8 \\ x &= 2 \end{aligned}$$

Check: $3 \log_{10} 2 = \log_{10} (2^3) = \log_{10} 8$

Consequently the solution is 2. □

EXAMPLE 3. Solve the equation $\log_8 (x + 9) - \log_8 (x - 1) = \frac{1}{3}$ for x.

Solution. This time we use (iii) and the definition of the logarithm to the base 8:

$$\begin{aligned} \log_8 (x + 9) - \log_8 (x - 1) &= \frac{1}{3} \\ \log_8 \frac{x + 9}{x - 1} &= \frac{1}{3} \\ \frac{x + 9}{x - 1} &= 8^{1/3} = 2 \\ x + 9 &= 2(x - 1) \\ x + 9 &= 2x - 2 \\ x &= 11 \end{aligned}$$

Check: $$\begin{aligned} \log_8 (11 + 9) - \log_8 (11 - 1) &= \log_8 20 - \log_8 10 \\ &= \log_8 \frac{20}{10} = \log_8 2 \\ &= \log_8 (8^{1/3}) = \frac{1}{3} \end{aligned}$$

Thus the solution of the given equation is 11. □

Change of Base

It is sometimes necessary to convert from logarithms in one base to logarithms in another base, that is, to write $\log_b x$ in terms of $\log_a x$ (where values of $\log_a x$ are presumably known). Since calculators usually have keys only for natural logarithms and common logarithms, this conversion may even be necessary when one wishes to use a calculator to make computations involving logarithms.

Suppose a and b are positive numbers different from 1 and we wish to express $\log_b x$ in terms of $\log_a x$. Let

$$r = \log_b x \quad \text{so that} \quad x = b^r$$

Since $a \neq 1$ by hypothesis, we can take logarithms to the base a of both sides of the latter equation and obtain

$$\log_a x = \log_a (b^r)$$

But by (iv) in the Laws of Logarithms,

$$\log_a (b^r) = r \log_a b \tag{3}$$

Since $b \neq 1$ by hypothesis, it follows that $\log_a b \neq 0$. Therefore we may divide both sides of (3) by $\log_a b$ to obtain

$$r = \frac{\log_a x}{\log_a \text{b}}$$

Substituting $\log_b x$ for r, we obtain the following formula.

Change of Base Formula

$$\log_b x = \frac{\log_a x}{\log_a b} \quad \text{for } x > 0 \tag{4}$$

As we observed earlier, most calculators have a key for the common logarithm of a given number. Now if $a = 10$, then (4) becomes

$$\log_b x = \frac{\log_{10} x}{\log_{10} b} \quad \text{for } x > 0 \tag{5}$$

EXAMPLE 4. Use a calculator to approximate the value of $\log_2 5$.

Solution. By (5),

$$\log_2 5 = \frac{\log_{10} 5}{\log_{10} 2} \tag{6}$$

Using (6), we find by calculator that

$$\log_2 5 \approx 2.321928095 \quad \square$$

If $x = a$, then the change of base formula (4) becomes

$$\log_b a = \frac{\log_a a}{\log_a b} \tag{7}$$

Since $\log_a a = 1$, (7) simplifies to

$$\boxed{\log_b a = \frac{1}{\log_a b}} \tag{8}$$

a formula that is occasionally of use.

The invention of logarithms is generally credited to John Napier (1550–1617), a Scotsman of noble heritage for whom mathematics was a hobby. His invention arose through his desire to associate ratios of pairs of numbers in a geometric progression with ratios of pairs of numbers in an arithmetic progression. Thus the name *logarithmus*, which means "ratio number" and has been anglicized to "logarithm." What resulted from Napier's creation was an association between quotients and differences, namely Law of Logarithms (iii). The base of Napier's logarithms happened to be related to e, and this is the reason why $\log_e x$ (that is, $\ln x$) is sometimes called a Napierian logarithm. Not only did Napier invent his special logarithms, but he made an extensive table of seven-place logarithms that was published in 1614.

EXERCISES 4.3

In Exercises 1–4, find the numerical value of the given expression.

1. $\log_3 (27^5)$ **2.** $\log_3 [(\frac{1}{27})^5]$

3. $\log_{16} (4^{7/9})$ **4.** $\log_4 [(\frac{1}{16})^6]$

In Exercises 5–8, approximate the numerical value of the given logarithm. Use the fact that $\log_{10} 4 \approx 0.6021$.

5. $\log_{10} 40$ **6.** $\log_{10} 400{,}000{,}000$

7. $\log_{10} \frac{1}{4}$ **8.** $\log_{10} 0.00000025$

In Exercises 9–12, approximate the numerical value of the given logarithm. Assume that $\log_{10} a = 2.7318$.

9. $\log_{10} (1000a)$ **10.** $\log_{10} \dfrac{a}{100}$

11. $\log_{10} (a^{10})$ **12.** $\log_{10} (a^{-5})$

In Exercises 13–18, use a calculator to approximate the numerical value of the given logarithm.

13. $\log_{10} 5$ **14.** $\log_{10} 217$

15. $\log_{10} 0.00849$ **16.** $\ln 3.06$

17. $\ln \frac{1}{53}$ **18.** $\ln 17{,}329.6$

In Exercises 19–24, rewrite the given expression as a single logarithm.

19. $\log_2 3 + \log_2 \frac{4}{3} + \log_2 \frac{5}{4}$

20. $\ln (3/e) + \ln \frac{1}{6} + \ln 2$

21. $\log_2 (xy^2) + \log_2 (y^{-3}) + \log_2 (x^5)$

***22.** $\ln 40 + \log_2 8$

***23.** $\log_{10} 2 + 3 \log_{\sqrt{10}} x - \log_{10} (2y)$

***24.** $\log_2 (9x) - \log_4 (7x^2) - \log_8 (3/x^2)$

In Exercises 25–30, express the given logarithm in terms of $\log_a x$, $\log_a y$, and $\log_a z$.

25. $\log_a \dfrac{xy}{z^3}$

26. $\log_a (x^2yz^3)$

27. $\log_a (z^3\sqrt{xy})$

28. $\log_a \dfrac{1}{xyz}$

29. $\log_a \sqrt{\dfrac{xy^5}{z^3}}$

30. $\log_a \sqrt[4]{\dfrac{x\sqrt{y}}{y^{1/3}}}$

In Exercises 31–36, solve the given equation for x.

31. $2 \log_{10} x = \log_{10} 16$

32. $4 \log_2 x = \log_2 81$

33. $-2 \log_{10} x = \log_{10} 4$

34. $\log_{10} (4x - 1) - \log_{10} \left(\dfrac{x}{5}\right) = 1$

35. $\log_2 (x + 1) + \log_2 (x - 1) = 3$

36. $\ln (2x - 1) - \ln (3x + 1) = -1$

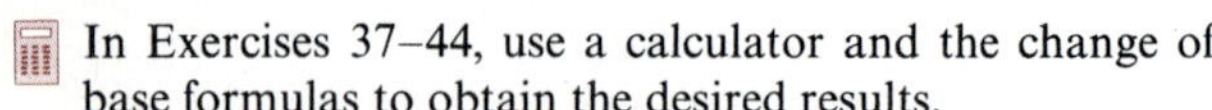

In Exercises 37–44, use a calculator and the change of base formulas to obtain the desired results.

37. $\log_2 5$

38. $\log_2 0.48$

39. $\log_7 14.3$

40. $\log_{13} \dfrac{419}{73}$

41. $\log_{1/3} \dfrac{13}{7}$

42. $\log_{9/10} 0.00346$

43. $\log_{.4} 5432$

44. $\log_{.02} 0.00749$

45. a. Show that $\log_2 x = \log_4 x^2$.
b. Show that $\log_a x = \log_{a^2} x^2$ for any positive numbers a and x such that $a \neq 1$.

46. Prove that $\ln (x + \sqrt{x^2 - 1}) = -\ln (x - \sqrt{x^2 - 1})$.

47. Prove that $\log_a (1/x) = \log_{1/a} x$.

48. a. Show that if $x > 0$ and x satisfies $\dfrac{\ln x}{x} < \dfrac{\ln 2}{2}$, then $x^2 < 2^x$.
b. Sketch the graphs of x^2 and 2^x simultaneously, and use the trace to determine approximately the intervals on which $x^2 < 2^x$. Round the numbers to the nearest hundredth.

49. Let $b = e^{-1/10^7}$. Show that $\log_b x = -10^7 \ln x$ for any $x > 0$. (The "logarithm" that John Napier defined in the seventeenth century was actually $\log_b x + 10^7 \ln 10^7$. Thus, although his logarithm was not the natural logarithm, it was very closely related to it.)

50. Prove Law (ii) by using Law (i) with y replaced by $1/x$.

51. Prove Law (iii) by using Laws (i) and (ii).

52. Prove Law (iv). (*Hint:* Let $r = \log_a x$. Using first the definition of the logarithm to the base a, next a law of exponents, and then the definition of the logarithm to the base a again , show in turn that $x = a^r$, $x^c = a^{cr}$, and $cr = \log_a (x^c)$, and complete the proof.)

53. a. Sketch the graphs of $\log_2 x$ and $\log_3 x$ simultaneously.
b. What does the Change of Base Formula (with $a = 2$ and $b = 3$) imply about the relationship between the two graphs?

54. a. Let $a > 0$ with $a \neq 1$. Prove that $\log_{1/a} x = -\log_a x$ for all $x > 0$.
b. Sketch the graphs of $\log_2 x$ and $\log_{1/2} x$ simultaneously.
c. What does the result in part (a) imply about the relationship of the graphs of $\log_2 x$ and $\log_{1/2} x$?

4.4 APPLICATIONS OF COMMON LOGARITHMS

This section is devoted to mathematical and scientific applications of common logarithms. The mathematical application concerns solutions of equations involving exponentials, and the scientific applications are represented by loudness of sound and magnitude of earthquakes, both of which employ common logarithms in their formulation. For simplicity we will write $\log x$ for $\log_{10} x$.

Solutions of Equations

For the examples below, we will use the fact that

$$x = y \quad \text{if and only if} \quad \log_a x = \log_a y$$

When we pass from an equation of the form $x = y$ to the equivalent equation $\log_a x = \log_a y$, we say that we ***take logarithms of both sides*** of the equation $x = y$.

EXAMPLE 1. Solve the equation $4^x = 17$ for x.

Solution. We take common logarithms of both sides and then use Law of Logarithms (iv):

$$\log(4^x) = \log 17$$

$$x \log 4 = \log 17$$

$$x = \frac{\log 17}{\log 4}$$

By calculator we find that

$$x \approx 2.043731421 \quad \square$$

EXAMPLE 2. Solve the equation $2^{x+1} = 3^{x^2-1}$ for x.

Solution. Taking common logarithms of both sides and then using (iv) of the Laws of Logarithms, we obtain

$$\log(2^{x+1}) = \log(3^{x^2-1})$$

$$(x + 1) \log 2 = (x^2 - 1) \log 3$$

$$(x + 1) \log 2 = (x + 1)[(x - 1) \log 3]$$

From this last equation we conclude that either $x = -1$ or

$$\log 2 = (x - 1) \log 3$$

Solving for $x - 1$ in the latter equation, we obtain

$$x - 1 = \frac{\log 2}{\log 3} \quad \text{so that} \quad x = \frac{\log 2}{\log 3} + 1$$

Thus the solutions of the given equation are -1 and $[(\log 2)/(\log 3)] + 1$ (which by calculator we find is approximately 1.630929754). $\square$

Loudness of Sound

We turn now to physical applications of common logarithms, represented by loudness of sound and magnitude of earthquakes. In each of these applications the primary reason for using common logarithms is to restrict the numbers involved to a more manageable range. For instance, if a certain quantity q can take any value from, say, 10^{-100} (which is miniscule) to 10^{100} (which is astronomical), then the value of $\log q$ lies between -100 and 100.

The ***intensity*** of a sound wave is the amount of energy that the wave carries through a unit area in a unit time, or equivalently, the amount of power per unit area. The unit we will use for the intensity of sound is one watt per square meter. The ***threshold of audibility*** of a sound wave is the minimal intensity the sound wave can have and still be audible to the normal human ear. The threshold of audibility of a sound wave varies according to the pitch, or frequency, of the wave. To simplify the discussion that follows, let us assume that each sound wave under consideration has the same frequency—100 hertz (cycles per second), producing a sound about an octave and a half below middle C on the piano.

Let I_0 be the intensity of such a sound wave at the threshold of audibility, approximately 10^{-12} watts per square meter. If x denotes the intensity of any sound wave of the same frequency, then the ***noise level*** (or ***loudness***) $L(x)$ of the sound wave is defined by

$$L(x) = 10 \log \frac{x}{I_0} \tag{1}$$

The units for $L(x)$ are ***decibels***, in honor of Alexander Graham Bell (1847–1922). Notice that if $x = I_0$, then (1) becomes

$$L(I_0) = 10 \log \frac{I_0}{I_0} = 10 \log 1 = 0$$

so that at the threshold of audibility the noise level is just 0.

EXAMPLE 3. Suppose a sound wave from a loud conversation is 25,000 times as intense as a sound wave from a whisper, whose noise level is 22 decibels. Determine the noise level of the loud conversation.

Solution. Let x and z, respectively, represent the intensities of sound waves from the loud conversation and the whisper. We wish to determine $L(x)$. By hypothesis,

$$\frac{x}{z} = 25{,}000, \quad \text{so that} \quad x = 25{,}000z$$

Using (1) and then the Law of Logarithms, we obtain

$$L(x) = L(25{,}000z) = 10 \log \frac{25{,}000z}{I_0}$$

$$= 10 \log 25{,}000 + 10 \log \frac{z}{I_0}$$

Since $L(z) = 22$ by hypothesis, (1) implies that

$$10 \log \frac{z}{I_0} = L(z) = 22$$

Using a calculator, we find that

$$L(x) = 10 \log 25{,}000 + 22 \approx 10(4.39790009) + 22 \approx 66$$

Consequently the noise level of the loud conversation is approximately 66 decibels. □

Earthquakes

During the past four decades several formulas have been proposed for converting seismographic readings into a unified scale that would represent the magnitude of an earthquake. The scale most commonly used is the ***Richter scale***, devised by Charles F. Richter, an American geologist. In order to describe measurement on the Richter scale, we first introduce a reference, or ***zero-level***, earthquake, which by definition is any earthquake whose largest seismic wave would measure 0.001 millimeter (1 micron) on a standard seismograph located 100 kilometers from the epicenter of the earthquake. The ***magnitude*** M of a given earthquake can be obtained by the formula

$$M = \log \frac{a}{a_0} = \log a - \log a_0 \tag{2}$$

where a is the amplitude measured on a standard seismograph of the largest seismic wave of the given earthquake, and a_0 is the amplitude on the same seismograph of the largest seismic wave of a zero-level earthquake with the same epicenter. Values of a_0 for various distances from the epicenter have been calculated, so one only needs to measure the number a in order to be able to assess the magnitude of a given earthquake.

Notice that the stronger the earthquake is, the greater the ratio a/a_0 of amplitudes of seismic waves is. Let us call a/a_0 the ***intensity*** of an earthquake.

EXAMPLE 4. The great San Francisco earthquake of 1906 measured 8.3 on the Richter scale. On October 17, 1989, San Francisco suffered a less intense earthquake that measured 7.1 on the Richter scale. How many times more intense was the 1906 earthquake than the 1989 earthquake?

Solution. Let a_1 and a_2, respectively, represent the amplitudes of the maximum seismic waves of the 1906 and 1989 earthquakes at, say 1000 kilometers from their respective epicenters. We must determine $(a_1/a_0)/(a_2/a_0)$, which equals a_1/a_2. By hypothesis,

$$8.3 = \log \frac{a_1}{a_0} \quad \text{and} \quad 7.1 = \log \frac{a_2}{a_0}$$

Therefore by the Laws of Logarithms,

$$1.2 = 8.3 - 7.1 = \log \frac{a_1}{a_0} - \log \frac{a_2}{a_0} = \log \frac{a_1/a_0}{a_2/a_0} = \log \frac{a_1}{a_2}$$

It follows that

$$\frac{a_1}{a_2} = 10^{1.2} \approx 15.84893192$$

This means that the 1906 earthquake was some 15.8 times as intense as the 1989 earthquake. □

Theoretically, the magnitude of an earthquake can be any real number. In practice, however, a standard seismograph can only record those earthquakes whose magnitudes exceed 2. On the other end of the scale, no magnitude has ever exceeded 9.0, and only half a dozen times has one been 8.5 or higher.

EXERCISES 4.4

In Exercises 1–12, solve the given equation. Leave your answer in terms of common logarithms.

1. $2^x = 5$
2. $3^{-x} = 17$
3. $5^{2x} = 51$
4. $7^{\sqrt{2}x} = 3$
5. $4^{x+2} = 5$
6. $(\frac{1}{4})^{x+2} = 5$
7. $5^{2x+3} = 3^{5x-2}$
8. $8^{4x-1} = 6^{1-x}$
9. $3^x = 2^{(x^2)}$
10. $3^x = 2^{(x^3)}$
11. $4^{x+2} = 5^{x^2-4}$
12. $2^{x-1} = 10^{x^2+x-2}$
13. Using (1), show that if $L(x_1)$ decibels corresponds to intensity x_1 and $L(x_2)$ corresponds to intensity x_2, then

$$L(x_2) - L(x_1) = 10 \log \frac{x_2}{x_1}$$

14. Determine the number of decibels that corresponds to each of the following intensities (in watts per square meter).
    a. 10^{-11} (refrigerator motor)
    b. 10^{-5} (conversation)
    c. 10^{-4} (electric typewriter)
15. Using the value of 10^{-12} for I_0, find the intensity of a sound at 50 decibels.
16. What is the ratio of the intensity of a given sound to that of one that is 100 decibels higher?
17. What is the difference in the noise levels of two sounds one of which is 100 times as intense as the other?
18. Suppose the cannon fired at the time of a football touchdown produces a sound intensity 1000 times as strong as a cheerleader's shouts. Determine how many more decibels the noise level from the cannon produces than the cheerleader's cry.
19. The human ear can just barely distinguish between two sounds if one is 0.6 decibels higher than the other. What is the ratio of the intensity of one sound to that of another sound which is lower than the first and is just barely distinguishable from the first sound?
20. Suppose soundproofing a room results in a noise

reduction of 30 decibels in the noise from a busy highway. What percentage of the original intensity is the reduced intensity?

21. The Alaska Good Friday earthquake of March 28, 1964, measured 8.5 on the Richter scale. Find the intensity a/a_0.

22. The magnitude of the Good Friday Alaska earthquake of 1964 is sometimes given as 8.4 and sometimes as 8.5. What is the ratio of the maximum amplitude of an earthquake of magnitude 8.4 to that of an earthquake of magnitude 8.5?

23. Of seven major earthquakes to hit Iran in recent times, two have measured approximately 6.9 on the Richter scale (those occurring in April 1972 and in March 1977). Find the ratio of the amplitudes of the largest waves of these earthquakes to that of the San Francisco earthquake of 1906, which measured 8.3 in magnitude.

24. The strongest earthquakes ever recorded occurred off the coast of Ecuador and Columbia in 1906, and in Japan in 1933. Each had magnitude 8.9. Find the ratio of the amplitude of the largest wave of such a quake to the corresponding amplitude of a zero-level quake.

25. Suppose a seismograph is located exactly 100 kilometers from the epicenter of an earthquake. Determine the magnitude of the earthquake if the largest amplitude of the seismic waves registered on the seismograph is

a. 1 micron b. 1 millimeter c. 1 centimeter

26. If an earthquake has magnitude 2, find the amplitude of its largest wave 100 kilometers from the epicenter.

27. Suppose the amplitude of the maximal seismic wave is doubled. By how much is the magnitude of the earthquake increased?

28. In chemistry the ***hydrogen potential*** (denoted pH) of a solution is defined by

$$\text{pH} = \log \frac{1}{[H^+]} = -\log [H^+]$$

where H^+ is the concentration of hydrogen ions in the solution in moles per liter. The pH of a solution ranges from 0 to 14. If the pH of a solution is less than 7, the solution is acidic and turns litmus paper red. In contrast, if the pH is greater than 7, it is alkaline and turns litmus paper blue. Distilled water has a pH of 7 and is neither acid nor alkaline. Determine the hydrogen ion concentration of pure water.

29. Find the pH of a soft drink whose hydrogen ion concentration is roughly 0.004 mole per liter. Is the drink highly acidic, or highly alkaline.?

30. Orange juice has a pH of approximately 3.5, and tomato juice has a pH of approximately 4.2. What is the ratio of hydrogen ion concentration in orange juice to that in tomato juice?

31. Wheat grows best in soil that has a pH of between 6 and 7.5, whereas oats grow best in soil that has a pH of between 5 and 6.2. Suppose the hydrogen ion concentration in the soil in a field is found to be 4.47×10^{-7} mole per liter. Is the soil better suited to wheat or to oats?

4.5 EXPONENTIAL GROWTH AND DECAY

In physical and mathematical applications the exponential function that is most often used is the exponential function with base e, that is, the function e^x. In particular, many physical quantities, such as the amount of a specific radioactive substance, depend on time according to an equation of the form

$$f(t) = Ae^{ct} \tag{1}$$

where A and c are constants. If we substitute 0 for t in (1), we have

$$f(0) = Ae^{c \cdot 0} = Ae^0 = A \cdot 1 = A$$

so that $A = f(0)$. Thus (1) becomes

$$f(t) = f(0)e^{ct} \tag{2}$$

Since $e^{ct} > 0$ for all t, it follows from (2) that $f(t)$ has the same sign as $f(0)$. In applications $f(t)$ is almost always positive. As a result, we will assume throughout this section that $f(t) > 0$ for all t.

If $c > 0$, then customarily one lets $c = k$, so that

$$\boxed{f(t) = f(0)e^{kt} \quad \text{with } k > 0} \tag{3}$$

In this case, f is an increasing function of t (Figure 4.12a), and we say that f ***grows exponentially***. Analogously, if $c < 0$, then customarily one lets

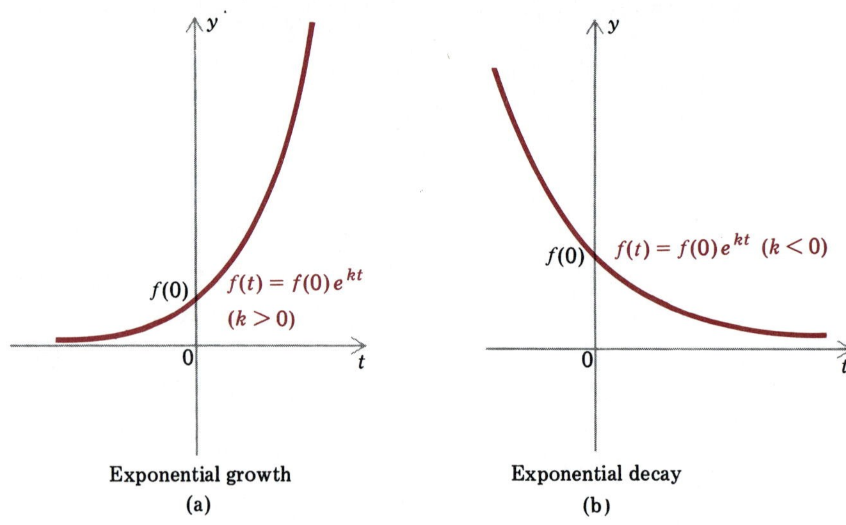

FIGURE 4.12

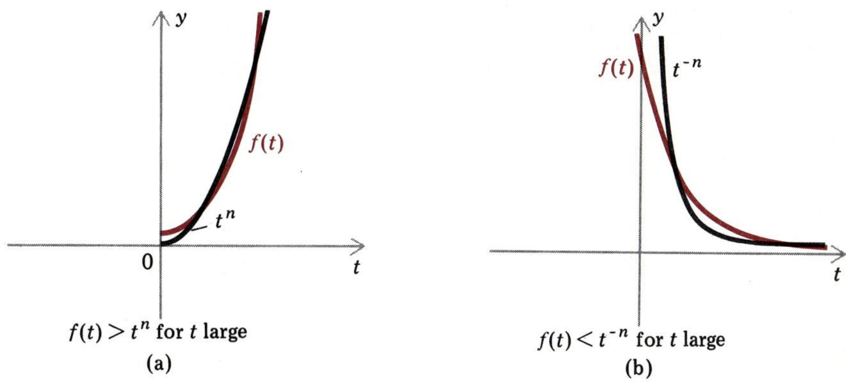

FIGURE 4.13

$c = -k$, so that

$$f(t) = f(0)e^{-kt} \quad \text{with } k > 0 \tag{4}$$

In this case, f is a decreasing function of t (Figure 4.12b), and we say that f ***decays exponentially***. If f grows exponentially, then for any positive integer n, $f(t) > t^n$ for all sufficiently large t (Figure 4.13a), and we say that $f(t)$ grows faster than any function of the form t^n. Similarly, if f decays exponentially, then for any positive integer n, $f(t) < t^{-n}$ for all sufficiently large t (Figure 4.13b), and we say that $f(t)$ decreases faster than any function of the form t^{-n}.

Solving Exponential Equations by Natural Logarithms

In the examples of exponential growth and decay that we will consider, it will be necessary to solve equations containing expressions like e^{kt}. Since

$$\ln (e^{kt}) = kt \tag{5}$$

the use of natural logarithms will simplify the solutions. This is illustrated in the next two examples.

EXAMPLE 1. Solve the equation $e^{5k} = 2$ for k.

Solution. We take the natural logarithm of each side, use (5), and then solve for k:

$$\ln (e^{5k}) = \ln 2$$

$$5k = \ln 2$$

$$k = \frac{1}{5} \ln 2$$

By calculator we find that $k \approx 0.1386294361$. □

EXAMPLE 2. Solve the equation $e^{(t/5)\ln 2} = 3$ for t.

Solution. We take the natural logarithm of each side, use (5), and solve for t:

$$\ln\left(e^{(t/5)\ln 2}\right) = \ln 3$$

$$\frac{t}{5}\ln 2 = \ln 3$$

$$t = 5\frac{\ln 3}{\ln 2}$$

By calculator we find that $t \approx 7.924812504$. □

Population Growth

When the number of organisms in a group is very large, the graph of the function representing the number in the group at any given time is very nearly a smooth curve. If the group has plenty of food and space and no enemies (such as disease or predators), the curve can be essentially the graph of an exponential function of the form of (3). This means that the population of the group can grow exponentially. Statistics show that various types of bacteria, insects, and rodents, and even certain human populations, have in the past grown exponentially for awhile—until conditions such as lack of food, overcrowding, and war changed the growth pattern.

EXAMPLE 3. Experiment shows that under optimal conditions and at a temperature of 29°C, the population of a colony of 1000 rice weevils can grow exponentially to about 2139 weevils in a week. Assuming that there are 1000 initially and that their population grows exponentially, find a formula for the number of weevils at time t.

Solution. Let $f(t)$ denote the number of weevils at time t, where t is measured in weeks. We take $t = 0$ to represent the initial time, which means that $f(0) = 1000$, the initial number of weevils. Then (3) becomes

$$f(t) = 1000e^{kt}$$

and we need to determine the numerical value of k. By assumption there are 2139 weevils after 1 week, so that $f(1) = 2139$. Thus

$$2139 = f(1) = 1000e^{k(1)} = 1000e^{k}$$

or equivalently,

$$e^{k} = \frac{2139}{1000} = 2.139$$

Taking natural logarithms of both sides yields

$$k = \ln 2.139$$

Therefore

$$f(t) = 1000e^{(\ln 2.139)t} \quad \square$$

EXAMPLE 4. From the data given in Example 3, determine the length of time in weeks for the population of weevils to double.

Solution. Since the initial population of weevils is 1000, what we seek is the value of t for which

$$f(t) = 2000$$

By the solution of Example 3, this means that we wish to find the value of t for which

$$2000 = 1000e^{(\ln 2.139)t}$$

or equivalently,

$$2 = e^{(\ln 2.139)t}$$

Taking natural logarithms of both sides, we obtain

$$\ln 2 = (\ln 2.139)t$$

so that

$$t = \frac{\ln 2}{\ln 2.139}$$

By calculator we find that

$$t \approx 0.9116298126$$

Consequently the population of weevils will double after approximately 0.9116298126 weeks, which is just over 6 days and 9 hours. $\square$

The length of time it takes a population growing exponentially to double is called the ***doubling time*** of the population. Thus under optimal conditions the doubling time for rice weevils is approximately 6 days and 9 hours.

Carbon Dating

Certain isotopes, such as carbon 14 (C^{14}) and uranium 238 (U^{238}), are radioactive, which means that they are unstable and in time decay, changing into other substances by the emission of various particles from their nuclei. For example, C^{14} decays into nitrogen 14 by emitting an electron (a beta particle), and U^{238} decays into thorium 234 by emitting a helium 4 nucleus (an alpha particle).

If $f(t)$ is the amount of a given radioactive element in a substance at time t, then by experiment it is known that f satisfies (4) for a suitable constant k depending on the isotope. Notice that the amount $f(t)$ decreases as t increases, and in fact, $f(t)$ eventually approaches 0, no matter how much of the element is initially present. The ***half-life*** of a radioactive element is the length of time necessary for half of any given amount of it to decay.

EXAMPLE 5. The half-life of C^{14} is approximately 5730 years. Using this figure, show that the amount $f(t)$ of C^{14} remaining after an elapsed time of t years is given by

$$f(t) = f(0)e^{-(1/5730)(\ln 2)t} \tag{6}$$

Solution. We need to determine the constant k so that

$$f(t) = f(0)e^{-kt} \quad \text{for } t > 0 \tag{7}$$

By assumption, half the original C^{14} remains after 5730 years, so that

$$f(5730) = \frac{1}{2}f(0)$$

However, by substituting 5730 for t in (7), we obtain

$$f(5730) = f(0)e^{-5730k}$$

From the last two equations it follows that

$$\frac{1}{2}f(0) = f(0)e^{-5730k}$$

or equivalently,

$$\frac{1}{2} = e^{-5730k}$$

Taking natural logarithms of both sides, we obtain

$$\ln\frac{1}{2} = -5730k$$

Therefore

$$k = -\frac{1}{5730} \ln \frac{1}{2} = \left(-\frac{1}{5730}\right)(-\ln 2) = \frac{1}{5730} \ln 2$$

Consequently the formula we seek is

$$f(t) = f(0)e^{-(1/5730)(\ln 2)t} \quad \square$$

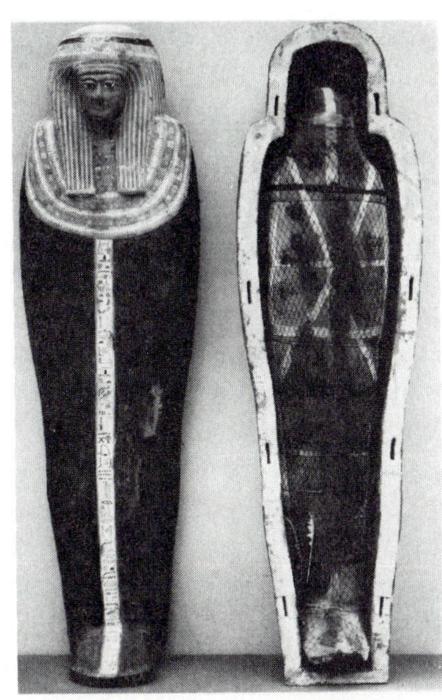

An ancient Egyptian mummy in its wood coffin. Carbon dating is often used in determining the age of mummies.

Every living object contains C^{14}. Because of the radioactivity of C^{14}, you might think that the amount of C^{14} in living objects would be ever decreasing. But to offset the decay of C^{14} in any living object, there is an increase of C^{14} in the atmosphere due to cosmic radiation. This C^{14} enters the carbon cycle of living plants and animals, with the effect that the amount per gram of C^{14} in living organisms remains constant. However, when an organism dies, it abruptly ceases to assimilate any more C^{14}, and consequently the amount of C^{14} per gram in the organism starts decreasing. This decrease in C^{14} after death is the basis for the process known as ***carbon dating*** of once-living organisms.

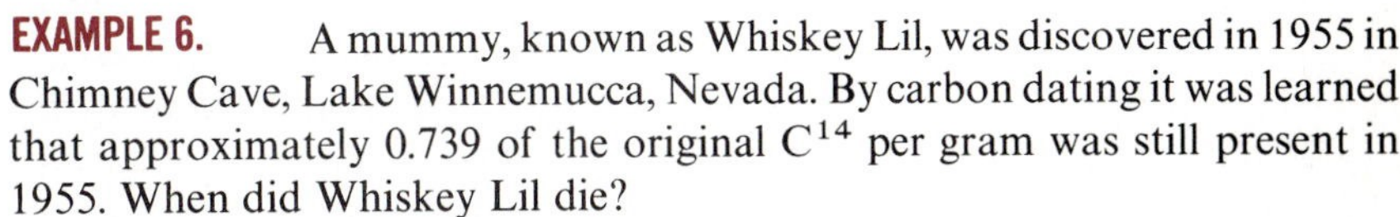

EXAMPLE 6. A mummy, known as Whiskey Lil, was discovered in 1955 in Chimney Cave, Lake Winnemucca, Nevada. By carbon dating it was learned that approximately 0.739 of the original C^{14} per gram was still present in 1955. When did Whiskey Lil die?

Solution. We let $t = 0$ at the time of death and let t_0 be the time elapsed between the death of Whiskey Lil and 1955. First we will determine the value of t_0, and then we will calculate when she died. By hypothesis,

$$f(t_0) = 0.739 f(0)$$

and by (6) with t replaced by t_0,

$$f(t_0) = f(0)e^{-(1/5730)(\ln 2)t_0}$$

Therefore

$$0.739 f(0) = f(0)e^{-(1/5730)(\ln 2)t_0}$$

or equivalently,

$$0.739 = e^{-(1/5730)(\ln 2)t_0}$$

Taking logarithms of both sides, we find that

$$\ln 0.739 = -\left(\frac{1}{5730} \ln 2\right) t_0$$

Solving for t_0, we obtain

$$t_0 = -\frac{5730 \ln 0.739}{\ln 2}$$

By calculator, we conclude that

$$t_0 \approx 2500$$

Therefore Whiskey Lil died about 2500 years before 1955 A.D., which means about 545 B.C. □

Since the mid-1950s, when Willard Libby proposed dating objects by analysis of electron emissions of C^{14}, carbon dating has been intimately connected with the questions of when the Americas were settled and what kinds of tools early cultures included. Bones, skulls, artifacts—archaeological finds of all kinds—were subjected to carbon dating analysis. Until recently the best estimates pointed to the first American settlements as having taken place around 11,000 B.C., but new evidence suggests, again with the help of carbon dating, that the first settlements were perhaps as much as 16,000 years earlier. (See Exercise 25.)

Yet there are limitations to the use of carbon dating. First, the outer limits of dates that can be ascertained by carbon dating are approximately 1500 A.D. and 50,000 B.C. Second, carbon dating gives approximate, rather than precise, dates; the error can be more than 10%.

EXERCISES 4.5

In Exercises 1–10, solve the given equation for k or t, whichever is appropriate. Leave your answer in terms of natural logarithms.

1. $e^{3k} = 4$

2. $e^{-4k} = \frac{1}{5}$

3. $7e^{-1.2k} = 3.5$

4. $3e^{(t/2)\ln 2} = 4$

5. $1.7e^{-(t/3)\ln 2} = 0.85$

6. $0.005e^{-(t/1.5)\ln 3} = 0.0004$

7. $e^{2t} - 3e^{t} + 2 = 0$
(*Hint:* This is a quadratic equation in e^t.)

8. $e^{4t} - e^{2t} - 6 = 0$

9. $e^{t} + e^{-t} = 2$

10. $2^{kt} = e^{-t}$

11. Suppose the number of organisms of a species is given by (3). Show that the doubling time is $(\ln 2)/k$. (*Hint:* Compute $f((\ln 2)/k)$.)

12. With the data given in Example 3, determine how long it would take for the population of 1000 rice weevils to reach
a. 3000 b. 4000

13. If the temperature is lowered to 23°C, then an initial population of 1000 rice weevils could ideally grow to approximately 1537 during a week's time.
a. Determine a formula for the number of rice weevils at a future time t.
b. Determine the doubling time of the weevils.

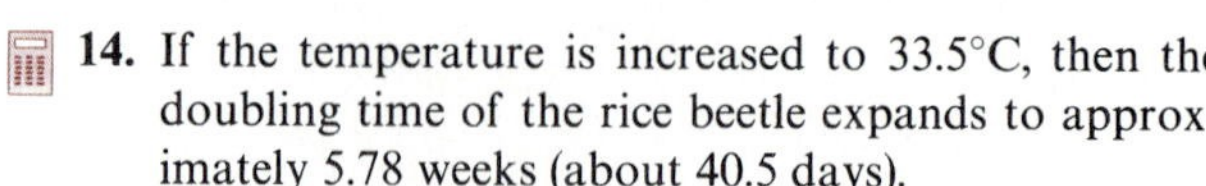

14. If the temperature is increased to 33.5°C, then the doubling time of the rice beetle expands to approximately 5.78 weeks (about 40.5 days).

a. Find the numerical value of k in the population formula.

b. Determine how long it would take for a population of 217 to increase to 864.

15. Experiment has shown that under ideal conditions and a constant temperature of 28.5°C the population of a certain type of flour beetle grows according to the formula

$$f(t) = f(0)e^{0.71t}$$

where t is measured in weeks. Does it take more than a week for the population to double? Explain your answer.

16. Using the information from Exercise 15, determine the length of time it would take for a population of 5000 flour beetles to grow to 17,000.

17. It is known that a population of one type of grain beetle can grow from 1000 to 1994 in a week if the temperature is set at 32.3°C, whereas it can grow from 2500 to 4465 in a week at a temperature of 29.1°C. Determine at which temperature the doubling time is less.

18. Suppose the number of bacteria in a culture grows exponentially and increases from 1 million to 3 million in 6 hours.

a. Find a formula for the number $f(t)$ of bacteria at time t.

b. How long would it take for the number of bacteria to quadruple?

19. The world population in 1962 was approximately 3.15 billion, and in 1971 it was approximately 3.706 billion. Assuming that the world population was growing exponentially during the interim and continued to do so, determine what the (approximate) world population would have been in 1990.

20. In 1930 the world's population was 2 billion, and in 1976 the population passed the 4 billion mark. Assuming that the population grew exponentially during the interim, determine the constant k.

21. Suppose a radioactive substance decays according to (4). Show that the half-life of the substance is $(\ln 2)/k$. (See Exercise 11.)

22. What percent of the original C^{14} content in a skull would remain now if the animal lived around

a. 1500 A.D.? b. 50,000 B.C.?

23. In the late 1960s a wrench made from bone was unearthed in southeast Arizona. The date of the bone was calculated to be approximately 11,200 B.C. What percentage of the original C^{14} remained?

24. About 1960, awls used in basketmaking and formed from antlers were found in an archaeological site in western Tennessee. At that time approximately 32.86% of the original C^{14} remained. How old was the antler?

25. During the beginning of the 1970s a toothed bone scraper found in northern Yukon, Canada, gave new evidence of human settlements in North America more ancient than had been supposed. The C^{14} content remaining in the scraper was calculated to be approximately 3% of the original C^{14}. Does this suggest that there were human settlements in the Yukon at least 25,000 years ago?

26. Some trees that were destroyed during the Fourth Ice Age now contain only 27% of the amount of C^{14} per gram that living trees of the same type now contain. When did the Fourth Ice Age occur?

27. The recently discovered Dead Sea Scrolls were determined by carbon dating to have been written about

33 A.D. What percentage of the original amount of C^{14} presently remains in the Scrolls?

28. Suppose you are told that a bone found in an archaelogical site is more than 17,000 years old. When you submit the bone to carbon dating analysis, you determine that approximately 20% of the original C^{14} remains. Are your findings compatible with the claim of the bone's age? Explain why or why not.

29. Determine the percentage of C^{14} in a skull that decays during any 100 year period.

30. The half-life of strontium 90 (Sr^{90}) is 19.9 years. How long is required for a given amount of Sr^{90} to decay to
a. 25% of its original amount?
b. 20% of its original amount?

31. The half-life of uranium 238 (U^{238}) is approximately 4.5 billion years. How long is required for 1% of a given quantity of U^{238} to decay?

32. Suppose the half-life of a radioactive isotope is h years. Show that the amount $f(t)$ of the isotope after t years is given by

$$f(t) = f(0)2^{-t/h}$$

33. Halley's Law states that the barometric pressure $p(x)$ in inches of mercury at x miles above sea level is given by

$$p(x) = 29.92e^{-0.2x} \quad \text{for} \quad x \geq 0$$

Find the barometric pressure
a. at sea level
b. 1 mile above sea level
c. atop Mt. Whitney (14,495 feet above sea level)

34. The Bouguer–Lambert Law states that as a light beam passes through water, the intensity of the beam x meters from the surface of the water is given by

$$f(x) = ce^{-1.4x}$$

where c is the intensity at the surface. Determine the ratio of the intensity 10 meters below the surface to the intensity at the surface—that is, find $f(10)/c$. (From your answer it will be apparent why most flora cannot survive at depths greater than 10 meters.)

35. Let $A(t)$ denote the area in square centimeters of the unhealed portion of a wound t days after it is sustained, and let $A(0)$ represent the area of the original wound. Then according to the Law of Healing, $A(t)$ is given by

$$A(t) = A(0)e^{-0.15t} \quad \text{for} \quad t \geq 0$$

Suppose that just after a collision, a basketball player is told he cannot play until 90% of a leg wound is healed. Would he be able to play in a tournament in two weeks?

4.6 COMPOUND AND CONTINUOUS INTEREST

Recall from Section 1.2 that if P dollars are deposited into an account earning simple interest at an annual rate r, then the amount of interest after n years is Prn, and therefore the amount A_n in the account after n years is given by the following formula:

SIMPLE INTEREST

$$A_n = P + Prn = P(1 + rn) \tag{1}$$

In particular, after 1 year the amount A_1 in the account is given by

$$A_1 = P(1 + r) \tag{2}$$

For example, if the rate were 6%, then $A_1 = P(1 + 0.06) = 1.06\,P$.

If instead of using simple interest, the bank compounds interest annually, then the amount in the account after 1 year is A_1, as before. But during the second year interest is earned on A_1 dollars (rather than on the original principal P). Thus after 2 years the amount A_2 in the account is given by

$$A_2 = A_1 + A_1 r = A_1(1 + r)$$

Substituting for A_1 from (2), we have

$$A_2 = P(1 + r)^2$$

During the third year interest is earned on A_2 dollars, and in the same way as above, we find that the amount A_3 in the account after 3 years is given by the formula

$$A_3 = A_2 + A_2 r = A_2(1 + r) = P(1 + r)^3$$

More generally, the formula for the amount A_n in the account after n years of compounding interest annually is given by the following formula:

INTEREST COMPOUNDED ANNUALLY

$$A_n = P(1 + r)^n \tag{3}$$

Nowadays most banks go a step farther and compound interest not annually, but quarterly, monthly, or even daily (so the interest period becomes 3 months, 1 month, or 1 day, respectively). If a bank offers 6% interest compounded monthly, then since there are 12 months in a year, the interest rate for each month is 6% divided by 12, or $\frac{1}{2}\%$, and if 6% interest is compounded daily, the rate for each day is $\frac{6}{365}\%$ (except in leap years). In general, if a bank offers compound interest at an annual rate r compounded k times a year, then the interest rate for each interest period is r/k, so that after the first interest period a deposit of P dollars is worth A_1 dollars, where

$$A_1 = P + P \cdot \frac{r}{k} = P\left(1 + \frac{r}{k}\right)$$

If A_n represents the amount in the account after n interest periods, if r is the annual interest rate, and if the interest is compounded k times a year, then A_n can be shown to be given by the following formula:

INTEREST COMPOUNDED k TIMES ANNUALLY

$$A_n = P\left(1 + \frac{r}{k}\right)^n \tag{4}$$

Notice that (4) agrees with (3) when $k = 1$, that is, when the interest is compounded once a year (annually).

EXAMPLE 1. Suppose that \$1000 is deposited into an account. How much money is in the account after 5 years if the account earns 6% interest compounded

a. annually? b. monthly? c. daily?

Solution.

a. We use (4) with $P = 1000$, $r = 0.06$, $k = 1$, and $n = 5$:

$$A_5 = 1000(1 + 0.06)^5 = 1000(1.06)^5 \approx 1338.23$$

where the final computation can be accomplished either by hand or by calculator. Thus \$1338.23 is in the account after 5 years.

b. Here we use (4) with $P = 1000$, $r = 0.06$, $k = 12$, and $n = 5(12) = 60$:

$$A_{60} = 1000\left(1 + \frac{0.06}{12}\right)^{60} = 1000(1.005)^{60}$$

Using a calculator, we find that $A_{60} \approx 1348.85$. Consequently the account contains approximately \$1348.85 after 5 years.

c. Again we use (4), but this time with $P = 1000$, $r = 0.06$, $k = 365$, and $n = 5(365) = 1825$:

$$A_{1825} = 1000\left(1 + \frac{0.06}{365}\right)^{1825}$$

Using a calculator, we find that $A_{1825} \approx 1349.83$. This means that after 5 years the account contains \$1349.83. □

Observe from the results of Example 1 that the shorter the interest period is (that is, the more times per year interest is compounded), the more money is in the account after 5 years. More generally, the shorter the interest period is, the more money there would be in the account after any given length of time.

Continuously Compounded Interest

As we mentioned above, the shorter the interest period is, the more money there is in the account after a fixed period of time. A natural question to ask is: What would happen to the money in the account in, say, one year, if the interest period shrinks to 0? Since (4) tells us that with k interest periods a year the amount in the account at the end of one year would be given by

$$A_k = P\left(1 + \frac{r}{k}\right)^k$$

the question just posed is equivalent to the following question: What happens to A_k as k grows without bound? Would A_k also grow without bound? Although we will not prove it, the answer is that A_k approaches a limiting value A as k increases without bound, and moreover,

$$A = Pe^r \tag{5}$$

When the interest for one year is computed by (5), we say that interest has been ***compounded continuously***. Some banks actually do offer interest compounded continuously on savings accounts.

Now let us obtain a formula for the amount in an account after t years when interest is compounded continuously at an annual rate r. By (5) the amount after one year is Pe^r. Therefore after 2 years the amount is $(Pe^r)e^r = Pe^{2r}$, and in general, the amount $A(t)$ in the account after t years is given by the following formula:

INTEREST COMPOUNDED CONTINUOUSLY

$$A(t) = Pe^{rt} \tag{6}$$

Formula (6) is valid for any positive value of t, integer or not.

EXAMPLE 2. Suppose that \$1000 is deposited into an account that pays 6% interest compounded continuously. How much is in the account after 5 years?

Solution. We use (6) with $P = 1000$, $r = 0.06$, and $t = 5$:

$$A(5) = 1000e^{(0.06)5} = 1000e^{0.3}$$

Using a calculator, we find that $A(5) \approx 1349.86$. This means that the account contains \$1349.86 after 5 years. □

Compounding continuously can be regarded as a kind of ultimate—it produces more than compounding annually, monthly, or even daily. However, it is worthwhile noticing that the amount attained by compounding \$1000 continuously for 5 years is a mere 3¢ more than that attained by compounding daily, and only \$1.01 more than that attained by compounding monthly.

EXERCISES 4.6

1. Suppose \$1000 is deposited into an account. How much money will be in the account after 5 years if the account earns 5% interest compounded
   a. annually? b. monthly? c. daily?

2. Suppose the interest in Exercise 1 were compounded continuously. How much would be in the account after 5 years? Compare your answer with those given in Exercise 1.

3. Suppose \$10,000 is deposited into an account. How much money will be in the account after 10 years if the account earns 9% interest compounded
   a. annually? b. monthly? c. daily?

4. Suppose the interest in Exercise 3 were compounded continuously. How much would be in the account after 10 years? Compare your answer with those given in Exercise 3.

5. How much money would you have to deposit into an account that earns 7% interest compounded semi-annually if you wished the account to contain \$1000 after 4 years?

6. How much money would you have to deposit into an account that earns 7% interest compounded continuously if you wish the account to contain \$1000 after 4 years? Compare your answer with that given in Exercise 5.

7. Suppose a benefactor plans to open an account that will be yours in 5 years. Which would you prefer—a \$2000 initial deposit that earns 6% simple interest, or a \$1500 initial deposit that earns 12% interest compounded monthly? Justify your answer.

8. Approximately how many months does it take for \$1000 to double in an account that earns
   a. 6% interest compounded monthly?
   b. 9% interest compounded monthly?

9. Approximately how long does it take for \$1000 to double in an account that earns interest compounded continuously at a rate of
   a. 6%? b. 9%?

10. Suppose \$5000 is put into a saving account compounded monthly, and after 3 years the account is worth \$7000. What is the interest rate?

11. It has been estimated that in the past, productivity in the United States has increased at a rate of 4% annually. Assuming that this is true and productivity is compounded annually, determine how much productivity increased during a 25-year span.

12. If inflation is 9% per year over a 6-year period and is compounded annually, what is the overall inflation during the period?

13. Suppose that on January 1, \$800,000 is deposited in an account earning 6% interest compounded daily. Determine the difference, if any, between the amounts in the account at the end of a regular year (with 365 days) and a leap year (with 366 days).

14. You have \$1000 with which to open a savings account. Which would be worth more at the end of a 10-year period—an account that draws 6% simple interest, or an account that earns 5% interest compounded continuously? Justify your answer.

15. In 1626, Manhattan was purchased from native Americans for \$24. What would the value be in the year 2000 if native Americans had been able to invest the \$24 at 5% interest compounded
    a. annually? b. continuously?

16. If the \$24 the native Americans received for Manhattan had been compounded continuously at a rate of 5%, during what year would it have been worth 1 billion dollars?

17. The rate r at which a savings account draws compound interest is sometimes called the ***nominal rate*** of the account. The ***effective rate*** E (or effective annual interest rate) of the account is the simple interest that would yield the same interest during the course of one year.

    a. Assume that interest is compounded k times a year. Show that E is given by

    $$E = \left(1 + \frac{r}{k}\right)^k - 1$$

    b. What is the effective rate for an account that draws interest at a rate of 7.5% compounded monthly?
    c. Find a formula for the nominal interest rate r in terms of E and k.

18. A loan worth P dollars is to be paid off in n equal installments of A dollars each, and the interest rate is r per pay period. The ***installment payment*** A is given by the formula

$$A = P\left(\frac{r(1+r)^n}{(1+r)^n - 1}\right)$$

Solve for n in terms of P, r, and A.

The formula for A given in Exercise 18 applies to Exercises 19–22.

19. Pat receives a loan of \$30,000, with an annual interest rate of 9% compounded annually, and the loan is to be paid off in 20 equal yearly installments.
a. Determine the annual payment.
b. Determine the total to be paid over the life of the loan.

20. Marian receives a loan of \$6000, with an annual interest of 11.5% compounded monthly, and is to pay off the loan in equal monthly installments over a 10-year period.
a. Determine the monthly payment.
b. Determine the total to be paid back over the life of the loan.

21. Suppose Marian can get a 30-year loan at 9% per year, and can afford to pay up to \$263 a month. Determine the maximum loan under these conditions.

22. Jack and Jill each take out a 5-year loan of \$8000, with an annual interest rate of 10%. Jack pays the loan back in equal annual installments, whereas Jill pays the loan back in equal monthly installments. Who pays back more over the life of the loan, Jack or Jill? What is the difference?

***23.** Suppose you negotiate a 3-year contract according to which you are paid \$20,000 the first year and receive a total salary of \$70,000 over the 3-year period. If the salary is treated as being compounded annually, what is the rate at which it is compounded?

KEY TERMS

exponential function
  base
  exponent
  natural exponential function
exponential growth
  doubling time
exponential decay
  half-life
logarithmic function
  base
  natural logarithm
  common logarithm
  characteristic
  mantissa
antilogarithm
linear interpolation
compound interest
continuously compounded interest

KEY FORMULAS

Laws of Exponents

$a^x a^y = a^{x+y}$

$(a^x)^y = a^{xy}$

$(ab)^x = a^x b^x$

$\left(\dfrac{a}{b}\right)^x = \dfrac{a^x}{b^x}$

$\dfrac{a^x}{a^y} = a^{x-y}$

$a^{-x} = \dfrac{1}{a^x} = \left(\dfrac{1}{a}\right)^x$

$a^0 = 1$

$1^x = 1$

Laws of Logarithms

$a^{\log_a x} = x$

$\log_a (a^x) = x$

$\log_a (xy) = \log_a x + \log_a y$

$\log_a \dfrac{1}{x} = -\log_a x$

$\log_a \dfrac{x}{y} = \log_a x - \log_a y$

$\log_a (x^c) = c \log_a x$

$\log_b x = \dfrac{\log_a x}{\log_a b}$

$\log_a 1 = 0$

$\log_a a = 1$

$A(t) = Pe^{rt}$ (interest compounded continuously)

REVIEW EXERCISES

In Exercises 1–4, sketch the graph of the given function.

1. $f(x) = 3^x$

2. $f(x) = (\frac{1}{2})^{x-1}$

3. $f(x) = \frac{1}{2} \log_2 x$

4. $f(x) = |\log_2 x|$

In Exercises 5–8, simplify the given expression.

5. $(\sqrt{2^{\sqrt{2}}})^{\sqrt{2}}$

6. $\dfrac{5^{1-\sqrt{2}} \cdot 25^{1/\sqrt{2}}}{5^{2+\sqrt{2}}}$

7. $\ln 12 - \ln 2 + 2 \ln e^3$

8. $\log_{10} 5 - \log_{10} \dfrac{50}{7}$

In Exercises 9–14, solve the given equation.

9. $7^{x-1} = (\sqrt{7})^{-2x^2}$

10. $2^{x^3} = 4^x$

11. $\log_2 x = -3$

12. $\log_x 5 = 2$

13. $\log_{10} (3 - 5x) - \log_{10} (1 - x) = 1$

14. $3 \log_2 \sqrt{x} = 2 \log_2 8$

In Exercises 15–20, solve the given equation. Leave your answer in terms of logarithms.

15. $2^{-x} = 7$

16. $4^{x-4} = 5^{3x+5}$

17. $2^x = 3^{(x^4)}$

18. $e^{2x} = \dfrac{1}{e^{(x^2)}}$

19. $e^{5k} = 4.37$

20. $e^{-3t \ln 2} = 0.37$

In Exercises 21–24, evaluate the given expression.

21. $\log_2 \frac{1}{16}$

22. $\log_{10} (10^{-5})$

23. $e^{\ln 5}$

24. $10^{\log_{10} 1.23}$

25. Solve the equation $y = \ln(x - \sqrt{x^2 - 1})$ for x.

26. Suppose $y = a + b \ln x$, where a and b are constants. Express x in terms of y.

27. Show that $\log_2 x = 2 \log_4 x$.

28. Show that $\log_2 x = 3 \log_8 x$.

29. Let $f(x) = e^{cx}$, where c is a constant. Suppose that the graph of f passes through the point $(\frac{1}{3}, e^2)$. Determine c.

30. Let $g(x) = ae^{cx}$, where a and c are constants. Suppose that the graph of g passes through the points $(0, \frac{1}{5})$ and $(-1, \frac{1}{5}e^2)$. Determine a and c.

31. The noise level of one sound is 20 decibels higher than that of a second sound. Find the ratio of the intensity of the first sound to that of the second.

32. If an earthquake measures 6.1 on the Richter scale, what will one that is 100 times as intense measure?

33. It has been shown experimentally that under optimal conditions, the population of a colony of brown rats could grow exponentially from 50 rats to approximately 11,070 during the course of a year.
 a. Determine the doubling time.
 b. How long would it take for the population to increase from 120 to 340 rats?

34. The total amount of timber in a young forest increases exponentially.
 a. If the annual growth rate k is 3%, how long does it take for the total amount of timber to double?
 b. How much sooner would the total amount of timber double if the rate of growth were 4%?

35. The radioactive isotope potassium 40 (K^{40}) has a half-life of approximately 91.3 billion years. If a rock formed 4 billion years ago (shortly after the earth came into existence) initially contained 1 gram of K^{40}, how much of it would the rock contain today?

36. Radioactive argon 39 (Ar^{39}) has a half-life of 4 minutes. How long would it take for 99% of a given amount of Ar^{39} to decay?

37. Until about 200 years ago, charcoal, which is derived from wood, was used in the smelting of iron, and thus it is possible to use carbon dating to determine the age of objects made long ago from iron alloys (such as steel). Fragments of a steel sword unearthed in Yugoslavia in 1970 were found to contain only 77.286% of the C^{14} the sword would have contained at the time it was produced. Determine the age of the sword.

38. Nails buried long ago at a Roman Legionary fortress in Scotland were found in 1960 to contain 79.948% of the C^{14} they would have contained when produced. Determine when the nails were produced. (According to historical accounts, the fort was built in 83 A.D. and dismantled in 87 A.D.)

39. Suppose $2000 is deposited into a savings account. How much money is in the account after 10 years if the account earns 8% compounded
 a. annually? b. monthly? c. daily?

40. Suppose $2000 is deposited into a savings account. How much money is in the account after 10 years if the account earns 8% compounded continuously?

41. How much money must be deposited into a savings account that earns 6% compounded continuously if the account is to contain $5000 after 8 years?

5 THE TRIGONOMETRIC FUNCTIONS

Next we turn to trigonometry and trigonometric functions. Trigonometry is related to the study of triangles, which were studied long ago by the Babylonians and ancient Greeks. In fact, the word trigonometry is derived from the Greek word for "the measurement of triangles." Today trigonometry and trigonometric functions are indispensable tools not only in mathematics, but also in many practical applications, especially those involving oscillations and rotations.

As the first of three chapters on trigonometry, Chapter 5 is devoted mainly to defining the trigonometric functions and studying their properties, with a special emphasis on their graphs. The last section of the chapter involves inverses of trigonometric functions.

Of the many ways the concepts in this chapter can be applied, we will discuss two in detail:

> In the first application, one can study the lengths of the sides and measures of the angles in an appropriate triangle in order to determine the height of the tallest known tree in the world, a mammoth redwood tree located in Humboldt Redwood State Park, California (see Example 1 of Section 5.3). For a second application, trigonometric functions discussed in this chapter can help in understanding sound waves. For example, they can be used to analyze the vibrations of the strings on a violin or guitar (see Section 5.6).

5.1 ANGLE MEASUREMENT

Before we can define the trigonometric functions, we must discuss angles and methods of measuring them. An ***angle*** in the plane is formed by two lines or line segments, called the ***sides*** of the angle, that have a common endpoint, called the ***vertex*** of the angle (Figure 5.1). We call one side the ***initial side*** of the angle and the other the ***terminal side***, and we identify the direction that indicates rotation from the initial side to the terminal side. In a diagram this is indicated by a curved arrow from the initial side to the terminal side (Figure 5.1).

In this chapter angles will generally be located in a plane equipped with a Cartesian coordinate system. Normally the vertex of such an angle will be the origin and the initial side will lie along the positive x axis (Figure 5.2). In this case we say that the angle is in ***standard position*** in the coordinate system.

There are two common units in which we measure angles: degrees and radians. We will employ both of these units. First we will study degrees, then radians.

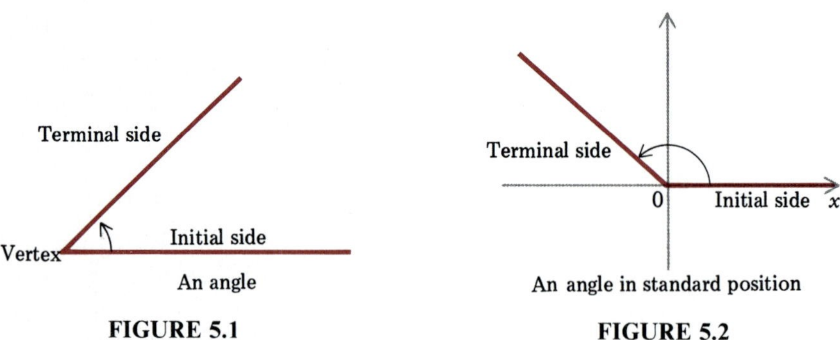

FIGURE 5.1

FIGURE 5.2

Degrees

One unit for measuring angles is the ***degree***, chosen in such a way that an angle formed by one complete revolution in the counterclockwise direction contains 360 degrees (written 360°), as Figure 5.3a indicates. Angles of 360°,

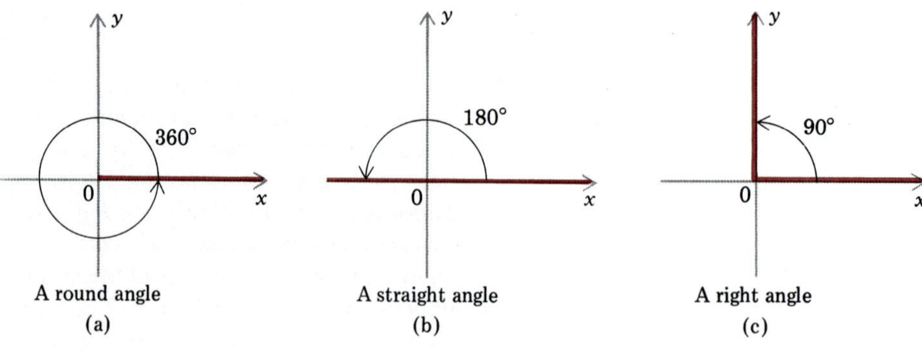

FIGURE 5.3

180°, and 90° have special names:

angle of 360°:	round angle
angle of 180°:	straight angle
angle of 90°:	right angle

(Figure 5.3a,b,c). We can draw certain other angles by using our knowledge of these angles.

EXAMPLE 1. Draw angles of 45°, 60°, and 270°.

Solution. To obtain an angle of 45° we bisect an angle of 90° (Figure 5.4a). For an angle of 60° we first mentally trisect an angle of 90° to obtain an angle of 30°, and then take 2 copies (Figure 5.4b). Finally, for an angle of 270° we can either take 3 copies of an angle of 90°, or we can add an angle of 90° to an angle of 180° (Figure 5.4c). □

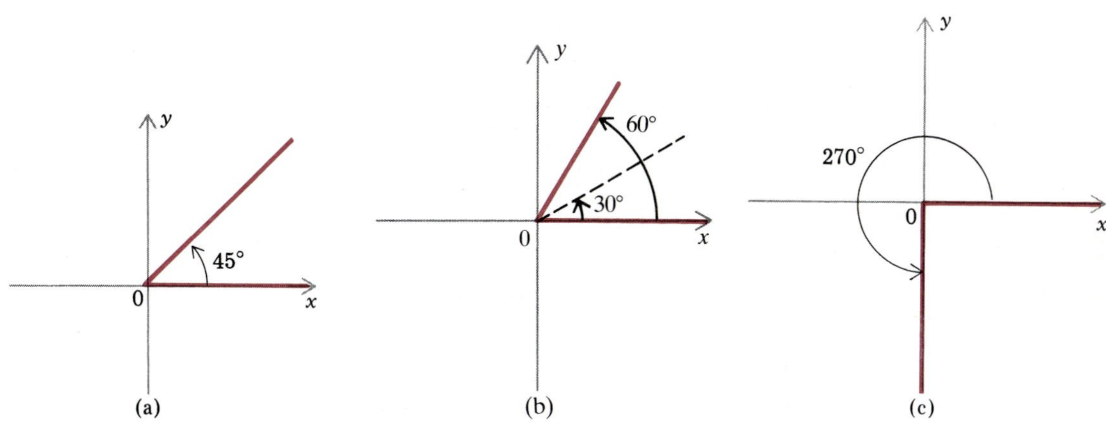

FIGURE 5.4

Since a 45° angle is formed by bisecting a 90° angle, it follows from the congruence of the triangles in Figure 5.5a that the terminal side of a 45° angle in standard position lies along the line $y = x$. Similarly, the terminal side of a 135° angle bisects the second quadrant, so it lies along the line $y = -x$ (Figure 5.5b).

If an angle is formed by rotating the initial side counterclockwise, then the angle measurement is positive; if an angle is formed by rotating the initial side clockwise, then the angle measurement is negative. We usually say that an angle is positive (or negative) if its degree measurement is positive (or negative).

Because angle measurement is positive or negative according to whether the rotation of the initial side is counterclockwise or clockwise, it follows that

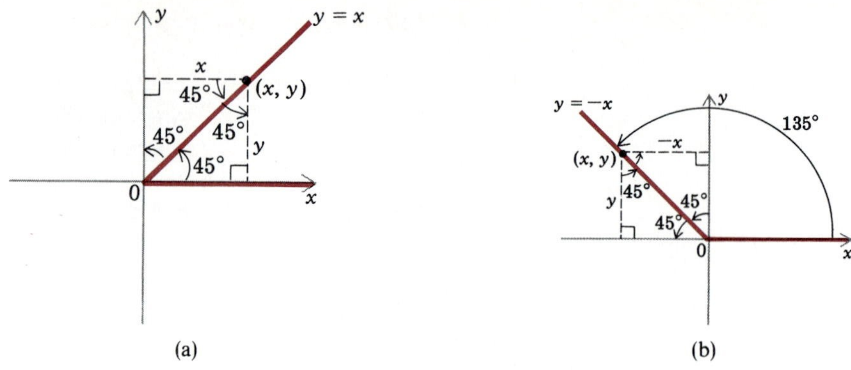

FIGURE 5.5

if we have drawn an angle in standard position, then we can draw the negative of the angle by rotating in the opposite direction (clockwise rather than counterclockwise, or counterclockwise rather than clockwise).

EXAMPLE 2. Draw angles of $-45°$ and $-270°$.

Solution. Figure 5.4a depicts an angle of 45°. Rotating in the clockwise direction, we obtain an angle of $-45°$ (Figure 5.6a). Similarly, if we rotate clockwise the angle of 270° in Figure 5.4c, we obtain an angle of $-270°$ (Figure 5.6b). □

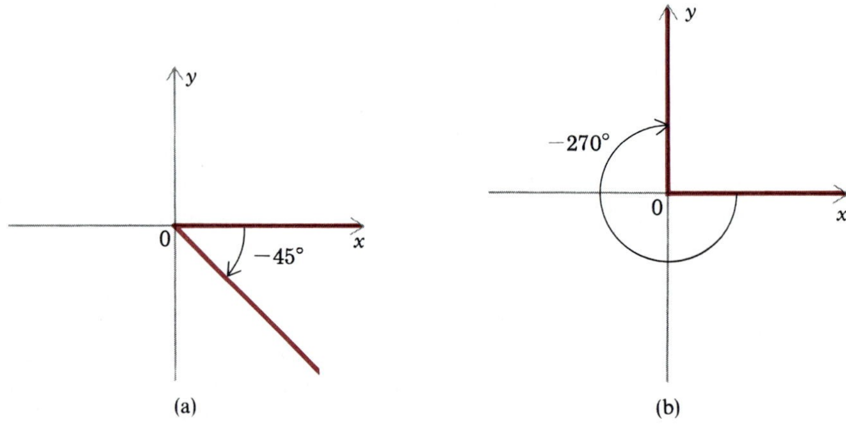

FIGURE 5.6

Before we present the next example, let us recall that a 360° angle corresponds to one complete revolution, so the initial and terminal sides of the angle coincide. Moreover, the sides of a 360° angle lie along the positive x axis if the angle is in standard position. This observation will help us draw angles of more than 360°.

EXAMPLE 3. Draw angles of 810° and −390°.

Solution. Since $810 = 2 \cdot 360 + 90$, an angle of 810° is formed by rotating the positive x axis counterclockwise through two complete revolutions and then continuing the rotation through an additional 90° (Figure 5.7). Since $-390 = -360 - 30$, and angle of −390° is formed by rotating the positive x axis clockwise through one revolution and then continuing the rotation through an additional 30° (Figure 5.8). □

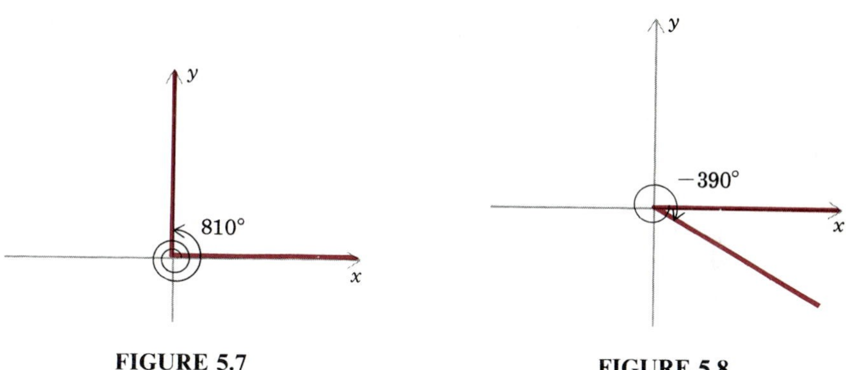

FIGURE 5.7

FIGURE 5.8

From Figures 5.3c and 5.7 it is apparent that angles in standard position of 90° and 810° have the same terminal sides. Such angles are called ***coterminal angles***. The only difference between two coterminal angles is that one of them can be formed by rotating the initial side through more complete revolutions than the other. Since each revolution corresponds to 360°, it follows that the degree measures of two coterminal angles differ by an integral multiple of 360°. For example, the coterminal angles of 90° and 810° have a difference of 720° (which is twice 360°).

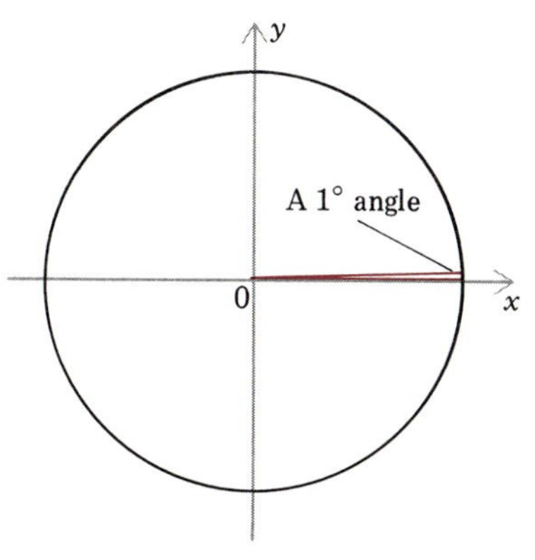

FIGURE 5.9

As one can see from Figure 5.9, an angle of 1° is essentially imperceptible in a small circle. However, suppose a circle is as large as the silhouette of the earth, some 24,881 miles in circumference. In this case, 1° subtends an arc of $\frac{24{,}881}{360}$ (approximately 69) miles, or about the length of Death Valley in California. In light of this it might not be surprising that degrees have been subdivided. Indeed, each degree is divided into 60 minutes, and each minute is divided into 60 seconds. A minute is denoted by the symbol ′, and a second by the symbol ″. Thus an angle of 34 degrees, 23 minutes, and 59 seconds is written 34°23′59″. For accurate location of stars and satellites in the sky, astronomers use not only degrees but also minutes and seconds. Longitude and latitude of points on earth are also described in terms of degrees and minutes (and seconds if greater precision is required).

Radians

The measure of angles most frequently used in mathematics is the radian, a unit much larger than the degree. In order to set the stage for the definition of radian, we first draw the circle $x^2 + y^2 = 1$ of radius 1, called the ***unit circle***.

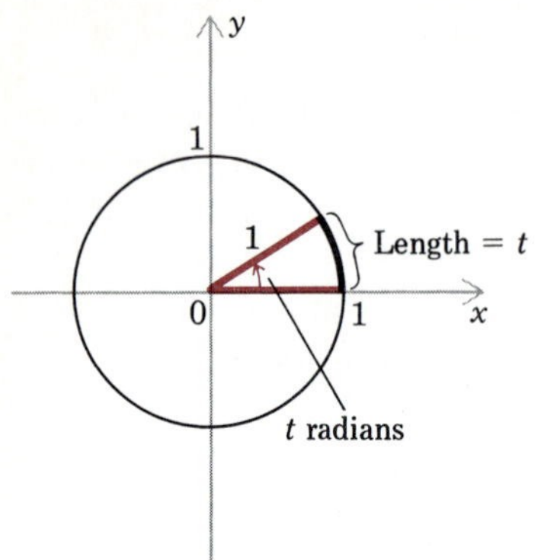

FIGURE 5.10

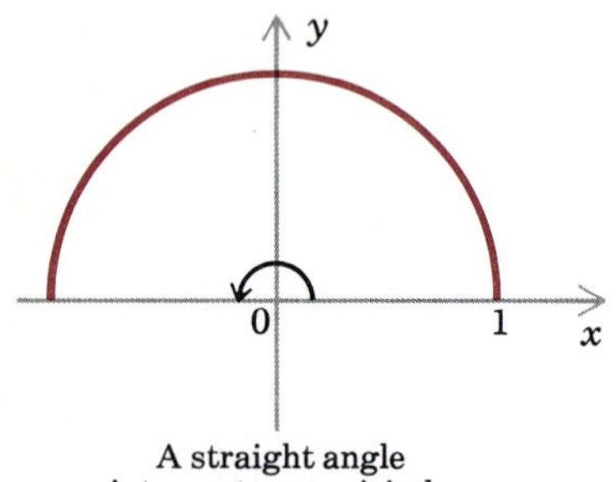

A straight angle intercepts a semicircle

FIGURE 5.11

Any positive angle in standard position intercepts an arc on the unit circle, and the length of the arc intercepted is proportional to the size of the angle. The radian is chosen so that the ***radian measure*** of a positive angle equals the length of the arc it intercepts on the unit circle (Figure 5.10).

Since a straight angle intercepts half of the unit circle (Figure 5.11), the radian measure of a straight angle is half the circumference of the unit circle, or $\frac{1}{2}(2\pi \cdot \text{radius}) = \frac{1}{2}(2\pi \cdot 1) = \pi$. Thus a straight angle contains π radians. Since such an angle also contains 180°, we have

$$180^\circ = \pi \text{ radians}$$

From this it follows that for any number θ,

$$\theta^\circ = \left(\frac{\theta}{180} \cdot 180\right)^\circ = \left(\frac{\theta}{180} \cdot \pi\right) \text{radians} = \left(\frac{\pi}{180} \cdot \theta\right) \text{radians}$$

where θ (the Greek letter *theta*) represents the degree measure of an angle. Likewise,

$$t \text{ radians} = \left(\frac{t}{\pi} \cdot \pi\right) \text{radians} = \left(\frac{t}{\pi} \cdot 180\right)^\circ = \left(\frac{180}{\pi} \cdot t\right)^\circ$$

where t represents the radian measure of an angle. The preceding two sets of equations yield formulas for converting between degrees and radians:

$$\boxed{\theta^\circ = \left(\frac{\pi}{180} \cdot \theta\right) \text{radians}} \tag{1}$$

and

$$\boxed{t \text{ radians} = \left(\frac{180}{\pi} \cdot t\right)^\circ} \tag{2}$$

In particular,

$$1 \text{ radian} = \left(\frac{180}{\pi}\right)^\circ \approx 57^\circ 17' 45''$$

EXAMPLE 4. Convert the following to radian measure.

a. 30° b. 60° c. 291°

Solution. In each case we use (1):

a. $30^\circ = \left(\frac{\pi}{180} \cdot 30\right) \text{radians} = \frac{\pi}{6} \text{ radians}$

b. $60^\circ = \left(\dfrac{\pi}{180} \cdot 60\right)$ radians $= \dfrac{\pi}{3}$ radians

c. $291^\circ = \left(\dfrac{\pi}{180} \cdot 291\right)$ radians $= \dfrac{97\pi}{60}$ radians □

EXAMPLE 5. Convert the following to degree measure.

a. $\dfrac{\pi}{4}$ radians b. $\dfrac{\pi}{12}$ radians c. $\dfrac{7\pi}{10}$ radians

Solution. In each case we use (2):

a. $\dfrac{\pi}{4}$ radians $= \left(\dfrac{180}{\pi} \cdot \dfrac{\pi}{4}\right)^\circ = 45^\circ$

b. $\dfrac{\pi}{12}$ radians $= \left(\dfrac{180}{\pi} \cdot \dfrac{\pi}{12}\right)^\circ = 15^\circ$

c. $\dfrac{7\pi}{10}$ radians $= \left(\dfrac{180}{\pi} \cdot \dfrac{7\pi}{10}\right)^\circ = 126^\circ$ □

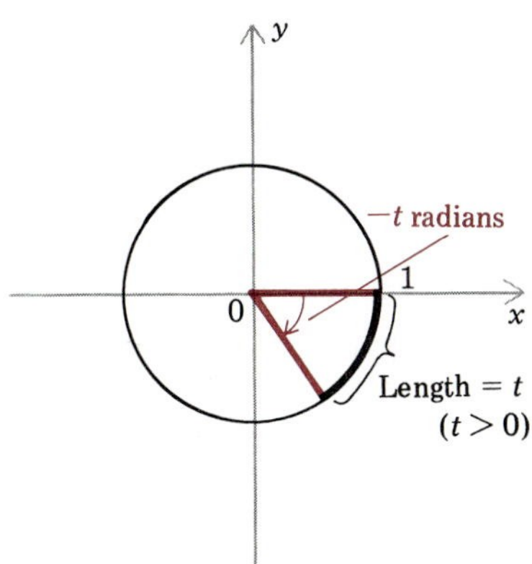

FIGURE 5.12

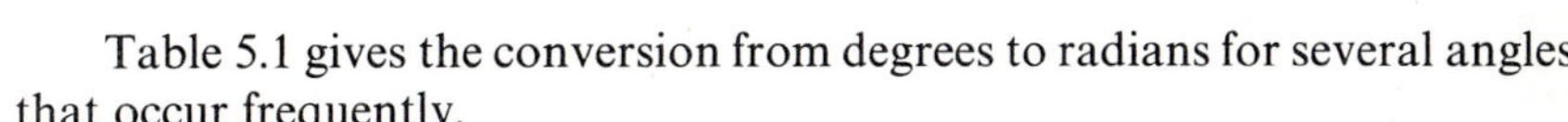

Table 5.1 gives the conversion from degrees to radians for several angles that occur frequently.

If an angle in standard position has negative measure in degrees, then the radian measure is also negative; it is the negative of the length of the arc subtended on the unit circle (Figure 5.12).

EXAMPLE 6. Convert -1110° to radian measure.

Solution. By the preceding remark, along with (1), we have

$$-1110^\circ = -\left(\frac{\pi}{180}\right)(1110) \text{ radians} = -\frac{37\pi}{6} \text{ radians} \quad \square$$

TABLE 5.1

Degrees	0°	30°	45°	60°	90°	120°	135°	150°	180°	210°	225°	240°	270°	300°	315°	330°	360°
Radians	0	$\frac{\pi}{6}$	$\frac{\pi}{4}$	$\frac{\pi}{3}$	$\frac{\pi}{2}$ (Right angle)	$\frac{2\pi}{3}$	$\frac{3\pi}{4}$	$\frac{5\pi}{6}$	π (Straight angle)	$\frac{7\pi}{6}$	$\frac{5\pi}{4}$	$\frac{4\pi}{3}$	$\frac{3\pi}{2}$	$\frac{5\pi}{3}$	$\frac{7\pi}{4}$	$\frac{11\pi}{6}$	2π (Round angle)

Radian Measure for Circles of Arbitrary Radius

If $t > 0$, then an angle of t radians subtends an arc of length t on the unit circle. If the unit circle is replaced by the circle $x^2 + y^2 = r^2$, with an arbitrary radius $r > 0$, then t can be related to the length of the arc on the new circle subtended by an angle of t radians. Since the circumference of the new circle $x^2 + y^2 = r^2$ is $2\pi r$, it follows that the length s of the arc subtended by an angle of t radians is given by the formula

$$s = \left(\frac{t}{2\pi}\right)(2\pi r) = rt$$

or as the formula often appears (with θ substituted for t),

$$s = r\theta \tag{3}$$

If $\theta > 2\pi$, then the arc subtended is to be interpreted as consisting of more than one complete circle, and the value of s will be greater than the circumference of the circle. For example, if $r = 3$ and $\theta = 5\pi/2$, then

$$s = 3\left(\frac{5\pi}{2}\right) = \frac{15\pi}{2}$$

EXAMPLE 7. Find the length s of the arc on a circle of radius 6 subtended by the central angle of the given radian measure.

a. $\dfrac{\pi}{6}$ b. $\dfrac{7\pi}{4}$ c. $\dfrac{43\pi}{3}$

Solution. Each part follows from (3).

a. $s = 6\left(\dfrac{\pi}{6}\right) = \pi$

b. $s = 6\left(\dfrac{7\pi}{4}\right) = \dfrac{21\pi}{2}$

c. $s = 6\left(\dfrac{43\pi}{3}\right) = 86\pi$ □

Now that we have discussed angle measurement and have introduced radians, we are prepared to define the trigonometric functions. We will take up that topic in the next section.

EXERCISES 5.1

In Exercises 1–16, draw the angle in standard position having the given degree measure.

1. 60° **2.** 120° **3.** 150°

4. 210° **5.** 225° **6.** 300°

7. −75° **8.** −120° **9.** −150°

10. −210° **11.** −240° **12.** −330°

13. .510° **14.** 750° **15.** −450°

16. 855°

In Exercises 17–32, draw the angle in standard position with the given radian measure.

17. $\frac{\pi}{6}$ **18.** $\frac{3\pi}{4}$ **19.** $\frac{7\pi}{6}$ **20.** $\frac{5\pi}{3}$

21. 2π **22.** $\frac{7\pi}{3}$ **23.** $\frac{9\pi}{2}$ **24.** $-\frac{\pi}{4}$

25. $-\frac{\pi}{3}$ **26.** $-\frac{\pi}{2}$ **27.** $-\frac{5\pi}{6}$ **28.** $-\frac{3\pi}{2}$

29. $-\frac{7\pi}{4}$ **30.** $-\frac{17\pi}{6}$ **31.** $-\frac{11\pi}{3}$ **32.** -3π

In Exercises 33–46, convert from degrees to radians.

33. 0° **34.** 15° **35.** −45°

36. −210° **37.** 390° **38.** 450°

39. −180° **40.** −300° **41.** 1470°

42. 20° **43.** −72° **44.** 1°

45. 4°30′ **46.** 15°15′

In Exercises 47–58, convert from radians to degrees.

47. $\frac{9\pi}{4}$ radians **48.** $\frac{7\pi}{3}$ radians

49. 6π radians **50.** $\frac{17\pi}{2}$ radians

51. 11π radians **52.** $-\frac{2\pi}{3}$ radians

53. $-\frac{13\pi}{4}$ radians **54.** $-\frac{16\pi}{3}$ radians

55. $\frac{7\pi}{18}$ radians **56.** $\frac{\pi}{90}$ radians

57. 1.8 radians **58.** −0.6 radians

In Exercises 59–68, determine the quadrant in which the terminal side of the angle in standard position with the given measure is located.

59. $\frac{4\pi}{7}$ radians **60.** $\frac{29\pi}{5}$ radians

61. 3.15π radians **62.** $-\frac{19\pi}{5}$ radians

63. $-\frac{37\pi}{6}$ radians **64.** -4.57π radians

65. 2 radians **66.** 721°

67. −809° **68.** −1261°

69. Identify those angles that are coterminal with an angle of $2\pi/3$ radians and those coterminal with an angle of $4\pi/3$ radians. Assume all angles are in standard position.

a. $\frac{16\pi}{3}$ b. $\frac{20\pi}{3}$ c. $-\frac{4\pi}{3}$

d. $-\frac{8\pi}{3}$ e. $-\frac{20\pi}{3}$ f. $-\frac{26\pi}{3}$

g. 240° h. −240° i. 840°

j. −840°

70. Find the length of the arc on a circle of radius 8 subtended by a central angle of the given radian measure.

a. $\frac{4\pi}{3}$ b. $\frac{17\pi}{6}$ c. $\frac{2\pi}{5}$

71. Find the length of the arc on a circle of the given radius subtended by a central angle of $\pi/8$ radians.

a. $r = 1$ b. $r = \frac{1}{3}$ c. $r = 4$

72. Find the radius of the circle on which a central angle of the given radian measure subtends an arc of length 5.

a. $\frac{2\pi}{3}$ b. $\frac{5\pi}{4}$ c. $\frac{13\pi}{6}$

73. An eighth of a right angle, sometimes called a ***point***, has seen use in navigation. Give the degree and the radian measures of a point.

74. Determine the number of radians in the angle formed by the hands of a clock at the following times.

a. 7 A.M. b. 4 P.M.

c. 1:30 P.M. d. 11:15 A.M.

75. The earth rotates 360° in a 24-hour period. Determine through how many radians it rotates in

a. 1 hour b. 1 minute c. 1 second

76. What is the length of an arc subtended by 1″ (that is, one second) on earth?

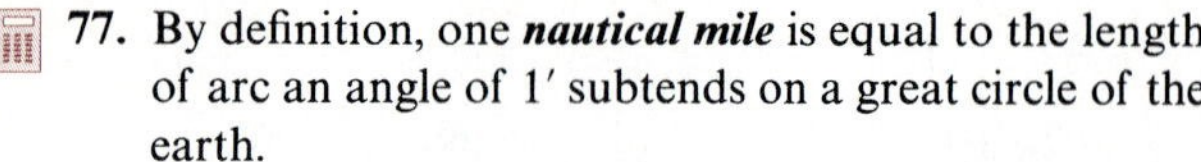

77. By definition, one ***nautical mile*** is equal to the length of arc an angle of 1′ subtends on a great circle of the earth.

a. Assuming that the radius of the earth is 3960 miles, approximate the length of one nautical mile in ordinary (that is, ***statute***) miles.

b. The distance between San Francisco and Honolulu is 2091 nautical miles. Approximate the number of statute miles separating the two cities.

78. An airplane is flying 900 feet directly above the summit of Pike's Peak. If the angle subtended by the airplane appears to an observer on the summit to be 5° (Figure 5.13), use (3) to approximate the length of the airplane.

79. Miami, Florida, and Pittsburgh, Pennsylvania, have approximately the same longitudes. Using the latitudes of Miami as 25°46′37″ and of Pittsburgh as 42°26′19″ and taking the radius of the earth to be 3960 miles, determine the approximate distance between Miami and Pittsburgh.

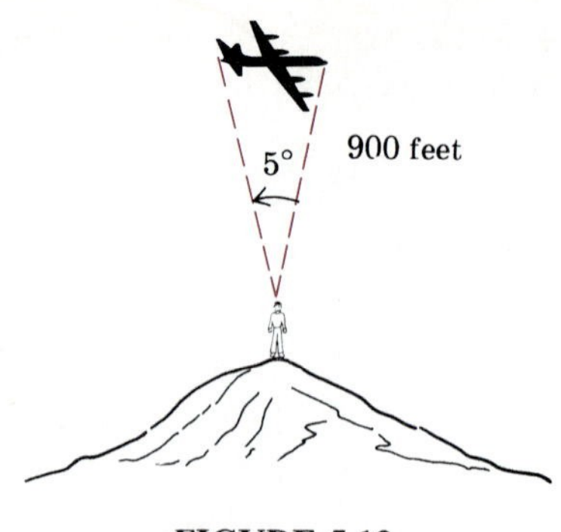

FIGURE 5.13

5.2 TRIGONOMETRIC FUNCTIONS OF ACUTE ANGLES

Having discussed angles measured in degrees and radians in Section 5.1, we are ready to define six basic trigonometric functions. For the present, when we write θ for an angle, it will not matter whether the angle is measured in degrees or in radians.

Let θ be an acute angle, as in Figure 5.14a, and form an associated right triangle as in Figure 5.14b, with legs of length a and b, and hypotenuse of length c. Using the customary abbreviations, we define the functions ***sine***, ***cosine***, ***tangent***, ***cotangent***, ***secant***, and ***cosecant*** at θ as follows:

$$\begin{aligned}
\sin\theta &= \frac{b}{c} = \frac{\text{length of opposite side}}{\text{length of hypotenuse}} \\
\cos\theta &= \frac{a}{c} = \frac{\text{length of adjacent side}}{\text{length of hypotenuse}} \\
\tan\theta &= \frac{b}{a} = \frac{\text{length of opposite side}}{\text{length of adjacent side}} \\
\cot\theta &= \frac{a}{b} = \frac{\text{length of adjacent side}}{\text{length of opposite side}} \\
\sec\theta &= \frac{c}{a} = \frac{\text{length of hypotenuse}}{\text{length of adjacent side}} \\
\csc\theta &= \frac{c}{b} = \frac{\text{length of hypotenuse}}{\text{length of opposite side}}
\end{aligned} \tag{1}$$

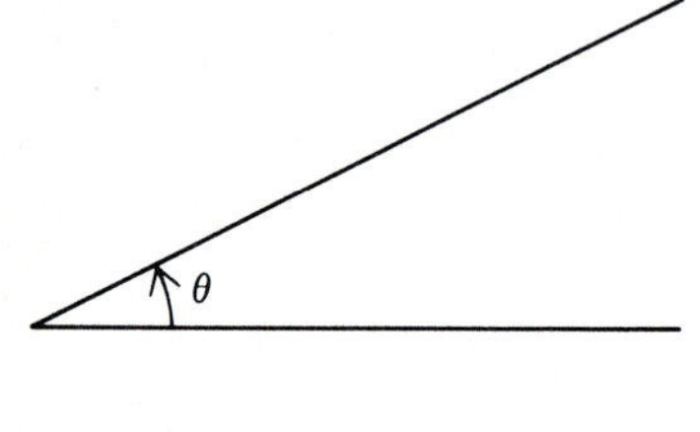

(a)

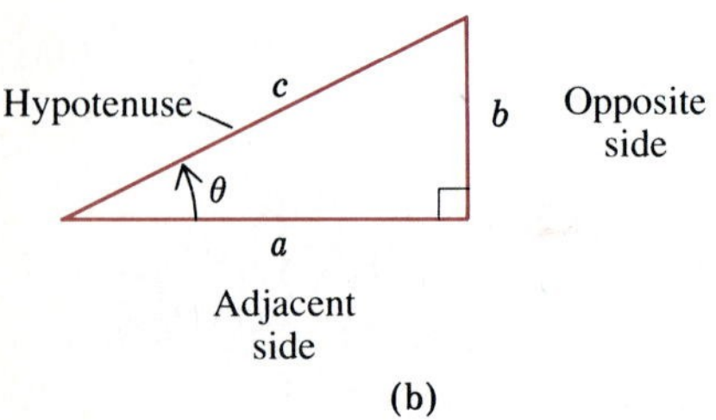

(b)

FIGURE 5.14

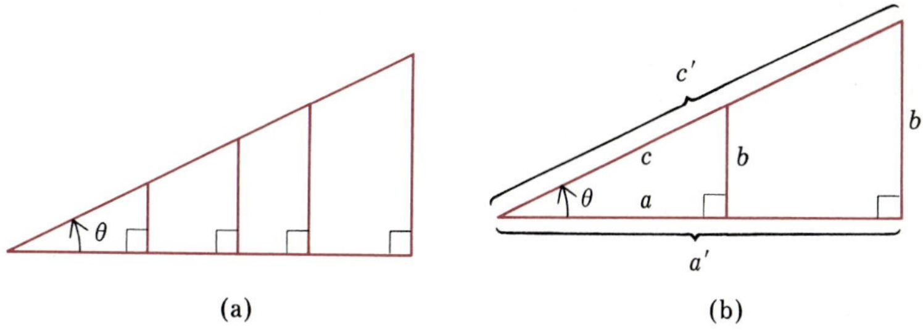

FIGURE 5.15

The angle θ appearing in the formulas above can lie in many right triangles (see Figure 5.15a). However, all such triangles are similar, so the ratios of the lengths of any two corresponding sides are identical. Thus for any two triangles as in Figure 5.15b, we have

$$\frac{b}{c} = \frac{b'}{c'}$$

Consequently $\sin \theta$, $\cos \theta$, $\tan \theta$, $\cot \theta$, $\sec \theta$, and $\csc \theta$ do not depend on the lengths of the sides of the right triangle, but *only* on the angle θ. Collectively the six functions defined for acute angles are referred to as the ***trigonometric functions***. Observe that each of the trigonometric values in (1) is nonnegative.

Special Values of the Trigonometric Functions

Next we find some values of the trigonometric functions for specific values of θ. In so doing we will repeatedly refer to the following results from geometry (see Figure 5.16):

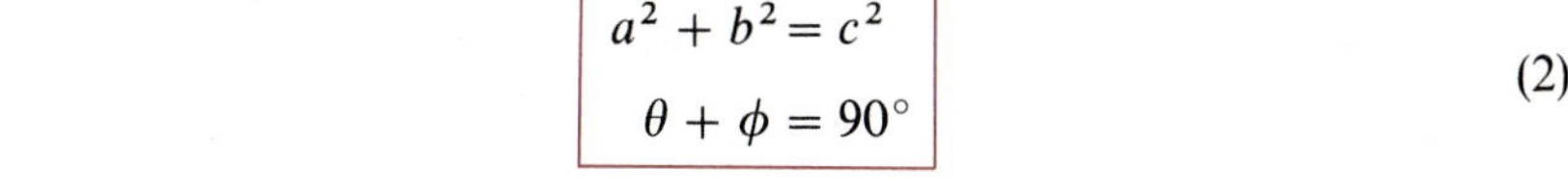

$$\boxed{\begin{aligned} a^2 + b^2 &= c^2 \\ \theta + \phi &= 90^\circ \end{aligned}} \qquad (2)$$

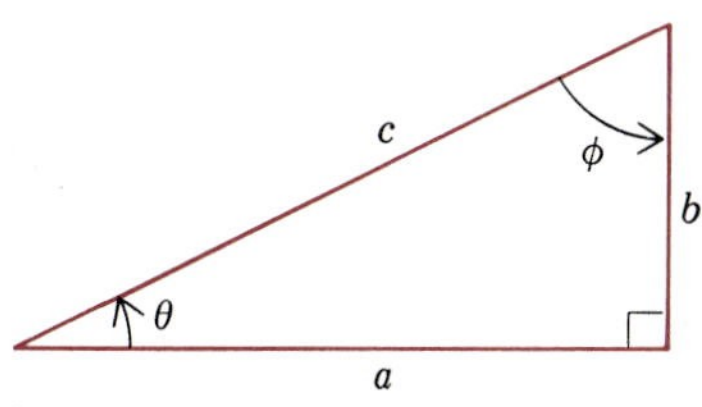

FIGURE 5.16

The first of these formulas is the Pythagorean Theorem; the second can be proved by recalling that the sum of the angles of any triangle is 180° and that a right triangle has a 90° angle.

First we will compute the values of the trigonometric functions for $\theta = 45^\circ$, which means that the other acute angle in the triangle is also 45°, so the triangle is isosceles.

EXAMPLE 1. Find $\sin 45^\circ$.

Solution. Since the value of $\sin 45^\circ$ is independent of the lengths of sides of the triangle, assume for simplicity that the legs each have length 1 (Figure 5.17a). Then by (2) we find that

$$c^2 = 1^2 + 1^2 = 2$$

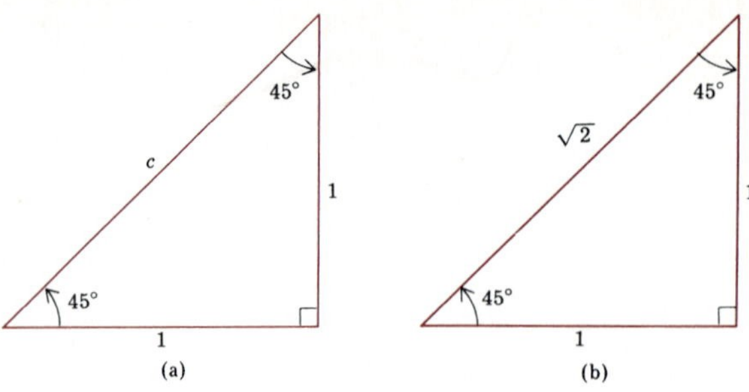

FIGURE 5.17

so that $c = \sqrt{2}$. It then follows from the definition of the sine function and the lengths shown in Figure 5.17b that $\sin 45° = 1/\sqrt{2} = \frac{1}{2}\sqrt{2}$. □

The values of the other trigonometric functions at 45° can also be found by using the defining formulas in (1), along with the lengths of the sides in Figure 5.17b. All these values are listed below:

$$\sin 45° = \frac{1}{2}\sqrt{2} \qquad \tan 45° = 1 \qquad \sec 45° = \sqrt{2}$$

$$\cos 45° = \frac{1}{2}\sqrt{2} \qquad \cot 45° = 1 \qquad \csc 45° = \sqrt{2}$$

Our next goal is to determine sin 60°.

EXAMPLE 2. Find sin 60°.

Solution. First we draw the triangle in Figure 5.18a, with hypotenuse of length 2. Next we place a congruent triangle beside it (Figure 5.18b).

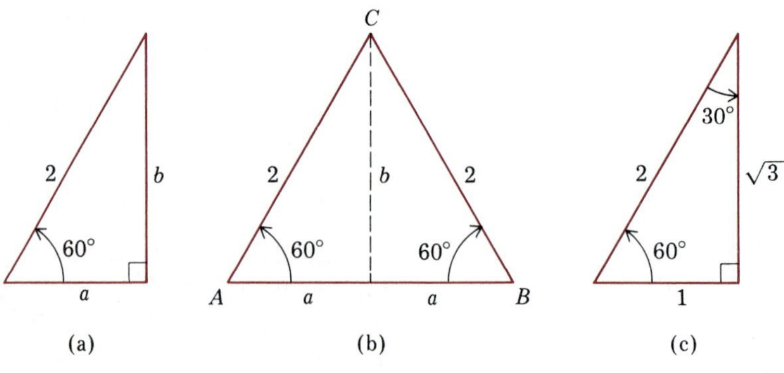

FIGURE 5.18

Because two right triangles are placed side by side, the resulting figure ABC is in fact a triangle. Moreover, since it has three angles of 60°, it is equilateral, each side having length 2. Therefore $2a = 2$, and $a = 1$. Consequently by the Pythagorean Theorem,

$$b^2 = 2^2 - a^2 = 4 - 1 = 3$$

so that $b = \sqrt{3}$. Thus we have the triangle shown in Figure 5.18c, so the definition of $\sin\theta$ yields

$$\sin 60° = \frac{\sqrt{3}}{2} = \frac{1}{2}\sqrt{3} \quad \square$$

Using the triangle in Figure 5.18c we also find that

$$\sin 30° = \frac{1}{2}$$

In fact, from Figure 5.18c we can find the values of all the trigonometric functions at $\theta = 30°$ and $\theta = 60°$. These values, along with the values that we already found at 45°, are listed in Table 5.2.

TABLE 5.2

θ (in degrees)	θ (in radians)	$\sin\theta$	$\cos\theta$	$\tan\theta$	$\cot\theta$	$\sec\theta$	$\csc\theta$
30°	$\frac{\pi}{6}$	$\frac{1}{2}$	$\frac{1}{2}\sqrt{3}$	$\frac{1}{3}\sqrt{3}$	$\sqrt{3}$	$\frac{2}{3}\sqrt{3}$	2
45°	$\frac{\pi}{4}$	$\frac{1}{2}\sqrt{2}$	$\frac{1}{2}\sqrt{2}$	1	1	$\sqrt{2}$	$\sqrt{2}$
60°	$\frac{\pi}{3}$	$\frac{1}{2}\sqrt{3}$	$\frac{1}{2}$	$\sqrt{3}$	$\frac{1}{3}\sqrt{3}$	2	$\frac{2}{3}\sqrt{3}$

Computing Trigonometric Values by Calculator

Table 5.2 gives the values of the trigonometric functions at 30°, 45°, and 60°. However, the values of the trigonometric functions for other acute angles are generally impossible to calculate. In such cases the best we can do is to approximate the values by using a calculator.

In general, calculators can approximate the values of the trigonometric functions when either degrees or radians are used. However, the procedure used in setting the calculator for degrees or radians varies from manufacturer to manufacturer. One must be aware of the functioning of the keys on any given calculator. After that, calculating trigonometric values proceeds something like this: To evaluate sin 40°, make certain that the calculator is in "degrees" mode, and next either key in sin and then 40, or 40 and then sin, according to the configuration of the calculator. The resulting display should read something like .6427876097, meaning that

$$\sin 40^\circ \approx 0.6427876097$$

To approximate cos 40° or tan 40°, we would proceed the same way, but press the key marked "cos" or "tan" instead of the one marked "sin." Although there generally are no keys representing cotangent, secant, or cosecant, we can still approximate their values on a calculator. For example, to approximate csc 40° we would approximate sin 40° as above, and then press the "$1/x$" key to obtain

$$\csc 40^\circ = \frac{1}{\sin 40^\circ} \approx 1.555723827$$

Evaluating an expression like sin 18°49′ on a calculator usually necessitates converting the minutes portion, 49′, into a decimal fraction of a degree, utilizing the fact that

$$1' = \left(\frac{1}{60}\right)^\circ$$

First we find that

$$49' = \left[49\left(\frac{1}{60}\right)\right]^\circ \approx (0.8166666667)^\circ$$

so that

$$18^\circ 49' \approx (18.81666667)^\circ$$

where two decimal places were lost in the last expression because the display on this particular calculator has only ten places. After we have 18°49′ in the calculator in decimal form, we press the key marked "sin" and find that

$$\sin 18^\circ 49' \approx 0.3225410508$$

If the angle measure includes seconds as well as minutes, we use the fact that

$$1'' = \left(\frac{1}{60}\right)' = \left(\frac{1}{3600}\right)^\circ$$

to help write the decimal expansion of the angle.

EXAMPLE 3. Use a calculator to approximate sin 64°13′28″.

Solution. From our remarks above,

$$64°13'28'' = 64° + \left[13\left(\frac{1}{60}\right)\right]^{\circ} + \left[28\left(\frac{1}{3600}\right)\right]^{\circ} \approx (64.22444444)^{\circ}$$

The calculator now yields

$$\sin 64°13'28'' \approx 0.9005043747 \quad \square$$

EXAMPLE 4. Use a calculator to approximate the value of $\cos \frac{\pi}{5}$.

Solution. First we set the calculator to compute in radians. Next we key in $\pi/5$ and then cos, or cos and then $\pi/5$, to obtain

$$\cos \frac{\pi}{5} \approx 0.8090169944 \quad \square$$

Using a Known Value to Find Other Values

In general, if we know the lengths of any two sides of the right triangle in Figure 5.19, then by using the Pythagorean Theorem we can determine the length of the third side and consequently the values of the six trigonometric functions—even if we do not know the value of θ. In our next example we show how to determine them.

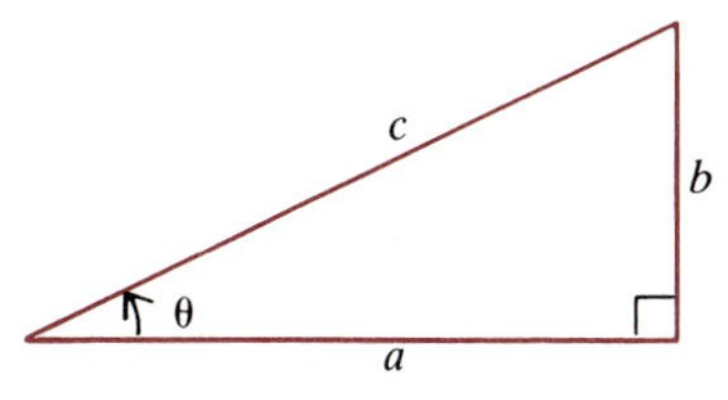

FIGURE 5.19

EXAMPLE 5. Suppose θ is an acute angle and $\tan \theta = \frac{2}{3}$. Find the values of the other five trigonometric functions at θ.

Solution. Since

$$\tan \theta = \frac{\text{length of opposite side}}{\text{length of adjacent side}} = \frac{2}{3}$$

we obtain the triangle in Figure 5.20a with $a = 3$ and $b = 2$. The Pythagorean Theorem tells us that $c^2 = 2^2 + 3^2 = 13$, so $c = \sqrt{13}$. Thus the triangle is as in Figure 5.20b, and we can compute the required values:

$$\sin \theta = \frac{2}{\sqrt{13}} = \frac{2}{13}\sqrt{13} \qquad \cos \theta = \frac{3}{\sqrt{13}} = \frac{3}{13}\sqrt{3}$$

$$\cot \theta = \frac{3}{2} \qquad \sec \theta = \frac{\sqrt{13}}{3} \qquad \csc \theta = \frac{\sqrt{13}}{2} \quad \square$$

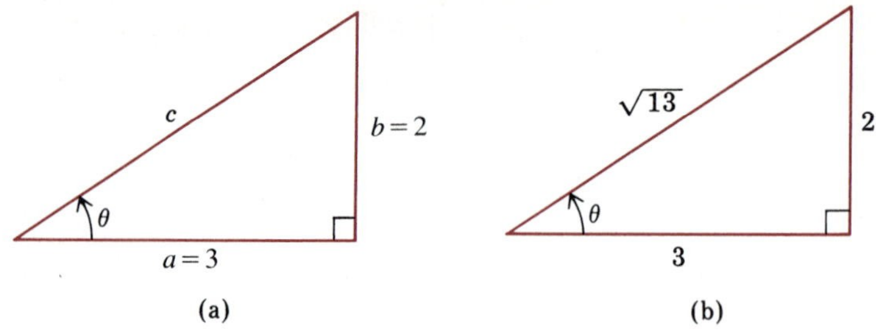

FIGURE 5.20

In the solution of Example 5 we could have selected another size triangle; the only stipulation was that the *ratio* of the lengths of the opposite and adjacent sides had to be $\frac{2}{3}$.

Relations among the Trigonometric Functions

Of the many relationships among the trigonometric functions, we note the following:

$$\tan\theta = \frac{\sin\theta}{\cos\theta} \qquad \cot\theta = \frac{\cos\theta}{\sin\theta} \qquad \cot\theta = \frac{1}{\tan\theta} \tag{3}$$

$$\sec\theta = \frac{1}{\cos\theta} \qquad \csc\theta = \frac{1}{\sin\theta} \tag{4}$$

Each of these formulas can be proved from the formulas in (1). For example, the first formula in (3) can be proved from (1) by observing that

$$\frac{\sin\theta}{\cos\theta} = \frac{b/c}{a/c} = \frac{b}{a} = \tan\theta$$

The proofs of the other formulas in (3) and (4) follow in the same general manner (see Exercises 62–64).

By applying the Pythagorean Theorem to the triangle in Figure 5.21 and then dividing by c^2, we obtain the fundamental formula relating the sine and cosine:

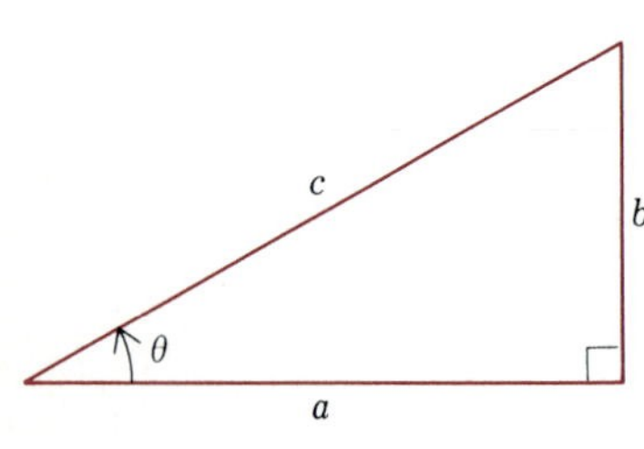

FIGURE 5.21

$$b^2 + a^2 = c^2$$

$$\left(\frac{b}{c}\right)^2 + \left(\frac{a}{c}\right)^2 = 1$$

$$\sin^2\theta + \cos^2\theta = 1 \tag{5}$$

Because of the intimate relationship between (5) and the Pythagorean Theorem, (5) is frequently called a ***Pythagorean Identity***. If we divide each side of (5) by $\cos^2\theta$, we obtain

$$\frac{\sin^2\theta}{\cos^2\theta} + \frac{\cos^2\theta}{\cos^2\theta} = \frac{1}{\cos^2\theta}$$

which by (3) and (4) simplifies to

$$\boxed{\tan^2\theta + 1 = \sec^2\theta} \tag{6}$$

Similarly, dividing both sides of (5) by $\sin^2\theta$ leads to the formula

$$\boxed{1 + \cot^2\theta = \csc^2\theta} \tag{7}$$

Exercise 65 asks for proof of this formula. Together (5)–(7) are the **Pythagorean Identities**; they are valid for any acute angle θ.

The preceding formulas provide a second method of finding the values of all the trigonometric functions of an acute angle when the value of one of the functions is known. Furthermore, it will not be necessary to draw a triangle.

EXAMPLE 6. Suppose θ is an acute angle and $\cos\theta = \frac{1}{4}$. Find the values of the other five trigonometric functions at θ.

Solution. From (5) we have

$$\sin^2\theta + \left(\frac{1}{4}\right)^2 = 1$$

so that

$$\sin^2\theta = 1 - \frac{1}{16} = \frac{15}{16}$$

Therefore

$$\sin\theta = \sqrt{\frac{15}{16}} = \frac{1}{4}\sqrt{15}$$

Then we use (3) and (4) to deduce that

$$\tan\theta = \frac{\sin\theta}{\cos\theta} = \frac{\frac{1}{4}\sqrt{15}}{1/4} = \sqrt{15}$$

$$\cot\theta = \frac{1}{\tan\theta} = \frac{1}{\sqrt{15}} = \frac{1}{15}\sqrt{15}$$

$$\sec\theta = \frac{1}{\cos\theta} = \frac{1}{1/4} = 4$$

$$\csc\theta = \frac{1}{\sin\theta} = \frac{1}{\frac{1}{4}\sqrt{15}} = \frac{4}{15}\sqrt{15} \quad \square$$

Snell's Law

As indicated in the introduction to the chapter, trigonometric functions are widely used in physics and engineering. Let us use the sine function to discuss briefly ***Snell's Law***, which is fundamental to the study of light.

Snell's Law states that when light passes from one medium (such as air) to a second medium (such as water), it is bent, or refracted, according to the equation

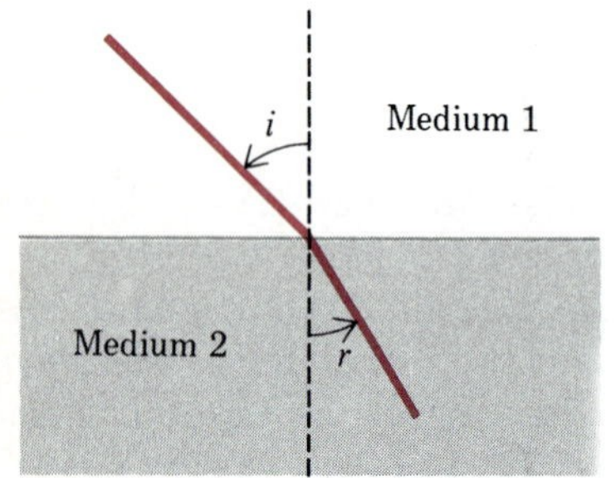

Snell's Law: $\sin i = \mu \sin r$

FIGURE 5.22

$$\sin i = \mu \sin r$$

where i is the angle of incidence, r is the angle of refraction, and μ is the index of refraction (Figure 5.22). The index of refraction depends on the color (or frequency) of the light and on the two media through which the light passes, but not on the angles of incidence or refraction. The index of refraction for a light ray in the visible spectrum passing from air into water varies from approximately 1.330 for red to 1.342 for violet.

EXERCISES 5.2

In Exercises 1–6, evaluate the expression.

1. $2\sin 45^\circ - 3\cos 30^\circ$

2. $5\cos 60^\circ + \sec 30^\circ - 1$

3. $\tan 45^\circ - \cot 45^\circ$

4. $\sin\frac{\pi}{4} + \cos\frac{\pi}{4} + \tan\frac{\pi}{4}$

5. $\csc\frac{\pi}{6} + \csc\frac{\pi}{3}$

6. $\cot\frac{\pi}{3} - \sec\frac{\pi}{6}$

7. Determine which of the following are equal.

a. $\tan 45^\circ$ b. $\csc 60^\circ$ c. $\csc 30^\circ$
d. $\frac{1}{2}\sec 60^\circ$ e. $\sin 30^\circ$ f. $\cos 30^\circ$
g. $\cos 60^\circ$ h. $2\sin 30^\circ$

8. Determine which of the following are equal.

a. $\sin\frac{\pi}{3}$ b. $\cos\frac{\pi}{3}$ c. $\cos\frac{\pi}{6}$
d. $2\tan\frac{\pi}{6}$ e. $\frac{1}{2}\tan\frac{\pi}{6}$ f. $\frac{1}{2}\tan\frac{\pi}{3}$
g. $\sec\frac{\pi}{6}$ h. $\csc\frac{\pi}{3}$

In Exercises 9–20, use the given values along with Figure 5.23 to determine the values of the six trigonometric functions at θ.

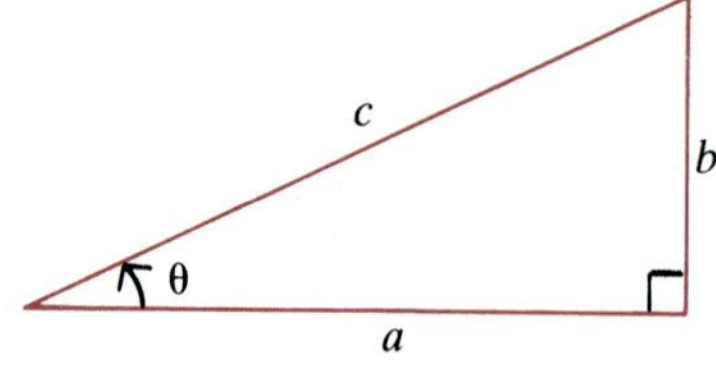

FIGURE 5.23

9. $a = 6, b = 1$

10. $a = 2, b = 4$

11. $a = 4, c = 5$

12. $a = 1, c = 6$

13. $b = 1.2, c = 1.3$

14. $b = 5, c = 13$

15. $a = 1.5, b = 1.5$

16. $a = 1, b = \sqrt{2}$

17. $a = 7, c = 25$

18. $a = 3, c = 6$

19. $b = \frac{1}{2}, c = 3$

20. $b = 0.3, c = 0.8$

In Exercises 21–32, use the given value to determine the values of the remaining five trigonometric functions at θ.

21. $\sin\theta = \dfrac{2}{5}$ **22.** $\sin\theta = 0.6$ **23.** $\cos\theta = \dfrac{3}{4}$

24. $\cos\theta = 0.2$ **25.** $\tan\theta = 2$ **26.** $\tan\theta = \dfrac{1}{\pi}$

27. $\cot\theta = 1.3$ **28.** $\cot\theta = \sqrt{2}$ **29.** $\sec\theta = 1.1$

30. $\sec\theta = 4$ **31.** $\csc\theta = \dfrac{5}{3}$ **32.** $\csc\theta = 1.2$

In Exercises 33–44, use a calculator to approximate the value of the given expression.

33. $\sin 8°$ **34.** $\sin 74°$ **35.** $\cos 27°$

36. $\cos 83°$ **37.** $\tan 14°$ **38.** $\tan 61°$

39. $\cot 41°$ **40.** $\cot 87°$ **41.** $\sec 1°$

42. $\sec 43°$ **43.** $\csc 54°$ **44.** $\csc 5°$

In Exercises 45–56, use a calculator to approximate the value of the given expression.

45. $\sin 38°20'$

46. $\sin 64°30'$

47. $\cos 16°10'$

48. $\cos 44°44'$

49. $\tan 57°17'$

50. $\sec 85°59'$

51. $\sin 63°21'49''$

52. $\cos 41°35'23''$

53. $\tan 88°59'59''$

54. $\cot 20°20'20''$

55. $\sec 1'37''$

56. $\csc 18°12'6''$

57. Use a calculator to approximate the value of each of the following expressions.
a. $\sin 0.1$ b. $\sin 0.0001$
c. $\sin 10^{-6}$ d. $\cos 0.1$
e. $\cos 0.001$ f. $\cos 0.00001$

58. Use a calculator to approximate the value of each of the following expressions.
a. $\tan(89.9)°$ b. $\tan(89.99)°$
c. $\tan(89.999)°$ d. $\tan(89.9999)°$

59. a. Try to evaluate $\tan 90°$ by calculator. Justify the calculator's response.
b. Find a positive number c such that the calculator displays 0 when $\sin c$ is keyed in.

60. For what values of θ in $(0, \pi/2)$ is $\tan\theta > 1$? [*Hint:* From (1), $\tan\theta = b/a$. Notice that $a = b$ if $\theta = \pi/4$. Now determine those θ in $(0, \pi/2)$ for which $b > a$.]

61. For what values of θ in $(0, \pi/2)$ is $\csc\theta < 2$?

62. a. Show that $\cot\theta = \dfrac{\cos\theta}{\sin\theta}$ for $0 < \theta < \pi/2$.
b. Show that $\cot\theta = \dfrac{1}{\tan\theta}$ for $0 < \theta < \pi/2$.

63. Show that $\sec\theta = \dfrac{1}{\cos\theta}$ for $0 < \theta < \pi/2$.

64. Show that $\csc\theta = \dfrac{1}{\sin\theta}$ for $0 < \theta < \pi/2$.

65. Show that $1 + \cot^2\theta = \csc^2\theta$ for $0 < \theta < \pi/2$.

66. Show that $\sin\theta = \cos\left(\dfrac{\pi}{2} - \theta\right)$ for $0 < \theta < \pi/2$.
(*Hint:* If one of the acute angles in a right triangle is θ, then the other is $\dfrac{\pi}{2} - \theta$.)

67. Show that $\cos\theta = \sin\left(\dfrac{\pi}{2} - \theta\right)$ for $0 < \theta < \pi/2$.

68. Show that $\tan\theta = \cot\left(\dfrac{\pi}{2} - \theta\right)$ for $0 < \theta < \pi/2$.

69. Show that $\sec\theta = \csc\left(\dfrac{\pi}{2} - \theta\right)$ for $0 < \theta < \pi/2$.

70. Can the angle of incidence of a ray passing from air to water be 45° and the angle of refraction be 30°? Explain your answer.

71. Using Exercise 66 but without using a calculator, compute
a. $\sin^2 16° + \sin^2 74°$
b. $\sin^2 10° + \sin^2 20° + \sin^2 30° + \sin^2 40° + \sin^2 50° + \sin^2 60° + \sin^2 70° + \sin^2 80°$

72. During a 24-hour period the ground under a Foucault pendulum located at θ latitude will rotate

$(360 \sin \theta)°$, so that the plane in which the pendulum swings appears to rotate through $(360 \sin \theta)°$. Through how many degrees does a Foucault pendulum in New York City (located at latitude 40°40′) appear to rotate during a 24-hour period?

73. In 1729 the English astronomer James Bradley found a way of determining the speed of the earth around the sun. He noticed that to view a star straight overhead by telescope, he had to tilt the telescope 20.49″. He concluded that the speed s in kilometers per second of the earth was given by

$$s = c \tan 20.49''$$

where c is the speed of light (approximately 299,792.5 kilometers per second). Use a calculator to determine s.

74. A yellow sodium light wave has an index of refraction of approximately 1.333. Using Snell's Law and a calculator, determine the angle of refraction if the angle of incidence is

a. 20° b. 53° c. 88°

75. Suppose a ramp c feet long is inclined at an angle of θ degrees. In foot-pounds the work W required to move an object weighing p pounds up the ramp is given by the formula

$$W = pc \sin \theta$$

Determine the work involved in pushing a 60-pound crate of apples up a 20-foot long ramp whose angle of inclination is 18°.

76. The perimeter of a regular polygon of n sides inscribed in a circle of radius r is given by

$$p_n(r) = 2nr \sin\left(\frac{180°}{n}\right)$$

Suppose the radius is 8 inches. Find the perimeter of a regular polygon with the given number of sides.

a. 5 b. 20 c. 180 d. 360

77. Suppose a circular curve in a highway has a radius of r feet and is banked at an angle θ (Figure 5.24). Assume further that the road is icy, so there is no friction. Then the maximum speed v in miles per hour that a car can travel on the curve without skidding

FIGURE 5.24

satisfies the formula

$$v^2 = \frac{1800}{121} r \tan \theta$$

a. Suppose the curve has a radius of 1000 feet and is banked at an angle of 10°. Find the maximum speed on the curve under icy conditions.
b. If the curve is regraded so that it is banked at an angle of 12°, how is the maximum speed on the curve under icy conditions affected?

78. Suppose a projectile is shot from ground level at an angle θ with respect to the horizontal and with an initial speed of v feet per second. If the ground is level, air resistance is negligible, and distances are measured in feet, then the maximum height H attained by the projectile and the range R of the projectile are given by

$$H = \frac{v^2 \sin^2 \theta}{64} \quad \text{and} \quad R = \frac{v^2 \sin 2\theta}{32}$$

(See Figure 5.25). Suppose a golf ball is hit at an angle of 30° with respect to the horizontal and has an initial speed of 100 feet per second. Find the maximum height and range.

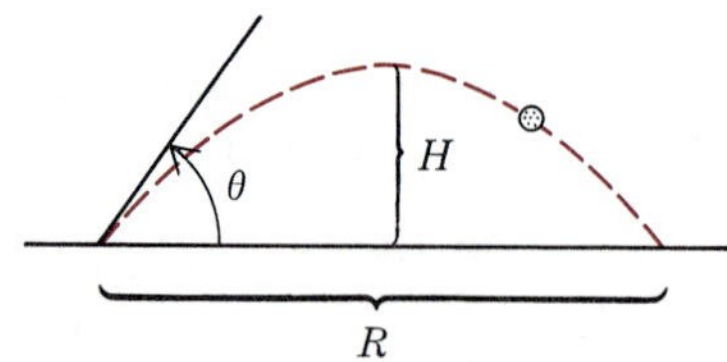

FIGURE 5.25

5.3 SOLVING RIGHT TRIANGLES

Trigonometry can be used in measuring the height of a redwood tree.

An almanac lists a 362-foot tall redwood located in Humboldt Redwood State Park, California, as the world's tallest tree. How was the tree's height first measured? We don't know, but this section contains methods of calculating such heights. More specifically, we will learn how to determine the lengths of the sides and the measures of the angles of a given right triangle when certain of them are known, a process called ***solving the triangle***. In Chapter 7 we will learn how to solve triangles that are not necessarily right triangles.

To compute the height of the redwood tree mentioned, assume that the tree stands essentially upright. Then the tree could be represented by one side of a right triangle with one vertex at a point A some distance away from the tree (Figure 5.26a). By measuring the distance between A and the tree, and the angle α in Figure 5.26a, one can then determine the height of the tree (that is, the length a of the appropriate leg of the triangle); one can even solve the triangle completely. In order to prepare for such calculations, we will give some preliminary notation and information.

Let the lengths of the legs of a given right triangle be denoted by a and b, the length of the hypotenuse by c, and the angles opposite a, b, and c by α, β, and γ, respectively (Figure 5.26b). In this section the angle γ will always be the right angle of the triangle, containing 90°, and therefore will need no further mention.

When we solve a right triangle, we will repeatedly use, as we did in Section 5.2, two results from geometry:

$$a^2 + b^2 = c^2 \tag{1}$$

$$\alpha + \beta = 90^\circ \tag{2}$$

We can solve a right triangle provided that either

(a) one acute angle and the length of one side are known, or
(b) the lengths of two sides are known.

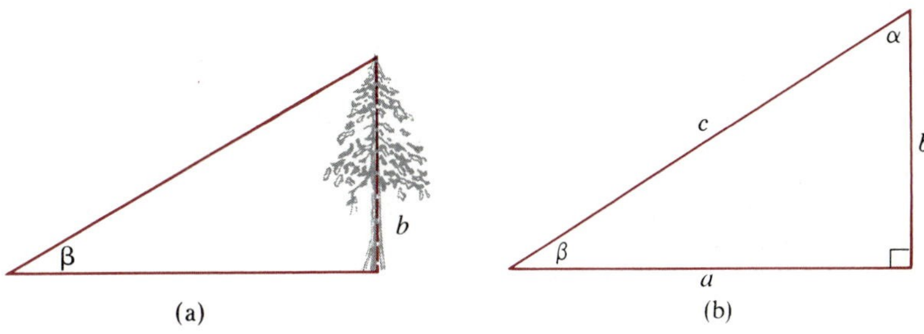

FIGURE 5.26

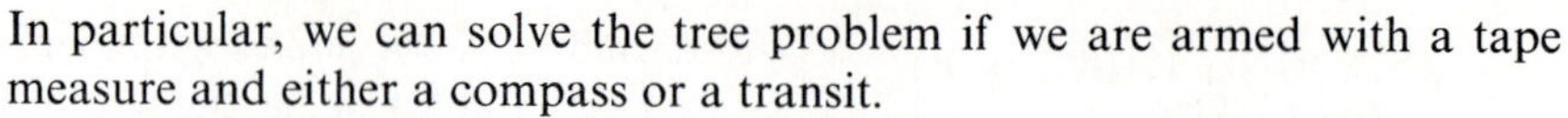

In particular, we can solve the tree problem if we are armed with a tape measure and either a compass or a transit.

In the examples and exercises of this section we will use a calculator and, for simplicity, round our final answers to four places.

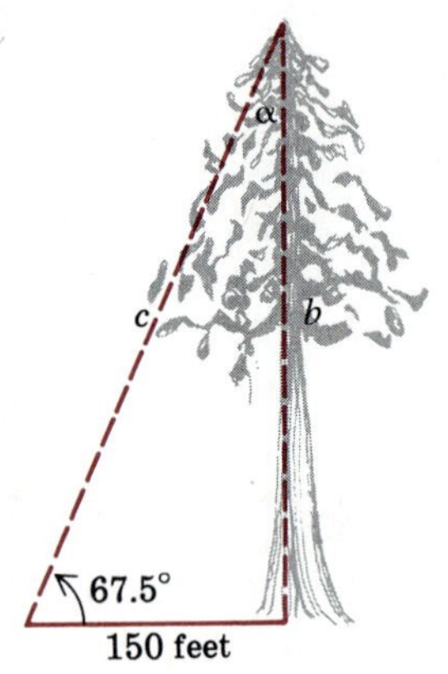

FIGURE 5.27

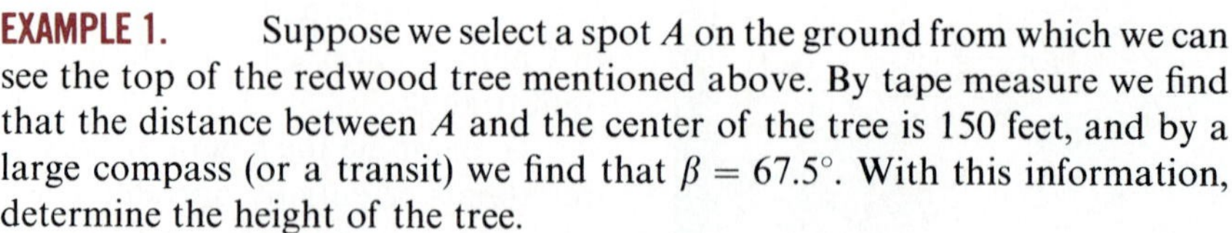

EXAMPLE 1. Suppose we select a spot A on the ground from which we can see the top of the redwood tree mentioned above. By tape measure we find that the distance between A and the center of the tree is 150 feet, and by a large compass (or a transit) we find that $\beta = 67.5°$. With this information, determine the height of the tree.

Solution. We must calculate b in Figure 5.27. But

$$\tan 67.5° = \frac{b}{150} \quad \text{so that} \quad b = 150 \tan 67.5°$$

By calculator we find that

$$b \approx 362.1$$

Consequently the height of the tree is approximately 362 feet, as the almanac says. □

The angle β, projecting upward from the horizontal to the top of the tree, is called the ***angle of elevation*** of the tree from the point A. It plays an important role in the determination of heights of trees and structures.

Once we know the tree's height, the horizontal distance from A to the center of the tree, and the angle of elevation, we can easily finish solving the corresponding triangle. We do this in Example 2.

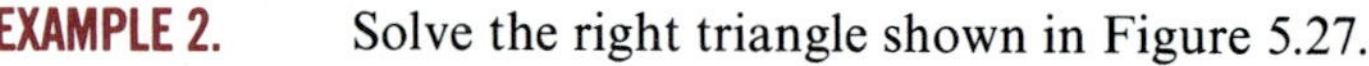

EXAMPLE 2. Solve the right triangle shown in Figure 5.27.

Solution. We already know that $a = 150$, $b \approx 362.1$, and $\beta = 67.5°$. As a result, we need only determine c and α. By (1),

$$c = \sqrt{a^2 + b^2} \approx \sqrt{(150)^2 + (362.1)^2} \approx 391.9$$

and by (2),

$$\alpha = 90° - 67.5° = 22.5° \quad \square$$

In Example 2 the length of one leg and one acute angle were given. In the next two examples we will solve a triangle when other information is provided.

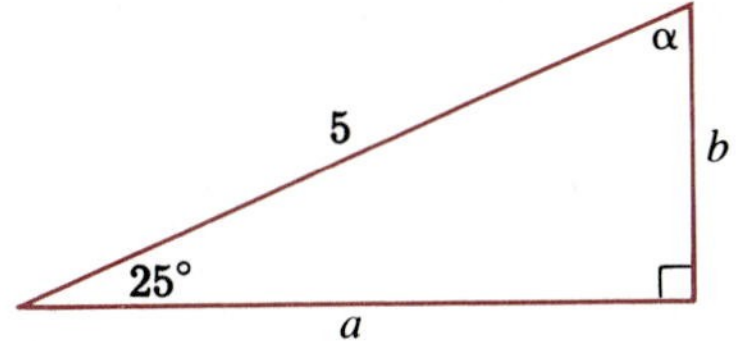

FIGURE 5.28

EXAMPLE 3. Solve the right triangle shown in Figure 5.28.

Solution. From the figure, $c = 5$ and $\beta = 25°$. We must calculate a, b, and α. From (2),

$$\alpha = 90° - 25° = 65°$$

Now we use the known values of β and c to determine a and b:

$$\sin 25° = \frac{b}{5}, \quad \text{so that} \quad b = 5 \sin 25°$$

and

$$\cos 25° = \frac{a}{5}, \quad \text{so that} \quad a = 5 \cos 25°$$

By calculator we find that

$$b \approx 2.113 \quad \text{and} \quad a \approx 4.532 \quad \square$$

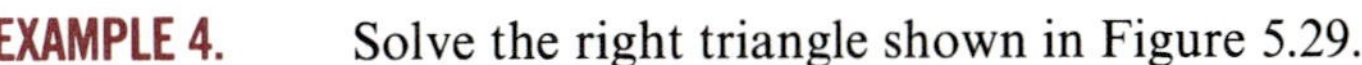

EXAMPLE 4. Solve the right triangle shown in Figure 5.29.

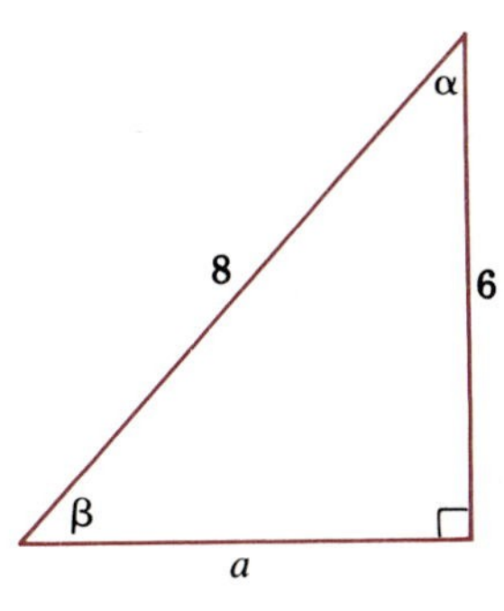

FIGURE 5.29

Solution. This time $b = 6$ and $c = 8$. We need to find a, α, and β. By (1),

$$a = \sqrt{8^2 - 6^2} = \sqrt{28} \approx 5.292$$

For β, we notice that

$$\sin \beta = \frac{6}{8} = 0.75$$

so that by calculator we have

$$\beta \approx 48.59° \ (\approx 48°35')$$

Consequently, it follows from (2) that

$$\alpha = 90° - \beta \approx 90° - 48.59° = 41.41° \ (\approx 41°25') \quad \square$$

We end with a more complicated example involving heights.

EXAMPLE 5. In order to estimate the height of a bridge, two sightings are made, one each at points A and B, which are 500 meters apart (Figure 5.30). How tall is the bridge?

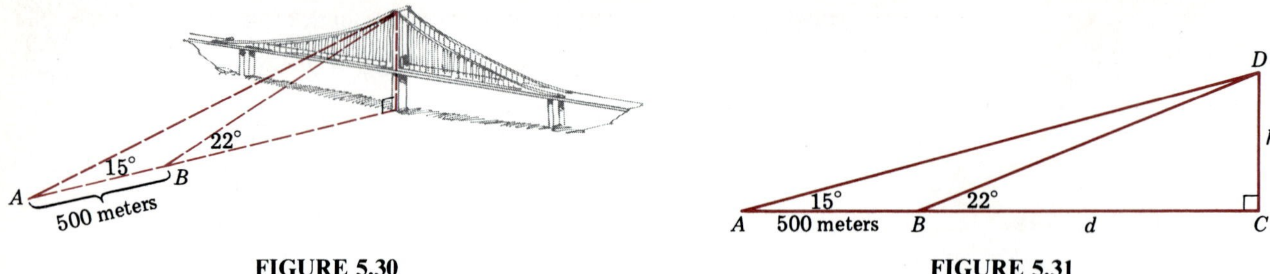

FIGURE 5.30

FIGURE 5.31

Solution. Figure 5.31 contains the information we use to solve the problem, including the letter h for the height we need to determine. From triangle ACD we obtain

$$\tan 15^\circ = \frac{h}{500 + d}, \quad \text{so that} \quad h = (500 + d) \tan 15^\circ \tag{3}$$

From triangle BCD we obtain

$$\cot 22^\circ = \frac{d}{h}, \quad \text{so that} \quad d = h \cot 22^\circ$$

Substituting for d in the second equation of (3) yields

$$h = (500 + h \cot 22^\circ) \tan 15^\circ = 500 \tan 15^\circ + h \cot 22^\circ \tan 15^\circ$$

Rearranging, we find that

$$h - h \cot 22^\circ \tan 15^\circ = 500 \tan 15^\circ$$

or

$$h(1 - \cot 22^\circ \tan 15^\circ) = 500 \tan 15^\circ$$

Thus

$$h = \frac{500 \tan 15^\circ}{1 - \cot 22^\circ \tan 15^\circ}$$

By calculator we find that

$$h \approx 397.8$$

Consequently the bridge is approximately 397.8 meters tall. □

EXERCISES 5.3

In Exercises 1–12, solve the right triangle with the given data pertaining to Figure 5.32.

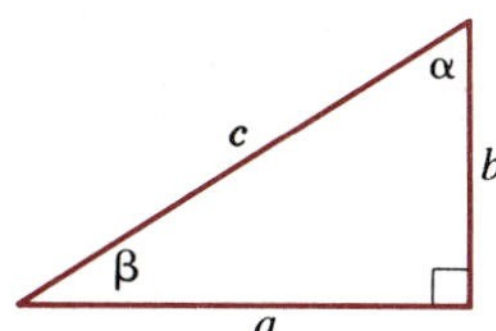

FIGURE 5.32

1. $\alpha = 15°, c = 4$

2. $\alpha = 15°, b = 4$

3. $\alpha = 40°, a = 2$

4. $\beta = 69.2°, a = 5$

5. $\beta = 38.4°, b = 3.2$

6. $\beta = 83.7°, c = 11.6$

7. $b = 15, c = 17$

8. $a = 60, b = 11$

9. $a = 3, c = \sqrt{10}$

10. $a = 2, b = 2.1$

11. $a = \frac{2}{5}, b = \frac{1}{3}$

12. $a = \frac{5}{7}, c = \frac{10}{3}$

13. Find the distance from point A to point B on the opposite side of the pond represented in Figure 5.33.

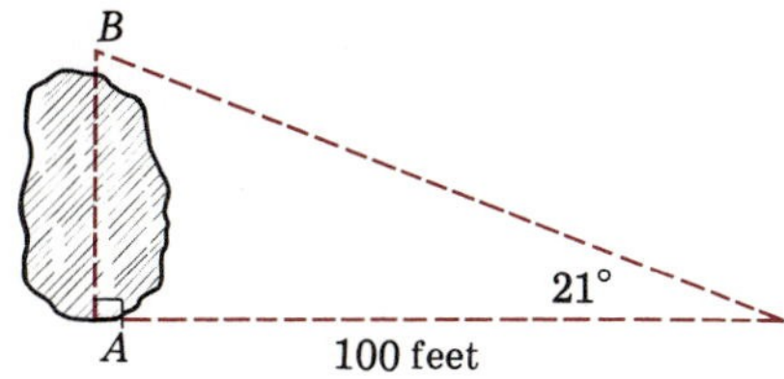

FIGURE 5.33

14. Find the height h of the flagpole atop the building pictured in Figure 5.34.

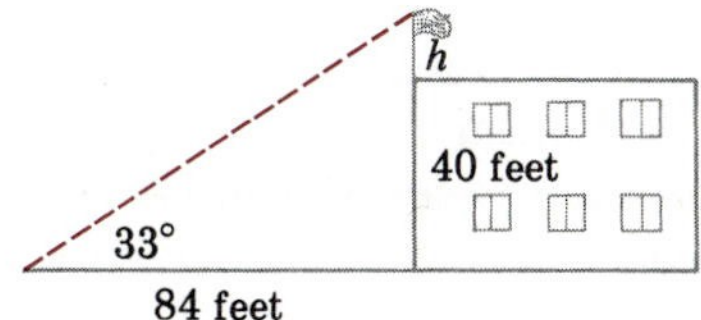

FIGURE 5.34

15. The top of a 12-foot ladder resting against a wall is 9 feet above the ground. Find the angle of elevation of the ladder.

16. A police car is located 80 feet from a street having a speed limit of 30 miles per hour (44 feet per second). Ten seconds after a truck passes the intersection, the line connecting the truck and the police car makes an angle of 10° with the street (Figure 5.35).

a. Has the truck been speeding?

b. If the truck had been traveling exactly 30 miles per hour, determine the angle the line connecting the truck and the police car would have made with the street 10 seconds after the truck passed the intersection.

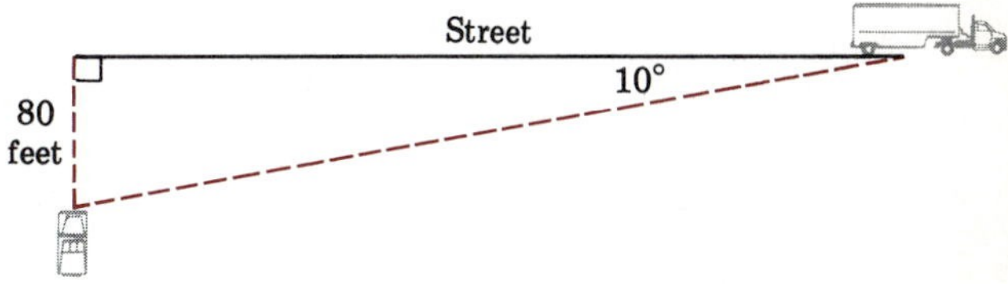

FIGURE 5.35

17. A jet is flying at 400 miles per hour. One minute after the jet passes directly over an airport, its angle of elevation from the airport is 35°. Determine the height of the jet.

18. A pleasure boat travels 10 miles due east from port and then 5 miles due south before running aground on a reef. The Coast Guard plans to send a rescue vessel along a straight line from port to the distressed boat. What angle should that straight line make with the line that points straight east from the port?

19. A surveyor takes a measurement 5 feet above ground and 100 feet from an apartment building (Figure 5.36). If the angle of elevation of the top of the building is 36°, how tall is the building?

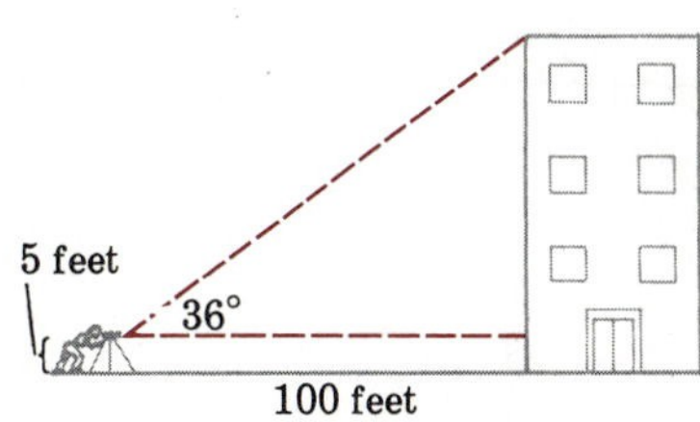

FIGURE 5.36

20. Points A and B lie directly opposite each other on a river and are invisible from each other because of an island in the river. Using the information in Figure 5.37, determine the distance between A and B.

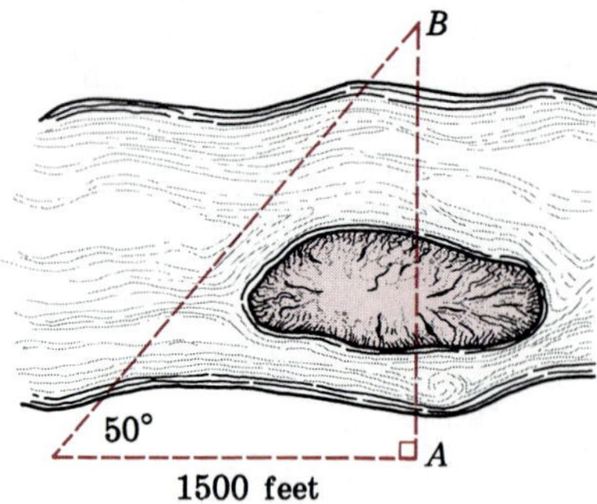

FIGURE 5.37

21. You are standing 4000 feet from the launch pad at Cape Kennedy. A few seconds after launch, the angle of elevation of the tail of the space shuttle is 50° and the angle of elevation of the nose is 52.05°. What is the length of the rocket?

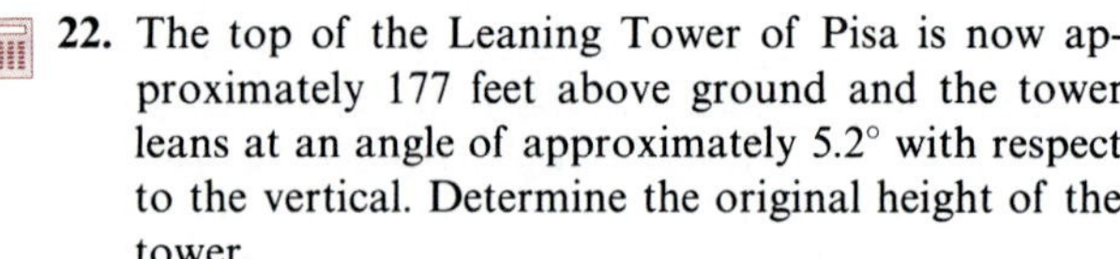

22. The top of the Leaning Tower of Pisa is now approximately 177 feet above ground and the tower leans at an angle of approximately 5.2° with respect to the vertical. Determine the original height of the tower.

23. Suppose it is decided to build a bridge 10 miles long from the south rim to the north rim of the Grand Canyon. Using the fact that the north rim is 1200 feet higher than the south rim, find the angle of elevation of the bridge.

24. The sun is approximately 93,000,000 miles from the earth and 865,000 miles in diameter, whereas the moon is approximately 240,000 miles from the earth and 2160 miles in diameter. Using this information, determine which appears larger from the earth. (*Hint:* Determine the angle each subtends in the eye of an observer.)

25. Originally the Great Pyramid of Cheops had a square base approximately 754 feet on a side. In order to determine its height, a visitor 800 feet away from the midpoint of one side noted that the angle of elevation of the pyramid from the visitor's feet was 22.27° (Figure 5.38). How tall was the pyramid?

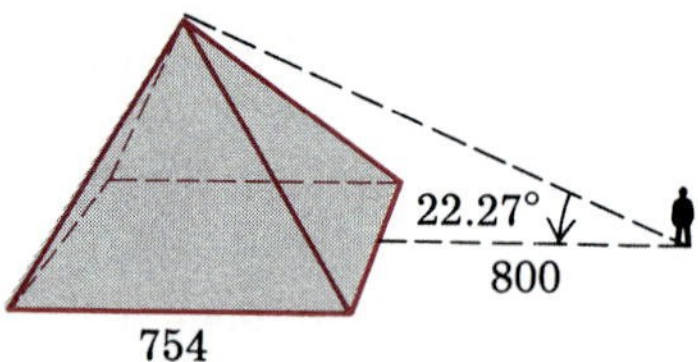

FIGURE 5.38

5.4 TRIGONOMETRIC FUNCTIONS OF ARBITRARY ANGLES

Until now we have discussed the trigonometric functions only for acute angles. However, since many applications involve angles that are not acute, it is desirable to define the trigonometric functions for arbitrary angles, acute or not. As in Section 5.2, the definitions in this section are valid whether the angles are measured in degrees or radians.

Let θ be any angle in standard position, and let $P(a, b)$ be an arbitrary point on the terminal side distinct from the origin (Figure 5.39). Furthermore, let $c = \sqrt{a^2 + b^2}$, the distance between the origin and P. Then the six trigonometric functions are defined by formulas like those in Section 5.2:

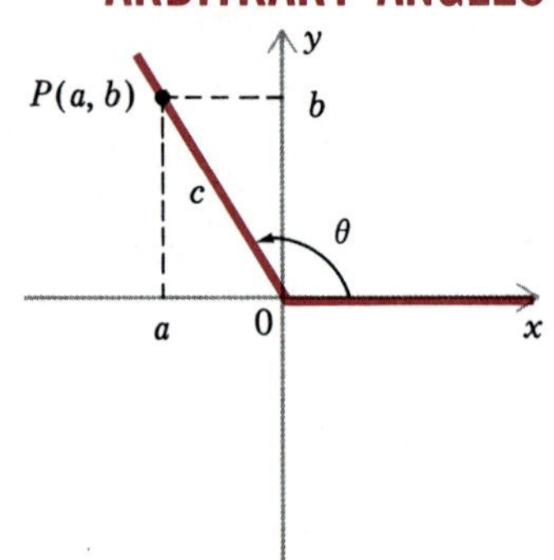

FIGURE 5.39

$$\sin\theta = \frac{b}{c} \qquad \tan\theta = \frac{b}{a} \qquad \sec\theta = \frac{c}{a}$$
$$\cos\theta = \frac{a}{c} \qquad \cot\theta = \frac{a}{b} \qquad \csc\theta = \frac{c}{b} \qquad (1)$$

As a result, the basic formulas we derived in Section 5.2 remain valid for any angle θ and can be proved the same way as for acute angles:

$$\tan\theta = \frac{\sin\theta}{\cos\theta} \qquad \cot\theta = \frac{\cos\theta}{\sin\theta} \qquad \cot\theta = \frac{1}{\tan\theta}$$

$$\sec\theta = \frac{1}{\cos\theta} \qquad \csc\theta = \frac{1}{\sin\theta}$$

$$\sin^2\theta + \cos^2\theta = 1 \qquad \tan^2\theta + 1 = \sec^2\theta \qquad 1 + \cot^2\theta = \csc^2\theta$$

If we know the values of a and b, we can find the trigonometric values at θ by means of (1).

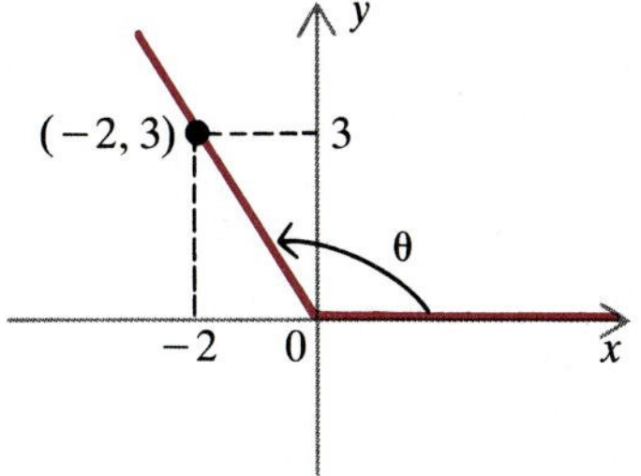

FIGURE 5.40

EXAMPLE 1. Consider the triangle and angle θ drawn in Figure 5.40. Determine the values of the 6 trigonometric functions at θ.

Solution. By assumption, $a = -2$ and $b = 3$. Therefore

$$c = \sqrt{(-2)^2 + 3^2} = \sqrt{13}$$

Consequently by (1),

$$\sin\theta = \frac{b}{c} = \frac{3}{\sqrt{13}} \qquad \cos\theta = \frac{a}{c} = \frac{-2}{\sqrt{13}} \qquad \tan\theta = \frac{b}{a} = \frac{3}{-2}$$

$$\cot\theta = \frac{a}{b} = \frac{-2}{3} \qquad \sec\theta = \frac{c}{a} = \frac{\sqrt{13}}{-2} \qquad \csc\theta = \frac{c}{b} = \frac{\sqrt{13}}{3} \qquad \square$$

Since either or both of the coordinates a and b in Figure 5.39 may be negative, it follows from (1) that each of the trigonometric functions can have negative values as well as positive values. In fact, in Example 1 four of the values turned out to be negative. Table 5.3 summarizes the signs of the various trigonometric functions, depending on the quadrant in which the terminal side of the angle lies.

TABLE 5.3

Quadrant in which the terminal side of θ lies	$\sin\theta$, $\csc\theta$	$\cos\theta$, $\sec\theta$	$\tan\theta$, $\cot\theta$
first	+	+	+
second	+	−	−
third	−	−	+
fourth	−	+	−

Next, we observe that $\sin\theta$ and $\cos\theta$ are defined for any angle θ. However, the remaining functions are not defined whenever their respective denominators equal 0. For example,

$$\sin 0 = 0 = \sin \pi$$

which means that $\cot\theta$ and $\csc\theta$ are not defined when $\theta = 0$ or $\theta = \pi$. Similarly,

$$\cos\frac{\pi}{2} = 0 = \cos\frac{3\pi}{2}$$

so that $\tan\theta$ and $\sec\theta$ are not defined when $\theta = \pi/2$ or $\theta = 3\pi/2$.

The Reference Angle

If an angle θ is in standard position but is not acute, one way of facilitating the computation of the various trigonometric values of θ involves associating with θ an acute angle called the reference angle. The ***reference angle*** of θ is the acute angle α formed by the terminal side of θ and either the positive or the negative x axis, whichever makes α acute (Figure 5.41a–d).

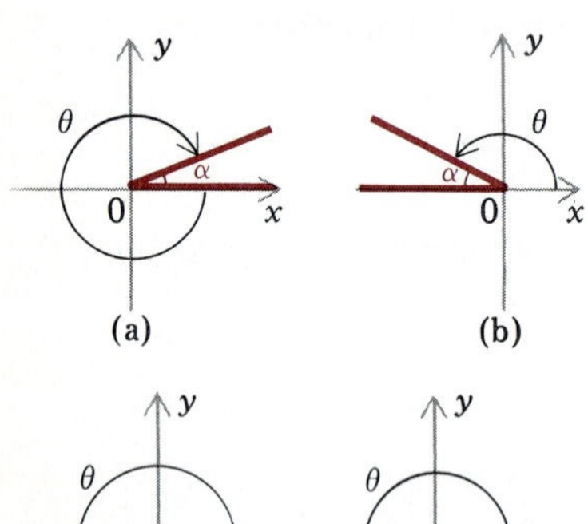

FIGURE 5.41

EXAMPLE 2. Find the reference angle α for each of the following angles.

a. $120°$ b. $-135°$ c. $\frac{5}{4}\pi$ d. $\frac{11}{6}\pi$

Solution.

a. Since the terminal side of $120°$ lies in the second quadrant (Figure 5.42a), the associated acute angle uses the negative x axis, so $\alpha = 180° - 120° = 60°$.
b. Since the terminal side of $-135°$ lies in the third quadrant (Figure 5.42b), the associated acute angle uses the negative x axis, so $\alpha = 180° - 135° = 45°$.
c. The terminal side of $\frac{5}{4}\pi$ lies in the third quadrant (Figure 5.42c), so the associated acute angle uses the negative x axis. Thus $\alpha = \frac{5}{4}\pi - \pi = \frac{1}{4}\pi$.
d. The terminal side of $\frac{11}{6}\pi$ lies in the fourth quadrant (Figure 5.42d), so the associated acute angle uses the positive x axis. Thus $\alpha = 2\pi - \frac{11}{6}\pi = \frac{1}{6}\pi$. □

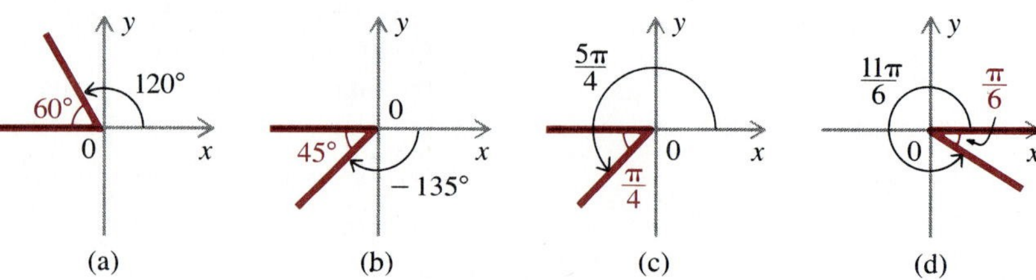

FIGURE 5.42

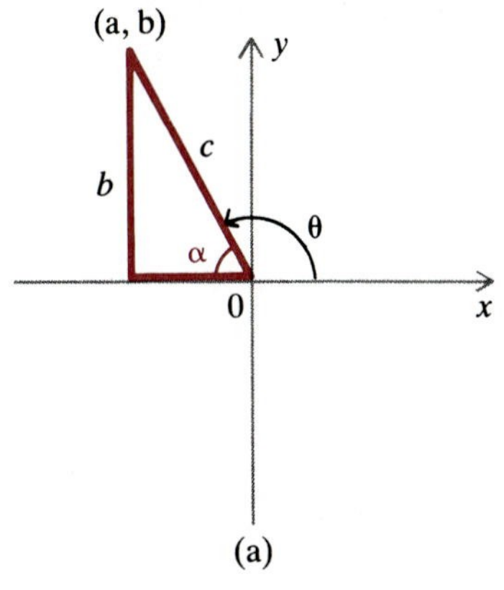

(a)

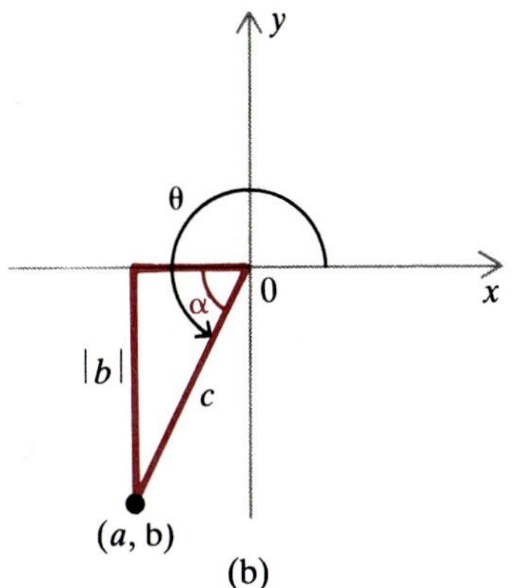

(b)

FIGURE 5.43

If a given angle θ has reference angle α, then

$$|\sin \theta| = \sin \alpha \tag{2}$$

Indeed, if the terminal side of θ lies in the first or second quadrant (as in Figure 5.43a), then

$$\sin \theta = \frac{b}{c} = \sin \alpha > 0$$

whereas if the terminal side of θ lies in the third or fourth quadrant (as in Figure 5.43b), then

$$\sin \theta = \frac{b}{c} < 0 \quad \text{and} \quad \sin \alpha = \frac{|b|}{c}$$

Either way,

$$|\sin \theta| = \sin \alpha$$

Formula (2) provides us with a procedure for determining the value of $\sin \theta$ for an arbitrary angle θ:

1. Determine the reference angle α.
2. Compute $\sin \alpha$.
3. Determine the sign of $\sin \theta$ (see Table 5.3).
4. If $\sin \theta \geq 0$, then $\sin \theta = \sin \alpha$.
   If $\sin \theta < 0$, then $\sin \theta = -\sin \alpha$.

A similar procedure applies to any other trigonometric function.

EXAMPLE 3. Compute $\sin 120°$ and $\cos 120°$.

Solution. By the solution of Example 2(a) we know that the reference angle is 60°. By Table 5.3 we know that $\sin 120° > 0$ and $\cos 120° < 0$. Using our knowledge of the values of the trigonometric functions at 60°, we find that

$$\sin 120° = \sin 60° = \frac{\sqrt{3}}{2} \quad \text{and} \quad \cos 120° = -\cos 60° = -\frac{1}{2} \quad \square$$

EXAMPLE 4. Compute the values of the six trigonometric functions at $11\pi/6$.

Solution. By the solution of Example 2(d) we know that the reference angle is $\frac{\pi}{6}$. By Table 5.3 we know that $\cos\frac{11\pi}{6}$ and $\sec\frac{11\pi}{6}$ are positive, whereas $\sin\frac{11\pi}{6}$, $\tan\frac{11\pi}{6}$, $\cot\frac{11\pi}{6}$, and $\csc\frac{11\pi}{6}$ are negative. Using our knowledge of the values of the trigonometric functions at $\frac{\pi}{6}$, we conclude that

$$\sin\frac{11\pi}{6} = -\sin\frac{\pi}{6} = -\frac{1}{2} \qquad \cos\frac{11\pi}{6} = \cos\frac{\pi}{6} = \frac{\sqrt{3}}{2}$$

$$\tan\frac{11\pi}{6} = -\tan\frac{\pi}{6} = -\frac{\sqrt{3}}{3} \qquad \cot\frac{11\pi}{6} = -\cot\frac{\pi}{6} = -\sqrt{3}$$

$$\sec\frac{11\pi}{6} = \sec\frac{\pi}{6} = \frac{2}{3}\sqrt{3} \qquad \csc\frac{11\pi}{6} = -\csc\frac{\pi}{6} = -2 \quad \square$$

Suppose one trigonometric value at an angle θ is given, along with the quadrant in which the terminal side of θ is located. Then the reference angle and an associated triangle provide the information needed to compute the values of the other trigonometric functions at θ.

EXAMPLE 5. Suppose $\sin\theta = -\frac{2}{7}$ and $\pi < \theta < \frac{3\pi}{2}$. Find the values of the remaining trigonometric functions at θ.

Solution. Let α be the reference angle for θ. Since the terminal side of θ lies in the third quadrant and $\sin\theta = -\frac{2}{7}$, we draw the triangle in Figure 5.44, with all lengths positive. By the Pythagorean Theorem,

$$7^2 = 2^2 + d^2, \quad \text{so that} \quad d = \sqrt{7^2 - 2^2} = \sqrt{45} = 3\sqrt{5}$$

From Table 5.3 we know that $\cos\theta < 0$, $\tan\theta > 0$, $\cot\theta > 0$, $\sec\theta < 0$, and $\csc\theta < 0$. Using the triangle in Figure 5.44, we conclude that

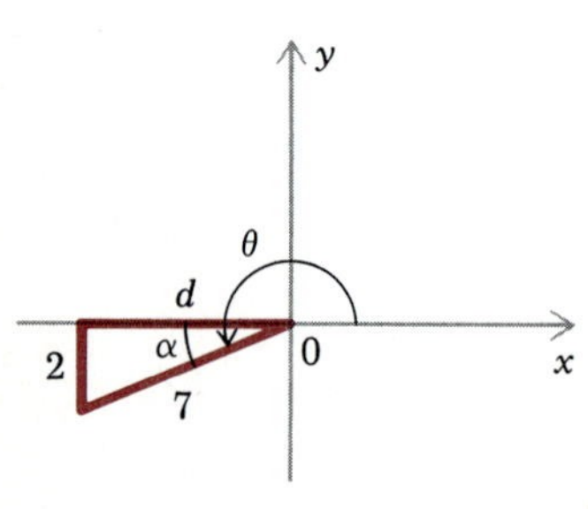

FIGURE 5.44

$$\cos\theta = -\cos\alpha = -\frac{d}{7} = -\frac{3\sqrt{5}}{7} \qquad \cot\theta = \cot\alpha = \frac{d}{2} = \frac{3\sqrt{5}}{2}$$

$$\tan\theta = \tan\alpha = \frac{2}{d} = \frac{2}{3\sqrt{5}} = \frac{2}{15}\sqrt{5} \qquad \sec\theta = -\sec\alpha = -\frac{7}{d} = -\frac{7\sqrt{5}}{15}$$

$$\csc\theta = -\csc\alpha = -\frac{7}{2} \quad \square$$

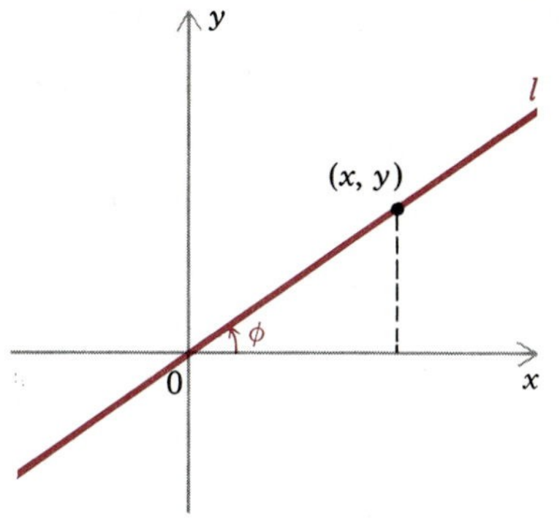

FIGURE 5.45

We have one more observation concerning the tangent function. Let l be a nonvertical line that passes through the origin and makes an angle θ with the positive x axis. If (x, y) is any point on l except the origin, then

$$\tan \theta = \frac{y}{x}$$

(Figure 5.45). Since l passes through the origin, it follows from the definition of the slope m of l that

$$m = \frac{y - 0}{x - 0} = \frac{y}{x}$$

Therefore

$$\tan \theta = \text{the slope of } l$$

EXERCISES 5.4

In Exercises 1–8, determine all the trigonometric functions that are defined for the given angle.

1. $360°$ **2.** $450°$ **3.** $-270°$

4. $-1080°$ **5.** $\frac{3\pi}{2}$ **6.** 10π

7. $-\frac{\pi}{2}$ **8.** $-\pi$

In Exercises 9–12, find the reference angle of the given angle.

9. $150°$ **10.** $-120°$ **11.** $\frac{9\pi}{4}$ **12.** $-\frac{\pi}{3}$

In Exercises 13–24, use the reference angle to find the values of the six trigonometric functions at the given angle.

13. $135°$ **14.** $210°$ **15.** $\frac{5\pi}{3}$

16. $\frac{11\pi}{6}$ **17.** $-120°$ **18.** $-225°$

19. $-\frac{3\pi}{4}$ **20.** $\frac{8\pi}{3}$ **21.** $750°$

22. $-945°$ **23.** $\frac{41\pi}{6}$ **24.** $-\frac{20\pi}{3}$

In Exercises 25–30, find the values of the remaining trigonometric functions at θ.

25. $\sin \theta = -\frac{3}{5}$ and $\pi < \theta < \frac{3\pi}{2}$

26. $\cos \theta = \frac{4\sqrt{3}}{7}$ and $-\frac{\pi}{2} < \theta < 0$

27. $\tan \theta = \sqrt{2}$ and $3\pi < \theta < \frac{7\pi}{2}$

28. $\cot \theta = -\sqrt{3}$ and $\frac{5\pi}{2} < \theta < 3\pi$

29. $\sec \theta = -\sqrt{3}$ and $-\frac{11\pi}{2} < \theta < -5\pi$

30. $\csc \theta = \frac{7}{4}$ and $4\pi < \theta < \frac{9\pi}{2}$

In Exercises 31–38, determine which statements are true and which are false. Justify your answer.

31. $\sin \frac{\pi}{4} \stackrel{?}{=} \frac{1}{2} \sin \frac{\pi}{2}$ **32.** $\cos 2\pi \stackrel{?}{=} (\cos \pi)^2$

33. $\cos 2\pi \stackrel{?}{=} 2 \cos \pi$ **34.** $\sin \frac{\pi}{6} + \sin \frac{\pi}{3} \stackrel{?}{=} \sin \frac{\pi}{2}$

35. $\sin \frac{\pi}{6} + \cos \frac{\pi}{3} \stackrel{?}{=} \sin \frac{\pi}{2}$

36. $\sin\frac{\pi}{6} + \left(\cos\frac{3\pi}{4}\right)^2 \stackrel{?}{=} \cos 4\pi$

37. $\frac{2}{3}\sin\frac{7\pi}{3}\cos\frac{11\pi}{6} \stackrel{?}{=} \left(\sin\frac{3\pi}{4}\right)^2$

38. $\sin^2 n\pi - \cos^2 n\pi \stackrel{?}{=} -1$

39. Find a formula for the distance between two points $(\cos s, \sin s)$ and $(\cos t, \sin t)$ on the unit circle. Simplify your answer.

5.5 GRAPHS OF THE TRIGONOMETRIC FUNCTIONS

In this section we will draw the graphs of the trigonometric functions. Throughout this section angles will be measured in radians.

Before actually sketching the graphs of the trigonometric functions, we will first show that

$$\sin(\theta + 2\pi) = \sin\theta \quad \cos(\theta + 2\pi) = \cos\theta$$

for any real number θ. To that end, let $P(\theta)$ be the point on the unit circle that lies on the terminal side of the angle of θ radians in standard position (Figure 5.46). The point $P(\theta)$ can be used in order to describe the trigonometric functions, and in fact sheds light on the etymology of "sine."* (The association of any number θ with the corresponding point $P(\theta)$ on the unit circle is sometimes called the ***wrapping function***.)

Since there are 2π radians in one complete revolution, it follows that $P(\theta) = P(\theta + 2\pi)$(Figure 5.48). Thus if $P(\theta) = (a, b)$, then $P(\theta + 2\pi) = (a, b)$,

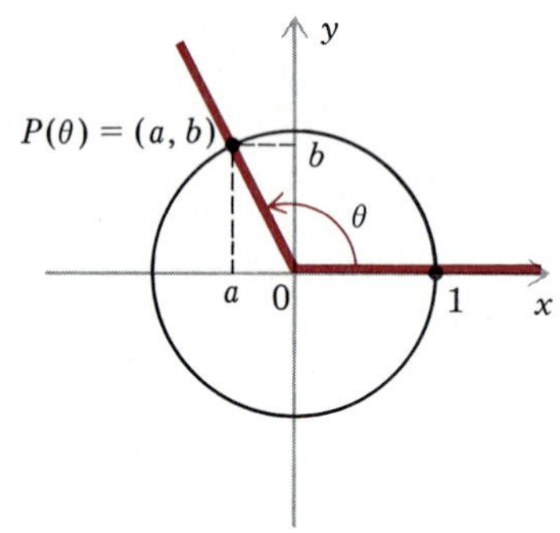

FIGURE 5.46

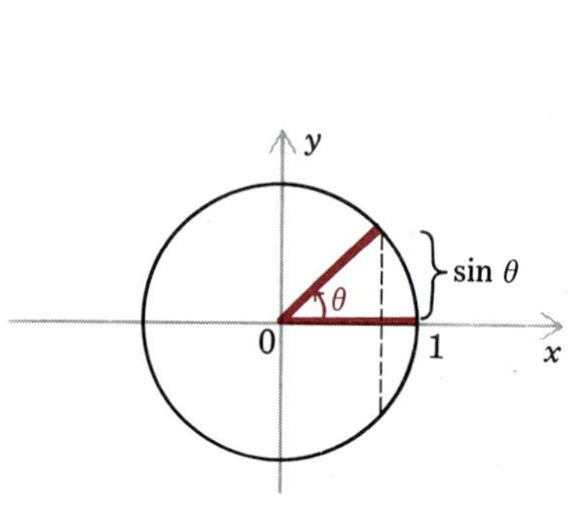

FIGURE 5.47

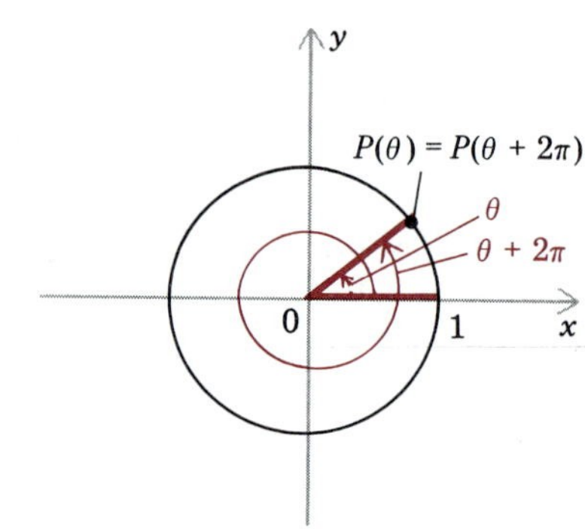

FIGURE 5.48

* Notice from Figure 5.47 that if $0 \leq \theta \leq \pi$, then $\sin\theta$ is the length of a half-chord, a fact that seems to have influenced the etymology of the word "sine." In the fifth century A.D. the Indian mathematician Aryabbatta called the sine of an angle by the name "ardha-jya," meaning "half-chord." Soon the word condensed to "jya," or "chord." The Arabs translated the word into their vowelless written languages as "jb." Later, as the Arabs once again began adding vowels to their language, "jb" somehow became "jaib," which already meant "bosom." Around the twelfth century, when mathematics was translated into Latin, the Arabic "jaib" was translated into "sinus," the Latin equivalent for "bosom," from which our word "sine" comes. It was in 1626 that the abbreviations "sin" and "cos" were introduced, by Albert Girard of the Netherlands.

which implies that

$$\sin\theta = b = \sin(\theta + 2\pi) \quad \text{and} \quad \cos\theta = a = \sin(\theta + 2\pi)$$

We conclude that

$$\begin{array}{|lll|}\hline \sin(\theta + 2\pi) = \sin\theta & \tan(\theta + 2\pi) = \tan\theta & \sec(\theta + 2\pi) = \sec\theta \\ \cos(\theta + 2\pi) = \cos\theta & \cot(\theta + 2\pi) = \cot\theta & \csc(\theta + 2\pi) = \csc\theta \\ \hline \end{array} \quad (1)$$

More generally, for any integer n we have $P(\theta + 2n\pi) = P(\theta)$, so that

$$\boxed{\sin(\theta + 2n\pi) = \sin\theta \quad \text{and} \quad \cos(\theta + 2n\pi) = \cos\theta} \quad (2)$$

with similar formulas for the other trigonometric functions. Thus

$$\sin\frac{25}{3}\pi = \sin\left(8\pi + \frac{1}{3}\pi\right) = \sin\frac{1}{3}\pi$$

From (2) and the fact that $\sin 0 = 0$, $\cos 0 = 1$, $\sin\pi = 0$, and $\cos\pi = -1$, we conclude that for any integer n,

$$\boxed{\sin n\pi = 0 \quad \text{and} \quad \begin{cases}\cos 2n\pi = 1 \\ \cos(2n + 1)\pi = -1\end{cases}}$$

Periodic Functions

It is a direct consequence of (1) that the trigonometric functions repeat their values every 2π units. Functions that repeat their values regularly have a special name.

DEFINITION 5.1 A function f is ***periodic*** if there is a positive number p such that $f(x + p) = f(x)$ for all x. If there is a smallest such number p, then p is called the ***period*** of f.

The formulas in (1) tell us that the trigonometric functions are periodic. It will follow later that the sine, cosine, secant, and cosecant have period 2π, whereas the tangent and the cotangent have period π.

None of the functions encountered in earlier chapters (except the constant functions) is periodic—neither the polynomials nor the rational functions nor the exponential functions nor the logarithmic functions. One of the reasons that the trigonometric functions are so important is their periodic nature. They are indispensable in the mathematical description of such periodic natural phenomena as tides, alternating current, and the oscillation of springs.

The Graphs of the Sine and Cosine Functions

Now we are ready to draw the graphs of the sine and cosine functions. Throughout the discussion, angles will be measured in radians, and we will use x instead of θ for the angle.

First we prepare Table 5.4 with special values of $\sin x$:

TABLE 5.4

x	0	$\frac{\pi}{6}$	$\frac{\pi}{4}$	$\frac{\pi}{3}$	$\frac{\pi}{2}$	$\frac{2\pi}{3}$	$\frac{3\pi}{4}$	$\frac{5\pi}{6}$	π	$\frac{7\pi}{6}$	$\frac{5\pi}{4}$	$\frac{4\pi}{3}$	$\frac{3\pi}{2}$	$\frac{5\pi}{3}$	$\frac{7\pi}{4}$	$\frac{11\pi}{6}$	2π
$\sin x$	0	$\frac{1}{2}$	$\frac{\sqrt{2}}{2}$	$\frac{\sqrt{3}}{2}$	1	$\frac{\sqrt{3}}{2}$	$\frac{\sqrt{2}}{2}$	$\frac{1}{2}$	0	$-\frac{1}{2}$	$-\frac{\sqrt{2}}{2}$	$-\frac{\sqrt{3}}{2}$	-1	$-\frac{\sqrt{3}}{2}$	$-\frac{\sqrt{2}}{2}$	$-\frac{1}{2}$	0

As usual, we plot the points obtained by the table and connect them smoothly. Finally, we use the fact that $\sin(x + 2n\pi) = \sin x$ to obtain the remainder of the graph (Figure 5.49).

Next we assemble Table 5.5 with special values of $\cos x$:

TABLE 5.5

x	0	$\frac{\pi}{6}$	$\frac{\pi}{4}$	$\frac{\pi}{3}$	$\frac{\pi}{2}$	$\frac{2\pi}{3}$	$\frac{3\pi}{4}$	$\frac{5\pi}{6}$	π	$\frac{7\pi}{6}$	$\frac{5\pi}{4}$	$\frac{4\pi}{3}$	$\frac{3\pi}{2}$	$\frac{5\pi}{3}$	$\frac{7\pi}{4}$	$\frac{11\pi}{6}$	2π
$\cos x$	1	$\frac{\sqrt{3}}{2}$	$\frac{\sqrt{2}}{2}$	$\frac{1}{2}$	0	$-\frac{1}{2}$	$-\frac{\sqrt{2}}{2}$	$-\frac{\sqrt{3}}{2}$	-1	$-\frac{\sqrt{3}}{2}$	$-\frac{\sqrt{2}}{2}$	$-\frac{1}{2}$	0	$\frac{1}{2}$	$\frac{\sqrt{2}}{2}$	$\frac{\sqrt{3}}{2}$	1

Then we plot the corresponding points, connect them and use the fact that $\cos(x + 2n\pi) = \cos x$ to complete the graph (Figure 5.50).

Observe from Figures 5.49 and 5.50 that the period of $\sin x$ and $\cos x$ is precisely 2π, a fact alluded to in our discussion of periodic functions.

From Figures 5.49 and 5.50 it appears that the graph of the sine function is symmetric with respect to the origin, that is, $\sin(-\theta) = -\sin \theta$. To show that this is true, we observe that if $P(\theta)$ is any point on the unit circle, then

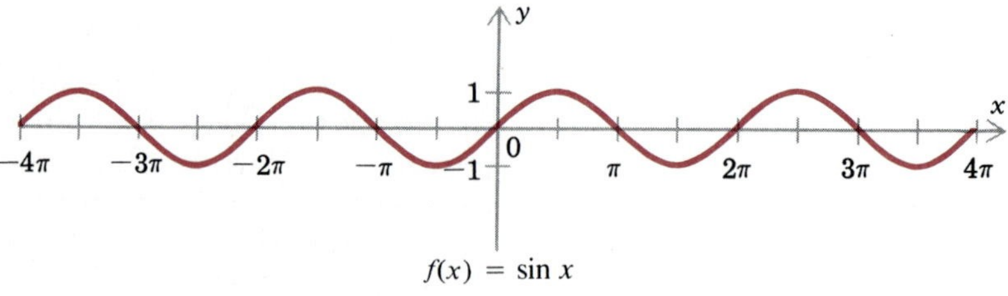

$f(x) = \sin x$

FIGURE 5.49

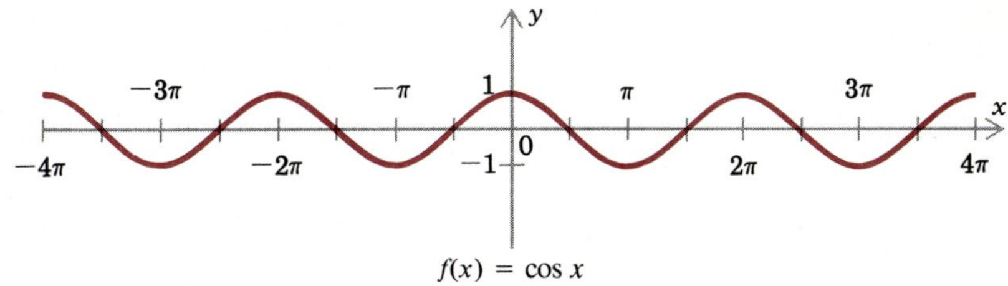

$f(x) = \cos x$

FIGURE 5.50

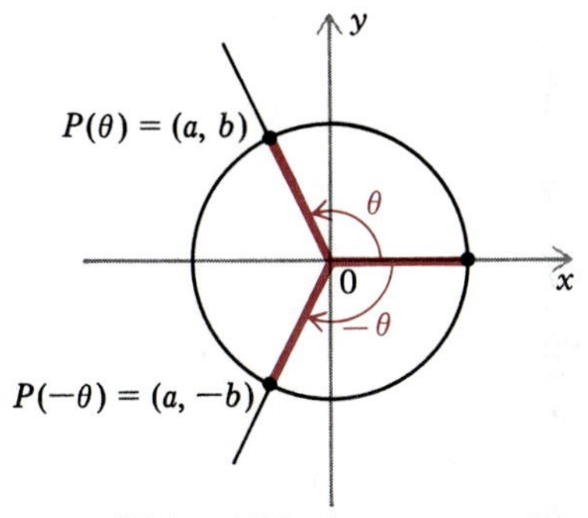

$P(\theta)$ and $P(-\theta)$ are symmetric with respect to the x axis

FIGURE 5.51

$P(\theta)$ and $P(-\theta)$ are symmetric with respect to the x axis (Figure 5.51). It follows that if $P(\theta) = (a, b)$, then $P(-\theta) = (a, -b)$. Therefore

$$\sin\theta = b \quad \text{and} \quad \sin(-\theta) = -b$$

so that

$$\boxed{\sin(-\theta) = -\sin\theta} \tag{3}$$

Similarly,

$$\cos\theta = a \quad \text{and} \quad \cos(-\theta) = a$$

so that

$$\cos(-\theta) = \cos\theta \tag{4}$$

In the same way,

$$\begin{aligned} \tan(-\theta) &= -\tan\theta \qquad & \sec(-\theta) &= \sec\theta \\ \cot(-\theta) &= -\cot\theta \qquad & \csc(-\theta) &= -\csc\theta \end{aligned} \tag{5}$$

In studying the graphs of the sine and cosine functions, it might appear that one graph is merely a shift of the other. In fact this observation is valid: the graph of the sine function can be obtained by shifting the graph of the cosine function $\pi/2$ units to the right (Figure 5.52). We will be able to verify this fact by the results of Section 6.2.

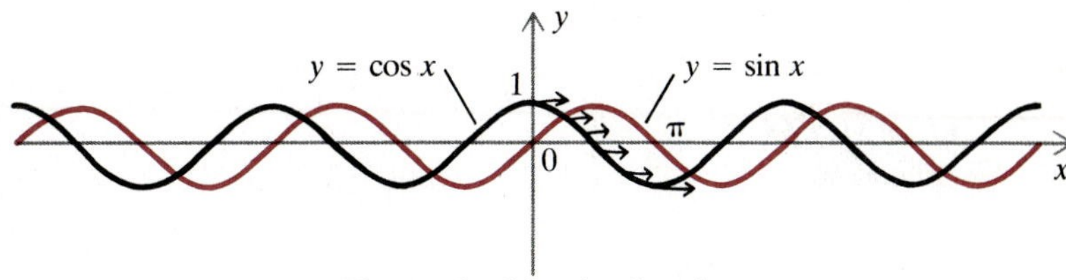

The graph of the sine function is the graph of the cosine function shifted to the right $\pi/2$ units

FIGURE 5.52

Knowledge of the graphs of the sine and cosine functions enables us to sketch the graphs of closely related functions.

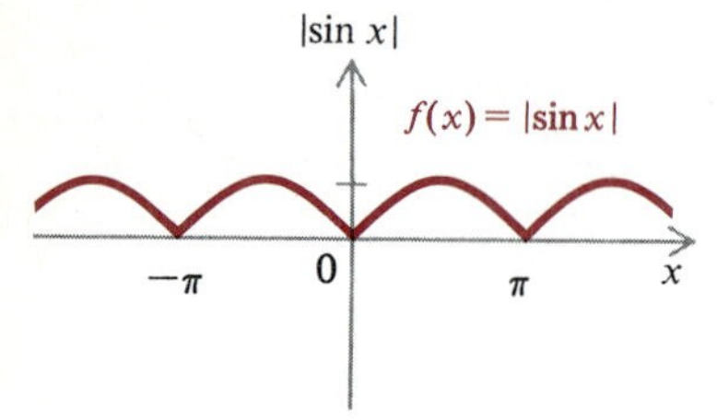

FIGURE 5.53

EXAMPLE 1. Let $f(x) = |\sin x|$. Sketch the graph of f.

Solution. Notice that if $\sin x \geq 0$, then $f(x) = \sin x$, whereas if $\sin x < 0$, then $f(x) = -\sin x$. Therefore we obtain the graph of f from the graph of $\sin x$ by leaving unaltered the portions that lie above the x axis and by reflecting through the x axis the portions that lie below the x axis (Figure 5.53). □

Before we sketch the next trigonometric graph, we recall that if

$$g(x) = f(x) + c \quad \text{and} \quad h(x) = f(x + c)$$

then the graph of g is the graph of f shifted upward c units if $c > 0$ and downward $|c|$ units if $c < 0$. Similarly, the graph of h is the graph of f shifted to the left c units if $c > 0$ and to the right $|c|$ units if $c < 0$ (see Section 2.5).

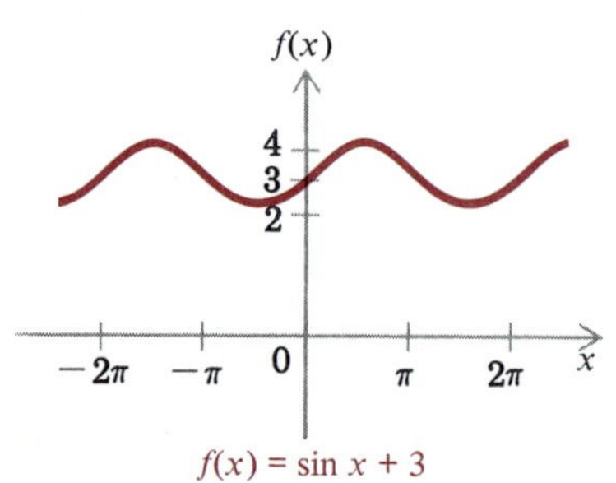

FIGURE 5.54

EXAMPLE 2. Sketch the graphs of the following functions, and determine their periods.

a. $f(x) = \sin x + 3$

b. $g(x) = \cos\left(x - \dfrac{4\pi}{3}\right)$

Solution.

a. The graph of f is the graph of $\sin x$, shifted upward 3 units (Figure 5.54). Since $\sin x$ has period 2π, we find that

$$f(x + 2\pi) = \sin(x + 2\pi) + 3 = \sin x + 3 = f(x)$$

so that the period of f is also 2π, as Figure 5.54 confirms.

b. The graph of g is the graph of $\cos x$, shifted to the right $4\pi/3$ units (Figure 5.55). The fact that $\cos x$ has period 2π implies that

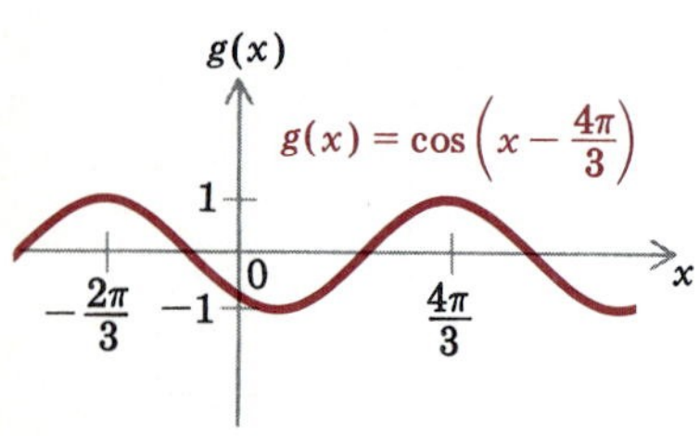

FIGURE 5.55

$$\begin{aligned} g(x + 2\pi) &= \cos\left[(x + 2\pi) - \frac{4\pi}{3}\right] \\ &= \cos\left[\left(x - \frac{4\pi}{3}\right) + 2\pi\right] = \cos\left(x - \frac{4\pi}{3}\right) = g(x) \end{aligned}$$

so that the period of g is also 2π, as Figure 5.55 suggests. □

As Example 2 illustrates, any function of the form

$$\sin x + c \quad \text{or} \quad \cos x + c$$

and any function of the form

$$\sin(x + a) \quad \text{or} \quad \cos(x + a)$$

have period 2π. Thus when the graphs of the sine and cosine functions are shifted vertically or horizontally, the period remains the same—namely, 2π.

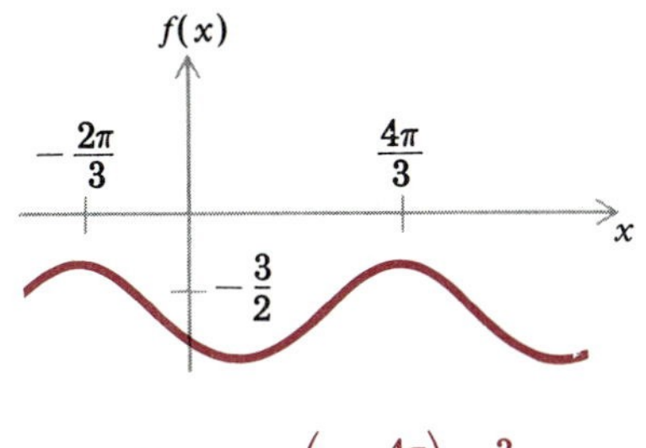

$f(x) = \cos\left(x - \frac{4\pi}{3}\right) - \frac{3}{2}$

FIGURE 5.56

EXAMPLE 3. Let $f(x) = \cos\left(x - \frac{4\pi}{3}\right) - \frac{3}{2}$. Sketch the graph of f.

Solution. The graph of f is obtained from the graph of $\cos x$ by shifting it to the right $4\pi/3$ units and then downward $\frac{3}{2}$ units, which yields the graph in Figure 5.56. □

The Graphs of the Remaining Trigonometric Functions

To prepare for sketching the graph of the tangent function, we will first prove that

$$\boxed{\tan(\theta + \pi) = \tan \theta} \tag{6}$$

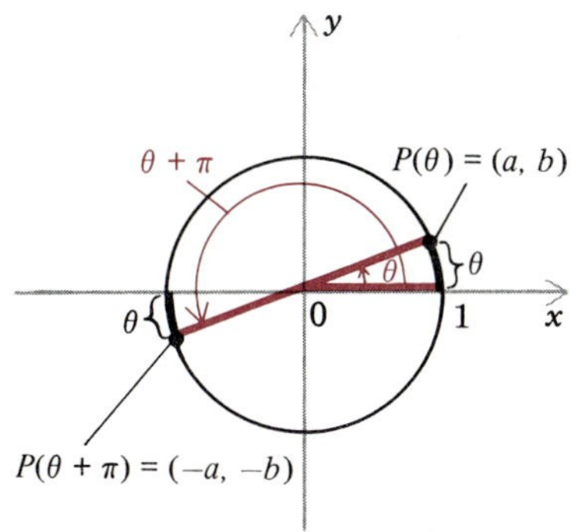

$P(\theta)$ and $P(\theta + \pi)$ are symmetric with respect to the origin

FIGURE 5.57

For this we let $P(\theta) = (a, b)$ once again denote the point on the unit circle corresponding to an angle of θ radians. Then $(-a, -b)$ is the point on the unit circle corresponding to an angle of $\theta + \pi$ radians (Figure 5.57). By the definition of the tangent function (see (1) in Section 5.4), it follows that

$$\tan(\theta + \pi) = \frac{-b}{-a} = \frac{b}{a} = \tan \theta$$

which proves (6).

Next we assemble Table 5.6, containing special values of $\tan x$:

TABLE 5.6

x	$-\frac{\pi}{3}$	$-\frac{\pi}{4}$	$-\frac{\pi}{6}$	0	$\frac{\pi}{6}$	$\frac{\pi}{4}$	$\frac{\pi}{3}$
$\tan x$	$-\sqrt{3}$	-1	$-\frac{\sqrt{3}}{3}$	0	$\frac{\sqrt{3}}{3}$	1	$\sqrt{3}$

Then we connect the corresponding points with a smooth curve, as in Figure 5.58a. Since $\pi/2$ is not in the domain of the tangent function, we investigate the values of $\tan x$ for which x is close to and less than $\pi/2$. Recall that

$$\tan x = \frac{\sin x}{\cos x}$$

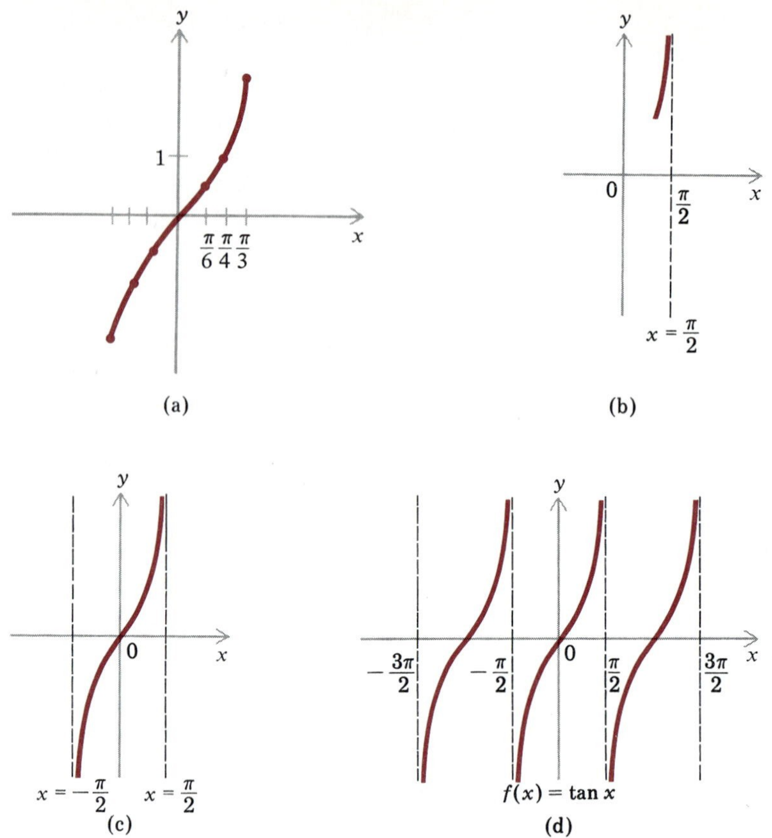

FIGURE 5.58

Now if x is close to $\pi/2$ and $0 < x < \pi/2$, then sin x is a positive number close to 1 and cos x is a positive number close to 0. Thus tan x is a large positive number. The closer x comes to $\pi/2$, the closer cos x comes to 0 and sin x to 1, and hence the larger tan x becomes. This is reflected in the graph drawn in Figure 5.58b. It follows that the vertical line $x = \pi/2$ is a vertical asymptote of the graph of tan x. Since $\tan(-x) = -\tan x$, the vertical line $x = -\pi/2$ is also a vertical asymptote of the graph, so we obtain Figure 5.58c. Finally, we use the fact that $\tan(x + \pi) = \tan x$ in order to complete the graph of tan x, shown in Figure 5.58d. Incidentally, Figure 5.58d indicates that the period of tan x is π, which is in fact true.

EXAMPLE 4. Sketch the graphs of the following functions.

a. $f(x) = \tan\left(x - \dfrac{\pi}{2}\right)$ b. $g(x) = \tan\left(x + \dfrac{\pi}{4}\right) - 1$

Solution.

a. The graph of f is obtained from the graph of $\tan x$ by shifting it to the right $\pi/2$ units. Thus we obtain the graph in Figure 5.59.

b. The graph of g is obtained from the graph of $\tan x$ by shifting it to the left $\pi/4$ units and then downward 1 unit. This yields the graph in Figure 5.60.

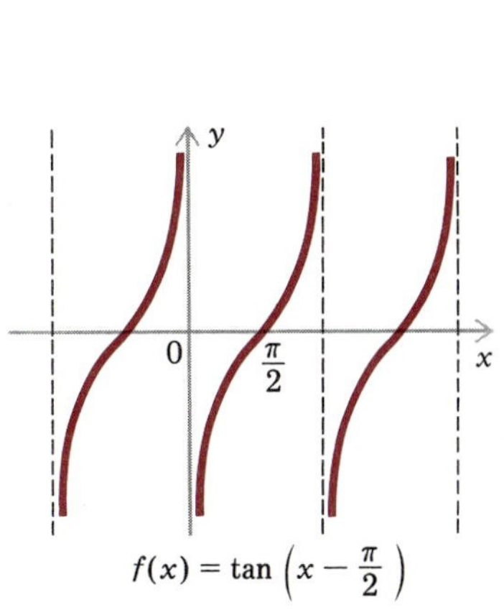

$f(x) = \tan\left(x - \frac{\pi}{2}\right)$

FIGURE 5.59

$g(x) = \tan\left(x + \frac{\pi}{4}\right) - 1$

FIGURE 5.60 □

We complete this section by drawing the graphs of the remaining trigonometric functions: $\cot x$, $\sec x$, and $\csc x$ (Figure 5.61a–c). Notice that any line of the form $x = n\pi$ is a vertical asymptote of the graphs of $\cot x$ and $\csc x$, and any line of the form $x = \frac{\pi}{2} + n\pi$ is a vertical asymptote of the graph of $\sec x$.

In the next section we will discuss graphs of other functions related to trigonometric functions.

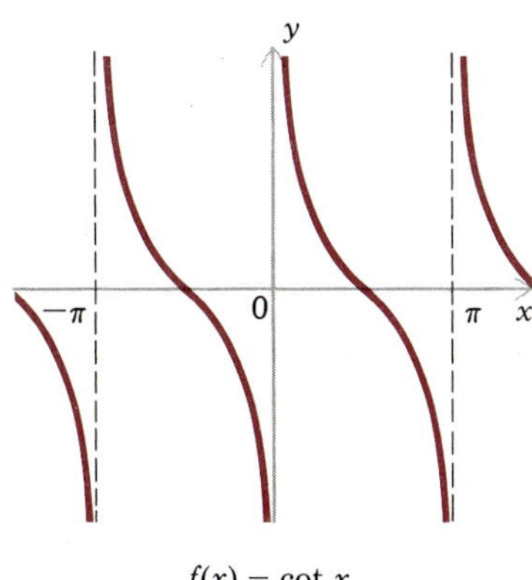

$f(x) = \cot x$
(a)

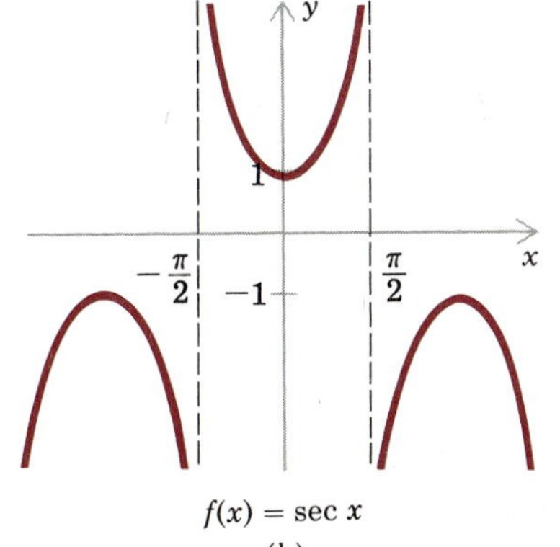

$f(x) = \sec x$
(b)

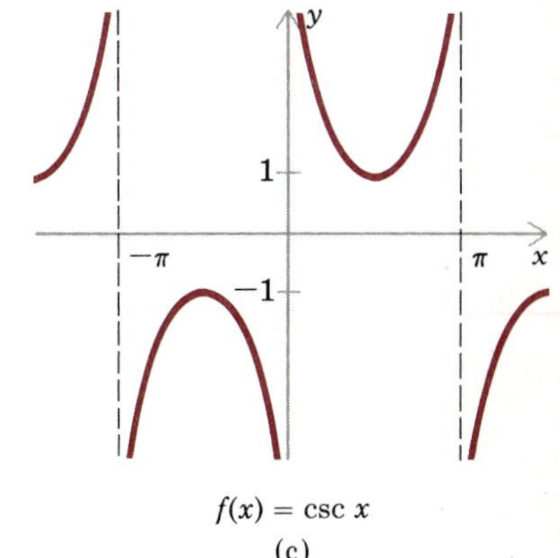

$f(x) = \csc x$
(c)

FIGURE 5.61

EXERCISES 5.5

In Exercises 1–12, sketch the graph of the function.

1. $\sin x + 2$

2. $\sin x - \frac{1}{2}$

3. $\cos x + \frac{3}{2}$

4. $\cos x - 1$

5. $\sin\left(x - \frac{\pi}{2}\right)$

6. $\sin(x + \pi)$

7. $\cos\left(x + \frac{3\pi}{4}\right)$

8. $\cos\left(x + \frac{4\pi}{3}\right)$

9. $\sin(-x)$

10. $\sin\left(\frac{3\pi}{2} - x\right)$

11. $\cos(-x)$

12. $\cos(\pi - x)$

In Exercises 13–20, sketch the graph of the function.

13. $\sin(x + \pi) - 2$

14. $\sin\left(x - \frac{\pi}{3}\right) - \frac{1}{3}$

15. $\sin\left(x + \frac{\pi}{4}\right) + 1$

16. $\cos\left(x - \frac{\pi}{2}\right) - 1$

17. $\cos\left(x + \frac{3\pi}{2}\right) + \frac{2}{3}$

18. $\cos\left(x + \frac{2\pi}{3}\right) + 3$

19. $|\cos x|$

20. $\left|\sin x + \frac{1}{2}\right|$

In Exercises 21–30, sketch the graph of the function.

21. $\tan\left(x + \frac{\pi}{2}\right)$

22. $\tan\left(x - \frac{\pi}{3}\right)$

23. $\cot(x - \pi)$

24. $\cot\left(x + \frac{\pi}{4}\right)$

25. $\sec\left(x - \frac{\pi}{6}\right)$

26. $1 + \sec x$

27. $\csc(x + \pi)$

28. $2 + \csc x$

29. $\tan\left(x - \frac{\pi}{3}\right) - 2$

30. $\sec\left(x + \frac{\pi}{4}\right) + 3$

The method of approximating a zero of a polynomial that was discussed in Section 3.9 is valid for any trigonometric function defined at every point in a given interval $[a, b]$. In Exercises 31–32, use a graphics calculator to approximate to within .01 the zero of the function in the given interval.

31. $f(x) = \cos x - x$; $[0, \pi/2]$

32. $f(x) = 2 + \tan 2x$; $[-\pi/5, 0]$

5.6 SINUSOIDAL GRAPHS

In Section 5.5 we analyzed the graphs of the trigonometric functions, along with horizontal and vertical shifts of these graphs. In this section we will discuss the graphs of ***sinusoidal functions***, which are functions of the form $a \sin b(x + c)$ or $a \cos b(x + c)$. These functions and their graphs are important in physical applications.

Vertical Expansions and Contractions

Let $a \neq 0$. For any real number x,

$$-|a| \leq a \sin x \leq |a|$$

Moreover, all numbers between $-|a|$ and $|a|$ are values of $a \sin x$. The number $|a|$ is called the ***amplitude*** of $a \sin x$. Notice that $a \sin x$ has a greater amplitude than $\sin x$ does if $|a| > 1$ and has a smaller amplitude than $\sin x$ does if $0 < |a| < 1$.

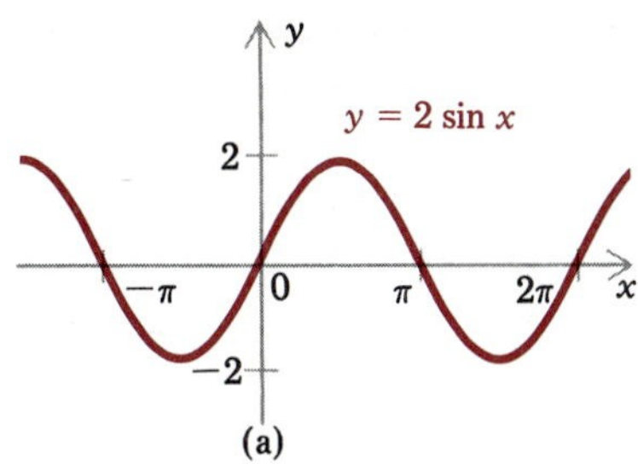

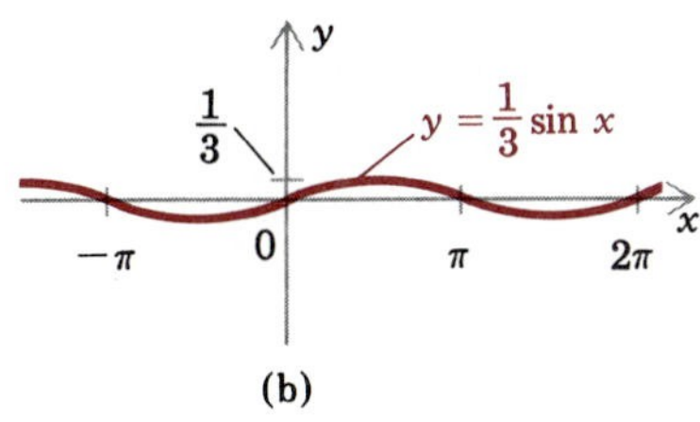

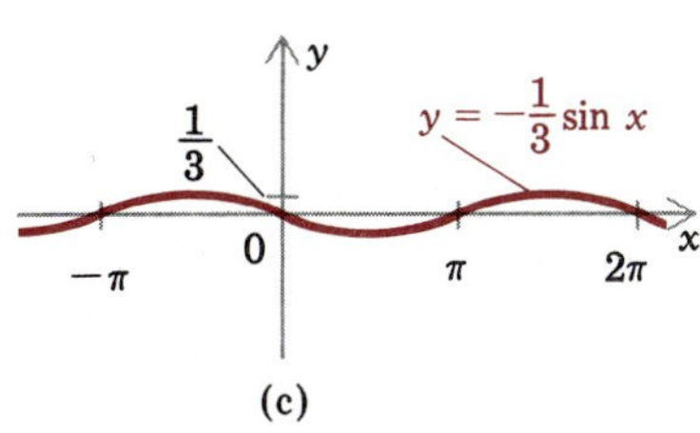

FIGURE 5.62

EXAMPLE 1. Sketch the graphs of the following functions.

a. $2 \sin x$ b. $\dfrac{1}{3} \sin x$ c. $-\dfrac{1}{3} \sin x$

Solution.

a. To obtain the graph of $2 \sin x$, we multiply the y coordinate of each point on the graph of $\sin x$ by 2 (Figure 5.62a).
b. To obtain the graph of $\frac{1}{3} \sin x$, we multiply the y coordinate of each point on the graph of $\sin x$ by $\frac{1}{3}$ (Figure 5.62b).
c. To obtain the graph of $-\frac{1}{3} \sin x$, we reflect the graph of $\frac{1}{3} \sin x$ through the x axis (Figure 5.62c). □

Similar remarks apply to the function $a \cos x$. In particular, the ***amplitude*** of $a \cos x$ is $|a|$.

EXAMPLE 2. Sketch the graphs of the following functions.

a. $\dfrac{3}{2} \cos x$ b. $\dfrac{1}{2} \cos x$ c. $-\dfrac{1}{2} \cos x$

Solution.

a. For the graph of $\frac{3}{2} \cos x$ we multiply the y coordinate of each point on the graph of $\cos x$ by $\frac{3}{2}$ (Figure 5.63a).
b. Here we multiply the y coordinate of each point on the graph of $\cos x$ by $\frac{1}{2}$ (Figure 5.63b).
c. For the graph of $-\frac{1}{2} \cos x$ we reflect the graph of $\frac{1}{2} \cos x$ through the x axis (Figure 5.63c). □

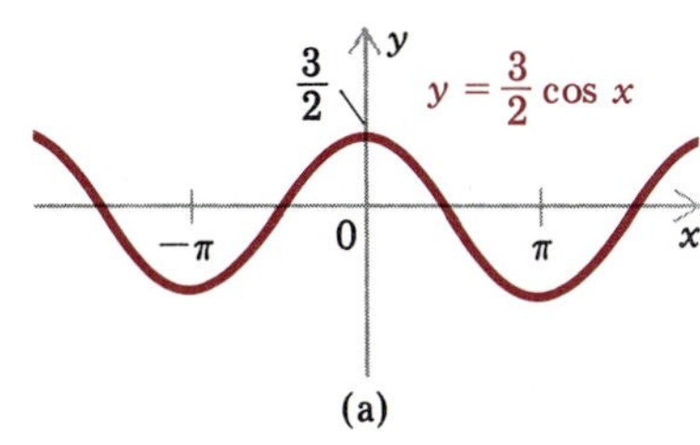

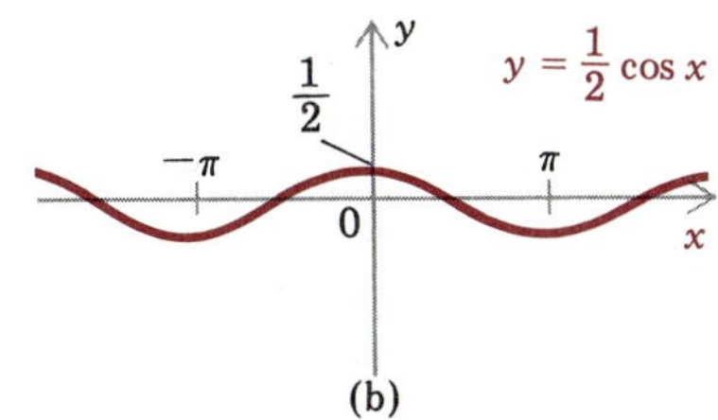

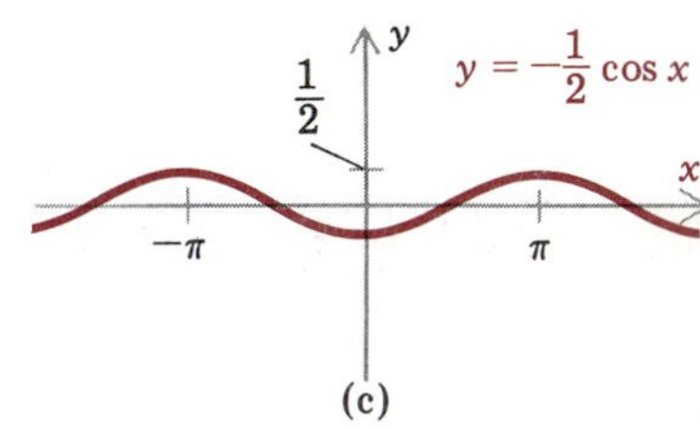

FIGURE 5.63

Notice that the graph of $2 \sin x$ in Figure 5.62a is obtained by "pulling" the graph of $\sin x$ away from the x axis so that it extends between the lines $y = -2$ and $y = 2$. We say that the graph has been expanded vertically. More generally, the graph of $a \sin x$ is a ***vertical expansion*** of the graph of $\sin x$ if $a > 1$. Thus the graph of $2 \sin x$ is a vertical expansion of the graph of $\sin x$. In contrast, from Figure 5.62b we see that the graph of $\frac{1}{3} \sin x$ is obtained from the graph of $\sin x$ by compressing it so that it fits between the lines $y = -\frac{1}{3}$ and $y = \frac{1}{3}$. We say that the graph has been contracted vertically. More generally, the graph of $a \sin x$ is a ***vertical contraction*** of the graph of $\sin x$ if $0 < a < 1$. As a result, the graph of $\frac{1}{3} \sin x$ is a vertical contraction of the graph of $\sin x$. Finally, if $a < 0$, then the graph of $a \sin x$ is obtained by a vertical expansion or contraction of the graph of $\sin x$, plus a reflection through the x axis, as part (c) of Example 1 suggests. Analogous remarks pertain to the graph of $a \cos x$.

Horizontal Expansions and Contractions

Let b be a fixed positive number. Then it can be shown that the function $\sin bx$ is periodic (see Definition 5.1) and its values are repeated every $2\pi/b$ units. After all,

$$\sin b\left(x + \frac{2\pi}{b}\right) = \sin(bx + 2\pi) = \sin bx$$

In fact, it turns out that the period of $\sin bx$ is exactly $2\pi/b$, so that the graph of $\sin bx$ is determined by its graph on $[0, 2\pi/b]$. Comparing the graphs of $\sin bx$ and $\sin x$, we see that if $b > 1$, then the graph of $\sin x$ must be contracted horizontally in order to obtain the graph of $\sin bx$, whereas if $0 < b < 1$, then the graph of $\sin x$ must be expanded horizontally to obtain the graph of $\sin bx$. In either case the graph is still contained in the region between the horizontal lines $y = -1$ and $y = 1$, just as the graph of $\sin x$ is. Similar comments apply to the graph of $\cos bx$ for $b > 0$.

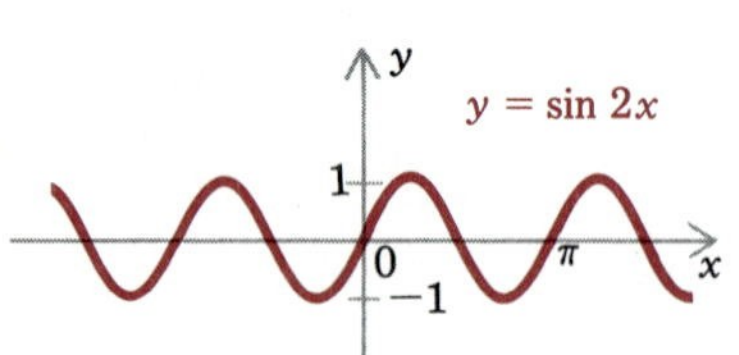

(a)

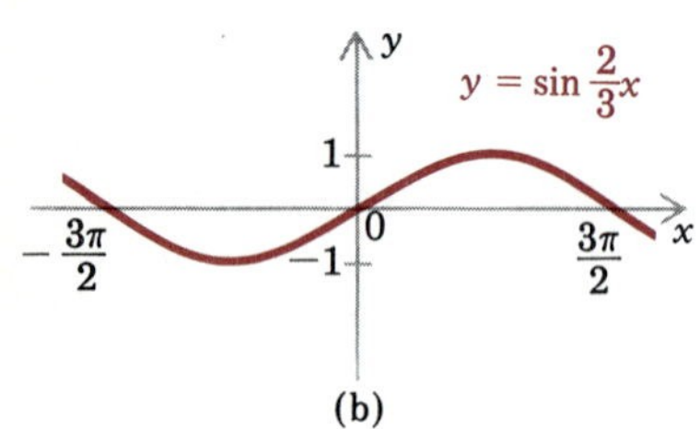

(b)

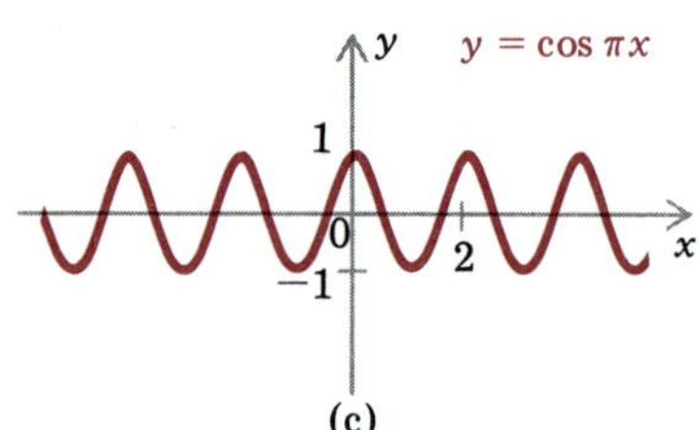

(c)

FIGURE 5.64

EXAMPLE 3. Sketch the graphs of the following functions.

a. $\sin 2x$ b. $\sin \frac{2}{3} x$ c. $\cos \pi x$

Solution.

a. In this case $b = 2$, so the period is $2\pi/2 = \pi$. Thus the graph of $\sin 2x$ is obtained by contracting the graph of $\sin x$ horizontally so that the portion between $x = 0$ and $x = 2\pi$ fits into the region between $x = 0$ and $x = \pi$ (Figure 5.64a).
b. This time $b = \frac{2}{3}$, so the period is $2\pi/(\frac{2}{3}) = 3\pi$. It follows that the graph of $\sin \frac{2}{3}x$ is obtained by expanding the graph of $\sin x$ horizontally so that the portion between $x = 0$ and $x = 2\pi$ fits into the region between $x = 0$ and $x = 3\pi$ (Figure 5.64b).
c. Now $b = \pi$, so the period is $2\pi/\pi = 2$. Therefore the graph of $\cos \pi x$ is obtained by contracting the graph of $\cos x$ horizontally so that the portion between $x = 0$ and $x = 2\pi$ fits into the region between $x = 0$ and $x = 2$ (Figure 5.64c). □

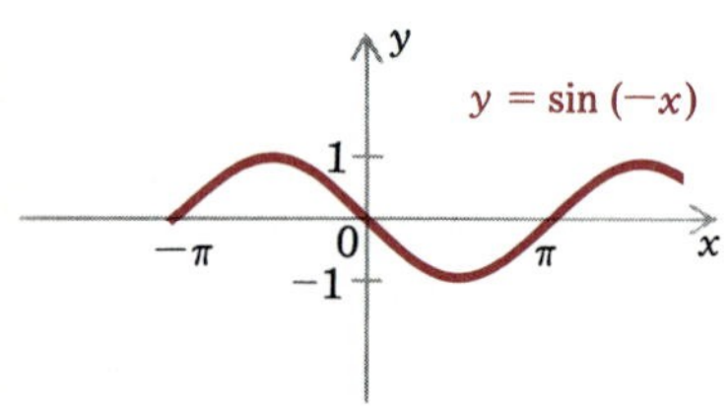

FIGURE 5.65

Since $\sin(-x) = -\sin x$, the graph of $\sin(-x)$ is just the reflection of the graph of $\sin x$ through the x axis (Figure 5.65). Therefore, to obtain the graph of $\sin bx$ when $b < 0$, one draws the graph of $\sin(|b|\, x)$ and then reflects it through the x axis. The period is $2\pi/|b|$.

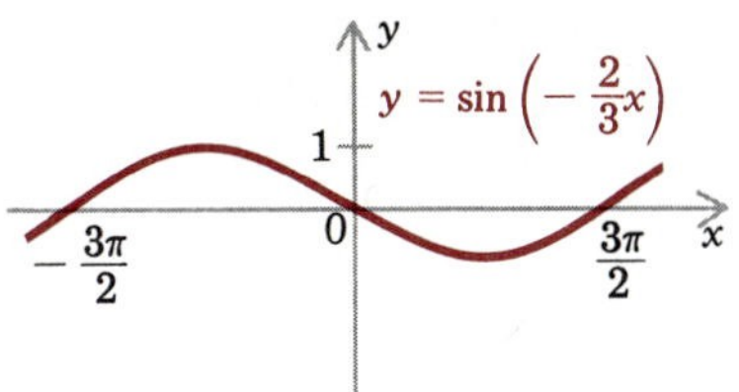

FIGURE 5.66

EXAMPLE 4. Sketch the graph of $\sin\left(-\dfrac{2}{3}x\right)$.

Solution. We recall the graph of $\sin \frac{2}{3}x$ from Figure 5.64b and reflect it through the x axis to obtain the graph we desire (Figure 5.66). □

Since $\cos(-x) = \cos x$, it follows that $\cos bx = \cos(-bx)$ for any constant b, so the graphs of $\cos bx$ and $\cos(-bx)$ are the same. Thus the graph of, say, $\cos(-\pi x)$ is the same as the graph of $\cos \pi x$.

Horizontal Expansions and Contractions Plus Shifts

Next we will discuss the graphs of $\sin b(x + c)$ and $\cos b(x + c)$, where $b \neq 0$ and c may be any number. The graph of $\sin b(x + c)$ is the same as the graph of $\sin bx$ shifted to the left c units if $c > 0$, and to the right $-c$ units if $c < 0$. In particular, $\sin b(x + c)$ and $\sin bx$ have the same period. Similarly, the graph of $\cos b(x + c)$ is the same as the graph of $\cos bx$ shifted $|c|$ units to the left or right. The number c is called the ***phase shift*** of $\sin b(x + c)$ and of $\cos b(x + c)$.

EXAMPLE 5. Sketch the graph of $\sin\left[-2\left(x - \dfrac{\pi}{3}\right)\right]$.

Solution. First we sketch the graph of $\sin 2x$ (see Figure 5.64a again). Then we reflect the graph through the x axis to obtain the graph of $\sin(-2x)$ (Figure 5.67a). Finally, we shift the resulting graph $\pi/3$ units to the right to obtain the graph of the given function (Figure 5.67b). □

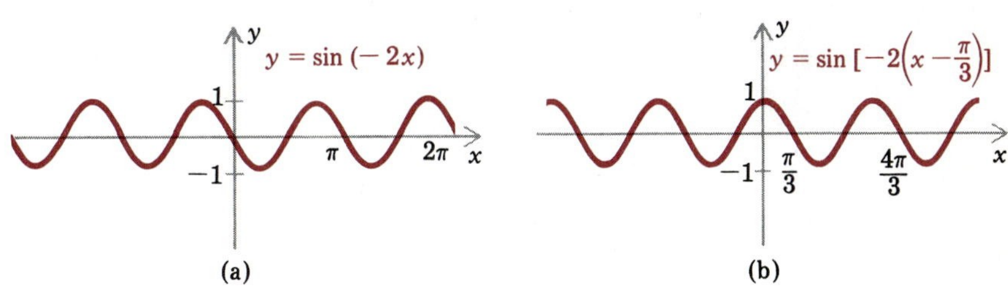

FIGURE 5.67

EXAMPLE 6. Sketch the graph of $\cos\left(-\dfrac{1}{2}x - \dfrac{\pi}{8}\right)$.

Solution. Notice that

$$\cos\left(-\frac{1}{2}x - \frac{\pi}{8}\right) = \cos\left[-\frac{1}{2}\left(x + \frac{\pi}{4}\right)\right]$$

and that the right side of this equation has the form $\cos b(x + c)$. Thus we draw the graph of $\cos(-\frac{1}{2}x)$, which is the same as that of $\cos \frac{1}{2}x$, and shift the graph $\pi/4$ units to the left (Figure 5.68). □

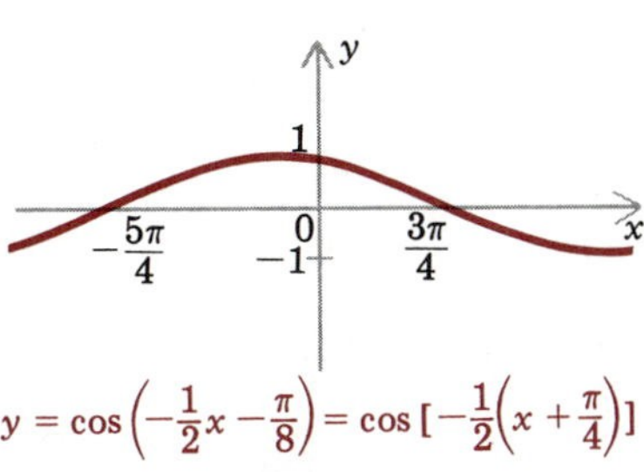

$y = \cos\left(-\frac{1}{2}x - \frac{\pi}{8}\right) = \cos\left[-\frac{1}{2}\left(x + \frac{\pi}{4}\right)\right]$

FIGURE 5.68

General Sinusoidal Graphs

Finally, we discuss graphs of the general sinusoidal functions, which are obtained from sin x and cos x by horizontal shifts as well as horizontal and vertical contractions and expansions. Such altered functions can be written in the form $a \sin b(x + c)$ or $a \cos b(x + c)$.

EXAMPLE 7. Sketch the graphs of the following functions.

a. $3 \sin\left(2x - \dfrac{4\pi}{3}\right)$ b. $\dfrac{1}{2} \sin 2\left(x - \dfrac{2\pi}{3}\right)$

Solution.

a. When written in the form $a \sin b(x + c)$, the function will become $3 \sin 2(x - 2\pi/3)$. First we sketch the graph of $\sin 2x$ (see Figure 5.64a again) and shift it $2\pi/3$ units to the right to obtain the graph of $\sin 2(x - 2\pi/3)$. Then we expand the latter graph vertically by a factor of 3 to obtain the graph of the given function (Figure 5.69a).

b. If instead of expanding the graph of $\sin 2(x - 2\pi/3)$, as we did in the solution of part(a), we contract it vertically by a factor of $\frac{1}{2}$ to obtain the desired graph (Figure 5.69b). □

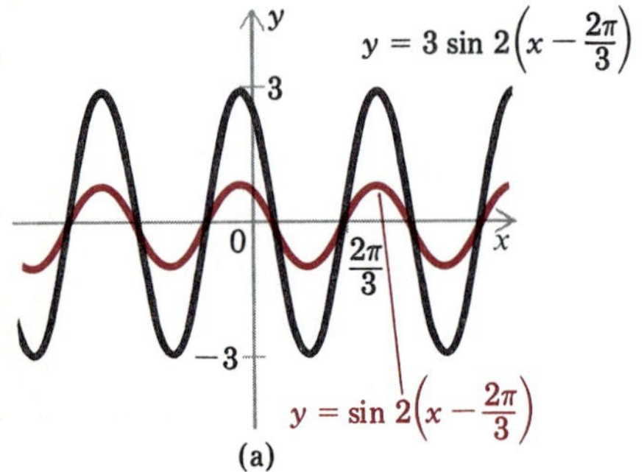

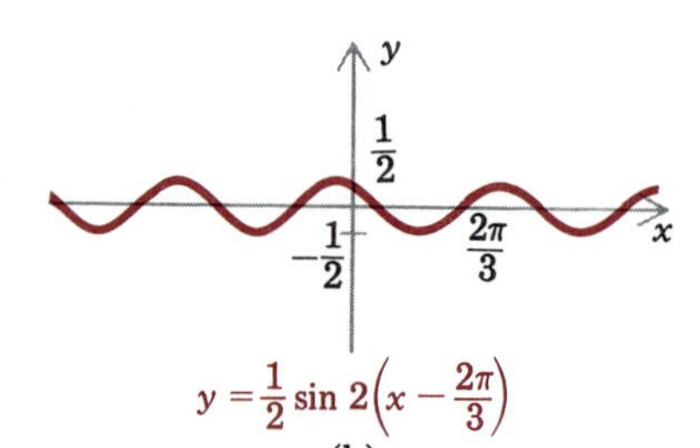

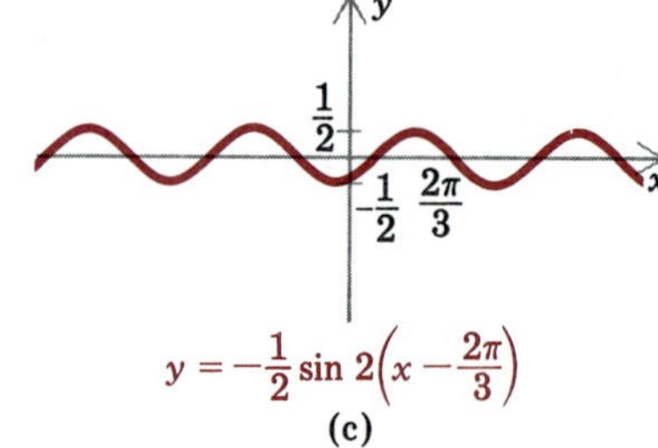

FIGURE 5.69

In those cases for which $a < 0$ rather than $a > 0$, we first determine the graph of $|a| \sin b(x + c)$ and then reflect it through the x axis in order to obtain the graph of $a \sin b(x + c)$. Thus the graph of $-\frac{1}{2} \sin 2(x - 2\pi/3)$ is as in Figure 5.69c.

The graph of $a \cos b(x + c)$ is obtained from the graph of $\cos b(x + c)$ in exactly the same manner. Thus to obtain the graph of $-\frac{4}{3} \cos(-\frac{1}{2}x - \pi/8)$, we would expand the graph of $\cos(-\frac{1}{2}x - \pi/8)$ in Figure 5.68 vertically by the factor $\frac{4}{3}$ and then reflect the graph through the x axis. The result appears in Figure 5.70.

$y = -\frac{4}{3} \cos\left(-\frac{1}{2}x - \frac{\pi}{8}\right)$

FIGURE 5.70

We call $|a|$ the ***amplitude*** of $a \sin b(x + c)$ and of $a \cos b(x + c)$.

Simple Harmonic Motion

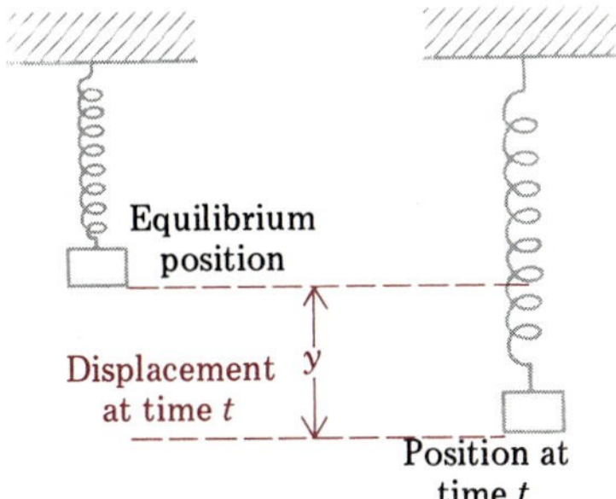

FIGURE 5.71

Many objects in the physical world vibrate, that is, move back and forth in a regular motion. Examples include the heart, waves in an ocean, the strings on a guitar, the vocal cords of a singer, the pendulum on a grandfather clock, a piston in the engine of a car, and the blade of a pneumatic hammer. Because any musical tone (in fact, any sound whatever) is produced by vibrations, the term ***harmonic motion*** is applied to any vibration, whether it produces a sound or not.

The displacement of a vibrating object from its equilibrium position is frequently given by an equation of the form

$$y = a \sin b(t + c)$$

where t represents time and y the displacement. Any harmonic motion that can be so described is called ***simple harmonic motion***. The motion of a bob attached to a spring (Figure 5.71) would be simple harmonic motion if there were no air resistance (for example, if the spring were located in a vacuum).

Another example of simple harmonic motion is provided by a vibrating string, such as a string on a musical instrument. Suppose the string lies along the x axis. Under certain conditions, when the string is plucked or bowed, the disturbance that arises is propagated along the string in a way that can be described by the equation

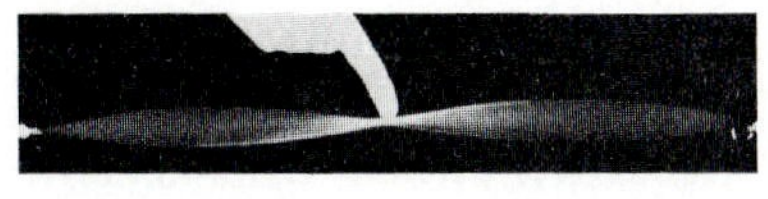

$$y = a \sin 2\pi\left(\frac{x}{\lambda} - \frac{t}{T}\right) \tag{1}$$

In this equation, y represents the displacement at time t of the point on the string corresponding to x; the letters a, T, and λ have physical interpretations described as follows:

First, the number $|a|$ is the ***amplitude*** of the vibration. It represents the maximum displacement of any point on the string from the x axis and is related to the loudness of the sound produced by the vibration.

Second, the number T is the ***period*** of the vibration. If x is held constant in (1), then the resulting function of t is sinusoidal with period T. Physically, T is the length of time it takes a given point on the string to make one complete vibration, and is related to the pitch of the sound produced.

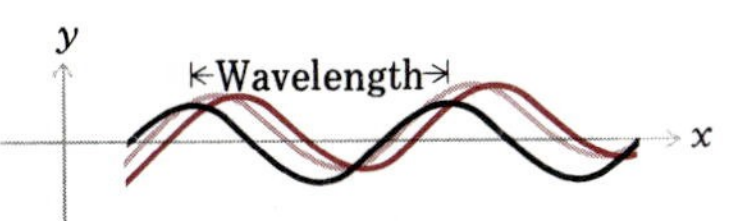

FIGURE 5.72

Finally, the number λ is called the ***wavelength*** of the vibration. If we hold t constant in (1), then the resulting function of x is sinusoidal with period λ. Physically, λ is the length between two successive points on the string that experience the same displacement at any given time t (Figure 5.72).

Notice that if we compare the graph of y as a function of x for successive values of t that are close together (Figure 5.72), we see that the points of maximal displacement move along the string. For this reason the disturbance is sometimes called a "traveling wave."

In Section 6.4 we will explain this topic further.

EXERCISES 5.6

In Exercises 1–12, sketch the graph of the given function. Note the amplitude, period, and phase angle of the function.

1. $3 \sin x$

2. $-\frac{2}{3} \sin x$

3. $-\cos x$

4. $3 \cos x$

5. $\sin \pi x$

6. $\sin(-\frac{1}{2}x)$

7. $\cos 2x$

8. $\cos(-\frac{2}{3}x)$

9. $2 \sin(x - \pi/3)$

10. $-\sin(-x - 3\pi/4)$

11. $3 \cos 2(x + \pi/3)$

12. $\frac{3}{2} \cos 4\pi(x - \frac{1}{2})$

In Exercises 13–15, write down a function with the given amplitude, period, and phase shift, and then sketch the graph of the function.

13. amplitude $= 4$, period $= 2\pi/3$, and phase shift $= \pi$

14. amplitude $= \frac{4}{3}$, period $= 3\pi$, and phase shift $= 5\pi/3$

15. amplitude $= \frac{3}{2}$, period $= \pi$, and phase shift $= \pi/3$

***16.** Assume that as a result of the tides, the depth d of the ocean above a certain barrier reef depends on time x according to the equation

$$d = d_0 + a \sin b\left(x - \frac{\pi}{2}\right)$$

where d_0 is the average depth, a is the difference between the depth at high tide and the average depth, and b is related to the tidal period. Assume that the tides have a period of 12.5 hours, and that the water is 10 feet above the reef at high tide but only 2 feet above the reef at low tide. If a certain boat needs at least 8 feet of water to cross the reef, how long during one tidal period is it safe for the boat to cross the reef?

5.7 INVERSE TRIGONOMETRIC FUNCTIONS

Each of the six basic trigonometric functions is periodic; consequently none has an inverse (see Section 2.7). However, it is possible to restrict the domain of each of these functions to a suitable interval so that the resulting function has an inverse. These inverse functions are called ***inverse trigonometric functions***; we will presently define and discuss them.

The Arcsine Function

If we restrict the domain of the sine function to $[-\pi/2, \pi/2]$, the resulting function f is increasing (Figure 5.73). Thus, by the comments at the end of Section 2.7, f has an inverse. Since the range of f is $[-1, 1]$, the domain of f^{-1} is also $[-1, 1]$. The inverse f^{-1} is called the ***arcsine function*** (or the inverse sine function). Its value at x is usually denoted by $\arcsin x$ (sometimes by $\text{Arcsin } x$ or $\sin^{-1} x$). Thus by (6) in Section 2.7

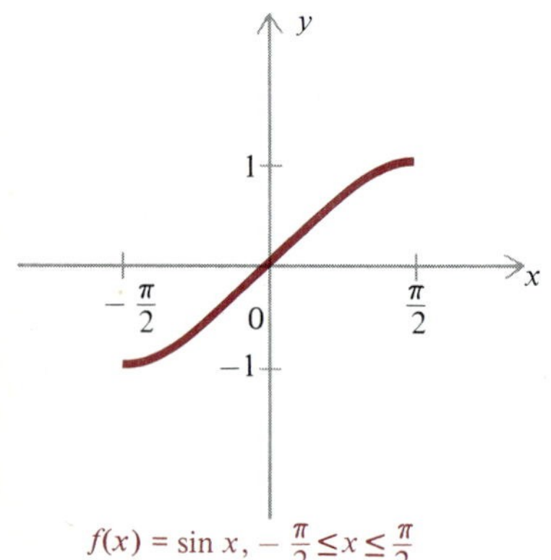

FIGURE 5.73

$$\arcsin x = y \quad \text{if and only if} \quad \sin y = x \quad \text{for } -1 \le x \le 1 \quad \text{and} \quad -\pi/2 \le y \le \pi/2 \tag{1}$$

By definition, if $-1 \le x \le 1$, then $\arcsin x$ is the unique number y between $-\pi/2$ and $\pi/2$ whose sine is x.

CAUTION: Even though we write $\sin^2 x$ for $(\sin x)^2$, we *never* write $\sin^{-1} x$ for $(\sin x)^{-1}$. To avoid any possible confusion we will use the notation $\arcsin x$ instead of $\sin^{-1} x$ for the arcsine function.

EXAMPLE 1. Evaluate the following expressions.

a. $\arcsin 0$ b. $\arcsin \dfrac{1}{2}$

c. $\arcsin 1$ d. $\arcsin\left(-\dfrac{\sqrt{3}}{2}\right)$

Solution.

a. By (1), $\arcsin 0 = y$ for the value of y in $[-\pi/2, \pi/2]$ such that $\sin y = 0$. But $\sin y = 0$ if $y = 0$. Thus $\arcsin 0 = 0$.

b. By (1), $\arcsin \frac{1}{2} = y$ for the value of y in $[-\pi/2, \pi/2]$ such that $\sin y = \frac{1}{2}$. But $\sin y = \frac{1}{2}$ if $y = \pi/6$. Thus $\arcsin \frac{1}{2} = \pi/6$.

c. By (1), $\arcsin 1 = y$ for the value of y in $[-\pi/2, \pi/2]$ such that $\sin y = 1$. But $\sin y = 1$ if $y = \pi/2$, so $\arcsin 1 = \pi/2$.

d. By (1), $\arcsin(-\sqrt{3}/2) = y$ for the value of y in $[-\pi/2, \pi/2]$ such that $\sin y = -\sqrt{3}/2$. But $\sin y = -\sqrt{3}/2$ if $y = -\pi/3$, so $\arcsin(-\sqrt{3}/2) = -\pi/3$. □

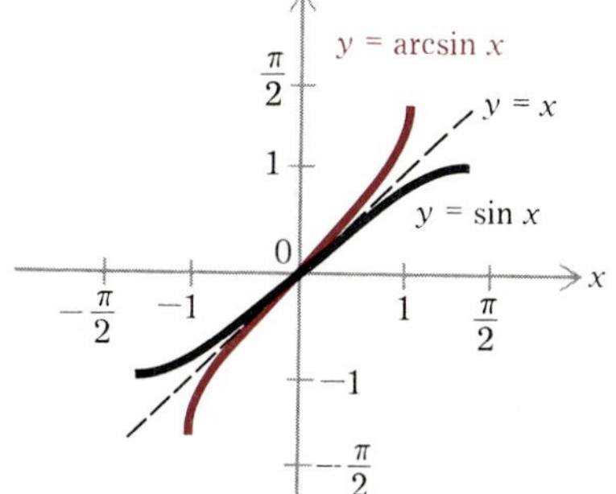

FIGURE 5.74

Since the graph of an inverse function can always be obtained from the graph of the original function by reflection through the line $y = x$, the graph of the arcsine function can be obtained from the graph of the restricted sine function in just this manner (Figure 5.74).

The fundamental formulas relating inverses have the following versions for the restricted sine function and the arcsine function:

$$\arcsin(\sin x) = x \quad \text{for} \quad -\frac{\pi}{2} \leq x \leq \frac{\pi}{2} \tag{2}$$

$$\sin(\arcsin x) = x \quad \text{for} \quad -1 \leq x \leq 1 \tag{3}$$

For example,

$$\arcsin\left(\sin \frac{\pi}{12}\right) = \frac{\pi}{12} \quad \text{and} \quad \sin\left(\arcsin \frac{1}{3}\right) = \frac{1}{3}$$

CAUTION: Formula (2) is not valid for values of x outside the interval $[-\pi/2, \pi/2]$. However, for such values of x we can still calculate $\arcsin(\sin x)$ in the following way. Find the number y in $[-\pi/2, \pi/2]$ such that

$$\sin x = \sin y$$

Then by (2),

$$\arcsin(\sin y) = y$$

so that

$$\arcsin(\sin x) = \arcsin(\sin y) = y$$

EXAMPLE 2. Evaluate $\arcsin(\sin 21\pi/5)$.

Solution. Notice that $21\pi/5$ is not in $[-\pi/2, \pi/2]$, so we cannot evaluate $\arcsin(\sin 21\pi/5)$ by using (2) directly. But since

$$\sin\frac{21\pi}{5} = \sin\left(4\pi + \frac{\pi}{5}\right) = \sin\frac{\pi}{5}$$

and $-\pi/2 < \pi/5 < \pi/2$, it follows from (2) with $x = \pi/5$ that

$$\arcsin\left(\sin\frac{21\pi}{5}\right) = \arcsin\left(\sin\frac{\pi}{5}\right) = \frac{\pi}{5} \quad \square$$

A well-chosen triangle will assist us in evaluating the expression given in the next example.

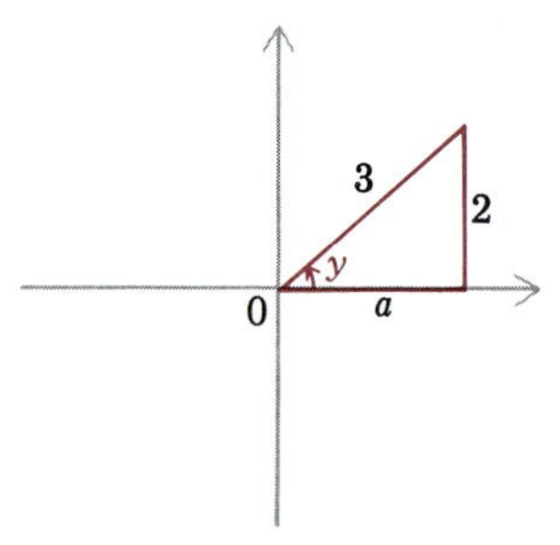

FIGURE 5.75

EXAMPLE 3. Evaluate $\tan(\arcsin \frac{2}{3})$.

Solution. We will evaluate $\tan(\arcsin \frac{2}{3})$ by evaluating $\tan y$, where y is the number in $[-\pi/2, \pi/2]$ such that $\arcsin \frac{2}{3} = y$, that is, $\sin y = \frac{2}{3}$. Since $\sin y > 0$, it follows that $0 < y < \pi/2$, and therefore $\tan y > 0$. By applying the Pythagorean Theorem to the triangle drawn in Figure 5.75, which is drawn so that $\sin y = \frac{2}{3}$, we find that

$$a = \sqrt{3^2 - 2^2} = \sqrt{9 - 4} = \sqrt{5}$$

Consequently

$$\tan y = \frac{2}{\sqrt{5}} = \frac{2}{5}\sqrt{5}$$

so that

$$\tan\left(\arcsin\frac{2}{3}\right) = \tan y = \frac{2}{5}\sqrt{5} \quad \square$$

A notable feature of Example 3 is that we did *not* need to evaluate $\arcsin \frac{2}{3}$ in order to evaluate $\tan(\arcsin \frac{2}{3})$.

The Arccosine Function

If we restrict the domain of the cosine function to $[0, \pi]$, the resulting function f is decreasing (Figure 5.76). Thus by the comments at the end of Section 2.7, f has an inverse. Since f has range $[-1, 1]$, it follows that the domain of f^{-1} is also $[-1, 1]$. The inverse f^{-1} is called the ***arccosine function*** (or inverse cosine function). Its value at x is denoted by $\arccos x$

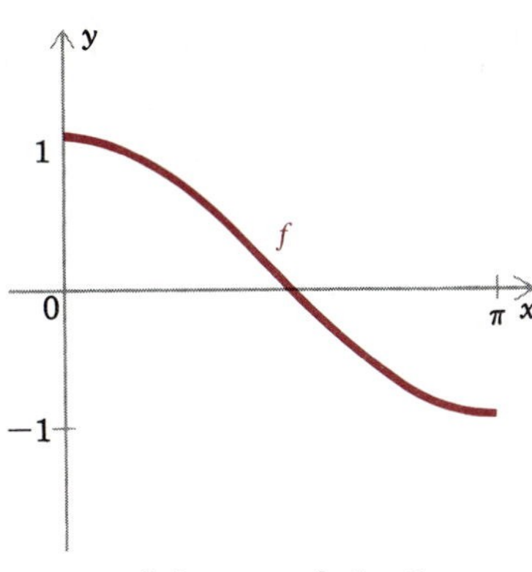

$f(x) = \cos x, 0 \leq x \leq \pi$

FIGURE 5.76

(sometimes by Arccos x or $\cos^{-1} x$). Thus by (6) in Section 2.7,

$$\arccos x = y \quad \text{if and only if} \quad \cos y = x \quad \text{for } -1 \leq x \leq 1 \quad \text{and} \quad 0 \leq y \leq \pi \tag{4}$$

By definition, if $-1 \leq x \leq 1$, then $\arccos x$ is the unique number y in $[0, \pi]$ such that $\cos y = x$.

EXAMPLE 4. Evaluate the following expressions.

a. $\arccos 1$ b. $\arccos\left(-\frac{1}{2}\right)$

Solution.

a. By (4), $\arccos 1 = y$ for the value of y in $[0, \pi]$ such that $\cos y = 1$. But $\cos y = 1$ if $y = 0$, so that $\arccos 1 = 0$.
b. By (4), $\arccos(-\frac{1}{2}) = y$ for the value of y in $[0, \pi]$ such that $\cos y = -\frac{1}{2}$. But $\cos y = -\frac{1}{2}$ if $y = 2\pi/3$, so $\arccos(-\frac{1}{2}) = 2\pi/3$. □

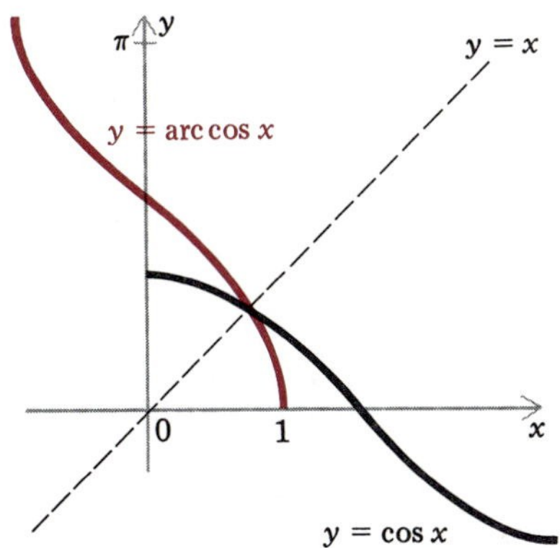

FIGURE 5.77

The graph of the arccosine function, which is obtained by reflecting the restricted cosine function through the line $y = x$, appears in Figure 5.77.

The fundamental relations between the restricted cosine function and the arccosine function are:

$$\arccos(\cos x) = x \quad \text{for} \quad 0 \leq x \leq \pi \tag{5}$$

$$\cos(\arccos x) = x \quad \text{for} \quad -1 \leq x \leq 1 \tag{6}$$

In particular,

$$\arccos\left[\cos\frac{2\pi}{3}\right] = \frac{2\pi}{3} \quad \text{and} \quad \cos\left(\arccos\frac{1}{\sqrt{5}}\right) = \frac{1}{\sqrt{5}}$$

If x lies outside $[0, \pi]$ we can evaluate $\arccos(\cos x)$ by finding the value of y in $[0, \pi]$ for which $\cos y = \cos x$. Then by (5),

$$\arccos(\cos x) = \arccos(\cos y) = y$$

EXAMPLE 5. Evaluate $\arccos\left(\cos\frac{20\pi}{9}\right)$.

Solution. We cannot evaluate $\arccos(\cos 20\pi/9)$ by invoking (5) with $x = 20\pi/9$ because $20\pi/9$ is not in $[0, \pi]$. But since $20\pi/9 = 2\pi/9 + 2\pi$ and $0 < 2\pi/9 < \pi$, we have

$$\cos\frac{20\pi}{9} = \cos\left(2\pi + \frac{2\pi}{9}\right) = \cos\frac{2\pi}{9}$$

Consequently by (5) with $x = 2\pi/9$, we conclude that

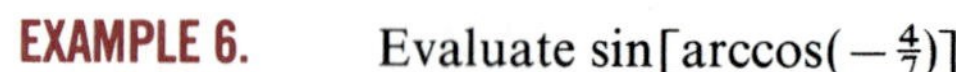

$$\arccos\left(\cos\frac{20\pi}{9}\right) = \arccos\left(\cos\frac{2\pi}{9}\right) = \frac{2\pi}{9} \quad \square$$

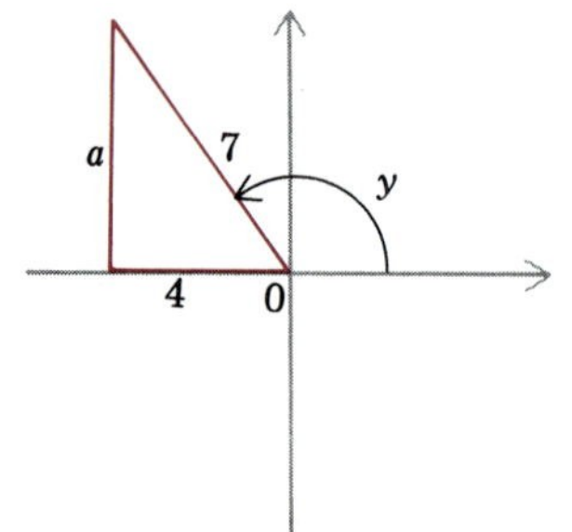

FIGURE 5.78

EXAMPLE 6. Evaluate $\sin[\arccos(-\frac{4}{7})]$.

Solution. We will evaluate $\sin[\arccos(-\frac{4}{7})]$ by evaluating $\sin y$, where y is the number in $[0, \pi]$ such that $\arccos(-\frac{4}{7}) = y$, that is, $\cos y = -\frac{4}{7}$. Since $\cos y < 0$, it follows that $\pi/2 < y < \pi$, so that $\sin y > 0$. By applying the Pythagorean Theorem to the triangle in Figure 5.78, for which $\cos y = -\frac{4}{7}$, we find that

$$a = \sqrt{7^2 - 4^2} = \sqrt{49 - 16} = \sqrt{33}$$

Therefore

$$\sin y = \frac{\sqrt{33}}{7}$$

so that

$$\sin\left[\arccos\left(-\frac{4}{7}\right)\right] = \sin y = \frac{\sqrt{33}}{7} \quad \square$$

The Arctangent Function

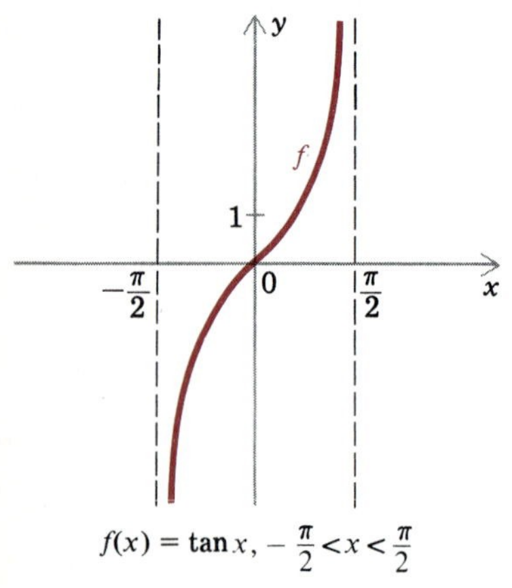

$f(x) = \tan x,\ -\frac{\pi}{2} < x < \frac{\pi}{2}$

FIGURE 5.79

If we restrict the domain of the tangent function to $(-\pi/2, \pi/2)$, the resulting function f is increasing (Figure 5.79). Thus, by the comments at the end of Section 2.7, f has an inverse. Since f has range $(-\infty, \infty)$, the inverse f^{-1} has domain $(-\infty, \infty)$. The inverse function f^{-1} is called the ***arctangent function*** (or inverse tangent function). Its value at x is denoted by $\arctan x$ (sometimes $\text{Arctan } x$ or $\tan^{-1} x$). Thus by (6) in Section 2.7,

$$\arctan x = y \quad \text{if and only if} \quad \tan y = x \quad \text{for any } x \text{ and for } -\frac{\pi}{2} < y < \frac{\pi}{2} \tag{7}$$

For any number x, $\arctan x$ is the unique number y in $(-\pi/2, \pi/2)$ such that $\tan y = x$.

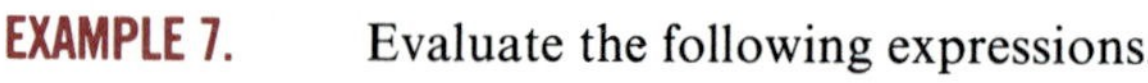

EXAMPLE 7. Evaluate the following expressions.

a. $\arctan 1$

b. $\arctan(-\sqrt{3})$

Solution.

a. By (7), $\arctan 1 = y$ for the value of y in $(-\pi/2, \pi/2)$ such that $\tan y = 1$. But $\tan \pi/4 = 1$, so that $\arctan 1 = \pi/4$.

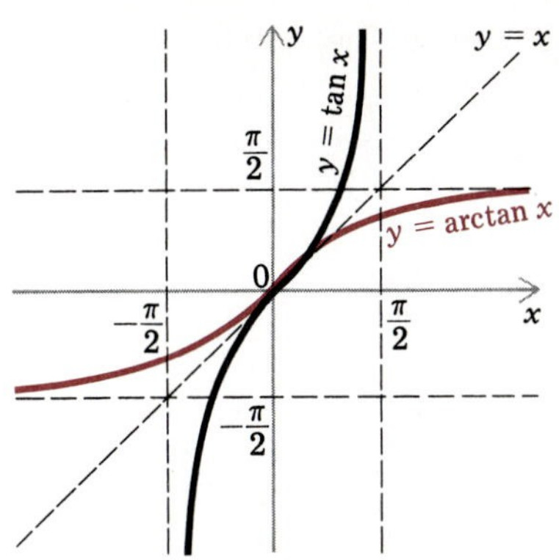

FIGURE 5.80

b. By (7), $\arctan(-\sqrt{3}) = y$ for the value of y in $(-\pi/2, \pi/2)$ such that $\tan y = -\sqrt{3}$. But $\tan(-\pi/3) = -\sqrt{3}$, so that $\arctan(-\sqrt{3}) = -\pi/3$. □

The graph of the arctangent function, which is obtained by reflecting the graph of the restricted tangent function through the line $y = x$, is shown in Figure 5.80.

We also have

$$\arctan(\tan x) = x \quad \text{for} \quad -\frac{\pi}{2} < x < \frac{\pi}{2} \tag{8}$$

$$\tan(\arctan x) = x \quad \text{for all } x \tag{9}$$

EXAMPLE 8. Evaluate $\sec(\arctan \frac{3}{5})$.

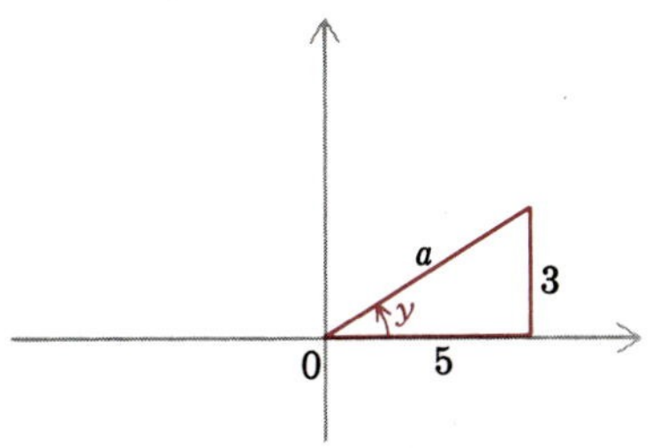

FIGURE 5.81

Solution. We will evaluate $\sec(\arctan \frac{3}{5})$ by evaluating $\sec y$ for the value of y in $(-\pi/2, \pi/2)$ such that $\arctan \frac{3}{5} = y$, that is, $\tan y = \frac{3}{5}$. Since $\tan y > 0$, we know that $0 < y < \pi/2$, so that $\sec y > 0$. Applying the Pythagorean Theorem to the triangle in Figure 5.81, for which $\tan y = \frac{3}{5}$, we find that

$$a = \sqrt{3^2 + 5^2} = \sqrt{9 + 25} = \sqrt{34}$$

Therefore
$$\sec y = \frac{\sqrt{34}}{5}$$

so that
$$\sec\left(\arctan \frac{3}{5}\right) = \sec y = \frac{\sqrt{34}}{5} \quad \square$$

EXAMPLE 9. Express $\sec(\arctan x)$ in terms of x, without the use of trigonometric functions.

Solution. First we write $\sec(\arctan x)$ as $\sec y$, where $y = \arctan x$. But $y = \arctan x$ means that $\tan y = x$ and y is in $(-\pi/2, \pi/2)$. Now x can be either positive or negative. However, either of the triangles that arise (see Figure 5.82)

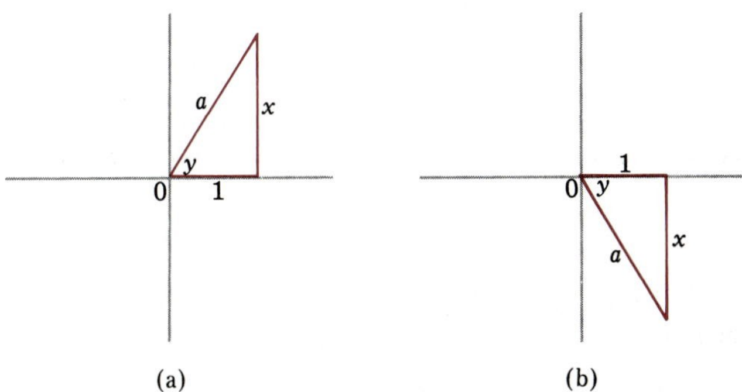

FIGURE 5.82

yields $\tan y = x$. For each triangle, the Pythagorean Theorem tells us that

$$a = \sqrt{1^2 + x^2} = \sqrt{1 + x^2}$$

Consequently

$$\sec(\arctan x) = \sec y = \frac{a}{1} = a = \sqrt{1 + x^2} \quad \square$$

The Remaining Inverse Trigonometric Functions

There are also inverses of suitable restrictions of the cotangent, secant, and cosecant functions. They are not used nearly as frequently as the three we have presented. For that reason we merely give their definitions.

$\operatorname{arccot} x = y$ if and only if $\cot y = x$

for any x and for $0 < y < \pi$

$\operatorname{arcsec} x = y$ if and only if $\sec y = x$

for $|x| \geq 1$ and for $0 \leq y \leq \pi$ except $y = \frac{\pi}{2}$

$\operatorname{arccsc} x = y$ if and only if $\csc y = x$

for $|x| \geq 1$ and for $0 < |y| \leq \frac{\pi}{2}$

Approximating Values of the Inverse Trigonometric Functions

If angles are measured in degrees rather than radians, then it is convenient to express the values of the inverse trigonometric functions in terms of degrees. In that case the values of the inverse sine and the inverse tangent functions would lie between $-90°$ and $90°$, rather than between $-\pi/2$ and $\pi/2$. Similarly, the values of the inverse cosine function would lie between $0°$ and $180°$.

EXAMPLE 10. Use a calculator to approximate arcsin 0.3456.

Solution. On our calculator, when we key in the inverse of sin and .3456, the display reads 20.21842629, which means that

$$\arcsin 0.3456 \approx (20.21842629)° \approx 20°13' \quad \square$$

EXERCISES 5.7

In Exercises 1–6, find the value of the given expression.

1. $\arcsin(-1)$

2. $\arcsin \sqrt{3}/2$

3. $\arccos 0$

4. $\arccos(-1)$

5. $\arctan 0$

6. $\arctan \sqrt{3}/3$

In Exercises 7–18, find the value of the given expression.

7. $\operatorname{arccot} 0$

8. $\operatorname{arccot} \sqrt{3}$

9. $\operatorname{arccot}(-1)$

10. $\operatorname{arccot}(-\sqrt{3}/3)$

11. $\operatorname{arcsec} 1$

12. $\operatorname{arcsec} \sqrt{2}$

13. $\operatorname{arcsec}(-2)$

14. $\operatorname{arcsec}(-\frac{2}{3}\sqrt{3})$

15. $\operatorname{arccsc} 1$

16. $\operatorname{arccsc} \frac{2}{3}\sqrt{3}$

17. $\operatorname{arccsc}(-1)$

18. $\operatorname{arccsc}(-2)$

In Exercises 19–30, use a calculator to estimate the value of the given expression.

19. $\arcsin 0.3$

20. $\arcsin(-0.84)$

21. $\arccos 0.54$

22. $\arccos(-0.17)$

23. $\arctan 0.11$

24. $\arctan(-7.81)$

25. $\operatorname{arccot} 0.94$

26. $\operatorname{arccot}(-55.5)$

27. $\operatorname{arcsec} 2.4$

28. $\operatorname{arcsec}(-1.5)$

29. $\operatorname{arccsc}(-4.7)$

30. $\operatorname{arccsc} 6.17$

In Exercises 31–42, determine the value of the given expression.

31. $\arcsin(\sin \pi/6)$

32. $\arcsin(\sin \frac{1}{5})$

33. $\arcsin(\sin 5\pi/3)$

34. $\arcsin[\sin(-7\pi/6)]$

35. $\arccos(\cos 2\pi/3)$

36. $\arccos(\cos \pi)$

37. $\arccos(\cos 19\pi/4)$

38. $\arccos[\cos(-8\pi/3)]$

39. $\arctan(\tan \pi/13)$

40. $\arctan[\tan(-5\pi/6)]$

41. $\operatorname{arccot}[\cot(-19\pi/6)]$

42. $\operatorname{arcsec}(\sec 23\pi/4)$

In Exercises 43–46, determine the value of the given expression.

43. $\arcsin(\cos \pi/3)$

44. $\arccos(\sin \pi/4)$

45. $\arctan(\cos \pi/2)$

46. $\arctan(\cos 8\pi)$

In Exercises 47–60, determine the value of the given expression.

47. $\sin(\arcsin \frac{1}{4})$

48. $\sin(\arccos \frac{1}{2})$

49. $\sin[\arctan(-\sqrt{3}/3)]$

50. $\cos(\arcsin 1)$

51. $\cos[\arctan(-1)]$

52. $\tan[\arccos(-1)]$

53. $\tan[\arcsin(-\sqrt{2}/2)]$

54. $\sin(\arccos \frac{1}{6})$

55. $\cos[\arctan(-\frac{4}{3})]$

56. $\cos(\operatorname{arccsc} \frac{9}{2})$

57. $\tan[\arcsin(-\frac{4}{5})]$

58. $\tan[\operatorname{arcsec}(-3)]$

59. $\sec[\arcsin(-\frac{1}{3})]$

60. $\csc[\arctan(-14)]$

In Exercises 61–64, determine the value of the given expression.

61. $\arcsin[\tan(\arccos \sqrt{2}/2)]$

62. $\arccos\{\sin[\arctan(-1)]\}$

63. $\arctan\{\frac{2}{3}\sin[\arccos(-\frac{1}{2})]\}$

64. $\arccos[\frac{1}{2}\tan(\arcsin \sqrt{3}/2)]$

In Exercises 65–70, verify the given equation.

65. $\sin(\arccos x) = \sqrt{1-x^2}$ (*Hint:* See Example 9, but here let $\arccos x = y$, so that $\cos y = x$. Then express $\sin y$ in terms of x.)

66. $\sin(\arctan x) = \dfrac{x}{\sqrt{1+x^2}}$

67. $\sec(\arctan x) = \sqrt{1+x^2}$

68. $\tan(\arcsin x) = \dfrac{x}{\sqrt{1-x^2}}$

69. $\cos(\arcsin x^2) = \sqrt{1-x^4}$

70. $\tan\left(\arccos \dfrac{1}{\sqrt{x}}\right) = \sqrt{x-1}$

In Exercises 71–73, sketch the graph of the function.

71. $|\arcsin x|$

72. $-2 + \arccos x$

73. $\arctan |x|$

In Exercises 74–78, verify the equation.

74. $\arcsin(-x) = -\arcsin x$ (*Hint:* Let $\arcsin(-x) = z$, and show that $\arcsin x = -z$.)

75. $\arctan(-x) = -\arctan x$

76. $\arcsin x = \arctan \dfrac{x}{\sqrt{1-x^2}}$

77. $\arcsin x + \arccos x = \pi/2$ (*Hint:* Let $\arcsin x = z$, and show that $\arccos x = \pi/2 - z$.)

78. $\arctan x + \operatorname{arccot} x = \pi/2$

79. a. C. L. Dodgson (better known as Lewis Carroll) proved that if a, b, and c are constants such that

$bc = 1 + a^2$, then

$$\arctan \frac{1}{a+b} + \arctan \frac{1}{a+c} = \arctan \frac{1}{a}$$

Taking $a = b = 1$ and $c = 2$, show that

$$\arctan \frac{1}{2} + \arctan \frac{1}{3} = \frac{\pi}{4}$$

b. Corroborate part (a) by computing

$$\frac{\pi}{4} - (\arctan \tfrac{1}{2} + \arctan \tfrac{1}{3})$$

on a calculator.

KEY TERMS

angle
- initial side
- terminal side
- vertex
- standard position
- coterminal angles

degree measure
radian measure

trigonometric functions
- sine function
- cosine function
- tangent function
- cotangent function
- secant function
- cosecant function

Pythagorean Identity

reference angle
periodic function
sinusoidal function
- phase shift
- amplitude

inverse trigonometric function
- arcsine function
- arccosine function
- arctangent function

KEY FORMULAS

$$\theta^\circ = \left(\frac{\pi}{180} \cdot \theta\right) \text{radians}$$

$$\theta \text{ radians} = \left(\frac{180}{\pi} \cdot \theta\right)^\circ$$

$$\sin \theta = \frac{b}{c} \qquad \cot \theta = \frac{\cos \theta}{\sin \theta} = \frac{a}{b}$$

$$\cos \theta = \frac{a}{c} \qquad \sec \theta = \frac{1}{\cos \theta} = \frac{c}{a}$$

$$\tan \theta = \frac{\sin \theta}{\cos \theta} = \frac{b}{a} \qquad \csc \theta = \frac{1}{\sin \theta} = \frac{c}{b}$$

$$\sin^2 \theta + \cos^2 \theta = 1$$

$$\tan^2 \theta + 1 = \sec^2 \theta$$

$$1 + \cot^2 \theta = \csc^2 \theta$$

For any integer n:

$$\sin(\theta + 2n\pi) = \sin \theta \qquad \cot(\theta + n\pi) = \cot \theta$$

$$\cos(\theta + 2n\pi) = \cos \theta \qquad \sec(\theta + 2n\pi) = \sec \theta$$

$$\tan(\theta + n\pi) = \tan \theta \qquad \csc(\theta + 2n\pi) = \csc \theta$$

$\arcsin x = y$ if and only if $\sin y = x$

$\arccos x = y$ if and only if $\cos y = x$

$\arctan x = y$ if and only if $\tan y = x$

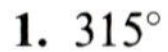

REVIEW EXERCISES

In Exercises 1–6, draw the angle in standard position having the given measure.

1. $315°$

2. $-390°$

3. $960°$

4. $\frac{8}{3}\pi$ radians

5. $-\frac{11}{4}\pi$ radians

6. 7π radians

In Exercises 7–22, determine the value of the expression.

7. $\sin(-19\pi/4)$

8. $\sin 13\pi/2$

9. $\cos 29\pi/6$

10. $\cos(-13\pi/3)$

11. $\tan 15\pi/4$

12. $\tan(-17\pi/6)$

13. $\cot 7\pi/3$

14. $\cot(-11\pi/4)$

15. $\sec 13\pi/6$

16. $\sec(-23\pi/4)$

17. $\csc(44\pi/3)$

18. $\csc(-23\pi/2)$

19. $\sin 930°$

20. $\cos(-585°)$

21. $\tan 1110°$

22. $\sec(-750°)$

In Exercises 23–28, find the values of the six trigonometric functions at θ.

23. $\sin\theta = -\frac{2}{5}$, $\pi < \theta < 3\pi/2$

24. $\cos\theta = 0.9$, $-2\pi < \theta < -3\pi/2$

25. $\tan\theta = -8$, $\pi/2 < \theta < \pi$

26. $\cot\theta = -\frac{2}{3}$, $-\pi/2 < \theta < 0$

27. $\sec\theta = 4$, $4\pi < \theta < 9\pi/2$

28. $\csc\theta = -\frac{4}{3}$, $-\pi < \theta < -\pi/2$

In Exercises 29–34, use a calculator to approximate the values of the given expressions.

29. $\sin 14°20'$

30. $\sin 62°50'$

31. $\cos 47°36'$

32. $\cos 185°15'$

33. $\tan 8°8'$

34. $\tan(-22°48')$

In Exercises 35–38, solve the right triangle with the given data pertaining to Figure 5.83.

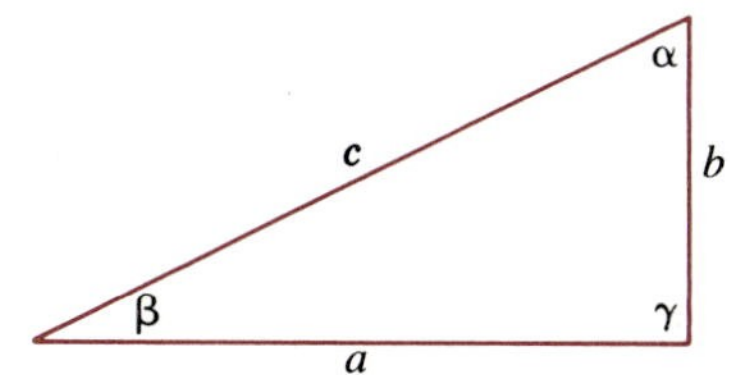

FIGURE 5.83

35. $\alpha = 10°$, $b = 3$

36. $\beta = 29.1°$, $c = 2.1$

37. $a = 6.7$, $b = 5$

38. $a = 4.9$, $c = 17$

39. Use the information in Figure 5.84 to determine the values of all six trigonometric functions at θ.

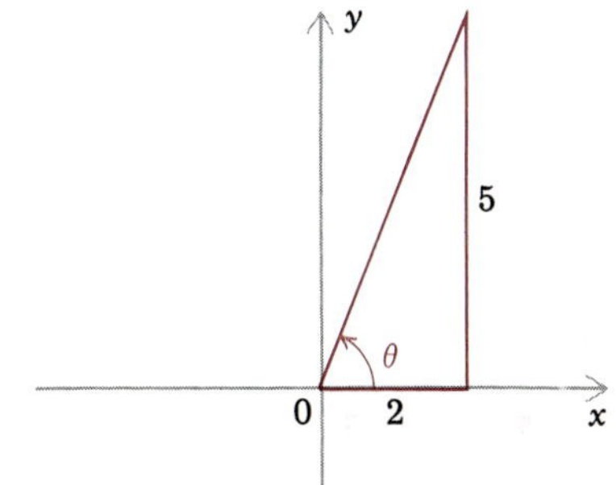

FIGURE 5.84

In Exercises 40–42, sketch the graph.

40. $1 + \sin x$

41. $\cos(x + \pi/6)$

42. $|\tan x|$

In Exercises 43–56, find the value of the given expression.

43. $\arccos \frac{1}{2}$

44. $\arcsin(-\frac{1}{2})$

45. $\arctan\sqrt{3}/3$

46. $\arctan(-1)$

47. $\arcsin(\sin 8\pi/3)$

48. $\arccos[\cos(-\pi/2)]$

49. $\arctan(\tan 7\pi/4)$

50. $\arctan[\tan(-1.7)]$

51. $\arcsin(\tan \pi/4)$

52. $\arccos(\sec \pi)$

53. $\arctan(2 \sin \pi/6)$

54. $\sin(\text{arccsc } 3)$

55. $\cos[\text{arccot}(-2)]$

56. $\tan[\text{arcsec}(-\pi/2)]$

57. For each of the numbers $x = 0.1, 0.01$, and 0.001 (in radians), calculate the value of the given expression. Then guess the limiting value of the expression as x approaches 0.

a. $\dfrac{\sin x}{x}$

b. $\dfrac{\cos x - 1}{x}$

c. $\dfrac{\tan x}{x}$

d. $x \tan^2 x$

58. For $x = 0.1, 0.01$, and 0.001 (in radians), compare the values of $\sin x$ and $x - \frac{1}{6}x^3$. (Often $x - \frac{1}{6}x^3$ is used as an approximate value of $\sin x$ for small values of x.)

59. For $x = 0.1, 0.01$, and 0.001 (in radians), compare the values of $\cos x$ and $1 - \frac{1}{2}x^2$. (Often $1 - \frac{1}{2}x^2$ is used as an approximate value of $\cos x$ for small values of x.)

60. Show that $\pi \sin \pi/6 + \arcsin 1 = \pi$.

61. Prove that $\sec(x + \pi) = -\sec x$.

62. Explain why the equation $\sin x \cos x = 1$ has no solution.

63. If the hour hand on a kitchen clock is 4 inches long and the minute hand is 6 inches long, how much farther does the tip of the minute hand travel during a 12-hour span than the tip of the hour hand?

64. When fully inflated, a tire on an automobile is 1 foot in radius. Through how many revolutions does the tire travel in a mile?

65. Chris sees a bolt of lightning strike a tower on the ground some 7000 feet away. The angle of elevation of the cloud is 40°. How high is the cloud?

66. At the time when the moon is half-moon, the moon angle in Figure 5.85 is 90°. Suppose the earth angle is calculated to be 89.85°. Determine how many times as far away from the earth the sun is as the moon.

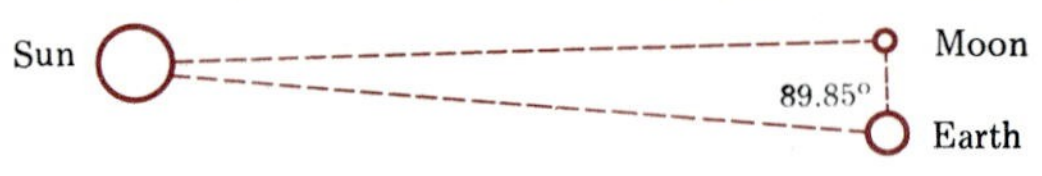

FIGURE 5.85

67. A ski lift is divided into two parts. The first part is 1600 feet long and makes an angle of elevation of 25° with the horizontal. The second part is 1100 feet long and makes an angle of elevation of 40° with the horizontal. Determine the straight-line distance covered by the two parts of the ski lift together.

68. A pilot is instructed by the control tower to rise from ground level to an altitude of 30,000 feet by the time it has flown 10 miles. If the pilot is to achieve the goal by flying along a straight line,

a. at what angle would it rise?

b. what ground distance would the plane travel during the interim?

69. About 200 B.C., Eratosthenes of Alexandria, Egypt, devised a way of approximating the circumference of the earth. From astronomical data available to him as head librarian in Alexandria he knew that at noon on a particular June day the sun would be directly overhead in Syene, Egypt. By direct measurement he found that at the same time the sun was approximately 7°12′ from the vertical in Alexandria (Figure 5.86). Using 490 miles as the distance between Syene and Alexandria, along with (3) of Section 5.1, determine an approximate value for the earth's circumference. Round your answer to the nearest mile.

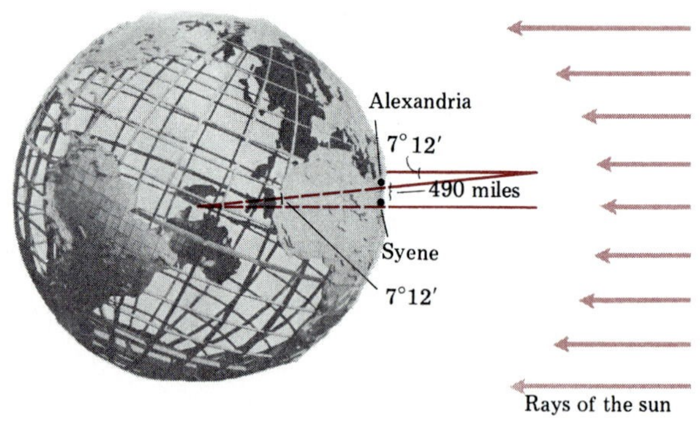

The measurement of the circumference of the earth by Eratosthenes

FIGURE 5.86

70. Suppose the angle of elevation of the sun is 30. Use Snell's Law (see Section 5.2) to find the angle θ at which a swimmer under water must look up in order to see the sun (Figure 5.87). Take the index of refraction μ to be 1.333.

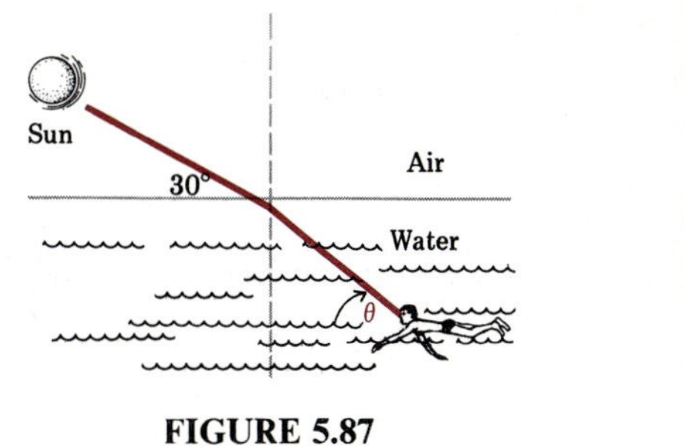

FIGURE 5.87

6 TRIGONOMETRIC IDENTITIES AND EQUATIONS

In Chapter 1 we studied equations involving polynomials, and in Chapter 4 we considered equations involving logarithms and exponentials. In this chapter we study equations involving trigonometric functions. Those equations are of two types. The first type consists of equations such as

$$\sin^2 x + \cos^2 x = 1 \quad \text{and} \quad \tan(x + y) = \frac{\tan x + \tan y}{1 - \tan x \tan y}$$

that are valid for all values of the variables for which every expression in the equation is defined. Such an equation is a ***trigonometric identity***, and the values of the variables for which a trigonometric identity is defined are its ***admissible values***. The initial four sections of the chapter will be devoted to trigonometric identities, which we will usually refer to just as identities. As an application of the trigonometric identities, we will show that identity (5) in Section 6.4 can be used to help us understand the nature of musical notes produced by stringed instruments such as pianos, violins, or guitars, and in particular the nature of overtones.

The other type of equation we will study in this chapter consists of ***conditional trigonometric equations***, which are valid only for isolated values of the variable. They are exemplified by

$$\cos 2x = \cos x \quad \text{and} \quad \tan^2 x - \sec x - 1 = 0$$

In the final section of the chapter we will discuss solutions of conditional trigonometric equations.

6.1 EXAMPLES OF TRIGONOMETRIC IDENTITIES

We have already encountered a number of trigonometric identities in Chapter 5. Those that will play a prominent role in this chapter are the following:

Defining Identities

$$\tan x = \frac{\sin x}{\cos x} \qquad \sec x = \frac{1}{\cos x} \qquad \cot x = \frac{1}{\tan x}$$

$$\cot x = \frac{\cos x}{\sin x} \qquad \csc x = \frac{1}{\sin x}$$

Symmetry Identities

$\sin(-x) = -\sin x$	$\cot(-x) = -\cot x$
$\cos(-x) = \cos x$	$\sec(-x) = \sec x$
$\tan(-x) = -\tan x$	$\csc(-x) = -\csc x$

Pythagorean Identities

$$\sin^2 x + \cos^2 x = 1$$

$$\tan^2 x + 1 = \sec^2 x$$

$$\cot^2 x + 1 = \csc^2 x$$

Each of the Defining Identities except $\cot x = 1/\tan x$ is actually the definition of the function on the left side of the equation, and this one is an immediate consequence of the definitions of $\tan x$ and $\cot x$. The Symmetry Identities follow readily from the definitions of the functions involved and are closely related to the graphs of the respective functions. Finally, the Pythagorean Identities first appeared in (5)–(7) of Section 5.2.

In each of the examples that follow we will be asked to verify a proposed identity, by which we mean to demonstrate that the given equation is valid for all admissible values of the variable. The verification usually will involve the use of identities listed above, together with algebraic manipulation.

A basic method we will employ in verifying identities involves altering only *one* side of the proposed identity and showing, perhaps in several steps, that this side is equal to the other side of the proposed identity. Normally we will pick the more complicated side for alteration. Examples 1–4 illustrate this method.

EXAMPLE 1. Verify $\cos^4 x - \sin^4 x = \cos^2 x - \sin^2 x$.

Solution. We alter the left side by factoring it and then using the first Pythagorean Identity, $\sin^2 x + \cos^2 x = 1$:

$$\begin{aligned}\cos^4 x - \sin^4 x &= (\cos^2 x - \sin^2 x)\overbrace{(\cos^2 x + \sin^2 x)}^{=1} \\ &= \cos^2 x - \sin^2 x\end{aligned}$$

We have thus obtained the right side of the proposed identity, so the solution is complete. □

EXAMPLE 2. Verify $\dfrac{1 - \tan t}{1 + \tan t} = \dfrac{\cot t - 1}{\cot t + 1}$.

Solution. We alter the right side by using the identity $\cot t = 1/\tan t$:

$$\begin{aligned}\frac{\cot t - 1}{\cot t + 1} &= \frac{\dfrac{1}{\tan t} - 1}{\dfrac{1}{\tan t} + 1} \\ &= \frac{\dfrac{1 - \tan t}{\tan t}}{\dfrac{1 + \tan t}{\tan t}} = \frac{1 - \tan t}{1 + \tan t}\end{aligned}$$

Since the final fraction on the right is the left side of the proposed identity, the solution is complete. □

EXAMPLE 3. Verify $\dfrac{\sec^4 x - 1}{\tan^2 x} = \tan^2 x + 2$.

Solution. Again we alter the left side, but this time we use the second Pythagorean Identity, $\tan^2 x + 1 = \sec^2 x$:

$$\begin{aligned}\frac{\sec^4 x - 1}{\tan^2 x} &= \frac{(\sec^2 x)^2 - 1}{\tan^2 x} \\ &= \frac{(\tan^2 x + 1)^2 - 1}{\tan^2 x} \\ &= \frac{(\tan^4 x + 2\tan^2 x + 1) - 1}{\tan^2 x} \\ &= \frac{\tan^4 x + 2\tan^2 x}{\tan^2 x} = \tan^2 x + 2\end{aligned}$$

Since we have obtained the right side of the proposed identity, the solution is finished. □

Another way of altering one side of an identity is to write all expressions in terms of sines and cosines and then rearrange as needed.

EXAMPLE 4. Verify $\dfrac{\cot x - \tan x}{\cos x + \sin x} = \dfrac{\cos x - \sin x}{\sin x \cos x}$.

Solution. We start by writing the expressions on the left side in terms of sine and cosine, and only then do we simplify:

$$\begin{aligned}
\frac{\cot x - \tan x}{\cos x + \sin x} &= \frac{\dfrac{\cos x}{\sin x} - \dfrac{\sin x}{\cos x}}{\cos x + \sin x} \\
&= \frac{\dfrac{\cos^2 x - \sin^2 x}{\sin x \cos x}}{\cos x + \sin x} \\
&= \frac{\cos^2 x - \sin^2 x}{\sin x \cos x\,(\cos x + \sin x)} \\
&= \frac{(\cos x - \sin x)(\cos x + \sin x)}{\sin x \cos x\,(\cos x + \sin x)} \\
&= \frac{\cos x - \sin x}{\sin x \cos x}
\end{aligned}$$

Again we have obtained the right side of the proposed identity, so the identity is verified. □

When a proposed identity contains fractions, the verification may be facilitated by clearing fractions.

EXAMPLE 5. Verify $\dfrac{1 - \sin x}{\cos x} = \dfrac{1}{\sec x + \tan x}$.

Solution. In order to clear fractions, we multiply both sides of the given identity by $(\cos x)(\sec x + \tan x)$:

$$\frac{1 - \sin x}{\cos x}(\cos x)(\sec x + \tan x) = \frac{1}{\sec x + \tan x}(\cos x)(\sec x + \tan x)$$

which reduces to

$$(1 - \sin x)(\sec x + \tan x) = \cos x \tag{1}$$

Now we alter the left side of (1) with the help of the Defining Identities and the first Pythagorean Identity written in the form $\cos^2 x = 1 - \sin^2 x$:

$$\begin{aligned}(1 - \sin x)(\sec x + \tan x) &= (1 - \sin x)\left(\frac{1}{\cos x} + \frac{\sin x}{\cos x}\right) \\ &= (1 - \sin x)\left(\frac{1 + \sin x}{\cos x}\right) \\ &= \frac{1 - \sin^2 x}{\cos x} \\ &= \frac{\cos^2 x}{\cos x} \\ &= \cos x\end{aligned}$$

Therefore (1), and hence the given identity, is verified. □

There are numerous ways of verifying proposed identities. Although one method alone will frequently not suffice, some of the more common and effective methods include:

1. **1.** Altering one side of the proposed identity by using one or more of the identities appearing at the outset of this section.
2. **2.** Writing all expressions of the proposed identity in terms of sines and cosines.
3. **3.** Clearing any fractions that appear in the proposed identity.

EXERCISES 6.1

In Exercises 1–6, verify the identity.

1. $\sin x \csc x = 1$

2. $\tan x \cot x = 1$

3. $\cot x = \csc x \cos x$

4. $\sin x = \dfrac{\tan x}{\sec x}$

5. $\tan x = \dfrac{\sec x}{\csc x}$

6. $\csc x = \cot x \sec x$

In Exercises 7–22, verify the identity.

7. $(\sec t - 1)(\sec t + 1) = \tan^2 t$

8. $\dfrac{\cos t - \sin t}{\cos t} = 1 - \tan t$

9. $\dfrac{\sin z}{\csc z} + \dfrac{\cos z}{\sec z} = 1$

10. $\tan^2 z + \sec^2 z = 2\sec^2 z - 1$

11. $\cot^2 z + \csc^2 z + 1 = 2\csc^2 z$

12. $\sec^4 z - \tan^4 z = \sec^2 z + \tan^2 z$

13. $\sin x(\csc x - \sin x) = \cos^2 x$

14. $(\cos x + \sin x)^2 + (\cos x - \sin x)^2 = 2$

15. $(\cos^2 x \sin x)^2 + (\sin^2 x \cos x)^2 = \sin^2 x \cos^2 x$

16. $(1 - \cos t)^2 + \sin^2 t = 2(1 - \cos t)$

17. $(\cos t - t\sin t)^2 + (\sin t + t\cos t)^2 = 1 + t^2$

18. $\cot^2 t + \cos^2 t = -\sin^2 t + \csc^2 t$

19. $(1 - \sin^2 t)\sec t = \cos t$

20. $4\sin^2 w + 9\cos^2 w = 4 + 5\cos^2 w$

21. $6\cos^2 w + 8\sin^2 w = 6 + 2\sin^2 w$

22. $(1 + \tan w)^2 = \sec^2 w + 2\tan w$

In Exercises 23–42, verify the identity.

23. $\dfrac{\sec^2 x}{\sec^2 x - 1} = \csc^2 x$

24. $\dfrac{\csc x}{\sec x} + \dfrac{\cos x}{\sin x} = 2\cot x$

25. $\dfrac{\sin x}{1 - \cos x} = \dfrac{1 + \cos x}{\sin x}$

26. $\dfrac{\tan x}{\sec x - 1} = \dfrac{\sec x + 1}{\tan x}$

27. $\dfrac{1}{\csc x + \cot x} = \csc x - \cot x$

28. $\dfrac{1 - \tan^2 x}{1 - \cot^2 x} = 1 - \sec^2 x$

29. $\dfrac{1}{\tan^2 x} - \dfrac{1}{\sec^2 x} = \dfrac{\cos^4 x}{\sin^2 x}$

30. $\dfrac{\cos x - \csc x}{\sec x - \sin x} + \cot x = 0$

31. $\dfrac{\sec x - \cos x}{\sec x + \cos x} = \dfrac{\sin^2 x}{1 + \cos^2 x}$

32. $\dfrac{\cos x}{\cos x - \sin x} = \dfrac{1}{1 - \tan x}$

33. $\dfrac{1}{\tan x + \sin x} = \dfrac{\cos x \csc x}{1 + \cos x}$

34. $\dfrac{\cos x}{1 + \sin x} + \dfrac{1 + \sin x}{\cos x} = 2\sec x$

35. $\dfrac{1}{1 + \sin x} + \dfrac{1}{1 - \sin x} = 2\sec^2 x$

36. $\dfrac{1 + \tan z}{\sec z} = \dfrac{\cot z + 1}{\csc z}$

37. $\dfrac{\sin^2 z}{\sec z - 1} = \cos^2 z + \cos z$

38. $\dfrac{1 + \sec z}{\sin z + \tan z} = \csc z$

39. $\sec^2 t \csc^2 t = \sec^2 t + \csc^2 t$

40. $\sec t - \cos t = \sin t \tan t$

41. $\tan t + \cot t = \sec t \csc t$

42. $\sec t + \csc t = (\tan t + \cot t)(\sin t + \cos t)$

In Exercises 43–56, verify the identity.

43. $\tan^2 x - \sin^2 x = \tan^2 x \sin^2 x$

44. $\dfrac{1}{\sec x - \cos x} = \csc x \cot x$

45. $\dfrac{1 + \sec x}{\sec x} = \dfrac{\sin^2 x}{1 - \cos x}$

46. $\sec x - \tan x = \dfrac{\cos x}{1 + \sin x}$

47. $\dfrac{1 - \cos x}{\csc x} = \dfrac{\sin^3 x}{1 + \cos x}$

48. $\dfrac{1}{1 + \cos x} = \csc^2 x - \csc x \cot x$

49. $(\tan x + \cot x)(\sec x - \cos x) = \sec x \tan x$

50. $\dfrac{\sin^3 x - \cos^3 x}{\sin x - \cos x} = 1 + \sin x \cos x$

51. $\dfrac{1 - \sin x}{1 + \sin x} = (\sec x - \tan x)^2$

52. $\dfrac{\cot x - \tan x}{\cot x + \tan x} = \cos^2 x - \sin^2 x$

53. $\log(\sec x + \tan x) + \log(\sec x - \tan x) = 0$

54. $\log(\csc x + \cot x) + \log(\csc x - \cot x) = 0$

55. $\dfrac{\sin x + \cos y}{\cos x + \sin y} = \dfrac{\cos x - \sin y}{\cos y - \sin x}$

56. $\dfrac{\tan x + \tan y}{\cot x + \cot y} = \dfrac{\tan x \tan y - 1}{1 - \cot x \cot y}$

In Exercises 57–62, show that the equation is *not* a trigonometric identity by finding an admissible value of x for which the equation is not valid.

57. $\cos x \stackrel{?}{=} 1 - \sin x$

58. $1 + \sec^2 x \stackrel{?}{=} \tan^2 x$

59. $\dfrac{\cos x}{1 + \sin x} \stackrel{?}{=} \dfrac{1 + \sin x}{\cos x}$

60. $(\sin x + \cos x)^2 \stackrel{?}{=} \sin^2 x + \cos^2 x$

61. $\tan x \stackrel{?}{=} \sqrt{\tan^2 x}$

62. $\sin x \stackrel{?}{=} \sqrt{1 - \cos^2 x}$

In Exercises 63–64, sketch the graph of the function.

63. $\dfrac{\cos x}{1 + \sin x} + \dfrac{1 + \sin x}{\cos x}$ (*Hint:* Use Exercise 34.)

64. $(1 + \tan x)^2 - \sec^2 x$

6.2 THE ADDITION FORMULAS

Sometimes we encounter trigonometric expressions such as $\sin(x + y)$, $\cos(x - y)$, and $\tan(x - y)$ that involve the sum or difference of two numbers x and y. The following identities relate such expressions to combinations of the more basic expressions $\sin x$, $\sin y$, $\cos x$, $\cos y$, $\tan x$, and $\tan y$:

Addition Formulas

$$\cos(x + y) = \cos x \cos y - \sin x \sin y \qquad (1)$$

$$\cos(x - y) = \cos x \cos y + \sin x \sin y \qquad (2)$$

$$\sin(x + y) = \sin x \cos y + \cos x \sin y \qquad (3)$$

$$\sin(x - y) = \sin x \cos y - \cos x \sin y \qquad (4)$$

$$\tan(x + y) = \frac{\tan x + \tan y}{1 - \tan x \tan y} \qquad (5)$$

$$\tan(x - y) = \frac{\tan x - \tan y}{1 + \tan x \tan y} \qquad (6)$$

The identities in (1)–(6) are called the Addition Formulas for trigonometric functions. (Even though three of the identities feature the difference $x - y$, one can think of $x - y$ as the sum $x + (-y)$.) We will prove (2) first, then (1), and finally (3)–(6).

The Addition Formulas for the Cosine

To prove (2), we will use the fact that two angles of the same measure subtend chords of the same length on the unit circle. Thus the length of the chord joining P and Q in Figure 6.1a is equal to the length of the chord joining $(1, 0)$ and R in Figure 6.1b, since both are chords of angles of measure $x - y$. Using

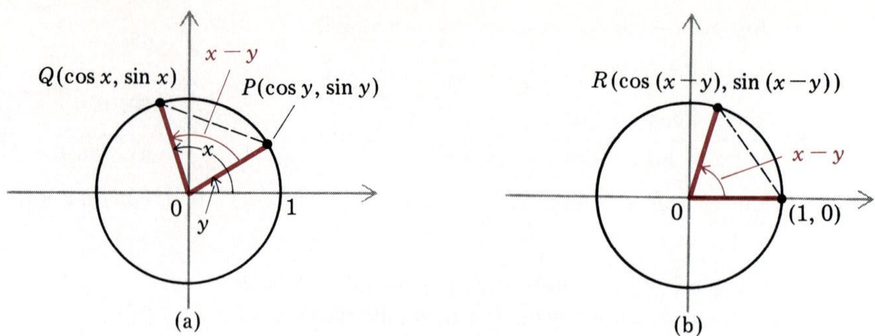

FIGURE 6.1

the coordinates of P, Q, and R appearing in Figure 6.1a, b, we find that

$$\overbrace{(\cos(x-y)-1)^2+(\sin(x-y)-0)^2}^{\text{square of distance between (1, 0) and } R} = \overbrace{(\cos x-\cos y)^2+(\sin x-\sin y)^2}^{\text{square of distance between } P \text{ and } Q}$$

Squaring out the binomials and using the first Pythagorean Identity of Section 6.1, we obtain

$$\begin{aligned} \cos^2(x-y) - 2\cos(x-y) + 1 + \sin^2(x-y) &\\ = \cos^2 x - 2\cos x\cos y + \cos^2 y &+ \sin^2 x - 2\sin x\sin y + \sin^2 y \\ 1 - 2\cos(x-y) + 1 &= 1 - 2\cos x\cos y + 1 - 2\sin x\sin y \\ -2\cos(x-y) &= -2\cos x\cos y - 2\sin x\sin y \\ \cos(x-y) &= \cos x\cos y + \sin x\sin y \end{aligned}$$

The last equation is (2).

Addition Formula (1) follows from (2) by substituting $-y$ for y in (2) and recalling the Symmetry Identities $\cos(-y) = \cos y$ and $\sin(-y) = -\sin y$:

$$\begin{aligned} \cos(x+y) &= \cos(x-(-y)) \\ &\overset{(2)}{=} \cos x\cos(-y) + \sin x\sin(-y) \\ &= \cos x\cos y - \sin x\sin y \end{aligned}$$

For a typical application of Addition Formulas (1) and (2), we will use the fact that $7\pi/12 = 3\pi/12 + 4\pi/12 = \pi/4 + \pi/3$ to write $\cos 7\pi/12$, whose value is not immediately recognizable, as $\cos(\pi/4 + \pi/3)$, and then apply (1).

EXAMPLE 1. Evaluate $\cos \dfrac{7\pi}{12}$.

Solution. Since

$$\frac{7\pi}{12} = \frac{3\pi}{12} + \frac{4\pi}{12} = \frac{\pi}{4} + \frac{\pi}{3}$$

we use (1) with $x = \pi/4$ and $y = \pi/3$:

$$\begin{aligned}\cos \frac{7\pi}{12} &= \cos\left(\frac{\pi}{4} + \frac{\pi}{3}\right)\\ &\overset{(1)}{=} \cos\frac{\pi}{4}\cos\frac{\pi}{3} - \sin\frac{\pi}{4}\sin\frac{\pi}{3}\\ &= \left(\frac{\sqrt{2}}{2}\right)\left(\frac{1}{2}\right) - \left(\frac{\sqrt{2}}{2}\right)\left(\frac{\sqrt{3}}{2}\right)\\ &= \frac{1}{4}(\sqrt{2} - \sqrt{6}) \quad \square\end{aligned}$$

EXAMPLE 2. Evaluate $\cos \dfrac{\pi}{12}$.

Solution. Notice that

$$\frac{\pi}{12} = \frac{4\pi}{12} - \frac{3\pi}{12} = \frac{\pi}{3} - \frac{\pi}{4}$$

Therefore we can apply (2) with $x = \pi/3$ and $y = \pi/4$ to obtain

$$\begin{aligned}\cos \frac{\pi}{12} &= \cos\left(\frac{\pi}{3} - \frac{\pi}{4}\right)\\ &\overset{(2)}{=} \cos\frac{\pi}{3}\cos\frac{\pi}{4} + \sin\frac{\pi}{3}\sin\frac{\pi}{4}\\ &= \left(\frac{1}{2}\right)\left(\frac{\sqrt{2}}{2}\right) + \left(\frac{\sqrt{3}}{2}\right)\left(\frac{\sqrt{2}}{2}\right)\\ &= \frac{1}{4}(\sqrt{2} + \sqrt{6}) \quad \square\end{aligned}$$

The Addition Formulas are valid when x and y represent degrees.

EXAMPLE 3. Evaluate $\cos 75°$.

Solution. Since $75° = 30° + 45°$, it follows from (1) that

$$\begin{aligned}\cos 75° &= \cos(30° + 45°)\\ &\overset{(1)}{=} \cos 30° \cos 45° - \sin 30° \sin 45°\\ &= \left(\frac{\sqrt{3}}{2}\right)\left(\frac{\sqrt{2}}{2}\right) - \left(\frac{1}{2}\right)\left(\frac{\sqrt{2}}{2}\right)\\ &= \frac{1}{4}(\sqrt{6} - \sqrt{2}) \quad \square\end{aligned}$$

We could also have evaluated $\cos 75°$ by first noticing that $75° = 120° - 45°$ and then using (2) instead of (1).

Next we will use Addition Formula (2) in order to prove two trigonometric identities relating sines and cosines.

EXAMPLE 4. Prove the following formulas.

a. $\sin x = \cos\left(\frac{\pi}{2} - x\right)$ b. $\cos x = \sin\left(\frac{\pi}{2} - x\right)$

Solution.

a. Replacing x by $\pi/2$ and y by x in (2), we obtain

$$\begin{aligned}\cos\left(\frac{\pi}{2} - x\right) &= \cos\frac{\pi}{2}\cos x + \sin\frac{\pi}{2}\sin x\\ &= 0\,(\cos x) + 1\,(\sin x)\\ &= \sin x\end{aligned}$$

b. Replacing x by $\pi/2 - x$ in part (a), we find that

$$\cos\left[\frac{\pi}{2} - \left(\frac{\pi}{2} - x\right)\right] = \sin\left(\frac{\pi}{2} - x\right)$$

or equivalently,

$$\cos x = \sin\left(\frac{\pi}{2} - x\right) \quad \square$$

The Addition Formulas for the Sine

Now we turn to the Addition Formulas for $\sin(x + y)$ and $\sin(x - y)$. In the proof of (3) we will use (2), along with the identities derived in Example 4:

$$
\begin{aligned}
\sin(x+y) &= \cos\left[\frac{\pi}{2}-(x+y)\right] \\
&= \cos\left[\left(\frac{\pi}{2}-x\right)-y\right] \\
&\overset{(2)}{=} \cos\left(\frac{\pi}{2}-x\right)\cos y + \sin\left(\frac{\pi}{2}-x\right)\sin y \\
&= \sin x \cos y + \cos x \sin y
\end{aligned}
$$

Formula (3) is thus verified. Formula (4) is verified by replacing y by $-y$ in (3) and using the Symmetry Identities (see Exercise 65).

EXAMPLE 5. Evaluate $\sin \dfrac{11\pi}{12}$.

Solution. We observe that

$$\frac{11\pi}{12} = \frac{8\pi}{12} + \frac{3\pi}{12} = \frac{2\pi}{3} + \frac{\pi}{4}$$

and then use (3):

$$
\begin{aligned}
\sin\frac{11\pi}{12} &= \sin\left(\frac{2\pi}{3}+\frac{\pi}{4}\right) \\
&\overset{(3)}{=} \sin\frac{2\pi}{3}\cos\frac{\pi}{4} + \cos\frac{2\pi}{3}\sin\frac{\pi}{4} \\
&= \left(\frac{\sqrt{3}}{2}\right)\left(\frac{\sqrt{2}}{2}\right) + \left(-\frac{1}{2}\right)\left(\frac{\sqrt{2}}{2}\right) \\
&= \frac{1}{4}(\sqrt{6}-\sqrt{2}) \quad \square
\end{aligned}
$$

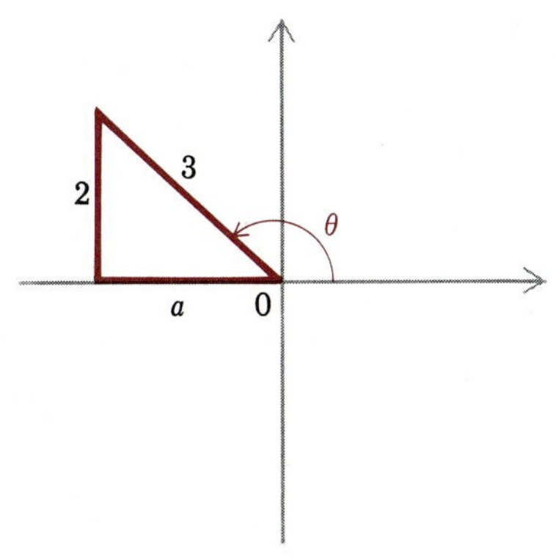

FIGURE 6.2

EXAMPLE 6. Assume that $\sin\theta° = \frac{2}{3}$ and $90 < \theta < 180$. Evaluate $\sin(60° - \theta°)$.

Solution. To find $\sin(60° - \theta°)$ we will use (4). But in order to do so we must find $\cos\theta°$. Since $\sin\theta° = \frac{2}{3}$ and $90 < \theta < 180$, we deduce that $\cos\theta° < 0$ and that the length a in Figure 6.2 is given by

$$a = \sqrt{3^2 - 2^2} = \sqrt{5}$$

Therefore

$$\cos \theta^\circ = -\frac{\sqrt{5}}{3}$$

Now we use (4) with $x = 60^\circ$ and $y = \theta^\circ$ to conclude that

$$\begin{aligned}\sin(60^\circ - \theta^\circ) &= \sin 60^\circ \cos \theta^\circ - \cos 60^\circ \sin \theta^\circ \\ &= \left(\frac{\sqrt{3}}{2}\right)\left(-\frac{\sqrt{5}}{3}\right) - \left(\frac{1}{2}\right)\left(\frac{2}{3}\right) \\ &= -\frac{\sqrt{15}}{6} - \frac{1}{3} \quad \square\end{aligned}$$

EXAMPLE 7. Prove the formula $\sin(\pi - x) = \sin x$.

Solution. Replacing x and y in (4) by π and x, we obtain

$$\begin{aligned}\sin(\pi - x) &= \sin \pi \cos x - \cos \pi \sin x \\ &= 0(\cos x) - (-1) \sin x \\ &= \sin x \quad \square\end{aligned}$$

The Addition Formula

$$\sin(x + y) = \sin x \cos y + \cos x \sin y \tag{7}$$

can help us sketch the graph of a function of the form $p \sin x + q \cos x$ when $p \neq 0$ and $q \neq 0$. First we will find appropriate values of a and c so that

$$p \sin x + q \cos x = a \sin(x + c) \quad \text{for all } x$$

Since we know by Section 5.6 how to sketch the graph of any function of the form $a \sin(x + c)$, once we know a and c we can graph $p \sin x + q \cos x$. Let us see how all this is accomplished in a concrete example.

EXAMPLE 8. Sketch the graph of $\sin x - \cos x$.

Solution. We will first determine numbers a and c so that

$$\sin x - \cos x = a \sin(x + c) \tag{8}$$

By (7) we can write (8) as

$$\begin{aligned}\sin x - \cos x &= a \sin x \cos c + a \cos x \sin c \\ &= (a \cos c) \sin x + (a \sin c) \cos x\end{aligned}$$

Evidently these equations will be valid for all values of x if

$$a \cos c = 1 \quad \text{and} \quad a \sin c = -1 \tag{9}$$

Now we use (9) to calculate the values of a and c. To determine a we use the Pythagorean Theorem and (9) to obtain

$$\begin{aligned} a^2 &= a^2(\cos^2 c + \sin^2 c) = (a \cos c)^2 + (a \sin c)^2 \\ &= 1^2 + (-1)^2 = 2 \end{aligned}$$

It follows that $a = \sqrt{2}$ or $a = -\sqrt{2}$. For simplicity we take $a = \sqrt{2}$. To find c we observe from (9) that

$$\frac{a \sin c}{a \cos c} = \frac{-1}{1}$$

so that

$$\tan c = -1$$

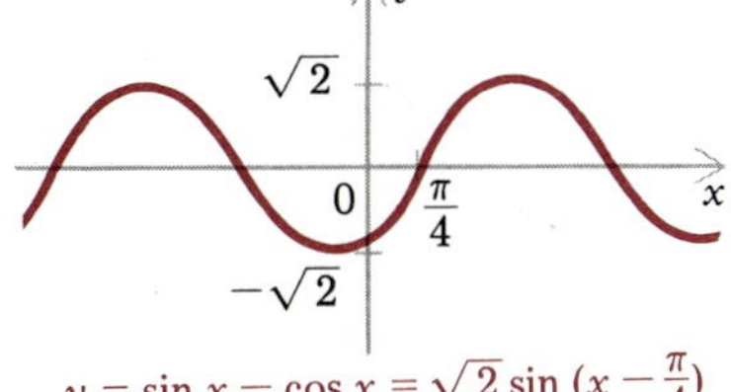

$y = \sin x - \cos x = \sqrt{2} \sin (x - \frac{\pi}{4})$

FIGURE 6.3

Thus we could take $c = 3\pi/4$ or $c = -\pi/4$. But from the two formulas in (9) and the fact that $a > 0$, we know that $\cos c > 0$ and $\sin c < 0$. This implies that $c = -\pi/4$. Substituting our values of a and c into (9), we conclude that

$$\sin x - \cos x = \sqrt{2} \sin \left(x - \frac{\pi}{4} \right)$$

Consequently the graph of $\sin x - \cos x$ is the same as the graph of $\sqrt{2} \sin(x - \pi/4)$, which appears in Figure 6.3. □

If we wish to determine the graph of $p \sin x + q \cos x$ for other values of p and q, we can follow the same procedure as we used in Example 8 in order to find values of a and c so that

$$p \sin x + q \cos x = a \sin (x + c) \tag{10}$$

It turns out that (10) is valid for all values of x if

$$p = a \cos c \text{ and } q = a \sin c \tag{11}$$

If we want a to be positive, then

$$a = \sqrt{p^2 + q^2} \quad \text{and} \quad \tan c = \frac{q}{p}$$

(provided $p \neq 0$). The given values of p and q, along with signs of $\sin c$ and $\cos c$ determined by (11), then yield the values of a and c for (10).

A physical interpretation of (10) is that any motion given by $p \sin x + q \cos x$, where x represents time, is simple harmonic motion, as described in Section 5.6.

The Addition Formulas for the Tangent

Addition Formula (5) for $\tan(x + y)$ is derived from (1) and (3) by using the fact that the tangent is the quotient of the sine and cosine and by dividing both numerator and denominator by $\cos x \cos y$:

$$\tan(x + y) = \frac{\sin(x + y)}{\cos(x + y)}$$

$$= \frac{\sin x \cos y + \cos x \sin y}{\cos x \cos y - \sin x \sin y}$$

$$= \frac{\dfrac{\sin x \cos y}{\cos x \cos y} + \dfrac{\cos x \sin y}{\cos x \cos y}}{\dfrac{\cos x \cos y}{\cos x \cos y} - \dfrac{\sin x \sin y}{\cos x \cos y}}$$

$$= \frac{\dfrac{\sin x}{\cos x} + \dfrac{\sin y}{\cos y}}{1 - \dfrac{\sin x}{\cos x}\dfrac{\sin y}{\cos y}} = \frac{\tan x + \tan y}{1 - \tan x \tan y}$$

Thus (5) is proved. Formula (6) can be proved from (5) by replacing y by $-y$ (see Exercise 66).

EXAMPLE 9. Evaluate $\tan \dfrac{11\pi}{12}$.

Solution. Since

$$\frac{11\pi}{12} = \frac{8\pi}{12} + \frac{3\pi}{12} = \frac{2\pi}{3} + \frac{\pi}{4}$$

we find that

$$\tan \frac{11\pi}{12} = \tan\left(\frac{2\pi}{3} + \frac{\pi}{4}\right) \overset{(5)}{=} \frac{\tan 2\pi/3 + \tan \pi/4}{1 - \tan 2\pi/3 \tan \pi/4}$$

$$= \frac{-\sqrt{3} + 1}{1 - (-\sqrt{3})(1)} = \frac{1 - \sqrt{3}}{1 + \sqrt{3}} \quad \square$$

EXAMPLE 10. Assume that $\sin x = \frac{3}{5}$ with $\pi/2 < x < \pi$, and $\cos y = -\frac{1}{4}$ with $\pi < y < 3\pi/2$. Evaluate $\tan(x - y)$.

Solution. We will use (6). But to do so we need to determine the values of tan x and tan y. From the hypotheses that $\sin x = \frac{3}{5}$ and $\pi/2 < x < \pi$ we deduce that $\tan x < 0$ and that the length a in Figure 6.4 is given by

$$a = \sqrt{5^2 - 3^2} = 4$$

Therefore

$$\tan x = -\frac{3}{4}$$

In a similar way, we deduce that $\tan y > 0$ and that the length a in Figure 6.5 is given by

$$a = \sqrt{4^2 - 1^2} = \sqrt{15}$$

Therefore

$$\tan y = \sqrt{15}$$

Consequently by (6),

$$\tan(x - y) = \frac{\tan x - \tan y}{1 + \tan x \tan y} = \frac{-\frac{3}{4} - \sqrt{15}}{1 + (-\frac{3}{4})(\sqrt{15})}$$

$$= \frac{-3 - 4\sqrt{15}}{4 - 3\sqrt{15}} \quad \square$$

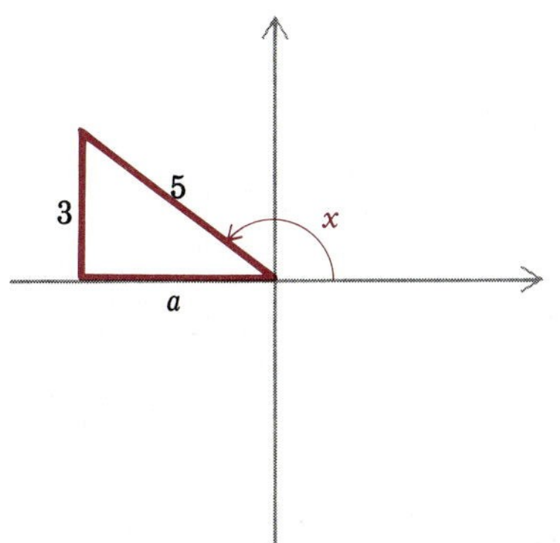

FIGURE 6.4

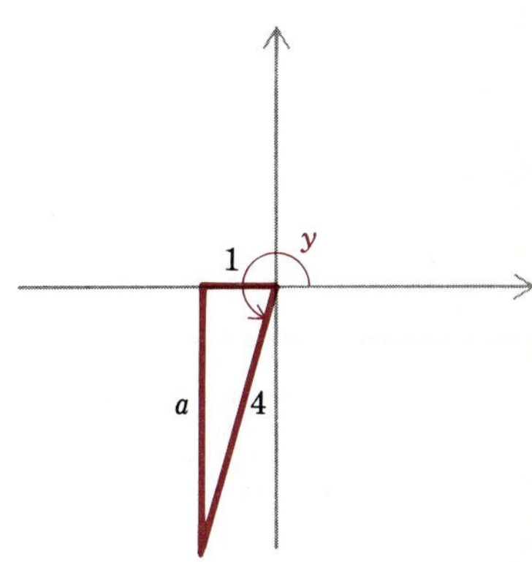

FIGURE 6.5

EXERCISES 6.2

In Exercises 1–12, evaluate the given expression.

1. $\sin \frac{\pi}{12}$

2. $\sin \frac{7\pi}{12}$

3. $\sin(-165^\circ)$

4. $\sin 465^\circ$

5. $\cos\left(-\frac{5\pi}{12}\right)$

6. $\cos \frac{13\pi}{12}$

7. $\cos 255^\circ$

8. $\cos 105^\circ$

9. $\tan \frac{\pi}{12}$

10. $\tan \frac{17\pi}{12}$

11. $\tan(-285^\circ)$

12. $\tan 465^\circ$

13. Suppose that $\sin x = -\frac{1}{4}$ with $\pi < x < 3\pi/2$. Find $\sin(\pi/6 + x)$.

14. Suppose that $\cos x = -\frac{4}{7}$ with $\pi/2 < x < \pi$. Find $\cos(\pi/4 - x)$.

15. Suppose that $\sin x = \frac{1}{5}$ with $0 < x < \pi/2$, and $\cos y = \frac{1}{4}$ with $3\pi/2 < y < 2\pi$. Find $\sin(x + y)$.

16. Suppose that $\tan x = \frac{1}{2}$ and $\sec y = 3$ with $0 < y < \pi/2$. Find $\tan(x + y)$.

17. Suppose that $\sin(x + y) = \frac{1}{3}$ with $0 < x + y < \pi/2$, and $\cos y = \frac{1}{4}$ with $0 < y < \pi/2$. Find $\sin x$.

18. Suppose that $\sin x = \frac{2}{3}$ with $0 < x < \pi/2$, and $\cos(x - y) = \frac{4}{5}$ with $0 < x - y < \pi/2$. Find $\cos y$.

In Exercises 19–30, use the Addition Formulas to verify the identity.

19. $\sin(\pi + x) = -\sin x$

20. $\cos(\pi + x) = -\cos x$

21. $\cos(\pi - x) = -\cos x$

22. $\tan(\pi - x) = -\tan x$

23. $\sin\left(\frac{\pi}{2} + x\right) = \cos x$

24. $\cos\left(\frac{\pi}{2} + x\right) = -\sin x$

25. $\sin\left(\frac{\pi}{2} + x\right) = \sin\left(\frac{\pi}{2} - x\right)$

26. $\cos\left(\frac{\pi}{2} + x\right) = -\cos\left(\frac{\pi}{2} - x\right)$

27. $\sin\left(\frac{\pi}{3} + x\right) = \frac{1}{2}(\sqrt{3} \cos x + \sin x)$

28. $\cos\left(\frac{\pi}{3} + x\right) = \frac{1}{2}(\cos x - \sqrt{3} \sin x)$

29. $\sin\left(\frac{\pi}{4} - x\right) = \frac{\sqrt{2}}{2}(\cos x - \sin x)$

30. $\cos\left(\frac{\pi}{4} + x\right) = \frac{\sqrt{2}}{2}(\cos x - \sin x)$

In Exercises 31–52, verify the identity.

31. $\sin(x + y) + \sin(x - y) = 2 \sin x \cos y$

32. $\cos(x + y) + \cos(x - y) = 2 \cos x \cos y$

33. $\sin(x + y) \sin(x - y) = \sin^2 x \cos^2 y - \cos^2 x \sin^2 y$

34. $\sin(x + y) \sin(x - y) = \sin^2 x - \sin^2 y$

35. $\cos(x + y) \cos(x - y) = \cos^2 x - \sin^2 y$

36. $(\cos x - \sin y)^2 + (\sin x + \cos y)^2 = 2 + 2 \sin(x - y)$

37. $(\cos x - \cos y)^2 + (\sin x + \sin y)^2 = 2 - 2 \cos(x + y)$

38. $\frac{\sin x}{\sin y} + \frac{\cos x}{\cos y} = \csc y \sec y \sin (x + y)$

39. $\frac{\sin(x + y)}{\cos x \cos y} = \tan x + \tan y$

40. $\frac{\cos(x - y)}{\cos x \sin y} = \cot y + \tan x$

41. $\frac{\cos(x + y)}{\sin(x - y)} = \frac{\cot x - \tan y}{1 - \cot x \tan y}$

42. $\frac{\sin(x + y)}{\cos(x - y)} = \frac{\cot x + \cot y}{\cot x \cot y + 1}$

43. $\frac{\sin(x + y)}{\sin(x - y)} = \frac{\tan x + \tan y}{\tan x - \tan y}$

44. $\frac{\cos(x + y)}{\cos(x - y)} = \frac{1 - \tan x \tan y}{1 + \tan x \tan y}$

45. $\sin^2 (\pi/4 + x) - \sin^2 (\pi/4 - x) = \sin 2x$

46. $\frac{\tan(\pi/4 - x)}{\tan(\pi/4 + x)} = \frac{(1 - \tan x)^2}{(1 + \tan x)^2}$

47. $\tan(x+y) = \dfrac{\cot x + \cot y}{\cot x \cot y - 1}$

48. $\tan(x-y) - \tan(y-x) = \dfrac{2(\tan x - \tan y)}{1 + \tan x \tan y}$

49. $\cot(x+y) = \dfrac{\cot x \cot y - 1}{\cot x + \cot y}$

50. $\cot(x-y) = \dfrac{\cot x \cot y + 1}{\cot y - \cot x}$

51. $\sec(x+y) = \dfrac{\csc x \csc y}{\cot x \cot y - 1}$

52. $\sec(x-y) = \dfrac{\sec x \sec y}{1 + \tan x \tan y}$

53. Verify that

$$\frac{\sin(x+h) - \sin x}{h} = \cos x \frac{\sin h}{h} - \sin x \frac{1 - \cos h}{h}$$

(This formula arises in calculus.)

54. Verify that

$$\frac{\cos(x+h) - \cos x}{h} = -\sin x \frac{\sin h}{h} - \cos x \frac{1 - \cos h}{h}$$

(This formula arises in calculus.)

In Exercises 55–58, show that the equation is *not* a trigonometric identity by finding admissible values of x and y for which the equation is not valid.

55. $\sin(x+y) \stackrel{?}{=} \sin x + \sin y$

56. $\cos(x+y) \stackrel{?}{=} \cos x + \cos y$

57. $\tan(x+y) \stackrel{?}{=} \tan x + \tan y$

58. $\sec(x+y) \stackrel{?}{=} \sec x + \sec y$

In Exercises 59–64, sketch the graph of the function.

59. $\sin x + \cos x$

60. $\sin x - \sqrt{3} \cos x$

61. $2\sqrt{3} \sin x + 6 \cos x$

62. $\cos x \cos 3x - \sin x \sin 3x$

63. $\cos x \cos 3x + \sin x \sin 3x$

64. $\cos \frac{3}{5}x \cos \frac{7}{5}x - \sin \frac{3}{5}x \sin \frac{7}{5}x$

65. Prove formula (4) from (3) by replacing y with $-y$.

66. Prove formula (6) from (5) by replacing y with $-y$.

6.3 MULTIPLE-ANGLE FORMULAS

From the Addition Formulas presented in the preceding section we will obtain formulas for $\sin 2x$, $\sin 3x$, $\sin \frac{1}{2}x$, and indeed the sines of many other multiples of x, in terms of $\sin x$ and $\cos x$. Similar results will appear for the cosine and tangent functions.

The Double-Angle Formulas

The following formulas are called Double-Angle Formulas:

Double-Angle Formulas

$$\sin 2x = 2 \sin x \cos x \tag{1}$$

$$\cos 2x = \cos^2 x - \sin^2 x \tag{2}$$

$$\tan 2x = \frac{2 \tan x}{1 - \tan^2 x} \tag{3}$$

Notice that the angle on the left side of each formula is $2x$, rather than x.

Formula (1) is proved by letting $y = x$ in the Addition Formula (3) of Section 6.2:

$$\sin 2x = \sin(x + x) = \sin x \cos x + \cos x \sin x = 2 \sin x \cos x$$

Formulas (2) and (3) are verified by similar methods (see Exercises 56 and 57).

EXAMPLE 1. Suppose that $\cos x = \frac{2}{3}$ and $3\pi/2 < x < 2\pi$. Evaluate the following expressions.

a. $\sin 2x$ b. $\cos 2x$ c. $\tan 2x$

Solution.

a. In order to use (1) we need to know the value of $\sin x$ as well as $\cos x$. Since $3\pi/2 < x < 2\pi$ and $\cos x = \frac{2}{3}$, it follows that $\sin x < 0$ and that the length a in Figure 6.6 is given by

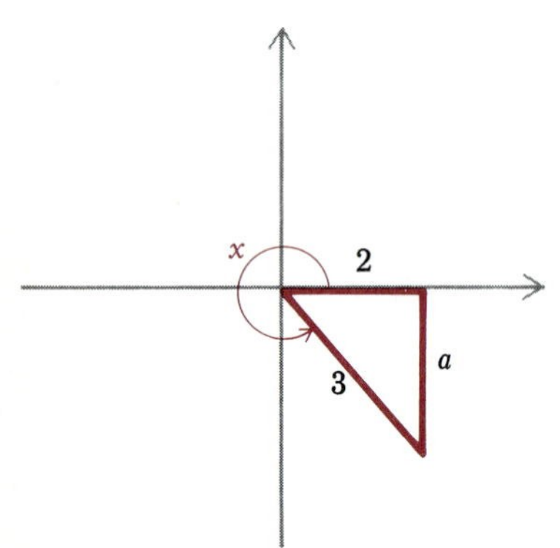

FIGURE 6.6

$$a = \sqrt{3^2 - 2^2} = \sqrt{5}$$

Therefore

$$\sin x = -\frac{\sqrt{5}}{3}$$

Consequently (1) implies that

$$\sin 2x = 2 \sin x \cos x = 2\left(-\frac{\sqrt{5}}{3}\right)\left(\frac{2}{3}\right) = -\frac{4}{9}\sqrt{5}$$

b. Using (2), the given value of $\cos x$, and the value of $\sin x$ determined in part (a), we have

$$\cos 2x = \cos^2 x - \sin^2 x = \left(\frac{2}{3}\right)^2 - \left(-\frac{\sqrt{5}}{3}\right)^2 = \frac{4}{9} - \frac{5}{9} = -\frac{1}{9}$$

c. The values we have of $\sin 2x$ and $\cos 2x$ yield

$$\tan 2x = \frac{\sin 2x}{\cos 2x} = \frac{-\frac{4}{9}\sqrt{5}}{-\frac{1}{9}} = 4\sqrt{5} \quad \square$$

EXAMPLE 2. Find a formula for $\sin 3x$ in terms of $\sin x$ and $\cos x$.

Solution. If we think of $3x$ as $2x + x$, then Addition Formula (3) in Section 6.2 yields

$$\sin 3x = \sin(2x + x) = \sin 2x \cos x + \cos 2x \sin x$$

By virtue of Double-Angle Formulas (1) and (2),

$$\begin{aligned}\sin 2x \cos x + \cos 2x \sin x &= (2 \sin x \cos x)\cos x + (\cos^2 x - \sin^2 x)\sin x \\ &= 2 \sin x \cos^2 x + \cos^2 x \sin x - \sin^3 x \\ &= 3 \sin x \cos^2 x - \sin^3 x\end{aligned}$$

Therefore

$$\sin 3x = 3 \sin x \cos^2 x - \sin^3 x \quad \square$$

Using the same method, one could find a formula for $\sin 4x$, or even $\sin nx$ for any positive integer n, in terms of $\sin x$ and $\cos x$. The same holds true for $\cos nx$ and $\tan nx$.

Formula (2) can be modified to yield the auxiliary formulas

$$\sin^2 x = \frac{1 - \cos 2x}{2} \tag{4}$$

$$\cos^2 x = \frac{1 + \cos 2x}{2} \tag{5}$$

which are of interest in their own right. To prove (5) we start with (2) and utilize the first Pythagorean Identity written in the form $\sin^2 x = 1 - \cos^2 x$:

$$\cos 2x = \cos^2 x - \sin^2 x = \cos^2 x - (1 - \cos^2 x)$$

so that

$$\cos 2x = 2 \cos^2 x - 1 \tag{6}$$

Therefore

$$\cos^2 x = \frac{1 + \cos 2x}{2}$$

which is (5). Formula (4) can be proved either from (2) or from (5) (see Exercise 58).

As a byproduct of the auxiliary formulas (4) and (5), the graphs of $\sin^2 x$ and $\cos^2 x$ are now very accessible, for they are simple modifications of the graph of $\cos 2x$ (see Figure 6.7a, b on the following page). Were we to graph $\sin^2 x$ (or $\cos^2 x$) by plotting points and using various properties of $\sin x$ (or $\cos x$), it would be much more tedious.

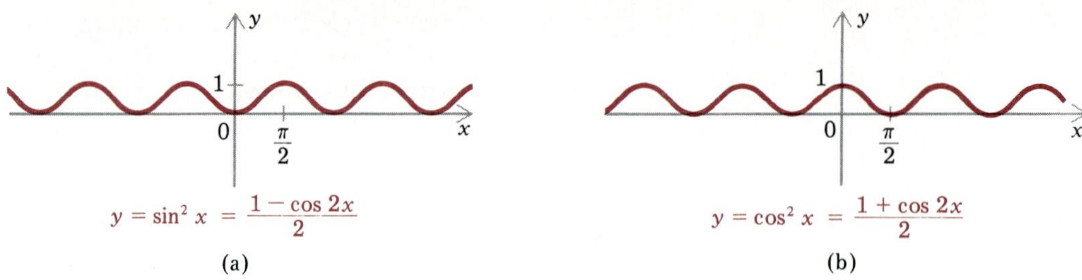

FIGURE 6.7

The Half-Angle Formulas

Formulas (4) and (5) give rise to formulas for $\sin \frac{1}{2}x$ and $\cos \frac{1}{2}x$. We first replace x by y and take square roots in (4) to obtain

$$\sin y = \sqrt{\frac{1 - \cos 2y}{2}} \quad \text{or} \quad -\sqrt{\frac{1 - \cos 2y}{2}}$$

depending on whether $\sin y \geq 0$ or $\sin y < 0$. Likewise, taking square roots in (5), we obtain

$$\cos y = \sqrt{\frac{1 + \cos 2y}{2}} \quad \text{or} \quad -\sqrt{\frac{1 + \cos 2y}{2}}$$

depending on whether $\cos y \geq 0$ or $\cos y < 0$. Then we let $y = \frac{1}{2}x$ in these formulas to obtain the so-called Half-Angle Formulas:

Half-Angle Formulas

$$\sin \frac{1}{2} x = \begin{cases} \sqrt{\dfrac{1 - \cos x}{2}} & \text{if } \sin \frac{1}{2} x \geq 0 \\ -\sqrt{\dfrac{1 - \cos x}{2}} & \text{if } \sin \frac{1}{2} x < 0 \end{cases} \tag{7}$$

$$\cos \frac{1}{2} x = \begin{cases} \sqrt{\dfrac{1 + \cos x}{2}} & \text{if } \cos \frac{1}{2} x \geq 0 \\ -\sqrt{\dfrac{1 + \cos x}{2}} & \text{if } \cos \frac{1}{2} x < 0 \end{cases} \tag{8}$$

CAUTION: Care must be taken in applying (7) or (8) in order to ensure that the correct sign is obtained. Recall that $\sin \frac{1}{2}x$ is nonnegative when the terminal side of the angle in standard position associated with $\frac{1}{2}x$ lies in the first or second quadrant and is negative otherwise. Similarly, $\cos \frac{1}{2}x$ is positive when the terminal side of the angle associated with $\frac{1}{2}x$ lies in the first or fourth quadrants and is negative otherwise.

EXAMPLE 3. Evaluate $\sin \dfrac{\pi}{8}$.

Solution. We have

$$\frac{\pi}{8} = \frac{1}{2} \cdot \frac{\pi}{4}$$

and $\cos \pi/4 = \sqrt{2}/2$. Since $0 < \pi/8 < \pi/2$, it follows that $\sin \pi/8 > 0$. Now we calculate $\sin \pi/8$ by using (7):

$$\sin \frac{\pi}{8} = \sin\left(\frac{1}{2} \cdot \frac{\pi}{4}\right) = \sqrt{\frac{1 - \cos \dfrac{\pi}{4}}{2}}$$

$$= \sqrt{\frac{1 - \dfrac{\sqrt{2}}{2}}{2}} = \sqrt{\frac{2 - \sqrt{2}}{4}} = \frac{1}{2}\sqrt{2 - \sqrt{2}} \quad \square$$

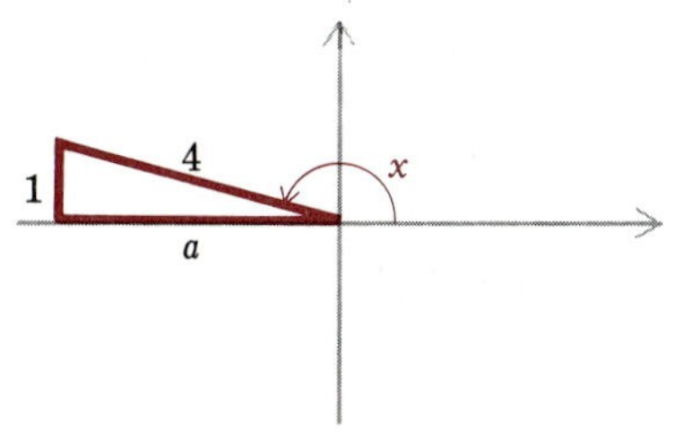

FIGURE 6.8

EXAMPLE 4. Suppose that $\sin x = \frac{1}{4}$ and $\pi/2 < x < \pi$. Evaluate $\cos \frac{1}{2}x$.

Solution. In order to use (8), we first need to find the value of $\cos x$. Since $\pi/2 < x < \pi$ and $\sin x = \frac{1}{4}$, it follows that $\cos x < 0$ and the length a in Figure 6.8 is given by

$$a = \sqrt{4^2 - 1^2} = \sqrt{15}$$

Therefore

$$\cos x = -\frac{\sqrt{15}}{4}$$

Because

$$\frac{\pi}{4} < \frac{1}{2}x < \frac{\pi}{2}$$

it follows that $\cos \frac{1}{2}x > 0$, so that by (8),

$$\cos \frac{1}{2}x = \sqrt{\frac{1 + \cos x}{2}} = \sqrt{\frac{1 - \dfrac{\sqrt{15}}{4}}{2}} = \sqrt{\frac{4 - \sqrt{15}}{8}} \quad \square$$

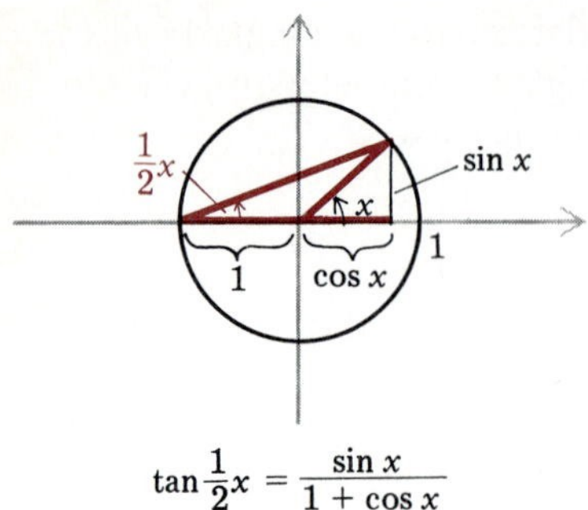

$\tan\frac{1}{2}x = \frac{\sin x}{1 + \cos x}$

FIGURE 6.9

To derive a half-angle formula for the tangent function, we use (1) and (5), with x replaced by y:

$$\tan y = \frac{\sin y}{\cos y} = \frac{\sin y}{\cos y}\,\frac{2\cos y}{2\cos y}$$

$$= \frac{2\sin y\cos y}{2\cos^2 y} = \frac{\sin 2y}{2\left(\dfrac{1+\cos 2y}{2}\right)} = \frac{\sin 2y}{1+\cos 2y}$$

Letting $\frac{1}{2}x$ replace y now yields the desired Half-Angle Formula for the tangent:

$$\boxed{\tan\frac{1}{2}x = \frac{\sin x}{1+\cos x}} \tag{9}$$

In case $0 < x < \pi/2$, formula (9) can be proved geometrically (Figure 6.9).

EXAMPLE 5. Evaluate tan 67.5°.

Solution. Observe that $67.5° = [\frac{1}{2}(135)]°$. Since

$$\sin 135° = \frac{\sqrt{2}}{2} \quad \text{and} \quad \cos 135° = -\frac{\sqrt{2}}{2}$$

it follows from (9) that

$$\tan 67.5° = \tan\left(\frac{1}{2}(135)\right)^{\circ} = \frac{\sin 135°}{1+\cos 135°}$$

$$= \frac{\dfrac{\sqrt{2}}{2}}{1-\dfrac{\sqrt{2}}{2}} = \frac{\sqrt{2}}{2-\sqrt{2}} \qquad \square$$

EXERCISES 6.3

In Exercises 1–6, find sin $2x$, cos $2x$, and tan $2x$ under the given conditions.

1. $\sin x = -\frac{1}{3}$ and $3\pi/2 < x < 2\pi$
2. $\cos x = \frac{3}{4}$ and $0 < x < \pi/2$
3. $\tan x = -\sqrt{2}$ and $\pi/2 < x < \pi$
4. $\cot x = \frac{2}{3}$ and $\pi < x < 3\pi/2$
5. $\sec x = -5$ and $\pi/2 < x < \pi$
6. $\csc x = 4$ and $0 < x < \pi/2$

In Exercises 7–12, find $\sin \frac{1}{2}x$, $\cos \frac{1}{2}x$, and $\tan \frac{1}{2}x$ under the given conditions.

7. $\sin x = \frac{1}{8}$ and $\pi/2 < x < \pi$

8. $\cos x = \frac{1}{4}$ and $0 < x < \pi/2$

9. $\tan x = \frac{1}{2}$ and $\pi < x < 3\pi/2$

10. $\cot x = 3$ and $2\pi < x < 5\pi/2$

11. $\sec x = -3$ and $\pi < x < 3\pi/2$

12. $\csc x = -4$ and $3\pi/2 < x < 2\pi$

In Exercises 13–24, use the Half-Angle Formulas to evaluate the expressions.

13. $\sin \dfrac{7\pi}{8}$ **14.** $\sin \dfrac{11\pi}{8}$

15. $\sin\left(-\dfrac{\pi}{12}\right)$ **16.** $\sin 105°$

17. $\cos \dfrac{5\pi}{8}$ **18.** $\cos(-247.5°)$

19. $\cos \dfrac{5\pi}{12}$ **20.** $\cos\left(-\dfrac{19\pi}{12}\right)$

21. $\tan \dfrac{5\pi}{8}$ **22.** $\tan \dfrac{9\pi}{8}$

23. $\tan\left(-\dfrac{13\pi}{12}\right)$ **24.** $\tan 75°$

25. Using the fact that $\cos \pi/8 = \frac{1}{2}\sqrt{2 + \sqrt{2}}$, evaluate the following.

a. $\sin \dfrac{\pi}{16}$ b. $\cos \dfrac{\pi}{16}$ c. $\tan \dfrac{\pi}{16}$

26. Using the fact that $\cos \pi/12 = \frac{1}{2}\sqrt{2 + \sqrt{3}}$, evaluate the following.

a. $\sin \dfrac{\pi}{24}$ b. $\cos \dfrac{\pi}{24}$ c. $\tan \dfrac{\pi}{24}$

In Exercises 27–38, verify the identity.

27. $\cos 2x + 2\sin^2 x = 1$

28. $\cos 2x = \cos^4 x - \sin^4 x$

29. $\sin 2x = \dfrac{2\sin x}{\sec x}$

30. $\cos 2x = \dfrac{\cot x - \tan x}{\cot x + \tan x}$

31. $\cot 2x = \frac{1}{2}(\cot x - \tan x)$

32. $\csc 2x = \frac{1}{2}(\cot x + \tan x)$

33. $\cot 2x = \dfrac{\cot^2 x - 1}{2\cot x}$

34. $\sec 2x = \dfrac{1}{2\cos^2 x - 1}$

35. $\csc 2x = \cot x - \cot 2x$

36. $\dfrac{\sec x}{1 + \tan x} = \dfrac{\cos x - \sin x}{\cos 2x}$

37. $\tan 2x - \sec 2x = \dfrac{\sin x - \cos x}{\sin x + \cos x}$

38. $\log(\cos x - \sin x) + \log(\cos x + \sin x) = \log \cos 2x$

In Exercises 39–42, verify the identity.

39. $\cos 3x = 4\cos^3 x - 3\cos x$

40. $\sin 4x = 8\sin x \cos^3 x - 4\sin x \cos x$

***41.** $\cos 4x = 8\cos^4 x - 8\cos^2 x + 1$

42. $\dfrac{\sin 3x}{\sin x} - \dfrac{\cos 3x}{\cos x} = 2$ (*Hint:* Use Example 2 and Exercise 39.)

In Exercises 43–51, verify the identity.

43. $\sec^2 \dfrac{1}{2}x = \dfrac{2}{1 + \cos x}$

44. $\csc^2 \dfrac{1}{2}x = \dfrac{2}{1 - \cos x}$

45. $\sec^2 \dfrac{1}{2}x = \dfrac{2\sec x}{\sec x + 1}$

46. $\sin^2 \dfrac{1}{2}x = \dfrac{\tan x - \sin x}{2\tan x}$

47. $\cos^2 \dfrac{1}{2}x = \dfrac{\sec x + 1}{2\sec x}$

48. $\tan \dfrac{1}{2}x = \dfrac{\tan x}{1 + \sec x}$

49. $\tan \dfrac{1}{2}x = \csc x - \cot x$

50. $\cot \dfrac{1}{2} x = \dfrac{\sin x}{1 - \cos x}$

51. $\cot^2 \dfrac{1}{2} x = \dfrac{\sec x + 1}{\sec x - 1}$

52. a. Verify that $\sin x = \dfrac{2 \tan \frac{1}{2}x}{1 + \tan^2 \frac{1}{2}x}$.

b. Verify that $\cos x = \dfrac{1 - \tan^2 \frac{1}{2}x}{1 + \tan^2 \frac{1}{2}x}$.

(The identities in this exercise sometimes appear in calculus.)

53. Let $f(x) = \sin 2x \cos 2x$. Sketch the graph of f.

54. Let $f(x) = \cos^2 \frac{1}{6}x - \sin^2 \frac{1}{6}x$. Sketch the graph of f.

55. Let $f(x) = \dfrac{4 \tan \frac{1}{2}x}{1 - \tan^2 \frac{1}{2} x}$. Sketch the graph of f.

56. Verify (2).

57. Verify (3).

58. a. Prove (4) from (2).
b. Prove (4) from (5).

6.4 THE SUM AND PRODUCT FORMULAS

In this section we will employ the Addition Formulas to prove the following identities, which are useful in calculus and in the mathematical description of certain aspects of wave propagation.

Product Formulas

$$\sin x \sin y = \frac{1}{2}[\cos(x - y) - \cos(x + y)] \quad (1)$$

$$\cos x \cos y = \frac{1}{2}[\cos(x + y) + \cos(x - y)] \quad (2)$$

$$\sin x \cos y = \frac{1}{2}[\sin(x + y) + \sin(x - y)] \quad (3)$$

Sum Formulas

$$\sin x + \sin y = 2 \sin \frac{1}{2}(x + y) \cos \frac{1}{2}(x - y) \quad (4)$$

$$\sin x - \sin y = 2 \sin \frac{1}{2}(x - y) \cos \frac{1}{2}(x + y) \quad (5)$$

$$\cos x + \cos y = 2 \cos \frac{1}{2}(x + y) \cos \frac{1}{2}(x - y) \quad (6)$$

$$\cos x - \cos y = -2 \sin \frac{1}{2}(x + y) \sin \frac{1}{2}(x - y) \quad (7)$$

Since (1)–(3) are all proved in the same way, we will prove (1) and leave (2) and (3) as exercises. By (1) and (2) of Section 6.2, we have

$$\begin{aligned}\cos(x - y) - \cos(x + y) \\ &= (\cos x \cos y + \sin x \sin y) - (\cos x \cos y - \sin x \sin y) \\ &= 2 \sin x \sin y\end{aligned}$$

Dividing both sides by 2 and switching the sides, we obtain formula (1).

Formulas (4)–(7) follow from (1)–(3). We will prove (7) and leave the others as exercises. By applying (1) with x replaced by $\frac{1}{2}(x + y)$ and y replaced by $\frac{1}{2}(x - y)$, we have

$$\begin{aligned}\sin\frac{1}{2}(x + y) \sin\frac{1}{2}(x - y) \\ &= \frac{1}{2}\left[\cos\left(\frac{1}{2}(x + y) - \frac{1}{2}(x - y)\right) - \cos\left(\frac{1}{2}(x + y) + \frac{1}{2}(x - y)\right)\right] \\ &= \frac{1}{2}(\cos y - \cos x)\end{aligned}$$

Multiplying both sides by -2 and reversing the order of the terms, we obtain (7).

EXAMPLE 1. Use formula (1) to express $\sin 3x \sin 2x$ in the form $a(\cos bx - \cos cx)$.

Solution. Applying (1) with x replaced by $3x$ and y replaced by $2x$, we have

$$\sin 3x \sin 2x = \frac{1}{2}[\cos(3x - 2x) - \cos(3x + 2x)] = \frac{1}{2}(\cos x - \cos 5x) \quad \square$$

EXAMPLE 2. Verify the identity

$$\frac{\sin x - \sin 3x}{\cos x + \cos 3x} = -\tan x$$

Solution. By (5) with y replaced by $3x$, we have

$$\begin{aligned}\sin x - \sin 3x &= 2 \sin\frac{1}{2}(x - 3x)\cos\frac{1}{2}(x + 3x) \\ &= 2 \sin(-x)\cos 2x \\ &= -2 \sin x \cos 2x\end{aligned}$$

and by (6) with y replaced by $3x$, we have

$$\begin{aligned}\cos x + \cos 3x &= 2\cos\frac{1}{2}(x + 3x)\cos\frac{1}{2}(x - 3x)\\ &= 2\cos 2x\cos(-x)\\ &= 2\cos 2x\cos x\end{aligned}$$

Therefore

$$\frac{\sin x - \sin 3x}{\cos x + \cos 3x} = \frac{-2\sin x\cos 2x}{2\cos 2x\cos x} = -\frac{\sin x}{\cos x} = -\tan x$$

as desired. □

Now we have completed our presentation of trigonometric identities and formulas. The more important ones are listed at the end of the chapter.

Musical Strings

One application of the Sum Formula occurs in the analysis of musical sounds produced by stringed instruments such as the violin or guitar. Recall from Section 5.6 that when the string is plucked or bowed, the displacement at time t of the point on the string corresponding to x is given by

$$y_1 = a\sin 2\pi\left(\frac{x}{\lambda} - \frac{t}{T}\right)$$

where $|a|$ denotes the amplitude, λ the wavelength, and T the period of the motion. The motion of the string produces a wave, moving along the string. As the wave reaches the end of the string, it is reflected back along the string with the same amplitude, wavelength, and period. Considered by itself, this reflected wave would yield a displacement y_2 given by

$$y_2 = -a\sin 2\pi\left(\frac{x}{\lambda} + \frac{t}{T}\right)$$

However, the reflected wave combines with the original wave to yield a combined wave. The superposition principle from physics tells us that the displacement y of a given point x for the combined wave equals the sum $y_1 + y_2$ of the displacements y_1 and y_2 for the original and reflected waves.

In order to analyze the combined wave further, we use the Sum Formula (5):

$$\begin{aligned}y = y_1 + y_2 &= a\sin 2\pi\left(\frac{x}{\lambda} - \frac{t}{T}\right) - a\sin 2\pi\left(\frac{x}{\lambda} + \frac{t}{T}\right)\\ &\overset{(5)}{=} 2a\sin\frac{1}{2}\left[2\pi\left(\frac{x}{\lambda} - \frac{t}{T}\right) - 2\pi\left(\frac{x}{\lambda} + \frac{t}{T}\right)\right]\cos\frac{1}{2}\left[2\pi\left(\frac{x}{\lambda} - \frac{t}{T}\right) + 2\pi\left(\frac{x}{\lambda} + \frac{t}{T}\right)\right]\\ &= -2a\cos\left(2\pi\frac{x}{\lambda}\right)\sin\left(2\pi\frac{t}{T}\right)\end{aligned}$$

Therefore

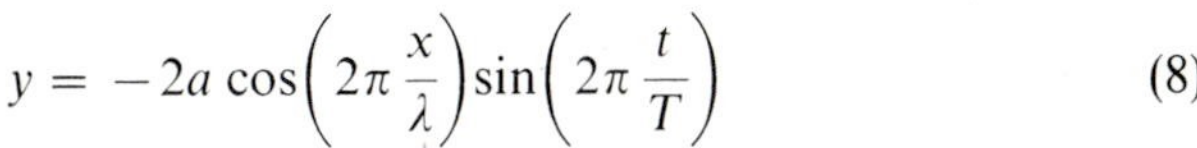

$$y = -2a\cos\left(2\pi\frac{x}{\lambda}\right)\sin\left(2\pi\frac{t}{T}\right) \tag{8}$$

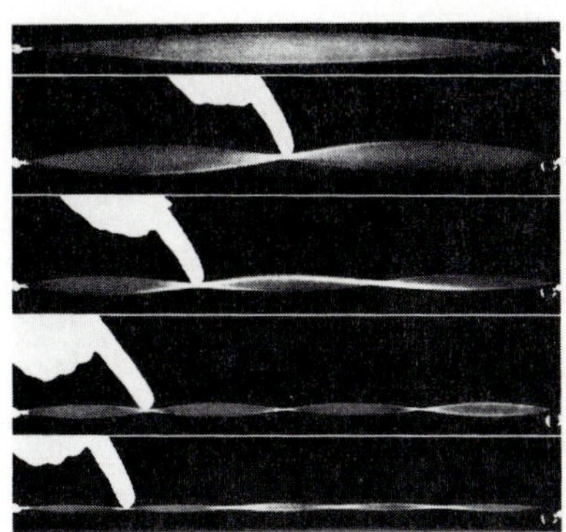

Notice that if $\cos(2\pi x/\lambda) = 0$ in (8), then $y = 0$, which can be interpreted to mean that the point x is stationary (because $y = 0$ for any value of t). Such points are called ***nodes***. In the photograph the nodes of each vibrating string are clearly visible. Because of the existence of nodes, the waves in the string appear to be stationary. This is the reason such waves are called ***stationary waves***, or ***standing waves***.

Because both ends of the string are fixed, and hence the ends are nodes themselves, there are exactly an integral number of "halfwaves" on the string. Thus the halfwave length $\lambda/2$ must divide the length L of the string:

$$L = n\left(\frac{\lambda}{2}\right)$$

so that

$$\lambda = \frac{2L}{n}$$

for some positive integer n. As a result, a given string with a predetermined length can vibrate only with certain wavelengths. The corresponding frequencies, which determine the sounds we hear, depend on various physical properties of the string, such as tension and mass per unit length. The frequency corresponding to a wave with two nodes (at the ends of the string) is called the ***first harmonic*** or ***fundamental***. If n is a positive integer, then the frequency corresponding to a wave with $n + 2$ nodes is called the ***nth harmonic*** or ***nth overtone***. The fundamental and first four overtones are shown in the photograph.

EXERCISES 6.4

In Exercises 1–8, write each expression as a sum or difference.

1. $\sin 2x \sin 4x$ **2.** $\sin 3x \sin 5x$

3. $\cos x \cos 3x$ **4.** $\cos 4x \cos 5x$

5. $\cos \frac{5}{3}x \cos \frac{1}{3}x$ **6.** $\sin \frac{1}{2}x \cos \frac{3}{2}x$

7. $\cos 7x \sin 3x$ **8.** $\sin 2x \cos 2x$

In Exercises 9–16, write each expression as a product.

9. $\sin 4x + \sin 2x$ **10.** $\sin x + \sin 3x$

11. $\sin 2x - \sin \frac{1}{2}x$ **12.** $\sin 5x - \sin 7x$

13. $\cos \frac{1}{2}x + \cos \frac{3}{2}x$ **14.** $\cos \frac{7}{5}x + \cos \frac{3}{5}x$

15. $\cos 7x - \cos 3x$ **16.** $\cos x - \cos 6x$

In Exercises 17–28, verify the identity.

17. $\dfrac{\sin x + \sin 3x}{\cos x + \cos 3x} = \tan 2x$

18. $\dfrac{\sin x + \sin 3x}{\cos x - \cos 3x} = \cot x$

19. $\dfrac{\sin x - \sin 3x}{\cos x - \cos 3x} = -\cot 2x$

20. $\dfrac{\sin x + \sin 3x}{\sin x - \sin 3x} = -\tan 2x \cot x$

21. $\dfrac{\sin 5x + \sin 3x}{\cos 3x - \cos 5x} = \cot x$

22. $\dfrac{\cos x - \cos 3x}{\cos x + \cos 3x} = \tan 2x \tan x$

23. $\dfrac{\sin 8x + \sin 3x}{\cos 8x + \cos 3x} = \tan \dfrac{11}{2} x$

24. $\dfrac{\sin 8x + \sin 3x}{\cos 8x - \cos 3x} = \cot \dfrac{5}{2} x$

25. $\dfrac{\sin 8x - \sin 3x}{\cos 8x + \cos 3x} = \tan \dfrac{5}{2} x$

26. $\dfrac{\sin 8x - \sin 3x}{\cos 8x - \cos 3x} = -\cot \dfrac{11}{2} x$

27. $\dfrac{\sin 8x + \sin 3x}{\sin 8x - \sin 3x} = \tan \dfrac{11}{2} x \cot \dfrac{5}{2} x$

28. $\dfrac{\cos 8x - \cos 3x}{\cos 8x + \cos 3x} = -\tan \dfrac{11}{2} x \tan \dfrac{5}{2} x$

In Exercises 29–34, verify the identity.

29. $\dfrac{\sin x + \sin y}{\cos x + \cos y} = \tan \dfrac{1}{2}(x + y)$

30. $\dfrac{\sin x + \sin y}{\cos x - \cos y} = -\cot \dfrac{1}{2}(x - y)$

31. $\dfrac{\sin x - \sin y}{\cos x + \cos y} = \tan \dfrac{1}{2}(x - y)$

32. $\dfrac{\sin x - \sin y}{\cos x - \cos y} = -\cot \dfrac{1}{2}(x + y)$

33. $\dfrac{\sin x + \sin y}{\sin x - \sin y} = \tan \dfrac{1}{2}(x + y) \cot \dfrac{1}{2}(x - y)$

34. $\dfrac{\cos x - \cos y}{\cos x + \cos y} = -\tan \dfrac{1}{2}(x + y) \tan \dfrac{1}{2}(x - y)$

35. Verify (2).

36. Verify (3).

37. Verify (4).

38. Verify (5).

39. Verify (6).

6.5 CONDITIONAL TRIGONOMETRIC EQUATIONS

Throughout this chapter we have studied trigonometric identities, that is, equations that are valid for all admissible values of the variables. In contrast, some trigonometric equations are valid only for special values of the variables. Such equations, which are called conditional trigonometric equations, are the topic of this section. As always, the goal will be to solve such an equation, by which we mean to find all values of the variable for which the equation is true. In this section we will use radian measure exclusively for the variables in trigonometric expressions.

We have already encountered simple conditional trigonometric equations such as $\sin x = 0$ and $\tan x = 0$. But as you will see, the equations can be much more complicated. Many of the techniques we will use to solve conditional trigonometric equations (such as factoring and using the quadratic formula) were used in solving equations in Chapter 1. Yet you should be aware of two major differences between conditional trigonometric equations and the conditional equations of Chapter 1. In the first place, trigonometric identities are available to aid us in the solutions. Second, and more important, the equations will in general have an infinite number of solutions, because the equations will involve trigonometric functions, each of which is periodic. Thus if x is a solution of an equation such as

$$\tan^2 x - \sec x - 1 = 0$$

then for any integer n, $x + 2n\pi$ is also a solution of the equation, because

$$\tan^2(x + 2n\pi) - \sec(x + 2n\pi) - 1 = \tan^2 x - \sec x - 1$$

We will generally determine the solutions that are in $[0, 2\pi)$ first, and from these we will deduce all solutions.

In each of the first three examples we will solve an equation involving a single trigonometric function.

EXAMPLE 1. Solve the equation $\sin x = \frac{1}{2}$.

Solution. The solutions of $\sin x = \frac{1}{2}$ belonging to $[0, 2\pi)$ are $\pi/6$ and $5\pi/6$. Therefore the total collection of solutions of $\sin x = \frac{1}{2}$ consists of $\pi/6 + 2n\pi$ and $5\pi/6 + 2n\pi$, for all integers n. □

EXAMPLE 2. Solve the equation $\cos 2x = -1$.

Solution. Let $z = 2x$, so that the given equation becomes $\cos z = -1$. The only solution of $\cos z = -1$ belonging to $[0, 2\pi)$ is π, so the total collection of solutions of $\cos z = -1$ consists of $\pi + 2n\pi = (2n + 1)\pi$. Substituting $2x$ back for z, we find that the solutions of the given equation $\cos 2x = -1$ are the numbers x such that

$$2x = (2n + 1)\pi$$

that is, the numbers of the form $(2n + 1)\pi/2$, for all integers n. □

EXAMPLE 3. Solve the equation $\tan \frac{1}{4}x = -\sqrt{3}/3$.

Solution. Let $z = \frac{1}{4}x$, so that the given equation becomes $\tan z = -\sqrt{3}/3$. Since the period of $\tan z$ is π, we will first find the solutions of $\tan z = -\sqrt{3}/3$ belonging to $(-\pi/2, \pi/2)$. Now the lone solution of $\tan z = -\sqrt{3}/3$ belonging to that interval is $-\pi/6$, so the total collection of solutions of $\tan z = -\sqrt{3}/3$ consists of $-\pi/6 + n\pi$ for all integers n. Since $z = \frac{1}{4}x$, the solutions of the given equation satisfy $\frac{1}{4}x = -\pi/6 + n\pi$ and thus are $4(-\pi/6 + n\pi) = -2\pi/3 + 4n\pi$, for all integers n. □

EXAMPLE 4. Solve the equation $2 \sin^2 x + 5 \sin x + 2 = 0$.

Solution. Let $z = \sin x$, so that the given equation becomes $2z^2 + 5z + 2 = 0$. Now the solutions of $2z^2 + 5z + 2 = 0$ can be obtained by factoring directly or by the quadratic formula. In the latter case we obtain

$$z = \frac{-5 \pm \sqrt{5^2 - 4(2)(2)}}{2(2)} = \frac{-5 \pm \sqrt{9}}{4} = \frac{-5 \pm 3}{4} = -\tfrac{1}{2} \quad \text{or} \quad -2$$

Since $z = \sin x$ and $|\sin x| \leq 1$, it follows that $\sin x = -\frac{1}{2}$. Thus the original equation is equivalent to the equation $\sin x = -\frac{1}{2}$, whose solutions in $[0, 2\pi)$

are $7\pi/6$ and $11\pi/6$. Consequently the solutions of the given equation are $7\pi/6 + 2n\pi$ and $11\pi/6 + 2n\pi$, for all integers n. □

The solutions of Examples 5–9 will utilize the various identities we have encountered in the preceding sections of Chapter 6. Generally speaking, in our solutions we will want to convert a given equation into an equivalent equation (with the same solutions) in which *only one* trigonometric function and its powers appear. Then we will solve that equation instead of the original one. The conversion will normally utilize identities such as those listed at the outset of Section 6.1, as well as the Addition, Double-Angle, and Half-Angle Formulas.

EXAMPLE 5. Solve the equation $\tan^2 x - \sec x - 1 = 0$.

Solution. Since $\tan^2 x = \sec^2 x - 1$ by the second Pythagorean Identity, we can convert the given equation to one having only powers of $\sec x$:

$$(\sec^2 x - 1) - \sec x - 1 = 0$$

or equivalently,

$$\sec^2 x - \sec x - 2 = 0$$

But this means that

$$(\sec x + 1)(\sec x - 2) = 0$$

so that $\sec x = -1$ or $\sec x = 2$. Now the only solution of the equation $\sec x = -1$ belonging to $[0, 2\pi)$ is π, and the only such solutions of the equation $\sec x = 2$ are $\pi/3$ and $5\pi/3$. Therefore the solutions of the given equation are the numbers of the form $\pi + 2n\pi$, $\pi/3 + 2n\pi$, or $5\pi/3 + 2n\pi$, for all integers n. □

EXAMPLE 6. Solve the equation $4 \sin x \cos x = -\sqrt{3}$.

Solution. Rather than attempting to write $\sin x$ in terms of $\cos x$ or $\cos x$ in terms of $\sin x$, we will utilize the Double-Angle Formula

$$\sin 2x = 2 \sin x \cos x$$

which converts the given equation into

$$2 \sin 2x = -\sqrt{3}$$

or equivalently,

$$\sin 2x = -\frac{\sqrt{3}}{2}$$

Letting $z = 2x$, we notice that the only solutions of $\sin z = -\sqrt{3}/2$ belonging to $[0, 2\pi)$ are $4\pi/3$ and $5\pi/3$. Therefore the collection of solutions of $\sin z = -\sqrt{3}/2$ consists of the numbers z such that

$$z = \frac{4\pi}{3} + 2n\pi \quad \text{or} \quad z = \frac{5\pi}{3} + 2n\pi$$

Substituting $2x$ back in for z, we find that the collection of solutions of $\sin 2x = -\sqrt{3}/2$ consists of the numbers x such that

$$2x = \frac{4\pi}{3} + 2n\pi \quad \text{or} \quad 2x = \frac{5\pi}{3} + 2n\pi$$

which means that the solutions of the given equation are the numbers of the form $2\pi/3 + n\pi$ or $5\pi/6 + n\pi$, for all integers n. □

EXAMPLE 7. Solve the equation $\cos 2x = \cos x$.

Solution. Notice that $\cos 2x$ and $\cos x$ are two distinct functions. First we use (6) of Section 6.3:

$$\cos 2x = 2\cos^2 x - 1$$

This allows us to rewrite the given equation as

$$2\cos^2 x - 1 = \cos x$$

which yields

$$2\cos^2 x - \cos x - 1 = 0$$

Factoring, we obtain

$$(2\cos x + 1)(\cos x - 1) = 0$$

so that

$$2\cos x + 1 = 0 \quad \text{or} \quad \cos x - 1 = 0$$

As a result,

$$\cos x = -\frac{1}{2} \quad \text{or} \quad \cos x = 1$$

Now the only solutions of the equation $\cos x = -\frac{1}{2}$ belonging to $[0, 2\pi)$ are $2\pi/3$ and $4\pi/3$, and the only such solution of the equation $\cos x = 1$ is 0. Consequently the solutions of the given equation are the numbers of the form $2n\pi$, $2\pi/3 + 2n\pi$, or $4\pi/3 + 2n\pi$, for all integers n. □

EXAMPLE 8. Solve the equation $\sin x = \frac{1}{2}\sqrt{3} \tan x$.

Solution. Here we again have two trigonometric functions—this time the sine and tangent functions. If we write $\tan x = (\sin x)/(\cos x)$, then the given equation becomes

$$\sin x = \frac{\sqrt{3}}{2} \frac{\sin x}{\cos x} \tag{1}$$

or equivalently,

$$\sin x - \frac{\sqrt{3}}{2} \frac{\sin x}{\cos x} = 0$$

Factoring out $\sin x$, we obtain

$$\sin x \left(1 - \frac{\sqrt{3}}{2} \frac{1}{\cos x}\right) = 0$$

This equation is satisfied if

$$\sin x = 0 \quad \text{or} \quad 1 - \frac{\sqrt{3}}{2} \frac{1}{\cos x} = 0$$

Now $\sin x = 0$ if and only if $x = n\pi$ for some integer n. Analogously, the equation

$$1 - \frac{\sqrt{3}}{2} \frac{1}{\cos x} = 0$$

is equivalent to

$$1 = \frac{\sqrt{3}}{2} \frac{1}{\cos x}, \quad \text{or} \quad \cos x = \frac{\sqrt{3}}{2}$$

We observe that the only solutions of the equation $\cos x = \sqrt{3}/2$ belonging to $[0, 2\pi)$ are $\pi/6$ and $11\pi/6$. Therefore the solutions of the given equation are the numbers of the form $n\pi$, $\pi/6 + 2n\pi$, or $11\pi/6 + 2n\pi$, for all integers n. □

CAUTION: Had we canceled $\sin x$ from (1) and solved the resulting equation for x, we could have lost the solutions of the form $n\pi$. We see once again that we *must not* cancel expressions that can be 0.

EXERCISES 6.5

In Exercises 1–18, solve the equation.

1. $\sin x = 1$

2. $\sin x = -\sqrt{3}/2$

3. $\cos x = -\frac{1}{2}$

4. $\cos x = \sqrt{2}/2$

5. $\tan x = \sqrt{3}$

6. $\tan x = -1$

7. $\cot x = \sqrt{3}$

8. $\sec x = -2$

9. $\csc x = \frac{2}{3}\sqrt{3}$

10. $\csc x = -1$

11. $\sin 3x = \frac{1}{2}$

12. $\cos \frac{1}{2}x = \sqrt{3}/2$

13. $\tan \frac{1}{5}x = 1$

14. $\tan 2x = \sqrt{3}$

15. $\sec 6x = \sqrt{2}$

16. $\sin^2 x = 1$

17. $\sin^2 x/2 = \frac{1}{4}$

18. $\cos^2 x = \frac{1}{2}$

In Exercises 19–30, solve the equation.

19. $\sin x = \sqrt{3} \cos x$

20. $\cos x = \dfrac{\sqrt{2}}{2} \cot x$

21. $\tan x \csc x = -\frac{2}{3}\sqrt{3}$

22. $\cos x = -\frac{1}{4} \csc x$

23. $\sin 2x = \sin x$

24. $\tan \frac{1}{2}x = \sin x$

25. $(\tan x + 1) \cot x = 0$

26. $\sin^2 x + \sin x = 0$

27. $\cos^2 x - \sin x \cos x = 0$

28. $\sin x \cos x - \cos x + \sin x - 1 = 0$

***29.** $2 \sin x \cos x - \sqrt{2} \sin x - \sqrt{2} \cos x + 1 = 0$

***30.** $4 \sin x \cos x - 2\sqrt{3} \sin x - 2 \cos x + \sqrt{3} = 0$

In Exercises 31–44, solve the equation.

31. $\sin^2 x + 5 \sin x + 4 = 0$

32. $4 \cos^2 x - 4 \cos x + 1 = 0$

33. $2 \cos^2 x - \cos x - 1 = 0$

34. $\sin^2 x + \frac{1}{2} \sin x - \frac{1}{2} = 0$

35. $\cos^2 x - \frac{3}{2} \cos x - 1 = 0$

36. $\tan^2 x - 1 = 0$

37. $\tan^2 x + \frac{2}{3}\sqrt{3} \tan x - 1 = 0$

38. $\sec^2 x + 3 \sec x + 2 = 0$

39. $\csc^5 x - 4 \csc x = 0$

40. $\sin^2 x = 2 \cos x + 2$

41. $\sin x + 2 \cos^2 x = 1$

42. $\csc^2 x - \cot x - 1 = 0$

43. $\sec^2 x - \frac{3}{2} \sec x - 1 = 0$

44. $\csc^2 x - 2 \csc x = 0$

In Exercises 45–50, find the solutions of the given equation that lie in the interval $[0°, 360°)$. Use a calculator where necessary to approximate the solutions.

45. $6 \sin^2 \theta° - \sin \theta° - 1 = 0$

46. $\sin^2 \theta° + \sin \theta° - 1 = 0$

47. $15 \cos^2 \theta° + 2 \cos \theta° - 1 = 0$

48. $\tan^2 \theta° + 5 \tan \theta° + 6 = 0$

49. $\sec^2 \theta° - 3 = 0$

50. $\csc^2 \theta° + 2 \csc \theta° - 8 = 0$

KEY TERMS

trigonometric identity
admissible values
conditional trigonometric equation
Pythagorean Identities

KEY IDENTITIES AND FORMULAS

Defining Identities

$$\tan x = \frac{\sin x}{\cos x} \qquad \csc x = \frac{1}{\sin x}$$

$$\cot x = \frac{\cos x}{\sin x} \qquad \cot x = \frac{1}{\tan x}$$

$$\sec x = \frac{1}{\cos x}$$

Symmetry Identities

$$\sin(-x) = -\sin x \qquad \cot(-x) = -\cot x$$

$$\cos(-x) = \cos x \qquad \sec(-x) = \sec x$$

$$\tan(-x) = -\tan x \qquad \csc(-x) = -\csc x$$

Pythagorean Identities

$$\sin^2 x + \cos^2 x = 1 \qquad \cot^2 x + 1 = \csc^2 x$$

$$\tan^2 x + 1 = \sec^2 x$$

Addition Formulas

$$\cos(x + y) = \cos x \cos y - \sin x \sin y$$

$$\cos(x - y) = \cos x \cos y + \sin x \sin y$$

$$\sin(x + y) = \sin x \cos y + \cos x \sin y$$

$$\sin(x - y) = \sin x \cos y - \cos x \sin y$$

$$\tan(x + y) = \frac{\tan x + \tan y}{1 - \tan x \tan y}$$

$$\tan(x - y) = \frac{\tan x - \tan y}{1 + \tan x \tan y}$$

Double-Angle Formulas

$$\sin 2x = 2 \sin x \cos x$$

$$\cos 2x = \cos^2 x - \sin^2 x$$

$$\tan 2x = \frac{2 \tan x}{1 - \tan^2 x}$$

REVIEW EXERCISES

In Exercises 1–32, verify the identity.

1. $\tan x + \cot x = \dfrac{1}{\cos x \sin x}$

2. $(\sin x + \cos x)^2 = 1 + \sin 2x$

3. $(\sin x + \cos x)^2 - (\sin x - \cos x)^2 = 2 \sin 2x$

4. $\sin^2 x = \dfrac{\tan^2 x}{1 + \tan^2 x}$

5. $\sec^2 t + \tan^4 t = \sec^4 t - \tan^2 t$

6. $\sin t = (1 + \cos t)(\csc t - \cot t)$

7. $\dfrac{\sec t}{\tan t + \cot t} = \sin t$

8. $(1 + \tan^2 x)(1 + \cot^2 x) = \sec^2 x \csc^2 x$

9. $\dfrac{\sec^2 x}{\sec 2x} = \dfrac{2 \tan x}{\tan 2x}$

10. $\dfrac{1}{\sec x - \tan x} - \dfrac{1}{\sec x + \tan x} = 2 \tan x$

11. $\dfrac{1}{\csc x - \cot x} - \dfrac{1}{\csc x + \cot x} = 2 \cot x$

12. $\log|1 + \sin x| + \log|1 - \sin x| - 2 \log|\cos x| = 0$

13. $\dfrac{\tan^4 x - 1}{\sin^4 x - \cos^4 x} = \sec^4 x$

14. $\csc^2 x = \dfrac{1 + 2 \cot^2 x}{1 + \cos^2 x}$

15. $(\cos x + \sin x)(\tan x + \cot x) = \sec x + \csc x$

16. $\sin x = \sin(x + 60°) + \sin(x - 60°)$

17. $\cos x = \cos(x + 60°) + \cos(x - 60°)$

18. $\tan^2 x = \dfrac{1 - \cos 2x}{1 + \cos 2x}$

19. $\cos\left(\dfrac{\pi}{2} + x - y\right) = \sin(y - x)$

20. $\tan\left(x + \dfrac{\pi}{4}\right) = \dfrac{1 + \tan x}{1 - \tan x}$

21. $\tan x \tan y = \dfrac{\tan x + \tan y}{\cot x + \cot y}$

22. $\sin x \cos(x - y) - \cos x \sin(x - y) = \sin y$

23. $\cos x \cos(x + y) + \sin x \sin(x + y) = \cos y$

24. $\dfrac{\sin(x - y)}{\sin x \sin y} = \cot y - \cot x$

25. $\cot x = \dfrac{1 + \cos x + \cos 2x}{\sin x + \sin 2x}$

26. $\sin x + \sin 3x = 2 \sin 2x \cos x$

27. $\tan \dfrac{x}{2} = \dfrac{1 - \cos x}{\sin x}$

28. $\dfrac{\sin 2x}{\sin x} - \dfrac{\cos 2x}{\cos x} = \sec x$

29. $\dfrac{\cos 2x}{\sin x} + \dfrac{\sin 2x}{\cos x} = \csc x$

30. $\sin 4x = 4 \sin x \cos x \cos 2x$

31. $\cos 4x = 1 - 8 \cos^2 x \sin^2 x$

32. $2 \cos^2 \dfrac{x}{2} - \cos x - 1 = 0$

In Exercises 33–36, evaluate the given expression.

33. $\sin 5\pi/8$

34. $\cos 165°$

35. $\sec 75°$

36. $\tan \pi/8$

37. Suppose that $\sin x = \frac{1}{7}$ with $\pi/2 < x < \pi$. Find $\cos(x - \pi/3)$.

38. Suppose that $\tan x = \frac{3}{5}$, and $\sin y = \frac{2}{5}$ with $0 < y < \pi/2$. Find $\tan(x + y)$.

39. Suppose $\sec x = 4$ with $-\pi/2 < x < 0$. Find $\sin 2x$.

40. Suppose $\cot x = 5$ with $\pi < x < 3\pi/2$. Find $\tan \frac{1}{2}x$.

41. Suppose $x + y + z = \pi$. Show that

$$\tan x + \tan y + \tan z = \tan x \tan y \tan z$$

(*Hint:* $z = \pi - (x + y)$.)

42. Prove that if $\sin x = 0$, then $\sin 2x = 0$.

43. a. If $\cos x = 0$, does it follow that $\cos 2x = 0$? Explain your answer.
    b. If $\cos x = 0$, does it follow that $\cos 3x = 0$? Explain your answer.

44. Show that $\tan^2 x + \sec^2 x \stackrel{?}{=} 1$ is not an identity by finding an admissible value of x for which the equation does not hold.

45. a. Use a Half-Angle Formula to compute $\cos \pi/12$.
    b. By comparing your answer to part (a) with the result of Example 2 in Section 6.2, conclude that

$$\sqrt{2 + \sqrt{3}} = \frac{\sqrt{2} + \sqrt{6}}{2}$$

46. a. Express $2 \cos 5x \sin 3x$ as a sum or a difference of functions of the form $\sin ax$.
    b. Express $\cos 2x - \cos 4x$ as a product of functions of the form $\sin ax$.

47. Let $f(x) = \sqrt{3} \sin x + \cos x$. Sketch the graph of f.

48. Let $f(x) = \sin 3x \cos x - \cos 3x \sin x$. Sketch the graph of f.

In Exercises 49–58, solve the given equation.

49. $\cos x = -\sqrt{3}/2$

50. $\tan^2 x = \frac{1}{3}$

51. $\cos^2 x = 1 + \sin^2 x$

52. $\cos^2 x + 2 \cos x + \sin^2 x = 0$

53. $2 \sin^2 x + \sin x = 1$

54. $\sec^2 x - \sec x - 2 = 0$

55. $\cos 2x = \sin x$

56. $2 \sin x - \sin 2x \cos x = 0$

57. $\tan^2 x - 2 \sec^2 x = -5$

*58. $\sec x + \tan x = 1$ (*Hint:* Multiply both sides by $\sec x - \tan x$.)

7 TRIGONOMETRY AND ITS APPLICATIONS

Since prehistoric times people have been preoccupied with calculating distances, whether it be finding the height of a tree, the distance across a lake, or the dimensions of one's own property. Each of the distances just mentioned can be ascertained by studying the angles and sides of a convenient triangle. The case in which the triangle is a right triangle was analyzed in Section 5.3. In the present chapter we turn to triangles that are not necessarily right triangles and learn how to determine the lengths of the sides and the measures of the angles of a given triangle when certain of them are already known. As before, this process is called ***solving the triangle***.

Sections 7.1 and 7.2 are devoted to solving triangles by means of two famous laws, the Law of Sines and the Law of Cosines. In Section 7.3 we derive two quite different formulas for the area of an arbitrary triangle. The chapter ends with introductions to polar coordinates and vectors, each of which makes use of trigonometric functions and triangles. Throughout the chapter we use a calculator and round our final answers to four places.

As an illustration of the applications of the mathematics in this chapter, consider the following:

> Suppose that forest rangers at two park stations eight miles apart spot smoke from a forest fire in the distance. By measuring certain angles and using the methods of this chapter, we can determine which ranger is closer to the fire (see Example 4 in Section 7.1).

7.1 THE LAW OF SINES

It is important not only in the study of trigonometry, but also in applications, that triangles can be solved whether or not they are right triangles. A triangle that is not a right triangle is said to be ***oblique***. An oblique triangle has either three acute angles or two acute and one obtuse angle (Figure 7.1).

In this and the next section we discuss the two main tools—the Law of Sines and the Law of Cosines—that are especially helpful in solving oblique triangles. We will use α, β, and γ for the angles of a triangle and a, b, c for the lengths of the sides opposite α, β, and γ, respectively.

Now we give the main result of this section.

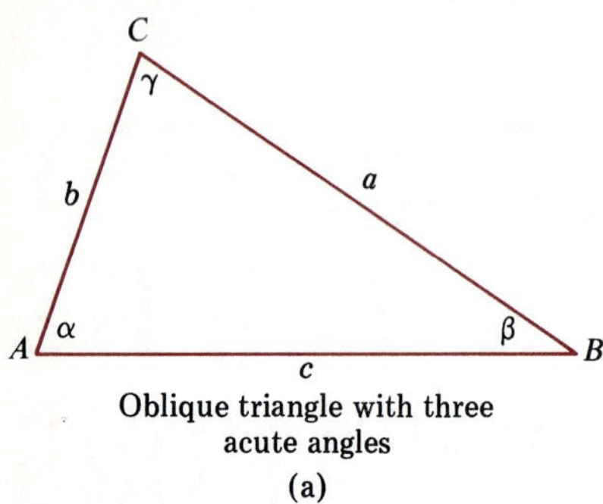

Oblique triangle with three acute angles
(a)

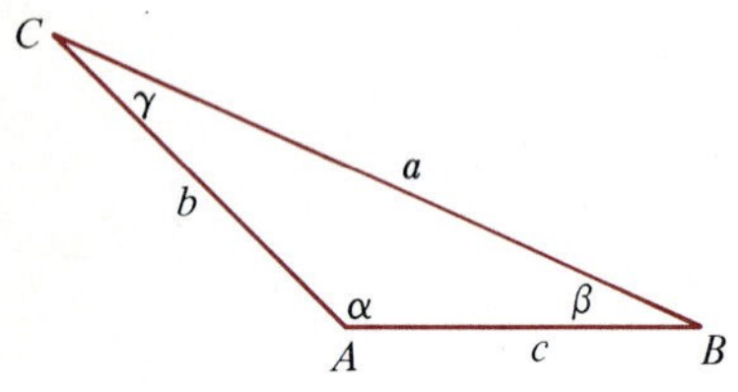

Oblique triangle with two acute angles and one obtuse angle
(b)

FIGURE 7.1

The Law of Sines

$$\frac{\sin \alpha}{a} = \frac{\sin \beta}{b} = \frac{\sin \gamma}{c} \tag{1}$$

To prove the Law of Sines, we first consider a triangle ABC with three acute angles and drop a perpendicular from C as in Figure 7.2. The perpendicular divides the given triangle into the smaller right triangles ADC and BCD. From triangle ADC,

$$\sin \alpha = \frac{h}{b}, \quad \text{so that} \quad h = b \sin \alpha \tag{2}$$

and from triangle BCD,

$$\sin \beta = \frac{h}{a}, \quad \text{so that} \quad h = a \sin \beta \tag{3}$$

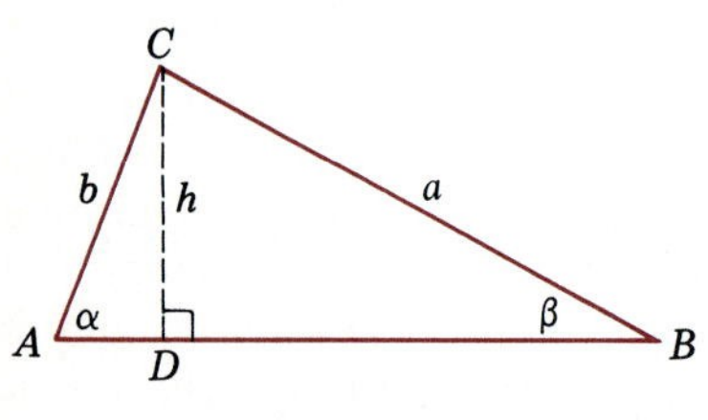

FIGURE 7.2

Equating the two values of h in (2) and (3), we have

$$b \sin \alpha = a \sin \beta$$

so that

$$\frac{\sin \alpha}{a} = \frac{\sin \beta}{b} \tag{4}$$

Similarly, it can be shown by dropping a perpendicular from A as in Figure 7.3 that

$$\frac{\sin \beta}{b} = \frac{\sin \gamma}{c} \tag{5}$$

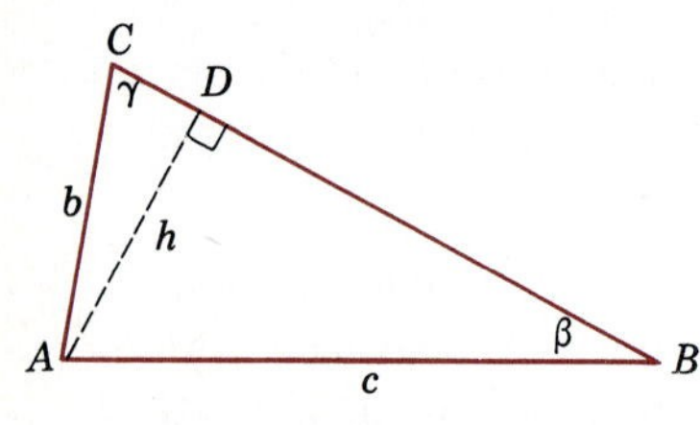

FIGURE 7.3

(see Exercise 29). Together (4) and (5) yield (1) for triangles with three acute angles.

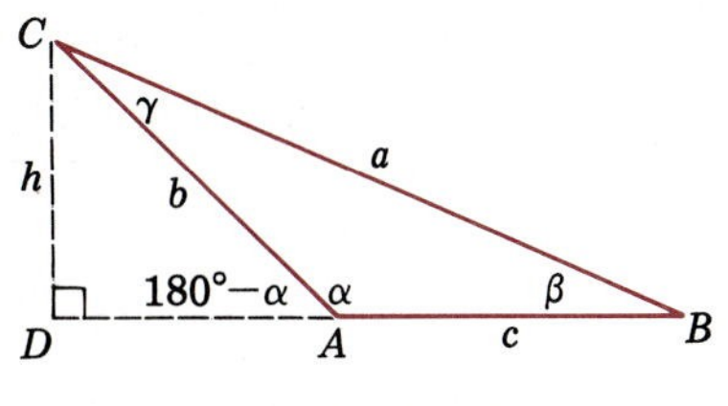

FIGURE 7.4

If a triangle ABC has an obtuse angle at α, we drop a perpendicular from C as in Figure 7.4. From triangle ACD,

$$\sin(180^\circ - \alpha) = \frac{h}{b}$$

so that

$$h = b \sin(180^\circ - \alpha) = b \sin \alpha \tag{6}$$

and from triangle BCD,

$$\sin \beta = \frac{h}{a}, \quad \text{so that} \quad h = a \sin \beta \tag{7}$$

Equating the values of h in (6) and (7) yields

$$b \sin \alpha = a \sin \beta$$

so that

$$\frac{\sin \alpha}{a} = \frac{\sin \beta}{b}$$

As before, one can show that

$$\frac{\sin \beta}{b} = \frac{\sin \gamma}{c}$$

so that (1) still holds for triangles with an obtuse angle. This completes the proof of the Law of Sines.

Solutions of Triangles Using the Law of Sines

The Law of Sines is applicable to solving triangles, provided that either

(a) two angles and the length of a side are known, or
(b) the lengths of two sides and the angle opposite one of these sides are known.

In the solution of a triangle by means of the Law of Sines we will use the fact that in any triangle,

$$\boxed{\alpha + \beta + \gamma = 180^\circ} \tag{8}$$

Triangles that fall under case (a) can be easier to solve than those falling under case (b). Example 1 illustrates case (a).

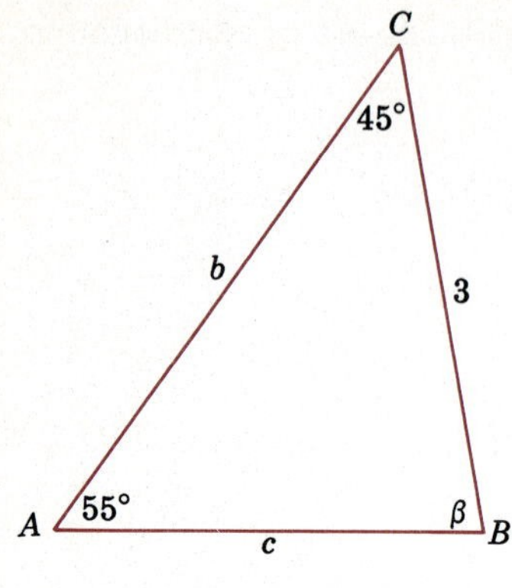

FIGURE 7.5

EXAMPLE 1. Solve the triangle shown in Figure 7.5.

Solution. From the figure we know that $a = 3$, $\alpha = 55°$, and $\gamma = 45°$. We need to determine b, c, and β. By (8),

$$\beta = 180° - 55° - 45° = 80°$$

so we have all the angles. With appropriate substitutions the Law of Sines becomes

$$\frac{\sin 55°}{3} = \frac{\sin 80°}{b} = \frac{\sin 45°}{c}$$

Thus

$$\frac{\sin 55°}{3} = \frac{\sin 80°}{b}, \quad \text{so that} \quad b = \frac{3 \sin 80°}{\sin 55°}$$

and

$$\frac{\sin 55°}{3} = \frac{\sin 45°}{c}, \quad \text{so that} \quad c = \frac{3 \sin 45°}{\sin 55°}$$

By calculator we find that

$$b \approx 3.607 \quad \text{and} \quad c \approx 2.590 \quad \square$$

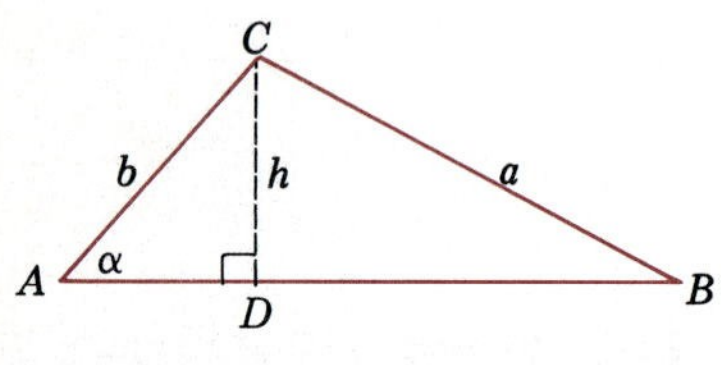

FIGURE 7.6

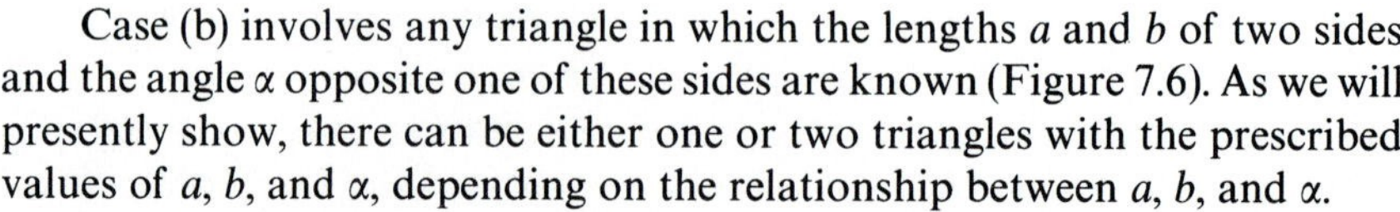

Case (b) involves any triangle in which the lengths a and b of two sides and the angle α opposite one of these sides are known (Figure 7.6). As we will presently show, there can be either one or two triangles with the prescribed values of a, b, and α, depending on the relationship between a, b, and α.

To determine whether the given values of a, b, and α yield one or two triangles, let h be the length of the perpendicular drawn from C to the opposite side (Figure 7.6). Since a is the hypotenuse of the right triangle BCD and h is a leg of the same triangle, it follows that $a \geq h$. Next, $h = b \sin \alpha$ because triangle ADC is a right triangle. Therefore

$$a \geq h = b \sin \alpha \tag{9}$$

so that either $a = b \sin \alpha$ or $a > b \sin \alpha$. These two possibilities, along with the possibilities $a \geq b$ or $b > a$, yield three cases, each of which determines uniquely the number of triangles that arise:

- **i.** $a \geq b$, in which case exactly one triangle exists (Figure 7.7a).
- **ii.** $b > a = b \sin \alpha$, in which case exactly one triangle exists, a right triangle with $a < b$ (Figure 7.7b).
- **iii.** $b > a > b \sin \alpha$, in which case two triangles exist, one with three acute angles and the other with an obtuse angle (triangles AB_1C and AB_2C, respectively, in Figure 7.7c).

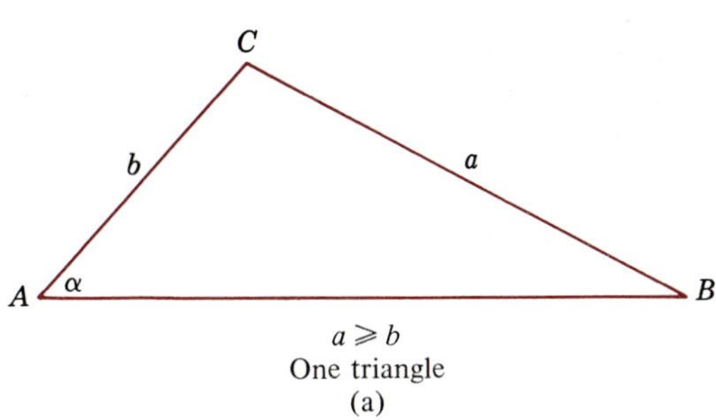

$a \geqslant b$
One triangle
(a)

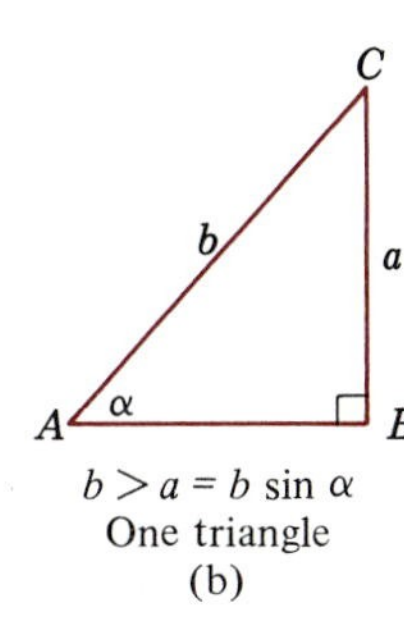

$b > a = b \sin \alpha$
One triangle
(b)

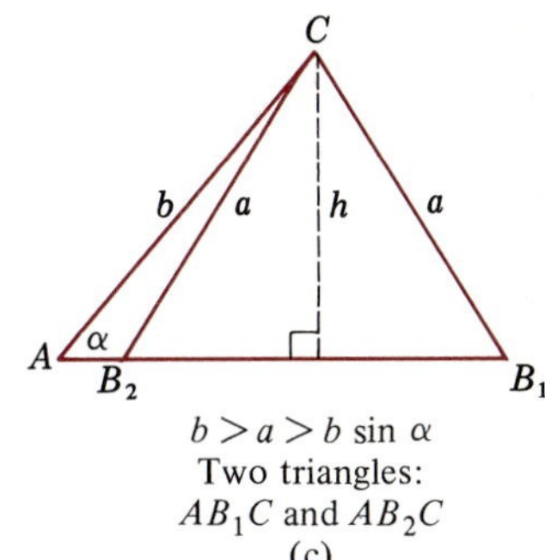

$b > a > b \sin \alpha$
Two triangles:
AB_1C and AB_2C
(c)

FIGURE 7.7

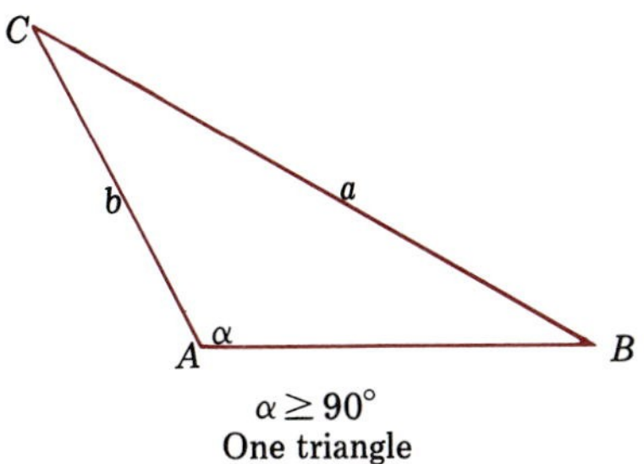
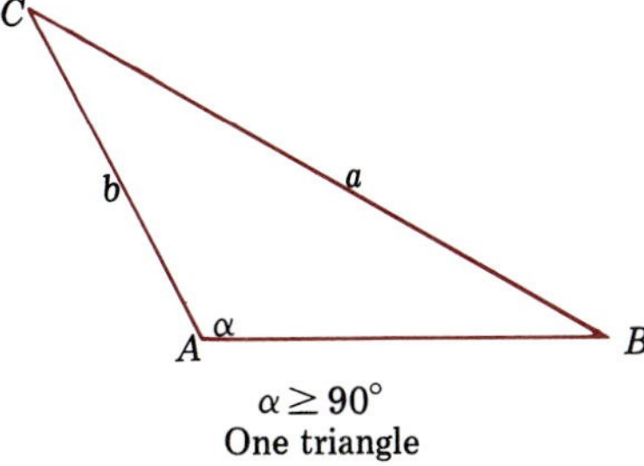

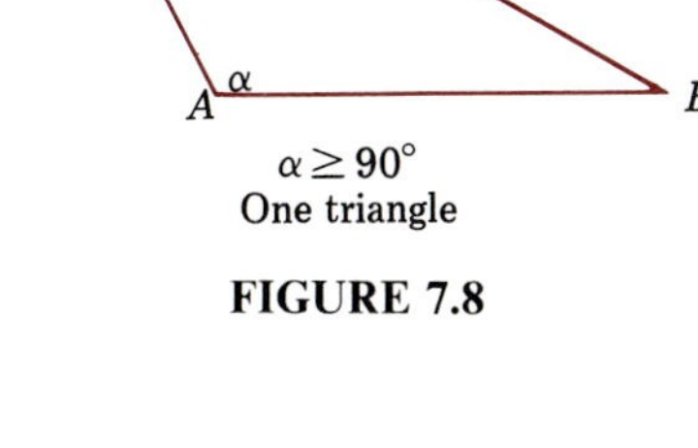

$\alpha \geq 90°$
One triangle

FIGURE 7.8

Notice that if $\alpha \geq 90°$, that is, if α is a right or obtuse angle, then $a \geq b$, so there is only one triangle (Figure 7.8). Notice also that under case **(iii)** there are two triangles; this case if often called the ***ambiguous case*** of the Law of Sines.

EXAMPLE 2. Solve the triangle in Figure 7.9.

Solution. From the figure we know that $\alpha = 110°$, $a = 6$ and $b = 2$. Since $a \geq b$, the triangle satisfies **(i)**, so there is only one triangle to solve. By the Law of Sines,

$$\frac{\sin 110°}{6} = \frac{\sin \beta}{2} = \frac{\sin \gamma}{c}$$

so that

$$\frac{\sin 110°}{6} = \frac{\sin \beta}{2} \quad \text{and} \quad \frac{\sin 110°}{6} = \frac{\sin \gamma}{c} \tag{10}$$

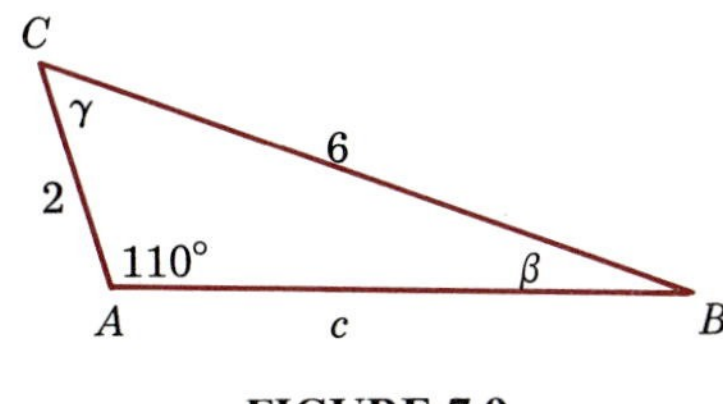

FIGURE 7.9

By calculator we find that

$$\sin \beta = \frac{1}{3} \sin 110° \approx 0.313$$

so that

$$\beta \approx 18.25° \; (\approx 18°15')$$

For γ we use (8):

$$\begin{aligned} \gamma &= 180° - \alpha - \beta \\ &\approx 180° - 110° - 18.25° = 51.75° \quad (\approx 51°45') \end{aligned}$$

To find c we use the fact that $\gamma \approx 51.75°$, and deduce from the second equation in (10) that

$$c \approx 6\frac{\sin 51.75^\circ}{\sin 110^\circ}$$

By calculator we find that

$$c \approx 5.014 \quad \square$$

The next example illustrates the ambiguous case of the Law of Sines.

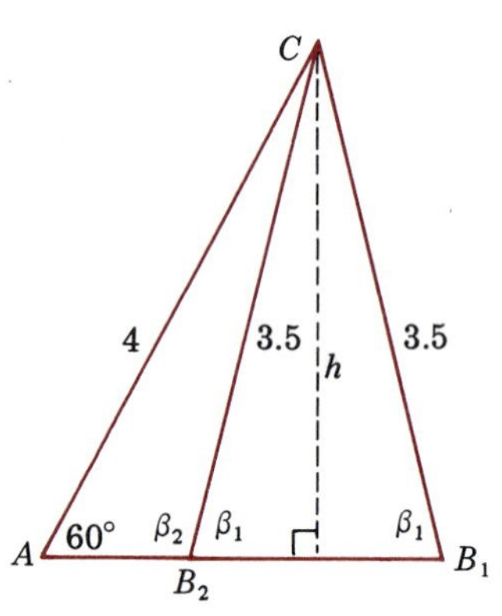

FIGURE 7.10

EXAMPLE 3. Suppose that $a = 3.5$, $b = 4$, and $\alpha = 60^\circ$. Show that two triangles result, and solve them.

Solution. We have

$$b \sin \alpha = 4 \sin 60^\circ = 4\left(\frac{\sqrt{3}}{2}\right) = 2\sqrt{3} \approx 3.464$$

Therefore $b > a > b \sin \alpha$, so that **(iii)** is satisfied, and consequently there are two triangles (AB_1C and AB_2C in Figure 7.10) that meet the given specifications.

We will first solve triangle AB_1C, all of whose angles are acute (Figure 7.11). This will entail calculating β_1, γ_1, and then c_1. For β_1, we use the Law of Sines to obtain

$$\frac{\sin 60^\circ}{3.5} = \frac{\sin \beta_1}{4}$$

so that

$$\sin \beta_1 = \frac{4 \sin 60^\circ}{3.5} = \frac{2\sqrt{3}}{3.5}$$

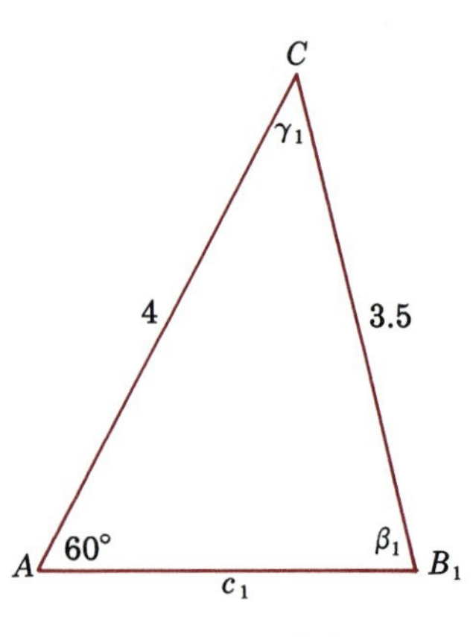

FIGURE 7.11

By calculator we find that

$$\beta_1 \approx 81.79^\circ \quad (\approx 81^\circ 47')$$

For γ_1 we use (8):

$$\begin{aligned}\gamma_1 &= 180^\circ - \alpha - \beta_1 \\ &\approx 180^\circ - 60^\circ - 81.79^\circ = 38.21^\circ \quad (\approx 38^\circ 13')\end{aligned}$$

To find c_1 we apply the Law of Sines a second time to triangle AB_1C and obtain

$$\frac{\sin 60^\circ}{3.5} = \frac{\sin \gamma_1}{c_1}$$

so that

$$c_1 = \frac{3.5 \sin \gamma_1}{\sin 60^\circ} \approx \frac{3.5 \sin 38.21^\circ}{\sin 60^\circ}$$

By calculator we find that

$$c_1 \approx 2.500$$

Thus we have solved triangle AB_1C of Figure 7.11.

For triangle AB_2C, shown in Figure 7.12, we must find β_2, γ_2, and c_2. The process is simpler than the process in the first part of the solution because triangle B_2B_1C in Figure 7.10 is isosceles and consequently we can find β_2 from β_1:

$$\beta_2 = 180^\circ - \beta_1 \approx 180^\circ - 81.79^\circ = 98.21^\circ \quad (\approx 98^\circ 13')$$

To find γ_2 we apply (8) again:

$$\begin{aligned} \gamma_2 &= 180^\circ - \alpha - \beta_2 \\ &\approx 180^\circ - 60^\circ - 98.21^\circ = 21.79^\circ \quad (\approx 21^\circ 47') \end{aligned}$$

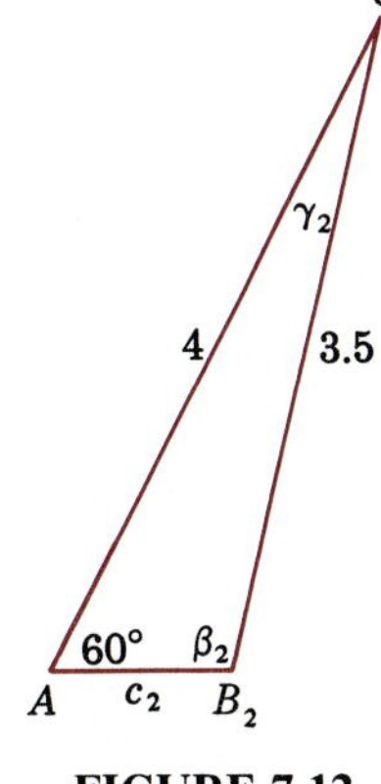

FIGURE 7.12

To determine c_2, we use the Law of Sines a final time:

$$\frac{\sin 60^\circ}{3.5} = \frac{\sin \gamma_2}{c_2}$$

so that

$$c_2 = \frac{3.5 \sin \gamma_2}{\sin 60^\circ} \approx \frac{3.5 \sin 21.79^\circ}{\sin 60^\circ}$$

By calculator we find that

$$c_2 \approx 1.500$$

Therefore triangle AB_2C is solved, and the solution of the example is complete. □

We note in passing that if we are given arbitrary positive values for a, b, and α, there may not exist any triangle with such prescribed values, since for any triangle we must have $a \geq b \sin \alpha$ by (9). (See Exercise 28.)

The Law of Sines is especially valuable in determining the distances between two given points and a third point.

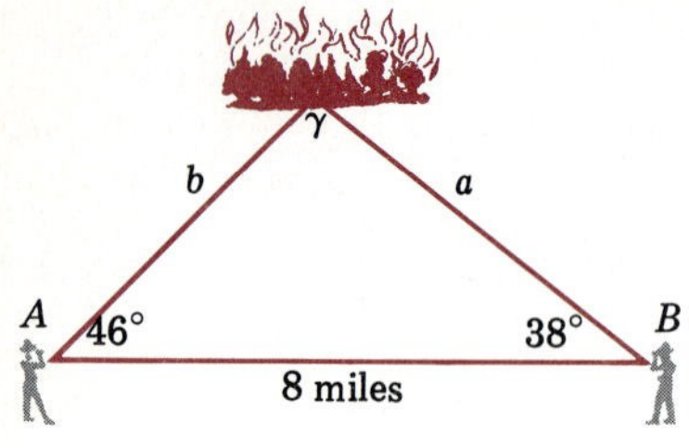

FIGURE 7.13

EXAMPLE 4. Forest rangers at stations A and B, which are 8 miles apart, observe a fire in the distance. By using the angles given in Figure 7.13, determine how far the fire is from each of the rangers.

Solution. We let $c = 8$, $\alpha = 46°$, and $\beta = 38°$. We need to determine the values of a and b and compare them. Before we can use the Law of Sines, we need the value of γ, which by (8) is given by

$$\gamma = 180° - \alpha - \beta = 180° - 46° - 38° = 96°$$

The Law of Sines then becomes

$$\frac{\sin 46°}{a} = \frac{\sin 38°}{b} = \frac{\sin 96°}{8}$$

In particular, this means that

$$a = \frac{8 \sin 46°}{\sin 96°} \quad \text{and} \quad b = \frac{8 \sin 38°}{\sin 96°}$$

By calculator we find that

$$a \approx 5.786 \quad \text{and} \quad b \approx 4.952$$

Therefore station B is approximately 5.8 miles from the fire and station A is approximately 5.0 miles from the fire, so station A is approximately 0.8 mile closer than station B is to the fire. □

It can be shown that if $\alpha < \beta$, then $a < b$. Consequently in any triangle the largest (smallest) side lies opposite the largest (smallest) angle. From this observation and the angles given in Example 4, we could tell immediately that station A is closer to the fire than station B.

EXERCISES 7.1

In Exercises 1–22, determine how many triangles there are satisfying the given data, and solve any triangles that exist.

1. $\alpha = 60°, \beta = 55°, b = 8$
2. $\alpha = 72°, \beta = 48°, c = 20$
3. $\beta = 29°, \gamma = 122°, c = 6$
4. $\beta = 34°, \gamma = 135°, c = 7$
5. $\beta = 34°, \gamma = 23°, a = 5$
6. $\alpha = 51.2°, \gamma = 65.7°, a = 8.3$
7. $\alpha = 23.8°, \gamma = 13.3°, b = 10.5$
8. $\alpha = 76.2°, \gamma = 27.6°, b = 13.9$
9. $\alpha = 27°24', \gamma = 61°9', c = 2.6$
10. $\beta = 53°43', \gamma = 87°29', b = 4.2$
11. $\alpha = 37°, a = 1.4, b = 2$
12. $\alpha = 50°, a = 3, b = 1$

13. $\alpha = 20°, a = 2.3, b = 7$

14. $\alpha = 20°, a = 2.4, b = 7$

15. $\alpha = 30°, a = 3, b = 6$

16. $\alpha = 33.5°, a = 4.9, c = 7.8$

17. $\beta = 74.9°, b = 5, c = 4.5$

18. $\beta = 14.4°, b = 1.2, c = 8.7$

19. $\gamma = 26.8°, b = 3.1, c = 2.5$

20. $\gamma = 27°17', a = 5.2, c = 3.6$

21. $\alpha = 110°45', a = 4.2, b = 2.6$

22. $\beta = 47.6°, b = \sqrt{2}, a = 1.5$

23. Two lifeguards, who are 500 yards apart at stations A and B, spot a swimmer in distress in the ocean. If the angles are as shown in Figure 7.14, which guard is in a better position to help the swimmer, and how much closer is that guard to the swimmer than the other guard?

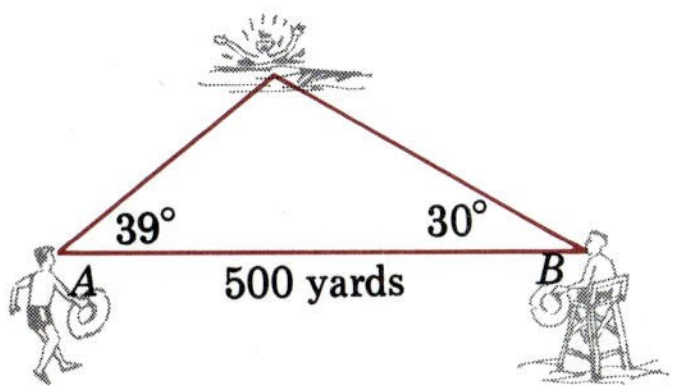

FIGURE 7.14

24. Two houses are located at points A and B, on opposite sides of a river (Figure 7.15). What is the distance between A and B?

25. The angle measurements shown in Figure 7.16 were taken on opposite sides of a tunnel through a mountain. Compute the height of the mountain.

26. The building in Figure 7.17 casts a shadow 200 feet long on the hill. How tall is side b of the building?

27. A ship leaves port in an easterly direction and later turns to the southeast. After traveling 10 miles in a southeasterly direction, the ship is 15 miles from port. How far did the ship travel in the easterly direction?

28. Show that there is no triangle with $a = 3, b = 5$, and $\alpha = 80°$.

29. By using Figure 7.3, prove that $(\sin \beta)/b = (\sin \gamma)/c$ if all three angles of the triangle are acute.

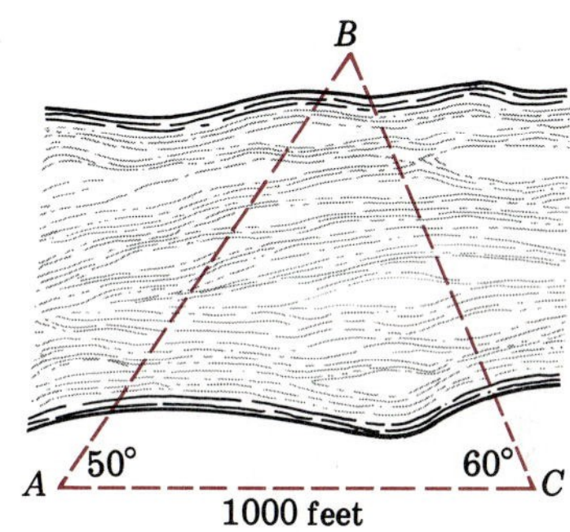

FIGURE 7.15

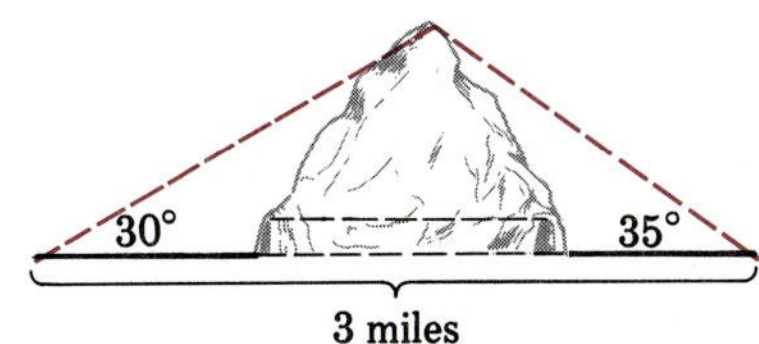

FIGURE 7.16

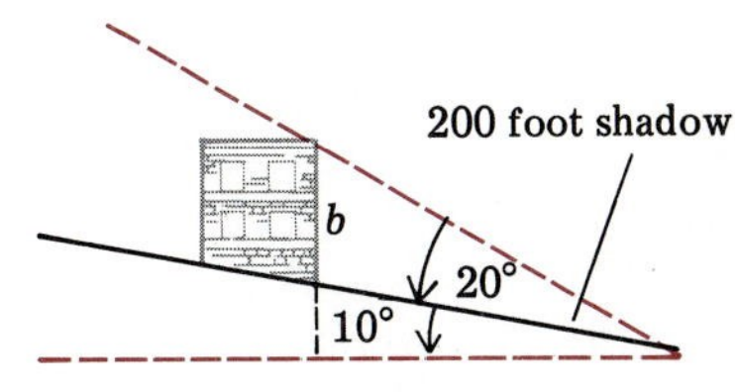

FIGURE 7.17

7.2 THE LAW OF COSINES

The companion of the Law of Sines is the Law of Cosines. It too can help us solve an arbitrary triangle when appropriate information about the sides and angles is given.

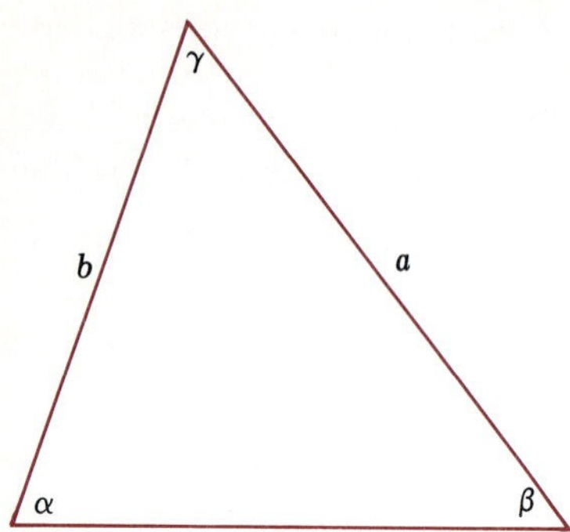

FIGURE 7.18

For an arbitrary triangle, the following three formulas, which differ only in the side of the triangle whose length is featured, comprise the Law of Cosines (see Figure 7.18):

Law of Cosines

$$a^2 = b^2 + c^2 - 2bc \cos \alpha \tag{1}$$

$$b^2 = c^2 + a^2 - 2ca \cos \beta \tag{2}$$

$$c^2 = a^2 + b^2 - 2ab \cos \gamma \tag{3}$$

Notice that if the angle in any of the formulas (1)–(3) is 90°, then the triangle is a right triangle, the cosine of the angle is 0, and the corresponding formula is just the Pythagorean Theorem. As a result, the Pythagorean Theorem is a special case of the Law of Cosines.

Since (1)–(3) can be proved by the same method, we will prove only (1). To that end, let a triangle be given. Since the shape and size of the triangle are independent of the location in a coordinate system, we let the triangle be placed as in Figure 7.19. The distance a between the points B and C is given by the distance formula:

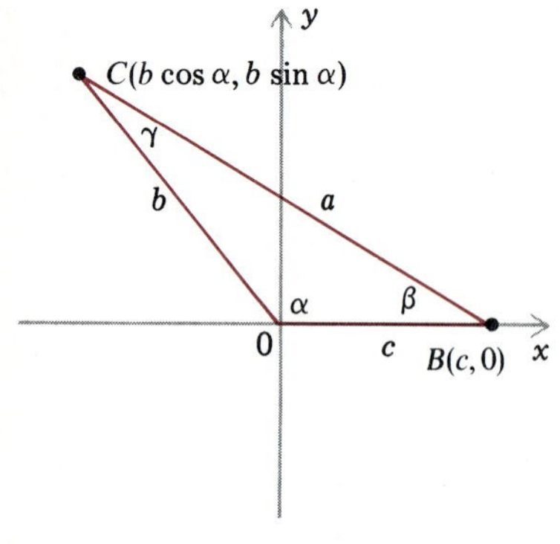

FIGURE 7.19

$$a = \sqrt{(b \cos \alpha - c)^2 + (b \sin \alpha - 0)^2}$$

Squaring both sides, rearranging, and using the fact that

$$\sin^2 \alpha + \cos^2 \alpha = 1$$

we find that

$$\begin{aligned} a^2 &= (b \cos \alpha - c)^2 + (b \sin \alpha - 0)^2 \\ &= b^2 \cos^2 \alpha - 2bc \cos \alpha + c^2 + b^2 \sin^2 \alpha \\ &= b^2(\cos^2 \alpha + \sin^2 \alpha) + c^2 - 2bc \cos \alpha \\ &= b^2 + c^2 - 2bc \cos \alpha \end{aligned}$$

This proves (1).

If we solve (1)–(3) for their respective cosine terms, we obtain the following formulas:

$$\cos \alpha = \frac{b^2 + c^2 - a^2}{2bc} \tag{4}$$

$$\cos \beta = \frac{c^2 + a^2 - b^2}{2ca} \tag{5}$$

$$\cos \gamma = \frac{a^2 + b^2 - c^2}{2ab} \tag{6}$$

These formulas will be helpful in the solutions of triangles later.

Solutions of Triangles Using the Law of Cosines

The Law of Cosines may be used to solve triangles, provided that either

(a) the lengths of the three sides are known (Figure 7.20a), or
(b) the lengths of two sides and the included angle are known (Figure 7.20b).

Our first two examples illustrate (a) and (b), in turn.

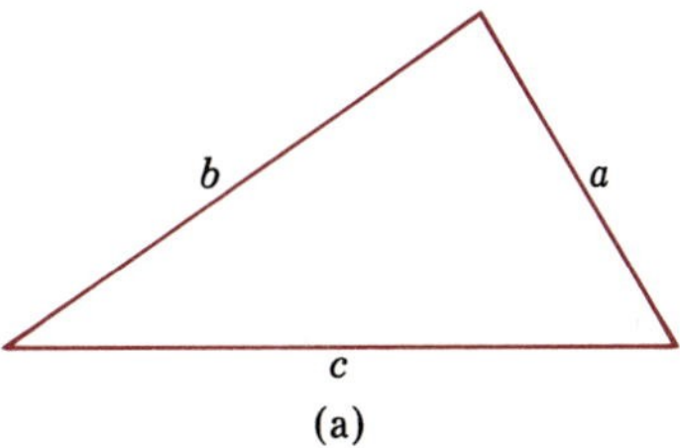

(a)

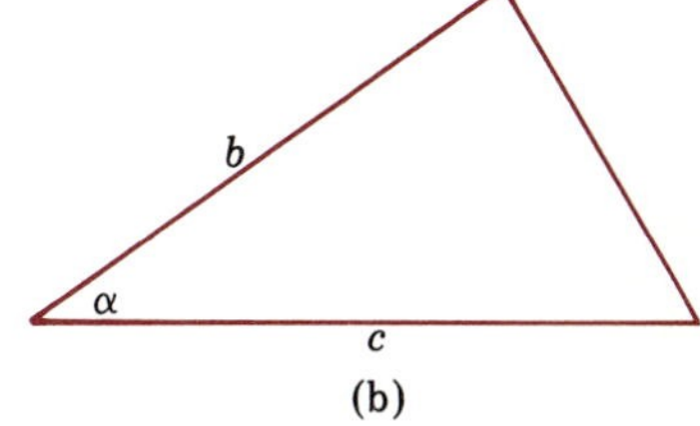

(b)

FIGURE 7.20

EXAMPLE 1. Solve the triangle shown in Figure 7.21.

FIGURE 7.21

Solution. We have $a = 4, b = 5$, and $c = 2$, and we must determine α, β, and γ. Using (4)–(6), we obtain

$$\cos \alpha = \frac{b^2 + c^2 - a^2}{2bc} = \frac{5^2 + 2^2 - 4^2}{2(5)(2)} = \frac{13}{20}$$
$$= 0.65$$
$$\cos \beta = \frac{c^2 + a^2 - b^2}{2ca} = \frac{2^2 + 4^2 - 5^2}{2(2)(4)} = \frac{-5}{16}$$
$$= -0.3125$$
$$\cos \gamma = \frac{a^2 + b^2 - c^2}{2ab} = \frac{4^2 + 5^2 - 2^2}{2(4)(5)} = \frac{37}{40}$$
$$= 0.925$$

By calculator we find that

$$\alpha \approx 49.46^\circ \; (\approx 49^\circ 28')$$

and

$$\gamma \approx 22.33 \; (\approx 22^\circ 20')$$

Consequently

$$\beta = 180^\circ - \alpha - \gamma \approx 180^\circ - 49.46^\circ - 22.33^\circ$$
$$= 108.21^\circ \; (\approx 108^\circ 13') \quad \square$$

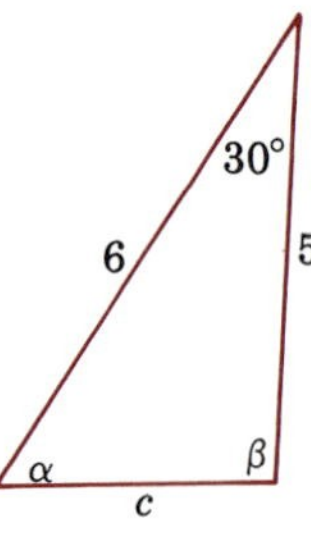

FIGURE 7.22

EXAMPLE 2. Solve the triangle shown in Figure 7.22.

Solution. Because $a = 5$, $b = 6$, and $\gamma = 30°$, we need only find c, α, and β. Applying (3), we deduce that

$$\begin{aligned} c^2 &= a^2 + b^2 - 2ab\cos\gamma \\ &= 5^2 + 6^2 - 2(5)(6)\cos 30° \\ &= 61 - 60\left(\frac{\sqrt{3}}{2}\right) \\ &= 61 - 30\sqrt{3} \end{aligned}$$

By calculator we find that $c \approx 3.006$. Next we calculate α and β. For α we use (4):

$$\begin{aligned} \cos\alpha &= \frac{b^2 + c^2 - a^2}{2bc} \approx \frac{6^2 + (61 - 30\sqrt{3}) - 5^2}{2(6)(3.006)} \\ &\approx 0.5555 \end{aligned}$$

By calculator we find that

$$\alpha \approx 56.25° \ (\approx 56°15')$$

To determine β, we can either use (5) or the values of α and γ. We choose the latter course because it entails less computation:

$$\beta = 180° - \alpha - \gamma \approx 180° - 56.25° - 30° \approx 93.75° \ (\approx 93°.45') \quad \square$$

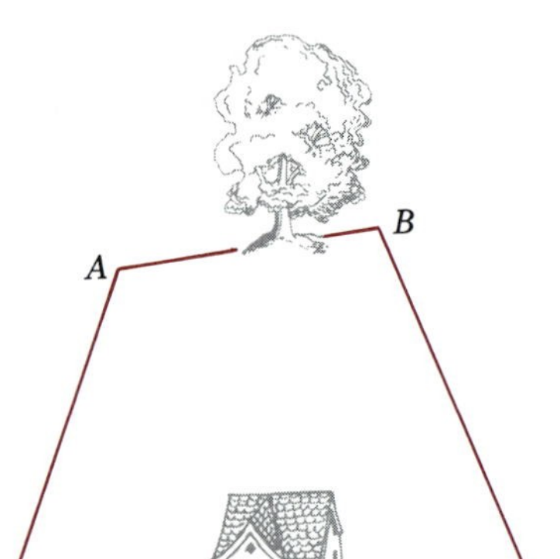

FIGURE 7.23

The Law of Cosines is especially pertinent in surveying. Suppose we wish to determine the distance between two corners A and B of our property (Figure 7.23). Unless there is an obstacle on the line joining A and B, a tape measure suffices. But let us assume that there is a mature oak tree stretched out over the line joining A and B, as in Figure 7.23. A tape measure no longer gives an accurate estimate of the length. What a surveyor would do then is to ***triangulate***, which means to locate a point C from which both A and B can be seen without obstruction, and use the triangle ABC to evaluate the length of AB (Figure 7.24). Measuring AC and BC is simple, and finding the measure of the angle at C is simple if one is armed with a transit, so finding the length of AB becomes straightforward, falling under the heading of case (b).

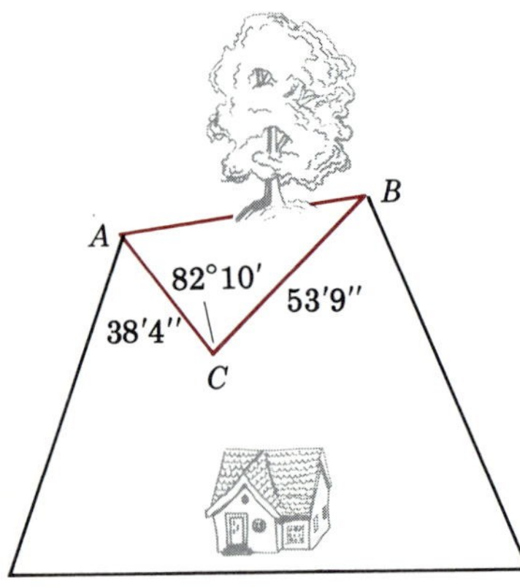

FIGURE 7.24

EXAMPLE 3. Suppose that a surveyor places a transit at point C in Figure 7.24 and gathers the data provided in the figure. Determine the distance c between A and B.

Solution. In order to use (3), we must have lengths in feet alone (rather than feet and inches). Thus we convert 53 feet 9 inches into 53.75 feet, and

38 feet 4 inches into approximately 38.33 feet. Now (3) applies, with $a = 53.75$, $b \approx 38.33$, $c =$ the length of the side to be determined, and $\gamma = 82°10'$. We find that

$$\begin{aligned} c^2 &= a^2 + b^2 - 2ab \cos \gamma \\ &= (53.75)^2 + (38.33)^2 - 2(53.75)(38.33)(\cos 82°10') \end{aligned}$$

By calculator we deduce that

$$c \approx 61.62$$

This means that the distance between the corners A and B of the property is approximately 61.62 feet, that is, 61 feet and 7 inches. □

Recapitulation

A triangle can be solved if we know one of the following:

i. the lengths of all three sides (Law of Cosines)
ii. the lengths of two sides and the included angle (Law of Cosines)
iii. the lengths of two sides and the angle opposite one of these sides, although the solution may not be unique (ambiguous case of the Law of Sines)
iv. two angles and the length of one side (Law of Sines)

The other general case, in which only the three angles are known, does not identify the triangle, as Figure 7.25 demonstrates.

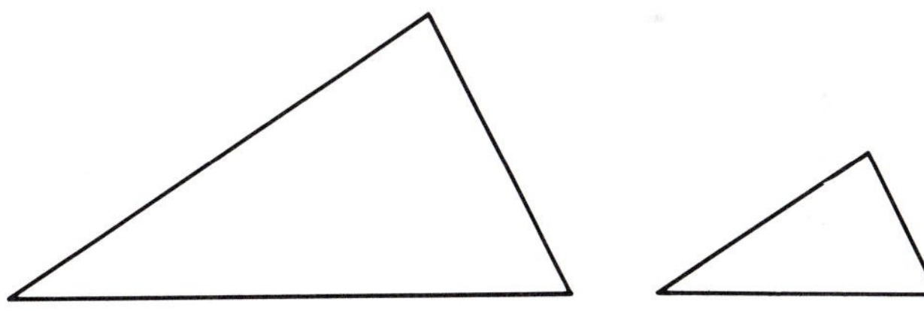

Two triangles with the same angles

FIGURE 7.25

EXERCISES 7.2

In Exercises 1–14, solve the triangle satisfying the given data.

1. $a = 6, b = 7, c = 9$

2. $a = 3, b = 12, c = 10$

3. $a = 3, b = 5, c = 7$

4. $a = 3, b = 5, c = 4.3$

5. $a = 2.3, b = 3.1, c = 3.7$

6. $a = 1, b = \frac{24}{7}, c = \frac{25}{7}$

7. $\alpha = 20°, b = 3, c = 4$

8. $\alpha = 64°, b = 7, c = 2.5$

9. $\beta = 11.7°, a = 3.2, c = 3.1$

10. $\beta = 130°, a = 5, c = 2$

11. $\gamma = 171.7°$, $a = 6.3$, $b = 4.7$

12. $\gamma = 42.8°$, $a = 14.3$, $b = 14.3$

13. $\alpha = 52°41'$, $b = 1.3$, $c = 5.4$

14. $\beta = 115°9'$, $a = 0.79$, $c = 0.61$

In Exercises 15–24 use the Law of Sines or the Law of Cosines to solve the triangle satisfying the given data.

15. $\beta = 58°30'$, $b = 5.7$, $c = 6.1$

16. $a = 0.54$, $b = 0.31$, $c = 0.33$

17. $\gamma = 99.9°$, $a = 5$, $c = 8.4$

18. $\alpha = 36°5'$, $a = 8.1$, $b = 11.4$

19. $\beta = 37°24'$, $a = 2.1$, $c = 4.8$

20. $\alpha = 96.4°$, $\gamma = 77.7°$, $a = 0.58$

21. $a = 17.1$, $b = 19.5$, $c = 10$

22. $\gamma = 14°3'$, $a = 6.4$, $b = 5.9$

23. $\beta = 151°7'$, $b = 0.3$, $c = 0.1$

24. $\beta = 6.3°$, $\gamma = 11.5°$, $c = 1.4$

25. Six dormitories are located at the vertices of a regular hexagon, as shown in Figure 7.26. How far is it from dormitory A to dormitory B?

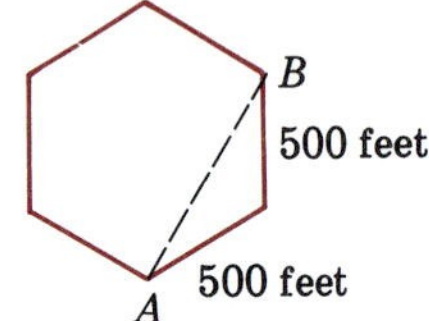

FIGURE 7.26

26. Points A and C on a gas line are separated by a fire station. Using the information in Figure 7.27, find the distance between A and C.

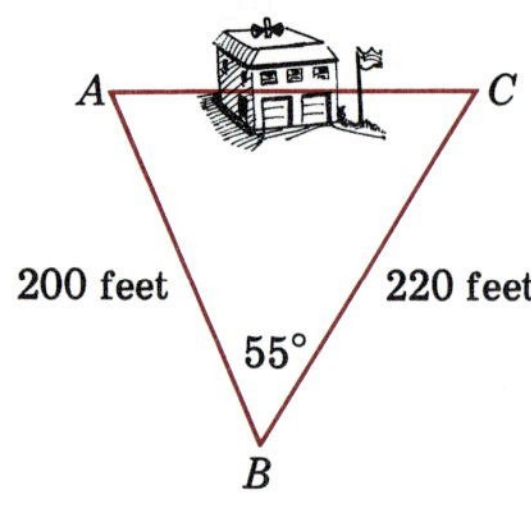

FIGURE 7.27

27. Two ships leave port at the same time, one traveling south at 20 miles per hour and the other traveling northwest at 30 miles per hour. How far apart are the ships after 30 minutes?

28. Two airplanes leave an airport at the same time. One flies north at 200 miles per hour; the other flies in a northeasterly direction at 300 miles per hour. After two hours the planes are 700 miles apart. How many degrees east from due north is the second plane flying?

29. Logs of length 10, 12, and 14 feet are placed in a triangular pattern around a campfire. What angles do the logs form?

30. In Yankee Stadium the distance between home plate and dead center field is 417 feet. Determine the distance from dead center field to third base. (*Hint:* The distance between home plate and third base is 90 feet.)

31. A carpenter plans to make a triangular frame, two sides having length 6 feet and 8 feet and forming an angle of 35° with each other. What must the length of the other side be?

32. One diagonal of a given parallelogram is 4 inches long, and the other is 7 inches long. They form an angle of 38° with one another. Determine the lengths of the sides of the parallelogram. (*Hint:* The diagonals of a parallelogram bisect each other.)

33. Find the angles of the triangle whose vertices are the points (0, 0), (3, 0), and (2, 1).

34. Find the angles of the triangle whose vertices are the points (−1, 3), (2, 2), and (0, 1).

35. A vertical flagpole is situated on a hillside that makes an angle of 15° with the horizontal. Two supporting wires are to be attached, as shown in Figure 7.28. How long should the wires be?

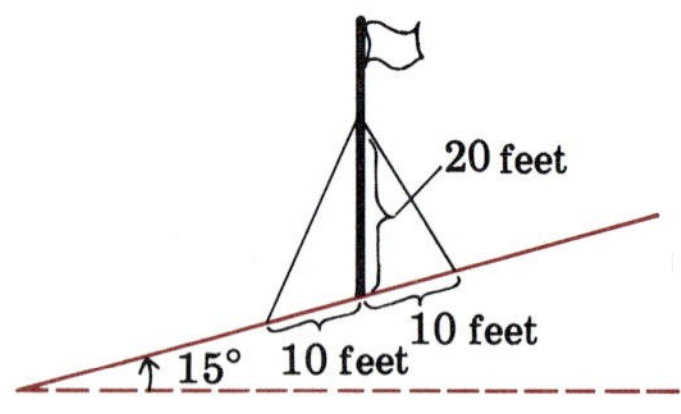

FIGURE 7.28

***36.** Ship A is moving away from port at a speed of 5 miles per hour in the direction 30° west of north, as indicated in Figure 7.29. Ship B is 8 miles due south of port and is moving at a speed of 7 miles per hour. How many degrees west of north should B travel in

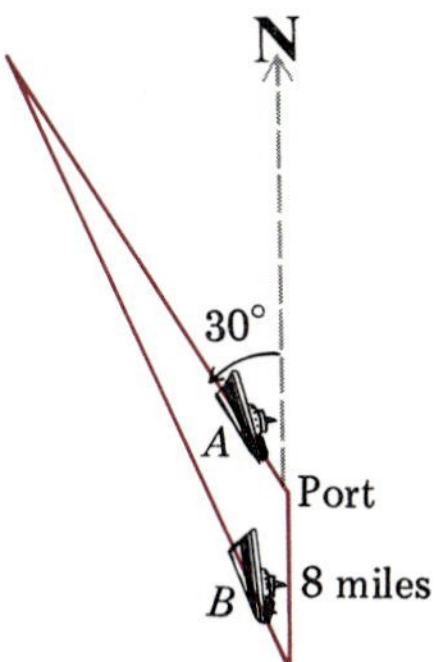

FIGURE 7.29

order to intercept A? (*Hint:* If t is the number of hours required for B to overtake A, then the triangle in Figure 7.29 has sides of length 8, $5t$, and $7t$ miles.)

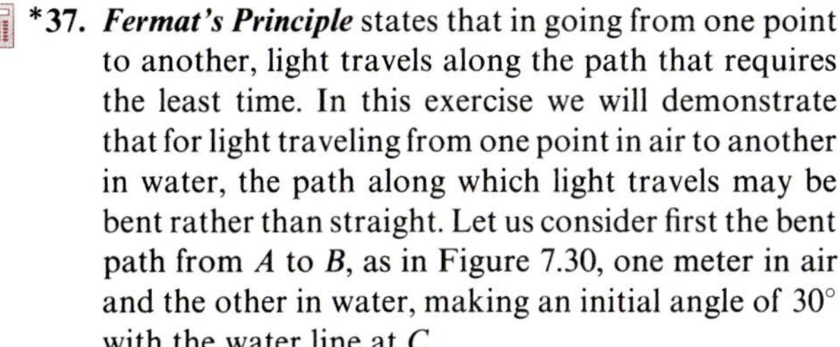

*37. ***Fermat's Principle*** states that in going from one point to another, light travels along the path that requires the least time. In this exercise we will demonstrate that for light traveling from one point in air to another in water, the path along which light travels may be bent rather than straight. Let us consider first the bent path from A to B, as in Figure 7.30, one meter in air and the other in water, making an initial angle of 30° with the water line at C.

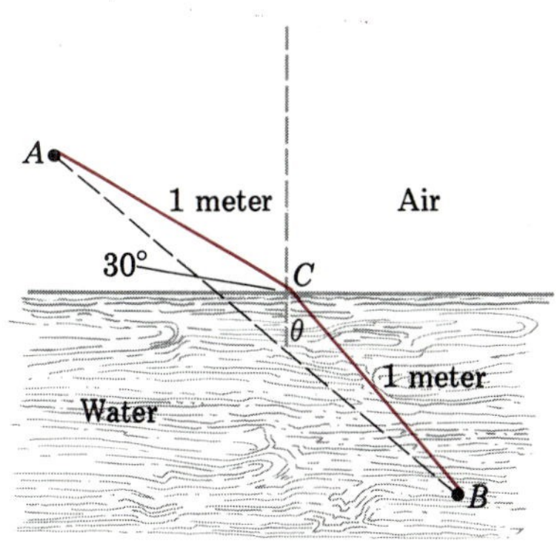

FIGURE 7.30

a. Using the speed c of light in air and the approximate speed $c/1.333$ of light in water, determine the time required for light to travel along the bent path ACB.

*b. Use Snell's Law (see Section 7.2) to determine the angle θ in Figure 7.30, next find angles A and B, and then use the Law of Cosines to determine the lengths of the segments of the straight line AB. From this information deduce the time required for light to travel in a straight line from A to B.

38. Using (1), (2), and (3), show that

$$\frac{a^2 + b^2 + c^2}{2abc} = \frac{\cos \alpha}{a} + \frac{\cos \beta}{b} + \frac{\cos \gamma}{c}$$

7.3 THE AREA OF A TRIANGLE

Let triangle ABC have base b, height h, and area $\mathscr{A}$ (Figure 7.31). The formula

$$\mathscr{A} = \frac{1}{2}bh \tag{1}$$

is the most familiar of several formulas for the area of the triangle. The formula can be verified by noticing that triangle I in Figure 7.32 is half of

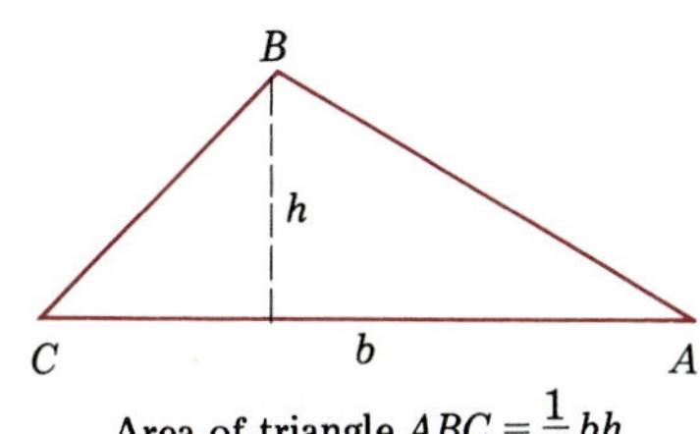

Area of triangle $ABC = \frac{1}{2}bh$

FIGURE 7.31

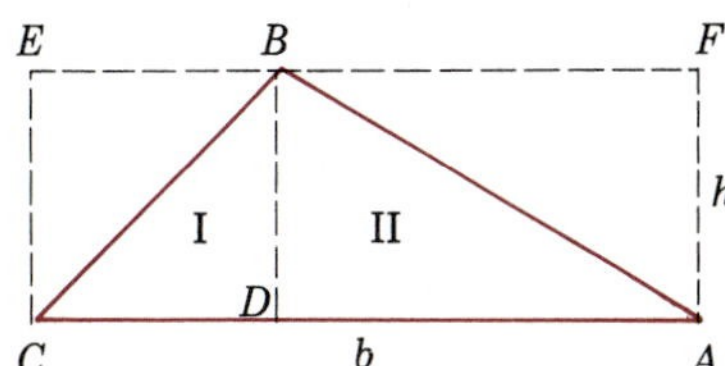

FIGURE 7.32

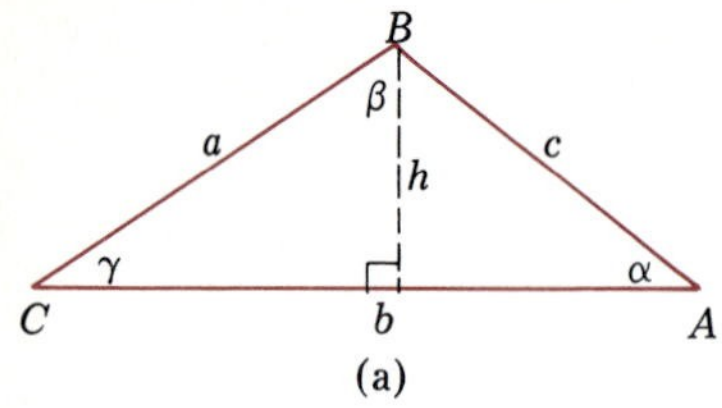

(a)

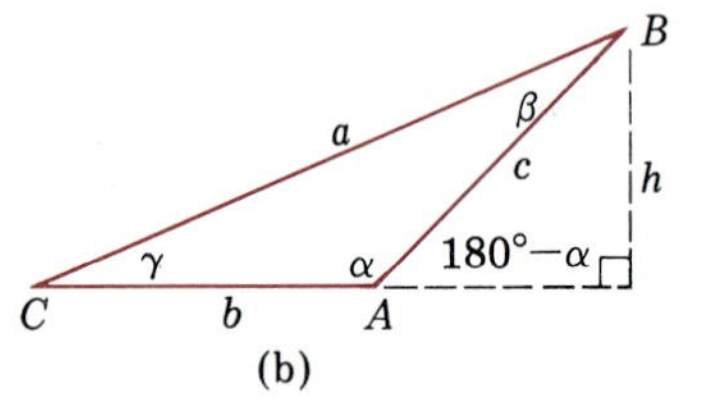

(b)

FIGURE 7.33

rectangle $CDBE$, and triangle II is half of rectangle $AFBD$, so that the area of triangle ABC is $\frac{1}{2}$ the area bh of rectangle $AFEC$.

If b and h are known, then computing the area of triangle ABC by (1) is straightforward. But if instead we know the lengths of two sides and the angle between them, or the lengths of all three sides of the triangle, and do not know h, then other formulas for the area are easier to apply. It is to these other formulas that we turn now.

First of all, suppose that for triangle ABC, b, c, and α are known (Figure 7.33a, b). By dropping a perpendicular from B, as in the figures, we see that in each case,

$$b = \text{base} \quad \text{and} \quad h = \text{height} = c \sin \alpha$$

(Note in Figure 7.33b that $h = c \sin(180° - \alpha) = c \sin \alpha$.) Therefore by (1) the area $\mathscr{A}$ of triangle ABC is given by

$$\mathscr{A} = \frac{1}{2}bh = \frac{1}{2}b(c \sin \alpha) = \frac{1}{2}bc \sin \alpha$$

By dropping perpendiculars from the other two vertices in turn, we obtain corresponding formulas involving $\sin \beta$ and $\sin \gamma$. All told, we have:

$$\mathscr{A} = \frac{1}{2}bc \sin \alpha \qquad (2)$$

$$\mathscr{A} = \frac{1}{2}ac \sin \beta \qquad (3)$$

$$\mathscr{A} = \frac{1}{2}ab \sin \gamma \qquad (4)$$

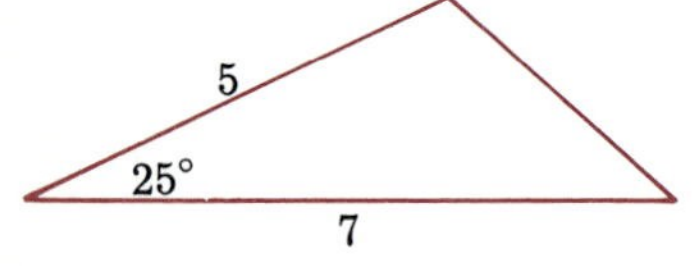

FIGURE 7.34

EXAMPLE 1. Find the area of the triangle shown in Figure 7.34.

Solution. Using (4) with $a = 5$, $b = 7$, and $\gamma = 25°$, we have

$$\mathscr{A} = \frac{1}{2}(5)(7) \sin 25°$$

By calculator we find that $\mathscr{A} \approx 7.396$. □

If we know the lengths a, b, and c of all three sides of triangle ABC, then we can compute the area by means of a formula called Heron's Formula. To derive this formula we let

$$s = \frac{1}{2}(a + b + c) \qquad (5)$$

and observe that

$$a + b - c = a + b + c - 2c = 2s - 2c$$

so that

$$a + b - c = 2(s - c) \tag{6}$$

Similarly, by interchanging the sides of the triangle, we have

$$b + c - a = 2(s - a) \tag{7}$$

and

$$c + a - b = 2(s - b) \tag{8}$$

From (4) above and the Law of Cosines, we find that

$$\begin{aligned}
16\mathscr{A}^2 &= 16\left(\frac{1}{2}ab \sin \gamma\right)^2 = 4a^2b^2 \sin^2 \gamma \\
&= 4a^2b^2(1 - \cos^2 \gamma) \overset{\substack{\text{Law of}\\ \text{Cosines}}}{=} 4a^2b^2\left[1 - \left(\frac{a^2 + b^2 - c^2}{2ab}\right)^2\right] \\
&= 4a^2b^2\left[1 - \frac{(a^2 + b^2 - c^2)^2}{4a^2b^2}\right] \\
&= 4a^2b^2 - (a^2 + b^2 - c^2)^2 \\
&= [2ab + (a^2 + b^2 - c^2)][2ab - (a^2 + b^2 - c^2)] \\
&= [(a^2 + 2ab + b^2) - c^2][c^2 - (a^2 - 2ab + b^2)] \\
&= [(a + b)^2 - c^2][c^2 - (a - b)^2] \\
&= [(a + b + c)(a + b - c)][(c + a - b)(c - a + b)]
\end{aligned}$$

Using (5)–(8), we can write the above in the form

$$\begin{aligned}
16\mathscr{A}^2 &= (2s)[2(s - c)][2(s - b)][2(s - a)] \\
&= 16s(s - a)(s - b)(s - c)
\end{aligned}$$

It follows that

$$\boxed{\mathscr{A} = \sqrt{s(s - a)(s - b)(s - c)}}$$

which is called ***Heron's Formula*** (alternatively known as ***Hero's Formula***), after the Greek mathematician Heron, who lived in the first century A.D.

(Actually, the formula was known to Archimedes, who lived four centuries earlier.)

Since s is determined by the lengths of the sides of the triangle, Heron's Formula yields the area of a triangle once the lengths of all three sides are known.

EXAMPLE 2. Find the area of a triangle having sides of length 3, 5, and 6.

Solution. If we let $a = 3$, $b = 5$, and $c = 6$, then

$$s = \frac{1}{2}(a + b + c) = \frac{1}{2}(3 + 5 + 6) = 7$$

so that by Heron's Formula,

$$\begin{aligned}\mathcal{A} &= \sqrt{7(7-3)(7-5)(7-6)}\\ &= \sqrt{56} = 2\sqrt{14} \quad \square\end{aligned}$$

EXERCISES 7.3

In Exercises 1–16, find the area of the triangle satisfying the given data.

1. $\alpha = 45°$, $b = 4$, $c = 3$
2. $\alpha = 27°$, $b = 8$, $c = 10$
3. $\alpha = 16°$, $b = 4.2$, $c = 4.2$
4. $\beta = 120°$, $a = 5$, $c = 10$
5. $\beta = 90°$, $a = 8$, $c = 2.5$
6. $\gamma = 153.2°$, $a = 6.3$, $b = 2.1$
7. $\gamma = 172°18'$, $a = 5.7$, $b = 8.1$
8. $\gamma = 2°$, $a = 9.3$, $b = 5.8$
9. $a = 3$, $b = 6$, $c = 7$
10. $a = 4$, $b = 3$, $c = 2$
11. $a = 1.5$, $b = 1$, $c = 1.5$
12. $a = 4$, $b = 10$, $c = 7$
13. $a = 2$, $b = 2$, $c = 2$
14. $a = 3.2$, $b = 4.5$, $c = 6.1$
15. $\alpha = 40°$, $\beta = 60°$, $c = 3.8$
16. $\beta = 108°$, $\gamma = 41°$, $a = 1.6$
17. Let the sides of a triangle have positive lengths a, b, and c. Prove that $s > a$. (Likewise, $s > b$ and $s > c$, so that $s(s-a)(s-b)(s-c) > 0$ for Heron's Formula.)
18. Suppose the coordinates of the vertices of a triangle are (x_1, y_1), (x_2, y_2), and (x_3, y_3), listed in the order in which they occur as we traverse the triangle counterclockwise. Then it can be shown that the area $\mathcal{A}$ of the triangle is

$$\mathcal{A} = \frac{1}{2}(x_1y_2 - y_1x_2) + \frac{1}{2}(x_2y_3 - y_2x_3) + \frac{1}{2}(x_3y_1 - y_3x_1)$$

Find the area $\mathcal{A}$ of the triangle having vertices
a. $(0, 0)$, $(2, 4)$, and $(-2, 2)$
b. $(-1, -2)$, $(2, 5)$, and $(5, 2)$

7.4 POLAR COORDINATES

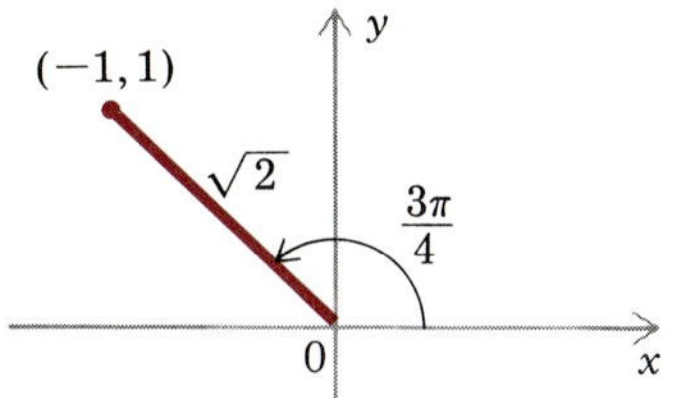

FIGURE 7.35

Until this point we have always described points in the plane by their rectangular (or Cartesian) coordinates (x, y). However, there is another commonly used system for describing points: the polar coordinate system.

To describe the polar coordinate system, consider the point P with rectangular coordinates $(-1, 1)$. The distance between P and the origin is

$$\sqrt{(-1-0)^2 + (1-0)^2}$$

which is equal to $\sqrt{2}$ (Figure 7.35). Moreover, the line joining P and the origin makes an angle of $3\pi/4$ radians with the positive x axis. The numbers $\sqrt{2}$ and $3\pi/4$, representing the distance and the angle, suffice to locate the point P: Just start at the origin and proceed $\sqrt{2}$ units along the line that makes an angle of $3\pi/4$ radians with the positive x axis. The point P is at the end of the journey.

More generally, if we start with a given rectangular coordinate system, then to any point P in the plane except the origin we can associate an ordered pair of polar coordinates (r, θ), where r is the distance between P and the origin, and where the line segment joining P and the origin makes an angle of θ radians with the positive x axis (Figure 7.36). Since there are infinitely many choices for θ (any two differing by a multiple of 2π), P has infinitely many sets of polar coordinates. For convenience we also consider the pair $(-r, \theta + \pi)$ to be a set of polar coordinates for P. (You may think of $(-r, \theta + \pi)$ as corresponding to the point reached by proceeding a distance r in the direction opposite that of the angle $\theta + \pi$ (Figure 7.37).) Therefore if (r, θ) is one set of polar coordinates of P, then any other set will have the form

$$\begin{aligned} &(r, \theta + 2n\pi), \quad \text{where } n \text{ is an integer} \\ &(-r, \theta + \pi + 2n\pi), \quad \text{where } n \text{ is an integer} \end{aligned} \tag{1}$$

To the origin we assign polar coordinates $(0, \theta)$, where θ may be any number. We have thus defined the ***polar coordinate system***. The origin is called the ***pole*** of the polar coordinate system.

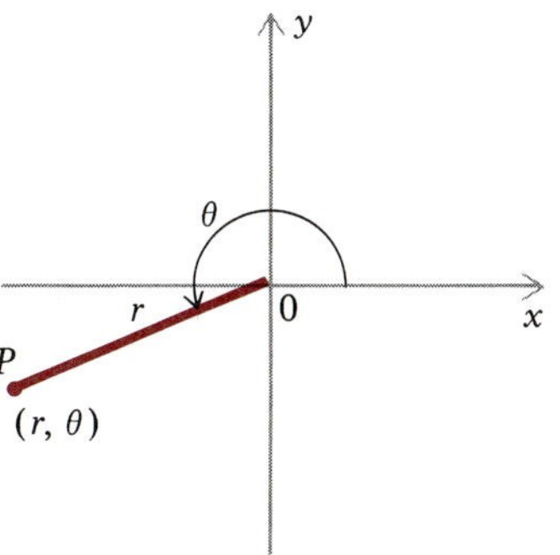

FIGURE 7.36

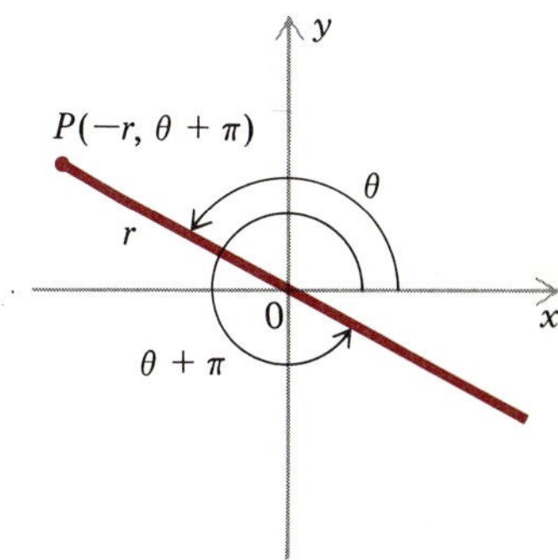

FIGURE 7.37

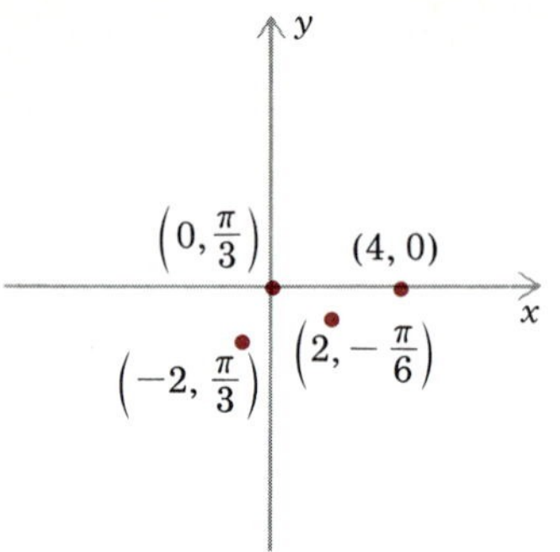

FIGURE 7.38

EXAMPLE 1. Plot the points having polar coordinates $(0, \pi/3)$, $(4, 0)$, $(-2, \pi/3)$, and $(2, -\pi/6)$.

Solution. The points are plotted in Figure 7.38. □

Graph paper for rectangular coordinates contains horizontal lines (on which y is constant) and vertical lines (on which x is constant) (Figure 7.39a). In contrast, on graph paper specially designed for polar coordinates, the polar coordinate r is constant on circles centered at the origin and the polar coordinate θ is constant on lines emanating from the origin. Figure 7.39b depicts polar coordinate graph paper.

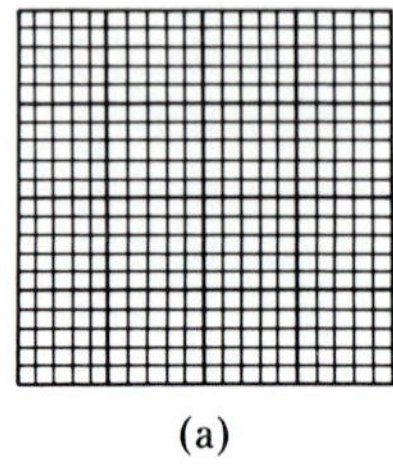

(a)

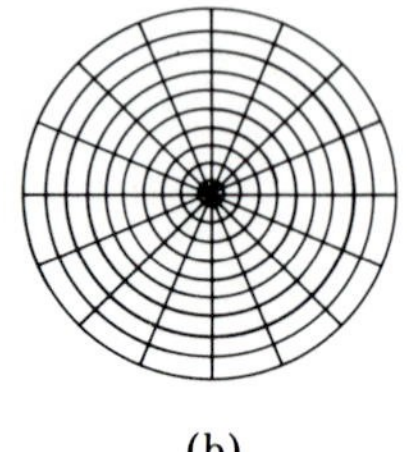

(b)

FIGURE 7.39

The dance a bee performs to communicate the location of a source of food seems to be related to polar coordinates. The orientation of the bee's body locates the direction of the food, and the intensity of the dance indicates the distance from the source.

Conversion between Cartesian and Polar Coordinates

Every point in the plane has both Cartesian and polar coordinates. We will now derive the formulas used for converting from one type of coordinates to the other.

Suppose a point P in the plane has polar coordinates (r, θ) and Cartesian coordinates (x, y). Then

$$\cos \theta = \frac{x}{r} \quad \text{and} \quad \sin \theta = \frac{y}{r}$$

so that

$$\boxed{x = r \cos \theta \quad \text{and} \quad y = r \sin \theta} \tag{2}$$

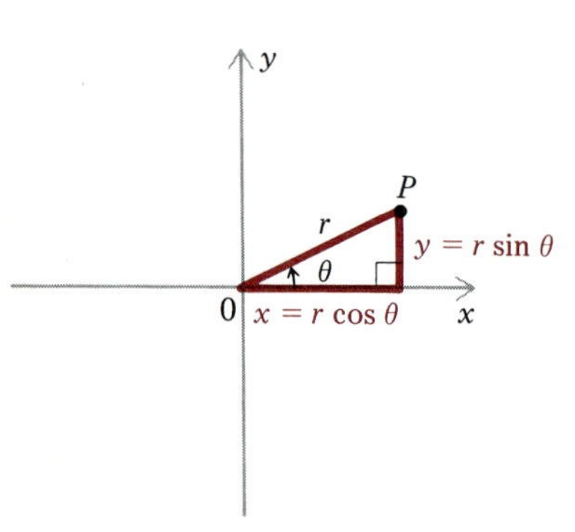

FIGURE 7.40

(Figure 7.40). It follows that

$$\boxed{r^2 = x^2 + y^2 \quad \text{and} \quad \tan \theta = \frac{y}{x} \quad \text{for} \quad x \neq 0} \tag{3}$$

From (2) we see that x and y are uniquely determined by r and θ.

EXAMPLE 2. Find the Cartesian coordinates of the point P having polar coordinates $(2, 35\pi/6)$.

Solution. From (2) we obtain

$$x = 2\cos\frac{35\pi}{6} = 2\cos\left(-\frac{\pi}{6}\right) = 2\cos\frac{\pi}{6} = 2\left(\frac{\sqrt{3}}{2}\right) = \sqrt{3}$$

and

$$y = 2\sin\frac{35\pi}{6} = 2\sin\left(-\frac{\pi}{6}\right) = -2\sin\frac{\pi}{6} = -2\left(\frac{1}{2}\right) = -1$$

Therefore the Cartesian coordinates of P are $(\sqrt{3}, -1)$. □

Although we cannot determine unique values for r and θ from the pair (x, y) of Cartesian coordinates by simply applying (3), it is possible to determine all sets of polar coordinates for the point with Cartesian coordinates (x, y).

EXAMPLE 3. Find all sets of polar coordinates for the point P having Cartesian coordinates $(-7, 7\sqrt{3})$.

Solution. First we will find the polar coordinates (r, θ) such that $r > 0$ and $0 \le \theta < 2\pi$. For that purpose we use (3), which tells us that

$$r^2 = (-7)^2 + (7\sqrt{3})^2 = 49 + 147 = 196$$

Because we desire that $r > 0$, we conclude that

$$r = \sqrt{196} = 14$$

We also know from (3) that

$$\tan\theta = \frac{7\sqrt{3}}{-7} = -\sqrt{3}$$

Therefore θ could be either $2\pi/3$ or $5\pi/3$. But since the given point $(-7, 7\sqrt{3})$ lies in the second quadrant, it follows that

$$\theta = \frac{2\pi}{3}$$

Thus one set of polar coordinates for the point with Cartesian coordinates $(-7, 7\sqrt{3})$ is $(14, 2\pi/3)$. Consequently by (1) any set of polar coordinates of P

must be of the form

$$\left(14, \frac{2\pi}{3} + 2n\pi\right) \quad \text{for any integer } n$$

or

$$\left(-14, \frac{5\pi}{3} + 2n\pi\right) \quad \text{for any integer } n \quad \square$$

CAUTION: The solution of Example 3 shows that when we convert from Cartesian to polar coordinates, it is not enough merely to choose values of r and θ that satisfy (3). We must be sure as well that the point with polar coordinates (r, θ) lies in the correct quadrant.

Polar Graphs and Equations

The ***polar graph*** of an equation involving polar coordinates (r, θ) is the set of all points in the plane having a set of polar coordinates satisfying the given equation.

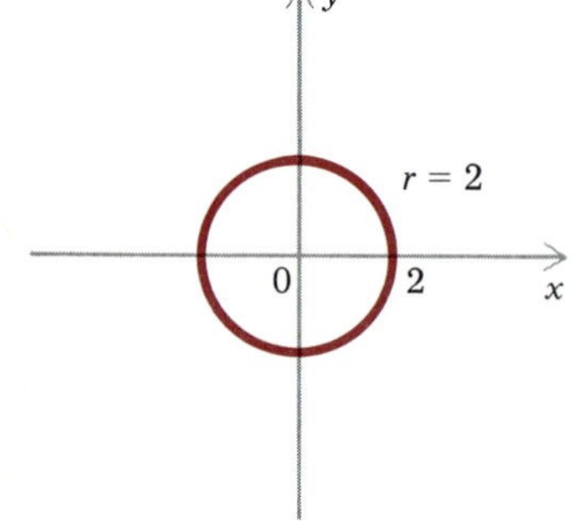

FIGURE 7.41

EXAMPLE 4. Sketch the polar graph of the equation $r = 2$.

Solution. Since every point that satisfies the equation $r = 2$ has distance 2 from the origin, evidently the graph is the circle with radius 2 and center at the origin (Figure 7.41). $\square$

More generally, if r_0 is an arbitrary positive constant, then the polar graph of the equation

$$r = r_0$$

is the circle with radius r_0 and center at the origin (Figure 7.42).

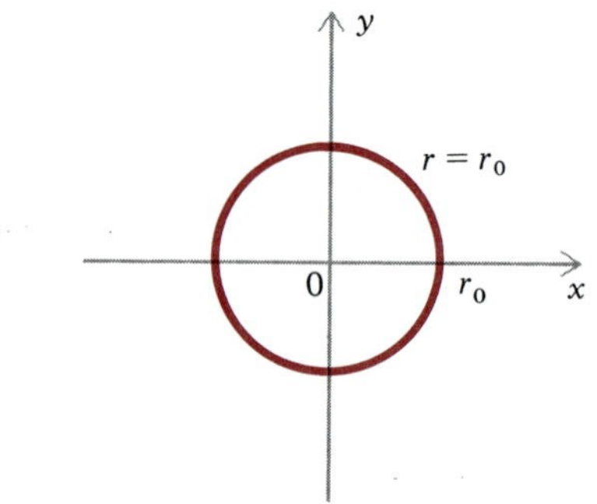

FIGURE 7.42

EXAMPLE 5. Sketch the polar graph of the equation $\theta = \pi/3$.

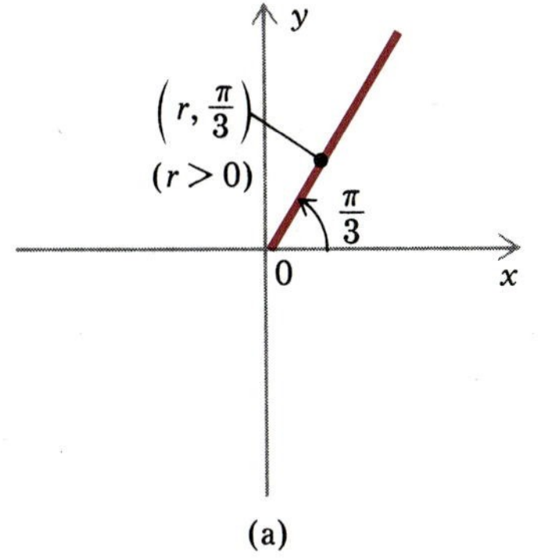

(a)

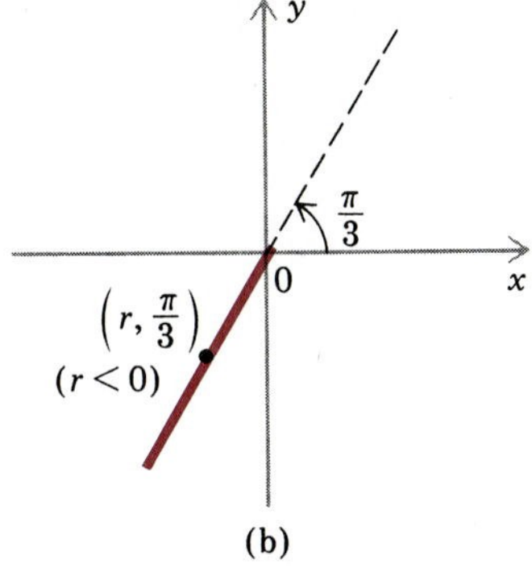

(b)

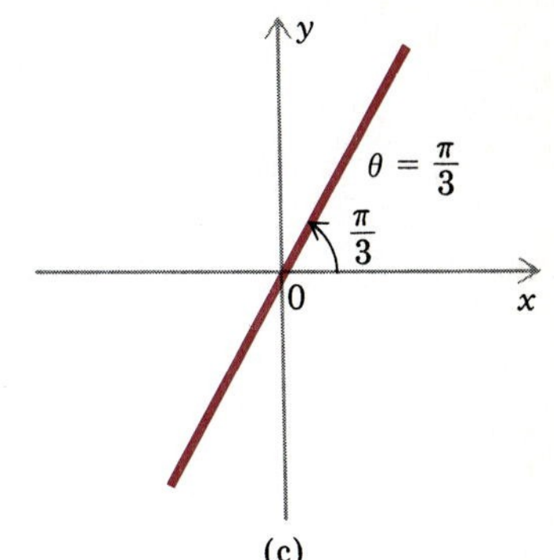

(c)

FIGURE 7.43

Solution. For any number r, the point with polar coordinates $(r, \pi/3)$ is on the graph of the given equation. If $r \geq 0$, then this point lies on the half-line shown in Figure 7.43a. But if $r < 0$, then the point corresponding to $(r, \pi/3)$ also has polar coordinates $(-r, \pi/3 + \pi)$, that is, $(-r, 4\pi/3)$, and since $-r > 0$, it lies on the solid half-line in Figure 7.43b. Therefore the polar graph of the equation $\theta = \pi/3$ is the line in Figure 7.43c. □

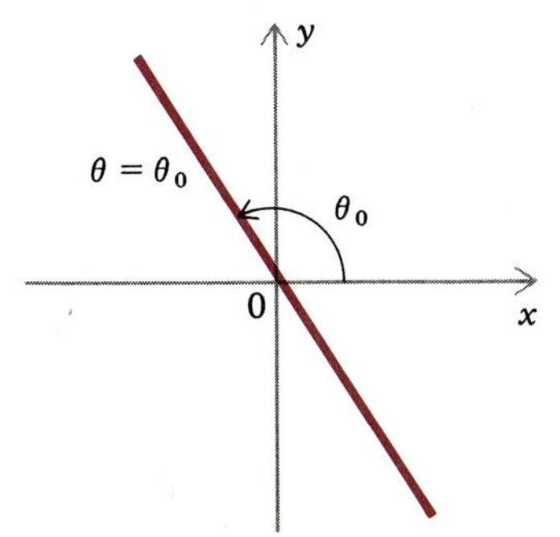

FIGURE 7.44

One can show in a similar way that for any number θ_0, the polar graph of the equation

$$\theta = \theta_0$$

is the line through the origin that makes an angle of θ_0 radians with the positive x axis (Figure 7.44).

Two other equations whose polar graphs are easy to determine are

$$r = a \cos \theta \tag{4}$$

and

$$r = a \sin \theta \tag{5}$$

To determine the graph of (4), we multiply both sides of (4) by r to obtain

$$r^2 = ar \cos \theta$$

Substituting for r^2 from (3), and for $r \cos \theta$ from (2), yields

$$x^2 + y^2 = ax$$

which means that

$$x^2 - ax + y^2 = 0$$

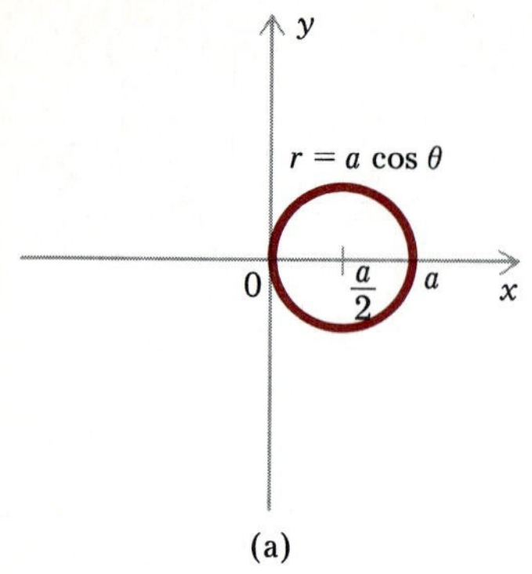

(a)

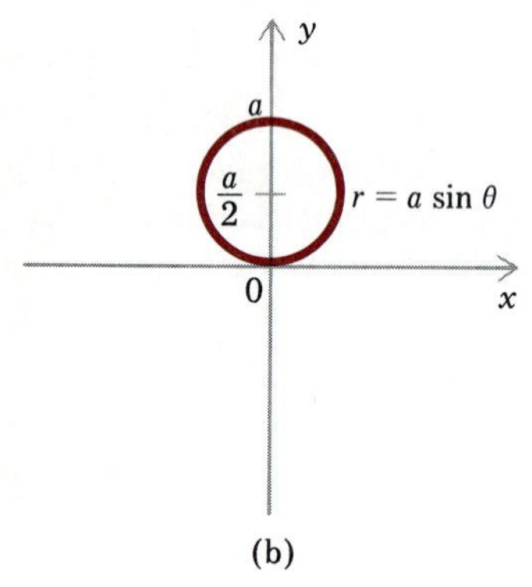

(b)

FIGURE 7.45

The graph of this equation becomes apparent once we complete the square:

$$\left(x^2 - ax + \frac{1}{4}a^2 - \frac{1}{4}a^2\right) + y^2 = 0$$

$$\left(x^2 - ax + \frac{1}{4}a^2\right) - \frac{1}{4}a^2 + y^2 = 0$$

$$\left(x - \frac{1}{2}a\right)^2 + y^2 = \frac{1}{4}a^2 = \left(\frac{1}{2}a\right)^2 \qquad (6)$$

Notice that (6) is an equation in rectangular coordinates of the circle with radius $\frac{1}{2}|a|$ and center at the point $(\frac{1}{2}a, 0)$ (Figure 7.45a), and therefore the circle is also the graph of (4). Similarly, it is possible to prove that the graph of (5) is the circle with radius $\frac{1}{2}|a|$ and center at the point $(0, \frac{1}{2}a)$ (Figure 7.45b).

EXAMPLE 6. Sketch the polar graph of the equation $r = 3 \cos \theta$.

Solution. Taking $a = 3$ in (4) and using Figure 7.45a as a guide, we obtain the graph sketched in Figure 7.46. □

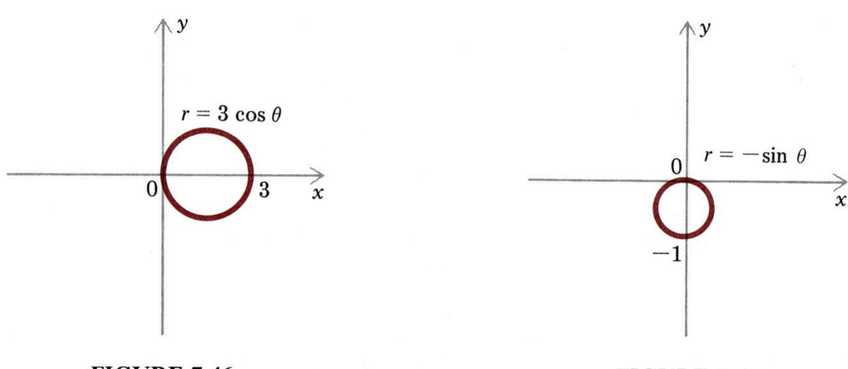

FIGURE 7.46

FIGURE 7.47

EXAMPLE 7. Sketch the polar graph of the equation $r = -\sin \theta$.

Solution. Taking $a = -1$ in (5) and using Figure 7.45b as a guide, we obtain the graph shown in Figure 7.47. □

EXERCISES 7.4

In Exercises 1–18, find the rectangular coordinates of the point having the given polar coordinates.

1. $(0, 0)$ **2.** $(0, \pi/6)$

3. $(-5, 0)$ **4.** $(4, 2\pi)$

5. $(3, \pi/2)$ **6.** $(\frac{1}{2}, -\pi/2)$

7. $(2.3, \pi)$ **8.** $(\sqrt{2}, \pi/4)$

9. $(3, \pi/3)$ **10.** $(-2, -\pi/6)$

11. $(1, -\pi/3)$

12. $(17, 3\pi/2)$

13. $(2, 7\pi)$

14. $(2\sqrt{2}, 5\pi/4)$

15. $(-4, -2\pi/3)$

16. $(-2, 23\pi/6)$

17. $(2, 1)$

18. $(-1, -4)$

In Exercises 19–30, find all sets of polar coordinates of the point having the given rectangular coordinates.

19. $(5, 0)$

20. $(0, 5)$

21. $(-5, 0)$

22. $(0, -5)$

23. $(\sqrt{2}, \sqrt{2})$

24. $(3\sqrt{2}, -3\sqrt{2})$

25. $(\sqrt{3}, 1)$

26. $(\frac{1}{3}\sqrt{3}, -1)$

27. $(-\sqrt{2}, \sqrt{6})$

28. $(-12, -4\sqrt{3})$

29. $(18, 10)$

30. $(-3, 4)$

In Exercises 31–47, sketch the polar graph of the given equation.

31. $r = 1$

32. $r = 4$

33. $r = -2$

34. $r = -\dfrac{1}{2}$

35. $\theta = 0$

36. $\theta = \dfrac{3\pi}{4}$

37. $\theta = -\dfrac{\pi}{6}$

38. $\theta = \dfrac{13\pi}{3}$

39. $r = \cos\theta$

40. $r = 5\cos\theta$

41. $r = -3\cos\theta$

42. $r = 4\sin\theta$

43. $r = 2\sin\theta$

44. $r = -\dfrac{1}{2}\sin\theta$

***45.** $r = \cos(\theta + \pi/2)$

***46.** $r = -\dfrac{1}{2}\sin(\theta + \pi/2)$

***47.** $r = \theta$

48. a. Use a graphics calculator to determine the number of "petals" in each of the following graphs.

i. $r = \sin 2\theta$ ii. $r = \sin 3\theta$

iii. $r = \sin 4\theta$ iv. $r = \sin 5\theta$

b. Using the results of part (a), predict the number of petals in the graph of $r = \sin n\theta$, where n is a positive integer.

In Exercises 49–52, use a graphics calculator to determine the relationship between the graphs of the given pairs of polar equations.

49. $r = 1 + \sin\theta,\ r = 1 + \cos\theta$

50. $r = 2 - \sin\theta,\ r = 2 + \sin\theta$

51. $r = \dfrac{1}{2} - \cos\theta,\ r = \dfrac{1}{2} + \sin\theta$

52. $r^2 = \cos 2\theta,\ r^2 = \sin 2\theta$

7.5 VECTORS

Many physical and mathematical quantities have only magnitude and thus can be described by numbers. Examples are cost, profit, speed, height, length, area, and volume. Other quantities have both magnitude and direction. Two notable examples are velocity and force. The velocity of a moving object involves not only the speed of the object but also the direction of motion. Similarly, a force acting on an object is partly determined by its magnitude and partly by the direction in which the force acts.

Quantities that involve both magnitude and direction are described mathematically by vectors. Geometrically, a vector is a directed line segment, often described by a line segment with an arrow at one end (Figure 7.48a). For example, the velocity vector of a moving object is the directed line segment that points in the direction of motion and has length equal to the speed of the object. Similarly, the vector that describes a force is the line segment that points in the direction in which the force acts and has length equal to the magnitude of the force.

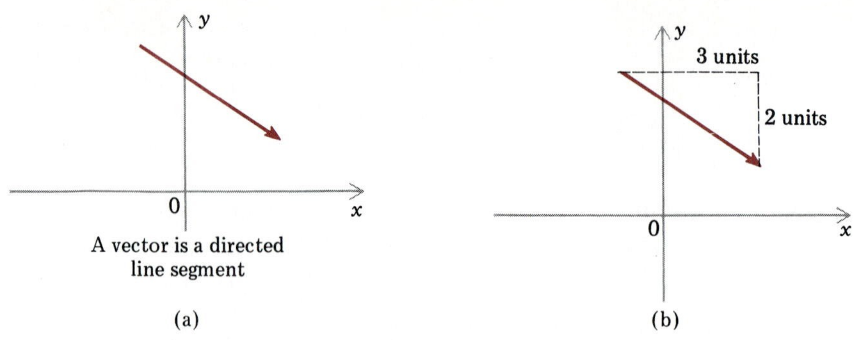

FIGURE 7.48

One simple way of describing a vector algebraically is to give the coordinates of the initial and terminal points of the directed line segment. But at least with velocity and force, one is normally concerned more with the direction and length of the line segment than with its initial and terminal points. Now observe that the vector appearing in Figure 7.48b represents a change of 3 units in the positive x direction and a change of 2 units in the negative y direction (which we will think of as a change of -2 units in the positive y direction). The numbers 3 and -2 identify the direction and length of the line segment, and hence the given vector can be described by the ordered pair $(3, -2)$. More generally, if a directed line segment has initial point (x_0, y_0) and terminal point (x_1, y_1) (Figure 7.49), then the corresponding vector represents a change of $x_1 - x_0$ units in the positive x direction and a change of $y_1 - y_0$ units in the positive y direction. We group these two numbers into the ordered pair $(x_1 - x_0, y_1 - y_0)$ and identify the vector with this ordered pair. In this vein, we make the following definition of a vector.

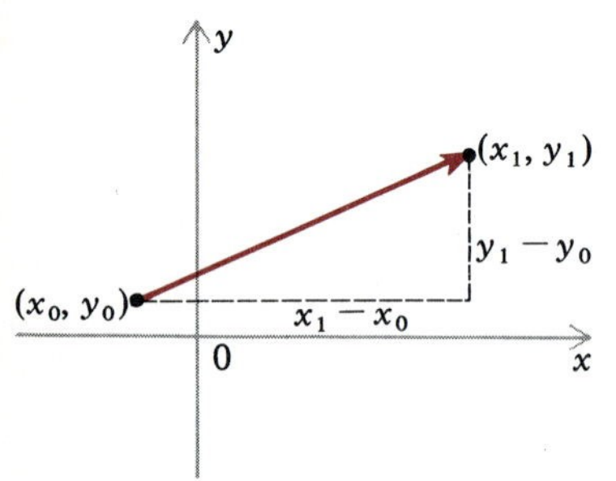

FIGURE 7.49

DEFINITION 7.1 A ***vector*** is an ordered pair (a_1, a_2) of numbers. The numbers a_1 and a_2 are called the ***components*** of the vector. The vector associated with the directed line segment with initial point $P = (x_0, y_0)$ and terminal point $Q = (x_1, y_1)$ is the vector $(x_1 - x_0, y_1 - y_0)$ and is denoted $\overrightarrow{PQ}$. Two vectors (a_1, a_2) and (b_1, b_2) are ***equal*** if $a_1 = b_1$ and $a_2 = b_2$.

EXAMPLE 1. Let $P = (2, -1)$, $Q = (5, 3)$, $R = (-1, 6)$, and $S = (2, 10)$. Show that $\overrightarrow{PQ}$ and $\overrightarrow{RS}$ are equal.

Solution. Applying Definition 7.1, we have

$$\overrightarrow{PQ} = (5 - 2, 3 - (-1)) = (3, 4)$$
$$\overrightarrow{RS} = (2 - (-1), 10 - 6) = (3, 4)$$

Thus $\overrightarrow{PQ} = \overrightarrow{RS}$. □

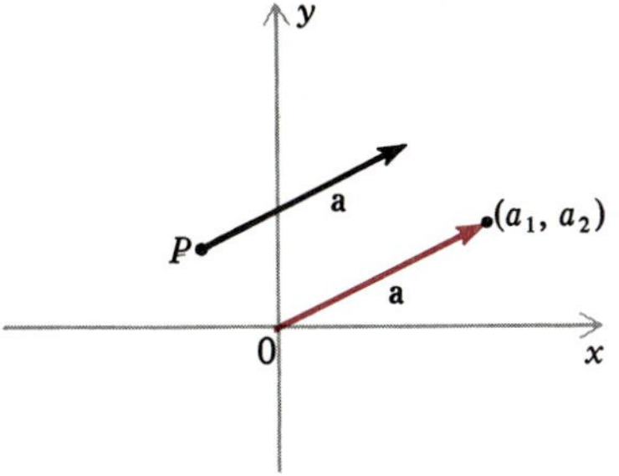

FIGURE 7.50

To distinguish vectors from numbers, which are sometimes called ***scalars***, we will denote vectors by lower-case boldface letters, such as **a**, **b**, and **c**, near the beginning of the alphabet. An exception is the ***zero vector*** (0, 0), which is denoted **0**. Since it is difficult to write a boldface letter by hand, vectors are usually written by hand by placing an arrow over a symbol or expression. Thus we would write $\vec{a}$ instead of **a**.

Each vector $\mathbf{a} = (a_1, a_2)$ can be associated with a directed line segment having an arbitrary initial point P (Figure 7.50), and different initial points give us different geometrical representations of the same vector. If P is the origin, then **a** is associated with the directed line segment from the origin to the point (a_1, a_2), or more simply, with the point (a_1, a_2). Thus we have another way of thinking of a vector—as a directed line segment starting at the origin.

A natural way to assign a length to a vector $\mathbf{a} = (a_1, a_2)$ is to assign it the length of the directed line segment from the origin to the point (a_1, a_2). In view of this, we define the length of a vector as follows.

Length (or Norm) of a Vector

$$\|\mathbf{a}\| = \sqrt{a_1^2 + a_2^2}$$

For example, if $\mathbf{a} = (-1, 2)$, then

$$\|\mathbf{a}\| = \sqrt{(-1)^2 + 2^2} = \sqrt{5}$$

If $P = (x_0, y_0)$ and $Q = (x_1, y_1)$ are any two given points, then $\overrightarrow{PQ} = (x_1 - x_0, y_1 - y_0)$, so that

$$\|\overrightarrow{PQ}\| = \sqrt{(x_1 - x_0)^2 + (y_1 - y_0)^2}$$

Referring to the distance formula, we see that $\|\overrightarrow{PQ}\|$ is just the distance between P and Q. For example, if P and Q are as in Example 1, then

$$\|\overrightarrow{PQ}\| = \sqrt{(5 - 2)^2 + (3 - (-1))^2} = \sqrt{25} = 5$$

A ***unit vector*** is a vector having length 1. There are two unit vectors that will simplify the remainder of our work with vectors:

$$\mathbf{i} = (1, 0) \quad \text{and} \quad \mathbf{j} = (0, 1)$$

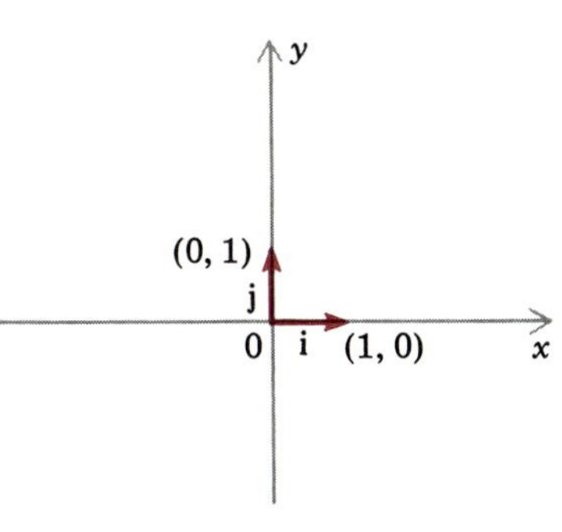

FIGURE 7.51

(Figure 7.51). Since

$$\|\mathbf{i}\| = \sqrt{1^2 + 0^2} = 1 \quad \text{and} \quad \|\mathbf{j}\| = \sqrt{0^2 + 1^2} = 1$$

i and **j** are indeed unit vectors.

Combination of Vectors

Let $\mathbf{a} = (a_1, a_2)$ and $\mathbf{b} = (b_1, b_2)$ be vectors, and let c be a number. Then the ***sum*** $\mathbf{a} + \mathbf{b}$, the ***difference*** $\mathbf{a} - \mathbf{b}$, and the ***scalar multiple*** $c\mathbf{a}$ are defined by

$$\mathbf{a} + \mathbf{b} = (a_1 + b_1, a_2 + b_2)$$
$$\mathbf{a} - \mathbf{b} = (a_1 - b_1, a_2 - b_2)$$
$$c\mathbf{a} = (ca_1, ca_2)$$

There are many laws governing these combinations. For example,

$$\mathbf{0} + \mathbf{a} = \mathbf{a} + \mathbf{0} = \mathbf{a} \qquad \mathbf{a} + \mathbf{b} = \mathbf{b} + \mathbf{a}$$
$$0\mathbf{a} = \mathbf{0} \qquad 1\mathbf{a} = \mathbf{a}$$
$$\mathbf{a} + (\mathbf{b} + \mathbf{c}) = (\mathbf{a} + \mathbf{b}) + \mathbf{c} \qquad c(\mathbf{a} + \mathbf{b}) = c\mathbf{a} + c\mathbf{b}$$
$$\mathbf{a} - \mathbf{b} = \mathbf{a} + (-1)\mathbf{b}$$

EXAMPLE 2. Let $\mathbf{a} = (2, 3)$ and $\mathbf{b} = (-1, 4)$. Find $\mathbf{a} + \mathbf{b}$, $\mathbf{a} - \mathbf{b}$, and $\frac{1}{2}\mathbf{a}$.

Solution. From the definitions we have

$$\mathbf{a} + \mathbf{b} = (2 + (-1), 3 + 4) = (1, 7)$$
$$\mathbf{a} - \mathbf{b} = (2 - (-1), 3 - 4) = (3, -1)$$
$$\frac{1}{2}\mathbf{a} = \left(\frac{1}{2}(2), \frac{1}{2}(3)\right) = \left(1, \frac{3}{2}\right) \quad \square$$

Using addition and scalar multiplication of vectors, we can express any vector $\mathbf{a} = (a_1, a_2)$ as a combination of the unit vectors $\mathbf{i}$ and $\mathbf{j}$. Indeed,

$$\begin{aligned} \mathbf{a} &= (a_1, a_2) = (a_1, 0) + (0, a_2) \\ &= a_1(1, 0) + a_2(0, 1) \\ &= a_1\mathbf{i} + a_2\mathbf{j} \end{aligned}$$

Thus

$$\mathbf{a} = a_1\mathbf{i} + a_2\mathbf{j} \tag{1}$$

For example,

$$(3, -4) = 3\mathbf{i} - 4\mathbf{j} \quad \text{and} \quad \left(-\frac{1}{3}, \sqrt{2}\right) = -\frac{1}{3}\mathbf{i} + \sqrt{2}\mathbf{j}$$

We will frequently write vectors in the form of (1). For that reason we now restate our notions related to vectors in the form of (1):

$$\begin{aligned} \|a_1\mathbf{i} + a_2\mathbf{j}\| &= \sqrt{a_1^2 + a_2^2} \\ (a_1\mathbf{i} + a_2\mathbf{j}) + (b_1\mathbf{i} + b_2\mathbf{j}) &= (a_1 + b_1)\mathbf{i} + (a_2 + b_2)\mathbf{j} \\ (a_1\mathbf{i} + a_2\mathbf{j}) - (b_1\mathbf{i} + b_2\mathbf{j}) &= (a_1 - b_1)\mathbf{i} + (a_2 - b_2)\mathbf{j} \\ c(a_1\mathbf{i} + a_2\mathbf{j}) &= ca_1\mathbf{i} + ca_2\mathbf{j} \end{aligned} \qquad (2)$$

EXAMPLE 3. Let $\mathbf{a} = 3\mathbf{i} - \frac{5}{2}\mathbf{j}$ and $\mathbf{b} = -2\mathbf{i} - \frac{1}{2}\mathbf{j}$. Find $\mathbf{a} + \mathbf{b}$, $5\mathbf{b}$ and $\|\mathbf{a}\|$.

Solution. By the formulas given above,

$$\mathbf{a} + \mathbf{b} = (3 + (-2))\mathbf{i} + \left(-\frac{5}{2} + \left(-\frac{1}{2}\right)\right)\mathbf{j} = \mathbf{i} - 3\mathbf{j}$$

$$5\mathbf{b} = 5(-2)\mathbf{i} + 5\left(-\frac{1}{2}\right)\mathbf{j} = -10\mathbf{i} - \frac{5}{2}\mathbf{j}$$

$$\|\mathbf{a}\| = \sqrt{(3)^2 + \left(-\frac{5}{2}\right)^2} = \sqrt{9 + \frac{25}{4}} = \sqrt{\frac{61}{4}} = \frac{1}{2}\sqrt{61} \quad \square$$

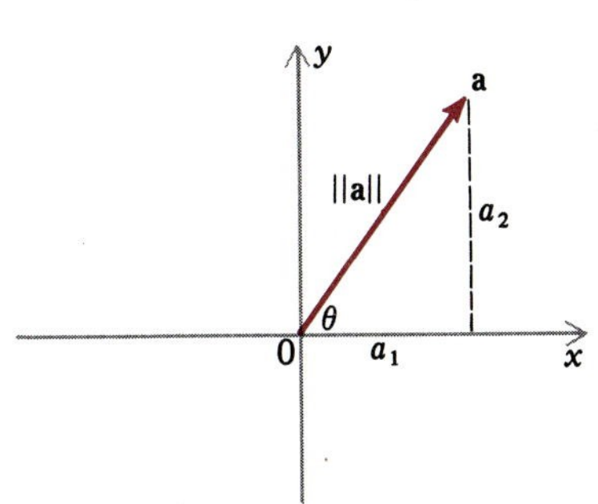

FIGURE 7.52

There is another way of representing vectors. To find it, we observe that the components a_1 and a_2 of the vector $\mathbf{a} = a_1\mathbf{i} + a_2\mathbf{j}$ are determined by the length of **a** and the angle θ that **a** makes with the positive x axis (Figure 7.52). Indeed, if $\mathbf{a} \neq \mathbf{0}$, then

$$\cos\theta = \frac{a_1}{\|\mathbf{a}\|} \quad \text{and} \quad \sin\theta = \frac{a_2}{\|\mathbf{a}\|}$$

so that

$$a_1 = \|\mathbf{a}\|\cos\theta \quad \text{and} \quad a_2 = \|\mathbf{a}\|\sin\theta$$

Therefore an alternative form of (1) is the following:

Polar Form of a Vector

$$\mathbf{a} = \|\mathbf{a}\|(\cos\theta\,\mathbf{i} + \sin\theta\,\mathbf{j}) \qquad (3)$$

EXAMPLE 4. Let $\mathbf{a} = -3\mathbf{i} + \sqrt{3}\mathbf{j}$. Express **a** in polar form.

Solution. To write **a** in polar form we need the length $\|\mathbf{a}\|$ and the angle θ. For $\|\mathbf{a}\|$ we use (2) to obtain

$$\|\mathbf{a}\| = \sqrt{(-3)^2 + (\sqrt{3})^2} = \sqrt{9 + 3} = 2\sqrt{3}$$

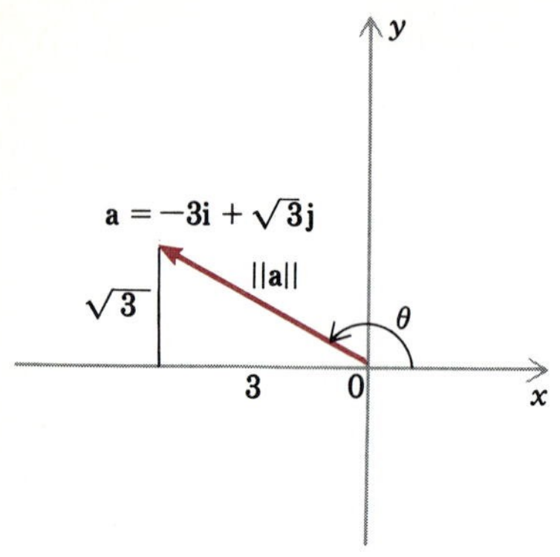

FIGURE 7.53

To find the angle θ, we draw the triangle shown in Figure 7.53 and observe that $\tan\theta < 0$ and therefore that

$$\tan\theta = -\frac{\sqrt{3}}{3}$$

Since the point corresponding to **a** lies in the second quadrant and $\tan\theta = -\sqrt{3}/3$, we take

$$\theta = \frac{5\pi}{6}$$

Thus the polar form of **a** is given by

$$\mathbf{a} = 2\sqrt{3}\left(\cos\frac{5\pi}{6}\mathbf{i} + \sin\frac{5\pi}{6}\mathbf{j}\right) \quad \square$$

Geometric Interpretations of Vector Operations

First we will interpret the sum $\mathbf{a} + \mathbf{b}$ of two vectors **a** and **b** geometrically. We begin by letting $\mathbf{a} = a_1\mathbf{i} + a_2\mathbf{j}$ and $\mathbf{b} = b_1\mathbf{i} + b_2\mathbf{j}$. We can think of **a**, **b**, and $\mathbf{a} + \mathbf{b}$ as the directed line segments from the origin to the points P, Q, and R having coordinates (a_1, a_2), (b_1, b_2), and $(a_1 + b_1, a_2 + b_2)$, respectively (Figure 7.54). Now observe from the figure that if the directed line segment representing **b** is placed so that its initial point is P, then its terminal point will be R. Thus the vector $\mathbf{a} + \mathbf{b}$ can be obtained geometrically by placing the initial point of **b** on the terminal point of **a**, and drawing the vector from the initial point of **a** to the terminal point of **b**. The figure also tells us that the two representations of **a** and **b** determine a parallelogram one of whose diagonals is $\mathbf{a} + \mathbf{b}$.

To analyze the vector $c\mathbf{a}$ geometrically, we let $\mathbf{a} = a_1\mathbf{i} + a_2\mathbf{j}$ be a vector and c any number. Then by the definition of $c\mathbf{a}$, we find that

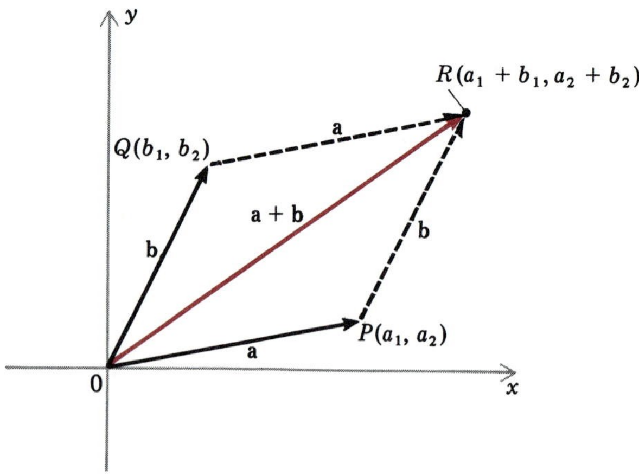

FIGURE 7.54

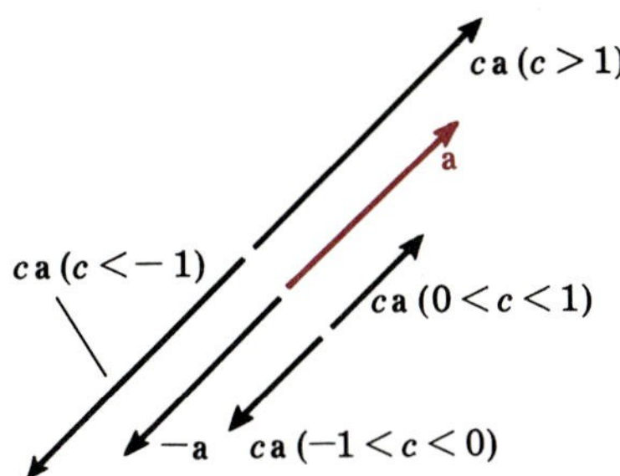

FIGURE 7.55

$$\|c\mathbf{a}\| = \sqrt{(ca_1)^2 + (ca_2)^2} = \sqrt{c^2(a_1^2 + a_2^2)}$$
$$= \sqrt{c^2}\,\sqrt{a_1^2 + a_2^2} = |c|\,\sqrt{a_1^2 + a_2^2} = |c|\,\|\mathbf{a}\|$$

Thus

$$\boxed{\|c\mathbf{a}\| = |c|\,\|\mathbf{a}\|}$$

It follows that when **a** is multiplied by a scalar c, the length of **a** is multiplied by $|c|$. If $c > 0$, the direction of **a** does not change (Figure 7.55). However, if $c < 0$, the direction of **a** is reversed (Figure 7.55). In particular, $-\mathbf{a}$ has the same length as **a** but points in the opposite direction.

For a geometric interpretation of the difference $\mathbf{a} - \mathbf{b}$ of two vectors **a** and **b**, we consider **a** and **b** as directed line segments with initial points at the origin. Next we observe that

$$\mathbf{a} = \mathbf{b} + (\mathbf{a} - \mathbf{b})$$

so that **a** is the sum of the vectors **b** and $\mathbf{a} - \mathbf{b}$. Thus it follows from our previous discussion of addition of vectors that if we place the initial point of $\mathbf{a} - \mathbf{b}$ at the terminal point of **b**, then the terminal point of $\mathbf{a} - \mathbf{b}$ will coincide with the terminal point of **a** (Figure 7.56). In other words, the vector $\mathbf{a} - \mathbf{b}$ is represented by the directed line segment that joins the terminal points of **a** and **b** and points toward **a**. The vector $\mathbf{a} - \mathbf{b}$ is one diagonal of the parallelogram determined by **a** and **b** (Figure 7.57), the other diagonal being $\mathbf{a} + \mathbf{b}$.

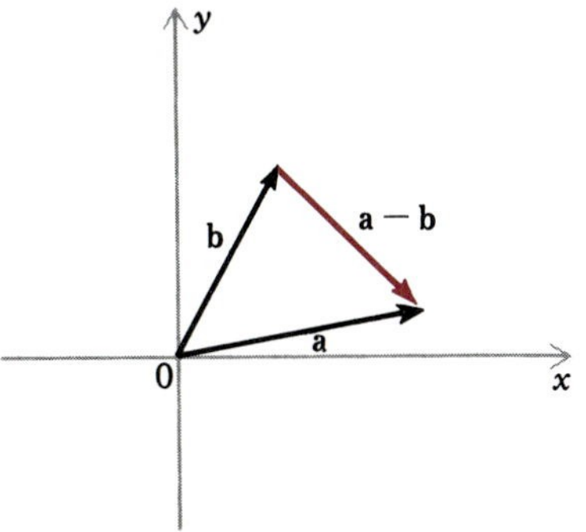

FIGURE 7.56

EXAMPLE 5. Let $\mathbf{a} = 2\mathbf{i} + \mathbf{j}$ and $\mathbf{b} = -\mathbf{i} + \mathbf{j}$. Construct $\mathbf{a} + \mathbf{b}$ and $\mathbf{a} - \mathbf{b}$ geometrically, as in Figure 7.57.

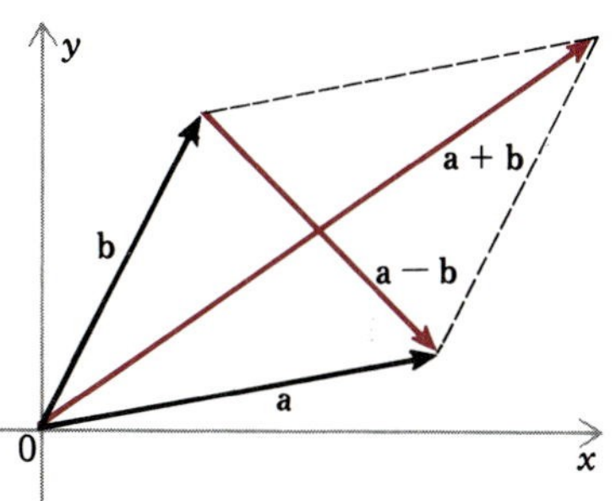

FIGURE 7.57

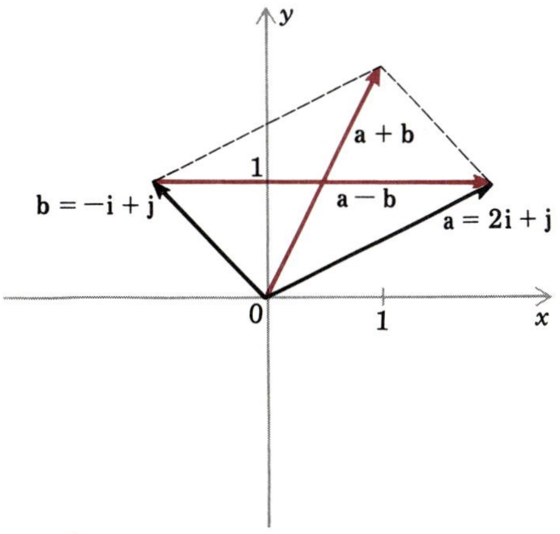

FIGURE 7.58

Solution. The constructions are shown in Figure 7.58. □

Application of Vector Addition

Many physical quantities combine according to vector addition; indeed, this is the reason we defined vector addition the way we did. For example, if several forces act simultaneously at the same point P on an object, then the object reacts as though a single force (called the ***resultant force***) equal to the vector sum of the several forces is acting on the object at P.

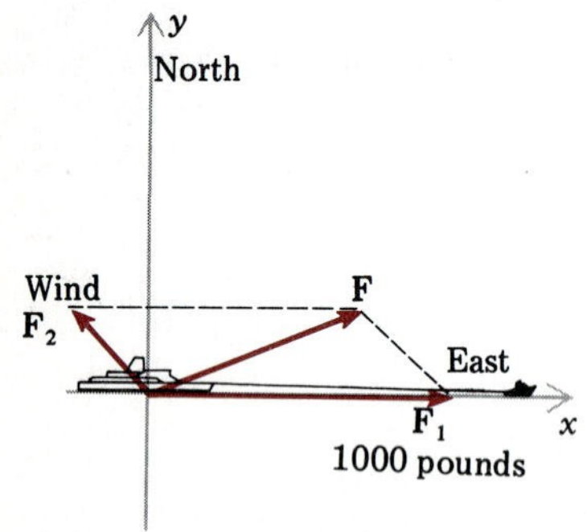

FIGURE 7.59

EXAMPLE 6. Suppose a tugboat exerts a force of 1000 pounds toward the east on a ship and a wind exerts a force of 400 pounds toward the northwest on the ship. Find the resultant force $\mathbf{F}$ and its magnitude.

Solution. We set up a coordinate system as in Figure 7.59. Our first goal is to find $\mathbf{F}$, which equals $\mathbf{F}_1 + \mathbf{F}_2$. Since $\|\mathbf{F}_1\| = 1000$ and $\mathbf{F}_1$ points along the positive x axis, it follows from (3) that

$$\mathbf{F}_1 = 1000\mathbf{i}$$

Since $\mathbf{F}_2$ points toward the northwest, it follows that $\mathbf{F}_2$ makes an angle of $3\pi/4$ with respect to the positive x axis. By hypothesis, $\|\mathbf{F}_2\| = 400$. Thus by (3),

$$\begin{aligned}\mathbf{F}_2 &= \|\mathbf{F}_2\|\left(\cos\frac{3\pi}{4}\mathbf{i} + \sin\frac{3\pi}{4}\mathbf{j}\right) = 400\left(-\frac{1}{2}\sqrt{2}\mathbf{i} + \frac{1}{2}\sqrt{2}\mathbf{j}\right)\\ &= 200\sqrt{2}(-\mathbf{i} + \mathbf{j})\end{aligned}$$

The resultant force $\mathbf{F}$ is $\mathbf{F}_1 + \mathbf{F}_2$, so that

$$\begin{aligned}\mathbf{F} = \mathbf{F}_1 + \mathbf{F}_2 &= 1000\mathbf{i} + 200\sqrt{2}(-\mathbf{i} + \mathbf{j})\\ &= (1000 - 200\sqrt{2})\mathbf{i} + 200\sqrt{2}\mathbf{j}\\ &= 200(5 - \sqrt{2})\mathbf{i} + 200\sqrt{2}\mathbf{j}\end{aligned}$$

By (2), the magnitude $\|\mathbf{F}\|$ of the resultant force is given by

$$\|\mathbf{F}\| = \sqrt{[200(5 - \sqrt{2})]^2 + (200\sqrt{2})^2}$$

By calculator we find that

$$\|\mathbf{F}\| \approx 770.9$$

Thus the magnitude of the resultant force is approximately 770.9 pounds. □

EXERCISES 7.5

In Exercises 1–6, find $\mathbf{a} + \mathbf{b}$, $\mathbf{a} - \mathbf{b}$, $2\mathbf{a} - 3\mathbf{b}$, and $-\frac{5}{2}\mathbf{a}$.

1. $\mathbf{a} = (3, 1)$, $\mathbf{b} = (-2, 4)$
2. $\mathbf{a} = (1, -4)$, $\mathbf{b} = (-2, -2)$
3. $\mathbf{a} = (-6, -3)$, $\mathbf{b} = (3, 6)$
4. $\mathbf{a} = (-5, -4)$, $\mathbf{b} = (-6, 1)$
5. $\mathbf{a} = (-2, 0)$, $\mathbf{b} = (0, 3)$
6. $\mathbf{a} = (0, 0)$, $\mathbf{b} = (-\sqrt{2}, 1)$

In Exercises 7–12, find $4\mathbf{a} - \mathbf{b}$ and $3\mathbf{a} + 2\mathbf{b}$.

7. $\mathbf{a} = 6\mathbf{i} + 2\mathbf{j}$, $\mathbf{b} = -\mathbf{i} - 2\mathbf{j}$

8. $\mathbf{a} = -3\mathbf{i} + \mathbf{j}$, $\mathbf{b} = \sqrt{3}\mathbf{i} - \sqrt{3}\mathbf{j}$

9. $\mathbf{a} = 3\mathbf{j}$, $\mathbf{b} = -5\mathbf{i} + 6\mathbf{j}$

10. $\mathbf{a} = -5\mathbf{i} - 2\mathbf{j}$, $\mathbf{b} = \mathbf{0}$

11. $\mathbf{a} = \mathbf{i} - \mathbf{j}$, $\mathbf{b} = \mathbf{i} + \mathbf{j}$

12. $\mathbf{a} = -\frac{7}{4}\mathbf{i} - \frac{4}{5}\mathbf{j}$, $\mathbf{b} = -\frac{2}{3}\mathbf{i} - \frac{6}{5}\mathbf{j}$

In Exercises 13–18, sketch vectors corresponding to $\mathbf{a}$, $\mathbf{b}$, $\mathbf{a} + \mathbf{b}$ and $\mathbf{a} - \mathbf{b}$.

13. $\mathbf{a} = (2, -1)$, $\mathbf{b} = (1, -1)$

14. $\mathbf{a} = (0, 3)$, $\mathbf{b} = (4, 0)$

15. $\mathbf{a} = (1, -3)$, $\mathbf{b} = (2, 2)$

16. $\mathbf{a} = -\mathbf{i} + \mathbf{j}$, $\mathbf{b} = 2\mathbf{i} - 3\mathbf{j}$

17. $\mathbf{a} = -\mathbf{j}$, $\mathbf{b} = 2\mathbf{i}$

18. $\mathbf{a} = 3\mathbf{i} + \mathbf{j}$, $\mathbf{b} = \mathbf{j} + 3\mathbf{i}$

In Exercises 19–24, determine the length of the vector.

19. $(3, -1)$ **20.** $(-\frac{1}{4}, -\frac{3}{4})$ **21.** $2\mathbf{i} - \mathbf{j}$

22. $-12\mathbf{i} + 5\mathbf{j}$ **23.** $\sqrt{2}\mathbf{i} + \sqrt{2}\mathbf{j}$ **24.** $-5\mathbf{j}$

In Exercises 25–30, determine whether the given vector is a unit vector.

25. $\mathbf{i} + \mathbf{j}$ **26.** $\frac{3}{5}\mathbf{i} + \frac{4}{5}\mathbf{j}$

27. $3\mathbf{i} - 2\mathbf{j}$ **28.** $-\frac{1}{2}\mathbf{i} - \frac{1}{2}\mathbf{j}$

29. $\frac{1}{2}\sqrt{2}\mathbf{i} - \frac{1}{2}\sqrt{2}\mathbf{j}$ **30.** $-\frac{1}{2}\mathbf{i} + \frac{3}{2}\mathbf{j}$

In Exercises 31–36, write the vector $\mathbf{a}$ in polar form.

31. $\mathbf{a} = (-1, -1)$ **32.** $\mathbf{a} = \sqrt{2}\mathbf{i}$

33. $\mathbf{a} = \mathbf{i} + \sqrt{3}\mathbf{j}$ **34.** $\mathbf{a} = 6\mathbf{i} - 2\sqrt{3}\mathbf{j}$

35. $\mathbf{a} = -\pi\mathbf{i} + \pi\mathbf{j}$ **36.** $\mathbf{a} = \sqrt{15}\mathbf{i} - \sqrt{5}\mathbf{j}$

37. Show that $\mathbf{a} - \mathbf{b} = \mathbf{a} + (-\mathbf{b})$.

38. Show that $-\mathbf{a} = (-1)\mathbf{a}$.

39. Suppose $\mathbf{a}$ and $\mathbf{b}$ are vectors such that $3\mathbf{i} + 4\mathbf{j} = 2\mathbf{a} - 3\mathbf{b} = -6\mathbf{a} + \mathbf{b}$. Find $\mathbf{a}$ and $\mathbf{b}$.

The ***dot product*** $\mathbf{a} \cdot \mathbf{b}$ of two vectors $\mathbf{a} = a_1\mathbf{i} + a_2\mathbf{j}$ and $\mathbf{b} = b_1\mathbf{i} + b_2\mathbf{j}$ is defined by

$$\mathbf{a} \cdot \mathbf{b} = a_1b_1 + a_2b_2$$

In Exercises 40–43, determine the dot product of $\mathbf{a}$ and $\mathbf{b}$.

40. $\mathbf{a} = 2\mathbf{i} - 3\mathbf{j}$, $\mathbf{b} = 4\mathbf{i} + \mathbf{j}$

41. $\mathbf{a} = \frac{5}{2}\mathbf{i} - \frac{1}{3}\mathbf{j}$, $\mathbf{b} = -6\mathbf{i} - 12\mathbf{j}$

42. $\mathbf{a} = \frac{1}{2}\sqrt{3}\mathbf{i} - \mathbf{j}$, $\mathbf{b} = \sqrt{3}\mathbf{i} + \frac{3}{2}\mathbf{j}$

43. $\mathbf{a} = \frac{1}{2}\mathbf{i}$, $\mathbf{b} = 5\mathbf{j}$

44. It can be proved that the dot product $\mathbf{a} \cdot \mathbf{b}$ of $\mathbf{a}$ and $\mathbf{b}$ is also given by

$$\mathbf{a} \cdot \mathbf{b} = \|\mathbf{a}\| \, \|\mathbf{b}\| \cos \theta \qquad (4)$$

where θ is the angle with initial side pointing in the direction of $\mathbf{a}$ and terminal side pointing in the direction of $\mathbf{b}$.

a. Use (4) to show that $\mathbf{a} \cdot \mathbf{b} = 0$ if and only if $\mathbf{a} = \mathbf{0}$, $\mathbf{b} = \mathbf{0}$, or $\mathbf{a}$ and $\mathbf{b}$ are perpendicular.

b. Show that $2\mathbf{i} - 3\mathbf{j}$ and $\frac{3}{2}\mathbf{i} + \mathbf{j}$ are perpendicular.

45. If a constant force $\mathbf{F}$ acts on an object that moves along a straight line from a point P_0 to a point P_1, then the work W done on the object by the force $\mathbf{F}$ is given by

$$W = \mathbf{F} \cdot \overrightarrow{P_0P_1}$$

Suppose a person pushes a box up a 30-foot ramp inclined at an angle of 30° by exerting a horizontal force $\mathbf{F}$ of magnitude 20 pounds (Figure 7.60). Determine the amount of work done on the box by $\mathbf{F}$.

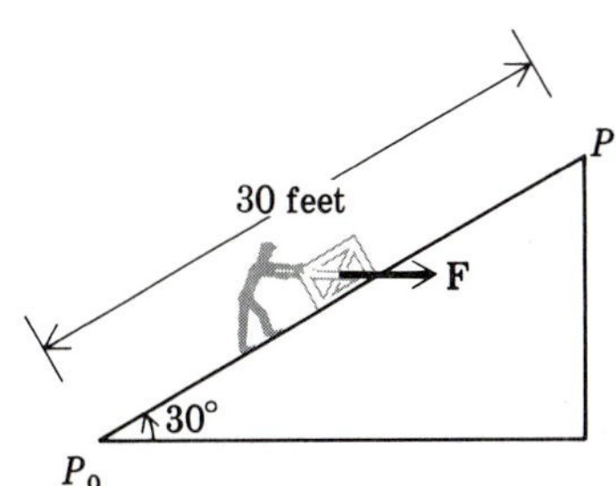

FIGURE 7.60

46. Suppose the tugboat in Example 6 exerts its force in the northeast direction. Find the resultant force $\mathbf{F}$ and its magnitude.

47. A plane is flying 300 mph (air speed), with its nose pointed 30° north from the east direction. If a wind blows at 30 mph from the west, find the velocity vector, and approximate the ground speed and the actual course of the plane.

48. If two children pull on a cart with forces given in Figure 7.61, determine the resultant force on the cart and the magnitude of the force.

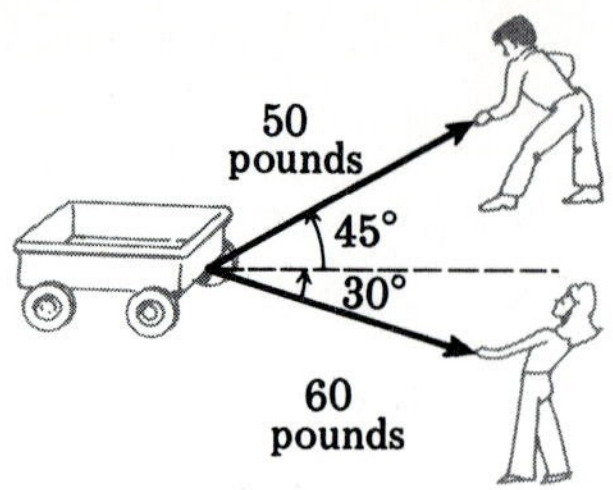

FIGURE 7.61

KEY TERMS

solving a triangle
polar coordinate system
polar graph

vector
- component
- length (or norm)
- unit vector

KEY FORMULAS

Law of Sines

$$\frac{\sin \alpha}{a} = \frac{\sin \beta}{b} = \frac{\sin \gamma}{c}$$

Law of Cosines

$$a^2 = b^2 + c^2 - 2bc \cos \alpha$$

$$b^2 = c^2 + a^2 - 2ca \cos \beta$$

$$c^2 = a^2 + b^2 - 2ab \cos \gamma$$

$$x = r \cos \theta \qquad y = r \sin \theta$$

$$r = x^2 + y^2 \qquad \tan \theta = \frac{y}{x} \quad \text{for} \quad x \neq 0$$

$$(a_1\mathbf{i} + a_2\mathbf{j}) + (b_1\mathbf{i} + b_2\mathbf{j}) = (a_1 + b_1)\mathbf{i} + (a_2 + b_2)\mathbf{j}$$

$$c(a_1\mathbf{i} + a_2\mathbf{j}) = ca_1\mathbf{i} + ca_2\mathbf{j}$$

$$\|\mathbf{a}\| = \sqrt{a_1^2 + a_2^2}$$

$$\mathbf{a} = \|\mathbf{a}\| (\cos \theta\, \mathbf{i} + \sin \theta\, \mathbf{j})$$

REVIEW EXERCISES

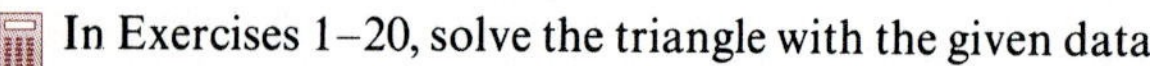

In Exercises 1–20, solve the triangle with the given data.

1. $a = 6.4$, $b = 6.4$, $c = 3.2$

2. $\gamma = 90°$, $a = 19.3$, $b = 16.8$

3. $\alpha = 32.1°$, $\beta = 50.5°$, $a = 32.1$

4. $\beta = 64°32'$, $a = 1.5$, $b = 1.4$

5. $\alpha = 13°27'$, $a = 16.2$, $b = 13.1$

6. $\alpha = 71°37'$, $b = 8.4$, $c = 9$

7. $\beta = 124°19'$, $a = 0.18$, $b = 0.23$

8. $a = 97, b = 65, c = 72$

9. $\beta = 80.3°, a = 10.3, b = 10.2$

10. $\alpha = 83°23', \gamma = 64°56', a = 0.7$

11. $a = 1.8, b = 7.2, c = 5.9$

12. $\beta = 80.8°, a = 0.35, c = 0.41$

13. $\beta = 132°46', \gamma = 26°12', b = 3.7$

14. $\beta = 134.6°, a = \sqrt{2}, c = 1.3$

15. $a = 0.8, b = 1.3, c = 1.5$

16. $\alpha = 7°11', a = 3.4, c = 6$

17. $\alpha = 60.7°, a = 10, c = 9.4$

18. $\alpha = 16.3°, \gamma = 100°, c = 16.1$

19. $\gamma = 90°, a = 90, b = 56$

20. $a = 4.8, b = 3.9, c = 3.6$

In Exercises 21–24, find the area $\mathscr{A}$ of the given triangle.

21. $\alpha = 83.2°, b = 7.1, c = 9.2$

22. $\gamma = 17°23', a = 0.6, b = 0.2$

23. $a = 1.4, b = 2.7, c = 3.5$

24. $a = 18.1, b = 9.9, c = 12.6$

In Exercises 25–28, find the rectangular coordinates of the point having the given polar coordinates.

25. $(\frac{3}{4}, 7\pi/4)$
26. $(\sqrt{3}, 17\pi/6)$

27. $(e, -2\pi/3)$
28. $(13, -9\pi/2)$

In Exercises 29–32, find all sets of polar coordinates of the point having the given rectangular coordinates.

29. $(\sqrt{5}, -\sqrt{15})$
30. $(9, 3\sqrt{3})$

31. $(0, \pi)$
32. $(-\sqrt{7}, -\sqrt{7})$

In Exercises 33–38, sketch the polar graph of the given equation.

33. $r = 3$
34. $r = -\frac{2}{3}$

35. $\theta = 7\pi/6$
36. $\theta = -23\pi/4$

37. $r = \frac{3}{2} \sin \theta$
38. $r = -2 \cos \theta$

In Exercises 39–42, find $-2\mathbf{a} + 4\mathbf{b}$.

39. $\mathbf{a} = (-1, -5), \mathbf{b} = (\frac{2}{3}, -3)$

40. $\mathbf{a} = (\sqrt{3}, 0), \mathbf{b} = (0, 1)$

41. $\mathbf{a} = \mathbf{i} + 2\mathbf{j}, \mathbf{b} = 2\mathbf{i} - \mathbf{j}$

42. $\mathbf{a} = -\sqrt{2}\mathbf{j}, \mathbf{b} = \frac{1}{2}\sqrt{2}\mathbf{i} + \sqrt{2}\mathbf{j}$

In Exercises 43–48, determine the length of **a**.

43. $\mathbf{a} = (6, -2\sqrt{3})$

44. $\mathbf{a} = (-\frac{2}{3}, \frac{2}{9}\sqrt{3})$

45. $\mathbf{a} = (0, \sqrt{3}\pi - 7\sqrt{2})$

46. $\mathbf{a} = 6\mathbf{i} - 5\mathbf{j}$

47. $\mathbf{a} = -\sqrt{2}\mathbf{i} - \mathbf{j}$

48. $\mathbf{a} = 7\mathbf{i} + 24\mathbf{j}$

In Exercises 49–52, write the vector **a** in polar form.

49. $\mathbf{a} = -\sqrt{21}\mathbf{i} + \sqrt{7}\mathbf{j}$
50. $\mathbf{a} = -e\mathbf{i} - e\mathbf{j}$

51. $\mathbf{a} = \frac{1}{10}\mathbf{i} + \frac{1}{10}\sqrt{3}\mathbf{j}$
52. $\mathbf{a} = 0.43\mathbf{i} - 0.43\mathbf{j}$

53. The lengths of the sides of a triangle are 4.2, 6.1, and 5.3. Determine the largest angle in the triangle.

54. A rectangle has sides of length 4.2 and 6.5 inches. Determine the angles the diagonals of the rectangle make with each other.

55. When the angle of elevation of the sun is 63°, the shadow of a sugar maple tree is 26 feet long. Determine the height of the maple tree.

56. The Washington Monument is 555 feet tall. What is the angle of elevation of the top when viewed from a distance of 1000 feet?

57. The Washington Monument is an obelisk whose base is 55 feet long on a side (Figure 7.62). The four walls of the monument, which are trapezoidal and almost vertical, gradually slant inward to a height of 500 feet, where they are but 34.5 feet on a side. (On top of the walls is a small pyramid.) Determine the angle of elevation of the walls.

FIGURE 7.62

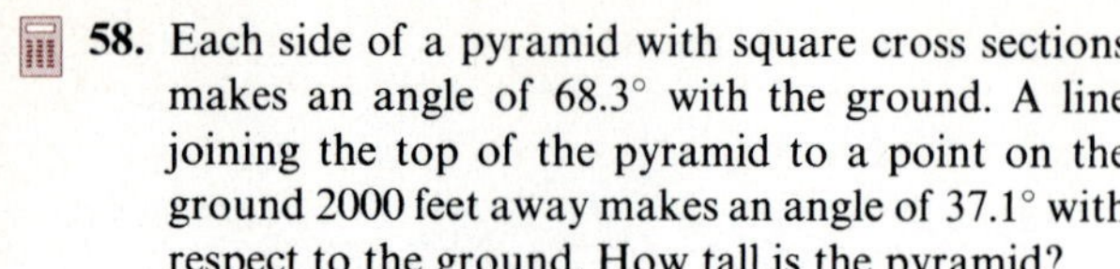

58. Each side of a pyramid with square cross sections makes an angle of 68.3° with the ground. A line joining the top of the pyramid to a point on the ground 2000 feet away makes an angle of 37.1° with respect to the ground. How tall is the pyramid?

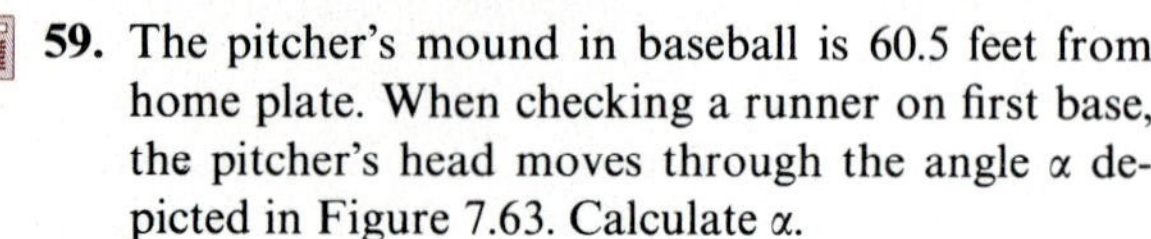

59. The pitcher's mound in baseball is 60.5 feet from home plate. When checking a runner on first base, the pitcher's head moves through the angle α depicted in Figure 7.63. Calculate α.

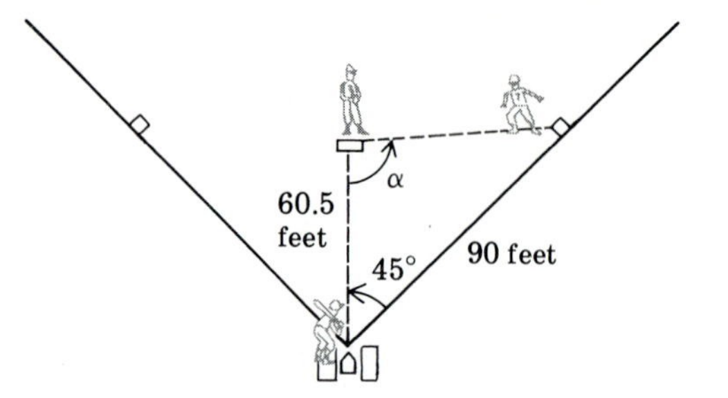

FIGURE 7.63

CUMULATIVE REVIEW II: CHAPTERS 4-7

In Exercises 1–4, evaluate the given expression.

1. $(\sqrt{3^{-4}})^{-1}$ **2.** $\dfrac{5^{1-\pi}5^{\pi+2}}{5^5}$

3. $\ln \dfrac{1}{e^{5/6}}$ **4.** $10^{-3\log_{10}5}$

In Exercises 5–12, solve the given equation.

5. $(7^x)^{2x-1} = (\sqrt{7})^{10x-8}$ **6.** $\log_2 x = -3$

7. $\log_x 5 = \dfrac{1}{2}$ **8.** $4 \ln \sqrt{2x+4} = \ln 3$

9. $1.5\, e^{0.02k} = 4$ **10.** $e^{\frac{t \ln 2}{2}} = 14$

11. $2^{2x+1} = 3^{x-1}$ **12.** $2^x = 5^{(x^2)}$

13. Simplify $\ln\left(\dfrac{1}{e^x}\right)^2$.

14. Rewrite as a single logarithm and simplify: $\log_5(x^2 - y^2) - \log_5(x+y) - \log_5 x$

15. Show that $\log \dfrac{1}{\sqrt{1-x^2}} = -\dfrac{1}{2}[\log(1-x) + \log(1+x)]$.

16. Find the base a if the graph of $y = a^x$ contains the point $(-2, 4)$.

17. Find the base a if the graph of $y = \log_a x$ contains the point $(4, 4)$.

In Exercises 18–19, sketch the graph of the given function.

18. $f(x) = \sqrt{3}(3^{x-\frac{3}{2}})$ **19.** $f(x) = \log_2(x+1)$

20. Find the length of arc on a circle of radius 2 subtended by a central angle of $(5\pi/6)$ radians.

In Exercises 21–26, find the value of the given expression.

21. $\sin \dfrac{41\pi}{6}$ **22.** $\tan 330°$ **23.** $\sec\left(-\dfrac{7\pi}{3}\right)$

24. $\csc \dfrac{5\pi}{4}$ **25.** $\cos 15°$ **26.** $\cot \dfrac{3\pi}{8}$

27. Find the values of the remaining 5 trigonometric functions at θ if $\tan\theta = \frac{1}{2}$ and $\pi < \theta < 3\pi/2$.

28. Find $\cos 2x$ if $\sin x = \frac{2}{3}$ and $\pi/2 < x < \pi$.

29. Find $\sin(x - \pi/4)$ if $\sin x = -\frac{3}{5}$ and $3\pi/2 < x < 2\pi$.

In Exercises 30–31, compute the value of the given expression.

30. $\cos[\arcsin(-\frac{1}{2})]$ **31.** $\arctan\left(\tan \dfrac{3\pi}{4}\right)$

32. Show that $\cot(\arcsin x) = \dfrac{\sqrt{1-x^2}}{x}$.

In Exercises 33–34, sketch the graph of the given function.

33. $f(x) = 1 + \sin\left(x - \dfrac{\pi}{4}\right)$

34. $f(x) = 2 - \cos 2\left(x + \dfrac{\pi}{2}\right)$

In Exercises 35–39, verify the given identity.

35. $(1 + \cos^2 x)(1 + \sin^2 x) = \dfrac{8 + \sin^2 2x}{4}$

36. $(\tan x + 1)^2 + (\tan x - 1)^2 = 2\sec^2 x$

37. $\dfrac{1}{4}\sin^2 2x = \dfrac{\sin^4 x - \cos^4 x}{\tan^2 x - \cot^2 x}$

38. $\dfrac{\sin 3x}{\sin x} - \dfrac{\cos 3x}{\cos x} = 2$ **39.** $\dfrac{\cos 5x + \cos x}{\cos x - \cos 5x} = \dfrac{\cot 2x}{\tan 3x}$

In Exercises 40–42, solve the given equation.

40. $\cos t = -\dfrac{\sqrt{3}}{2}$ **41.** $\cos x = \cot x$

42. $\sin^2 x - \dfrac{1}{2}\sin x - \dfrac{1}{2} = 0$

In Exercises 43–46, solve the triangle with the given data.

43. $a = 4, b = 5, \alpha = 40°$ **44.** $b = 5, c = 3, \alpha = 40°$

45. $a = 2, b = 5, c = 6.2$

46. $a = 4.2, c = 6.1, \gamma = 90°$

47. Find the area $\mathscr{A}$ of the triangle whose sides have lengths 4, 8, and 10.

48. Let $\mathbf{a} = 2\mathbf{i} - 4\mathbf{j}$ and $\mathbf{b} = -5\mathbf{i} - \mathbf{j}$.

a. Find $\dfrac{1}{2}\mathbf{a} - 2\mathbf{b}$.

b. Determine which is the longer vector, **a** or **b**.

49. Find all sets of polar coordinates of the point with rectangular coordinates $(6, -6\sqrt{3})$.

50. Sketch the polar graph of $r = \frac{3}{2}\cos\theta$.

51. Suppose the intensity of one sound wave is 10% greater than that of a second sound wave. Are the two sounds distinguishable to the human ear? (See Exercise 19 in Section 4.4.)

52. Earthquakes measuring 3.4 on the Richter scale preceded the major eruption of Mount St. Helens in the spring of 1980. Find the ratio of the maximal intensity of a seismic wave of such an earthquake to that of the earthquake on the California–Mexico border which occurred the same spring and which registered 6 on the Richter scale.

53. If the population of a species grows exponentially and doubles in three weeks, how long does it take for the population to multiply 10-fold?

54. At time $t = 0$ an electrical condenser with an initial charge of q_0 amperes begins to discharge. After t seconds its charge $q(t)$ is given by $q(t) = q_0 e^{-kt}$. Determine how long it takes for 95% of the charge to leave. Express your answer in terms of k.

55. The atmospheric pressure decays exponentially as a function of altitude. The pressure at sea level is known to be 14.7 pounds per square inch and is half as much at an altitude of 18,000 feet. Determine the atmospheric pressure at the top of Mt. Washington, in New Hampshire, whose altitude is 6288 feet.

56. The half-life of carbon 11 is approximately 20 minutes. If a gram of carbon 11 is present at 11 A.M., how much will be present one hour later?

57. A long-playing record makes $33\frac{1}{3}$ revolutions per minute. Through how many degrees does a point on the record travel in

a. 1 second? b. 30 seconds?

58. Approximately how far does a point 4 inches from the center of the record mentioned in Exercise 57 travel in one minute?

59. The floor of a rectangular swimming pool is inclined lengthwise at a 9° angle from the horizontal. If the swimming pool is 100 meters long at the surface, determine the length of the floor of the pool.

60. According to one model, the relationship between the speed v of a ball just before it is batted, and the speed v' just after it is batted, is given by the formula $v' \sin\phi = v \sin\theta$, where θ and ϕ are, respectively, the angles of incidence and reflection with respect to the normal to the bat (see Figure II.1). When a baseball is batted along a foul line, we have $\theta + \phi = \pi/4$. Show that in this case,

$$v' = \frac{v\sqrt{2}}{\cot\theta - 1}$$

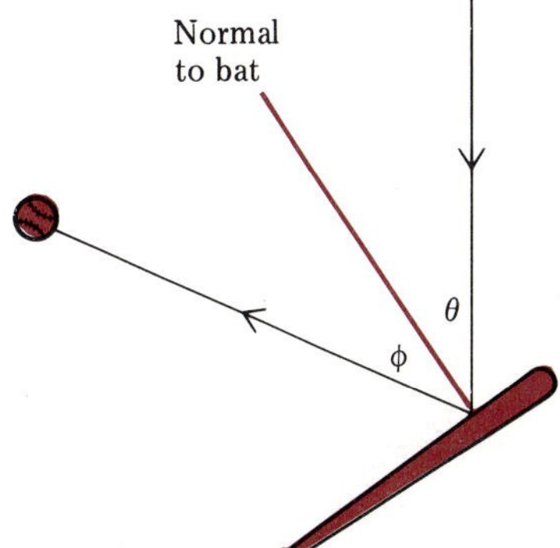

FIGURE II.1

61. A force of 6 pounds in the northeast direction, along with a force of $4\sqrt{2}$ pounds in the south direction, are applied to an object. Find the magnitude and approximate direction of the resulting force F.

8 SYSTEMS OF EQUATIONS AND INEQUALITIES

In Chapter 1 we studied methods of solving equations having a single unknown or variable (usually denoted by x). However, some problems involve more than one unknown. Suppose, for example, that we wish to find two numbers x and y whose sum is 10 and whose product is 21. Then there are two unknowns, which we will denote x and y. Moreover, since the sum of the two numbers is 10 and their product is 21, it follows that x and y are related by the two equations

$$x + y = 10$$
$$xy = 21$$

Thus our problem is to find numbers x and y that satisfy both of the above equations. In other problems there may be more than two unknowns and more than two equations relating them. In each such case we would wish to find values of the unknowns that satisfy all the equations relating the unknowns. The collection of equations relating the unknowns is called a ***system of equations*** and will be a central topic of this chapter.

As an application of systems of equations, we will solve the following problem (see Example 6 in Section 8.1).

> Suppose that at the beginning of the swimming season it is desired to heat a 200-cubic-yard swimming pool by mixing heated water with tap water. If the temperature of the heated water is 182°F and the tap water is 54°F, how much heated water must be used to achieve a temperature of 70°F?

The outline of the chapter is as follows: In Sections 8.1 through 8.3 we examine several methods of solving systems of equations. Sections 8.4 through 8.7 involve a discussion of matrices and determinants and their use in solving systems of equations. Section 8.8 is devoted to systems of inequalities and their solutions, and the final section of the chapter concerns linear programming, which utilizes both systems of equations and systems of inequalities.

8.1 SYSTEMS OF EQUATIONS IN TWO VARIABLES

By a system of equations in two variables x and y we mean a collection of equations that contain the variables x and y and no others. Examples are

$$\begin{cases} x + y = 10 \\ xy = 21 \end{cases} \tag{1}$$

and

$$\begin{cases} 2x + y = 1 \\ -3x + 2y = 9 \end{cases} \tag{2}$$

As is done above, we will use a brace to group together the equations in a given system.

A ***solution*** of a system of equations in two variables x and y is an ordered pair (a, b) such that if x is replaced by a and y is replaced by b in all equations, then all the equations are valid. For example, (3, 7) is a solution of (1) because if we replace x by 3 and y by 7, then (1) becomes

$$3 + 7 = 10$$
$$3 \cdot 7 = 21$$

which is a system of valid equations. Likewise (7, 3) is a solution of (1), as you can check. What may be less obvious, but will be shown a little later, is the fact that (3, 7) and (7, 3) are the only solutions of (1). The process of finding all solutions of a system of equations is called ***solving*** the system.

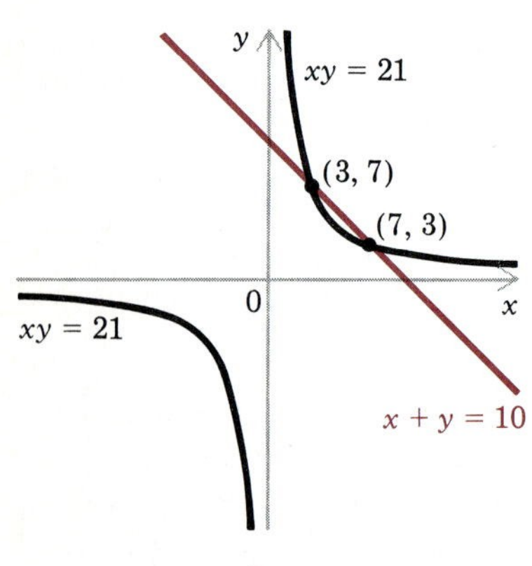

FIGURE 8.1

CAUTION: It is crucial to note that *each* equation in a system must be satisfied by a pair (a, b) for (a, b) to be a solution of the system. For example, (2, 8) is a solution of $x + y = 10$ in (1) because $2 + 8 = 10$, but since $2 \cdot 8 = 16$, (2, 8) is *not* a solution of $xy = 21$, and therefore (2, 8) is *not* a solution of the system in (1).

One can interpret the solutions of a system of equations in two variables graphically. For a specific example, let us consider (1) again. The graphs of the two equations are shown in Figure 8.1. Notice that the solutions (3, 7) and

(7, 3) are precisely the points at which the graphs of the two equations in (1) intersect. More generally, we observe that a point (a, b) is on the graph of each equation in a system if and only if each equation in the system is valid when x is replaced by a and y is replaced by b. Consequently (a, b) is a solution of the system if and only if (a, b) lies on the graph of each equation in the system. When the graphs of the equations in a system are easy to draw and happen to intersect at obvious points in the plane, it is possible to use graphs to detect solutions of the system. However, generally other methods are preferable because they are more likely to lead to precise solutions.

The Method of Substitution

Let us return to the system

$$\begin{cases} x + y = 10 \\ xy = 21 \end{cases} \tag{3}$$

and determine all solutions of (3). Solving for y in terms of x in the first equation of (3), we obtain

$$y = -x + 10 \tag{4}$$

Next we substitute $-x + 10$ for y in the second equation of (3), which leads to the following chain of equivalent equations:

$$x(-x + 10) = 21$$

$$-x^2 + 10x - 21 = 0$$

$$x^2 - 10x + 21 = 0$$

The left side of the last equation factors to give us

$$(x - 3)(x - 7) = 0$$

It follows that $x = 3$ or $x = 7$. If $x = 3$, then by (4),

$$y = -3 + 10 = 7$$

so that (3, 7) is a solution. If $x = 7$, then again by (4),

$$y = -7 + 10 = 3$$

so that (7, 3) is a solution. Thus the only possible solutions of (3) are (3, 7) and (7, 3). We noted earlier that they are indeed solutions, so (3) is solved.

The method we have just used is called the ***substitution method*** of solving a system of equations, because after solving for one variable in one equation, we substituted for that variable in the other equation. To guide us in using the method, we list the steps to be followed:

Substitution Method

i. Solve one of the equations for y in terms of x (or for x in terms of y).
ii. Substitute for y (or x) in the other equation in order to obtain an equation in the single variable x (or y).
iii. Find all solutions of the equation obtained in (ii).
iv. Use the equation obtained in (i) and the values for x (or y) obtained in (iii) in order to find the corresponding values of y (or x).
v. Check every proposed solution by substituting in *each* equation of the given system.

EXAMPLE 1. Use the substitution method to solve the system

$$\begin{cases} 2x + y = 1 \\ -3x + 2y = 9 \end{cases}$$

Solution. Solving for y in the first equation, we obtain

$$y = -2x + 1 \tag{5}$$

Substituting $-2x + 1$ for y in the second equation yields

$$\begin{aligned} -3x + 2(-2x + 1) &= 9 \\ -3x - 4x + 2 &= 9 \\ -7x &= 7 \\ x &= -1 \end{aligned}$$

From (5) we see that if $x = -1$, then

$$y = -2(-1) + 1 = 2 + 1 = 3$$

Thus the proposed solution is $(-1, 3)$.

Check: $$\begin{cases} 2(-1) + 3 = -2 + 3 = 1 \\ -3(-1) + 2(3) = 3 + 6 = 9 \end{cases}$$

Thus the only solution of the given system is $(-1, 3)$. □

EXAMPLE 2. Use the substitution method to solve the system

$$\begin{cases} xy + 2y^2 = 8 \\ x - 2y = 4 \end{cases}$$

Solution. Solving for x in the second equation yields

$$x = 2y + 4 \tag{6}$$

By substituting $2y + 4$ for x in the first equation, we obtain

$$\begin{aligned} (2y + 4)y + 2y^2 &= 8 \\ 2y^2 + 4y + 2y^2 &= 8 \\ 4y^2 + 4y - 8 &= 0 \\ y^2 + y - 2 &= 0 \\ (y + 2)(y - 1) &= 0 \end{aligned}$$

Therefore $y = -2$ or $y = 1$. If $y = -2$, then by (6),

$$x = 2(-2) + 4 = 0$$

which means that $(0, -2)$ is a proposed solution. If $y = 1$, then again by (6),

$$x = 2(1) + 4 = 6$$

so that $(6, 1)$ is a proposed solution.

Check: $\begin{cases} (0)(-2) + 2(-2)^2 = 8 \\ 0 - 2(-2) = 4 \end{cases}$ and $\begin{cases} (6)(1) + 2(1)^2 = 8 \\ 6 - 2(1) = 4 \end{cases}$

Consequently the solutions of the given system are $(0, -2)$ and $(6, 1)$. □

EXAMPLE 3. Use the substitution method to solve the system

$$\begin{cases} \dfrac{y}{x - 1} = x + 1 \\ x^2 + y^2 = 1 \end{cases}$$

Solution. Multiplying both sides of the first equation by $x - 1$, we obtain

$$y = (x - 1)(x + 1) = x^2 - 1 \tag{7}$$

By substituting $x^2 - 1$ for y in the second equation, we find that

$$\begin{aligned} x^2 + (x^2 - 1)^2 &= 1 \\ x^2 + (x^4 - 2x^2 + 1) &= 1 \\ x^4 - x^2 &= 0 \\ x^2(x - 1)(x + 1) &= 0 \end{aligned}$$

Thus $x = 0$, $x = 1$, or $x = -1$. From (7) we see that $y = -1$ if $x = 0$, $y = 0$ if $x = 1$, and $y = 0$ if $x = -1$. Therefore there are three proposed solutions: $(0, -1)$, $(1, 0)$, and $(-1, 0)$. A routine check shows that $(0, -1)$ and $(-1, 0)$ actually are solutions of the given system. However, when we check the proposed solution $(1, 0)$ by substituting 1 for x and 0 for y in the left side of the first equation, we obtain the meaningless expression $0/(1 - 1)$. We conclude that there are only two solutions: $(0, -1)$ and $(-1, 0)$. □

A proposed solution of a system that is not a solution is called an ***extraneous solution***. Thus $(1, 0)$ is an extraneous solution of the system in Example 3. Such solutions often arise when we square both sides of an equation or multiply both sides by an expression that may be 0.

So far we have solved systems that had one or two solutions. In general, there can be any number of solutions: none, any finite number, or infinitely many. The following example illustrates a system with no solutions.

EXAMPLE 4. Show that the system

$$\begin{cases} x - 2y = 3 \\ 2x - 4y = 7 \end{cases}$$

has no solutions.

Solution. If (x, y) is a solution of the system, then from the first equation we have

$$x = 2y + 3$$

Substituting $2y + 3$ for x in the second equation yields

$$\begin{aligned} 2(2y + 3) - 4y &= 7 \\ 4y + 6 - 4y &= 7 \\ 6 &= 7 \end{aligned}$$

which is impossible. Thus no solution exists. □

A system of equations that has no solutions is called ***inconsistent***. Thus the system in Example 4 is inconsistent. Any system that has solutions is by contrast ***consistent***. The systems given in Examples 1–3 exemplify consistent systems.

It is also possible to prove geometrically that the system in Example 4 is inconsistent. Observe that $x - 2y = 3$ and $2x - 4y = 7$ are equations of lines. The slope–intercept form of $x - 2y = 3$ is

$$y = \frac{1}{2}x - \frac{3}{2}$$

and the slope–intercept form of $2x - 4y = 7$ is

$$y = \frac{1}{2}x - \frac{7}{4}$$

It follows that the lines have the same slope, $\frac{1}{2}$, but different y intercepts. Thus the two lines are parallel and do not intersect. Since any solution of the system must be on both lines, there are no solutions of the system in Example 4.

Linear Systems

In general, any equation of the form $ax + by = c$, where a, b, and c are constants with either $a \neq 0$ or $b \neq 0$, is called a ***linear equation***, since its graph is a line. A system containing only linear equations is called a ***linear system***.

Now consider a linear system containing two equations in two variables. The graph of each equation is a straight line. Since two lines are identical, intersect at a single point, or do not intersect at all, it follows that a linear system has either infinitely many solutions, one solution, or no solutions. Because the nonlinear system appearing in Example 2 has two solutions, we know that nonlinear systems are not limited to infinitely many, one, or no solutions.

Our next example involves a linear system with infinitely many solutions.

EXAMPLE 5. Show that the system

$$\begin{cases} 2x - 5y = 1 \\ 8x - 20y = 4 \end{cases} \tag{8}$$

has infinitely many solutions.

Solution. If we multiply both sides of the first equation by 4, it becomes

$$8x - 20y = 4$$

which is the same as the second equation. Solving for y, we obtain

$$y = \frac{1}{20}(8x - 4) = \frac{1}{5}(2x - 1)$$

Thus (a, b) is a proposed solution of the given system if and only if

$$b = \frac{1}{5}(2a - 1)$$

This means that any ordered pair of the form $(a, \frac{1}{5}(2a - 1))$ is a proposed solution, where a can be any number.

$$\textit{Check:}\quad \begin{cases} 2a - 5\left[\dfrac{1}{5}(2a-1)\right] = 2a - 2a + 1 = 1 \\ 8a - 20\left[\dfrac{1}{5}(2a-1)\right] = 8a - 8a + 4 = 4 \end{cases}$$

Therefore the given system has infinitely many solutions. □

Particular solutions of the system in Example 5 can be found by assigning particular values to a. For example, the solutions arising from the values 0, 1, 3, and -7 for a are $(0, -\frac{1}{5})$, $(1, \frac{1}{5})$, $(3, 1)$, and $(-7, -3)$, respectively.

An Applied Problem

We conclude this section with an applied problem that involves the mixing of two liquids.

EXAMPLE 6. Suppose that at the beginning of the swimming season it is desired to heat a 200-cubic-yard swimming pool by mixing heated water with tap water. If the temperature of the heated water is 182°F and the tap water is 54°F, how much heated water must be used to achieve a temperature of 70°F?

Solution. Let

x = the amount in hundreds of cubic yards of tap water at 54°F to be used
y = the amount in hundreds of cubic yards of preheated water at 182°F to be used

By assumption, 200 cubic yards are needed, so that one equation relating x and y is

$$x + y = 2 \tag{9}$$

Next, since the proportion of tap water in the pool is $x/2$ and the proportion of heated water is $y/2$, the temperature of the pool is $(x/2)54 + (y/2)182$. Thus the assumption that the temperature of the pool is to be 70°F means that

$$\left(\frac{x}{2}\right)54 + \left(\frac{y}{2}\right)182 = 70$$

which simplifies to

$$27x + 91y = 70 \tag{10}$$

Equations (9) and (10) form the system

$$\begin{cases} x + \quad y = 2 \\ 27x + 91y = 70 \end{cases} \tag{11}$$

Solving for x in the first equation of (11), we have

$$x = 2 - y \tag{12}$$

Substituting $2 - y$ for x in the second equation of (11) and solving for y, we obtain

$$\begin{aligned} 27(2 - y) + 91y &= 70 \\ 54 - 27y + 91y &= 70 \\ 64y &= 16 \\ y &= \frac{1}{4} \end{aligned}$$

From (12) we conclude that

$$x = 2 - \frac{1}{4} = \frac{7}{4}$$

A routine check shows that $(7/4, 1/4)$ is the solution of (11). Consequently, $1/4$ hundred cubic yards, that is, 25 cubic yards, of preheated water are needed. □

In closing we mention that systems of equations in two variables have a long history. In fact, systems such as the one in (3) have appeared in one guise or another since antiquity. For example, a tablet dating from around 300 B.C. shows problems related to finding the dimensions of a rectangle with unit area and given semiperimeter s. If the sides of the rectangle are denoted by x and y, then these conditions on the area and semiperimeter give rise to the system

$$\begin{cases} xy = 1 \\ x + y = s \end{cases}$$

This system is a close relative of the system in (3) and can be solved the same way (see Exercise 31).

EXERCISES 8.1

In Exercises 1–4, verify that the given ordered pair is a solution of the given system.

1. $\begin{cases} 2x + 3y = 8 \\ -5x + \frac{1}{2}y = -4 \end{cases}$ $(1, 2)$

2. $\begin{cases} -\frac{2}{3}x + 3y = 1 \\ 17x - 6y = -2 \end{cases}$ $(0, \frac{1}{3})$

3. $\begin{cases} 2xy = 12 \\ \dfrac{1}{y} - \dfrac{1}{x} = \dfrac{1}{6} \end{cases}$ $(-2, -3)$

4. $\begin{cases} \dfrac{y}{x-1} = x + 1 \\ x^2 + y^2 = 1 \end{cases}$ $(0, -1)$

In Exercises 5–30, find all solutions of the given system of equations.

5. $\begin{cases} 3x - 5y = 13 \\ y = -2 \end{cases}$

6. $\begin{cases} x - y = 0 \\ 2x - 5y = 9 \end{cases}$

7. $\begin{cases} 2y - 3z = 3 \\ 3y + z = 10 \end{cases}$

8. $\begin{cases} z - 3y = 11 \\ 3y - 7z = -5 \end{cases}$

9. $\begin{cases} 7x - 5y = -6 \\ 15y - 21x = 18 \end{cases}$

10. $\begin{cases} 2x - 3y = 5 \\ 4x - 6y = 10 \end{cases}$

11. $\begin{cases} x = 15 - 4y \\ x = -13 + 3y \end{cases}$

12. $\begin{cases} x - 3y = 9 \\ \frac{1}{3}x - y = 27 \end{cases}$

13. $\begin{cases} 4y - z = 4.7 \\ 3y + 2z = 9.3 \end{cases}$

14. $\begin{cases} 7y + 5z = -3 \\ y + 2z = -2.1 \end{cases}$

15. $\begin{cases} \frac{1}{2}x - \frac{1}{3}y = 0 \\ 3x - 2y = 1 \end{cases}$

16. $\begin{cases} 2x + 3y - 5 = 0 \\ x - 2y - 4 = 0 \end{cases}$

17. $\begin{cases} 3x - 2y + 9 = 0 \\ y - 1.5x - 6 = 0 \end{cases}$

18. $\begin{cases} 5x - y - 7 = 0 \\ 15x = 3y + 21 \end{cases}$

19. $\begin{cases} 3x - y = 0 \\ xy = 3 \end{cases}$

20. $\begin{cases} x + 2y = 8 \\ xy = 6 \end{cases}$

21. $\begin{cases} x^2 + 4y^2 = 32 \\ x + 2y = 8 \end{cases}$

22. $\begin{cases} y^2 + z^2 = 25 \\ y - 2z = 0 \end{cases}$

23. $\begin{cases} x^2 - 3xy + 2y^2 = 27 \\ 2x + y = 0 \end{cases}$

24. $\begin{cases} x + y = 1 \\ \dfrac{1}{x} + \dfrac{1}{y} = 4 \end{cases}$

25. $\begin{cases} 6y - x - 1 = 0 \\ y = x^2 \end{cases}$

26. $\begin{cases} y = e^x \\ 0 = e^{2x} - 2y - 8 \end{cases}$

27. $\begin{cases} y = 2^x \\ 6 = 4^x - y \end{cases}$

28. $\begin{cases} x^2 + 2y^2 = 24 \\ \log_2 x + \log_2 y = 3 \end{cases}$

29. $\begin{cases} x + y = a - b \\ x - y = a + b \end{cases}$ where a and b are constants

30. $\begin{cases} x + 2y = a + b \\ 2x - 3y = 4a - 2b \end{cases}$ where a and b are constants

31. Let s be a given positive number, and consider the system

$$\begin{cases} xy = 1 \\ x + y = s \end{cases}$$

a. Solve the system under the condition that $s \geq 2$.
b. Show that the system has no solution if $0 < s < 2$.

32. Solve the system

$$\begin{cases} \dfrac{1}{x^2} - \dfrac{1}{y^2} = 20 \\ \dfrac{1}{x} - \dfrac{1}{y} = 2 \end{cases}$$

(*Hint:* Let $u = 1/x$ and $v = 1/y$, solve the resulting system in u and v, and then find the corresponding values of x and y.)

33. Consider the system

$$\begin{cases} x^2 + 3y^2 = a \\ 1 + x = y \end{cases}$$

Determine the values of a for which the system has
a. exactly two solutions
b. exactly one solution
c. no solutions

34. Determine the values of a and b for which the line with equation $ax + by = 1$ passes through the points $(2, -1)$ and $(-3, 3)$.

35. Determine the values of a and b for which the line with equation $ax + by = 1$ has y intercept $\frac{7}{5}$ and passes through the point $(-4, -1)$.

36. The sum of two numbers is 11, and their product is 28. Find the numbers.

37. The difference of two numbers is 6, and their product is 216. Find the numbers.

38. The difference of two numbers is 4, and the sum of their squares is 136. Find the numbers.

39. The area of a right triangle is 20, and the length of the hypotenuse is $4\sqrt{5}$. Find the lengths of the legs of the triangle.

40. The lengths of the sides of a triangle are 4, 8, and 10. Find the length of the altitude directed toward the longest side.

41. The perimeter of a rectangular room is 82 feet, and the area is 408 square feet. Find the dimensions of the room.

42. A Metroliner train travels the 216 miles from New York to Washington, D.C., at an average speed of 24 miles per hour faster than a local train. If the local train takes 90 minutes longer than the Metroliner, find the average speed of the Metroliner.

43. In making yogurt one places the ingredients in water at 110° Fahrenheit. How much tap water at 60° Fahrenheit and how much heated water at 140° Fahrenheit should be mixed to obtain 3 quarts of water at 110° Fahrenheit?

44. A person has a total of $800 in two savings accounts, one yielding 6% simple interest per year and the other $7\frac{1}{2}\%$ simple interest per year. If the total interest earned during a single year is $51.75, how much is invested in the 6% savings account?

45. In 8 years a girl will be $\frac{5}{8}$ as old as her father was 8 years ago, whereas in 4 years she will be $\frac{4}{9}$ as old as he was 4 years ago. What is the present age of each?

46. When a 4-foot-wide sidewalk is built along one end and one side of a rectangular plot of ground, the ground area is reduced from 7200 square feet to 6536 square feet. Find the original dimensions of the plot.

47. A smaller cube sits on top of a larger cube, which sits on the ground. If the combined height is 6 centimeters and the uncovered portions have a combined surface area of 96 square centimeters, find the lengths of the sides of the cubes.

8.2 SOLUTIONS OF SYSTEMS BY THE ELIMINATION METHOD

In addition to the substitution method discussed in Section 8.1, there is another important method for solving systems of equations: the elimination method. It is especially effective for solving linear systems with more than two equations and two variables, and for understanding methods we will discuss later in the chapter.

The elimination method involves manipulating the equations in a system in ways that do not affect the solutions of the system. Two systems that have the same solutions are called ***equivalent***, and the strategy will be to replace a given system by an equivalent system whose solutions are more easily identified.

One way to manipulate a system is to multiply an equation in the system by a nonzero constant, by which we mean to multiply each side of the equation by that constant. For example, multiplying the second equation of the system

$$\begin{cases} -2x + 3y = 10 \\ x - y = 6 \end{cases}$$

by 2 yields the equivalent system

$$\begin{cases} -2x + 3y = 10 \\ 2x - 2y = 12 \end{cases} \tag{1}$$

Since multiplying an equation by a nonzero number does not alter the solutions of the equation, this manipulation leads to an equivalent system.

A second manipulation of a system involves replacing one of the equations in the system by the sum (or difference) of the equations in the system. For example, taking the sum of the equations in (1) gives us

$$(-2x + 3y) + (2x - 2y) = 10 + 12$$

or more simply,

$$y = 22$$

Replacing the second equation in (1) by the simplified form $y = 22$ of the sum yields

$$\begin{cases} -2x + 3y = 10 \\ \qquad\quad y = 22 \end{cases} \tag{2}$$

The system in (2), obtained from the system in (1) by the addition manipulation, has the same solution as does the system in (1). Thus the two systems are equivalent. Observe that x does not appear in the second equation of (2). Thus in passing from (1) to (2) we have eliminated x from the second equation.

The method of solving a system by utilizing the two manipulations—multiplying an equation by a constant, and substituting the sum (or difference) of two equations for one of the equations—in order to eliminate one of the variables is called the ***elimination method***.

EXAMPLE 1. Use the elimination method to solve the system

$$\begin{cases} 2x + 3y = -4 \\ \ \ x - 2y = \ \ \ 5 \end{cases}$$

Solution. First we multiply the second equation by -2:

$$\begin{cases} \ \ 2x + 3y = -4 \\ -2x + 4y = -10 \qquad -2 \times \text{second equation} \end{cases}$$

Next we replace the second equation by its sum with the first equation:

$$\begin{cases} 2x + 3y = -4 \\ \qquad 7y = -14 \qquad \text{first equation} + \text{second equation} \end{cases}$$

This last system can be solved easily, since x has been eliminated from its second equation. Indeed, the second equation implies that $y = -2$, so by substituting -2 for y in the first equation we obtain

$$\begin{aligned} 2x + 3(-2) &= -4 \\ 2x &= -4 + 6 = 2 \\ x &= 1 \end{aligned}$$

Thus the proposed solution is $(1, -2)$.

Check: $\begin{cases} 2(1) + 3(-2) = -4 \\ \quad 1 - 2(-2) = \ \ 5 \end{cases}$

We conclude that $(1, -2)$ is the solution of the given system. □

Let us observe that if the variables appearing in the system are changed but the coefficients of the variables are unaltered, then the solutions of the system remain the same. Thus the solution of the system

$$\begin{cases} 2y + 3z = -4 \\ \ \ y - 2z = \ \ 5 \end{cases} \tag{3}$$

obtained from the system in Example 1 by replacing x by y and y by z, is the ordered pair $(1, -2)$. We will use this fact later.

In attempting to use the elimination method to solve a linear system, it is possible to obtain a new system that has two identical equations. In that case there are infinitely many solutions, as the following example shows.

EXAMPLE 2. Use the elimination method to solve the system

$$\begin{cases} \ \ 2x - \ y = \ \ 4 \\ -6x + 3y = -12 \end{cases}$$

Solution. If we multiply the second equation by $-\frac{1}{3}$, we obtain the system

$$\begin{cases} 2x - y = 4 \\ 2x - y = 4 \qquad -\tfrac{1}{3} \times \text{second equation} \end{cases}$$

It follows that

$$y = 2x - 4$$

Thus (a, b) is a proposed solution of the given system if and only if

$$b = 2a - 4$$

Check:
$$\begin{cases} 2a - (2a - 4) = 2a - 2a + 4 = 4 \\ -6a + 3(2a - 4) = -6a + 6a - 12 = -12 \end{cases}$$

We conclude that for any number a, $(a, 2a - 4)$ is a solution of the given system. □

The elimination method can also identify inconsistent systems, that is, systems with no solutions.

EXAMPLE 3. Use the elimination method to show that the system

$$\begin{cases} 4x - 2y = 8 \\ -6x + 3y = 3 \end{cases}$$

is inconsistent.

Solution. Multiplying the first equation by 3 and the second by -2, we obtain the system

$$\begin{cases} 12x - 6y = 24 & 3 \times \text{first equation} \\ 12x - 6y = -6 & -2 \times \text{second equation} \end{cases}$$

If we subtract the second equation from the first, we obtain $0 = 30$, which is impossible. Consequently the system has no solutions and hence is inconsistent. □

EXERCISES 8.2

In Exercises 1–14, use the elimination method to find all solutions of the given system of equations.

1. $\begin{cases} 2x + 3y = 3 \\ x + y = 2 \end{cases}$

2. $\begin{cases} 2x + y = 4 \\ 3x + y = 9 \end{cases}$

3. $\begin{cases} 7x - 5y = -6 \\ x + 5y = 22 \end{cases}$

4. $\begin{cases} 5x + 7y = 2 \\ 3x + 2y = -1 \end{cases}$

5. $\begin{cases} -5u + 3v = 1 \\ 9u - 10v = 12 \end{cases}$

6. $\begin{cases} \frac{2}{3}u - \frac{4}{3}v = 0 \\ \frac{3}{5}u + \frac{14}{5}v = 4 \end{cases}$

7. $\begin{cases} \frac{1}{2}x - 5y = 5 \\ -3x + 30y = -30 \end{cases}$

8. $\begin{cases} 5x - 2y = 1.4 \\ 2x + 3y = 9.3 \end{cases}$

9. $\begin{cases} 4x - 3y = 6 \\ -2x + 1.5y = 5 \end{cases}$

10. $\begin{cases} \dfrac{5x - 3y}{2} - 6 = 0 \\ \dfrac{x + y}{4} - 1 = 0 \end{cases}$

11. $\begin{cases} 6x - 4y = 7 \\ -4x + \frac{8}{3}y = -\frac{14}{3} \end{cases}$

12. $\begin{cases} 14x + 10y - 18 = 0 \\ 42 - 35x - 25y = 0 \end{cases}$

13. $\begin{cases} \dfrac{x}{a} + \dfrac{y}{b} = 6 \\ \dfrac{x}{a} - \dfrac{y}{b} = 2 \end{cases}$ where a and b are constants

14. $\begin{cases} ax + by = a^2 + b^2 \\ bx - ay = a^2 + b^2 \end{cases}$ where a and b are constants

15. Show that for each real number a, the system

$$\begin{cases} 3x + 2y = 7 \\ 4x - 2y = a \end{cases}$$

has exactly one solution.

16. Consider the system

$$\begin{cases} 6x - 4y = 8 \\ -9x + 6y = a \end{cases}$$

Find the values of a for which the system has
a. an infinite number of solutions
b. no solutions

17. Find the values of a and b for which the line with equation $ax + by = 6$ passes through the points $(10, 6)$ and $(-2, -3)$.

18. Let $f(x) = ax^2 + bx + 4$. Find the values of a and b for which $(1, 6)$ and $(-2, 18)$ lie on the graph of f.

19. Two students have \$50 together. One student loses a \$5 bet to the other, and then both have the same amount of money. How much money did each student have before the bet?

20. When a passenger jet travels with a constant air speed (the speed of the plane with respect to the air) from Seattle to Chicago, it has a 50 mph tailwind and makes the trip in 3 hours and 36 minutes. On the return flight it flies with the same air speed, has a headwind of 50 mph, and makes the trip in $4\frac{1}{2}$ hours. Find the air speed of the jet and the distance flown from Seattle to Chicago.

21. How many ounces of 24-carat (pure) gold and how many ounces of 14-carat gold must be fused together to obtain a 5-ounce bar of 18-carat gold?

22. An investor splits \$1000 between two savings accounts that earn simple interest at the same rate. At the end of 3 years one account contains \$472, and after 5 years the other contains \$780. Find the interest rate and the amount invested in each account.

23. A theater that has a seating capacity of 200 charges \$3 for adults and \$1.50 for children. If the receipts for a given sold-out performance are \$510, how many children's tickets were sold?

8.3 LINEAR SYSTEMS WITH MORE THAN TWO VARIABLES

Until now we have discussed and solved systems of equations with two variables. However, a system can have any number of variables. For example, the system

$$\begin{cases} x - 2y + z = -1 \\ 2x - 3y + 4z = -5 \\ 3x - 4y + 2z = 1 \end{cases} \tag{1}$$

has three variables, and the system

$$\begin{cases} w + 2x - y + 3z = 0 \\ 2w + y - z = 3 \\ -w + x - y + \frac{1}{2}z = -1 \\ 9w + 3x - 4y - z = 2 \end{cases} \tag{2}$$

has four variables. In this section we will restrict our attention to systems having three variables. However, the methods developed here can be extended to systems having four or more variables.

By a ***solution*** of a system of equations in three variables x, y, and z we mean an ordered triple (a, b, c) of numbers such that if x is replaced by a, y by b, and z by c in all equations of the system, all the resulting equations are valid. Thus $(3, 1, -2)$ is a solution of (1) because when we let $x = 3$, $y = 1$, and $z = -2$ in (1) we obtain the system

$$\begin{cases} 3 - 2(1) + (-2) = 3 - 2 - 2 = -1 \\ 2(3) - 3(1) + 4(-2) = 6 - 3 - 8 = -5 \\ 3(3) - 4(1) + 2(-2) = 9 - 4 - 4 = 1 \end{cases} \tag{3}$$

all of whose equations are valid.

An equation in three variables of the form

$$ax + by + cz = d$$

where a, b, c, and d are constants and either $a \neq 0$, $b \neq 0$, or $c \neq 0$, is called a ***linear equation***, and a system of linear equations is called a ***linear system***. The system in (1) is an example of a linear system; in this section we will solve only linear systems with three equations and three variables. One feature of linear systems with three equations and three variables is that there are either infinitely many solutions, one solution, or no solutions. We have ***solved*** the system once we have determined all its solutions.

The methods of substitution and elimination that we presented for systems having two variables have counterparts for systems with three variables.

The Substitution Method

Although the method of substitution is not as conveniently applied for systems with three variables as it is for systems of two variables, the basic idea is the same. We use one of the equations to express one of the variables in terms of the other two variables. Then we substitute for the chosen variable in the other equations. That yields a system of two equations and two variables, which we solve by the methods of the preceding two sections.

EXAMPLE 1. Use the substitution method to solve the system

$$\begin{cases} x - 2y + z = -1 \\ 2x - 3y + 4z = -5 \\ 3x - 4y + 2z = 1 \end{cases} \tag{4}$$

Solution. We use the first equation to express x in terms of y and z:

$$x = 2y - z - 1 \tag{5}$$

Now we substitute $2y - z - 1$ for x in the second and third equations of (4), obtaining the system

$$\begin{cases} 2(2y - z - 1) - 3y + 4z = -5 \\ 3(2y - z - 1) - 4y + 2z = 1 \end{cases}$$

which condenses to

$$\begin{cases} y + 2z = -3 \\ 2y - z = 4 \end{cases} \tag{6}$$

Next we solve the system in (6) by substitution. We use the first equation in (6) to express y in terms of z:

$$y = -2z - 3 \tag{7}$$

Substituting $-2z - 3$ for y in the second equation of (6) gives us

$$\begin{aligned} 2(-2z - 3) - z &= 4 \\ -5z &= 10 \\ z &= -2 \end{aligned}$$

To find y we use (7) with the known value -2 of z:

$$y = -2(-2) - 3 = 1$$

Then to find x we use (5), substituting 1 for y and -2 for z:

$$x = 2(1) - (-2) - 1 = 3$$

Thus the only possible solution is $(3, 1, -2)$. That it is in fact a solution was shown in (3). □

The Elimination Method

Consider the system

$$\begin{cases} 3x + 5y + 4z = 9 \\ 3x + 3y + z = 13 \\ \phantom{3x + {}} y - 2z = 5 \end{cases} \tag{8}$$

One way of solving this system involves subtracting the second equation from the first and substituting the result for the second equation to obtain

$$\begin{cases} 3x + 5y + 4z = 9 \\ \phantom{3x + {}} 2y + 3z = -4 \qquad \text{first equation} - \text{second equation} \\ \phantom{3x + 2{}} y - 2z = 5 \end{cases} \tag{9}$$

The latter two equations of (9) form a system in the two variables y and z. By the elimination method the system

$$\begin{cases} 2y + 3z = -4 \\ y - 2z = 5 \end{cases}$$

is equivalent to the system

$$\begin{cases} 2y + 3z = -4 \\ \phantom{2y + {}} 7z = -14 \end{cases}$$

(See Example 1 and the system in (3), both in Section 8.2.) Therefore (9) is equivalent to the system

$$\begin{cases} 3x + 5y + 4z = 9 \\ \phantom{3x + {}} 2y + 3z = -4 \\ \phantom{3x + 2y + {}} 7z = -14 \end{cases}$$

This system is easy to solve. First we find that $z = -2$ from the last equation. Then, substituting -2 for z in the middle equation, we find y:

$$\begin{aligned} 2y + 3(-2) &= -4 \\ 2y &= -4 + 6 = 2 \\ y &= 1 \end{aligned}$$

Finally, we substitute 1 for y and -2 for z in the first equation to obtain x:

$$\begin{aligned} 3x + 5(1) + 4(-2) &= 9 \\ 3x &= 12 \\ x &= 4 \end{aligned}$$

The proposed solution of the system in (8) is therefore $(4, 1, -2)$. A routine check bears out the validity of the solution. (See Exercise 1.)

Our normal strategy for solving systems of three equations in three variables by the elimination method involves replacing a given system by an equivalent system in which one equation has one variable, a second equation has two variables, and the remaining equation has three variables. The manipulations that we use in passing from one system to an equivalent one are similar to those used for two variables:

> i. Multiplying an equation in the system by a nonzero constant.
> ii. Replacing an equation in the system by the result of adding the equation to, or subtracting it from, another equation in the system.

As you will see below, we frequently combine (i) and (ii).

EXAMPLE 2. Use the elimination method to solve the system

$$\begin{cases} x + 3y + 4z = 2 \\ 2x - y - 5z = 3 \\ 3x + 2y - 2z = -1 \end{cases} \tag{10}$$

Solution. We begin by multiplying the first equation by 2 and then subtracting the second equation from the result:

$$\begin{array}{rll} 2x + 6y + 8z = 4 & & 2 \times \text{first equation} \\ 2x - y - 5z = 3 & & \text{second equation} \\ \hline 7y + 13z = 1 & & \end{array}$$

Replacing the second equation in (10) by $7y + 13z = 1$ yields the equivalent system

$$\begin{cases} x + 3y + 4z = 2 \\ 7y + 13z = 1 \\ 3x + 2y - 2z = -1 \end{cases} \tag{11}$$

Notice that x has been eliminated from the second equation. Next we multiply the first equation in (11) by 3 and subtract the third equation from the result:

$$\begin{array}{rll} 3x + 9y + 12z = 6 & & 3 \times \text{first equation} \\ 3x + 2y - 2z = -1 & & \text{third equation} \\ \hline 7y + 14z = 7 & & \end{array}$$

Replacing the third equation in (11) by $7y + 14z = 7$, we obtain the system

$$\begin{cases} x + 3y + 4z = 2 \\ 7y + 13z = 1 \\ 7y + 14z = 7 \end{cases} \tag{12}$$

Subtracting the second equation in (12) from the third, and substituting the result for the third equation in (12), we obtain the system

$$\begin{cases} x + 3y + 4z = 2 \\ \phantom{x + {}} 7y + 13z = 1 \\ z = 6 \end{cases} \tag{13}$$

This system has one equation in one variable, one in two variables, and one in three variables, so it is easy to substitute in the second equation to find y, and then to substitute in the first equation to find x. Indeed, substituting 6 for z in the second equation of (13), we find that

$$7y + 13(6) = 1$$
$$7y = 1 - 78 = -77$$
$$y = -11$$

Now we substitute -11 for y and 6 for z in the first equation of (13) and solve for x:

$$x + 3(-11) + 4(6) = 2$$
$$x = 11$$

Therefore $(11, -11, 6)$ is the proposed solution of (10).

Check:
$$\begin{cases} 11 + 3(-11) + 4(6) = 2 \\ 2(11) - (-11) - 5(6) = 3 \\ 3(11) + 2(-11) - 2(6) = -1 \end{cases}$$

We conclude that the solution of the given system is $(11, -11, 6)$. □

In the next example we illustrate how the method of elimination can be used to find all solutions of a system with infinitely many solutions.

EXAMPLE 3. Use the method of elimination to solve the system

$$\begin{cases} x - 2y - 3z = 1 \\ 2x + y + 4z = 2 \\ 8x - 6y - 4z = 8 \end{cases} \tag{14}$$

Solution. We begin by multiplying the first equation by 2 and subtracting the resulting equation from the second equation:

$$\begin{array}{rl} 2x + y + 4z = 2 & \text{second equation} \\ 2x - 4y - 6z = 2 & 2 \times \text{first equation} \\ \hline 5y + 10z = 0 & \end{array}$$

Replacing the second equation in (14) by $5y + 10z = 0$, we obtain the system

$$\begin{cases} x - 2y - 3z = 1 \\ 5y + 10z = 0 \\ 8x - 6y - 4z = 8 \end{cases} \tag{15}$$

Next we multiply the first equation in (15) by 8 and subtract the resulting equation from the third equation:

$$\begin{array}{rl} 8x - 6y - 4z = 8 & \text{third equation} \\ 8x - 16y - 24z = 8 & 8 \times \text{first equation} \\ \hline 10y + 20z = 0 & \end{array}$$

Replacing the third equation in (15) by $10y + 20z = 0$ yields the system

$$\begin{cases} x - 2y - 3z = 1 \\ 5y + 10z = 0 \\ 10y + 20z = 0 \end{cases} \tag{16}$$

Multiplying the third equation in (16) by $\frac{1}{2}$ brings us to the system

$$\begin{cases} x - 2y - 3z = 1 \\ 5y + 10z = 0 \\ 5y + 10z = 0 \end{cases} \tag{17}$$

Subtracting the third equation in (17) from the second and substituting the result for the third equation, we obtain the system

$$\begin{cases} x - 2y - 3z = 1 \\ 5y + 10z = 0 \\ 0 = 0 \end{cases} \tag{18}$$

The third equation yields no value for x, y, or z. However, if we solve for y in the second equation in (18), we obtain

$$5y = -10z$$

$$y = -2z \tag{19}$$

Now we substitute $-2z$ for y in the first equation in (18) and solve for x:

$$\begin{aligned} x - 2(-2z) - 3z &= 1 \\ x + z &= 1 \\ x &= 1 - z \end{aligned} \tag{20}$$

It follows that (a, b, c) is a proposed solution of the given system if and only if $b = -2c$ (from (19)) and $a = 1 - c$ (from (20)). Consequently there are infinitely many proposed solutions, given by $(1 - c, -2c, c)$ for any number c. In Exercise 4 you are asked to check that for every value of c, $(1 - c, -2c, c)$ satisfies all three equations in the given system. □

EXERCISES 8.3

In Exercises 1–4, verify that the given ordered triple is a solution of the given system.

1. $\begin{cases} 3x + 5y + 4z = 9 \\ 3x + 3y + z = 13 \\ y - 2z = 5 \end{cases}$ $(4, 1, -2)$

2. $\begin{cases} 4x - 2y + 3z = 2 \\ -2x + \frac{1}{2}y - \frac{1}{2}z = 1 \\ 7x - 13y + 2z = -3 \end{cases}$ $(-1, 0, 2)$

3. $\begin{cases} 2x + y - 6z = 1 \\ x + y - 4z = 1 \\ -4x - 3y + z = -\frac{5}{6} \end{cases}$ $(-\frac{1}{3}, \frac{2}{3}, -\frac{1}{6})$

4. $\begin{cases} x - 2y - 3z = 1 \\ 2x + y + 4z = 2 \\ 8x - 6y - 4z = 8 \end{cases}$ $(1 - c, -2c, c)$ for any number c

In Exercises 5–24, find all solutions of the given system.

5. $\begin{cases} 3x - 2y + z = 10 \\ 5y - z = 2 \\ 3z = -6 \end{cases}$

6. $\begin{cases} -2x + 3y + 5z = -1 \\ 6y - z = -11 \\ -2y + 4z = 0 \end{cases}$

7. $\begin{cases} x - y + z = 5 \\ 2x + 3y - z = -4 \\ x + y + 2z = 6 \end{cases}$

8. $\begin{cases} 2x - y + 3z = 1 \\ -4x + y - 2z = 4 \\ x - y - 2z = -2 \end{cases}$

9. $\begin{cases} x - y - z = 0 \\ x + 3y - z = 0 \\ x + y - z = 0 \end{cases}$

10. $\begin{cases} 4u - 3v - 3w = 22 \\ 2u - v - 6w = 15 \\ u + 3v - 2w = 3 \end{cases}$

11. $\begin{cases} u + 2v + 3w = 5 \\ 2u - 8v - 14w = 5 \\ 2u - 2v - 4w = 7 \end{cases}$

12. $\begin{cases} 3x - 4y + z = 0 \\ 7x - z = 0 \\ -9x + 12y - 3z = 0 \end{cases}$

13. $\begin{cases} 6x - 3y + 7z = 0 \\ 4x + 5y + 9z = 0 \\ 13x - 11y - 4z = 0 \end{cases}$

14. $\begin{cases} 2x - 3y + 4z + 12 = 0 \\ -4x + y - z - 7 = 0 \\ 3x + y - z = 0 \end{cases}$

15. $\begin{cases} 4x - 5y = 11 \\ 2x + z = 7 \\ 2y + z = 1 \end{cases}$

16. $\begin{cases} 2x - 3y + 5z = 4 \\ x + 5y - 2z = 5 \\ 5x - y + 8z = 6 \end{cases}$

17. $\begin{cases} 2x - y - 7 = 0 \\ 3y - 4z - 5 = 0 \\ -x + 2z + 7 = 0 \end{cases}$

18. $\begin{cases} x + 5y - z = 11 \\ 2x - y = 7 \\ 3x + y + z = 11 \end{cases}$

19. $\begin{cases} x + y = 0 \\ 2x - 4z = 3 \\ x - y + z = -2 \end{cases}$

20. $\begin{cases} x + 2y - z = -9 \\ -x + y + z = 3 \\ 2x + 3y + z = 8 \end{cases}$

21. $\begin{cases} x + y + z = 0 \\ x - y - z = 1 \\ 2x - 2y + 4z = 3 \end{cases}$

22. $\begin{cases} \frac{1}{2}x - \frac{1}{3}y - \frac{1}{6}z = \frac{5}{6} \\ \frac{1}{3}x + \frac{1}{6}y + z = -\frac{3}{2} \\ -\frac{2}{3}x - \frac{1}{2}y + \frac{1}{3}z = -\frac{1}{3} \end{cases}$

23. $\begin{cases} 2.4x - 1.6y - 0.8z = 4.8 \\ 1.2x + 1.4y - 2z = -0.6 \\ -x + 0.8y - 1.8z = -3.2 \end{cases}$

24. $\begin{cases} 40x - 50y - 10z = -21 \\ -5x - 3y + 20z = 13 \\ -10x + 20y + 30z = 31 \end{cases}$

25. Let $f(x) = ax^2 + bx + c$. Find the values of a, b, and c for which the graph of f contains the points $(-1, 10)$, $(1, 6)$, and $(2, 13)$.

26. Let $f(x) = ae^x + be^{-x} + c$. Find the values of a, b, and c for which the graph of f contains the points $(0, 1)$, $(\ln 2, 6)$, and $(\ln 4, 13)$.

27. The sum of three numbers is 58. The sum of the smaller two numbers is two more than the largest number, and the largest number is double the smallest number. Find the three numbers.

28. One angle in a certain triangle is $\frac{1}{3}$ a second angle in the triangle, and the third angle in the triangle is twice the sum of the other two. Find the angles. (*Hint:* The sum of the angles in any triangle is equal to 180°.)

29. A collection of quarters, dimes, and nickels is worth \$4.55 and has 44 coins in it. If there are 6 more nickels than dimes in the collection, find the number of quarters in the collection.

30. Peter, Paul, and Mary have a combined total of \$200. If Peter gives \$10 to Paul, they will have the same amount of money, and Mary will have twice as much as either Peter or Paul. How much money does each have before the transaction?

31. A university theater charges \$8 for adults other than students, \$3 for students, and \$1 for preschool children. Suppose that 580 people attend a performance of *My Fair Lady*, including 8 times as many students as children. If the receipts total \$2760, determine the number of students attending the performance.

32. A concert hall has 2100 seats. The orchestra seats cost \$8 each, whereas the seats in the first and second balcony cost \$6 and \$3, respectively. For a full house the hall grosses \$13,800. If all the seats in both balconies were sold for \$5 each, then the gross for a full house would be \$14,100. Find the total number of balcony seats.

8.4 SOLUTIONS OF SYSTEMS BY MATRICES

The elimination method of solving systems of equations can be reformulated so as to reduce the amount of writing necessary. The reformulation involves a bracketed rectangular array of numbers such as

$$\begin{bmatrix} 2 & -1 & 4 \\ 3 & 1 & -14 \end{bmatrix} \tag{1}$$

or

$$\begin{bmatrix} 3 & -1 & 2 \\ 1 & 0 & 3 \\ 6 & -4 & 2 \end{bmatrix} \tag{2}$$

called a matrix.

DEFINITION 8.1 A ***matrix*** is a rectangular array

$$\begin{bmatrix} a_{11} & a_{12} & \cdots & a_{1n} \\ a_{21} & a_{22} & \cdots & a_{2n} \\ \cdot & \cdot & & \cdot \\ \cdot & \cdot & & \cdot \\ \cdot & \cdot & & \cdot \\ a_{m1} & a_{m2} & \cdots & a_{mn} \end{bmatrix} \tag{3}$$

where $a_{11}, a_{12}, \ldots, a_{1n}; a_{21}, a_{22}, \ldots, a_{2n}; \ldots; a_{m1}, a_{m2}, \ldots, a_{mn}$ are numbers, called the ***entries*** of the matrix.

The matrix in (3) is called an $m \times n$ matrix (read "m by n matrix") to emphasize the fact that there are m rows and n columns of numbers in the matrix. The matrix in (1) is a 2×3 matrix, because there are 2 rows and 3 columns. For that matrix we have

$$a_{11} = 2, \quad a_{12} = -1, \quad a_{13} = 4, \quad a_{21} = 3, \quad a_{22} = 1, \quad a_{23} = -14$$

Similarly, the matrix in (2) is a 3×3 matrix, because there are 3 rows and 3 columns. A 1×1 matrix has the form $[a]$, where a is a real number. Thus a 1×1 matrix is essentially the same as a number. An $n \times n$ matrix is called a ***square matrix of order n***. The matrix in (2) is a square matrix of order 3.

Now consider the system

$$\begin{cases} 2x - y = 4 \\ 3x + y = -14 \end{cases} \tag{4}$$

There are two matrices associated with the system. First there is the ***coefficient matrix***

$$\begin{bmatrix} 2 & -1 \\ 3 & 1 \end{bmatrix}$$

consisting of the coefficients of x and y. Notice that the coefficients 2 and 3 of x occupy the first column, and the coefficients -1 and 1 of y the second column.

The second matrix related to a system of equations is the ***augmented matrix***, which for (4) is

$$\left[\begin{array}{cc|c} 2 & -1 & 4 \\ 3 & 1 & -14 \end{array}\right]$$

The augmented matrix is the matrix obtained by attaching to the coefficient matrix the constants to the right of the equality signs in the system. The vertical dashed line serves to separate the coefficient matrix from the column of constants. The rows of the augmented matrix correspond to the equations in the system, in their given order.

For the system

$$\begin{cases} x - y - 2z = -2 \\ 4x + 5z = \frac{1}{2} \\ 7x + 15y - z = 0 \end{cases} \tag{5}$$

the coefficient matrix and the augmented matrix are

$$\begin{bmatrix} 1 & -1 & -2 \\ 4 & 0 & 5 \\ 7 & 15 & -1 \end{bmatrix} \quad \text{and} \quad \left[\begin{array}{ccc:c} 1 & -1 & -2 & -2 \\ 4 & 0 & 5 & \frac{1}{2} \\ 7 & 15 & -1 & 0 \end{array}\right] \tag{6}$$

respectively. Notice that the second equation in the system (5) is equivalent to

$$4x + 0y + 5z = \frac{1}{2}$$

and as a result, 0 represents the coefficient of y in the second row of the matrices in (6).

Equivalent Matrices

The solutions of a given system of equations are unaffected by the order in which the equations appear. Similarly, the solutions of a system are unaffected by multiplying an equation by a nonzero constant or by replacing an equation by the sum of it and another equation. These alterations in a system of equations correspond to the first three of the following alterations in the rows of the associated augmented matrix:

> i. Interchanging two rows
> ii. Replacing a row by a nonzero multiple of that row
> iii. Replacing a row by the sum of it and another row
> iv. Replacing a row by the sum of it and a nonzero multiple of another row

Operations (i)–(iv) are sometimes called ***elementary row operations*** on a matrix. Frequently two of the four operations are performed in a single step. In fact, (iv) is a combination of (ii) and (iii), as is the operation of substituting for one row the difference of it and another row.

We say that two matrices are ***equivalent*** if one of them can be derived from the other by one or more of the operations (i)–(iv). We will use the symbol $\leftrightarrow$ between two matrices to indicate that the matrices are equivalent.

From the definition of equivalence, if two matrices are equivalent, then they have the same number of rows and the same number of columns. Moreover, if matrices A, B, and C have the property that $A \leftrightarrow B$ and $B \leftrightarrow C$, then $A \leftrightarrow C$.

In applying elementary row operations to a matrix, we will denote the top row by R_1, the next row by R_2, and so forth. If we write, say, $R_2 \rightarrow R_1 + (-3)R_2$, we mean that we are replacing the second row by the new row obtained by multiplying the second row by -3 and adding the result to the first row.

EXAMPLE 1. Show that

$$\begin{bmatrix} 3 & -1 & 0 & -3 \\ 1 & -\frac{1}{3} & 3 & -2 \\ -6 & -4 & 2 & 0 \end{bmatrix} \leftrightarrow \begin{bmatrix} 3 & -1 & 0 & -3 \\ 0 & -3 & 1 & -3 \\ 0 & 0 & -9 & 3 \end{bmatrix}$$

Solution. We proceed as follows:

$$\begin{bmatrix} 3 & -1 & 0 & -3 \\ 1 & -\frac{1}{3} & 3 & -2 \\ -6 & -4 & 2 & 0 \end{bmatrix} \xleftrightarrow{R_2 \rightarrow R_1 + (-3)R_2} \begin{bmatrix} 3 & -1 & 0 & -3 \\ 0 & 0 & -9 & 3 \\ -6 & -4 & 2 & 0 \end{bmatrix} \xleftrightarrow{R_3 \rightarrow R_1 + \frac{1}{2}R_3}$$

$$\begin{bmatrix} 3 & -1 & 0 & -3 \\ 0 & 0 & -9 & 3 \\ 0 & -3 & 1 & -3 \end{bmatrix} \xleftrightarrow[R_3 \rightarrow R_2]{R_2 \rightarrow R_3} \begin{bmatrix} 3 & -1 & 0 & -3 \\ 0 & -3 & 1 & -3 \\ 0 & 0 & -9 & 3 \end{bmatrix}$$

Thus the two given matrices are equivalent. □

Solutions of Systems by Augmented Matrices

Recall that every system of equations has an associated augmented matrix. Each elementary row operation performed on the augmented matrix corresponds to an operation on the given system that yields a system with the same solutions. As a result, two systems of equations that have equivalent augmented matrices have the same solutions.

EXAMPLE 2. Use augmented matrices to solve the system

$$\begin{cases} 3x - y = -3 \\ x - \frac{1}{3}y + 3z = -2 \\ -6x - 4y + 2z = 0 \end{cases} \tag{7}$$

Solution. The associated augmented matrix is

$$\left[\begin{array}{ccc|c} 3 & -1 & 0 & -3 \\ 1 & -\frac{1}{3} & 3 & -2 \\ -6 & -4 & 2 & 0 \end{array}\right]$$

By Example 1,

$$\left[\begin{array}{rrr|r} 3 & -1 & 0 & -3 \\ 1 & -\frac{1}{3} & 3 & -2 \\ -6 & -4 & 2 & 0 \end{array}\right] \leftrightarrow \left[\begin{array}{rrrr} 3 & -1 & 0 & -3 \\ 0 & -3 & 1 & -3 \\ 0 & 0 & -9 & 3 \end{array}\right]$$

But the latter matrix is the augmented matrix for the system

$$\begin{cases} 3x - \quad y \qquad = -3 \\ \qquad -3y + \ z = -3 \\ \qquad\qquad -9z = \ \ 3 \end{cases} \tag{8}$$

By the comments preceding the example, the solutions of (8) are exactly the solutions of (7). From (8),

$$z = \frac{3}{-9} = -\frac{1}{3}$$

To find y we substitute $-\frac{1}{3}$ for z in the second equation of (8) and solve for y:

$$\begin{aligned} -3y + \left(-\frac{1}{3}\right) &= -3 \\ -3y &= -3 + \frac{1}{3} \\ &= -\frac{8}{3} \\ y &= \frac{8}{9} \end{aligned}$$

To find x we substitute $\frac{8}{9}$ for y in the first equation of (8) and solve for x:

$$\begin{aligned} 3x - \frac{8}{9} &= -3 \\ 3x &= -3 + \frac{8}{9} = -\frac{19}{9} \\ x &= -\frac{19}{27} \end{aligned}$$

Thus $\left(-\frac{19}{27}, \frac{8}{9}, -\frac{1}{3}\right)$ is a proposed solution of (8) and hence of (7).

Check:
$$\begin{cases} 3\left(-\dfrac{19}{27}\right) - \dfrac{8}{9} = -3 \\ -\dfrac{19}{27} - \dfrac{1}{3}\left(\dfrac{8}{9}\right) + 3\left(-\dfrac{1}{3}\right) = -2 \\ -6\left(-\dfrac{19}{27}\right) - 4\left(\dfrac{8}{9}\right) + 2\left(-\dfrac{1}{3}\right) = 0 \end{cases}$$

Consequently $(-\frac{19}{27}, \frac{8}{9}, -\frac{1}{3})$ is the solution of the given system. □

We remark that the third augmented matrix appearing in the solution of Example 2 has zeros everywhere below the diagonal 3, −3, −9 starting at the upper left corner. For this reason, the final equation of the associated system in (8) has the lone variable z. In finding solutions of systems by means of augmented matrices, the goal is to replace the augmented matrix of a given system by an equivalent matrix with zeros below the designated diagonal, whether the augmented matrix is a 3×4 or a 2×3 matrix.

EXAMPLE 3. Use augmented matrices to show that the system

$$\begin{cases} x + y + z = -1 \\ x + 2y - z = -6 \\ 2x + y + 4z = 3 \end{cases} \tag{9}$$

has infinitely many solutions.

Solution. The augmented matrix is

$$\left[\begin{array}{ccc|c} 1 & 1 & 1 & -1 \\ 1 & 2 & -1 & -6 \\ 2 & 1 & 4 & 3 \end{array}\right]$$

We apply elementary row operations as follows:

$$\left[\begin{array}{ccc|c} 1 & 1 & 1 & -1 \\ 1 & 2 & -1 & -6 \\ 2 & 1 & 4 & 3 \end{array}\right] \xleftrightarrow{R_2 \to R_2 + (-1)R_1} \left[\begin{array}{ccc|c} 1 & 1 & 1 & -1 \\ 0 & 1 & -2 & -5 \\ 2 & 1 & 4 & 3 \end{array}\right] \xleftrightarrow{R_3 \to R_3 + (-2)R_1}$$

$$\left[\begin{array}{ccc|c} 1 & 1 & 1 & -1 \\ 0 & 1 & -2 & -5 \\ 0 & -1 & 2 & 5 \end{array}\right] \xleftrightarrow{R_3 \to R_3 + R_2} \left[\begin{array}{ccc|c} 1 & 1 & 1 & -1 \\ 0 & 1 & -2 & -5 \\ 0 & 0 & 0 & 0 \end{array}\right]$$

The system associated with the last matrix above is

$$\begin{cases} x + y + z = -1 \\ \phantom{x+{}} y - 2z = -5 \\ 0 = 0 \end{cases} \tag{10}$$

From the second equation in (10) it follows that

$$y = 2z - 5$$

Substituting $2z - 5$ for y in the first equation in (10), we obtain

$$x + (2z - 5) + z = -1$$

or equivalently,

$$x = -3z + 4$$

Consequently (a, b, c) is a solution of(10), and hence of (9), only if $b = 2c - 5$ and $a = -3c + 4$. Thus the solutions of (9) are $(-3c + 4, 2c - 5, c)$ for any number c. You can check that each triple of that form actually satisfies the given system. □

EXERCISES 8.4

In Exercises 1–6, write the coefficient matrix and the augmented matrix of the given system.

1. $\begin{cases} 2x - 4y = 7 \\ -5x + y = 6 \end{cases}$

2. $\begin{cases} -x - y = 0 \\ 4x + \frac{1}{2}y = 3 \end{cases}$

3. $\begin{cases} x + y = 2 \\ y = 1 \end{cases}$

4. $\begin{cases} x - y - z = 1 \\ 2x + 4y - 5z = -3 \\ -7x + y + z = -2 \end{cases}$

5. $\begin{cases} \frac{1}{3}x - \frac{1}{4}y + \frac{1}{2}z = 1 \\ \frac{2}{5}x + z = 3 \\ x - \frac{1}{2}y = 4 \end{cases}$

6. $\begin{cases} x - y = 0 \\ x + z = 1 \\ 2y - z = 3 \end{cases}$

In Exercises 7–12, show that the given matrices are equivalent, giving all reasons.

7. $\begin{bmatrix} 1 & 3 \\ 4 & 2 \end{bmatrix}$ and $\begin{bmatrix} 4 & 2 \\ 1 & 3 \end{bmatrix}$

8. $\begin{bmatrix} 2 & -6 & 4 \\ 6 & -3 & 7 \end{bmatrix}$ and $\begin{bmatrix} 14 & -12 & 18 \\ 6 & -3 & 7 \end{bmatrix}$

9. $\begin{bmatrix} 1 & 2 & 3 \\ -2 & 1 & 4 \\ 3 & -1 & 0 \end{bmatrix}$ and $\begin{bmatrix} 1 & 2 & 3 \\ 0 & 1 & 2 \\ 0 & 0 & 5 \end{bmatrix}$

10. $\begin{bmatrix} -1 & 2 & 0 & 4 \\ 3 & 1 & -2 & -1 \\ 1 & 5 & -3 & 0 \end{bmatrix}$ and $\begin{bmatrix} -1 & 2 & 0 & 4 \\ 0 & 7 & -2 & 11 \\ 0 & 0 & 1 & 7 \end{bmatrix}$

11. $\begin{bmatrix} -1 & 1 & -1 & 1 \\ 2 & 3 & 0 & -1 \\ 1 & 1 & -2 & 4 \end{bmatrix}$ and $\begin{bmatrix} -1 & 1 & -1 & 1 \\ 0 & 5 & -2 & 1 \\ 0 & 0 & -11 & 23 \end{bmatrix}$

12. $\begin{bmatrix} 1 & -1 & 5 & 11 \\ 2 & 0 & -1 & 7 \\ 3 & 1 & 1 & 11 \end{bmatrix}$ and $\begin{bmatrix} 1 & -1 & 5 & 11 \\ 0 & 2 & -11 & -15 \\ 0 & 0 & 8 & 8 \end{bmatrix}$

In Exercises 13–24, use augmented matrices to solve the given system.

13. $\begin{cases} x + y = 0 \\ 2x - 3y = 5 \end{cases}$

14. $\begin{cases} -3x - 4y = 13 \\ -x + 4y = -1 \end{cases}$

15. $\begin{cases} 4x + 10y = 19 \\ -6x + 12y = -15 \end{cases}$

16. $\begin{cases} x - \frac{1}{2}y = 3 \\ \frac{1}{4}x + y = \frac{3}{4} \end{cases}$

17. $\begin{cases} 0.1x + 0.3y = 0.18 \\ 1.7x + 2.4y = 1.98 \end{cases}$

18. $\begin{cases} x - y - z = 2 \\ 2x + y - z = 3 \\ x + y + z = 0 \end{cases}$

19. $\begin{cases} x + 3y - 5z = -20 \\ 2x - 5y + 3z = 10 \\ -x + 6y - 5z = -19 \end{cases}$

20. $\begin{cases} x - 2y + 3z = 0 \\ 4x - 9y + z = 0 \\ 5x + 2y - z = 0 \end{cases}$

21. $\begin{cases} x - y = -3 \\ 2x + z = 5 \\ y - 2z = 3 \end{cases}$

22. $\begin{cases} x + y - z = 12 \\ y - 2z = 8 \\ 3y + z = 3 \end{cases}$

23. $\begin{cases} x + y - 4z = 0 \\ 2x + y - 3z = 2 \\ -3x - y + 2z = -4 \end{cases}$

24. $\begin{cases} x - y - z = -3 \\ 3x - 2y - 6z = -9 \\ -5x + 7y - z = 15 \end{cases}$

In Exercises 25–26, use augmented matrices to show that the given system is inconsistent.

25. $\begin{cases} 4x - y - 2z = 4 \\ x - y - \frac{1}{2}z = 1 \\ 2x - y - z = 8 \end{cases}$

26. $\begin{cases} 6x - 12y + z = 5 \\ 9x - 18y + 2z = 3 \\ 2x - 4y + 3z = 2 \end{cases}$

8.5 DETERMINANTS

In the preceding section we discussed solutions of systems of linear equations by augmented matrices. If the given system happens to have the same number of equations as variables, so that the system's coefficient matrix is a square matrix, then the system can frequently be solved by another method that utilizes a special number associated with the coefficient matrix, called the determinant of the matrix. Since determinants are of interest in other contexts as well, we devote the present section to determinants and their properties. In the following section we will see how to solve systems by determinants.

In general, the determinant of a square matrix

$$\begin{bmatrix} a_{11} & a_{12} & \cdots & a_{1n} \\ a_{21} & a_{22} & \cdots & a_{2n} \\ \vdots & \vdots & & \vdots \\ a_{n1} & a_{n2} & \cdots & a_{nn} \end{bmatrix}$$

is a number denoted by

$$\begin{vmatrix} a_{11} & a_{12} & \cdots & a_{1n} \\ a_{21} & a_{22} & \cdots & a_{2n} \\ \cdot & \cdot & & \cdot \\ \cdot & \cdot & & \cdot \\ \cdot & \cdot & & \cdot \\ a_{n1} & a_{n2} & \cdots & a_{nn} \end{vmatrix}$$

Normally, computation of the determinant varies according to the order of the matrix. For a 1×1 matrix $[a_{11}]$, the ***determinant*** is defined to be a_{11}. Thus the determinant of a 1×1 matrix is just the number in the matrix. For example, the determinant of the matrix $[-2]$ is -2.

For a 2×2 matrix

$$\begin{bmatrix} a_{11} & a_{12} \\ a_{21} & a_{22} \end{bmatrix}$$

the ***determinant*** is defined by

$$\begin{vmatrix} a_{11} & a_{12} \\ a_{21} & a_{22} \end{vmatrix} = a_{11}a_{22} - a_{21}a_{12} \tag{1}$$

which is the difference of the products of the numbers in the two diagonals.

EXAMPLE 1. Evaluate $\begin{vmatrix} 2 & -4 \\ -5 & 3 \end{vmatrix}$.

Solution. By (1),

$$\begin{vmatrix} 2 & -4 \\ -5 & 3 \end{vmatrix} = (2)(3) - (-5)(-4) = 6 - 20 = -14 \quad \square$$

For a 3×3 matrix A, where

$$A = \begin{bmatrix} a_{11} & a_{12} & a_{13} \\ a_{21} & a_{22} & a_{23} \\ a_{31} & a_{32} & a_{33} \end{bmatrix}$$

the ***determinant*** of A, denoted by det A, is defined as follows:

$$\det A = \begin{vmatrix} a_{11} & a_{12} & a_{13} \\ a_{21} & a_{22} & a_{23} \\ a_{31} & a_{32} & a_{33} \end{vmatrix} = a_{11}a_{22}a_{33} + a_{12}a_{23}a_{31} + a_{13}a_{21}a_{32} - a_{31}a_{22}a_{13} - a_{32}a_{23}a_{11} - a_{33}a_{21}a_{12} \tag{2}$$

A way to remember the determinant given in (2) is by the diagram

$$\begin{array}{ccc|cc} + & + & + & & \\ a_{11} & a_{12} & a_{13} & a_{11} & a_{12} \\ a_{21} & a_{22} & a_{23} & a_{21} & a_{22} \\ a_{31} & a_{32} & a_{33} & a_{31} & a_{32} \\ - & - & - & & \end{array}$$

EXAMPLE 2. Evaluate $\begin{vmatrix} 1 & 2 & -1 \\ 4 & 3 & -4 \\ -2 & 0 & -3 \end{vmatrix}$.

Solution. By (2),

$$\begin{aligned} \begin{vmatrix} 1 & 2 & -1 \\ 4 & 3 & -4 \\ -2 & 0 & -3 \end{vmatrix} &= (1)(3)(-3) + (2)(-4)(-2) + (-1)(4)(0) \\ &\quad - (-2)(3)(-1) - (0)(-4)(1) - (-3)(4)(2) \\ &= -9 + 16 + 0 - 6 - 0 + 24 \\ &= 25 \quad \square \end{aligned}$$

For square matrices of order 4 and above, the definition of the determinant is more complicated. As a prelude to the definition, we introduce the notion of a minor of a matrix.

DEFINITION 8.2 Let A be an $n \times n$ matrix. For $1 \leq i \leq n$ and $1 \leq j \leq n$ the ***minor*** M_{ij} of A is the determinant of the matrix formed by deleting the ith row and jth column.

EXAMPLE 3. Determine the minor M_{23} of

$$\begin{bmatrix} 1 & 2 & -1 \\ 4 & 3 & -4 \\ -2 & 0 & -3 \end{bmatrix}$$

Solution. By definition, to obtain M_{23} we first delete the second row and third column:

$$\begin{bmatrix} 1 & 2 & -1 \\ 4 & 3 & -4 \\ -2 & 0 & -3 \end{bmatrix}$$

which yields the matrix

$$\begin{bmatrix} 1 & 2 \\ -2 & 0 \end{bmatrix}$$

Then M_{23} is the determinant of this matrix, which means that

$$M_{23} = \begin{vmatrix} 1 & 2 \\ -2 & 0 \end{vmatrix} = 0 - (-4) = 4 \quad \square$$

EXAMPLE 4. Find the minor M_{43} of

$$\begin{bmatrix} 2 & 3 & 0 & 1 \\ 0 & -1 & 2 & 4 \\ -2 & 2 & 3 & 0 \\ 3 & 5 & 1 & -2 \end{bmatrix}$$

Solution. By definition, for M_{43} we delete the fourth row and third column to obtain

$$M_{43} = \begin{vmatrix} 2 & 3 & 1 \\ 0 & -1 & 4 \\ -2 & 2 & 0 \end{vmatrix} = 0 - 24 + 0 - 2 - 16 - 0 = -42 \quad \square$$

The definition in (2) of the determinant of a 3×3 matrix can be reformulated in terms of minors. More precisely,

$$\begin{vmatrix} a_{11} & a_{12} & a_{13} \\ a_{21} & a_{22} & a_{23} \\ a_{31} & a_{32} & a_{33} \end{vmatrix} = a_{11}M_{11} - a_{12}M_{12} + a_{13}M_{13} \tag{3}$$

Notice that the sum in (3) contains the term $-a_{12}M_{12}$, with the minus sign. If we let the ***cofactor*** A_{ij} of a_{ij} be defined by

$$A_{ij} = (-1)^{i+j}M_{ij}$$

then we find, for instance, that

$$\begin{aligned} A_{11} &= (-1)^{1+1}M_{11} = (-1)^2M_{11} = M_{11} \\ A_{12} &= (-1)^{1+2}M_{12} = (-1)^3M_{12} = -M_{12} \\ A_{13} &= (-1)^{1+3}M_{13} = (-1)^4M_{13} = M_{13} \end{aligned}$$

Therefore by (3) the determinant can be written in terms of cofactors:

$$\begin{vmatrix} a_{11} & a_{12} & a_{13} \\ a_{21} & a_{22} & a_{23} \\ a_{31} & a_{32} & a_{33} \end{vmatrix} = a_{11}A_{11} + a_{12}A_{12} + a_{13}A_{13} \tag{4}$$

Formula (4) tells us that the determinant of a 3×3 matrix is obtained by multiplying the entries in the first row of the matrix by their respective cofactors and taking the sum of the resulting products. This leads us to the formal definition of the determinant for any square matrix of order greater than or equal to 3.

DEFINITION 8.3 The ***determinant*** of a given square matrix of order greater than or equal to 3 is the number obtained as follows: Select a row or column of the matrix. Multiply each entry in the selected row or column by its cofactor and take the sum of the resulting products.

Notice that because of (4), the definition of determinant given in Definition 8.3 is consistent with the definition of the determinant of a 3×3 matrix given in (2).

According to Definition 8.3, the evaluation of the determinant of a square matrix of order 4 is based on calculating four cofactors, each of which involves the determinant of a square matrix of order 3. Computation of the determinant of a square matrix of order 5 involves evaluating five cofactors, each associated with a square matrix of order 4. As you can well imagine, as the order of the square matrix increases from 4 to 5 and even higher, the computations needed in order to evaluate the determinant soon become unwieldy.

In using Definition 8.3 to calculate the determinant of a square matrix, one begins with the selection of a row or column. If the pth row of the nth-order square matrix is chosen, then

$$\begin{vmatrix} a_{11} & a_{12} & \cdots & a_{1n} \\ a_{21} & a_{22} & \cdots & a_{2n} \\ \cdot & \cdot & & \cdot \\ \cdot & \cdot & & \cdot \\ \cdot & \cdot & & \cdot \\ a_{n1} & a_{n2} & \cdots & a_{nn} \end{vmatrix} = a_{p1}A_{p1} + a_{p2}A_{p2} + \cdots + a_{pn}A_{pn}$$

whereas if the pth column of the matrix is chosen, then

$$\begin{vmatrix} a_{11} & a_{12} & \cdots & a_{1n} \\ a_{21} & a_{22} & \cdots & a_{2n} \\ \cdot & \cdot & & \cdot \\ \cdot & \cdot & & \cdot \\ \cdot & \cdot & & \cdot \\ a_{n1} & a_{n2} & \cdots & a_{nn} \end{vmatrix} = a_{1p}A_{1p} + a_{2p}A_{2p} + \cdots + a_{np}A_{np}$$

It can be proved that the same number is obtained for the value of the determinant, no matter which row or column is used.

EXAMPLE 5. Use Definition 8.3 to evaluate the determinant

$$\begin{vmatrix} -3 & 2 & 1 \\ -5 & 0 & 6 \\ -2 & -1 & 3 \end{vmatrix}$$

Solution. Because of the presence of 0 in the second row, we select the second row. We find that

$$\begin{aligned} \begin{vmatrix} -3 & 2 & 1 \\ -5 & 0 & 6 \\ -2 & -1 & 3 \end{vmatrix} &= (-5)A_{21} + (0)A_{22} + 6A_{23} \\ &= (-5)(-1)^{2+1}\begin{vmatrix} 2 & 1 \\ -1 & 3 \end{vmatrix} + 0 + (6)(-1)^{2+3}\begin{vmatrix} -3 & 2 \\ -2 & -1 \end{vmatrix} \\ &= 5[6-(-1)] - 6[3-(-4)] \\ &= 35 - 42 = -7 \quad \square \end{aligned}$$

As the solution of Example 5 indicates, if an entry in a row or column is 0 (such as a_{22} in Example 5), the product of the entry and its cofactor is 0. Thus when we compute a determinant by means of cofactors and in so doing select a row (or column), there is no need to include a term for any entry in the row (or column) that is 0. Therefore it can help to select a row or column containing zero entries.

The following properties of determinants aid in their evaluation.

i. If all entries of a single row or column are 0, then the determinant is 0.
ii. If two rows or two columns are identical, then the determinant is 0.
iii. Interchanging two rows or two columns alters the determinant by a factor of -1.
iv. If a nonzero multiple of one row (or column) is added to a second row (or column), then the determinant is unaltered.
v. If all the entries of a single row (or column) are multiplied by a number c, then the determinant of the new matrix is c times the determinant of the original matrix.

EXAMPLE 6. Evaluate

$$\begin{vmatrix} 3 & 8 & 5 & 8 \\ 1 & 6 & -4 & 7 \\ 0 & 3 & 9 & -2 \\ 1 & 6 & -4 & 7 \end{vmatrix}$$

Solution. Since the second and fourth rows are identical, the determinant is 0 by (ii). □

One procedure for evaluating a determinant is to use (iv) in order to make all, or all but one, of the entries in a particular row or column equal to 0, and then to evaluate the determinant by using the cofactors of that row or column.

EXAMPLE 7. Evaluate

$$\begin{vmatrix} 2 & -1 & 0 & 3 \\ 1 & 2 & 5 & -4 \\ 3 & -1 & 2 & 0 \\ -1 & 3 & 2 & 5 \end{vmatrix}$$

Solution. If we multiply the second column by 2 and add the resulting column to the first column, we obtain the determinant

$$\begin{vmatrix} 0 & -1 & 0 & 3 \\ 5 & 2 & 5 & -4 \\ 1 & -1 & 2 & 0 \\ 5 & 3 & 2 & 5 \end{vmatrix}$$

Next we multiply the second column by 3 and add the resulting column to the fourth column, obtaining

$$\begin{vmatrix} 0 & -1 & 0 & 0 \\ 5 & 2 & 5 & 2 \\ 1 & -1 & 2 & -3 \\ 5 & 3 & 2 & 14 \end{vmatrix} \tag{5}$$

By (iv) neither of the two operations we have performed changes the value of the determinant. However, since all but one entry in the first row of the determinant in (5) are 0, we can readily evaluate the determinant by using cofactors of the first row:

$$\begin{vmatrix} 0 & -1 & 0 & 0 \\ 5 & 2 & 5 & 2 \\ 1 & -1 & 2 & -3 \\ 5 & 3 & 2 & 14 \end{vmatrix} = (0)A_{11} + (-1)A_{12} + (0)A_{13} + (0)A_{14}$$

$$= (-1)(-1)^{1+2}\begin{vmatrix} 5 & 5 & 2 \\ 1 & 2 & -3 \\ 5 & 2 & 14 \end{vmatrix}$$

$$= 140 - 75 + 4 - 20 - (-30) - 70$$

$$= 9 \quad \square$$

EXERCISES 8.5

In Exercises 1–2, find the minors and cofactors of the entries of the second row of the given matrix.

1. $\begin{bmatrix} -4 & 0 & 2 \\ 3 & 6 & -1 \\ 2 & -5 & 7 \end{bmatrix}$

2. $\begin{bmatrix} -1 & -3 & 1 & 2 \\ -2 & 0 & -1 & 1 \\ 3 & 2 & 0 & 4 \\ 0 & -3 & 1 & -2 \end{bmatrix}$

In Exercises 3–18, evaluate the given determinant.

3. $\begin{vmatrix} 2 & 0 \\ 3 & 4 \end{vmatrix}$

4. $\begin{vmatrix} 4 & -1 \\ -2 & -3 \end{vmatrix}$

5. $\begin{vmatrix} 6 & 0 \\ 3 & 0 \end{vmatrix}$

6. $\begin{vmatrix} a & b \\ -b & a \end{vmatrix}$

7. $\begin{vmatrix} 3 & -1 & 0 \\ 2 & 4 & 1 \\ -2 & 5 & -3 \end{vmatrix}$

8. $\begin{vmatrix} 1 & -1 & 1 \\ 2 & 2 & -2 \\ 3 & 3 & 3 \end{vmatrix}$

9. $\begin{vmatrix} -5 & -4 & 2 \\ -3 & 6 & 4 \\ -2 & -3 & 1 \end{vmatrix}$

10. $\begin{vmatrix} 6 & 9 & -1 \\ 4 & 6 & 4 \\ 2 & 3 & 7 \end{vmatrix}$

11. $\begin{vmatrix} 3 & 2 & -5 \\ 4 & -1 & -2 \\ 5 & -3 & 1 \end{vmatrix}$

12. $\begin{vmatrix} 3 & 0 & 0 \\ 0 & -2 & 0 \\ 0 & 0 & 5 \end{vmatrix}$

13. $\begin{vmatrix} -1 & 1 & 1 \\ 1 & -1 & 1 \\ 1 & 1 & -1 \end{vmatrix}$

14. $\begin{vmatrix} a & r & s \\ 0 & b & t \\ 0 & 0 & c \end{vmatrix}$

15. $\begin{vmatrix} 0 & 1 & 0 & 2 \\ 4 & 3 & -1 & 2 \\ 6 & 2 & 0 & 3 \\ 1 & -1 & 4 & -5 \end{vmatrix}$

16. $\begin{vmatrix} 1 & -2 & -4 & 5 \\ 2 & -4 & -6 & 10 \\ 1 & 3 & 0 & 2 \\ -1 & 7 & 1 & -3 \end{vmatrix}$

17. $\begin{vmatrix} 4 & 1 & -1 & -2 \\ -1 & 1 & 1 & 2 \\ -2 & 0 & 1 & 2 \\ -2 & 1 & 1 & 3 \end{vmatrix}$

18. $\begin{vmatrix} 1 & -2 & -3 & -4 \\ 0 & 2 & -3 & -4 \\ 0 & 0 & 3 & -4 \\ 0 & 0 & 0 & 4 \end{vmatrix}$

In Exercises 19–22, use properties (i)–(v) to explain why each equation is valid.

19. $\begin{vmatrix} 5 & -4 & 3 \\ 6 & 2 & 4 \\ 8 & 1 & -1 \end{vmatrix} = -\begin{vmatrix} 8 & 1 & -1 \\ 6 & 2 & 4 \\ 5 & -4 & 3 \end{vmatrix}$

20. $\begin{vmatrix} 2 & -4 & 7 \\ 5 & 1 & 0 \\ 3 & -6 & 5 \end{vmatrix} = \begin{vmatrix} 5 & 1 & 0 \\ 3 & -6 & 5 \\ 2 & -4 & 7 \end{vmatrix}$

21. $\begin{vmatrix} 3 & 5 & 8 \\ -2 & -3 & 4 \\ 1 & 7 & -2 \end{vmatrix} = -\frac{1}{4}\begin{vmatrix} 3 & 5 & 8 \\ 8 & 12 & -16 \\ 1 & 7 & -2 \end{vmatrix}$

22. $\begin{vmatrix} 2 & 3 & -1 \\ 5 & 1 & 4 \\ -6 & -9 & 3 \end{vmatrix} = 0$

23. Find all values of x satisfying

$$\begin{vmatrix} x & 1 & 0 \\ -1 & 3 & 4 \\ 3 & 1 & x \end{vmatrix} = 30$$

24. Show that

$$\begin{vmatrix} \cos\theta & -r\sin\theta \\ \sin\theta & r\cos\theta \end{vmatrix} = r$$

25. Show that

$$\begin{vmatrix} \sin\phi\cos\theta & \rho\cos\phi\cos\theta & -\rho\sin\phi\sin\theta \\ \sin\phi\sin\theta & \rho\cos\phi\sin\theta & \rho\sin\phi\cos\theta \\ \cos\phi & -\rho\sin\phi & 0 \end{vmatrix} = \rho^2\sin\phi$$

26. Show that an equation of the straight line passing through the distinct points (a_1, b_1) and (a_2, b_2) is given by

$$\begin{vmatrix} 1 & x & y \\ 1 & a_1 & b_1 \\ 1 & a_2 & b_2 \end{vmatrix} = 0$$

27. Show that

$$\begin{vmatrix} 1 & a & a^2 \\ 1 & b & b^2 \\ 1 & c & c^2 \end{vmatrix} = (a-b)(b-c)(c-a)$$

28. It can be shown that the area $\mathscr{A}$ of the triangle in the plane with vertices $(0, 0)$, (a, b), and (c, d) is given by

$$\mathscr{A} = \left|\frac{1}{2}\begin{vmatrix} a & b \\ c & d \end{vmatrix}\right|$$

Find the area of the triangle whose vertices are

a. (0, 0), (2, 1), (−1, 3) b. (0, 0), (−2, −2), (5, 7)

29. Suppose that three sides of a parallelepiped no two of which are parallel are represented by the vectors $\mathbf{a} = (a_1, a_2, a_3)$, $\mathbf{b} = (b_1, b_2, b_3)$, and $\mathbf{c} = (c_1, c_2, c_3)$. Then it can be shown that the volume V of the parallelepiped is given by

$$V = \left\|\begin{matrix} a_1 & a_2 & a_3 \\ b_1 & b_2 & b_3 \\ c_1 & c_2 & c_3 \end{matrix}\right\|$$

Find the volume of the parallelepiped determined by the vectors

a. $\mathbf{a} = (2, 1, 0)$, $\mathbf{b} = (-1, -4, 3)$, $\mathbf{c} = (0, -2, 1)$

b. $\mathbf{a} = (1, 3, 4)$, $\mathbf{b} = (2, 2, -1)$, $\mathbf{c} = (4, 0, -1)$

8.6 SOLUTIONS OF SYSTEMS BY DETERMINANTS: CRAMER'S RULE

Having learned to evaluate determinants, we are ready to use them in solving linear systems of equations. In order to do so, we will confine our attention to systems that have the same number of equations and variables.

Let us begin by solving the linear system

$$\begin{cases} a_1x + b_1y = c_1 \\ a_2x + b_2y = c_2 \end{cases} \tag{1}$$

for x and y, with the assumption that the determinant D of the coefficient matrix is not 0, that is, that

$$D = \begin{vmatrix} a_1 & b_1 \\ a_2 & b_2 \end{vmatrix} = a_1b_2 - a_2b_1 \neq 0 \tag{2}$$

To solve for x, we multiply the first equation in (1) by b_2 and the second equation by b_1, so the coefficient of y in each will be the same:

$$\begin{aligned} a_1b_2x + b_1b_2y &= c_1b_2 \\ a_2b_1x + b_1b_2y &= c_2b_1 \end{aligned} \tag{3}$$

Subtracting the second equation in (3) from the first yields

$$(a_1b_2 - a_2b_1)x = c_1b_2 - c_2b_1$$

Using (2), we find that

$$x = \frac{c_1b_2 - c_2b_1}{a_1b_2 - a_2b_1}$$

or in determinant form,

$$x = \frac{\begin{vmatrix} c_1 & b_1 \\ c_2 & b_2 \end{vmatrix}}{\begin{vmatrix} a_1 & b_1 \\ a_2 & b_2 \end{vmatrix}} = \frac{\begin{vmatrix} c_1 & b_1 \\ c_2 & b_2 \end{vmatrix}}{D} \tag{4}$$

Similarly, we find that

$$y = \frac{\begin{vmatrix} a_1 & c_1 \\ a_2 & c_2 \end{vmatrix}}{\begin{vmatrix} a_1 & b_1 \\ a_2 & b_2 \end{vmatrix}} = \frac{\begin{vmatrix} a_1 & c_1 \\ a_2 & c_2 \end{vmatrix}}{D} \tag{5}$$

(see Exercise 19). If we let the numerator of (4) be D_x and the numerator of (5) be D_y, so that

$$D_x = \begin{vmatrix} c_1 & b_1 \\ c_2 & b_2 \end{vmatrix} \quad \text{and} \quad D_y = \begin{vmatrix} a_1 & c_1 \\ a_2 & c_2 \end{vmatrix}$$

then (4) and (5) can be rewritten in the condensed form

$$\boxed{x = \frac{D_x}{D} \quad \text{and} \quad y = \frac{D_y}{D}} \tag{6}$$

Notice that D_x is obtained from D by replacing the first column with the entries c_1 and c_2, and D_y comes from D by replacing the second column with c_1, c_2.

Formulas (4) and (5) give the unique solution of the system in (1), provided that $D \neq 0$. Together, the formulas are known as ***Cramer's Rule*** after the Swiss mathematician Gabriel Cramer (1704–1752).

Although Cramer's Rule always provides us with a solution for the system of equations, we might make errors in the computations leading to the solution. As a result, it is important to check the proposed solution, as we have done until now in the chapter. However, to save space we will omit the checks from now on.

EXAMPLE 1. Solve the linear system

$$\begin{cases} 3x - 5y = -2 \\ x + 2y = 4 \end{cases}$$

by means of determinants.

Solution. Since

$$D = \begin{vmatrix} 3 & -5 \\ 1 & 2 \end{vmatrix} = 6 - (-5) = 11$$

(2) is satisfied, and Cramer's Rule applies. Thus by (4),

$$x = \frac{\begin{vmatrix} -2 & -5 \\ 4 & 2 \end{vmatrix}}{D} = \frac{-4 - (-20)}{11} = \frac{16}{11}$$

and by (5),

$$y = \frac{\begin{vmatrix} 3 & -2 \\ 1 & 4 \end{vmatrix}}{D} = \frac{12 - (-2)}{11} = \frac{14}{11}$$

Consequently the solution of the given system is $(\frac{16}{11}, \frac{14}{11})$. □

Next we present the three-variable version of Cramer's Rule. Consider the system

$$\begin{cases} a_1x + b_1y + c_1z = d_1 \\ a_2x + b_2y + c_2z = d_2 \\ a_3x + b_3y + c_3z = d_3 \end{cases} \tag{7}$$

and assume that the determinant D of the coefficient matrix is nonzero, that is,

$$D = \begin{vmatrix} a_1 & b_1 & c_1 \\ a_2 & b_2 & c_2 \\ a_3 & b_3 & c_3 \end{vmatrix} \neq 0$$

Then there is a unique solution of the system in (7) given by

$$x = \frac{\begin{vmatrix} d_1 & b_1 & c_1 \\ d_2 & b_2 & c_2 \\ d_3 & b_3 & c_3 \end{vmatrix}}{D}, \qquad y = \frac{\begin{vmatrix} a_1 & d_1 & c_1 \\ a_2 & d_2 & c_2 \\ a_3 & d_3 & c_3 \end{vmatrix}}{D}, \qquad z = \frac{\begin{vmatrix} a_1 & b_1 & d_1 \\ a_2 & b_2 & d_2 \\ a_3 & b_3 & d_3 \end{vmatrix}}{D} \tag{8}$$

This result is also known as Cramer's Rule.

If we let

$$D_x = \begin{vmatrix} d_1 & b_1 & c_1 \\ d_2 & b_2 & c_2 \\ d_3 & b_3 & c_3 \end{vmatrix}, \qquad D_y = \begin{vmatrix} a_1 & d_1 & c_1 \\ a_2 & d_2 & c_2 \\ a_3 & d_3 & c_3 \end{vmatrix}, \qquad D_z = \begin{vmatrix} a_1 & b_1 & d_1 \\ a_2 & b_2 & d_2 \\ a_3 & b_3 & d_3 \end{vmatrix}$$

obtained from D by replacing the respective columns of D by the column consisting of d_1, d_2, d_3, then the formulas in (8) can be written in the condensed form

$$\boxed{x = \frac{D_x}{D}, \qquad y = \frac{D_y}{D}, \quad \text{and} \quad z = \frac{D_z}{D}} \tag{9}$$

This corresponds to the simplified formulas for x and y in (6).

EXAMPLE 2. Solve the linear system

$$\begin{cases} 2x - 3y - z = 0 \\ -5x + y - 4z = 1 \\ 3x + z = -2 \end{cases}$$

by means of determinants.

Solution. Since

$$D = \begin{vmatrix} 2 & -3 & -1 \\ -5 & 1 & -4 \\ 3 & 0 & 1 \end{vmatrix} = 2 + 36 + 0 - (-3) - 0 - 15 = 26$$

Cramer's Rule applies, so that by (8),

$$x = \frac{\begin{vmatrix} 0 & -3 & -1 \\ 1 & 1 & -4 \\ -2 & 0 & 1 \end{vmatrix}}{D} = \frac{0 - 24 + 0 - 2 - 0 - (-3)}{26} = -\frac{23}{26}$$

$$y = \frac{\begin{vmatrix} 2 & 0 & -1 \\ -5 & 1 & -4 \\ 3 & -2 & 1 \end{vmatrix}}{D} = \frac{2 + 0 - 10 - (-3) - 16 - 0}{26} = -\frac{21}{26}$$

$$z = \frac{\begin{vmatrix} 2 & -3 & 0 \\ -5 & 1 & 1 \\ 3 & 0 & -2 \end{vmatrix}}{D} = \frac{-4 - 9 + 0 - 0 - 0 - (-30)}{26} = \frac{17}{26}$$

Thus $(-\frac{23}{26}, -\frac{21}{26}, \frac{17}{26})$ is the solution of the given system. □

An interesting consequence of Cramer's Rule concerns systems of the form

$$\begin{cases} a_1x + b_1y + c_1z = 0 \\ a_2x + b_2y + c_2z = 0 \\ a_3x + b_3y + c_3z = 0 \end{cases}$$

all of whose equations have 0 on the right side. Such a system is called ***homogeneous***. If the determinant D of a homogeneous system is not 0, then the unique solution is (0, 0, 0), since $d_1 = d_2 = d_3 = 0$ in (8). For example, for the solutions of the linear system

$$\begin{cases} 6x - 4y + 3z = 0 \\ -5x + y - 3z = 0 \\ 7x - y - 8z = 0 \end{cases}$$

we can first calculate that $D \neq 0$ and then conclude that the unique solution of the system is (0, 0, 0). An analogous result holds for homogeneous systems of two equations.

If $D = 0$ for a given linear system of equations, then Cramer's Rule does not apply. In that case either the system is inconsistent or there are infinitely many solutions, which are best found by augmented matrices.

Finally, we mention that for a linear system of n equations with variables $x_1, x_2, \ldots, x_n$, if the determinant D of the coefficient matrix is not 0, then the analogue of Cramer's Rule in (9) remains valid. In particular, $(x_1, x_2, \ldots, x_n)$ is a solution of the system if and only if

$$x_1 = \frac{D_{x_1}}{D}, \quad x_2 = \frac{D_{x_2}}{D}, \ldots, \quad x_n = \frac{D_{x_n}}{D}$$

But the solutions are increasingly complicated to evaluate as n increases. In fact, if $n \geq 4$, then solving such a system by Cramer's Rule is more complicated than solving by augmented matrices, and Cramer's Rule is generally not used.

EXERCISES 8.6

1–4. Use Cramer's Rule to solve the systems in Exercises 5–8 of Section 8.1.

5–9. Use Cramer's Rule to solve the systems in Exercises 1–5 of Section 8.2.

10–14. Use Cramer's Rule to solve the systems in Exercises 17–21 of Section 8.3.

In Exercises 15–18, use Cramer's Rule to solve the given system.

15. $$\begin{cases} 2w + x - y + 2z = -3 \\ 3x + 2y + z = 0 \\ -w + x + y - 2z = 0 \\ 3w + 2x - y + 4z = -4 \end{cases}$$

16. $$\begin{cases} 2w + 2x + y - z = 0 \\ w + 4x + y + 2z = 3 \\ -w + 2x + 2y - z = -2 \\ 3w - 2x + 4y + 2z = 1 \end{cases}$$

17. $$\begin{cases} w - x + y - 2z = 0 \\ 2w + 2x + 2y - z = 5 \\ -3w + x - z = -4 \\ x + y + z = 7 \end{cases}$$

18. $$\begin{cases} 2w + x + y - z = -4 \\ 4w - x + 5y + 2z = -6 \\ -2w + 2x - z = 2 \\ -6w - 3x + 3y + 2z = 4 \end{cases}$$

19. Prove (5). (*Hint:* Use the same procedure we used to prove (4).)

8.7 PRODUCTS OF MATRICES

This section is devoted mainly to products of matrices and the relation of such products to systems of equations. Before we study multiplication of matrices, let us make the following definitions.

(a) Let A and B be $m \times n$ matrices. Then $A = B$ if and only if the corresponding entries of A and B are equal. Thus

$$\begin{bmatrix} a & b \\ c & d \end{bmatrix} = \begin{bmatrix} e & f \\ g & h \end{bmatrix} \quad \text{if and only if} \quad \begin{cases} a = e, b = f, c = g, \\ \text{and} \quad d = h \end{cases}$$

(b) Let A and B be $m \times n$ matrices. Then the ***sum*** $A + B$ is the $m \times n$ matrix formed by adding the corresponding entries of A and B together. Thus

$$\begin{bmatrix} 2 & 3 & 1 \\ 4 & -5 & 0 \end{bmatrix} + \begin{bmatrix} 7 & -3 & -6 \\ 1 & 2 & 8 \end{bmatrix} = \begin{bmatrix} 2+7 & 3+(-3) & 1+(-6) \\ 4+1 & -5+2 & 0+8 \end{bmatrix}$$
$$= \begin{bmatrix} 9 & 0 & -5 \\ 5 & -3 & 8 \end{bmatrix}$$

In contrast, the sum of

$$\begin{bmatrix} 2 & 3 & 1 \\ 4 & -5 & 0 \end{bmatrix} \quad \text{and} \quad \begin{bmatrix} 7 & 1 \\ -3 & 2 \\ -6 & 8 \end{bmatrix}$$

is undefined because the matrices do not have the same number of rows (nor the same number of columns, for that matter).

(c) Let A be a matrix and c a number. Then the ***constant multiple*** cA is the matrix each of whose entries is c times the corresponding entry in A. Thus

$$(-3)\begin{bmatrix} 2 & \pi & -4 \\ 6 & -2 & 1 \\ -1 & 5 & -\sqrt{2} \end{bmatrix} = \begin{bmatrix} (-3)2 & (-3)\pi & (-3)(-4) \\ (-3)6 & (-3)(-2) & (-3)(1) \\ (-3)(-1) & (-3)(5) & (-3)(-\sqrt{2}) \end{bmatrix}$$
$$= \begin{bmatrix} -6 & -3\pi & 12 \\ -18 & 6 & -3 \\ 3 & -15 & 3\sqrt{2} \end{bmatrix}$$

You might well anticipate that the product AB of two matrices A and B would involve products of corresponding entries, to reflect the way we add matrices. However, from the standpoint of applications of matrices, another definition of the product AB is more useful. It is restricted to those matrices A and B such that A has the same number of columns as B has rows.

To define the product, let A be an $m \times n$ matrix and B an $n \times p$ matrix. Then we define the ***product*** AB to be the $m \times p$ matrix C whose entry c_{ij} in the ith row and jth column is given by

$$c_{ij} = a_{i1}b_{1j} + a_{i2}b_{2j} + \cdots + a_{in}b_{nj}$$

In other words, c_{ij} is obtained by multiplying the ith row of A entrywise with the jth column of B and then adding.

EXAMPLE 1. Let $A = \begin{bmatrix} 3 & -1 & 4 \\ 2 & 0 & 1 \end{bmatrix}$ and $B = \begin{bmatrix} 1 & 0 & -3 & -2 \\ -2 & 4 & 5 & -1 \\ 3 & -1 & 0 & 6 \end{bmatrix}$.

Calculate AB.

Solution. Since A is a 2×3 matrix and B a 3×4 matrix, the product AB is defined and is a 2×4 matrix, which we designate by C. Now

$$AB = \begin{bmatrix} 3 & -1 & 4 \\ 2 & 0 & 1 \end{bmatrix} \begin{bmatrix} 1 & 0 & -3 & -2 \\ -2 & 4 & 5 & -1 \\ 3 & -1 & 0 & 6 \end{bmatrix}$$

$$= C = \begin{bmatrix} c_{11} & c_{12} & c_{13} & c_{14} \\ c_{21} & c_{22} & c_{23} & c_{24} \end{bmatrix}$$

where

$$c_{11} = (3)(1) + (-1)(-2) + (4)(3) = 17$$

$$c_{12} = (3)(0) + (-1)(4) + (4)(-1) = -8$$

$$c_{13} = (3)(-3) + (-1)(5) + (4)(0) = -14$$

$$c_{14} = (3)(-2) + (-1)(-1) + (4)(6) = 19$$

$$c_{21} = (2)(1) + (0)(-2) + (1)(3) = 5$$

$$c_{22} = (2)(0) + (0)(4) + (1)(-1) = -1$$

$$c_{23} = (2)(-3) + (0)(5) + (1)(0) = -6$$

$$c_{24} = (2)(-2) + (0)(-1) + (1)(6) = 2$$

Thus

$$AB = \begin{bmatrix} 17 & -8 & -14 & 19 \\ 5 & -1 & -6 & 2 \end{bmatrix} \quad \square$$

CAUTION: Notice that although the product AB is defined for the matrices A and B of Example 1, the product BA is *not* defined. The reason is that the number of columns of B (which is 4) is not equal to the number of rows of A (which is 2). In general, even if both the products AB and BA of two matrices A and B are defined, AB and BA need not be equal. (See Exercises 19–22.)

In general, if the product AB of two matrices A and B is defined, then the relationship among the numbers of columns and rows of A, B, and the product $C = AB$ is as follows:

$$\underset{m \times n}{A} \times \underset{n \times p}{B} = \underset{m \times p}{C}$$

EXAMPLE 2. Let $A = \begin{bmatrix} 1 & 4 \\ -3 & -2 \\ 5 & 0 \end{bmatrix}$ and $X = \begin{bmatrix} -1 \\ 6 \end{bmatrix}$. Find AX.

Solution. Since A is a 3×2 matrix and X is a 2×1 matrix, AX is a 3×1 matrix, whose entries are computed as follows:

$$AX = \begin{bmatrix} 1 & 4 \\ -3 & -2 \\ 5 & 0 \end{bmatrix}\begin{bmatrix} -1 \\ 6 \end{bmatrix} = \begin{bmatrix} (1)(-1) + (4)(6) \\ (-3)(-1) + (-2)(6) \\ (5)(-1) + (0)(6) \end{bmatrix} = \begin{bmatrix} 23 \\ -9 \\ -5 \end{bmatrix} \quad \square$$

If A, B, and C are $m \times n$, $n \times p$, and $p \times r$ matrices, respectively, then AB is an $m \times p$ matrix, so $(AB)C$ is defined and is an $m \times r$ matrix. Similarly, BC is an $n \times r$ matrix, so $A(BC)$ is defined and is an $m \times r$ matrix. What may not be so evident, but is nevertheless true, is that

$$\boxed{(AB)C = A(BC)} \tag{1}$$

so it does not matter whether we first multiply A with B, or B with C. (See Exercises 23 and 24.) We will use this fact later.

Identity Matrices

For any positive integer n, let I_n be the square matrix of order n which has 1's down the diagonal starting at the upper left corner, and all other entries 0. This matrix is called the ***nth-order identity matrix***. For example,

$$I_2 = \begin{bmatrix} 1 & 0 \\ 0 & 1 \end{bmatrix} \quad \text{and} \quad I_3 = \begin{bmatrix} 1 & 0 & 0 \\ 0 & 1 & 0 \\ 0 & 0 & 1 \end{bmatrix}$$

EXAMPLE 3. Let $A = \begin{bmatrix} -1 & 4 \\ 2 & -3 \end{bmatrix}$. Show that $AI_2 = I_2A = A$.

Solution. By definition,

$$AI_2 = \begin{bmatrix} -1 & 4 \\ 2 & -3 \end{bmatrix}\begin{bmatrix} 1 & 0 \\ 0 & 1 \end{bmatrix} = \begin{bmatrix} (-1)(1) + (4)(0) & (-1)(0) + (4)(1) \\ (2)(1) + (-3)(0) & (2)(0) + (-3)(1) \end{bmatrix}$$
$$= \begin{bmatrix} -1 & 4 \\ 2 & -3 \end{bmatrix} = A$$

and

$$I_2A = \begin{bmatrix} 1 & 0 \\ 0 & 1 \end{bmatrix}\begin{bmatrix} -1 & 4 \\ 2 & -3 \end{bmatrix} = \begin{bmatrix} (1)(-1)+(0)(2) & (1)(4)+(0)(-3) \\ (0)(-1)+(1)(2) & (0)(4)+(1)(-3) \end{bmatrix}$$

$$= \begin{bmatrix} -1 & 4 \\ 2 & -3 \end{bmatrix} = A \quad \square$$

If A is any square matrix of order n, then it can be shown that

$$\boxed{AI_n = I_nA = A} \tag{2}$$

Example 3 lends credence to this fact, which says that the nth-order identity matrix has the property that the product of it and any square matrix A of order n is A. In this way I_n resembles the number 1 in multiplication, because for any real number a, we have $a \cdot 1 = 1 \cdot a = a$.

Inverse Matrices

If A is a square matrix of order n, then A ***has an inverse*** if and only if there is a square matrix B of order n such that

$$AB = I_n$$

In that case B is unique, and we also have $BA = I_n$. The matrix B is called the ***inverse*** of A and is written A^{-1}. Thus

$$\boxed{AA^{-1} = I_n \quad \text{and} \quad A^{-1}A = I_n}$$

Two questions immediately arise:

1. Which square matrices have inverses?
2. Assuming that a given square matrix has an inverse, how does one determine the entries of the inverse?

The answer to the first question is contained in the following theorem, given without proof.

THEOREM 8.4 *Let A be a square matrix. Then A has an inverse if and only if the determinant of A is not 0.*

To simplify the notation below, we will denote the determinant of a square matrix A by det A.

EXAMPLE 4. Let $A = \begin{bmatrix} 3 & 0 \\ 1 & 2 \end{bmatrix}$. Show that A has an inverse.

Solution. Since

$$\det A = \begin{vmatrix} 3 & 0 \\ 1 & 2 \end{vmatrix} = 6 - 0 = 6$$

Theorem 8.4 assures us that A has an inverse. $\square$

It is possible to give a formula for the inverse of a matrix. As you might imagine, the formula is simple for 2×2 matrices. If

$$A = \begin{bmatrix} a & b \\ c & d \end{bmatrix}$$

and $\det A \neq 0$, then

$$A^{-1} = \frac{1}{\det A}\begin{bmatrix} d & -b \\ -c & a \end{bmatrix} \tag{3}$$

as you can check by showing that $AA^{-1} = I_2$.

EXAMPLE 5. Let $A = \begin{bmatrix} 3 & 0 \\ 1 & 2 \end{bmatrix}$. Find A^{-1}.

Solution. By Example 4 we know that $\det A = 6$, and thus A^{-1} exists. Then (3) with $a = 3$, $b = 0$, $c = 1$, and $d = 2$, implies that A^{-1} is given by

$$A^{-1} = \frac{1}{6}\begin{bmatrix} 2 & 0 \\ -1 & 3 \end{bmatrix} \quad \square$$

If $n \geq 3$ and A is an $n \times n$ matrix that has an inverse, then the formula that yields A^{-1} is more cumbersome to apply. Instead of giving such a formula, we will present a procedure that not only tells whether the inverse exists but also yields the inverse when it does exist. The procedure is as follows:

i. Write the $n \times 2n$ matrix formed by adjoining I_n to the right of A:

$$\begin{bmatrix} a_{11} & a_{12} & \cdots & a_{1n} & 1 & 0 & 0 & \cdots & 0 \\ a_{21} & a_{22} & \cdots & a_{2n} & 0 & 1 & 0 & \cdots & 0 \\ \cdot & \cdot & & \cdot & \cdot & \cdot & \cdot & & \cdot \\ \cdot & \cdot & & \cdot & \cdot & \cdot & \cdot & & \cdot \\ \cdot & \cdot & & \cdot & \cdot & \cdot & \cdot & & \cdot \\ a_{n1} & a_{n2} & \cdots & a_{nn} & 0 & 0 & 0 & \cdots & 1 \end{bmatrix}$$

ii. Perform elementary row operations on the matrix in (i), with the goal of bringing the matrix into the form

$$\begin{bmatrix} 1 & 0 & 0 & \cdots & 0 & b_{11} & b_{12} & \cdots & b_{1n} \\ 0 & 1 & 0 & \cdots & 0 & b_{21} & b_{22} & \cdots & b_{2n} \\ \cdot & \cdot & \cdot & & \cdot & \cdot & \cdot & & \cdot \\ \cdot & \cdot & \cdot & & \cdot & \cdot & \cdot & & \cdot \\ \cdot & \cdot & \cdot & & \cdot & \cdot & \cdot & & \cdot \\ 0 & 0 & 0 & \cdots & 1 & b_{n1} & b_{n2} & \cdots & b_{nn} \end{bmatrix}$$

Notice that this matrix has the entries of I_n on the left, and the entries of a matrix that we will call B on the right.

iii. If it is possible to obtain the matrix appearing in step (ii), then not only does A have an inverse, but also $A^{-1} = B$. Conversely, if it is not possible to attain the matrix appearing in step (ii), then A has no inverse.

EXAMPLE 6. Let

$$A = \begin{bmatrix} 1 & 2 & -1 \\ 4 & 3 & -4 \\ -2 & 0 & -3 \end{bmatrix}$$

Show that A has an inverse, and find A^{-1}.

Solution. Using the procedure (i)–(iii) given above, we first form the matrix

$$\begin{bmatrix} 1 & 2 & -1 & 1 & 0 & 0 \\ 4 & 3 & -4 & 0 & 1 & 0 \\ -2 & 0 & -3 & 0 & 0 & 1 \end{bmatrix}$$

Next we perform elementary row operations to bring the left side of the matrix into the form given in step (ii):

$$\begin{bmatrix} 1 & 2 & -1 & 1 & 0 & 0 \\ 4 & 3 & -4 & 0 & 1 & 0 \\ -2 & 0 & -3 & 0 & 0 & 1 \end{bmatrix} \overset{R_2 \to R_2 + (-4)R_1}{\longleftrightarrow} \begin{bmatrix} 1 & 2 & -1 & 1 & 0 & 0 \\ 0 & -5 & 0 & -4 & 1 & 0 \\ -2 & 0 & -3 & 0 & 0 & 1 \end{bmatrix} \overset{R_3 \to R_3 + 2R_1}{\longleftrightarrow}$$

$$\begin{bmatrix} 1 & 2 & -1 & 1 & 0 & 0 \\ 0 & -5 & 0 & -4 & 1 & 0 \\ 0 & 4 & -5 & 2 & 0 & 1 \end{bmatrix} \overset{R_2 \to (-\frac{1}{5})R_2}{\longleftrightarrow} \begin{bmatrix} 1 & 2 & -1 & 1 & 0 & 0 \\ 0 & 1 & 0 & \frac{4}{5} & -\frac{1}{5} & 0 \\ 0 & 4 & -5 & 2 & 0 & 1 \end{bmatrix} \overset{R_1 \to R_1 + (-2)R_2}{\longleftrightarrow}$$

$$\begin{bmatrix} 1 & 0 & -1 & -\frac{3}{5} & \frac{2}{5} & 0 \\ 0 & 1 & 0 & \frac{4}{5} & -\frac{1}{5} & 0 \\ 0 & 4 & -5 & 2 & 0 & 1 \end{bmatrix} \overset{R_3 \to R_3 + (-4)R_2}{\longleftrightarrow} \begin{bmatrix} 1 & 0 & -1 & -\frac{3}{5} & \frac{2}{5} & 0 \\ 0 & 1 & 0 & \frac{4}{5} & -\frac{1}{5} & 0 \\ 0 & 0 & -5 & -\frac{6}{5} & \frac{4}{5} & 1 \end{bmatrix} \overset{R_3 \to (-\frac{1}{5})R_3}{\longleftrightarrow}$$

$$\begin{bmatrix} 1 & 0 & -1 & -\frac{3}{5} & \frac{2}{5} & 0 \\ 0 & 1 & 0 & \frac{4}{5} & -\frac{1}{5} & 0 \\ 0 & 0 & 1 & \frac{6}{25} & -\frac{4}{25} & -\frac{1}{5} \end{bmatrix} \overset{R_1 \to R_1 + R_3}{\longleftrightarrow} \begin{bmatrix} 1 & 0 & 0 & -\frac{9}{25} & \frac{6}{25} & -\frac{1}{5} \\ 0 & 1 & 0 & \frac{4}{5} & -\frac{1}{5} & 0 \\ 0 & 0 & 1 & \frac{6}{25} & -\frac{4}{25} & -\frac{1}{5} \end{bmatrix}$$

Therefore by (iii), A has an inverse, and

$$A^{-1} = \begin{bmatrix} -\frac{9}{25} & \frac{6}{25} & -\frac{1}{5} \\ \frac{4}{5} & -\frac{1}{5} & 0 \\ \frac{6}{25} & -\frac{4}{25} & -\frac{1}{5} \end{bmatrix} = \frac{1}{25}\begin{bmatrix} -9 & 6 & -5 \\ 20 & -5 & 0 \\ 6 & -4 & -5 \end{bmatrix}$$

You can check this result by showing that $AA^{-1} = I_3$. □

Solutions of Linear Systems Using Inverses

We return to linear systems of equations and present one final method of solving them. The method works when the system has the same number of equations as variables, provided that the coefficient matrix has an inverse. For simplicity our discussion will involve systems consisting of three equations in three variables.

Consider the system

$$\begin{cases} a_1x + b_1y + c_1z = k_1 \\ a_2x + b_2y + c_2z = k_2 \\ a_3x + b_3y + c_3z = k_3 \end{cases} \tag{4}$$

Let

$$A = \begin{bmatrix} a_1 & b_1 & c_1 \\ a_2 & b_2 & c_2 \\ a_3 & b_3 & c_3 \end{bmatrix}, \quad X = \begin{bmatrix} x \\ y \\ z \end{bmatrix}, \quad \text{and} \quad K = \begin{bmatrix} k_1 \\ k_2 \\ k_3 \end{bmatrix}$$

Then

$$AX = \begin{bmatrix} a_1 & b_1 & c_1 \\ a_2 & b_2 & c_2 \\ a_3 & b_3 & c_3 \end{bmatrix}\begin{bmatrix} x \\ y \\ z \end{bmatrix} = \begin{bmatrix} a_1x + b_1y + c_1z \\ a_2x + b_2y + c_2z \\ a_3x + b_3y + c_3z \end{bmatrix}$$

By the definition of the equality of two matrices,

$$AX = K \tag{5}$$

if and only if

$$\begin{bmatrix} a_1x + b_1y + c_1z \\ a_2x + b_2y + c_2z \\ a_3x + b_3y + c_3z \end{bmatrix} = \begin{bmatrix} k_1 \\ k_2 \\ k_3 \end{bmatrix}$$

which means that the corresponding entries of AX and K are the same, that is, the equations in (4) are satisfied. Therefore (4) and (5) are equivalent. This means that determining values of x, y, and z such that (4) is satisfied is equivalent to finding a matrix X such that (5) is satisfied. In other words, solving the system of equations in (4) is equivalent to solving the matrix equation in (5) for the matrix X.

How can we solve (5) for X? Suppose that the determinant of the coefficient matrix A is nonzero. Then A^{-1} exists by Theorem 8.4, so that (5) is equivalent to

$$A^{-1}(AX) = A^{-1}K \tag{6}$$

Using (1), (2), and the definition of inverse, we conclude that

$$A^{-1}(AX) = (A^{-1}A)X = I_3X = X$$

Therefore (6) becomes

$$\boxed{X = A^{-1}K} \tag{7}$$

In other words, if $\det A \neq 0$, then in order to solve (5) for X, we need only compute the product $A^{-1}K$. And since

$$X = \begin{bmatrix} x \\ y \\ z \end{bmatrix}$$

once we know X, we know the solution of the system (4).

EXAMPLE 7. Use (7) to solve the linear system of equations

$$\begin{cases} x + 2y - z = 6 \\ 4x + 3y - 4z = -1 \\ -2x \qquad - 3z = 3 \end{cases}$$

Solution. As our discussion preceding this example indicated, the given system of equations is equivalent to the matrix equation

$$AX = K \tag{8}$$

where

$$A = \begin{bmatrix} 1 & 2 & -1 \\ 4 & 3 & -4 \\ -2 & 0 & -3 \end{bmatrix}, \quad X = \begin{bmatrix} x \\ y \\ z \end{bmatrix}, \quad \text{and} \quad K = \begin{bmatrix} 6 \\ -1 \\ 3 \end{bmatrix}$$

By Example 6 we know that A^{-1} exists and that

$$A^{-1} = \frac{1}{25}\begin{bmatrix} -9 & 6 & -5 \\ 20 & -5 & 0 \\ 6 & -4 & -5 \end{bmatrix}$$

It then follows from (7) that the solution X of (8) is given by

$$X = A^{-1}K$$

We find that

$$\begin{bmatrix} x \\ y \\ z \end{bmatrix} = X = A^{-1}K = \frac{1}{25}\begin{bmatrix} -9 & 6 & -5 \\ 20 & -5 & 0 \\ 6 & -4 & -5 \end{bmatrix}\begin{bmatrix} 6 \\ -1 \\ 3 \end{bmatrix}$$

$$= \frac{1}{25}\begin{bmatrix} (-9)(6) + (6)(-1) + (-5)(3) \\ (20)(6) + (-5)(-1) + (0)(3) \\ (6)(6) + (-4)(-1) + (-5)(3) \end{bmatrix}$$

$$= \frac{1}{25}\begin{bmatrix} -75 \\ 125 \\ 25 \end{bmatrix} = \begin{bmatrix} -3 \\ 5 \\ 1 \end{bmatrix}$$

By the definition of equality of matrices, this means that $x = -3$, $y = 5$, and $z = 1$. Consequently $(-3, 5, 1)$ is the proposed solution of the given system of equations.

Check:
$$\begin{cases} -3 + 2(5) - 1 = 6 \\ 4(-3) + 3(5) - 4(1) = -1 \\ -2(-3) - 3(1) = 3 \end{cases}$$

We conclude that $(-3, 5, 1)$ is the solution of the given system. □

EXERCISES 8.7

In Exercises 1–4, perform the indicated operation.

1. $\begin{bmatrix} 2 & 1 & 3 \\ -4 & 1 & 5 \end{bmatrix} + \begin{bmatrix} 3 & -3 & 0 \\ 1 & 4 & 2 \end{bmatrix}$

2. $\begin{bmatrix} -1 & 3 & 5 \\ 2 & 6 & -4 \\ -3 & 1 & 2 \end{bmatrix} + \begin{bmatrix} 6 & -1 & 5 \\ 1 & -3 & 2 \\ -4 & 5 & 1 \end{bmatrix}$

3. $-3\begin{bmatrix} 2 & -4 \\ -5 & 12 \end{bmatrix}$

4. $\frac{1}{2}\begin{bmatrix} 2 & 4 & 0 \\ -3 & 6 & -2 \\ 14 & -8 & 10 \end{bmatrix}$

In Exercises 5–18, calculate the given product.

5. $\begin{bmatrix} 1 & 2 \\ 4 & 3 \end{bmatrix}\begin{bmatrix} -2 & -1 \\ 0 & 3 \end{bmatrix}$

6. $\begin{bmatrix} 0 & 1 \\ -5 & 6 \end{bmatrix}\begin{bmatrix} 3 & -4 \\ 4 & 7 \end{bmatrix}$

7. $\begin{bmatrix} \frac{1}{2} & 2 \\ -1 & 0 \end{bmatrix}\begin{bmatrix} 4 & -1 \\ \frac{1}{2} & 1 \end{bmatrix}$

8. $\begin{bmatrix} 1 & 2 \\ -3 & -4 \end{bmatrix}\begin{bmatrix} 3 \\ -1 \end{bmatrix}$

9. $\begin{bmatrix} -4 & 1 \\ 3 & 6 \end{bmatrix}\begin{bmatrix} 1 \\ -1 \end{bmatrix}$

10. $[2 \quad -5]\begin{bmatrix} 6 \\ -1 \end{bmatrix}$

11. $\begin{bmatrix} 2 \\ -1 \end{bmatrix}[3 \quad -4]$

12. $\begin{bmatrix} 3 & 1 & -1 \\ 0 & 2 & -3 \\ 4 & 0 & -2 \end{bmatrix}\begin{bmatrix} 0 & -1 & 2 \\ 2 & -3 & 4 \\ 3 & 0 & 2 \end{bmatrix}$

13. $\begin{bmatrix} 1 & 0 & 0 \\ 0 & 2 & 0 \\ 0 & 0 & 3 \end{bmatrix}\begin{bmatrix} 4 & -2 & 5 \\ 9 & -1 & -7 \\ -3 & 6 & 3 \end{bmatrix}$

14. $\begin{bmatrix} -1 & 1 & 2 \\ 3 & 0 & -2 \\ -4 & \frac{1}{2} & 0 \end{bmatrix}\begin{bmatrix} -1 & 2 & 0 \\ -2 & 4 & 6 \\ 7 & -6 & -\frac{1}{2} \end{bmatrix}$

15. $\begin{bmatrix} 2 & -3 & 1 \\ -6 & 9 & -2 \\ -3 & 5 & -1 \end{bmatrix}\begin{bmatrix} -1 & -2 & 3 \\ 0 & -1 & 2 \\ 3 & 1 & 0 \end{bmatrix}$

16. $\begin{bmatrix} 3 & -4 \\ 2 & 5 \\ 1 & 6 \end{bmatrix}\begin{bmatrix} 1 \\ -2 \end{bmatrix}$

17. $[-3 \quad -6 \quad 2]\begin{bmatrix} -1 \\ 3 \\ 4 \end{bmatrix}$

18. $\begin{bmatrix} -5 \\ 1 \\ -3 \end{bmatrix}[2 \quad -1]$

In Exercises 19–22, calculate AB and BA.

19. $A = \begin{bmatrix} -2 & 0 \\ -1 & 3 \end{bmatrix}$, $B = \begin{bmatrix} 4 & 3 \\ -1 & 0 \end{bmatrix}$

20. $A = \begin{bmatrix} -7 & 3 \\ 3 & 1 \end{bmatrix}$, $B = \begin{bmatrix} -1 & 0 \\ 0 & -1 \end{bmatrix}$

21. $A = \begin{bmatrix} 2 & 3 & -1 \\ 4 & -3 & 5 \\ 1 & -1 & 0 \end{bmatrix}$, $B = \begin{bmatrix} 1 & 2 & 0 \\ -1 & 0 & 3 \\ 4 & -1 & 2 \end{bmatrix}$

22. $A = \begin{bmatrix} \frac{1}{2} & -\frac{1}{3} & \frac{1}{4} \\ 2 & \frac{1}{2} & \frac{1}{6} \\ 1 & -\frac{1}{4} & \frac{2}{9} \end{bmatrix}$, $B = \begin{bmatrix} 4 & 0 & 0 \\ 0 & 4 & 0 \\ 0 & 0 & 4 \end{bmatrix}$

In Exercises 23–24, calculate $(AB)C$ and $A(BC)$, and thereby support (1).

23. $A = \begin{bmatrix} 0 & -2 \\ 3 & -1 \end{bmatrix}$, $B = \begin{bmatrix} 3 & -1 \\ 1 & 0 \end{bmatrix}$, $C = \begin{bmatrix} 4 & 0 \\ -1 & -3 \end{bmatrix}$

24. $A = \begin{bmatrix} 3 & 0 & 1 \\ 2 & -1 & -3 \\ 0 & 4 & -1 \end{bmatrix}$, $B = \begin{bmatrix} 1 & -1 & 0 \\ 0 & -1 & 2 \\ 0 & 1 & 0 \end{bmatrix}$,

$C = \begin{bmatrix} 1 & 2 & -3 \\ -2 & 0 & 1 \\ -1 & 2 & 0 \end{bmatrix}$

In Exercises 25–40, determine whether the given matrix has an inverse. If it does, compute the inverse.

25. $\begin{bmatrix} 0 & 1 \\ 1 & 0 \end{bmatrix}$

26. $\begin{bmatrix} 3 & -2 \\ 1 & 0 \end{bmatrix}$

27. $\begin{bmatrix} 4 & -6 \\ 2 & -3 \end{bmatrix}$

28. $\begin{bmatrix} -2 & 2 \\ 1 & 1 \end{bmatrix}$

29. $\begin{bmatrix} 6 & 1 \\ -3 & 2 \end{bmatrix}$

30. $\begin{bmatrix} -6 & 9 \\ 1 & -\frac{3}{2} \end{bmatrix}$

31. $\begin{bmatrix} a & 0 \\ c & b \end{bmatrix}$ where $ab \neq 0$

32. $\begin{bmatrix} 1 & -1 & 0 \\ 0 & -1 & 1 \\ 1 & 2 & -1 \end{bmatrix}$

33. $\begin{bmatrix} 1 & 0 & 0 \\ 0 & 1 & 0 \\ 0 & 0 & 1 \end{bmatrix}$

34. $\begin{bmatrix} 0 & 0 & 1 \\ 0 & 1 & 0 \\ 1 & 0 & 0 \end{bmatrix}$

35. $\begin{bmatrix} 2 & 0 & 4 \\ 3 & -1 & 4 \\ 2 & 1 & 6 \end{bmatrix}$

36. $\begin{bmatrix} 2 & -3 & 1 \\ 4 & -1 & -2 \\ -1 & 0 & 2 \end{bmatrix}$

37. $\begin{bmatrix} -1 & -2 & 3 \\ 0 & -1 & 2 \\ 3 & 1 & 0 \end{bmatrix}$

38. $\begin{bmatrix} 2 & -1 & -2 \\ 0 & -3 & 1 \\ 4 & -14 & 0 \end{bmatrix}$

39. $\begin{bmatrix} 0 & 4 & 2 \\ 1 & -3 & -1 \\ -4 & 5 & -2 \end{bmatrix}$

40. $\begin{bmatrix} a & 0 & 0 \\ d & b & 0 \\ e & f & c \end{bmatrix}$ where $abc \neq 0$

In Exercises 41–48, solve the given system by using the inverse of the coefficient matrix.

41. $\begin{cases} 3x + 4y = -11 \\ 2x - y = 0 \end{cases}$

42. $\begin{cases} -2x + 3y = 13 \\ x + 5y = 39 \end{cases}$

43. $\begin{cases} -4x - 2y = 22 \\ 3x - 5y = -23 \end{cases}$

44. $\begin{cases} 12x + y = 0 \\ 9x - 8y = 0 \end{cases}$

45. $\begin{cases} 2x + 4y - z = -4 \\ x - y + 3z = 8 \\ 3x + 3y - 4z = -8 \end{cases}$

46. $\begin{cases} 2x - 4y + z = -9 \\ -3x - y - 2z = 0 \\ 4x + 3y + z = 5 \end{cases}$

47. $\begin{cases} 3x - 7y + 5z = 7 \\ 2x + y - 5z = -3 \\ -x - y + z = \frac{1}{2} \end{cases}$

48. $\begin{cases} 9x + 2y - 7z = 0 \\ -3x + 2y + 6z = 0 \\ 5x - 3y - 4z = 0 \end{cases}$

49. Show that $\begin{bmatrix} 0 & 1 \\ 0 & 1 \end{bmatrix}\begin{bmatrix} 1 & 0 \\ 0 & 0 \end{bmatrix} = \begin{bmatrix} 0 & 0 \\ 0 & 0 \end{bmatrix}$.

50. Let A and B be 2×2 matrices. Show that $\det(AB) = (\det A)(\det B)$.

8.8 SYSTEMS OF INEQUALITIES

In Sections 1.6 and 1.7 we solved single inequalities. Now we return to inequalities and solve systems of two or more inequalities, such as

$$\begin{cases} x + y \leq 1 \\ 2x - y + 1 \geq 0 \end{cases} \tag{1}$$

and

$$\begin{cases} x^2 + y^2 \leq 4 \\ x^2 \geq 3y \\ x \geq 1 \end{cases} \tag{2}$$

An ordered pair (x, y) is a ***solution*** of a given system of inequalities if and only if (x, y) satisfies each inequality in the system individually. Thus $(1, -2)$ is a solution of (1) because

$$1 + (-2) \leq 1 \quad \text{and} \quad 2(1) - (-2) + 1 \geq 0$$

With a system of two or more inequalities, the set of solutions is most easily presented on a graph, and this is the way we will give the solutions. The solutions normally form a region whose boundary is composed of line segments or curves. An essential part of graphing the set of solutions will involve determining the points at which such pieces of the boundary meet.

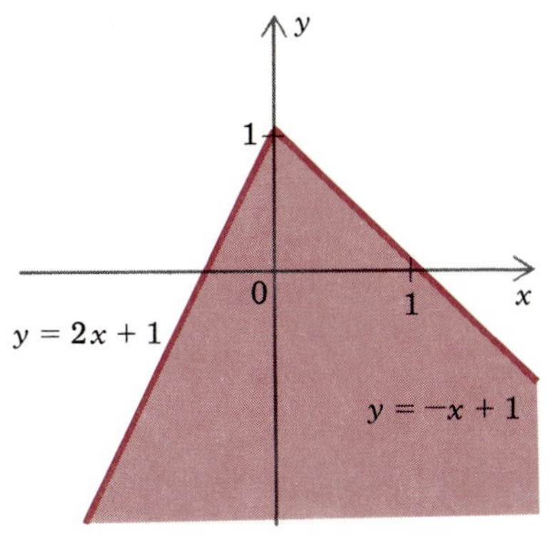

FIGURE 8.2

EXAMPLE 1. Sketch the solutions of the system

$$\begin{cases} y \le -x + 1 \\ y \le 2x + 1 \end{cases}$$

Solution. A pair (x, y) is a solution of the first inequality if (x, y) lies below or on the line $y = -x + 1$ (with slope -1 and y intercept 1). Likewise, (x, y) is a solution of the second inequality if (x, y) lies below or on the line $y = 2x + 1$ (with slope 2 and y intercept 1). The solutions of the system consist of those (x, y) satisfying both conditions, and such (x, y) are shaded in Figure 8.2. Because the y intercepts of the lines forming the boundary are both 1, evidently the lines meet at $(0, 1)$. □

To aid in solving other systems of inequalities, we will say that two inequalities are ***equivalent*** if they have the same solutions. For example, the inequalities

$$y \le -x + 1 \quad \text{and} \quad x + y \le 1$$

are equivalent, because the second can be obtained by adding x to each side of the first. Similarly,

$$y \le 2x + 1 \quad \text{and} \quad 2x - y + 1 \ge 0$$

are equivalent because the second can be obtained by subtracting y from each side of the first. In addition, two systems of inequalities are ***equivalent*** if they have the same solutions. For example, the systems

$$\begin{cases} y \le -x + 1 \\ y \le 2x + 1 \end{cases} \quad \text{and} \quad \begin{cases} x + y \le 1 \\ 2x - y + 1 \ge 0 \end{cases}$$

are equivalent, by our preceding comments.

If the boundary of the set of solutions of a system of inequalities consists of straight lines, it is usually convenient, when possible, to write the inequalities in the system in the equivalent standard form

$$y \le ax + b \quad \text{or} \quad y \ge ax + b \tag{3}$$

because of their close relationship with the slope-intercept form of lines. We will use this procedure in the following example.

EXAMPLE 2. Sketch the solutions of the system

$$\begin{cases} x \geq 0 \\ y \geq 0 \\ 2y + x \leq 8 \\ x + y \leq 5 \\ 2x + y \leq 8 \end{cases}$$

and determine the points of intersection of the boundary lines.

Solution. The given system is equivalent to the system

$$\begin{cases} x \geq 0 \\ y \geq 0 \\ y \leq -\dfrac{1}{2}x + 4 \\ y \leq -x + 5 \\ y \leq -2x + 8 \end{cases}$$

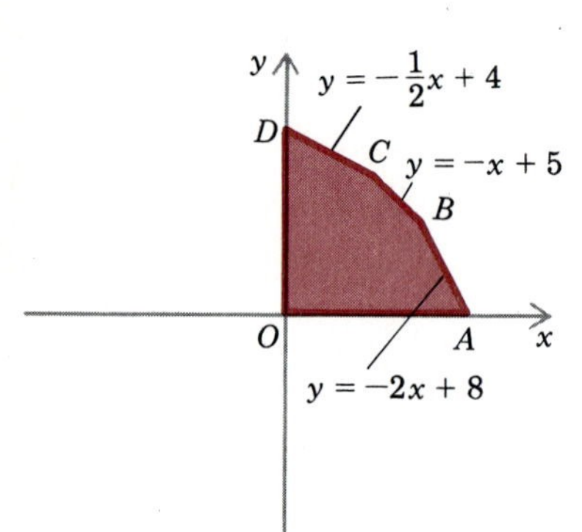

FIGURE 8.3

Observe that (x, y) satisfies $x \geq 0$ and $y \geq 0$ if and only if (x, y) lies in the first quadrant. For that reason, we can restrict our attention to the first quadrant when graphing the remaining inequalities. The graph of $y \leq -\frac{1}{2}x + 4$ consists of all points below or on the line $y = -\frac{1}{2}x + 4$ (with slope $-\frac{1}{2}$ and y intercept 4). The graph of $y \leq -x + 5$ consists of all points below or on the line $y = -x + 5$ (with slope -1 and y intercept 5). Finally, the graph of $y \leq -2x + 8$ consists of all points below or on the line $y = -2x + 8$ (with slope -2 and y intercept 8). Thus the solutions of the given system consist of all points in the first quadrant that lie on or below each of the three above-mentioned lines (Figure 8.3). Finally we determine the coordinates of the points O, A, B, C, and D shown in Figure 8.3. We know that O is the origin, $(0, 0)$, and D is the point $(0, 4)$ at which the line $y = -\frac{1}{2}x + 4$ intersects the y axis. Similarly, A is the point at which the line $y = -2x + 8$ intersects the x axis. We find the coordinates of A by setting y equal to 0 in the equation $y = -2x + 8$:

$$\begin{aligned} 0 &= -2x + 8 \\ 2x &= 8 \\ x &= 4 \end{aligned}$$

Thus A is the point $(4, 0)$. Next, B is the point of intersection of the lines $y = -x + 5$ and $y = -2x + 8$, which (as you can check) is the point $(3, 2)$.

Finally, C is the point of intersection of the lines $y = -\frac{1}{2}x + 4$ and $y = -x + 5$, which you can check is the point (2, 3). This completes the solution of the example. □

EXAMPLE 3. Sketch the solutions of the system

$$\begin{cases} x^2 + y^2 \le 4 \\ (x-2)^2 + (y+2)^2 \ge 4 \end{cases}$$

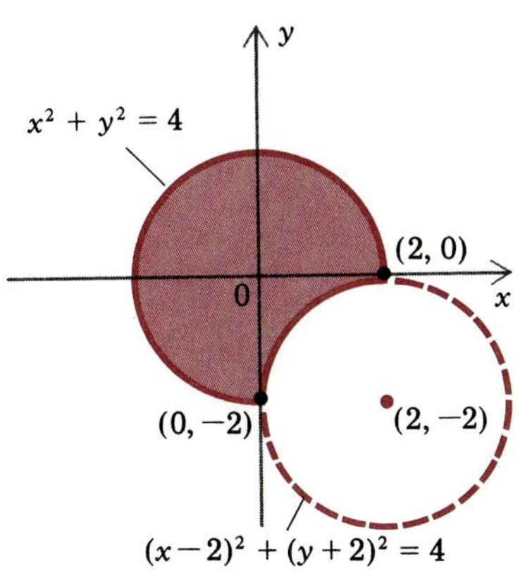

FIGURE 8.4

Solution. The graph of the equation $x^2 + y^2 = 4$ is the circle centered at the origin with radius 2, so the solutions of $x^2 + y^2 \le 4$ consist of all points (x, y) inside or on the circle. Similarly, the graph of the equation $(x-2)^2 + (y+2)^2 = 4$ is the circle centered at the point $(2, -2)$ with radius 2, so the solutions of $(x-2)^2 + (y+2)^2 \ge 4$ consist of all points (x, y) outside or on the circle. The solutions of the given system consist of those (x, y) satisfying both inequalities and form the region shaded in Figure 8.4. As indicated in Figure 8.4, the two circles meet at the points (2, 0) and (0, −2). You can check that this is true by solving the system consisting of the equations of the two circles. □

EXERCISES 8.8

In Exercises 1–6, determine whether the given ordered pair is a solution of the given system of inequalities.

1. $\begin{cases} 2x + y \le 3 \\ 3x - 4y \le 0 \end{cases}$ $(-1, 2)$

2. $\begin{cases} 4x - 3y \ge -1 \\ 2x - 3y \le 20 \end{cases}$ $(3, -4)$

3. $\begin{cases} 2x - 9y \le 0 \\ 3x - 2y \le 0 \end{cases}$ $(\frac{1}{2}, \frac{1}{3})$

4. $\begin{cases} x \ge 0 \\ y \ge 0 \\ 2x - y \ge 0 \\ 2x + 3y \le 50 \end{cases}$ $(6, 11)$

5. $\begin{cases} x \ge 0 \\ y \le 5 \\ 2x - y \le 10 \\ 5x + 6y \le 96 \end{cases}$ $(3, 8)$

6. $\begin{cases} x \ge 0 \\ y \le 5 \\ 7x + 8y \le 25 \\ 9x + 10y \le 30 \end{cases}$ $(0, 3)$

In Exercises 7–10, sketch the solutions of the given system of inequalities without calculating the points of intersection of the boundary lines.

7. $\begin{cases} y \le -2x + 4 \\ y \ge x + 2 \end{cases}$

8. $\begin{cases} 4x + 2y \le 5 \\ -3x + 2y \le 6 \end{cases}$

9. $\begin{cases} x \ge 0 \\ y \ge 0 \\ 2x + y \le 6 \\ 4x + y \le 8 \end{cases}$

10. $\begin{cases} x \geq 0 \\ y \geq 0 \\ y - \frac{1}{2}x \leq 3 \\ 2x - y \leq 4 \end{cases}$

In Exercises 11–22, sketch the solutions of the given system of inequalities, and find the points of intersection of the boundary lines.

11. $\begin{cases} x \geq 0 \\ x \leq 3 \\ y \geq 0 \\ x + y \leq 5 \end{cases}$

12. $\begin{cases} y \geq 0 \\ y - 2x \leq 3 \\ x + y \leq 6 \end{cases}$

13. $\begin{cases} x \geq 0 \\ y \geq 0 \\ y \leq 7 \\ 2y \geq x - 4 \end{cases}$

14. $\begin{cases} y \geq 0 \\ x \leq 3 \\ y - x \leq 1 \\ x \geq 2y - 3 \end{cases}$

15. $\begin{cases} y \leq 2x + 3 \\ x \leq 2y - 2 \\ y + 2x \leq 11 \end{cases}$

16. $\begin{cases} y \geq 0 \\ x - y \geq 1 \\ 2y \leq x \\ 2x + y \leq 10 \end{cases}$

17. $\begin{cases} x \geq 0 \\ y \geq 0 \\ 2y + x \leq 20 \\ x \leq -y + 12 \\ 2x + y \leq 20 \end{cases}$

18. $\begin{cases} x \geq 1 \\ x \leq 15 \\ y \geq 2 \\ y \leq 7 \\ x + 3y \leq 27 \end{cases}$

19. $\begin{cases} x^2 + y^2 \leq 9 \\ (x + 3)^2 + (y + 3)^2 \geq 9 \end{cases}$

20. $\begin{cases} x^2 + y^2 \leq 16 \\ x^2 + (y - 2)^2 \geq 1 \end{cases}$

21. $\begin{cases} y \geq x^2 \\ y \leq 1 - x^2 \end{cases}$

22. $\begin{cases} y \leq x + 2 \\ y \leq 4x \\ y \geq x^2 \end{cases}$

8.9 LINEAR PROGRAMMING

In the preceding sections of the chapter we have studied both systems of equations and systems of inequalities. We conclude the chapter with a discussion of linear programming, a topic that centers around both kinds of systems. Basically linear programming involves finding the maximum or minimum value of a linear expression, which in two variables has the form

$$ax + by$$

where the variables x and y are subject to a set of conditions or constraints given in the form of linear inequalities. The linear expression $ax + by$ might represent the profit resulting from allocating resources so that x units of one product and y units of another product are manufactured. In that case the linear inequalities might represent constraints on the resources used in production. The problem would be to allocate or "program" resources in such a way as to maximize the profit. This is the origin of the term "linear programming."

For example, consider the problem of maximizing the expression $3x + y$ subject to the linear inequalities

$$\begin{cases} x \geq 0 \\ y \geq 0 \\ 2x + y \leq 4 \\ x + 2y \leq 5 \end{cases} \tag{1}$$

Perhaps x represents the amount in thousands of gallons of yogurt and y the amount in thousands of gallons of ice cream that a company sells daily. If the profit in thousands of dollars per thousand gallons of yogurt is 3 and the profit in thousands of dollars per thousand gallons of ice cream is 1, then $3x + y$ would represent the daily profit in thousands of dollars on the sale of x thousand gallons of yogurt and y thousand gallons of ice cream. The first two constraints in (1), which are $x \geq 0$ and $y \geq 0$, say that only a nonnegative amount of the foods can be sold, which is reasonable. The next constraint, $2x + y \leq 4$, might represent restrictions on the available labor, and the final constraint, $x + 2y \leq 5$, might represent restrictions on plant equipment. After all this information is provided, the basic problem is to determine the values of x and y that maximize the profit $3x + y$ subject to the constraints in (1). After we develop the necessary tools, we will solve this problem in Example 1.

In this last 30 years linear programming has found widespread application in science and industry. Normally there are far more than two variables involved in linear programming problems that arise. However, in order to keep the calculations manageable, we will restrict our attention to those linear programming problems that involve only two variables.

Assume now that we have a linear programming problem such as the one discussed earlier, so that we wish to find the maximum or minimum value of a linear expression $ax + by$ subject to a set of linear inequalities (the constraints). Any ordered pair (x, y) that satisfies all the constraints is called a ***feasible solution*** of the problem. The set of all feasible solutions is called the ***feasible set***. A feasible solution that yields the desired maximum or minimum value of $ax + by$ is called an ***optimal solution*** of the problem. For the problem of maximizing $3x + y$ subject to the constraints in (1), the points $(1, \frac{1}{2})$ and $(1, 2)$ are feasible solutions, since they each satisfy all constraints:

$$\begin{cases} 1 \geq 0 \\ \dfrac{1}{2} \geq 0 \\ 2(1) + \dfrac{1}{2} \leq 4 \\ 1 + 2\left(\dfrac{1}{2}\right) \leq 5 \end{cases} \quad \text{and} \quad \begin{cases} 1 \geq 0 \\ 2 \geq 0 \\ 2(1) + 2 \leq 4 \\ 1 + 2(2) \leq 5 \end{cases}$$

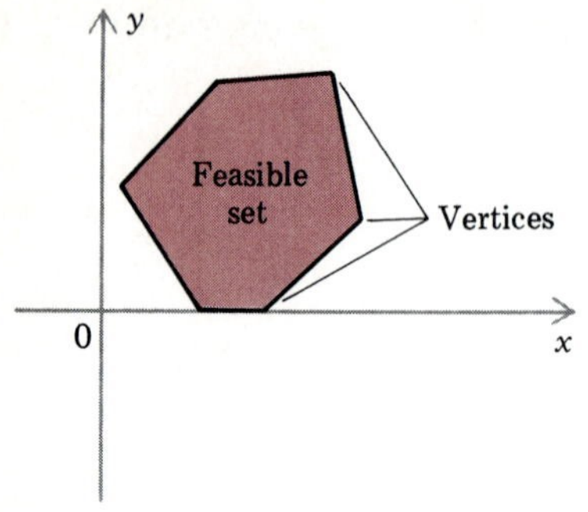

FIGURE 8.5

For the linear expression $3x + y$ the points $(1, \frac{1}{2})$ and $(1, 2)$ yield

$$3(1) + \frac{1}{2} = \frac{7}{2} \quad \text{and} \quad 3(1) + 2 = 5$$

respectively. We conclude that $(1, \frac{1}{2})$ does not yield the maximum value of $3x + y$ and hence is *not* an optimal solution. It will follow from the solution of Example 1 that (1, 2) is also not an optimal solution.

We can sketch the feasible set for a given linear programming problem by sketching the solutions of the system of linear inequalities that form the constraints. We do this by the methods of Section 8.8. Since all of the constraints are linear inequalities, the boundaries of the feasible set will be lines or line segments (Figure 8.5). The point at which two boundary lines intersect is called a ***vertex*** of the feasible set. The following theorem, which we will not prove, assures us that if optimal solutions exist, then at least one of them can be found among the vertices of the feasible set.

DEFINITION 8.5
Fundamental Theorem of Linear Programming

Assume that a linear programming problem has an optimal solution. Then at least one vertex of the feasible set is an optimal solution.

Since the feasible set may well have only a few vertices, the Fundamental Theorem of Linear Programming can reduce the work needed to find an optimal solution of a linear programming problem. The procedure is as follows:

i. Determine the feasible set by graphing the linear inequalities that form the constraints.
ii. Find the vertices of the feasible set by solving two at a time the equations of the boundary lines of the feasible set.
iii. Compute the value of $ax + by$ at each of the vertices.
iv. Select the largest value computed in step (iii) if $ax + by$ is to be maximized, and select the smallest value if $ax + by$ is to be minimized.

Before presenting examples and problems, we mention that a given linear programming problem may or may not have an optimal solution. However, all the problems in this book actually do have optimal solutions.

EXAMPLE 1. Find the maximum and minimum values of $3x + y$ subject to the constraints

$$\begin{cases} x \geq 0 \\ y \geq 0 \\ 2x + y \leq 4 \\ x + 2y \leq 5 \end{cases} \tag{2}$$

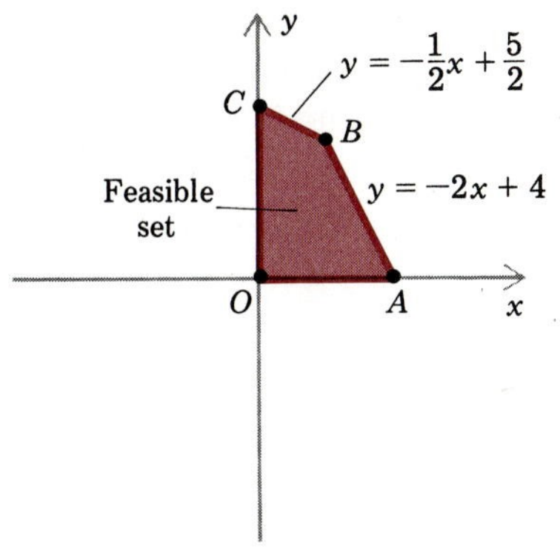

FIGURE 8.6

Solution. First we will determine the feasible set by graphing the inequalities in (2). Notice that (2) is equivalent to the system

$$\begin{cases} x \geq 0 \\ y \geq 0 \\ y \leq -2x + 4 \\ y \leq -\dfrac{1}{2}x + \dfrac{5}{2} \end{cases}$$

The inequalities $x \geq 0$ and $y \geq 0$ restrict the feasible set to the first quadrant. Thus the feasible set is the set of all points in the first quadrant that lie on or below each of the lines $y = -2x + 4$ (with slope -2 and y intercept 4) and $y = -\frac{1}{2}x + \frac{5}{2}$ (with slope $-\frac{1}{2}$ and y intercept $\frac{5}{2}$). The feasible set is sketched in Figure 8.6. Next we will determine the vertices O, A, B, and C of the feasible set. We know that O is the origin, and C is the point $(0, \frac{5}{2})$ at which the line $y = -\frac{1}{2}x + \frac{5}{2}$ crosses the y axis. Similarly, A is the point at which the line $y = -2x + 4$ crosses the x axis. To find the coordinates of A we substitute 0 for y in the equation $y = -2x + 4$:

$$\begin{aligned} 0 &= -2x + 4 \\ 2x &= 4 \\ x &= 2 \end{aligned}$$

Thus A is the point $(2, 0)$. Since B is the point of intersection of the lines $y = -2x + 4$ and $y = -\frac{1}{2}x + \frac{5}{2}$, we find the coordinates of B by solving the system

$$\begin{cases} y = -2x + 4 \\ y = -\dfrac{1}{2}x + \dfrac{5}{2} \end{cases} \tag{3}$$

Substituting $-\frac{1}{2}x + \frac{5}{2}$ for y in the first equation of (3) yields

$$\begin{aligned} -\frac{1}{2}x + \frac{5}{2} &= -2x + 4 \\ \frac{3}{2}x &= \frac{3}{2} \\ x &= 1 \end{aligned}$$

Substituting 1 for x in the first equation of (3), we find that

$$y = -2(1) + 4 = 2$$

Therefore the solution of (3) is (1, 2), which tells us that B is the point (1, 2). In order to find the maximum value of $3x + y$ among the feasible solutions, we compute the values of $3x + y$ at the vertices O, A, B, and C:

(x, y)	$O = (0, 0)$	$A = (2, 0)$	$B = (1, 2)$	$C = (0, \frac{5}{2})$
$3x + y$	0	6	5	$\frac{5}{2}$

From the table we see that the maximum value of $3x + y$ at the vertices is 6, corresponding to (2, 0). We conclude from the Fundamental Theorem of Linear Programming that (2, 0) is an optimal solution and that 6 is the maximum value of $3x + y$ subject to the constraints listed in (2). Likewise, we see from the table that the minimum value of $3x + y$ at the vertices is 0, corresponding to (0, 0). We conclude that (0, 0) is an optimal solution and that 0 is the minimum value of $3x + y$ subject to the constraints listed in (2). □

EXAMPLE 2. A small business produces tables and chairs. Suppose the material for each table costs \$75 and the material for each chair costs \$50, and that the business can spend a maximum of \$1200 per day for material. Suppose also that 3 hours of labor are required to assemble each table and 5 hours to assemble each chair, and that up to 75 hours of labor are available daily at a cost of \$5 per hour. Finally, let us assume that the delivery truck can deliver at most 18 items (tables or chairs) per day. If each table is sold for \$150 and each chair for \$125, determine the maximum daily profit and how many tables and chairs should be produced daily for maximum profit.

Solution. We will translate the given problem into a linear programming problem. Ultimately we wish to determine how many tables and how many chairs to produce daily, so let x be the number of tables and y the number of chairs to be produced daily. Now the material for each table costs \$75, and the labor costs \$15 (3 hours at \$5 per hour). Thus the total cost of producing each table is \$90. Likewise, the material for each chair costs \$50, and the labor costs \$25 (5 hours at \$5 per hour). Thus the total cost of producing each chair is \$75. Therefore the profit from each table sold for \$150 is \$60 (because $150 - 90 = 60$), and the profit for each chair sold for \$125 is \$50 (because $125 - 75 = 50$). Consequently if x tables and y chairs are produced and sold daily, then the daily profit is

$$60x + 50y$$

dollars. Since the business wishes to maximize its daily profit, we have now identified the linear expression to be maximized. Next we will determine the constraints. Recall first that the material for each table costs \$75 and the material for each chair costs \$50. Thus if x tables and y chairs are produced daily, then the daily cost of material is $75x + 50y$ dollars. Since by hypothesis the business can pay a maximum of \$1200 per day for material, we must have

$$75x + 50y \leq 1200$$

which reduces to

$$3x + 2y \leq 48$$

Next recall that each table requires 3 hours to assemble and each chair 5 hours. Therefore with x tables and y chairs produced daily, the number of hours required daily is $3x + 5y$. Since the number of hours available daily is by hypothesis limited to 75, we have

$$3x + 5y \leq 75$$

Recall also from the hypothesis that at most 18 items (tables or chairs) can be delivered a day. Thus

$$x + y \leq 18$$

Finally, we have

$$x \geq 0 \quad \text{and} \quad y \geq 0$$

since the number of items produced cannot be negative.

Now we can restate the problem as a linear programming problem: Find the maximum value of $60x + 50y$ subject to the constraints

$$\begin{cases} x \geq 0 \\ y \geq 0 \\ 3x + 2y \leq 48 \\ x + y \leq 18 \\ 3x + 5y \leq 75 \end{cases} \tag{4}$$

The first thing we do as we begin to solve the problem is to rewrite the system in (4) in the equivalent form

$$\begin{cases} x \geq 0 \\ y \geq 0 \\ y \leq -\frac{3}{2}x + 24 \\ y \leq -x + 18 \\ y \leq -\frac{3}{5}x + 15 \end{cases}$$

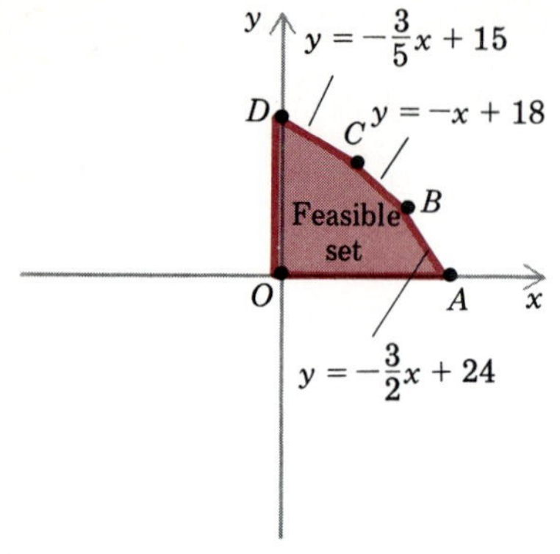

FIGURE 8.7

The feasible set for this problem is shown in Figure 8.7. There are 5 vertices: O, A, B, C, and D. To determine their coordinates, we notice that O is the origin, A is the point at which the line $y = -\frac{3}{2}x + 24$ intersects the x axis, B is the point of intersection of the lines $y = -x + 18$ and $y = -\frac{3}{2}x + 24$, C is the point of intersection of the lines $y = -\frac{3}{5}x + 15$ and $y = -x + 18$, and D is the point at which the line $y = -\frac{3}{5}x + 15$ intersects the y axis. You can verify that

$$O = (0, 0), \quad A = (16, 0), \quad B = (12, 6), \quad C = (7.5, 10.5), \quad D = (0, 15)$$

To complete the solution of the problem we evaluate $60x + 50y$ at the 5 vertices:

(x, y)	$60x + 50y$
$O = (0, 0)$	0
$A = (16, 0)$	960
$B = (12, 6)$	1020
$C = (7.5, 10.5)$	975
$D = (0, 15)$	750

From the table we see that the maximum value of $60x + 50y$ at the vertices is 1020, and occurs at (12, 6). It follows from the Fundamental Theorem of Linear Programming that the maximum value of $60x + 50y$ subject to the constraints in (4) is 1020, corresponding to the optimal solution (12, 6). We conclude that 12 tables and 6 chairs should be produced daily to maximize profit, and that the maximum daily profit is \$1020. □

EXERCISES 8.9

In Exercises 1–10, find the maximum and minimum values of the given linear expression subject to the given constraints, and find points at which those values are assumed.

1. $-4x + y$ subject to

$$\begin{cases} x \geq -1 \\ x \leq 3 \\ y \geq -2 \\ y \leq 1 \end{cases}$$

2. $3x - 2y$ subject to

$$\begin{cases} x \geq 0 \\ y \geq 0 \\ x \leq 5 \\ y \leq x \end{cases}$$

3. $\frac{1}{2}x + \frac{1}{3}y$ subject to

$$\begin{cases} x \geq 0 \\ y \geq 0 \\ y \leq 6 \\ 3x + y \leq 18 \end{cases}$$

4. $0.01x + 0.25y$ subject to

$$\begin{cases} x \geq 0 \\ y \geq 0 \\ 2x + 3y \leq 24 \\ 2x + y \leq 12 \end{cases}$$

5. $4y - 5x$ subject to

$$\begin{cases} y \geq 2 \\ 2x - y \leq 2 \\ 2y - x \leq 8 \end{cases}$$

6. $4y - 5x$ subject to

$$\begin{cases} x \leq 8 \\ y \geq 0 \\ 2x - y \geq 2 \\ 2y - x \leq 8 \end{cases}$$

7. $2x + \frac{1}{3}y$ subject to

$$\begin{cases} x \leq 6 \\ 3x + 2y \geq 18 \\ y \leq 2x + 9 \end{cases}$$

8. $\frac{1}{6}x + \frac{1}{4}y$ subject to

$$\begin{cases} x \geq 0 \\ y \geq 0 \\ y \leq 4 \\ x + 2y \geq 2 \\ 2x + 3y \leq 22 \end{cases}$$

9. $2x + 4y$ subject to

$$\begin{cases} x \geq 0 \\ y \geq 0 \\ 2x + 5y \leq 50 \\ 2x + 3y \leq 34 \\ 3x + y \leq 30 \end{cases}$$

10. $4x + 5y$ subject to

$$\begin{cases} y \geq 0 \\ y \leq 6 \\ x + 3y \geq 10 \\ 3y - x \leq 8 \\ 2x + y \leq 36 \end{cases}$$

11. A sidewalk vendor sells submarines and soft drinks from a cart. Each carton of submarines weighs 5 pounds, takes up 1 cubic foot of space in the cart, and yields a profit of \$8 to the vendor. Each case of soft drinks weights 20 pounds, takes up 0.3 cubic foot of space in the cart and yields a profit of \$3 to the vendor. The cart can carry up to 225 pounds and has a capacity of 8 cubic feet. Assuming that the vendor can sell whatever the cart can hold, find the maximum profit that the vendor can reap from one cartload.

12. A grocer plans to buy up to \$5000 worth of peanuts and raisins and sell the mixture for \$5 per pound. Peanuts cost \$1.25 per pound, and raisins cost \$2.50 per pound. The weight of the peanuts in the mixture should be at least three times and at most six times the weight of the raisins. How many pounds of peanuts and how many pounds of raisins should the grocer buy in order to maximize the profit?

13. A corporate executive plans to invest up to \$60,000 in municipal and corporate bonds for the next year. The municipal bonds yield 9% interest per year, and the corporate bonds 12%. If the executive wishes to invest at least \$10,000 in corporate bonds and at least twice as much money in municipal bonds as in corporate bonds, find the maximum amount of interest that can be earned during the year.

14. A builder can acquire up to 125 lots on which to build either modern or colonial houses. Each modern house requires an investment of \$30,000 and yields a profit of \$10,000, whereas each colonial house requires an investment of \$20,000 and yields a profit of \$8000. If the builder can invest up to \$3,000,000 and wishes the number of colonial houses built to be at most 4 times the number of modern houses, determine how many modern houses should be built to yield the maximum profit.

15. Smith's orchard produces peaches and apples. On the average, one bushel of peaches can be picked in 20 minutes and one bushel of apples in 30 minutes. Suppose that the labor force can pick up to 240 hours each day, that Smith's delivery truck can deliver at most 500 bushels daily to market, and that each day Smith must retain 40 bushels of fruit for sales on the premises. If there is a \$2 profit on each bushel of peaches and \$2.50 on each bushel of apples, determine the number of bushels of peaches that must be harvested daily to achieve the maximum daily profit, and compute that profit.

16. Suppose the profit per bushel of apples drops to \$2 in Exercise 15. Show that there would be multiple ways of obtaining a maximum profit.

17. A garden shop has 1100 pounds of Kentucky Bluegrass seed and 900 pounds of Kentucky Tall Fescue seed. In each bag of Premium seed the shop puts 4 pounds of Bluegrass and no Fescue seed, whereas in each bag of Shady seed it puts one pound of Bluegrass and 3 pounds of Fescue seed. If the profit on each bag of Premium seed is \$3 and on each bag of Shady seed is \$2, determine the maximum profit the shop can earn from the sale of Premium and Shady seed.

18. A sporting goods manufacturer produces sleeping bags and tents. Each sleeping bag requires $\frac{2}{3}$ hour of cutting, 1 hour of machine sewing, and $\frac{1}{2}$ hour of hand sewing, and brings a profit of \$15. Each tent requires 1 hour of cutting, 3 hours of machine sewing, and $\frac{1}{2}$ hour of hand sewing, and brings a profit of \$25. Each day the shop can provide 36 hours of labor for cutting, 90 hours for machine sewing, and 24 hours for hand sewing. Determine the number of sleeping bags that should be produced each day to bring in the maximum daily profit.

KEY TERMS

system of equations	determinant
solution of a system of equations	minor
method of substitution	cofactor
method of elimination	identity matrix
linear system	inverse of a matrix
matrix	system of inequalities
square matrix of order n	solution of a system of inequalities
coefficient matrix	equivalent systems of inequalities
augmented matrix	linear programming
equivalent matrices	feasible solution
elementary row operation	optimal solution

KEY FORMULAS

Cramer's Rule
Fundamental Theorem of Linear Programming

REVIEW EXERCISES

In Exercises 1–14, find all solutions of the given system of equations.

1. $\begin{cases} x + y = 7 \\ -2x - 7y = -4 \end{cases}$

2. $\begin{cases} 2x + 3y = 0 \\ -4x + 6y = -4 \end{cases}$

3. $\begin{cases} 6x - 12y = 1 \\ -9x + 18y = 7 \end{cases}$

4. $\begin{cases} x - 3y = -5 \\ -2x + 6y = 10 \end{cases}$

5. $\begin{cases} x - 2y = 0 \\ -3x + 6y = 0 \end{cases}$

6. $\begin{cases} x^2 + y^2 = 10 \\ 9x^2 + 2y^2 = 27 \end{cases}$

7. $\begin{cases} 2xy = -1 \\ 4x - y = 3 \end{cases}$

8. $\begin{cases} e^{x+y} = 1 \\ 2x - y = 3 \end{cases}$

9. $\begin{cases} 2x - xy = 0 \\ 3x + xy = 5 \end{cases}$

10. $\begin{cases} rs - 5s + r = 5 \\ 2s - r = 3 \end{cases}$

11. $\begin{cases} 3x - 2y - z = 3 \\ 2x + 5y + 2z = 0 \\ 5x + 7y + 2z = -1 \end{cases}$

12. $\begin{cases} 4x - 7y + 11z = 5 \\ 6x + 5y - 9z = -21 \\ 5x - y + z = 0 \end{cases}$

13. $\begin{cases} x - y - z = 1 \\ 2x + y - 8z = -1 \\ -3x + 2y + 5z = -2 \end{cases}$

14. $\begin{cases} 3x - 5y + 6z = 1 \\ 6x + 7y - 12z = 6 \\ -12x + 2y - 6z = -7 \end{cases}$

In Exercises 15–18, solve the given system by using augmented matrices.

15. $\begin{cases} x - 2y + 2z = 1 \\ 3x + 2y - z = -6 \\ 4x - 3y + 3z = -6 \end{cases}$

16. $\begin{cases} 2x - 3y - z = 0 \\ 7x - 3y + z = -2 \\ 4x - y + z = -1 \end{cases}$

17. $\begin{cases} 2x + 3y + 2z = 0 \\ 3x + 5y - 5z = 5 \\ -2x + y - z = 1 \end{cases}$

18. $\begin{cases} x - y + z = 3 \\ -3x + 2y + 3z = -7 \\ -5x + 3y + 7z = -11 \end{cases}$

In Exercises 19–24, evaluate the given determinant.

19. $\begin{vmatrix} 3 & -5 \\ 2 & 4 \end{vmatrix}$

20. $\begin{vmatrix} -4 & -5 \\ -7 & -13 \end{vmatrix}$

21. $\begin{vmatrix} -3 & 4 & 2 \\ 1 & 0 & 6 \\ 3 & 2 & -5 \end{vmatrix}$

22. $\begin{vmatrix} 2 & -4 & -5 \\ 0 & 1 & 1 \\ -3 & 5 & 6 \end{vmatrix}$

23. $\begin{vmatrix} 2 & -1 & 0 & -2 \\ 1 & 0 & -3 & -1 \\ 4 & -1 & 2 & 1 \\ -2 & 2 & 1 & 0 \end{vmatrix}$

24. $\begin{vmatrix} -2 & -1 & -1 & -3 \\ 1 & 0 & -2 & 0 \\ 3 & 2 & -6 & 4 \\ 3 & 2 & 1 & 5 \end{vmatrix}$

In Exercises 25–28, use Cramer's Rule to solve the given system.

25. $\begin{cases} 3x - 2y = 1 \\ -5x + 4y = 3 \end{cases}$

26. $\begin{cases} 2x + 16y = 1 \\ 3x + 26y = 2 \end{cases}$

27. $\begin{cases} 3x + 2y + 2z = 9 \\ 2x - 2y + 3z = 1 \\ 4x - 2y + 4z = 0 \end{cases}$

28. $\begin{cases} 3x + 2y + 2z = 2 \\ -4x - 3y - 2z = 4 \\ -2x + 3y + 3z = 3 \end{cases}$

In Exercises 29–30, perform the indicated operations.

29. $\begin{bmatrix} 3 & -1 & 5 \\ 2 & 1 & 4 \\ -7 & 2 & 0 \end{bmatrix} + 3\begin{bmatrix} -1 & -4 & 2 \\ 0 & -5 & 8 \\ 7 & 3 & 0 \end{bmatrix}$

30. $\begin{bmatrix} 4 & 3 & -2 \\ 6 & -5 & 7 \\ 0 & -4 & \frac{1}{2} \end{bmatrix}\begin{bmatrix} -4 & 5 & 6 \\ 2 & 4 & -3 \\ 0 & -2 & 2 \end{bmatrix}$

In Exercises 31–36, determine whether the given matrix has an inverse. If it does, compute the inverse.

31. $\begin{bmatrix} 2 & -1 \\ 0 & 4 \end{bmatrix}$

32. $\begin{bmatrix} 4 & 3 \\ 8 & 6 \end{bmatrix}$

33. $\begin{bmatrix} 1 & -1 \\ 1 & 1 \end{bmatrix}$

34. $\begin{bmatrix} 2 & -3 & -1 \\ 4 & 0 & 5 \\ -8 & 6 & -3 \end{bmatrix}$

35. $\begin{bmatrix} 1 & -1 & 2 \\ 2 & 1 & 3 \\ 2 & -2 & 5 \end{bmatrix}$

36. $\begin{bmatrix} 5 & -2 & 4 \\ 8 & -1 & 7 \\ 6 & -2 & 5 \end{bmatrix}$

In Exercises 37–40, solve the given system by using the inverse of the coefficient matrix.

37. $\begin{cases} 3x - 2y = -1 \\ 2x - 4y = 26 \end{cases}$

38. $\begin{cases} 6x + 3y = 1 \\ 4x + 5y = \frac{8}{3} \end{cases}$

39. $\begin{cases} 2x + 5y + z = 9 \\ -3x + 3y - 4z = -3 \\ -8x + 6y - 10z = -10 \end{cases}$

40. $\begin{cases} x + 2y + 3z = -3 \\ 2x + 7y + 5z = -\frac{17}{2} \\ -x - 5y - z = 5 \end{cases}$

In Exercises 41–44, sketch the solutions of the given system of inequalities, and find the points of intersection of the boundary lines.

41. $\begin{cases} x \geq 0 \\ y \geq 0 \\ y + \frac{4}{3}x \leq 1 \\ y + \frac{2}{3}x \leq \frac{2}{3} \end{cases}$

42. $\begin{cases} x \leq 5 \\ y \geq 0 \\ 2y - x \leq 7 \\ 3x \leq 19 - 2y \end{cases}$

43. $\begin{cases} x \geq 0 \\ y \geq 0 \\ x + y \leq 10 \\ y \leq -2(x - 6) \\ 4x \leq 20 - y \end{cases}$

44. $\begin{cases} y \geq 2x \\ 4x + y \geq 0 \\ 3x \leq 5 - y \\ y \leq x + 5 \end{cases}$

In Exercises 45–48, find the maximum and minimum values of the given linear expression subject to the given constraints, and find the points at which those values are assumed.

45. $2x + 4y$; the region in Exercise 41.

46. $-x + 2y$; the region in Exercise 42.

47. $3x + y$; the region in Exercise 43.

48. $4y - 5x$; the region in Exercise 44.

49. Let $A = \begin{bmatrix} 0 & 1 \\ 0 & 1 \end{bmatrix}$. Show that $A^2 = A$.

50. Let $A = \begin{bmatrix} 1 & 1 \\ 0 & 1 \end{bmatrix}$. Find a formula for A^n, for any positive integer n.

51. For which values of λ does the system

$$\begin{cases} \lambda x + 3y = 0 \\ 27x + \lambda y = 0 \end{cases}$$

have a solution other than the trivial solution (0, 0)?

52. For what values of a and b does the line $ax + by = 5$ contain the points (15, 10) and (−5, −5)?

53. If 6 oranges and 12 lemons together cost $3.30, and 12 oranges and 6 lemons together cost $3.00, how much does a single orange cost?

54. If Joe were 4 inches taller, he would be $\frac{2}{3}$ as tall as his mother. If he were 7 inches shorter, he would be only half as tall as his mother. How tall is his mother?

55. A swimming club deposited a $20,000 surplus into two separate savings accounts earning simple interest for a period for one year. One account earned 8% interest and the other 10%. If the total interest earned during the year was $1920, how much was deposited in each account?

56. An advertising company must have at least 1000 letters addressed within four hours. Its own staff can address 100 letters per hour at a cost of $40 per hour, and a professional typing company can address 200 letters per hour at a cost of $120 per hour. For how many hours should the advertising company employ the typing company if it wishes to minimize its total cost for addressing letters?

57. A company plans to introduce a new line of storm doors and storm windows made out of glass and aluminum. Because of limitations on experienced labor available at its glass-cutting and aluminum-cutting buildings, the company must limit production. The labor required for each door and window and the profit from each are shown in Table 8.1. Determine how many storm doors and how many storm windows should be produced daily in order to maximize the total profit.

TABLE 8.1

	Storm door	Storm window	Available daily
Time for cutting glass	2 minutes	1 minute	1000 minutes
Time for cutting aluminum	4 minutes	1 minute	1600 minutes
Profit	$7	$2	

9 COMPLEX NUMBERS

In Section 1.4 we introduced complex numbers. These are numbers of the form $a + bi$, where a and b are real numbers and i is a nonreal number satisfying

$$i^2 = -1$$

Our main goal in Section 1.4 was to be able to solve any quadratic equation

$$ax^2 + bx + c = 0$$

where a, b, and c are arbitrary real numbers with $a \neq 0$. Complex numbers are also important in solving higher-degree polynomial equations, such as

$$x^6 + 64 = 0 \qquad \text{and} \qquad x^4 - 2x^3 + 5x^2 - 8x + 4 = 0$$

as we will see in Chapters 9 and 10.

In the present chapter we define the quotient of complex numbers and associate complex numbers with points in a plane (just as real numbers are identified with points on a line). We end with a discussion of the nth roots of complex numbers. The importance of complex numbers lies in their having become indispensable in the analysis of problems in various sciences, especially physics and electrical engineering.

9.1 CONJUGATES AND DIVISION OF COMPLEX NUMBERS

Recall from Section 1.4 that the conjugate of a complex number $a + bi$ is the complex number $a - bi$, which is often denoted by $\overline{a + bi}$. For example,

$$\overline{2 + 7i} = 2 - 7i \quad \text{and} \quad \overline{4 - \frac{1}{3}i} = 4 + \frac{1}{3}i$$

It follows from the definition of the conjugate that a real number is equal to its conjugate. Thus

$$\overline{-5} = \overline{-5 + 0i} = -5 - 0i = -5$$

It also follows that the conjugate of a pure imaginary number is its negative. For example,

$$\overline{6i} = -6i$$

The conjugate is important because of its role in the definition of the multiplicative inverse and the quotient of complex numbers. To define the multiplicative inverse, we observe first that

$$(a + bi)(a - bi) = [a \cdot a - b(-b)] + [a(-b) + ba]i$$

which condenses to

$$\boxed{(a + bi)(a - bi) = a^2 + b^2} \tag{1}$$

Thus the product of any complex number and its conjugate is a nonnegative real number. For example,

$$(6 + i)(6 - i) = 6^2 + 1^2 = 37$$

Next we will give meaning to the complex fraction

$$\frac{a + bi}{c + di}$$

under the stipulation that $c + di \neq 0$. The hypothesis that $c + di \neq 0$ implies that $c^2 + d^2 \neq 0$. Therefore by (1),

$$\begin{aligned}(c + di)\left[\frac{1}{c^2 + d^2}(c - di)\right] &= \frac{1}{c^2 + d^2}[(c + di)(c - di)] \\ &= \frac{1}{c^2 + d^2}(c^2 + d^2) = 1 \end{aligned} \tag{2}$$

Thus $c + di$ has the multiplicative inverse

$$\frac{1}{c^2 + d^2}(c - di)$$

which we denote by $1/(c + di)$ (or $(c + di)^{-1}$). Consequently,

$$\boxed{\frac{1}{c + di} = \frac{1}{c^2 + d^2}(c - di)}$$

One way of evaluating $1/(c + di)$ as a complex number is to multiply it by 1, written as

$$\frac{c - di}{c - di}$$

This yields

$$\frac{1}{c + di} = \frac{1}{c + di}\frac{c - di}{c - di} = \frac{1}{(c + di)(c - di)}(c - di) = \frac{1}{c^2 + d^2}(c - di) \quad (3)$$

Now the ***quotient*** $(a + bi)/(c + di)$ is defined to be the product of $a + bi$ and $1/(c + di)$:

$$\frac{a + bi}{c + di} = (a + bi)\frac{1}{c + di} \quad (4)$$

Usually the easiest way of expressing a quotient $(a + bi)/(c + di)$ in the form $x + yi$ is to use the technique employed in (3) of multiplying it by 1 in the form $(c - di)/(c - di)$:

$$\boxed{\frac{a + bi}{c + di} = \frac{a + bi}{c + di}\frac{c - di}{c - di} = \frac{1}{c^2 + d^2}(a + bi)(c + di)} \quad (5)$$

Then we multiply out $(a + bi)(c + di)$ and condense the result into the form $x + yi$. The next two examples illustrate this technique.

EXAMPLE 1. Write the following expressions in the form $x + yi$.

a. $\dfrac{1}{4 + i}$ b. $(1 - i)^{-1}$

Solution. We solve both parts by (5):

a. $\dfrac{1}{4 + i} = \dfrac{1}{4 + i} \cdot \dfrac{4 - i}{4 - i} = \dfrac{1}{4^2 + 1^2}(4 - i) = \dfrac{1}{17}(4 - i) = \dfrac{4}{17} - \dfrac{1}{17}i$

b. $(1-i)^{-1} = \dfrac{1}{1-i} = \dfrac{1}{1-i} \cdot \dfrac{1+i}{1+i} = \dfrac{1}{1^2+1^2}(1+i) = \dfrac{1}{2}(1+i)$

$= \dfrac{1}{2} + \dfrac{1}{2}i$ □

EXAMPLE 2. Write $\dfrac{5+i}{-2-3i}$ in the form $x + yi$.

Solution. By (5),

$$\frac{5+i}{-2-3i} = \frac{5+i}{-2-3i} \cdot \frac{-2+3i}{-2+3i} = \frac{-10+15i-2i-3}{4+9}$$

$$= \frac{-13+13i}{13} = -1+i \quad \square$$

If a complex number is denoted by z, then the conjugate is denoted by $\bar{z}$. For later reference we list some of the properties of conjugation, most of which can be proved directly from the definition of conjugate.

i. $\bar{\bar{z}} = z$

ii. $\overline{z+w} = \bar{z} + \bar{w}$

iii. $\overline{zw} = \bar{z}\,\bar{w}$

iv. $\overline{z^n} = (\bar{z})^n$ for any positive integer n.

v. If $z = a + bi$, then $a = \dfrac{1}{2}(z + \bar{z})$ and $b = \dfrac{1}{2i}(z - \bar{z})$.

vi. $\bar{z} = z$ if and only if z is a real number.

It follows from (i) that z is the conjugate of $\bar{z}$, so that z and $\bar{z}$ are conjugates of one another. As a consequence of (v),

$$z + \bar{z} \text{ is real} \quad \text{for any complex number } z \tag{6}$$

It is easy to see that this is true by looking at an example. Suppose that

$$z = 2 + 3i$$

Then

$$z + \bar{z} = (2+3i) + (2-3i) = (2+2) + (3-3)i = 4$$

EXERCISES 9.1

In Exercises 1–6, find the conjugate of the given complex number.

1. $7 + 5i$
2. $-3 - \frac{1}{2}i$
3. $6 - 19i$
4. $-13i$
5. 23
6. -23

In Exercises 7–26, write the given expression in the form $a + bi$.

7. $\overline{6 - 2i}$
8. $\overline{-5i}$
9. $\overline{5 - \frac{1}{2}i}$
10. $\overline{\frac{1}{2} + \sqrt{2}i}$
11. $(1 + i)^{-1}$
12. $\dfrac{1}{4 - 5i}$
13. i^{-1}
14. $(-2 - i)^{-1}$
15. $\dfrac{1}{\overline{3 - 6i}}$
16. $\dfrac{5}{3 - i}$
17. $\dfrac{1 + i}{1 - i}$
18. $\dfrac{2i}{3 + 4i}$
19. $\dfrac{2 - 3i}{-5i}$
20. $\dfrac{-1 + 3i}{5 - 2i}$
21. $\dfrac{(-1 - 2i)^2}{4 - 3i}$
22. $\dfrac{1 - i}{(1 - 2i)^2}$
23. $(5 + 12i)\overline{(5 + 12i)}$
24. $\overline{(3 - 2i)}\overline{(-1 + 4i)}$
25. $2 + i + \dfrac{1}{2 + i}$
26. $\left(\dfrac{\overline{2 - 3i}}{1 - \sqrt{2}i}\right)$

27. Let θ be any real number for which $\tan\theta$ is defined. Show that

$$\frac{1 + i\tan\theta}{1 - i\tan\theta} = \cos 2\theta + i\sin 2\theta$$

(*Hint:* Write $\tan\theta = (\sin\theta)/(\cos\theta)$.)

28. Write the following in the form $a + bi$.
    a. $\dfrac{(1 + i)^5}{(1 - i)^3}$
    b. $\dfrac{(1 - i)^3 - 1}{(1 + i)^3 + 1}$

29. a. Show that $3 - 4i$ is a solution of the equation $x^2 - 6x + 25 = 0$.
    b. Without using the quadratic formula, determine the other solution of the equation in part (a).

30. a. Show that $-2 - 5i$ is a solution of the equation $x^2 + 4x + 29 = 0$.
    b. Without using the quadratic formula, determine the other solution of the equation in part (a).

31. Show that $\overline{z + w} = \bar{z} + \bar{w}$.

32. Show that if $z = a + bi$, then

$$a = \frac{1}{2}(z + \bar{z}) \quad \text{and} \quad b = \frac{1}{2i}(z - \bar{z})$$

33. Prove that $\bar{z} = z$ if and only if z is a real number.

34. Show that if $z^3 = z$, then z is real. (*Hint:* Factor $z^3 - z$.)

35. Determine the complex numbers z such that $z^5 - z = 0$. (*Hint:* Factor $z^5 - z$.)

*36. Find all complex numbers z such that $\bar{z} = z^2$. (Hint: Write $z = a + bi$ and determine the possible values for a and b.)

*37. Show that if both $z + w$ and zw are real, then either $w = \bar{z}$ or both z and w are real numbers.

9.2 THE COMPLEX PLANE

The complex number $a + bi$ can be identified with the ordered pair (a, b) of real numbers, which itself is identified with a point in the (Cartesian) plane. It follows that any complex number can be represented by a point in the plane

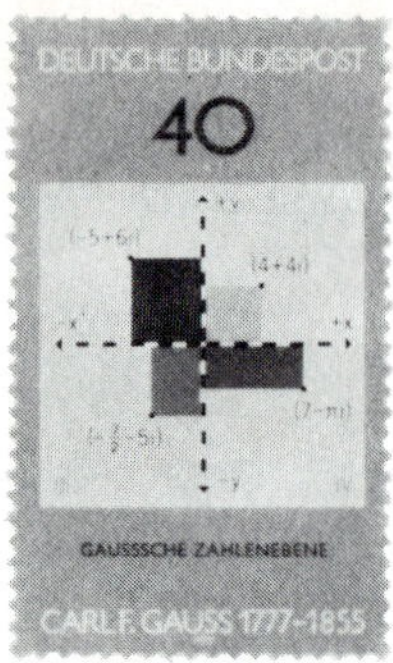

This West German postage stamp, featuring the complex plane, commemorated the bicentennial of Gauss's birth.

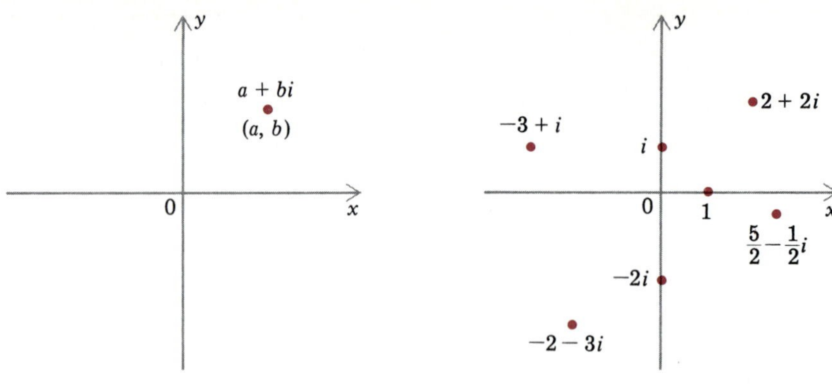

FIGURE 9.1

FIGURE 9.2

(Figure 9.1). In this representation, the first coordinate of the point is the real part of the complex number, and the second coordinate is the imaginary part of the complex number. We use this identification to represent several complex numbers in Figure 9.2. The horizontal axis, which contains the complex numbers that are real, is called the ***real axis***. Analogously, the vertical axis contains the pure imaginary numbers and is called the ***imaginary axis***. When the plane is identified with complex numbers, it is referred to as the ***complex plane***. The complex plane is also called "the Argand diagram," after the French–Swiss mathematician Jean Robert Argand (1768–1822), and "the Gaussian plane," after the German mathematician Carl Friedrich Gauss (1777–1855).

Absolute Value of a Complex Number

Let $a + bi$ be any complex number. Its ***absolute value***, or ***modulus***, written $|a + bi|$, is defined as the distance between the point (a, b) and the origin, namely, $\sqrt{a^2 + b^2}$:

$$|a + bi| = \sqrt{a^2 + b^2} \tag{1}$$

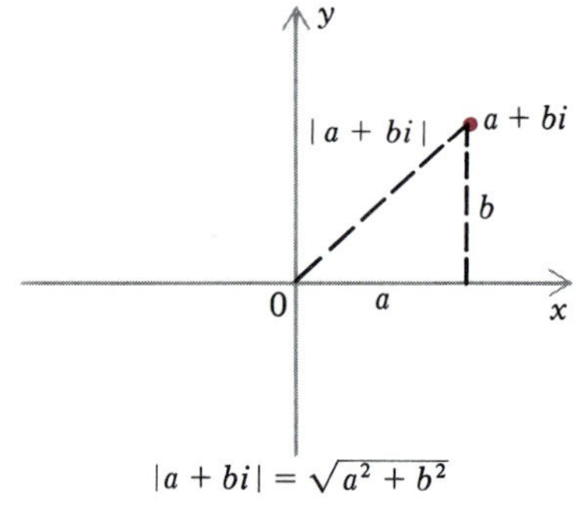

$|a + bi| = \sqrt{a^2 + b^2}$

FIGURE 9.3

(Figure 9.3). Thus if z is any complex number, then $|z|$ is the distance between z and the origin.

EXAMPLE 1. Compute the absolute values of the following complex numbers.

a. i b. $2 - 5i$

Solution.

a. Since $i = 0 + 1 \cdot i$, (1) implies that

$$|i| = \sqrt{0^2 + 1^2} = \sqrt{1^2} = 1 \tag{2}$$

b. By (1) we have

$$|2 - 5i| = \sqrt{2^2 + (-5)^2} = \sqrt{4 + 25} = \sqrt{29} \quad \square$$

The notion of absolute value of a complex number is consistent with absolute value as defined for real numbers in Section 0.2. Indeed, if $a + bi$ is real, so that $b = 0$, then

$$|a + bi| = |a + 0 \cdot i| = \sqrt{a^2 + 0^2} = \sqrt{a^2} = |a|$$

This means, for example, that

$$|-9| = 9$$

whether we think of -9 as a real number or as a complex number.

The following properties of absolute value hold for any complex numbers z and w and can be proved from the definition of absolute value:

$\lvert z\rvert \geq 0$	$\lvert zw\rvert = \lvert z\rvert\,\lvert w\rvert$
$\lvert z\rvert = 0$ if and only if $z = 0$	$\left\lvert \frac{z}{w} \right\rvert = \frac{\lvert z\rvert}{\lvert w\rvert}$
$\lvert z\rvert = \lvert -z\rvert$	$\lvert z + w\rvert \leq \lvert z\rvert + \lvert w\rvert$
$\lvert z - w\rvert = \lvert w - z\rvert$	

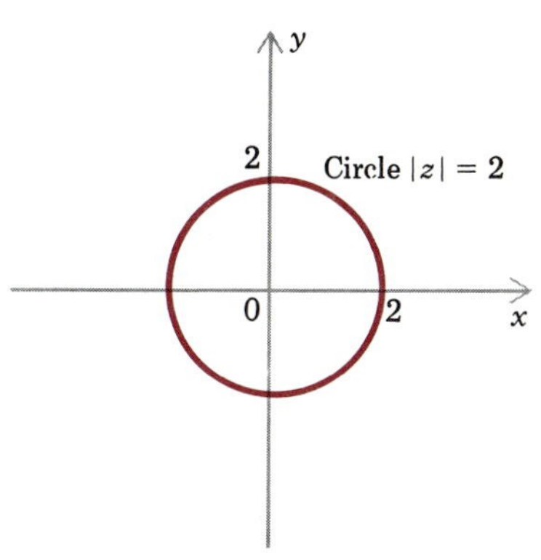

FIGURE 9.4

EXAMPLE 2. Sketch the graph of $|z| = 2$.

Solution. Because $|z|$ is the distance between z and the origin, $|z| = 2$ means that the distance between z and the origin is 2. Therefore the collection of all such complex numbers z forms the circle of radius 2 centered at the origin. It is sketched in Figure 9.4. □

Now let $z = x + yi$ and $z_0 = a + bi$ be two complex numbers. Then

$$\begin{aligned} |z - z_0| &= |(x + yi) - (a + bi)| \\ &= |(x - a) + (y - b)i| = \sqrt{(x - a)^2 + (y - b)^2} \end{aligned}$$

From this and the distance formula it follows that $|z - z_0|$ is the distance between the points (x, y) and (a, b) that represent z and z_0, respectively. Therefore if r is a positive real number and z_0 a fixed complex number, then the graph of the equation

$$|z - z_0| = r$$

is the circle of radius r with center at $a + bi$.

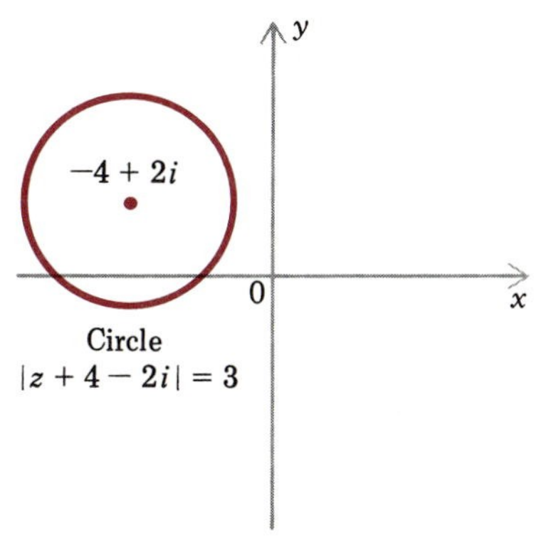

FIGURE 9.5

EXAMPLE 3. Sketch the graph of $|z + 4 - 2i| = 3$.

Solution. The graph is a circle of radius 3. Since

$$z + 4 - 2i = z - (-4 + 2i)$$

we find that the center of the circle is $-4 + 2i$. Now we are in a position to draw the graph (Figure 9.5). □

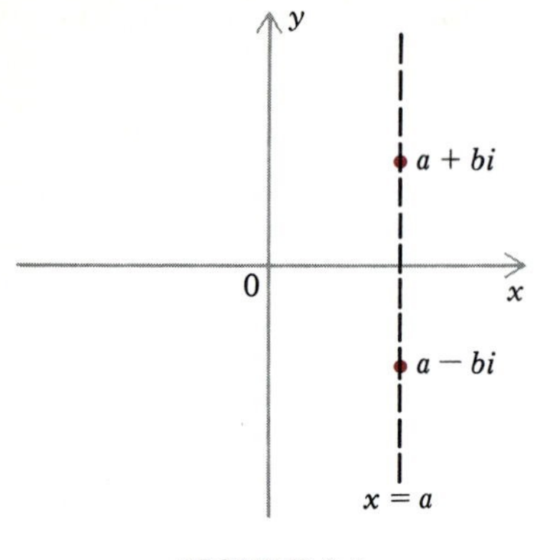

FIGURE 9.6

Conjugates have a geometric interpretation. Indeed, the real parts of $a + bi$ and $a - bi$ are the same (namely, a), so the two complex numbers both lie on the vertical line $x = a$. Since their imaginary parts are negatives of one another, the two conjugates are symmetric to one another with respect to the real axis (Figure 9.6). Therefore

$$|\bar{z}| = |z|$$

a fact that can be easily verified by algebraic considerations (see Exercise 47). Another important connection between the conjugate and the absolute value is the formula

$$\boxed{|z|^2 = z\bar{z}} \qquad (3)$$

To prove this formula, let $z = a + bi$. Then

$$z\bar{z} = (a + bi)(a - bi) = a^2 + b^2 = |z|^2$$

Polar Form of Complex Numbers

To obtain another representation of a complex number $z = a + bi$, we let $r = |z|$, which is the distance between z and the origin, so that

$$r = \sqrt{a^2 + b^2} \qquad (4)$$

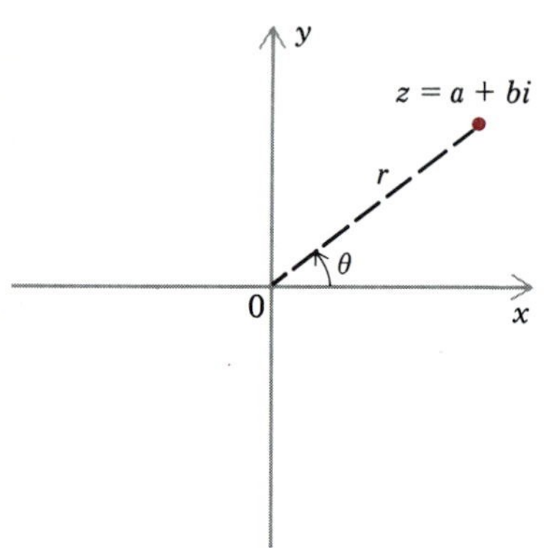

The polar form of $a + bi$ is $r(\cos\theta + i\sin\theta)$

FIGURE 9.7

Next we let θ be an angle that the line joining the origin and the point corresponding to z makes with the positive x axis. We call θ an ***argument*** of z (Figure 9.7). It follows that

$$a = r\cos\theta \quad \text{and} \quad b = r\sin\theta \qquad (5)$$

so that

$$z = a + bi = (r\cos\theta) + (r\sin\theta)i$$

Thus we can write z in polar (or trigonometric) form as follows:

Polar Form

$$z = r(\cos\theta + i\sin\theta) \qquad (6)$$

The number r is determined by z, but there are infinitely many choices of θ, any two of them differing by an integer multiple of 2π. However, there is a unique value of θ satisfying $0 \le \theta < 2\pi$. This value is called the ***principal argument*** of z.

In order to write a nonzero complex number $a + bi$ in polar form we need to determine r and an argument θ. Equation (4) yields r directly. The method for finding an argument θ depends on whether or not $a = 0$. If $a \neq 0$, then (5) implies that

$$\boxed{\frac{b}{a} = \frac{r \sin \theta}{r \cos \theta} = \tan \theta}$$

Thus $a + bi$ determines the value of $\tan \theta$, and by knowing the value of $\tan \theta$ and the quadrant in which $a + bi$ lies we can find an argument θ. If $a = 0$, then $a + bi$ lies on the y axis, so $\theta = \frac{\pi}{2}$ or $\theta = \frac{3\pi}{2}$, depending on whether $a + bi$ lies above or below the x axis.

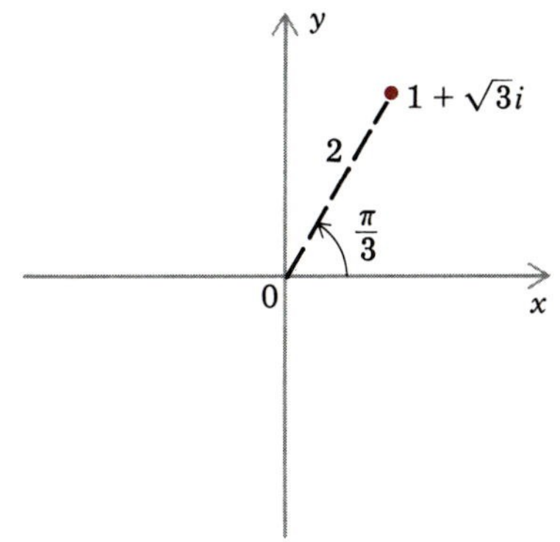

FIGURE 9.8

EXAMPLE 4. Find the principal argument and the corresponding polar form of

a. $1 + \sqrt{3}i$ b. $-6 + 6i$ c. -1

Solution.

a. For r and θ we have

$$r = |1 + \sqrt{3}i| = \sqrt{(1)^2 + (\sqrt{3})^2} = \sqrt{4} = 2$$

$$\tan \theta = \frac{\sqrt{3}}{1} = \sqrt{3}$$

Since $1 + \sqrt{3}i$ lies in the first quadrant (Figure 9.8) and $\tan \theta = \sqrt{3}$, we find that the principal argument is $\pi/3$. Therefore by (6),

$$1 + \sqrt{3}i = 2\left(\cos \frac{\pi}{3} + i \sin \frac{\pi}{3}\right)$$

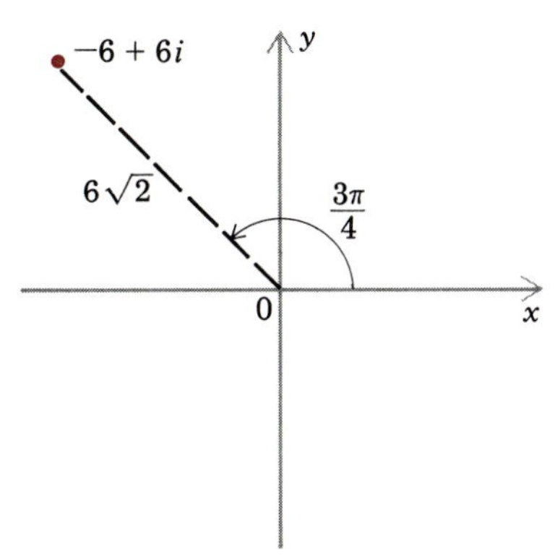

FIGURE 9.9

b. We have

$$r = |-6 + 6i| = \sqrt{(-6)^2 + (6)^2} = 6\sqrt{2}$$

$$\tan \theta = \frac{6}{-6} = -1$$

Since $-6 + 6i$ lies in the second quadrant (Figure 9.9), it follows that the principal argument is $3\pi/4$. Therefore by (6),

$$-6 + 6i = 6\sqrt{2}\left(\cos \frac{3\pi}{4} + i \sin \frac{3\pi}{4}\right)$$

c. We have $|-1| = 1$. Since -1 lies on the negative x axis, the principal argument is π, so that by (6),

$$-1 = 1(\cos \pi + i \sin \pi) \quad \square$$

The polar form is especially convenient for calculating products and quotients of complex numbers.

THEOREM 9.1 *Let $z_1 = r_1(\cos\theta_1 + i\sin\theta_1)$ and $z_2 = r_2(\cos\theta_2 + i\sin\theta_2)$. Then*

a. $z_1z_2 = r_1r_2[\cos(\theta_1+\theta_2) + i\sin(\theta_1+\theta_2)]$ (7)

b. $\dfrac{z_1}{z_2} = \dfrac{r_1}{r_2}[\cos(\theta_1-\theta_2) + i\sin(\theta_1-\theta_2)]$, if $r_2 \neq 0$ (8)

Proof. To prove (a) we use the Addition Formulas for sine and cosine from Section 6.2:

$$\begin{aligned} z_1z_2 &= [r_1(\cos\theta_1 + i\sin\theta_1)][r_2(\cos\theta_2 + i\sin\theta_2)] \\ &= r_1r_2[(\cos\theta_1\cos\theta_2 - \sin\theta_1\sin\theta_2) \\ &\quad + i(\sin\theta_1\cos\theta_2 + \cos\theta_1\sin\theta_2)] \\ &= r_1r_2[\cos(\theta_1+\theta_2) + i\sin(\theta_1+\theta_2)] \end{aligned}$$

To prove (b) we first notice that

$$\begin{aligned} \frac{1}{z_2} &= \frac{1}{r_2(\cos\theta_2 + i\sin\theta_2)} = \frac{1}{r_2(\cos\theta_2 + i\sin\theta_2)}\,\frac{\cos\theta_2 - i\sin\theta_2}{(\cos\theta_2 - i\sin\theta_2)} \\ &= \frac{\cos(-\theta_2) + i\sin(-\theta_2)}{r_2(\cos^2\theta_2 + \sin^2\theta_2)} \\ &= \frac{1}{r_2}[\cos(-\theta_2) + i\sin(-\theta_2)] \end{aligned}$$

Then the result of (a) yields

$$\begin{aligned} \frac{z_1}{z_2} &= z_1\left(\frac{1}{z_2}\right) = [r_1(\cos\theta_1 + i\sin\theta_1)]\left[\frac{1}{r_2}[\cos(-\theta_2) + i\sin(-\theta_2)]\right] \\ &= \frac{r_1}{r_2}[\cos(\theta_1-\theta_2) + i\sin(\theta_1-\theta_2)] \quad \blacksquare \end{aligned}$$

Thus to multiply two complex numbers, simply multiply their moduli and add their arguments. Similarly, to divide two complex numbers, divide their moduli and subtract their arguments.

EXAMPLE 5. Let $z_1 = -6 + 6i$ and $z_2 = 1 + \sqrt{3}i$. Find polar forms of z_1z_2 and z_1/z_2.

Solution. From parts (b) and (a) of Example 4 we know that

$$z_1 = 6\sqrt{2}\left(\cos\frac{3\pi}{4} + i\sin\frac{3\pi}{4}\right) \quad \text{and} \quad z_2 = 2\left(\cos\frac{\pi}{3} + i\sin\frac{\pi}{3}\right)$$

Therefore by (7),

$$\begin{aligned} z_1 z_2 &= (6\sqrt{2})(2)\left[\cos\left(\frac{3\pi}{4} + \frac{\pi}{3}\right) + i\sin\left(\frac{3\pi}{4} + \frac{\pi}{3}\right)\right] \\ &= 12\sqrt{2}\left(\cos\frac{13\pi}{12} + i\sin\frac{13\pi}{12}\right) \end{aligned}$$

and by (8),

$$\begin{aligned} \frac{z_1}{z_2} &= \frac{6\sqrt{2}}{2}\left[\cos\left(\frac{3\pi}{4} - \frac{\pi}{3}\right) + i\sin\left(\frac{3\pi}{4} - \frac{\pi}{3}\right)\right] \\ &= 3\sqrt{2}\left(\cos\frac{5\pi}{12} + i\sin\frac{5\pi}{12}\right) \quad \square \end{aligned}$$

EXERCISES 9.2

In Exercises 1–13, compute the absolute value of the given number.

1. $-i$ **2.** -12 **3.** i^{13}

4. $3 + 4i$ **5.** $3 - 4i$ **6.** $-3 + 4i$

7. $1 - i$ **8.** $-5 - 12i$

9. $(1 + i)(-3 - 4i)$ **10.** $\dfrac{1}{2 - i}$

11. $\dfrac{5 - i}{-2 + 4i}$ **12.** $-3i(4 + i)(5 - 2i)^2$

13. $\dfrac{(1 - 2i)(3 - i)}{4i(3 + 4i)}$

14. Plot the points in the complex plane that correspond to the following complex numbers.
a. $1 + i$ b. $3 - 4i$ c. $-2 + i$
d. $-\frac{1}{2} - \frac{3}{2}i$ e. -3 f. $2i$
g. $-3i$ h. 0

In Exercises 15–22, sketch the graph of the given equation.

15. $|z| = 1$ **16.** $|z| = 3$

17. $|z| = \sqrt{3}$ **18.** $|z| = \frac{1}{2}$

19. $|z - 1| = 1$ **20.** $|z - 5 - i| = 2$

21. $|z + 2 - 4i| = \frac{3}{2}$ **22.** $|z + 3 + 5i| = 4$

In Exercises 23–32, find the polar form of the given number.

23. $1 + i$ **24.** $1 - i$

25. $-1 + i$ **26.** $-1 - i$

27. -1 **28.** $-i$

29. $-7 + 7i$ **30.** $\sqrt{3} + i$

31. $\frac{1}{2} - \frac{\sqrt{3}}{2}i$ **32.** $-\frac{\sqrt{3}}{2} - \frac{1}{2}i$

In Exercises 33–36, calculate $z_1 z_2$ and z_1/z_2. (Leave the results in polar form.)

33. $z_1 = \cos \pi/4 + i \sin \pi/4,\quad z_2 = \cos \pi/6 + i \sin \pi/6$

34. $z_1 = \cos \pi/3 + i \sin \pi/3,\quad z_2 = \cos 2\pi/3 + i \sin 2\pi/3$

35. $z_1 = 2(\cos \pi/12 + i \sin \pi/12)$,
$z_2 = 4(\cos \pi/24 + i \sin \pi/24)$

36. $z_1 = 3(\cos \pi/12 - i \sin \pi/12)$,
$z_2 = 7(\cos \pi/24 + i \sin \pi/24)$

In Exercises 37–44, find the principal argument and the corresponding polar form of the given number.

37. $(\sqrt{3} + i)(7 + 7i)$

38. $\dfrac{\sqrt{3} + i}{7 + 7i}$

39. $(2\sqrt{3} - 2i)\left(-\dfrac{1}{2} + \dfrac{1}{2}i\right)$

40. $\dfrac{-5}{1 - \sqrt{3}i}$

41. $(-\sqrt{2} + \sqrt{6}i)(-1 - \sqrt{3}i)$

42. $i(i + 1)(-\sqrt{3} + i)$

43. $\dfrac{(2 - 2i)(\sqrt{3} - i)}{3 + 3\sqrt{3}i}$

44. $(1 - i)(2\sqrt{3} - 2i)(-4 - 4\sqrt{3}i)$

45. Show that $|z| = |-z|$.

46. Show that $|z - w| = |w - z|$.

47. Show that $|\bar{z}| = |z|$ by writing $z = a + bi$ and computing $|z|$ and $|\bar{z}|$.

48. If $\bar{z} = 1/z$, what can we say about $|z|$?

49. Let $z \neq 0$. Describe the relationship between the points in the complex plane that correspond to

a. z and iz b. z and $\dfrac{1}{\sqrt{2}}(1 + i)z$

50. Let z and w be any complex numbers. Show that

$$|z - w|^2 + |z + w|^2 = 2(|z|^2 + |w|^2)$$

(*Hint:* Let $z = a + bi$ and $w = c + di$. Express each side of the given equation in terms of a, b, c, and d.)

***51.** Suppose that z and w are distinct complex numbers with $|z| = 1$. Show that

$$\left|\frac{z - w}{1 - \bar{z}w}\right| = 1$$

(*Hint:* Multiply both numerator and denominator of the fraction by z.)

52. The ***complex exponential function*** is defined by

$$e^z = e^x(\cos y + i \sin y)$$

where $z = x + yi$.

a. Compute e^0, $e^{i\pi/2}$, and $e^{i\pi}$.
b. Show that $e^{z + 2\pi i} = e^z$. (It follows that e^z is periodic. The period is $2\pi i$.)
c. Show that $|e^z| = e^x$.
d. Let y be a real number. Prove ***Euler's Formula***

$$e^{iy} = \cos y + i \sin y$$

53. The ***complex logarithmic function*** is defined by

$$\log z = \ln r + i\theta$$

where $z = r(\cos \theta + i \sin \theta)$ and $0 \leq \theta < 2\pi$.

a. Compute $\log i$, $\log(-1)$, and $\log(-i)$.
b. Show that $e^{\log z} = z$.

9.3 DE MOIVRE'S THEOREM AND *n*th ROOTS OF COMPLEX NUMBERS

Let $z = r(\cos \theta + i \sin \theta)$ be any complex number. If we apply (7) of Section 9.2 with both z_1 and z_2 replaced by z, we find that

$$z^2 = r^2(\cos 2\theta + i \sin 2\theta)$$

Thus the modulus of z^2 is the square of the modulus of z, and twice the given

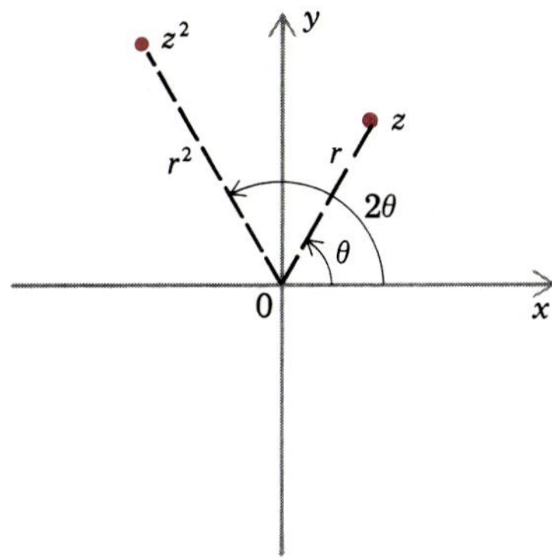

FIGURE 9.10

argument of z is an argument of z^2 (Figure 9.10). Similarly, one can prove that

$$z^3 = r^3(\cos 3\theta + i \sin 3\theta)$$

More generally, we have the following theorem, named after the French mathematician Abraham De Moivre (1667–1754).

THEOREM 9.2
De Moivre's Theorem

$[r(\cos \theta + i \sin \theta)]^n = r^n(\cos n\theta + i \sin n\theta)$ *for any integer* $n > 0$

Thus the modulus of the nth power of z is the nth power of the modulus of z, and n times the given argument of z is an argument of z^n.

EXAMPLE 1. Write $(1 + \sqrt{3}i)^7$ in the form $a + bi$.

Solution. We could multiply out $(1 + \sqrt{3}i)^7$, but that would be tedious. Alternatively, we could write $1 + \sqrt{3}i$ in polar form and use De Moivre's Theorem. We will do the latter. By Example 4(a) of Section 9.2,

$$1 + \sqrt{3}i = 2\left(\cos \frac{\pi}{3} + i \sin \frac{\pi}{3}\right)$$

so that by De Moivre's Theorem,

$$\begin{aligned}(1 + \sqrt{3}i)^7 &= \left[2\left(\cos \frac{\pi}{3} + i \sin \frac{\pi}{3}\right)\right]^7 = 2^7\left(\cos \frac{7\pi}{3} + i \sin \frac{7\pi}{3}\right)\\ &= 128\left(\cos \frac{\pi}{3} + i \sin \frac{\pi}{3}\right) = 128\left(\frac{1}{2} + \frac{\sqrt{3}}{2}i\right)\\ &= 64 + 64\sqrt{3}i \quad \square\end{aligned}$$

*n*th Roots

Let z be a complex number and n a positive integer. An nth ***root*** of z is a complex number w such that

$$w^n = z \tag{1}$$

If $z = 0$, then the only nth root of z is 0. If $z \neq 0$, we can use De Moivre's Theorem to find the polar form of any nth root of z. We set out to do this now.

Suppose $z \neq 0$ and $z = r(\cos\theta + i\sin\theta)$ in polar form. Assume further that z has an nth root w, which is written in polar form as

$$w = s(\cos\phi + i\sin\phi) \tag{2}$$

Then by De Moivre's Theorem,

$$w^n = s^n(\cos n\phi + i\sin n\phi)$$

so that by (1),

$$\overbrace{s^n(\cos n\phi + i\sin n\phi)}^{w^n} = \overbrace{r(\cos\theta + i\sin\theta)}^{z} \tag{3}$$

The modulus of the left side of (3) is s^n and the modulus of the right side is r, so that $s^n = r$. Since r and s are positive real numbers, it follows that

$$s = r^{1/n} \tag{4}$$

Moreover, because $s^n = r$, we can divide the left side of (3) by s^n and the right side by r to obtain

$$\cos n\phi + i\sin n\phi = \cos\theta + i\sin\theta$$

Equating the real and imaginary parts of the two sides of this equation gives us

$$\cos n\phi = \cos\theta \quad \text{and} \quad \sin n\phi = \sin\theta$$

Since the sine and cosine both have period 2π, it follows that

$$n\phi = \theta + 2\pi k$$

or equivalently,

$$\phi = \frac{\theta + 2\pi k}{n} \quad \text{for some integer } k \tag{5}$$

Using (4) and (5) to substitute for s and ϕ in (2), we conclude that if w is an nth root of z, then the polar form of w can be written as

$$w = r^{1/n}\left(\cos\frac{\theta + 2\pi k}{n} + i\sin\frac{\theta + 2\pi k}{n}\right) \quad \text{for some integer } k \tag{6}$$

Recall that we obtained (6) under the assumption that w was an nth root of z. Now we will show that the number w in (6) is indeed an nth root of z, for any integer k. Indeed, by (6) and De Moivre's Theorem,

$$\begin{aligned} w^n &= \left[r^{1/n}\left(\cos\frac{\theta + 2\pi k}{n} + i\sin\frac{\theta + 2\pi k}{n}\right)\right]^n \\ &= (r^{1/n})^n\left[\cos n\left(\frac{\theta + 2\pi k}{n}\right) + i\sin n\left(\frac{\theta + 2\pi k}{n}\right)\right] \\ &= r[\cos(\theta + 2\pi k) + i\sin(\theta + 2\pi k)] \\ &= r(\cos\theta + i\sin\theta) = z \end{aligned}$$

Thus w is an nth root of z.

It might appear that there would be infinitely many nth roots of z, since there are infinitely many choices for k. However, the fact that the sine and cosine both have period 2π implies that if k_1 and k_2 are two integers that differ by a multiple of n, so that $(k_1 - k_2)/n$ is an integer, then

$$\frac{\theta + 2\pi k_1}{n} - \frac{\theta + 2\pi k_2}{n} = 2\pi\frac{(k_1 - k_2)}{n} = 2\pi\ (\text{integer})$$

and hence the numbers on the right side of (6) corresponding to k_1 and k_2 are the same. Thus we obtain all nth roots of z by restricting k in (6) to be one of the integers $0, 1, 2, \ldots, n - 1$, since any integer differs from one of these by a multiple of n. Moreover, the values of w given by (6) that correspond to $k = 0, 1, 2, \ldots, n - 1$ are all distinct because the values of $(\theta + 2\pi k)/n$ corresponding to any two such values of k differ by less than 2π.

The net result is that there are exactly n distinct nth roots of any nonzero complex number z:

> If $z = r(\cos\theta + i\sin\theta)$, then $w^n = z$ if and only if
>
> $$w = r^{1/n}\left(\cos\frac{\theta + 2\pi k}{n} + i\sin\frac{\theta + 2\pi k}{n}\right) \quad \text{for } k = 0, 1, 2, \ldots, n - 1 \tag{7}$$

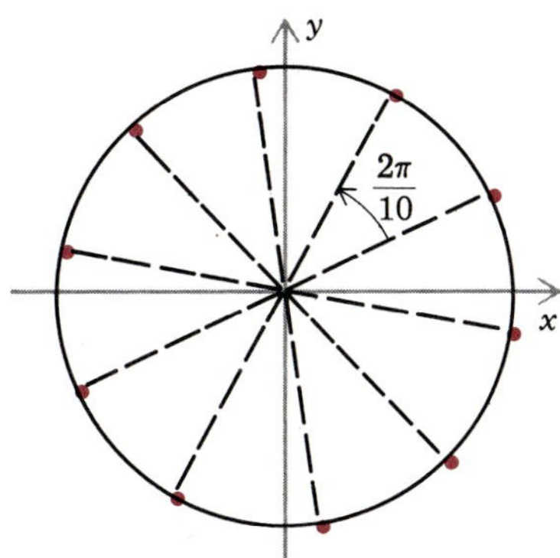

Uniform spacing of the tenth roots of a complex number about a circle

FIGURE 9.11

Observe that all nth roots of z have the same modulus $r^{1/n}$, but the arguments of the successive nth roots corresponding to $k = 0, 1, 2, \ldots, n - 1$ differ by $2\pi/n$. This means that the nth roots of z all lie on the circle with center 0 and radius $r^{1/n}$. Moreover, the nth roots are uniformly spaced on the circle (Figure 9.11). Thus in order to draw all n of the nth roots we need only locate one of them, say the one whose argument is θ/n, and then place the remaining $n-1$ of the nth roots around the circle every $2\pi/n$ radians.

EXAMPLE 2. Find the three third roots of -1, and sketch them on the unit circle.

Solution. By Example 4(c) of Section 9.2, the polar form of -1 is given by

$$-1 = 1(\cos \pi + i \sin \pi)$$

It follows from (7) with $n = 3$, $r = 1$, and $\theta = \pi$ that the third roots of -1 are given by

$$\cos\left(\frac{\pi + 2\pi k}{3}\right) + i \sin\left(\frac{\pi + 2\pi k}{3}\right) \quad \text{for } k = 0, 1, 2$$

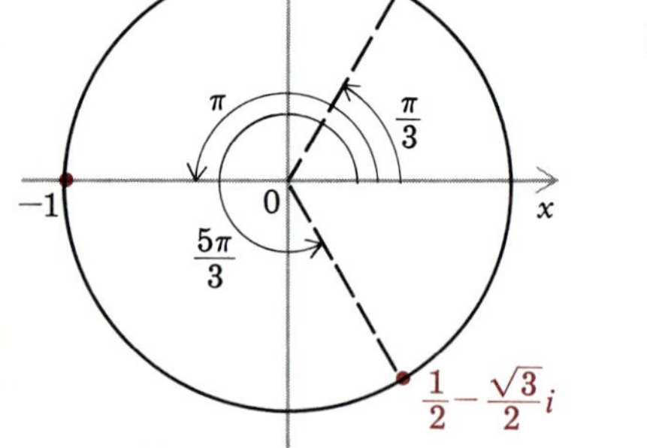

The third roots of -1

FIGURE 9.12

Individually, the three third roots are

$$\cos\frac{\pi}{3} + i \sin\frac{\pi}{3} = \frac{1}{2} + \frac{\sqrt{3}}{2}i \quad (\text{for } k = 0)$$

$$\cos\frac{3\pi}{3} + i \sin\frac{3\pi}{3} = -1 \quad (\text{for } k = 1)$$

$$\cos\frac{5\pi}{3} + i \sin\frac{5\pi}{3} = \frac{1}{2} - \frac{\sqrt{3}}{2}i \quad (\text{for } k = 2)$$

These roots are sketched in Figure 9.12. □

EXAMPLE 3. Find the two second (or square) roots of $1 - i$, and sketch them on the appropriate circle.

Solution. Let $z = 1 - i$. First we will find the polar form of z. For r and θ we have

$$r = |1 - i| = \sqrt{(1)^2 + (-1)^2} = \sqrt{2}$$

and

$$\tan \theta = \frac{-1}{1} = -1$$

Since $1 - i$ lies in the fourth quadrant and $\tan \theta = -1$, we take $\theta = 7\pi/4$.

Thus

$$1 - i = \sqrt{2}\left(\cos\frac{7\pi}{4} + i\sin\frac{7\pi}{4}\right)$$

Now (7) implies that the two square roots of $1 - i$ are given by

$$(\sqrt{2})^{1/2}\left[\cos\frac{1}{2}\left(\frac{7\pi}{4} + 2\pi k\right) + i\sin\frac{1}{2}\left(\frac{7\pi}{4} + 2\pi k\right)\right] \quad \text{for } k = 0, 1$$

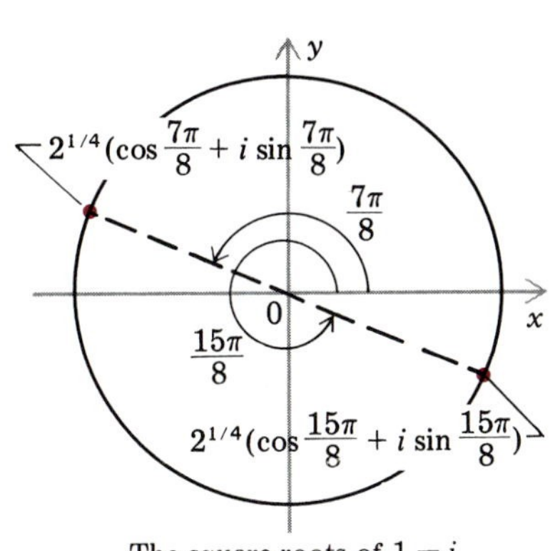

The square roots of $1 - i$

FIGURE 9.13

Individually, the square roots are

$$2^{1/4}\left(\cos\frac{7\pi}{8} + i\sin\frac{7\pi}{8}\right) \quad (\text{for } k = 0)$$

and

$$2^{1/4}\left(\cos\frac{15\pi}{8} + i\sin\frac{15\pi}{8}\right) \quad (\text{for } k = 1)$$

The square roots lie on the circle having center at the origin and radius $2^{1/4}$. They are shown in Figure 9.13. □

The various roots of a complex number can serve as the solutions of an equation. In particular, if c is any complex number and n a positive integer, then the solutions of the equation

$$z^n - c = 0, \quad \text{or equivalently,} \quad z^n = c$$

are the nth roots of c.

EXAMPLE 4. Find all solutions of the equation $z^6 + 64 = 0$.

Solution. By the comments above (with $c = -64$), the solutions of $z^6 + 64 = 0$ are the sixth roots of -64. Since

$$-64 = 64\,(\cos\pi + i\sin\pi)$$

formula (7) with $n = 6$ implies that the six sixth roots are given by

$$(64)^{1/6}\left[\cos\left(\frac{\pi + 2\pi k}{6}\right) + i\sin\left(\frac{\pi + 2\pi k}{6}\right)\right], \quad \text{for } k = 0, 1, 2, 3, 4, 5$$

or more simply,

$$2\left[\cos\left(\frac{\pi}{6}+\frac{\pi k}{3}\right)+i\sin\left(\frac{\pi}{6}+\frac{\pi k}{3}\right)\right],\quad \text{for } k=0,1,2,3,4,5$$

This means that the six solutions of the given equation are

$$2\left(\cos\frac{\pi}{6}+i\sin\frac{\pi}{6}\right)=2\left(\frac{1}{2}\sqrt{3}+\frac{1}{2}i\right)=\sqrt{3}+i \qquad (\text{for } k=0)$$

$$2\left(\cos\frac{\pi}{2}+i\sin\frac{\pi}{2}\right)=2i \qquad (\text{for } k=1)$$

$$2\left(\cos\frac{5\pi}{6}+i\sin\frac{5\pi}{6}\right)=2\left(-\frac{1}{2}\sqrt{3}+\frac{1}{2}i\right)=-\sqrt{3}+i \qquad (\text{for } k=2)$$

$$2\left(\cos\frac{7\pi}{6}+i\sin\frac{7\pi}{6}\right)=2\left(-\frac{1}{2}\sqrt{3}-\frac{1}{2}i\right)=-\sqrt{3}-i \qquad (\text{for } k=3)$$

$$2\left(\cos\frac{3\pi}{2}+i\sin\frac{3\pi}{2}\right)=-2i \qquad (\text{for } k=4)$$

$$2\left(\cos\frac{11\pi}{6}+i\sin\frac{11\pi}{6}\right)=2\left(\frac{1}{2}\sqrt{3}-\frac{1}{2}i\right)=\sqrt{3}-i \qquad (\text{for } k=5) \quad \square$$

Roots of Unity

The solutions of the equation $z^n = 1$, that is, the nth roots of 1, are of special importance in advanced mathematics. These are called the ***nth roots of unity***. They comprise n points in the complex plane and lie equally spaced on the unit circle centered at the origin. They include 1 itself (Figure 9.14).

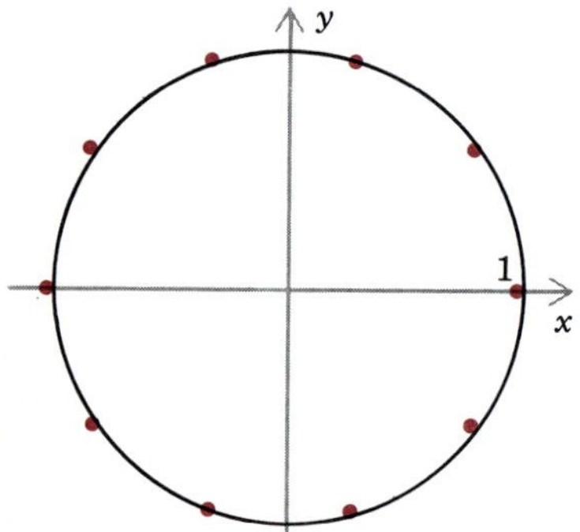

The tenth roots of unity

FIGURE 9.14

EXAMPLE 5. Find the four fourth roots of unity, and sketch them on the unit circle.

Solution. Since

$$1 = 1(\cos 0 + i\sin 0)$$

we conclude from (7) that the four fourth roots of unity are given by

$$1^{1/4}\left[\cos\left(\frac{0+2\pi k}{4}\right)+i\sin\left(\frac{0+2\pi k}{4}\right)\right] \quad \text{for } k=0,1,2,3$$

or more simply by

$$\cos\frac{\pi k}{2}+i\sin\frac{\pi k}{2} \quad \text{for } k=0,1,2,3$$

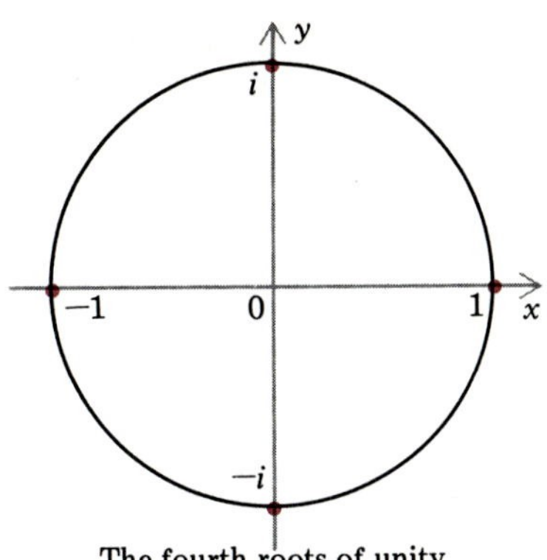

The fourth roots of unity

FIGURE 9.15

Therefore the roots are

$$1 \qquad (\text{for } k = 0)$$

$$\cos\frac{\pi}{2} + i\sin\frac{\pi}{2} = i \qquad (\text{for } k = 1)$$

$$\cos\pi + i\sin\pi = -1 \qquad (\text{for } k = 2)$$

$$\cos\frac{3\pi}{2} + i\sin\frac{3\pi}{2} = -i \qquad (\text{for } k = 3)$$

They are portrayed in Figure 9.15. □

EXERCISES 9.3

In Exercises 1–10, calculate the given power. Express your answers in the form $a + bi$.

1. $[2(\cos \pi/12 + i \sin \pi/12)]^6$

2. $[-(\cos \pi/5 + i \sin \pi/5)]^5$

3. $[3(\cos 2\pi/9 + i \sin 2\pi/9)]^3$

4. $[\cos \pi/8 - i \sin \pi/8]^6$

5. $(1 - i)^8$ **6.** $(-2 - 2i)^5$

7. $(1 + \sqrt{3}i)^{-5}$ **8.** $(\sqrt{3} - i)^7$

9. $(-\sqrt{5} + \sqrt{5}i)^6$ **10.** $(\sqrt{2} + \sqrt{6}i)^4$

In Exercises 11–18, determine the nth roots of the given number. (Leave the results in polar form.)

11. $8(\cos \pi/12 + i \sin \pi/12)$; $n = 3$

12. $2(\cos \pi/5 + i \sin \pi/5)$; $n = 4$

13. $10[\cos(-\pi/8) + i \sin(-\pi/8)]$; $n = 5$

14. i; $n = 4$ **15.** $-i$; $n = 6$

16. $1 + i$; $n = 2$ **17.** $1 - \sqrt{3}i$; $n = 3$

18. $-4 - 4i$; $n = 8$

In Exercises 19–22, determine the nth roots of the given number, and express each root in the form $a + bi$.

19. $9(\cos \pi/3 + i \sin \pi/3)$; $n = 2$

20. i; $n = 2$ **21.** 1; $n = 3$ **22.** -1; $n = 4$

In Exercises 23–32, determine the nth roots of the given number, and sketch them on the appropriate circle.

23. $8(\cos 3\pi/4 + i \sin 3\pi/4)$; $n = 3$

24. $256(\cos \pi + i \sin \pi)$; $n = 4$

25. $32[\cos(-5\pi/6) + i \sin(-5\pi/6)]$; $n = 5$

26. -1; $n = 2$ **27.** $-i$; $n = 2$ **28.** i; $n = 3$

29. $-i$; $n = 4$ **30.** 1; $n = 5$ **31.** 1; $n = 8$

32. $8 - 8i$; $n = 3$

In Exercises 33–38, solve the given equation. Express your solutions in the form $a + bi$.

33. $z^3 = 8$ **34.** $z^4 = -16$

35. $z^6 = 8$ **36.** $z^3 = 8i$

37. $z^4 + \frac{1}{2} - \frac{\sqrt{3}}{2}i = 0$

38. $z^4 + \frac{81}{2} + \frac{81}{2}\sqrt{3}i = 0$

39. a. Compute $(\cos\theta + i \sin\theta)^2$ by multiplying out, and by De Moivre's Theorem.

b. Use the result of part (a) to derive formulas for $\sin 2\theta$ and $\cos 2\theta$.

40. a. Compute $(\cos\theta + i\sin\theta)^3$ by multiplying out, and by De Moivre's Theorem.
b. Use the result of part (a) to derive formulas for $\sin 3\theta$ and $\cos 3\theta$.

41. Suppose one fourth root of a given complex number is $-1 + i$. Find the other three fourth roots.

42. Suppose one sixth root of a given complex number is $\sqrt{3} + i$. Find the other five sixth roots.

43. Suppose z_1 and z_2 have modulus 1 and have the same squares. What is the relationship between the arguments of z_1 and z_2? (*Hint:* Express z_1 and z_2 in polar form.)

KEY TERMS

conjugate
complex plane
absolute value (modulus) of a complex number
argument
polar form of a complex number
nth root
nth root of unity

KEY FORMULAS

$$\overline{a + bi} = a - bi$$
$$|a + bi| = \sqrt{a^2 + b^2}$$
$$[r(\cos\theta + i\sin\theta)]^n = r^n[\cos(n\theta) + i\sin(n\theta)]$$

KEY THEOREM

De Moivre's Theorem

REVIEW EXERCISES

In Exercises 1–12, perform the indicated operation, and express your answer in the form $a + bi$.

1. $\overline{-3 + 4i}$

2. $\overline{-0.4 - 0.01i}$

3. $\overline{(1 - i)^5}$

4. $\dfrac{1}{-5i - 1}$

5. $\dfrac{-i}{2i - 3}$

6. $\dfrac{4 - 2i}{1 + 5i}$

7. $(\sqrt{2} - \sqrt{3}i)\dfrac{1}{\overline{\sqrt{2} + \sqrt{3}i}}$

8. $\overline{\left(\dfrac{-2 - i}{7 + 3i}\right)}$

9. $|8 - 6i|$

10. $\left|\dfrac{i}{3i - 1}\right|$

11. $\left|\dfrac{(-1 - 2i)(3 + i)}{2i(3 - i)}\right|$

12. Evaluate the given square root.

a. $\sqrt{-36}$ b. $\sqrt{-(4-2i)(4+2i)}$ c. $\sqrt{i^{-14}}$

In Exercises 13–16, find all solutions (if any) of the given equation.

13. $\frac{1}{2} - \frac{1}{x} + \frac{3}{x^2} = 0$

14. $x + i = \frac{1}{x - i}$

15. $z^3 = -i$

16. $z^6 = 27$

In Exercises 17–22, find the principal argument and the corresponding polar form of the given number.

17. i

18. $5 - 5i$

19. $-2\sqrt{3} + 2i$

20. $(6 - 6i)(-1 - \sqrt{3}i)$

21. $\frac{(\sqrt{2} + \sqrt{6}i)(\sqrt{3} - i)}{-2 - 2i}$

22. $\frac{\cos \pi/5 + i \sin \pi/5}{\cos 2\pi/7 - i \sin 2\pi/7}$

In Exercises 23–26, sketch the graph of the given equation.

23. $|z| = \frac{2}{3}$

24. $|z - 2| = \frac{1}{2}$

25. $|z - 1 - i| = 3$

26. $|z + 4 + 6i| = \frac{5}{2}$

In Exercises 27–32, determine the nth roots of the given number. Leave your answers in polar form.

27. $4(\cos \pi/3 + i \sin \pi/3)$; $n = 4$

28. i; $n = 6$

29. $1 - i$; $n = 3$

30. $-\sqrt{3} - 3i$; $n = 8$

31. 1; $n = 9$

32. -1; $n = 5$

33. Describe geometrically the complex numbers $z = x + iy$ that satisfy the equation $\bar{z} = -iz$.

10 ROOTS OF POLYNOMIALS

In Section 1.4 we introduced the imaginary number i, which is a solution of the equation

$$x^2 + 1 = 0$$

More generally, we saw that any quadratic equation

$$ax^2 + bx + c = 0$$

has (possibly complex) solutions if a, b, and c are real numbers with $a \neq 0$. Now we turn to general polynomial equations, that is, equations of the form

$$a_n x^n + a_{n-1} x^{n-1} + \cdots + a_1 x + a_0 = 0 \tag{1}$$

where the coefficients $a_n, a_{n-1}, \ldots, a_1, a_0$ are (real or complex) constants. Any solution of (1) is called a ***root*** (or ***zero***) of the associated polynomial $a_n x^n + a_{n-1} x^{n-1} + \cdots + a_1 x + a_0$. For example, 2 is a root of the polynomial $3x^2 - 2x - 8$ because $3(2)^2 - 2(2) - 8 = 0$. The results appearing in this chapter can be given either in terms of solutions of polynomial equations or in terms of roots of polynomials. Because it is simpler to give the results in terms of roots of polynomials, the main subject of this chapter is roots of polynomials.

A high point of the chapter is the theorem that states that every nonconstant polynomial has a (possibly complex) root, even if we allow the coefficients to be complex numbers. This result, known as the Fundamental Theorem of Algebra, is one of the deepest and most remarkable theorems in mathematics and was first proved by the brilliant German mathematician, Carl Friedrich Gauss, at the age of 20. Although the theorem does not tell us what the roots are, we will devise methods for actually finding roots in case the coefficients of the polynomial are integers.

10.1 DIVISION OF POLYNOMIALS

In Section 3.4 we studied rational functions, which are quotients of polynomials. Now we will see what the result is when we actually "divide" one polynomial by another.

Recall that a polynomial has the form

$$a_n x^n + a_{n-1} x^{n-1} + \cdots + a_1 x + a_0$$

where $a_n, a_{n-1}, \ldots, a_1$, and a_0 are constants that henceforth are allowed to be either real or complex numbers. As before, if $a_n \neq 0$, then n is the degree of the polynomial.

The Division Algorithm

Let $f(x)$ and $g(x)$ be polynomials. We say that $g(x)$ ***divides*** $f(x)$ (or that $g(x)$ is a ***factor*** of $f(x)$, or that $f(x)$ is ***divisible*** by $g(x)$) if there is a polynomial $q(x)$ such that

$$f(x) = g(x)q(x)$$

For example, since

$$x^2 - 5x + 4 = (x - 1)(x - 4) \tag{2}$$

it follows that $x^2 - 5x + 4$ is divisible both by $x - 1$ and by $x - 4$. Equivalently, we can say that $x - 1$ and $x - 4$ are factors of $x^2 - 5x + 4$.

Equation (2) is equivalent to the equation

$$\frac{x^2 - 5x + 4}{x - 1} = x - 4$$

and the division of $x^2 - 5x + 4$ by $x - 1$ can be carried out much as one divides integers by long division:

$$\begin{array}{r|l} & x - 4 \\ \hline x - 1 & x^2 - 5x + 4 \\ & \underline{x^2 - x} \\ & - 4x + 4 \\ & \underline{- 4x + 4} \\ & 0 \end{array}$$

This method of dividing one polynomial by another applies to other polynomials as well.

Now assume that $f(x)$ and $g(x)$ are polynomials such that $g(x) \neq 0$, and assume that $f(x)$ is not necessarily divisible by $g(x)$. We can perform a long division patterned after our division above. We illustrate by dividing $2x^4 - 5x^3 + 5x^2 - 3x + 4$ by $x^2 - x$:

$$
\begin{array}{r|l}
 & 2x^2 - 3x + 2 \\
\hline
x^2 - x & 2x^4 - 5x^3 + 5x^2 - 3x + 4 \\
 & \underline{2x^4 - 2x^3} \\
 & \quad - 3x^3 + 5x^2 \\
 & \quad \underline{- 3x^3 + 3x^2} \\
 & \qquad\qquad 2x^2 - 3x \\
 & \qquad\qquad \underline{2x^2 - 2x} \\
 & \qquad\qquad\qquad - x + 4
\end{array}
$$

The result of this division can be expressed by the equation

$$2x^4 - 5x^3 + 5x^2 - 3x + 4 = (x^2 - x)(2x^2 - 3x + 2) + (-x + 4) \qquad (3)$$

The polynomial $2x^2 - 3x + 2$ is called the quotient of $2x^4 - 5x^3 + 5x^2 - 3x + 4$ by the divisor $x^2 - x$, and $-x + 4$ is called the remainder. Observe that the degree of the remainder $-x + 4$ is less than the degree of the divisor $x^2 - x$. In general, when we use long division to divide a polynomial $f(x)$ by a nonzero polynomial $g(x)$, we continue the procedure until the remainder is either 0 or has degree less than the degree of the divisor $g(x)$. It is possible to prove that such a division is always possible. We now state the result formally, but without proof.

THEOREM 10.1
The Division Algorithm

Let $f(x)$ and $g(x)$ be polynomials with $g(x) \neq 0$. Then there are unique polynomials $q(x)$ and $r(x)$ with either $r(x) = 0$ or $\deg r(x) < \deg g(x)$ such that

$$f(x) = g(x)q(x) + r(x)$$

The polynomial $q(x)$ is called the ***quotient***, and the polynomial $r(x)$ is called the ***remainder***. When we find $q(x)$ and $r(x)$, we say that we ***divide $f(x)$ by $g(x)$***. If $f(x)$ and $g(x)$ have real coefficients, then so do $q(x)$ and $r(x)$.

EXAMPLE 1. Find the quotient $q(x)$ and remainder $r(x)$ when $2x^4 - 5x^3 + 5x^2 - 3x + 4$ is divided by $x^2 - x$.

Solution. Let

$$f(x) = 2x^4 - 5x^3 + 5x^2 - 3x + 4 \quad \text{and} \quad g(x) = x^2 - x$$

From (3) we see that

$$q(x) = 2x^2 - 3x + 2 \quad \text{and} \quad r(x) = -x + 4 \quad \square$$

It can happen that either $q(x)$ or $r(x)$ is 0. For example, if $f(x) = 2x$ and $g(x) = x^2 - 1$, then

$$\overbrace{2x}^{f(x)} = \overbrace{(x^2 - 1)}^{g(x)}\ \overbrace{0}^{q(x)} + \overbrace{2x}^{r(x)}$$

so in this case, $q(x) = 0$. In general, if the degree of $f(x)$ is less than the degree of $g(x)$, then dividing $f(x)$ by $g(x)$ yields the quotient $q(x) = 0$. To show that $r(x)$ can be 0, let $f(x) = x^3 - 1$ and $g(x) = x - 1$. Then

$$\overbrace{x^3 - 1}^{f(x)} = \overbrace{(x - 1)}^{g(x)}\overbrace{(x^2 + x + 1)}^{q(x)} + \overbrace{0}^{r(x)}$$

It follows that $r(x) = 0$. The remainder is 0 because $g(x)$ is a factor of $f(x)$.

Synthetic Division

As we will see later, one frequently needs to divide a polynomial by a second polynomial of the special form $x - c$, where c is a constant. A method called synthetic division normally reduces the writing involved in such a long division. Consider dividing $2x^4 - 10x^3 + 38x - 10$ by $x - 3$, with the long division performed as above:

$$\begin{array}{r|rrrrr}
 & 2x^3 & -\ 4x^2 & -\ 12x & +\ 2 & \\
\hline
x - 3 & 2x^4 & -\ 10x^3 & & +\ 38x & -\ 10 \\
 & 2x^4 & -\ 6x^3 & & & \\
\cline{2-3}
 & & -4x^3 & & & \\
 & & -4x^3 & +\ 12x^2 & & \\
\cline{3-4}
 & & & -12x^2 & +\ 38x & \\
 & & & -12x^2 & +\ 36x & \\
\cline{4-5}
 & & & & 2x & -\ 10 \\
 & & & & 2x & -\ 6 \\
\cline{5-6}
 & & & & & -4
\end{array}$$

The result is

$$2x^4 - 10x^3 + 38x - 10 = (x - 3)\overbrace{(2x^3 - 4x^2 - 12x + 2)}^{\text{quotient}}\ \overbrace{-\ 4}^{\text{remainder}}$$

Observe first of all that the numbers in color in each vertical column are duplicates of one another. Notice next that from the dividend downward, like powers of the variable x appear in vertical columns. Because of these two observations, we can condense the division as follows by eliminating the repetitions, collapsing the lower lines all onto a second line, and putting the quotient at the bottom along with the remainder:

$$\begin{array}{r|rrrrr} -3 & 2 & -10 & 0 & 38 & -10 \\ & & -6 & 12 & 36 & -6 \\ \hline & 2 & -4 & -12 & 2 & \underline{|-4} \end{array}$$

In the display we have separated the divisor number, -3, from the dividend numbers by the symbol ⅃, and the remainder -4 from the quotient numbers by the symbol ∟. Moreover, we placed a 0 in the first row because the polynomial $2x^4 - 10x^3 + 38x - 10$, with no x^2 term, is equal to the polynomial $2x^4 - 10x^3 + 0x^2 + 38x - 10$. Notice also that the coefficients in the bottom row of the display correspond to powers of x that are one less than the powers to which the coefficients in the top row of the display correspond.

There is one further simplification we can make. Notice that each number (except the first) in the third row of the condensed display is obtained from the numbers directly above it by *subtracting* the number in the second row from the number in the first row. We can achieve the same third row by replacing each number in the second row by its negative and *adding* the resulting number to the number directly above it in the first row. For later convenience we also replace -3 by its negative, the number 3. (More generally, if the divisor is $x - c$, we would put c at the left of the display.) When we make these replacements, our condensed display becomes

$$\begin{array}{r|rrrrr} 3 & 2 & -10 & 0 & 38 & -10 \\ & & 6 & -12 & -36 & 6 \\ \hline & 2 & -4 & -12 & 2 & \underline{|-4} \end{array}$$

This final display is an example of ***synthetic division***, wherein powers of the variable have disappeared and the entire division appears in three rows.

Following Example 2, the method of synthetic division for dividing a polynomial $f(x)$ by $x - c$ is outlined and illustrated with the division of $2x^4 - 2x^3 - 7x^2 - 57$ by $x - 3$. The solutions of Examples 2–4 are based on the outline.

EXAMPLE 2. Use synthetic division to find the quotient and remainder when $3x^4 - 8x^3 + x^2 - 5x + 9$ is divided by $x - 2$.

Solution. Following the procedure outlined, we have:

$$\begin{array}{r|rrrrr} 2 & 3 & -8 & 1 & -5 & 9 \\ & & 6 & -4 & -6 & -22 \\ \hline & 3 & -2 & -3 & -11 & \underline{-13|} \end{array}$$

Therefore the quotient is $3x^3 - 2x^2 - 3x - 11$, and the remainder is -13. □

Method of Synthetic Division

1. Write the coefficients of $f(x)$ in a horizontal line. Include a 0 for each missing power of x.

$$\begin{array}{rrrrr} 2 & -2 & -7 & 0 & -57 \end{array}$$

2. Write the number c to the left of the coefficients of $f(x)$. Separate c from the coefficients by the symbol $\lrcorner$, and draw a horizontal line as indicated, well below the coefficients.

$$\begin{array}{r|rrrrr} 3 & 2 & -2 & -7 & 0 & -57 \\ & & & & & \\ \hline \end{array}$$

3. Bring down the first coefficient of $f(x)$ as indicated.

$$\begin{array}{r|rrrrr} 3 & 2 & -2 & -7 & 0 & -57 \\ & \downarrow & & & & \\ \hline & 2 & & & & \end{array}$$

4. Multiply the number below the line by c, and place the result under the second coefficient of $f(x)$.

$$\begin{array}{r|rrrrr} 3 & 2 & -2 & -7 & 0 & -57 \\ & & 6 & & & \\ \hline & 2 & & & & \end{array}$$

5. Add the numbers in the second column and place the result under the line.

$$\begin{array}{r|rrrrr} 3 & 2 & -2 & -7 & 0 & -57 \\ & & 6 & & & \\ \hline & 2 & 4 & & & \end{array}$$

6. Multiply the number below the line in the second column by c and place the result below the third coefficient of $f(x)$.

$$\begin{array}{r|rrrrr} 3 & 2 & -2 & -7 & 0 & -57 \\ & & 6 & 12 & & \\ \hline & 2 & 4 & & & \end{array}$$

7. Add the numbers in the third column and place the result below the line.

$$\begin{array}{r|rrrrr} 3 & 2 & -2 & -7 & 0 & -57 \\ & & 6 & 12 & & \\ \hline & 2 & 4 & 5 & & \end{array}$$

8. Continue in this fashion until the last column has been completed. Place the symbol $\llcorner$ around the last number of the bottom row to isolate the remainder.

$$\begin{array}{r|rrrrr} 3 & 2 & -2 & -7 & 0 & -57 \\ & & 6 & 12 & 15 & 45 \\ \hline & 2 & 4 & 5 & 15 & \llcorner -12 \end{array}$$

9. Read off the result. The last number on the right below the line is the remainder. The other numbers are the coefficients of the quotient. The degree of the quotient is one less than the degree of $f(x)$.

Quotient: $2x^3+4x^2+5x+15$

Remainder: -12

Result of Synthetic Division:

$$2x^4 - 2x^3 - 7x^2 - 57 = (x-3)(2x^3+4x^2+5x+15) - 12$$

EXAMPLE 3. Use synthetic division to show that $-x^3 - x^2 + 7x - 20$ is divisible by $x + 4$.

Solution. In performing the synthetic division we use the fact that $x + 4 = x - (-4)$. Then we have

$$\begin{array}{r|rrrr} -4 & -1 & -1 & 7 & -20 \\ & & 4 & -12 & 20 \\ \hline & -1 & 3 & -5 & \boxed{0} \end{array}$$

From the division we see that

$$\begin{aligned} -x^3 - x^2 + 7x - 20 &= (x + 4)\overbrace{(-x^2 + 3x - 5)}^{\text{quotient}} + \overbrace{0}^{\text{remainder}} \\ &= (x + 4)(-x^2 + 3x - 5) \end{aligned}$$

so that $-x^3 - x^2 + 7x - 20$ is divisible by $x + 4$. □

In our final example, we use synthetic division with polynomials whose coefficients include complex as well as real numbers.

EXAMPLE 4. Divide $ix^3 + 3x^2 + (-4 + i)x + 1 + 9i$ by $x - i$. Use synthetic division.

Solution. Following the same procedure as above, we have

$$\begin{array}{r|rrrr} i & i & 3 & -4 + i & 1 + 9i \\ & & -1 & 2i & -3 - 4i \\ \hline & i & 2 & -4 + 3i & \boxed{-2 + 5i} \end{array}$$

The quotient is $ix^2 + 2x - 4 + 3i$, and the remainder is $-2 + 5i$. □

CAUTION: Long division can be used with a divisor of any degree. However, synthetic division can only be used when the divisor is a polynomial of the form $x - c$.

EXERCISES 10.1

In Exercises 1–6, perform the indicated long division.

1. $x^2 + 1\overline{)2x^4 + 3x^3 - 4x + 7}$
2. $3x - 2\overline{)-9x^3 + 12x^2 + 2x + 12}$
3. $x^2 + 1\overline{)x^4 + 1}$
4. $x^2 + \sqrt{2}x + 1\overline{)x^4 + 1}$
5. $x^3 + x + 1\overline{)3x^4 - 2x^3 + 3x^2 + x - 1}$
6. $x^2 + 3\overline{)x^3 + 4x^2 - 7}$

In Exercises 7–18, use synthetic division to find the quotient and the remainder when the first polynomial is divided by the second.

7. $3x^2 - 2x + 7; x + 1$ **8.** $-x^3 + 3x^2 - 2x; x - 2$

9. $2x^3 + 3x^2 + 5; x - \frac{1}{2}$

10. $x^3 + 3x^2 - 9x + 5; x + 5$

11. $6x^3 - x^2 + 2x + 1; x + \frac{1}{3}$

12. $2x^4 - 3x^2 + 2x - 5; x - 3$

13. $5x^5 + 16x^4 - 15x^3 + x^2 + 19x; x + 4$

14. $x^7 - x^3 + x; x - 1$

15. $x^7 - 1; x - 1$

16. $x^{10} - 1; x - 1$

17. $x^5 - ix + 1; x - i$

18. $x^4 + (1 - 3i)x^3 + (-5 - i)x^2 + (-6 + 7i)x - 1 + 8i; x + 1 - i$

In Exercises 19–30, use synthetic division to show that the first polynomial is a factor of the second.

19. $x - 3; 2x^4 - 6x^3 - 3x^2 + 13x - 12$

20. $x + 2; x^6 + 6x^3 + 7x - 2$

21. $x - 1; 3x^5 - 4x^3 - 6x + 7$

22. $x + 1; 4x^4 + 3x^3 - 2x^2 + 2x + 3$

23. $x - 5; x^3 - 8x^2 + 17x - 10$

24. $x - \frac{1}{2}; 8x^3 - 4x^2 - 2x + 1$

25. $x + 6; 2x^3 + 9x^2 - 17x + 6$

26. $x + 2; x^5 + 2x^4 - x^3 - 2x^2 + 2x + 4$

27. $x + i; x^3 + 2x^2 + x + 2$

28. $x - i; x^4 + 3x^3 + 3x^2 + 3x + 2$

29. $x - i; x^5 + x^3$

30. $x - \frac{\sqrt{2}}{2} - \frac{\sqrt{2}}{2}i; x^2 - \sqrt{2}x + 1$

31. Use synthetic division to find the values of k for which $x - 2$ is a factor of $x^3 + k^2x^2 - 3kx + k^2 - 7$.

32. Use synthetic division to find the values of k for which $x + 2$ is a factor of $x^4 + k^2x + k - 10$.

10.2 THE REMAINDER AND FACTOR THEOREMS

When a polynomial $f(x)$ is divided by a first-degree polynomial $x - c$, the Division Algorithm yields

$$f(x) = (x - c)q(x) + r(x) \tag{1}$$

where $q(x)$ and $r(x)$ are polynomials and where either $r(x) = 0$ or $\deg r(x) < 1 = \deg(x - c)$. Let us show that the remainder $r(x)$ is the number $f(c)$. The fact that $r(x) = 0$ or $\deg r(x) < 1$ implies that $r(x)$ is a constant polynomial. It follows that for some constant b, $r(x) = b$, so that (1) becomes

$$f(x) = (x - c)q(x) + b \tag{2}$$

Replacing x by c in (2), we obtain

$$f(c) = \overbrace{(c - c)}^{0}q(c) + b = b$$

so that we can substitute $f(c)$ for b in (2) to obtain

$$f(x) = (x - c)q(x) + f(c) \tag{3}$$

This proves the following theorem.

THEOREM 10.2
The Remainder Theorem

If a polynomial $f(x)$ is divided by $x - c$, then the remainder is $f(c)$.

The Remainder Theorem is valid whether c is a real number or a complex number.

EXAMPLE 1. Let $f(x) = 3x^4 - 8x^3 + x^2 - 5x + 9$. Find the remainder when $f(x)$ is divided by $x - 2$.

Solution. Taking $c = 2$ in the Remainder Theorem, we find that the remainder is $f(2)$. But

$$f(2) = 3(2)^4 - 8(2)^3 + (2)^2 - 5(2) + 9 = -13$$

Therefore the remainder is -13. □

Now we have found the remainder when $3x^4 - 8x^3 + x^2 - 5x + 9$ is divided by $x - 2$ in two different ways: by synthetic division (Example 2 in Section 10.1) and by the Remainder Theorem (Example 1 above). In each case the remainder is -13.

The most important consequence of the Remainder Theorem is the Factor Theorem, which we mentioned (and used) in Section 3.3.

THEOREM 10.3
The Factor Theorem

Let $f(x)$ be any polynomial. Then $x - c$ is a factor of $f(x)$ if and only if $f(c) = 0$.

Proof. Suppose that $x - c$ is a factor of $f(x)$. Then

$$f(x) = (x - c)g(x)$$

for some polynomial $g(x)$. It follows that

$$f(c) = (c - c)g(c) = 0 \cdot g(c) = 0$$

Thus if $x - c$ is a factor of $f(x)$, then $f(c) = 0$. Conversely, assume that $f(c) = 0$. By (3),

$$f(x) = (x - c)q(x) + \overbrace{f(c)}^{0} = (x - c)q(x) \tag{4}$$

for some polynomial $q(x)$. Thus $x - c$ is a factor of $f(x)$. ■

EXAMPLE 2. Show that $x - 1$ is a factor of $x^3 + x^2 - x - 1$, and then factor $x^3 + x^2 - x - 1$ into linear factors.

Solution. Let

$$f(x) = x^3 + x^2 - x - 1$$

Since

$$f(1) = 1^3 + 1^2 - 1 - 1 = 0$$

the Factor Theorem implies that $x - 1$ is a factor of $f(x)$. We use synthetic division to divide $f(x)$, that is, $x^3 + x^2 - x - 1$, by $x - 1$.

$$\begin{array}{r|rrrr} 1 & 1 & 1 & -1 & -1 \\ & & 1 & 2 & 1 \\ \hline & 1 & 2 & 1 & \boxed{0} \end{array}$$

Thus

$$\begin{aligned} x^3 + x^2 - x - 1 &= (x - 1)(x^2 + 2x + 1) \\ &= (x - 1)(x + 1)^2 \quad \square \end{aligned}$$

EXAMPLE 3. Show that $x + 2$ is not a factor of $-3x^3 + 2x + 4$.

Solution. Let

$$f(x) = -3x^3 + 2x + 4$$

By the Factor Theorem, $x + 2$, which is the same as $x - (-2)$, is a factor of $f(x)$ if and only if $f(-2) = 0$. But

$$f(-2) = (-3)(-2)^3 + 2(-2) + 4 = 24$$

so that $f(-2) \neq 0$. Therefore $x + 2$ is not a factor of $-3x^3 + 2x + 4$. $\square$

Recall that a number c is a root (or zero) of a polynomial $f(x)$ if $f(c) = 0$. The Factor Theorem implies that c is a root of $f(x)$ if and only if $x - c$ is a factor of $f(x)$.

EXAMPLE 4. Find a polynomial of degree 3 that has roots 1, 2, and -3.

Solution. By the Factor Theorem, any polynomial that has roots 1, 2, and -3 must be divisible by $x - 1$, $x - 2$, and $x - (-3) = x + 3$. One such polynomial is $(x - 1)(x - 2)(x + 3)$. Since

$$(x - 1)(x - 2)(x + 3) = (x^2 - 3x + 2)(x + 3) = x^3 - 7x + 6$$

it follows that the polynomial $x^3 - 7x + 6$ has the required properties. $\square$

There are other polynomials of degree 3 that have roots 1, 2 and -3. But they all have the form $c(x - 1)(x - 2)(x - 3)$ for an appropriate nonzero constant c.

Factors of $x^2 + a^2$

Let a be a nonzero real number, and let

$$f(x) = x^2 + a^2$$

If c is any real number, then

$$f(c) = c^2 + a^2 \neq 0$$

so that by the Factor Theorem, $f(x)$ has no factor of the form $x - c$ with c real. However, $f(x)$ does have the factors $x - ai$ and $x + ai$, since

$$f(ai) = (ai)^2 + a^2 = -a^2 + a^2 = 0$$

and $$f(-ai) = (-ai)^2 + a^2 = -a^2 + a^2 = 0$$

Consequently for any nonzero real number a,

$$\boxed{x^2 + a^2 = (x - ai)(x + ai)}$$

For example,

$$x^2 + 1 = (x - i)(x + i)$$
$$x^2 + 16 = (x - 4i)(x + 4i)$$
$$x^2 + 5 = (x - \sqrt{5}i)(x + \sqrt{5}i)$$

EXERCISES 10.2

In Exercises 1–10, use the Remainder Theorem to find the remainder when the first polynomial is divided by the second.

1. $x^3 - 3x^2 + 2x + 1;\ x - 1$

2. $3x^4 - 6x^2 + 7;\ x + 1$

3. $\frac{1}{2}x^6 - 2x^3 - x + 7;\ x - 2$

4. $-x^5 + 7;\ x + 2$

5. $3x^4 - 4x^2 + 3x - 6;\ x - \sqrt{2}$

6. $-17x^3 + \pi x^2 - \sqrt{2}x + \frac{17}{32};\ x$

7. $x^4 + a^2;\ x + a$

8. $x^4 - 2x^2 + 1 + i;\ x - i$

9. $x^2 + x + 1 - 3i;\ x - 1 + i$

10. $x^{24} + x^{23} - 1;\ x - \frac{\sqrt{2}}{2} - \frac{\sqrt{2}}{2}i$

In Exercises 11–20, verify that the linear polynomial is a factor of the other polynomial, and then factor the other polynomial into linear and constant factors.

11. $x^3 - 6x^2 + 12x - 8;\ x - 2$

12. $x^3 + 2x^2 - 13x + 10;\ x - 1$

13. $x^3 + 3x^2 - 4;\ x + 2$

14. $x^3 + 4x^2 - 7x - 10;\ x + 5$

15. $2x^3 + 3x^2 - 8x + 3;\ x - \frac{1}{2}$

16. $x^3 - 3x^2 + 4x - 12;\ x - 3$

17. $x^4 - 2x^3 + 9x^2 - 18x;\ x - 2$

18. $x^3 + 3ix^2 - 3x - i;\ x + i$

19. $2x^4 - 7ix^3 - 7x^2 + 2ix;\ x - \frac{1}{2}i$

20. $x^3 + (1 + i)x^2 + (2 + i)x + 2;\ x - i$

In Exercises 21–28, show that the given number is a root of the given polynomial.

21. $2x^3 - 4x^2 - 3x + 3;\ -1$

22. $x^4 - 2x^3 + x^2 - 10x - 6; 3$

23. $x^3 + x^2 + x + \frac{3}{8}; -\frac{1}{2}$

24. $x^7 + 3x^4 - x; 0$

25. $\sqrt{2}x^5 - \frac{\sqrt{2}}{2}x^3 - x^2 - 4;\ \sqrt{2}$

26. $2x^3 - 5x^2 + 8x - 20; 2i$

27. $x^4 + 3ix^3 - 2x^2 - 6ix; -3i$

28. $2x^3 + 3x^2 - 2ix - 8i + 2; i + 1$

In Exercises 29–40, find a polynomial having the given roots and no others.

29. $3, -2$ **30.** $-1, 6$

31. $0, 13, -1$ **32.** $-1, 1, 2$

33. $-2, 1, 4$ **34.** $0, \frac{1}{2}, 1, -2$

35. $1, -1, 2, -2$ **36.** $1, 2, 3, 4$

37. $5i, -5i$ **38.** $3 + 4i, 3 - 4i$

39. $\sqrt{2} + 2i, \sqrt{2} - 2i$ **40.** $1 + \sqrt{3}i, 1 - \sqrt{3}i$

In Exercises 41–48, factor the given polynomial into linear and constant factors.

41. $x^2 + 4$ **42.** $x^2 + 25$

43. $x^2 + 7$ **44.** $x^2 + 17$

45. $2x^2 + 2$ **46.** $3x^2 + 5$

47. $x^3 + x$ **48.** $-3x^3 - 21x$

49. For what value of k is $kx^3 + 3x^2 - 2x + 4$ divisible by $x - 2$?

50. For what values of k is $kx^2 + k^2x + 12$ divisible by $x + 3$?

51. Let n be a positive integer and a a real number.
a. Show that $x - a$ is a factor of $x^n - a^n$. (*Hint:* Use the Factor Theorem.)
b. Show that if n is even, then $x + a$ is not a factor of $x^n + a^n$.
c. Show that if n is odd, then $x + a$ is a factor of $x^n + a^n$.

***52.** Suppose w is a root of the polynomial $x^3 - x + 1$. Show that w is a root of the polynomial $-x^6 + 2x^4 - x^2 + 1$. (*Hint:* From the fact that $w^3 = w - 1$, find expressions for w^6 and w^4.)

10.3 THE FUNDAMENTAL THEOREM OF ALGEBRA

The Factor Theorem provides a simple test for determining whether a polynomial $f(x)$ is divisible by $x - c$: Just find out whether $f(c) = 0$, that is, whether c is a root of $f(x)$. However, the Factor Theorem does not indicate which polynomials actually have roots, nor does it give any clue to finding roots when they exist. A deep theorem called the Fundamental Theorem of Algebra solves all mystery concerning which polynomials have roots—it says that *every* nonconstant polynomial with real or complex coefficients has a (possibly complex) root. The theorem was first proved rigorously by Carl Friedrich Gauss at the age of 20. Since all known proofs of the Fundamental Theorem of Algebra involve concepts not covered in this book, we will state the theorem but not prove it.

THEOREM 10.4
Fundamental Theorem of Algebra

Every nonconstant polynomial has at least one complex root.

From the Fundamental Theorem of Algebra we know that even a

complicated polynomial such as

$$x^8 - 4x^5 - \sqrt{2}x^3 - 5x + 1 \tag{1}$$

has a root, although the theorem does not suggest how to find a root. Using the Fundamental Theorem, we can even conclude that in a certain sense the polynomial in (1) has 8 roots, and more generally, any polynomial of positive degree n has n roots.

In order to prove this last result, we will need the fact that for any nonzero polynomial $g(x)$,

$$\text{degree of } (x - c)g(x) = 1 + \text{degree of } g(x) \tag{2}$$

After all, if

$$g(x) = a_m x^m + a_{m-1}x^{m-1} + \cdots + a_1 x + a_0 \quad \text{with} \quad a_m \neq 0$$

then

$$\begin{aligned}(x - c)g(x) &= (x - c)(a_m x^m + a_{m-1}x^{m-1} + \cdots + a_1 x + a_0) \\ &= a_m x^{m+1} + (a_{m-1} - ca_m)x^m + \cdots + (a_0 - ca_1)x + (-ca_0)\end{aligned}$$

so that the degree of $(x - c)g(x)$ is $m + 1$, whereas the degree of $g(x)$ is m.

Now we are ready to show that if $f(x)$ is a nonconstant polynomial of degree n, then $f(x)$ can be written as a product of a constant and n linear factors. By the Fundamental Theorem of Algebra, $f(x)$ has a root c_1, so the Factor Theorem tells us that $x - c_1$ is a factor of $f(x)$, that is, there is a polynomial $g_1(x)$ such that

$$f(x) = (x - c_1)g_1(x) \tag{3}$$

Since $f(x)$ is not the zero polynomial, neither is $g_1(x)$. By (3),

$$\text{degree of } (x - c_1)g_1(x) = \text{degree of } f(x) = n$$

so that by (2),

$$\text{degree of } g_1(x) = n - 1$$

If $n - 1 = 0$, then $g_1(x)$ is a constant polynomial. But if $n - 1 > 0$, we can apply the same procedure to $g_1(x)$ as we did to $f(x)$, and conclude that there is a nonzero polynomial $g_2(x)$ such that

$$g_1(x) = (x - c_2)g_2(x) \tag{4}$$

where c_2 is a root of $g_1(x)$ and where $g_2(x)$ has degree $n - 2$. Combining (3) and (4), we have

$$f(x) = (x - c_1)(x - c_2)g_2(x)$$

If we continue this process, we find that there are complex numbers c_1, $c_2, \ldots, c_n$ and a nonzero polynomial $g_n(x)$ such that

$$f(x) = (x - c_1)(x - c_2)\cdots(x - c_n)g_n(x)$$

where the degree of $g_n(x)$ is $n - n = 0$. But then

$$g_n(x) = b$$

for some nonzero constant b, which leads us to the following corollary of the Fundamental Theorem of Algebra.

COROLLARY 10.5 *Let $f(x)$ be a nonconstant polynomial of degree n. Then there are complex numbers $b, c_1, c_2, \ldots, c_n$ such that $b \neq 0$ and such that*

$$f(x) = b(x - c_1)(x - c_2)\cdots(x - c_n) \tag{5}$$

The numbers $c_1, c_2, \ldots, c_n$ are the roots of $f(x)$, and b is the leading coefficient of $f(x)$.

CAUTION: The numbers $c_1, c_2, \ldots, c_n$ in (5) need not be distinct from one another. Indeed, if

$$f(x) = x^2 - 2x + 1$$

then

$$f(x) = (x - 1)^2$$

which means that $c_1 = 1 = c_2$.

The import of Corollary 10.5 is that any nonconstant polynomial $f(x)$ can be written as a product of a constant and linear polynomials. This fact is expressed by saying that the polynomial $f(x)$ ***factors completely***.

The roots mentioned in Corollary 10.5 are the *only* roots of $f(x)$. The reason is that if c is a number different from each of the numbers $c_1, c_2, \ldots, c_n$, then because $b, c - c_1, c - c_2, \ldots, c - c_n$ are all different from 0, we have

$$f(c) = b(c - c_1)(c - c_2)\cdots(c - c_n) \neq 0$$

Therefore c cannot be a root of $f(x)$. This yields the next theorem.

THEOREM 10.6 *A polynomial of degree n has at most n distinct roots.*

One consequence of Theorem 10.6 is that the complete factorization of $f(x)$ in (5) is unique except for the order in which we write the roots $c_1, c_2, \ldots, c_n$ of $f(x)$.

EXAMPLE 1. Factor the polynomial $3x^5 + 2x^4 + x^3$ completely, and show that it has no nonzero real roots.

Solution. Notice that

$$3x^5 + 2x^4 + x^3 = x^3(3x^2 + 2x + 1) \tag{6}$$

so 0 is a root of the given polynomial. Next we will find the roots of $3x^2 + 2x + 1$, which means finding the solutions of the equation

$$3x^2 + 2x + 1 = 0 \tag{7}$$

By the quadratic formula, the solutions of (7) are given by

$$x = \frac{-2 \pm \sqrt{2^2 - 4(3)(1)}}{2(3)} = \frac{-2 \pm \sqrt{-8}}{6} = \frac{-2 \pm \sqrt{8}i}{6} = -\frac{1}{3} \pm \frac{1}{3}\sqrt{2}i$$

Therefore the roots of $3x^2 + 2x + 1$ are $-\frac{1}{3} + \frac{1}{3}\sqrt{2}i$ and $-\frac{1}{3} - \frac{1}{3}\sqrt{2}i$. Since the leading coefficient is 3, it follows from Corollary 10.5 that (6) can be written in the form

$$3x^5 + 2x^4 + x^3 = 3x^3\left[x - \left(-\frac{1}{3} + \frac{1}{3}\sqrt{2}i\right)\right]\left[x - \left(-\frac{1}{3} - \frac{1}{3}\sqrt{2}i\right)\right]$$

It is now clear that the given polynomial has no nonzero real roots. □

Next we let z be a fixed complex number and n a positive integer. Theorem 10.6 tells us that there are at most n distinct roots of the polynomial $x^n - z$. This is consistent with the results of Section 9.3, where we concluded in effect that if $z \neq 0$, then there are exactly n distinct nth roots of $x^n - z$, namely the n nth roots of z. For example, if $z = 16$, then there are four roots of $x^4 - 16$, which are the fourth roots of 16: 2, $2i$, -2, and $-2i$.

In general, the roots $c_1, c_2, \ldots, c_n$ of a nonconstant polynomial $f(x)$ need not be distinct from one another. If at least two of the roots are the same, then of course $f(x)$ has fewer than n distinct roots. A root c_k appearing m times in $b(x - c_1)(x - c_2) \cdots (x - c_n)$ is called a root of $f(x)$ of ***multiplicity m***. For example, if

$$f(x) = -\frac{3}{4}(x - 9)^3(x - i)^7(x - 6)$$

then $f(x)$ has degree 11 (because $3 + 7 + 1 = 11$) and has roots 9, i and 6. The root 9 has multiplicity 3, the root i has multiplicity 7, and the root 6 has multiplicity 1. If we count each root m times, where m is its multiplicity, then f has 11 roots, and in this sense every polynomial of positive degree n has n roots.

EXAMPLE 2. Let $f(x) = x^4 - 2x^3 + 5x^2 - 8x + 4$. Given that 1 is a root of multiplicity 2 of $f(x)$, factor $f(x)$ completely.

Solution. Since 1 is a root of $f(x)$ of multiplicity 2, we know by Corollary 10.5 that $(x - 1)^2$ divides $f(x)$. Next we use synthetic division to divide $f(x)$ by $x - 1$:

$$\begin{array}{r|rrrrr} 1 & 1 & -2 & 5 & -8 & 4 \\ & & 1 & -1 & 4 & -4 \\ \hline & 1 & -1 & 4 & -4 & \boxed{0} \end{array}$$

Thus

$$x^4 - 2x^3 + 5x^2 - 8x + 4 = (x - 1)(x^3 - x^2 + 4x - 4)$$

Now we use synthetic division again, this time to divide $x^3 - x^2 + 4x - 4$ by $x - 1$:

$$\begin{array}{r|rrrr} 1 & 1 & -1 & 4 & -4 \\ & & 1 & 0 & 4 \\ \hline & 1 & 0 & 4 & \boxed{0} \end{array}$$

Therefore

$$x^3 - x^2 + 4x - 4 = (x - 1)(x^2 + 4)$$

Consequently

$$x^4 - 2x^3 + 5x^2 - 8x + 4 = (x - 1)(x - 1)(x^2 + 4) = (x - 1)^2(x^2 + 4)$$

Since

$$x^2 + 4 = (x + 2i)(x - 2i)$$

we conclude that

$$x^4 - 2x^3 + 5x^2 - 8x + 4 = (x - 1)^2(x + 2i)(x - 2i)$$

which is the complete factorization of $f(x)$. □

We close by mentioning that one interpretation of the Fundamental Theorem of Algebra is that any equation of the form

$$a_n x^n + a_{n-1} x^{n-1} + \cdots + a_1 x + a_0 = 0$$

where $n \geq 1$, has a solution among the complex numbers. The corresponding interpretation of Theorem 10.6 is that such an equation has at most n distinct solutions.

EXERCISES 10.3

In Exercises 1–8, find the roots of the polynomial, along with their multiplicities.

1. $(x - 1)^2(2x + 1)^3$

2. $(x + \frac{1}{2})(x - \frac{4}{3})(x + \pi)^4$

3. $-3x^2(2x^2 + 6)^2$

4. $(3x - 5)(2x^2 + 5x - 3)$

5. $(x^2 - 9)(x^2 + 3)$

6. $(6x^2 + 7x - 3)^2$

7. $-\frac{1}{2}x^2(2x + 1)^2(3x^2 + 1)^3$

8. $17i[x(7x - 3)^2]^3(4x^2 + 1)$

In Exercises 9–16, factor the polynomial completely.

9. $x^3 - 3x^2 + 2x$

10. $x^3 + 3x^2 + 3x + 1$

11. $x^4 - 10x^2 + 24$

12. $2x^5 - 7x^4 - 4x^3$

13. $(x^2 + 9)^2$

14. $x^4 + 5x^2 - 36$

15. $-x^2 + 4x - 13$

16. $x^3 - ix^2 + 5x - 5i$

17. Show that $5x^2 - 2x + 9$ has no real roots.

18. Show that $(x^2 + 4)(3x^2 - x + 1)$ has no real roots.

19. Show that $-2x^3 + 2x^2 - 3x$ has exactly one real root, and find it.

20. Show that -1 is a root of multiplicity 2 of the polynomial $2x^3 + 5x^2 + 4x + 1$, and then factor the polynomial completely.

21. Show that 2 is a root of multiplicity 3 of the polynomial $x^5 - 6x^4 + 13x^3 - 14x^2 + 12x - 8$, and then factor the polynomial completely.

22. Show that $1 + i$ is a root of multiplicity 2 of the polynomial $x^3 - (1 + 2i)x^2 - 2x + 2i$, and then factor the polynomial completely.

23. Show that $-i$ is a root of multiplicity 4 of the polynomial $x^5 + 2ix^4 + 2x^3 + 8ix^2 - 7x - 2i$, and then factor the polynomial completely.

In Exercises 24–28, write a polynomial with the given roots and multiplicities, and no other roots. Leave the polynomial in factored form.

24. $2 + 3i$ with multiplicity 1, $3 - i$ with multiplicity 1, -4 with multiplicity 1.

25. 3 with multiplicity 4, $1 - i$ with multiplicity 1.

26. $\sqrt{2}$ with multiplicity 1, -4 with multiplicity 2, $3i$ with multiplicity 3.

27. $1 + i$ with multiplicity 2, $1 - i$ with multiplicity 2, -6 with multiplicity 3.

28. $2 - 4i$ with multiplicity 3, $-i$ with multiplicity 2, 0 with multiplicity 5.

29. Show that -3 is a root of multiplicity 2 of the polynomial $x^4 + 2x^3 - 11x^2 - 12x + 36$. Find the other root.

30. Show that 1 is a root of multiplicity 4 of the polynomial $x^5 - x^4 - 6x^3 + 14x^2 - 11x + 3$. Find the other root.

31. Show that -1 is a root of multiplicity 5 of the polynomial $x^6 + 3x^5 - 10x^3 - 15x^2 - 9x - 2$. Find the other root.

10.4 POLYNOMIALS WITH REAL COEFFICIENTS

Any finite set of complex numbers can serve as the roots of a polynomial with complex coefficients. For example, a polynomial whose roots are -3, $2i$, and $\frac{1}{2} - 4i$ is given by

$$f(x) = (x + 3)(x - 2i)\left[x - \left(\frac{1}{2} - 4i\right)\right]$$

However, if we insist that each coefficient $a_n, a_{n-1}, \ldots, a_1, a_0$ of the polynomial

$$f(x) = a_n x^n + a_{n-1}x^{n-1} + \cdots + a_1 x + a_0$$

be a *real* number, then it turns out that the conjugate of any nonreal root of $f(x)$ must also be a root of the polynomial. This fact was already observed in Section 1.4 for quadratic polynomials. We will now state and prove the result for polynomials of arbitrary degree.

THEOREM 10.7 *Let $f(x)$ be a polynomial with real coefficients. If c is a root of $f(x)$, then so is the conjugate $\bar{c}$.*

Proof. Let

$$f(x) = a_n x^n + a_{n-1}x^{n-1} + \cdots + a_1 x + a_0$$

where $a_n, a_{n-1}, \ldots, a_1, a_0$ are real numbers. Because the coefficients are real, each is equal to its conjugate. Thus for any complex number c,

$$\begin{aligned}
f(\bar{c}) &= a_n(\bar{c})^n + a_{n-1}(\bar{c})^{n-1} + \cdots + a_1\bar{c} + a_0 \\
&= a_n(\overline{c^n}) + a_{n-1}(\overline{c^{n-1}}) + \cdots + a_1\bar{c} + a_0 \\
&= \overline{a_n}(\overline{c^n}) + \overline{a_{n-1}}(\overline{c^{n-1}}) + \cdots + \overline{a_1}\bar{c} + \overline{a_0} \\
&= \overline{a_n c^n} + \overline{a_{n-1}c^{n-1}} + \cdots + \overline{a_1 c} + \overline{a_0} \\
&= \overline{a_n c^n + a_{n-1}c^{n-1} + \cdots + a_1 c + a_0} \\
&= \overline{f(c)}
\end{aligned}$$

Now suppose that c is a root of $f(x)$, so that $f(c) = 0$. By our preceding calculations, this means that

$$f(\bar{c}) = \overline{f(c)} = \bar{0} = 0$$

so that $\bar{c}$ is also a root of $f(x)$, which is what we wished to prove. ■

CAUTION: We must emphasize that the conjugate of a root of a given polynomial need not also be a root if one or more of the coefficients are not real (see Exercise 24).

Let $f(x)$ be a polynomial with real coefficients, and let c be a root of $f(x)$. If c is a real number, then $\bar{c} = c$, so that Theorem 10.7 tells us nothing new about $f(x)$. In contrast, if c is a root of $f(x)$ and is not a real number, then Theorem 10.7 tells us that the conjugate $\bar{c}$ is a root of $f(x)$, and of course in this case $\bar{c} \neq c$. Thus nonreal roots of any polynomial with real coefficients come in conjugate pairs.

EXAMPLE 1. Find a polynomial that has real coefficients, has roots 5 and $1 - 3i$, and has the value 100 for $x = 0$.

Solution. If $1 - 3i$ is a root of a polynomial $f(x)$ with real coefficients, then by Theorem 10.7 so is its conjugate $1 + 3i$. Since the remaining prescribed root is 5, we will let $f(x)$ be given by

$$f(x) = b(x - 5)[x - (1 - 3i)][x - (1 + 3i)]$$

where b is to be determined. Since

$$\begin{aligned}[x - (1 - 3i)][x - (1 + 3i)] &= [(x - 1) + 3i][(x - 1) - 3i] \\ &= (x - 1)^2 - (3i)^2 \\ &= (x^2 - 2x + 1) + 9 \\ &= x^2 - 2x + 10\end{aligned}$$

$f(x)$ is given by

$$f(x) = b(x - 5)(x^2 - 2x + 10) = b(x^3 - 7x^2 + 20x - 50) \tag{1}$$

The number b must be chosen so that $f(0) = 100$. By (1), it follows that

$$f(0) = -50b$$

Therefore $100 = -50b$, so that $b = -2$. Consequently

$$\begin{aligned}f(x) &= -2(x^3 - 7x^2 + 20x - 50) \\ &= -2x^3 + 14x^2 - 40x + 100\end{aligned}$$

Consequently, $-2x^3 + 14x^2 - 40x + 100$ is a polynomial having the required properties. □

EXAMPLE 2. Let $f(x) = x^4 + 3x^3 + 3x^2 + 3x + 2$. Given that i is a root of $f(x)$, factor $f(x)$ completely.

Solution. Since all the coefficients of $f(x)$ are real, Theorem 10.7 applies. Thus the fact that i is a root of $f(x)$ implies that its conjugate $-i$ is also a root. Now Corollary 10.5 tells us that $f(x)$ is divisible by $x - i$ and by $x - (-i) = x + i$. To complete the factorization of $f(x)$, we use synthetic division to divide first by $x - i$ and then by $x + i$:

$$\begin{array}{r|rrrrr} i & 1 & 3 & 3 & 3 & 2 \\ & & i & -1+3i & -3+2i & -2 \\ \hline & 1 & 3+i & 2+3i & 2i & \underline{|0} \end{array}$$

Therefore

$$x^4 + 3x^3 + 3x^2 + 3x + 2 = (x - i)[x^3 + (3 + i)x^2 + (2 + 3i)x + 2i]$$

Dividing the bracketed term by $x + i$ yields

$$\begin{array}{r|rrrr} -i & 1 & 3+i & 2+3i & 2i \\ & & -i & -3i & -2i \\ \hline & 1 & 3 & 2 & \underline{|0} \end{array}$$

Consequently

$$x^3 + (3 + i)x^2 + (2 + 3i)x + 2i = (x + i)(x^2 + 3x + 2)$$

so that

$$x^4 + 3x^3 + 3x^2 + 3x + 2 = (x - i)(x + i)(x^2 + 3x + 2)$$

Since

$$x^2 + 3x + 2 = (x + 1)(x + 2)$$

$f(x)$ is completely factored as follows:

$$x^4 + 3x^3 + 3x^2 + 3x + 2 = (x - i)(x + i)(x + 1)(x + 2) \quad \square$$

It is sometimes desirable to factor a polynomial with real coefficients so that the factors are also polynomials with real coefficients. Although it is not always possible to accomplish this if we insist that the factors have degree 1 (see the result of Example 2), it is possible if we allow the factors to have degree 1 or 2.

THEOREM 10.8 *Every nonconstant polynomial with real coefficients can be expressed as a product of linear and quadratic polynomials with real coefficients.*

Proof. Let $f(x)$ be a nonconstant polynomial of degree n. By Corollary 10.5,

$$f(x) = b(x - c_1)(x - c_2)\ldots(x - c_n) \tag{2}$$

where $c_1, c_2, \ldots, c_n$ are the roots of f (repeated with multiplicity), and where b is the leading coefficient of $f(x)$ and hence is real. Consider $x - c_1$. If c_1 is real, then $x - c_1$ is a linear polynomial with real coefficients. However, if c_1 is not real, then by Theorem 10.7 we know that $\overline{c_1}$ is also a root of $f(x)$, and hence c_1 is one of the numbers $c_2, c_3, \ldots, c_n$. In this case $x - \overline{c_1}$ is a factor of $f(x)$. Multiplying $x - c_1$ and $x - \overline{c_1}$ together, we obtain

$$(x - c_1)(x - \overline{c_1}) = x^2 - (c_1 + \overline{c_1})x + c_1\overline{c_1} \tag{3}$$

Now recall from (6) of Section 9.1 that $c_1 + \overline{c_1}$ is real, and from (3) of Section 9.2 that $c_1\overline{c_1} = |c_1|^2$, which is also real. Consequently the polynomial in (3) is a quadratic polynomial with real coefficients. The same analysis applies to each of the factors $x - c_2, x - c_3, \ldots, x - c_n$. Thus each of the factors in (2) is a real number or a linear polynomial with real coefficients or combines with a second factor in (2) to form a quadratic polynomial with real coefficients. This proves the theorem. ■

EXAMPLE 3. Let $f(x) = -\frac{1}{2}x^3 + 3x^2 - 2x + \sqrt{3}$. Show that $f(x)$ has a real root.

Solution. Since $f(x)$ has degree 3, Theorem 10.8 implies that $f(x)$ can be expressed either as the product of three linear polynomials with real coefficients or as the product of one linear and one quadratic polynomial, each with real coefficients. In either case, $f(x)$ has a linear factor $x - c$, where c is a real number. It then follows from the Factor Theorem that $f(c) = 0$, so c is a real root of $f(x)$. □

By similar reasoning it is possible to prove that any polynomial having real coefficients and odd degree has at least one real root.

EXERCISES 10.4

In Exercises 1–12, find a polynomial $f(x)$ with real coefficients that has the given roots and prescribed value.

1. root: $1 - 2i$; $f(0) = 10$

2. root: $-3 + i$; $f(i) = 3 + 2i$

3. root: $1 + i$; $f(0) = -3$

4. roots: $0, -2, -i$; $f(0) = 0$

5. roots: $0, -2, -i$; $f(1) = 8$

6. roots: $i, -1, 1$; $f(2) = -3$

7. roots: i, i^2, i^3, i^4; $f(-2) = \sqrt{2}$

8. roots: $3, 4 - \sqrt{3}i$; $f(0) = 57$

9. roots: $2, 2 + \sqrt{3}i$; $f(1) = -28$

10. roots: $1, 2i, 3i$; $f(-1) = 100$

11. roots: $2 - i, 4 - 3i$; $f(2) = \frac{13}{4}$

12. roots: $2, -5, i, \frac{1}{3}i$; $f(0) = -10$

13. Find a polynomial with real coefficients for which -1 is a root of multiplicity 2 and i is a root of multiplicity 3. Express your answer as a product of linear and quadratic polynomials.

14. Find a polynomial with real coefficients for which $\frac{1}{2}$ is a root of multiplicity 3, $1 + 2i$ is a root of multiplicity 2, and i is a root of multiplicity 1. Express your answer as a product of linear and quadratic polynomials.

15. Given that $1 + i$ is a root of the polynomial $2x^3 - 5x^2 + 6x - 2$, factor the polynomial completely.

16. Given that i is a root of the polynomial $2x^4 - x^3 + x^2 - x - 1$, factor the polynomial completely.

17. Given that $-2 + 3i$ is a root of the polynomial $x^3 + 3x^2 + 9x - 13$, factor the polynomial completely.

18. Given that i is a root of multiplicity 2 of the polynomial $2x^5 - x^4 + 4x^3 - 2x^2 + 2x - 1$, factor the polynomial completely.

19. Given that $2 - 3i$ is a root of the polynomial $2x^4 - 8x^3 + 27x^2 - 4x + 13$, factor the polynomial completely.

20. a. Factor $x^4 + 1$ into quadratic factors with real coefficients. (*Hint:* Use the fact that $x^4 + 1 = (x^2 + 1)^2 - (\sqrt{2}x)^2$, and then factor the right side.)

b. Factor $x^4 + 1$ into linear factors. (*Hint:* Use part (a) and the quadratic formula.)

21. Given that $2 + i$ and $1 - i$ are roots of the polynomial $2x^4 - 12x^3 + 30x^2 - 36x + 20$, factor the polynomial

a. into a constant and linear factors

b. into a constant and quadratic factors with real coefficients.

22. Show that the polynomial $-7x^5 + \frac{1}{2}x^3 + \pi x - \sqrt{2}$ has at least one real root.

23. Show that the polynomial $4x^{19} + 2x^{12} - 1$ has at least one real root.

24. Let $f(x) = x^2 - 2x + 1 + 2i$. Show that $2 - i$ is a root of $f(x)$, but its conjugate $2 + i$ is not. Does this contradict Theorem 10.7? Explain.

25. Let $a_6x^6 + a_5x^5 + a_4x^4 + a_3x^3 + a_2x^2 + a_1x + a_0$ be a polynomial with real coefficients and six distinct roots. Show that if one of the roots is real, then at least two of the roots are real.

10.5 ROOTS OF POLYNOMIALS WITH INTEGER COEFFICIENTS

From the Fundamental Theorem of Algebra we know that every nonconstant polynomial has roots, but the theorem gives no hint for finding the roots. Some information about the roots is gained if the coefficients of the polynomial are real: Complex roots come in pairs, that is, the conjugate of a complex root is itself a root. As we saw in the preceding section, this result can be a help in determining the remaining roots of a polynomial one of whose nonreal roots is known, but it does not tell us anything about how to find an initial root of such a polynomial.

In the present section we will address the question of determining integer and, more generally, rational roots of a polynomial whose coefficients are integers. The proofs of the theorems are omitted.

Integer Roots

Our discussion begins with integer roots. We say that an integer c ***divides*** a second integer a, or is a ***divisor*** of a, if there is an integer b such that $a = bc$. Thus the divisors of 12 are

$$1, \quad -1, \quad 2, \quad -2, \quad 3, \quad -3, \quad 4, \quad -4, \quad 6, \quad -6, \quad 12, \quad \text{and} \quad -12$$

whereas the divisors of -7 are

$$1, \quad -1, \quad 7, \quad \text{and} \quad -7$$

Now we are ready to state the following theorem, which we mentioned (and used) in Section 3.3.

THEOREM 10.9 *Let $f(x) = a_n x^n + a_{n-1}x^{n-1} + \cdots + a_1 x + a_0$ be a polynomial with integer coefficients. If c is an integer root of $f(x)$, then c is a divisor of a_0.*

Since there are only finitely many divisors of a given nonzero integer, Theorem 10.9 provides us with an effective way of determining any integer roots of a polynomial whose coefficients are integers.

EXAMPLE 1. Let $f(x) = x^3 + 4x^2 + x - 6$. Find the integer roots of $f(x)$.

Solution. In the framework of Theorem 10.9, $a_0 = -6$. Since the divisors of -6 are $1, -1, 2, -2, 3, -3, 6$, and -6, Theorem 10.9 tells us that any integer roots of $f(x)$ must be one of these numbers. Now by substituting one by one the divisors of -6 for x, we find that

$$\begin{aligned} f(1) &= 0 & f(3) &= 60 \\ f(-1) &= -4 & f(-3) &= 0 \\ f(2) &= 20 & f(6) &= 360 \\ f(-2) &= 0 & f(-6) &= -84 \end{aligned}$$

Consequently the integer roots of $f(x)$ are $1, -2$, and -3. □

EXAMPLE 2. Factor $-6x^3 - 7x^2 + 1$ completely.

Solution. Let

$$f(x) = -6x^3 - 7x^2 + 1$$

To start the factorization of $f(x)$, let us look for a factor $x - c$ with c an integer. Since the constant term is 1, Theorem 10.9 implies that the only possible integer values of c are 1 and -1. But

$$f(1) = -12 \quad \text{and} \quad f(-1) = 0$$

so it follows from the Factor Theorem that $x + 1$ is a factor and $x - 1$ is not. Next we divide $f(x)$ by $x + 1$ by means of synthetic division:

$$\begin{array}{r|rrrr} -1 & -6 & -7 & 0 & 1 \\ & & 6 & 1 & -1 \\ \hline & -6 & -1 & 1 & \boxed{0} \end{array}$$

As a result,

$$-6x^3 - 7x^2 + 1 = (x + 1)(-6x^2 - x + 1)$$

Either by the quadratic formula or by direct factorization we find that

$$-6x^2 - x + 1 = (2x + 1)(-3x + 1)$$

Consequently the required factorization is given by

$$-6x^3 - 7x^2 + 1 = (x + 1)(2x + 1)(-3x + 1) \quad \square$$

CAUTION: It is possible, of course, for the method just used of seeking integer roots of a given polynomial to turn up no roots at all. For example, consider

$$f(x) = 8x^3 - 4x^2 - 2x + 1 \tag{1}$$

The coefficient of the constant term is 1, so by Theorem 10.9 the only possible integer roots of $f(x)$ are 1 and -1. But

$$f(1) = 3 \quad \text{and} \quad f(-1) = -9$$

so that there are no integer roots of $f(x)$. In Example 3 we will show that the only roots are $\frac{1}{2}$ and $-\frac{1}{2}$.

Rational Roots

Our next theorem gives us a method of searching out rational roots of polynomials.

THEOREM 10.10 *Let $f(x) = a_n x^n + a_{n-1} x^{n-1} + \cdots + a_1 x + a_0$ be a polynomial with integer coefficients. Let $c = p/q$, where p and q are integers without a common factor other than 1 and -1. If c is a root of $f(x)$, then p divides a_0 and q divides a_n.*

Notice that if c is an integer in Theorem 10.10, so that $p = c$ and $q = 1$, then the theorem says that c divides a_0, which is the result of Theorem 10.9. Thus Theorem 10.10 generalizes Theorem 10.9.

We will now use Theorem 10.10 to determine the roots of the polynomial given in (1).

EXAMPLE 3. Let $f(x) = 8x^3 - 4x^2 - 2x + 1$. Find the roots of $f(x)$, and then factor $f(x)$ completely.

Solution. We look for roots of $f(x)$ of the form $c = p/q$. By Theorem 10.10, p must divide 1 and q must divide 8. Thus p must be 1 or -1, and q must be 1, -1, 2, -2, 4, -4, 8, or -8. Therefore c can be 1, -1, $\frac{1}{2}$, $-\frac{1}{2}$, $\frac{1}{4}$, $-\frac{1}{4}$, $\frac{1}{8}$, or $-\frac{1}{8}$. Since

$$
\begin{aligned}
f(1) &= 3 & f\left(\frac{1}{4}\right) &= \frac{3}{8} \\
f(-1) &= -9 & f\left(-\frac{1}{4}\right) &= \frac{9}{8} \\
f\left(\frac{1}{2}\right) &= 0 & f\left(\frac{1}{8}\right) &= \frac{45}{64} \\
f\left(-\frac{1}{2}\right) &= 0 & f\left(-\frac{1}{8}\right) &= \frac{75}{64}
\end{aligned}
$$

it follows that $\frac{1}{2}$ and $-\frac{1}{2}$ are the only rational roots of $f(x)$. In order to factor $f(x)$ completely, we use synthetic division to divide $f(x)$ by $x - \frac{1}{2}$:

$$
\begin{array}{r|rrrr}
\frac{1}{2} & 8 & -4 & -2 & 1 \\
 & & 4 & 0 & -1 \\
\hline
 & 8 & 0 & -2 & \boxed{0}
\end{array}
$$

Therefore

$$f(x) = (x - \tfrac{1}{2})(8x^2 - 2)$$

Next we notice that

$$8x^2 - 2 = 2(4x^2 - 1) = 2(2x - 1)(2x + 1)$$

so the complete factorization of $f(x)$ is given by

$$f(x) = 2\left(x - \frac{1}{2}\right)(2x - 1)(2x + 1) = 8\left(x - \frac{1}{2}\right)^2\left(x + \frac{1}{2}\right)$$

From the factorization we see that the rational roots $\frac{1}{2}$ and $-\frac{1}{2}$ are the only roots of $f(x)$. □

When we employ Theorem 10.10, we may well obtain a large collection of candidates for a root—even larger than the collection of eight candidates we had in Example 3. The following result, which concerns intervals in which roots of a given polynomial must lie, can allow us to ignore many of the candidates and focus on a select few among which we hope to find a root.

THEOREM 10.11
Intermediate Value Theorem for Polynomials

Let $f(x)$ be a polynomial with real coefficients, and let a and b be any two distinct numbers. If $f(a) > 0$ and $f(b) < 0$, then there is a root of $f(x)$ between a and b.

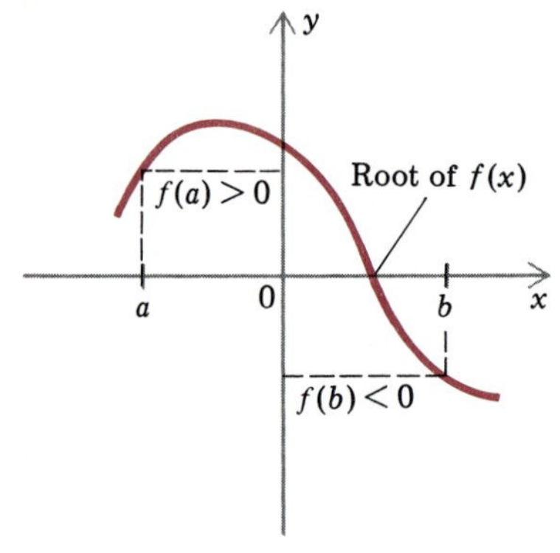

FIGURE 10.1

Figure 10.1 gives a pictorial idea of Theorem 10.11.

EXAMPLE 4. Let $f(x) = 12x^3 - 20x^2 - x + 6$. Factor $f(x)$ completely.

Solution. If $x - c$ is to be a factor of $f(x)$, where $c = p/q$, then by Theorem 10.10 we know that p must be a factor of 6 and q a factor of 12. Thus p must be

$$1, \quad -1, \quad 2, \quad -2, \quad 3, \quad -3, \quad 6, \quad \text{or} \quad -6$$

and q must be

$$1, \quad -1, \quad 2, \quad -2, \quad 3, \quad -3, \quad 4, \quad -4, \quad 6, \quad -6, \quad 12, \quad \text{or} \quad -12$$

Consequently the only possible values of c are

$$1, \quad -1, \quad \frac{1}{2}, \quad -\frac{1}{2}, \quad \frac{1}{3}, \quad -\frac{1}{3}, \quad \frac{1}{4}, \quad -\frac{1}{4}, \quad \frac{1}{6}, \quad -\frac{1}{6}, \quad \frac{1}{12}, \quad -\frac{1}{12},$$

$$2, \quad -2, \quad \frac{2}{3}, \quad -\frac{2}{3}, \quad 3, \quad -3, \quad \frac{3}{2}, \quad -\frac{3}{2}, \quad \frac{3}{4}, \quad -\frac{3}{4}, \quad 6, \quad \text{or} \quad -6$$

It seems an enormous task to test all these 24 values of c. However, if we look at, say, $c = -1$ and $c = 0$, we find that

$$f(-1) = -25 < 0 \quad \text{and} \quad f(0) = 6 > 0$$

Therefore by the Intermediate Value Theorem for Polynomials it follows that there is a root of $f(x)$ between -1 and 0, which cuts from 24 to 7 the collection from which we hope to find a rational root:

$$c = -\frac{1}{2}, \quad -\frac{1}{3}, \quad -\frac{1}{4}, \quad -\frac{1}{6}, \quad -\frac{1}{12}, \quad -\frac{2}{3}, \quad \text{or} \quad -\frac{3}{4}$$

When we test $-\frac{1}{2}$, we find that $f(-\frac{1}{2}) = 0$, so that $x + \frac{1}{2}$ is a factor of $f(x)$. Now we use synthetic division to divide $f(x)$ by $x + \frac{1}{2}$:

$$\begin{array}{r|rrrr} -\frac{1}{2} & 12 & -20 & -1 & 6 \\ & & -6 & 13 & -6 \\ \hline & 12 & -26 & 12 & \boxed{0} \end{array}$$

Thus

$$12x^3 - 20x^2 - x + 6 = \left(x + \frac{1}{2}\right)(12x^2 - 26x + 12) \tag{2}$$

Next we find that

$$12x^2 - 26x + 12 = 2(6x^2 - 13x + 6)$$

By the quadratic formula,

$$6x^2 - 13x + 6 = 0$$

if and only if

$$x = \frac{-(-13) \pm \sqrt{(-13)^2 - 4(6)(6)}}{2(6)} = \frac{13 \pm \sqrt{25}}{12} = \frac{13 \pm 5}{12}$$
$$= \frac{3}{2} \quad \text{or} \quad \frac{2}{3}$$

It follows that $x - \frac{3}{2}$ and $x - \frac{2}{3}$ are factors of $6x^2 - 13x + 6$, so that

$$6x^2 - 13x + 6 = 6\left(x - \frac{3}{2}\right)\left(x - \frac{2}{3}\right)$$

Consequently by (2) the desired factorization of the given polynomial is

$$12x^3 - 20x^2 - x + 6 = 12\left(x + \frac{1}{2}\right)\left(x - \frac{3}{2}\right)\left(x - \frac{2}{3}\right) \quad \square$$

Irrational Roots

Is there any method of finding irrational roots of polynomials? First of all, think about the difficulty of expressing irrational numbers. We already know that irrational numbers do not have repeating decimal expansions. Thus unless we have a special symbol (such as π, e, or $\sqrt{2}$) for an irrational number, about the best we could hope to do is to find the first several digits in its decimal expansion or to find a rational number that is "close" to it.

Consequently irrational roots can usually only be approximated. The methods used in such approximations can be found in more advanced mathematics texts.

EXERCISES 10.5

In Exercises 1–8, determine the divisors of the given number.

1. 2 **2.** 4 **3.** -5 **4.** -1

5. 8 **6.** 11 **7.** -24 **8.** 36

In Exercises 9–16, find all integer roots of the given polynomial.

9. $x^3 + 4x^2 - 7x - 10$

10. $2x^3 - 3x^2 - 11x + 6$

11. $2x^3 + 3x^2 - 17x + 12$

12. $x^3 - 2x^2 - 5x + 6$

13. $x^4 + 3x^3 - 3x - 1$

14. $x^4 + x^3 - x^2 - x - 6$

15. $x^4 - 5x^3 - 10x^2 + 20x + 24$

16. $x^4 + 2x^3 + 2x^2 + x$

In Exercises 17–30, use Theorems 10.9 and 10.10 to factor the given polynomial completely.

17. $2x^2 + 5x - 3$

18. $3x^3 + 6x^2 - 4x - 8$

19. $x^3 - 5x^2 + x - 5$

20. $x^3 + x - 2$

21. $x^3 - 9x^2 + 26x - 24$

22. $2x^3 + x^2 - 5x + 2$

23. $4x^3 + 8x^2 - 11x - 15$

24. $12x^3 + 23x^2 - 3x - 2$

25. $2x^4 - 2x^3 + 4x^2 + 2x - 6$

26. $4x^4 + 8x^3 + 7x^2 + 6x + 3$

27. $x^5 + 3x^4 + x^3 - x^2 - 4$

28. $12x^3 + 28x^2 - 7x - 5$

29. $24x^4 + 2x^3 - 5x^2 - x$

30. $16x^4 - 20x^3 - 56x^2 + 15x$

In Exercises 31–40, show that the given polynomial has no rational roots.

31. $x^3 + x + 3$ **32.** $2x^3 - 17x^2 + 12x - 2$

33. $x^5 - x - 1$

34. $2x^6 - 1$

35. $2x^7 - 3$

36. $3x^4 + 3x^2 + 4x + 4$

37. $x^5 - x^4 + x^3 - x^2 - x - 1$

38. $2x^6 + 4x^5 - 3x^3 + 5x - 1$

39. $-2x^7 - 4x^4 + x - 1$

40. $x^8 + x^7 + 4$

In Exercises 41–44, show that the given number is not rational.

41. $\sqrt{2}$ (*Hint:* Let $f(x) = x^2 - 2$ and use Theorem 10.10.)

42. $\sqrt{3}$ **43.** $2^{1/3}$ **44.** $8^{1/4}$

In Exercises 45–48, show that the polynomial has a root in the given interval.

45. $x^4 - x^3 + x^2 - x - 1$; $[-1, 1]$

46. $3x^4 + 8x^3 - 4x + 5$; $[-2, 0]$

47. $x^5 + x^3 - 8$; $[-1, 2]$

48. $x^3 + 3x^2 - 1$; $[-\frac{1}{2}, 1]$

49. Let $f(x) = 4x^4 - 9x^2 + 1$.

a. Use the Intermediate Value Theorem for Polynomials to determine in which of the unit intervals determined by $-3, -2, -1, 0, 1, 2, 3$ a root of $f(x)$ must occur.

b. Using the result of part (a) and Theorem 10.6, show that $f(x) > 0$ for $x > 3$.

50. Determine the solutions of the equation $4x^3 - 3x + 1 = 0$. (This equation is related to the problem of trisecting an angle.)

51. The problem of showing that a 60° angle cannot be trisected with a straightedge and compass can be reduced to showing that the equation $8x^3 - 6x - 1 = 0$ has no rational solutions. Show that indeed the equation has no rational solutions.

***52.** Occasionally one wishes to know whether a polynomial with integer coefficients can be factored into two other polynomials with positive degree that also have integer coefficients. The following result, due to Eisenstein, a student of Gauss, gives a condition that guarantees that the polynomial cannot be so factored.

Eisenstein's Criterion: Let

$$f(x) = a_n x^n + a_{n-1}x^{n-1} + \cdots + a_1 x + a_0$$

be a polynomial with integer coefficients. If there is a prime number p such that

i. p divides each of the numbers

$$a_0, a_1, a_2, \ldots, a_{n-1}$$

ii. p does not divide a_n

iii. p^2 does not divide a_0

then $f(x)$ cannot be written as the product of two polynomials each with integer coefficients and positive degree.

Use Eisenstein's Criterion to show that none of the following polynomials can be factored into two polynomials with integer coefficients and positive degree.

a. $x^3 + 3x^2 + 9x + 6$ (*Hint:* Take $p = 3$.)
b. $4x^4 - 15x^3 + 20x^2 + 1000x - 10$
c. $5x^3 + 21x^2 - 14x + 7$
d. $7x^5 + 543x^4 - 219x^3 + 72x^2 - 63x + 240$

10.6 PARTIAL FRACTIONS

In Section 0.7 we combined rational expressions to obtain one rational expression in lowest terms. For example, we took the sum

$$\frac{2x-1}{x^2+4x-5} + \frac{3}{x^2} - \frac{x-3}{x^2+5x}$$

found the least common denominator, and then combined the fractions to obtain

$$\frac{x^3+6x^2+9x-15}{x^2(x+5)(x-1)}$$

In calculus there are times when we need to reverse the procedure, splitting up a single rational expression like

$$\frac{2}{x^2-1}$$

into a sum of several component rational expressions:

$$\frac{2}{x^2-1} = \frac{1}{x-1} - \frac{1}{x+1}$$

Such a splitting up of a rational expression is called a ***partial fraction***

decomposition of the original expression, and the component fractions so obtained are ***partial fractions***.

Let $P(x)/Q(x)$ denote a given rational expression, where $P(x)$ is a polynomial and $Q(x)$ is a nonzero polynomial. In order to use the ideas of the preceding sections we will also assume throughout our discussion that $P(x)/Q(x)$ has real coefficients.

Our method of obtaining a partial fraction decomposition of $P(x)/Q(x)$ hinges on the degree of $P(x)$ being less than the degree of $Q(x)$. If this condition is not satisfied, then before decomposing the given expression we must divide $P(x)$ by $Q(x)$; afterward we decompose what remains. For example, the rational expressions

$$\frac{-x^4 + x^3 + x - 1}{x^3 + x} \quad \text{and} \quad \frac{x^4 - 1}{x^4 - x}$$

must both be divided because the degree of the numerator (which in each case is 4) is *not* less than that of the denominator. Division yields

$$\frac{-x^4 + x^3 + x - 1}{x^3 + x} = -x + 1 + \frac{x^2 - 1}{x^3 + x}$$

and

$$\frac{x^4 - 1}{x^4 - x} = 1 + \frac{x - 1}{x^4 - x}$$

Each of the expressions $(x^2 - 1)/(x^3 + x)$ and $(x - 1)/(x^4 - x)$ is in a form appropriate for decomposing into partial fractions.

From now on we will assume that any necessary division has already occurred, so that for our given rational expression $P(x)/Q(x)$, $\deg P(x) < \deg Q(x)$. Next we factor both numerator and denominator of $P(x)/Q(x)$ into expressions of the form

$$\text{constants}, \quad (ax + b)^m, \quad \text{and} \quad (ax^2 + bx + c)^n$$

where m and n are positive integers, $a \neq 0$, and $ax^2 + bx + c$ has no linear factors with real coefficients. (That such a factorization is possible was proved in Section 10.4.) Then we cancel any common factors.

EXAMPLE 1. Factor $\dfrac{x^3 - x}{x^4 - x}$ and cancel any common factors.

Solution. We obtain

$$\frac{x^3 - x}{x^4 - x} = \frac{x(x^2 - 1)}{x(x^3 - 1)} = \frac{x(x - 1)(x + 1)}{x(x - 1)(x^2 + x + 1)} = \frac{x + 1}{x^2 + x + 1} \qquad \square$$

Once the fraction $P(x)/Q(x)$ is factored in the above fashion and common factors are canceled, the resulting fraction $R(x)/S(x)$ will be decomposed into partial fractions. By this we mean that $R(x)/S(x)$ will be written as the sum of certain rational functions each of whose denominators contains only powers of a single factor of $S(x)$. In particular, for every expression $(ax + b)^m$ appearing in $S(x)$ we include a sum of the form

$$\frac{A_1}{ax + b} + \frac{A_2}{(ax + b)^2} + \cdots + \frac{A_m}{(ax + b)^m} \tag{1}$$

where the constants $A_1, A_2, \cdots, A_m$ are to be determined. If $m > 1$ then we say that $ax + b$ is a ***repeated factor*** of $S(x)$. For example, if $S(x) = x(x + 1)^2(x - 7)^3(2x + 1)$, then $x + 1$ and $x - 7$ are repeated factors of $S(x)$, whereas x and $2x + 1$ are not repeated.

In the same vein, for every expression in $S(x)$ of the form $(ax^2 + bx + c)^n$ we include a sum of the form

$$\frac{B_1x + C_1}{ax^2 + bx + c} + \frac{B_2x + C_2}{(ax^2 + bx + c)^2} + \cdots + \frac{B_nx + C_n}{(ax^2 + bx + c)^n} \tag{2}$$

where the constants $B_1, B_2, \cdots B_n$, and $C_1, C_2, \cdots C_n$ are to be determined. Again, if $n > 1$ we say that $ax^2 + bx + c$ is a repeated factor of $S(x)$.

This is a lot to manage. To simplify the discussion we first consider the partial fraction decomposition of $R(x)/S(x)$ under the condition that $S(x)$ has no repeated factors. Then we will turn to the decomposition when $S(x)$ has repeated factors.

No Repeated Factors

If $ax + b$ is a linear factor of $S(x)$ that is not repeated, then $m = 1$ and the expression in (1) reduces to

$$\frac{A}{bx + c} \tag{3}$$

Thus for every nonrepeated linear factor of $S(x)$ we include an expression in the form of (3). Similarly, for every nonrepeated quadratic factor $ax^2 + bx + c$ of $S(x)$ we include an expression of the form

$$\frac{Bx + C}{ax^2 + bx + c} \tag{4}$$

EXAMPLE 2. Express $\dfrac{x + 1}{x^2 - 4x}$ as a sum of partial fractions.

Solution. First we factor the denominator:

$$\frac{x+1}{x^2-4x} = \frac{x+1}{x(x-4)}$$

Since each of the factors x and $x - 4$ in the denominator is not repeated, the preceding discussion implies that we include expressions of the form A_1/x and $A_2/(x-4)$. Thus we write

$$\frac{x+1}{x^2-4x} = \frac{A_1}{x} + \frac{A_2}{x-4} \tag{5}$$

where A_1 and A_2 are to be determined. To calculate A_1 and A_2 we first clear fractions by multiplying both sides of (5) by $x(x-4)$. This yields

$$x + 1 = A_1(x-4) + A_2 x \tag{6}$$

This equation must hold for all values of x, so in particular it must hold for $x = 4$ and $x = 0$. Substituting 4 for x in (6), we obtain

$$4 + 1 = A_1\overbrace{(4-4)}^{0} + A_2 \cdot 4$$

so that

$$A_2 = \frac{5}{4}$$

Substituting 0 for x in (6), we obtain

$$0 + 1 = A_1(0-4) + A_2 \cdot 0$$

so that

$$A_1 = -\frac{1}{4}$$

Consequently (5) becomes

$$\frac{x+1}{x^2-4x} = \frac{-1/4}{x} + \frac{5/4}{x-4}$$

which is the desired partial fraction decomposition. □

EXAMPLE 3. Express $\dfrac{x^2 - 1}{x^3 + x}$ as a sum of partial fractions.

Solution. As in Example 2, we first factor:

$$\frac{x^2 - 1}{x^3 + x} = \frac{(x - 1)(x + 1)}{x(x^2 + 1)}$$

We notice first that there is no cancellation. Then we observe that the denominator contains the nonrepeated linear factor x and the nonrepeated quadratic factor $x^2 + 1$. Thus we use (3) and (4) once each, which yields

$$\frac{x^2 - 1}{x^3 + x} = \frac{A}{x} + \frac{Bx + C}{x^2 + 1} \tag{7}$$

where A, B, and C are to be determined. To calculate A, B, and C we first clear fractions by multiplying both sides of (7) by $x(x^2 + 1)$. The result is

$$x^2 - 1 = A(x^2 + 1) + (Bx + C)x$$

Combining like powers of x, we obtain

$$x^2 - 1 = Ax^2 + A + Bx^2 + Cx$$

so that

$$x^2 - 1 = (A + B)x^2 + Cx + A \tag{8}$$

Since both sides of (8) must be equal for all values of x, the coefficients of like powers of x on both sides must be equal, which implies that

$$A + B = 1 \qquad C = 0 \quad \text{and} \quad A = -1$$

It follows that $B = 1 - A = 1 - (-1) = 2$. Consequently (7) becomes

$$\frac{x^2 - 1}{x^3 + x} = \frac{-1}{x} + \frac{2x}{x^2 + 1} \qquad \square$$

Notice that in the solution of Example 2 we substituted special values of x in order to determine A_1 and A_2, whereas in Example 3 we equated coefficients of like powers of x in order to determine A, B, and C. One can combine both methods or use either of them, whichever is easier for a given rational expression.

Repeated Factors

Now we consider examples of partial fraction decompositions involving repeated factors.

EXAMPLE 4. Express $\dfrac{5x^2 - 24x + 25}{(x-2)^2(2x-3)}$ as a sum of partial fractions.

Solution. For the numerator the factorization is

$$5x^2 - 24x + 25 = 5\left(x - \frac{12}{5} - \frac{1}{5}\sqrt{19}\right)\left(x - \frac{12}{5} + \frac{1}{5}\sqrt{19}\right)$$

There is no cancellation with the denominator, which is already factored, with a linear factor $2x - 3$ and a repeated factor $(x-2)^2$. Thus we will use (1) with $m = 1$ for $2x - 3$, and (1) with $m = 2$ for $(x-2)^2$. This yields

$$\frac{5x^2 - 24x + 25}{(x-2)^2(2x-3)} = \frac{A}{2x-3} + \frac{B}{x-2} + \frac{C}{(x-2)^2} \tag{9}$$

where A, B, and C are to be determined. The process begins as we clear fractions by multiplying both sides by $(x-2)^2(2x-3)$. This yields

$$5x^2 - 24x + 25 = A(x-2)^2 + B(x-2)(2x-3) + C(2x-3) \tag{10}$$

Substituting 2 for x in (10), we obtain

$$-3 = 0 + 0 + C$$

so that $C = -3$. Substituting $\frac{3}{2}$ for x in (10), we obtain

$$\frac{45}{4} - \frac{72}{2} + 25 = \frac{A}{4} + 0 + 0$$

so that

$$A = 45 - 144 + 100 = 1$$

Although 0 is not a root of $x - 2$ or $2x - 3$, we can evaluate the remaining constant, B, by substituting 0 for x in (10). Using the fact that $A = 1$ and $C = -3$, we obtain

$$\begin{aligned} 25 &= 4A + 6B - 3C \\ &= 4 + 6B + 9 \end{aligned}$$

so that $6B = 12$, and hence $B = 2$. We conclude that

$$\frac{5x^2 - 24x + 25}{(x-2)^2(2x-3)} = \frac{1}{2x-3} + \frac{2}{x-2} - \frac{3}{(x-2)^2} \qquad \square$$

CAUTION: Observe carefully that in following our procedure we included the term $\dfrac{C}{(x-2)^2}$, rather than the term $\dfrac{Cx+D}{(x-2)^2}$, in (9).

EXAMPLE 5. Express $\dfrac{8x^2 - 5x + 32}{x(x^2+4)^2}$ as a sum of partial fractions.

Solution. The numerator cannot be factored into linear factors with real coefficients, so there is no cancellation. Since the denominator contains the nonrepeated linear factor x and the repeated quadratic factor $(x^2 + 4)^2$, we use (1) with $m = 1$ for x, and (2) with $n = 2$ for $(x^2 + 4)^2$. This yields

$$\frac{8x^2 - 5x + 32}{x(x^2+4)^2} = \frac{A}{x} + \frac{Bx + C}{x^2 + 4} + \frac{Dx + E}{(x^2+4)^2}$$

To clear fractions we multiply both sides by $x(x^2 + 4)^2$, obtaining

$$8x^2 - 5x + 32 = A(x^2+4)^2 + (Bx + C)x(x^2 + 4) + (Dx + E)x \tag{11}$$

Substituting 0 for x in (11) yields

$$32 = 16A + 0 + 0$$

so that

$$A = 2$$

We now replace A by 2 in (11), then expand the products in the result, and combine terms containing like powers of x:

$$\begin{aligned} 8x^2 - 5x + 32 &= 2(x^4 + 8x^2 + 16) + Bx^4 + Cx^3 + 4Bx^2 + 4Cx + Dx^2 + Ex \\ &= (2 + B)x^4 + Cx^3 + (16 + 4B + D)x^2 + (4C + E)x + 32 \end{aligned}$$

Equating the coefficients of like powers of x on both sides, we obtain

$$\begin{aligned} 2 + B &= 0 \\ C &= 0 \\ 16 + 4B + D &= 8 \\ 4C + E &= -5 \end{aligned} \tag{12}$$

From the first and second equations in (12) we see that

$$B = -2 \quad \text{and} \quad C = 0$$

If we substitute -2 for B in the third equation of (12), we find that

$$16 - 8 + D = 8$$

so that

$$D = 0$$

Substituting 0 for C in the fourth equation in (12) yields

$$E = -5$$

We conclude that

$$\frac{8x^2 - 5x + 32}{x(x^2 + 4)^2} = \frac{2}{x} - \frac{2x}{x^2 + 4} - \frac{5}{(x^2 + 4)^2} \quad \square$$

EXERCISES 10.6

In Exercises 1–21, express the given rational function as a sum of partial fractions.

1. $\dfrac{x-2}{x(x+1)}$
2. $\dfrac{20}{(-3x+2)(3x+2)}$
3. $\dfrac{21x}{(4x-2)(-x+4)}$
4. $\dfrac{x+3}{x(x-1)(x+1)}$
5. $\dfrac{-2x^2+16x+17}{(2x-1)(x-4)(x+3)}$
6. $\dfrac{3x-1}{x^2-1}$
7. $\dfrac{x+1/2}{x^2+x}$
8. $\dfrac{4x^2+3x-3}{x^3-x}$
9. $\dfrac{-6}{x(x^2+3)}$
10. $\dfrac{4x+5}{(2x+1)(x^2+x+1)}$
11. $\dfrac{7x^2}{(2x^2+3)(5x^2+4)}$
12. $\dfrac{x^2-3x-2}{(x^2+1)(4x^2+1)}$
13. $\dfrac{1}{x^3+x}$
14. $\dfrac{3x^3+5x^2-5x+5}{x^4-1}$
15. $\dfrac{2\sqrt{2}}{x^4+1}$

    (*Hint:* Use Exercise 20(a) of Section 10.4.)
16. $\dfrac{4x^2-4}{x^2(2x+3)}$
17. $\dfrac{-x^2-4x+7}{(x-1)(x+1)^2}$
18. $\dfrac{13x^2-3x+6}{x^2(8x^2+3)}$
19. $\dfrac{x^2}{(x^2+2)^2}$
20. $\dfrac{3x^2+3x+1}{x(x^2+1)^2}$
21. $\dfrac{4x^4-4x^3+3x^2-2x+1}{x^2(2x^2+1)^2}$

KEY TERMS

factor	multiplicity of a root
synthetic division	integer root
complete factorization	rational root

KEY THEOREMS

Division Algorithm
Remainder Theorem
Factor Theorem
Fundamental Theorem of Algebra
Intermediate Value Theorem for Polynomials

REVIEW EXERCISES

In Exercises 1–2, perform the indicated long division.

1. $x^2 - x - 1 \overline{)x^4 - 2x + 1}$

2. $x^3 + 1 \overline{)x^7 + x^6 - x^4 - x^3 + 2}$

In Exercises 3–8, use synthetic division to find the quotient and remainder when the first polynomial is divided by the second.

3. $-x^2 - 5x - 10; x + 3$

4. $2x^3 + 5x^2 - 7x + 7; x - \frac{1}{2}$

5. $x^3 + 2x + 3i; x + i$

6. $x^4 + 2x^3 - 2x^2 - 3x + 2; x + 2$

7. $x^6 - 3x^5 - (1 + i)x^4 + (1 - i)x^3 + 6ix^2 + 4x + 3; x - 3$

8. $x^8 + 1; x + 1$

In Exercises 9–12, use synthetic division to show that the first polynomial is a factor of the second.

9. $x + 3; 2x^3 + 5x^2 - 2x + 3$

10. $x + \frac{1}{2}; 2x^4 + 5x^3 + 2x^2 - x - \frac{1}{2}$

11. $x - 2i; 3x^5 + (-4 - 6i)x^4 + 7ix^3 + (-4 + i)x^2 + (3 + 4i)x - 2i$

12. $x + i; x^7 - i$

In Exercises 13–16, use the Remainder Theorem to find the remainder when the first polynomial is divided by the second.

13. $-x^4 + 5x^3 + 3x^2 - 4; x + 1$

14. $3x^6 - x^3 - 2x + 1; x - \sqrt{3}$

15. $ix^4 - 2x^3 - 3x - 2i; x + i$

16. $x^{19} + x^{16} - 1; x + \frac{\sqrt{2}}{2} - \frac{\sqrt{2}}{2}i$

In Exercises 17–22, verify that the linear polynomial is a factor of the other polynomial, and then factor the other polynomial into linear and constant factors.

17. $x^3 - 2x^2 - 5x + 6; x + 2$

18. $x^4 - 4x^3 - 2x^2 + 8x; x - 4$

19. $3x^4 - 8x^3 + 6x^2 - 1; x + \frac{1}{3}$

20. $x^3 + ix^2 + (i - 3)x + 2 - 2i; x - 1 + i$

21. $x^4 - x^2 - \sqrt{2}x;\ x - \sqrt{2}$

22. $x^3 - 3x^2 + x - 3;\ x + i$

In Exercises 23–26, factor the given polynomial into linear factors.

23. $x^3 - 4x^2 + 8x$

24. $x^3 + x^2 - x - 1$

25. $x^4 - x^2 - 12$

26. $x^4 + 2x^2 - 63$

In Exercises 27–30, show that the given number is a root of the polynomial. Then factor the polynomial completely.

27. $x^3 + 7x^2 + 19x + 21;\ -3$

28. $2x^3 - x^2 - x - 3;\ \frac{3}{2}$

***29.** $x^3 - 2ix^2 + (i - 2)x + i + 1;\ i$

***30.** $x^8 - 16;\ 1 - i$

In Exercises 31–34, find the roots of the polynomial, along with their multiplicities.

31. $(x - 2i)^2(x^2 + 49)^4(x - 3)^7$

32. $(3 - 4x)^2(2x^2 + 4x + 3)^5$

33. $x^3(\sqrt{6}x - \sqrt{2})^4(9x^2 + 1)^2$

34. $(x - 1)(x + 2)(x^3 - x^2 + x - 1)$

In Exercises 35–36, write a polynomial with the given roots and multiplicities. Leave the polynomial in factored form.

35. $3 - 5i$ with multiplicity 3, $2i$ with multiplicity 4, 0 with multiplicity 1.

36. $\sqrt{3} - 2i$ with multiplicity 10, -9 with multiplicity 7, $3 - \sqrt{3} + 4i$ with multiplicity 6.

37. Show that 3 is a root of multiplicity 3 of the polynomial $x^4 - 11x^3 + 45x^2 - 81x + 54$, and then find the other root.

38. Show that i is a root of multiplicity 4 of the polynomial $x^6 - 4ix^5 + 10x^4 - 60ix^3 - 95x^2 + 64ix + 16$, and then find the other roots.

In Exercises 39–40, find a polynomial $f(x)$ with real coefficients that has the given roots and prescribed value.

39. roots: $2, -1 - 3i$; $f(1) = -26$

40. roots: $i - 1, 2i - 4$; $f(2) = 80$

41. Given that $2 - i$ is a root of multiplicity 2 of the polynomial $x^4 - 8x^3 + 26x^2 - 40x + 25$, factor the polynomial completely.

42. Given that -1 is a root of multiplicity 2 and i is a root of multiplicity 1 of the polynomial $x^5 - 2x^4 - 6x^3 - 6x^2 - 7x - 4$, factor the polynomial completely.

43. Given that $1 + i$ and $-1 - 2i$ are roots of the polynomial $x^4 + 3x^2 - 6x + 10$, factor the polynomial into quadratic factors with real coefficients.

44. Show that the polynomial $3x^{11} - 4x^8 + 3x^5 - 1$ has at least one real root.

In Exercises 45–48, find all integer roots, if any, of the given polynomial.

45. $x^5 + 4x^3 + x^2 + 4$

46. $2x^4 - x^3 + x^2 + x$

47. $x^4 - x^3 - 6x^2 - 6x - 72$

48. $12x^3 + 20x^2 - x - 6$

In Exercises 49–54, use Theorems 10.9 and 10.10 to factor the given polynomial completely.

49. $2x^3 - 4x^2 - x + 2$

50. $8x^3 - 12x^2 + 12x - 18$

51. $x^5 - x^4 - x^3 - x^2 - 2x$

52. $x^6 - 1$

53. $16x^3 + 28x^2 - 28x + 5$

54. $18x^4 + 9x^3 - 8x^2 - 4x$

In Exercises 55–58, show that the given polynomial has no rational roots.

55. $2x^4 + 3x^3 - x^2 - 1$

56. $x^5 - 3x + 4$

57. $3x^6 - 2$

58. $2x^8 - 1$

In Exercises 59–60, show that the polynomial has a root in the given interval.

59. $x^3 - 4x^2 + 4;\ [-1, 0]$

60. $x^4 + 2x^3 - 4x^2 - 4x + \frac{1}{2};\ [-3, -1]$

61. Show that $x - 1$ is a factor of the polynomial $529x^{1345} - 284x^{612} - 245$.

62. Determine the relationship that must hold between a and b in order that $x + 1$ be a factor of the polynomial $x^3 + ax^2 - x - b$.

In Exercises 63–68, express the given rational function as a sum of partial fractions.

63. $\dfrac{4}{3x^2 - 8x + 4}$

64. $\dfrac{5x^2 + 2x - 1}{x(x - 1)(2x + 1)}$

65. $\dfrac{x^2 - x - 1}{(x + 2)(x^2 + 1)}$

66. $\dfrac{7x^2 - x + 5}{x^3 + x}$

67. $\dfrac{4 - 7x^2}{x^4 - 16}$

68. $\dfrac{3x^3 + x^2 + x}{(x^2 + x + 1)(x^2 - x + 1)}$

11 DISCRETE ALGEBRA

The title of this final chapter refers to the fact that each of the topics to be discussed is related to the discrete collection of natural numbers. We open the chapter with a presentation of the axiom of mathematical induction, which makes it possible to prove statements concerning the natural numbers whose proofs are otherwise unmanageable. Section 11.2 introduces sequences, that is, functions whose domains are sets of integers. Most notable among sequences are arithmetic sequences and geometric sequences, whose properties we discuss in Sections 11.3 and 11.4.

In Section 11.5 we turn to permutations and combinations, which involve counting the number of elements in various finite sets which occur in everyday life.

> For example, suppose that a 7-member committee sets out to pick 3 of its members to be officers: first the chairperson, next the vice-chairperson, and third the secretary. In how many different ways can the set of officers be picked? The answer to this question appears in Section 11.5.

We use the ideas of Section 11.5 in the final section of Chapter 11, where we discuss the Binomial Theorem, which provides us with an expansion of $(x + y)^n$ for any arbitrary positive integer n.

11.1 MATHEMATICAL INDUCTION

To set the stage for this axiom, let us consider sums of consecutive odd positive integers:

$$\begin{aligned} 1 &= 1 \\ 1 + 3 &= 4 \\ 1 + 3 + 5 &= 9 \\ 1 + 3 + 5 + 7 &= 16 \\ 1 + 3 + 5 + 7 + 9 &= 25 \end{aligned} \tag{1}$$

From the equations above, can you guess a formula for the sum

$$1 + 3 + 5 + \cdots + (2n - 1)$$

of the smallest n odd positive integers? If we notice that the right sides of the respective equations in (1) are 1^2, 2^2, 3^2, 4^2, and 5^2, then it becomes apparent that each of those equations has the form

$$1 + 3 + 5 + \cdots + (2n - 1) = n^2 \tag{2}$$

This suggests that (2) should be valid for *any* positive integer n. But how can we prove that a formula such as (2) is valid for every positive integer n? For a specific value, such as $n = 11$, we could compute the value of each side of (2) and check that the two values are the same. But it would be impossible to check (2) in this way for every positive integer n, since that would require infinitely many calculations. This is where the new axiom plays its role.

Informally, the axiom of mathematical induction states that we can prove a formula such as (2) for every positive integer n by verifying two statements:

i. The formula is true for $n = 1$.
ii. For any given positive integer k, if the formula is true for $n = k$, then it is also true for $n = k + 1$.

Let us verify (i) and (ii) for the formula in (2). In the first place, for $n = 1$ we have $2n - 1 = 1$, so that (2) becomes $1 = 1^2$, which is true. This verifies (i). Now to verify (ii), we suppose that any positive integer k is given and that the formula is true for $n = k$, that is,

$$1 + 3 + 5 + \cdots + (2k - 1) = k^2 \tag{3}$$

From this we will prove that the formula is true for $n = k + 1$; that is, we will prove that

$$1 + 3 + 5 + \cdots + (2k - 1) + [2(k + 1) - 1] = (k + 1)^2$$

which can be rewritten as

$$1 + 3 + 5 + \cdots + (2k - 1) + (2k + 1) = (k + 1)^2 \tag{4}$$

We accomplish this by adding $2k + 1$ to both sides of (3) and then combining terms to obtain

$$1 + 3 + 5 + \cdots + (2k - 1) + (2k + 1) = k^2 + (2k + 1) = (k + 1)^2$$

which is (4). Thus (ii) is verified.

We have proved (i) and (ii) for the formula in (2), so our version of mathematical induction above would imply that (2) is true for every positive integer n.

Now let us state the axiom of mathematical induction formally.

The Axiom of Mathematical Induction

For each positive integer n, let a statement (or formula or equation) $S(n)$ be given. Suppose that

i. $S(1)$ is true.
ii. For any positive integer k, if $S(k)$ is true, then $S(k + 1)$ is true.

Then $S(n)$ is true for every positive integer n.

Step (ii) is frequently called the ***inductive step***.

The ideas involved in mathematical induction might be illuminated by an analogy utilizing dominoes. Suppose an infinite string of dominoes is lined up (Figure 11.1a), and identify $S(n)$ as the statement "the nth domino falls over." In order for us to conclude that all dominoes will fall over, we need only see the initial domino fall over, and be assured that once an arbitrary kth domino falls, the neighboring $(k + 1)$st domino also falls (Figure 11.1b).

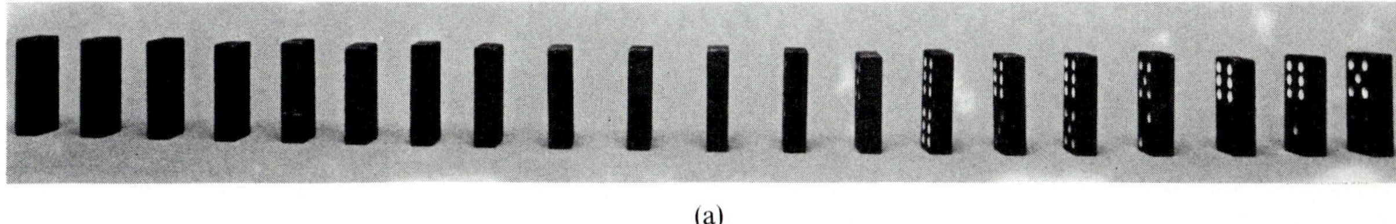

(a)

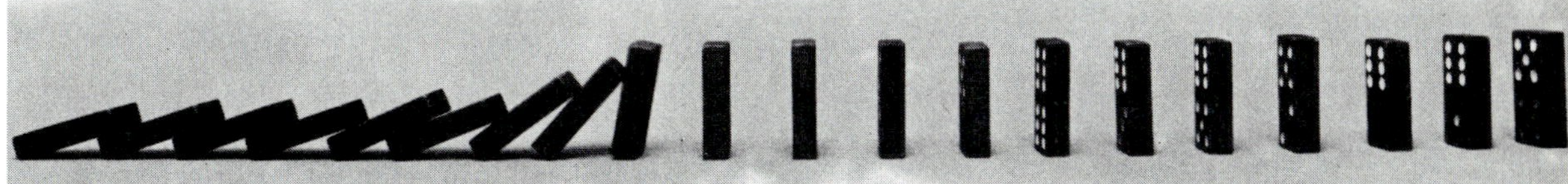

(b)

FIGURE 11.1

Mathematical induction is often used to prove that formulas are true for all positive integers. We will illustrate its use now.

EXAMPLE 1. Use mathematical induction to prove that for any positive integer n,

$$1 + 2 + 3 + \cdots + n = \frac{1}{2}n(n + 1) \tag{5}$$

Solution. Let $S(n)$ be the equation in (5). We must verify (i) and (ii) in the axiom of mathematical induction. First, $S(1)$ is the equation

$$1 = \frac{1}{2}(1)(1 + 1)$$

which is true. Therefore (i) is verified. To verify (ii), we assume that $S(k)$ is true, that is,

$$1 + 2 + 3 + \cdots + k = \frac{1}{2}k(k + 1) \tag{6}$$

From $S(k)$ we will prove that $S(k + 1)$ is true, that is,

$$1 + 2 + 3 + \cdots + k + (k + 1) = \frac{1}{2}(k + 1)(k + 2)$$

To accomplish this, we add $k + 1$ to both sides of (6) and rearrange the right side:

$$\begin{aligned} 1 + 2 + 3 + \cdots + k + (k + 1) &= \frac{1}{2}k(k + 1) + (k + 1) \\ &= (k + 1)\left(\frac{1}{2}k + 1\right) \\ &= \frac{1}{2}(k + 1)(k + 2) \end{aligned}$$

Thus from $S(k)$ we have obtained $S(k + 1)$, thereby verifying (ii). Consequently mathematical induction implies that (5) is valid for any positive integer n. □

CAUTION: Let us emphasize that in applying mathematical induction correctly, we must verify that *both* conditions (i) and (ii) hold. If we neglect one or the other of the conditions, we can be led to false statements (see Exercises 17 and 18). Moreover, in the verification of (ii), we do not prove that $S(k+1)$ is true. We prove only that *if* $S(k)$ is true, *then* $S(k+1)$ is also true.

EXAMPLE 2. Use mathematical induction to prove that for any positive integer n,

$$x^n - 1 = (x-1)(1 + x + x^2 + \cdots + x^{n-1}) \tag{7}$$

Solution. In order to verify (i) and (ii) of the mathematical induction axiom, we let the equation in (7) be $S(n)$. Then $S(1)$ is the equation

$$x^1 - 1 = (x-1)(1)$$

Since $x^1 = x$, $S(1)$ is true. Thus (i) is verified. For (ii) we let k be any given positive integer and assume that $S(k)$ is valid, that is,

$$x^k - 1 = (x-1)(1 + x + x^2 + \cdots + x^{k-1}) \tag{8}$$

From $S(k)$ we will prove that $S(k+1)$ is valid, that is,

$$x^{k+1} - 1 = (x-1)(1 + x + x^2 + \cdots + x^{k-1} + x^k) \tag{9}$$

Applying (8), we alter the right side of (9) as follows:

$$\begin{aligned}
(x-1)&(1 + x + x^2 + \cdots + x^{k-1} + x^k) \\
&= (x-1)[(1 + x + x^2 + \cdots + x^{k-1}) + x^k] \\
&= \underbrace{(x-1)(1 + x + x^2 + \cdots + x^{k-1})}_{\Downarrow (8)} + (x-1)x^k \\
&= \qquad\qquad (x^k - 1) \qquad\qquad + (x-1)x^k \\
&= x^k - 1 + x^{k+1} - x^k \\
&= x^{k+1} - 1
\end{aligned}$$

Thus from $S(k)$ we have obtained $S(k+1)$, so (ii) is verified. Consequently mathematical induction implies that (7) is valid for all positive integers n. □

Mathematical induction can also be used in order to prove that inequalities are valid.

EXAMPLE 3. Use mathematical induction to prove that for any positive integer n,

$$\text{if } |x| < 1, \quad \text{then} \quad |x^n| < 1 \tag{10}$$

Solution. Here we let $S(n)$ be the statement in (10). To verify (i) in the axiom, we notice first that $S(1)$ is the statement

$$\text{if } |x| < 1, \quad \text{then} \quad |x^1| < 1$$

Since $x^1 = x$, $S(1)$ is true, so (i) is verified. To verify (ii), we let k be any given positive integer and assume that $S(k)$ is valid, which means that

$$\text{if } |x| < 1, \quad \text{then} \quad |x^k| < 1 \tag{11}$$

From $S(k)$ we will prove that $S(k + 1)$ is valid, which means proving that

$$\text{if } |x| < 1, \quad \text{then} \quad |x^{k+1}| < 1 \tag{12}$$

Using properties of the absolute value and then (11), we find that

$$\text{if } |x| < 1, \quad \text{then} \quad |x^{k+1}| = |x^k x| = |x^k||x| < 1 \cdot 1 = 1 \tag{13}$$

Thus from $S(k)$ we have obtained $S(k + 1)$, thereby verifying (ii). By mathematical induction, (10) is true for all positive integers n. □

Extended Mathematical Induction

Occasionally we wish to prove that a statement or formula $S(n)$ is true for all integer values of n greater than or equal to some integer m, where $m \neq 1$. There is an extended version of mathematical induction that can be used to accomplish this. The inductive step (ii) remains essentially unchanged, and in step (i) we simply replace $S(1)$ by $S(m)$.

Axiom of Extended Mathematical Induction

Let m be an integer. For each integer $n \geq m$, let a statement (or formula or equation) $S(n)$ be given. Suppose that

i. $S(m)$ is true.
ii. For each integer $k \geq m$, if $S(k)$ is true, then $S(k + 1)$ is true.

Then $S(n)$ is true for every integer $n \geq m$.

To illustrate extended mathematical induction, we prove one of the Laws of Logarithms that was presented without proof in Chapter 4.

EXAMPLE 4. Let a be a positive number. Prove that

$$\ln a^n = n \ln a \quad \text{for any integer } n \geq 0 \tag{14}$$

Solution. Let $S(n)$ be the statement in (14). Here we have $m = 0$, so $S(m)$ is the statement

$$\ln a^0 = 0 \ln a \tag{15}$$

Since $a^0 = 1$ and $\ln 1 = 0$, (15) is true, and hence $S(m)$ is true. This verifies (i). Now we let k be an integer greater than or equal to 0 and suppose that $S(k)$ is valid, that is,

$$\ln a^k = k \ln a \tag{16}$$

From this we will prove that $S(k + 1)$ is valid, that is,

$$\ln a^{k+1} = (k + 1) \ln a$$

Using the formula $\ln (bc) = \ln b + \ln c$ with $b = a^k$ and $c = a$, along with (16), we have

$$\begin{aligned} \ln a^{k+1} &= \ln (a^k \cdot a) = \ln a^k + \ln a \\ &= k \ln a + \ln a = (k + 1) \ln a \end{aligned}$$

which is $S(k + 1)$. Thus from $S(k)$ we have obtained $S(k + 1)$, thereby verifying (ii). By extended mathematical induction, (14) is true for all nonnegative integers n. □

EXERCISES 11.1

In Exercises 1–16, use mathematical induction to prove the given formula for every positive integer n.

1. $2 + 4 + 6 + \cdots + 2n = n(n + 1)$
2. $1 + 4 + 7 + \cdots + (3n - 2) = \frac{1}{2}n(3n - 1)$
3. $1 + 5 + 9 + \cdots + (4n - 3) = n(2n - 1)$
4. $1 + 2 + 2^2 + \cdots + 2^{n-1} = 2^n - 1$
5. $1 + 3 + 3^2 + \cdots + 3^{n-1} = \frac{1}{2}(3^n - 1)$
6. $1 + 4 + 4^2 + \cdots + 4^{n-1} = \frac{1}{3}(4^n - 1)$
7. $\frac{1}{2} + \frac{1}{2^2} + \frac{1}{2^3} + \cdots + \frac{1}{2^n} = 1 - \frac{1}{2^n}$
8. $\frac{1}{3} + \frac{1}{3^2} + \frac{1}{3^3} + \cdots + \frac{1}{3^n} = \frac{1}{2}\left(1 - \frac{1}{3^n}\right)$
9. $\frac{1}{1 \cdot 2} + \frac{1}{2 \cdot 3} + \frac{1}{3 \cdot 4} + \cdots + \frac{1}{n(n + 1)} = \frac{n}{n + 1}$
10. $\frac{1}{1 \cdot 3} + \frac{1}{3 \cdot 5} + \frac{1}{5 \cdot 7} + \cdots + \frac{1}{(2n - 1)(2n + 1)}$
    $= \frac{n}{2n + 1}$

*11. $1^2 + 2^2 + 3^2 + \cdots + n^2 = \frac{1}{6}n(n + 1)(2n + 1)$

*12. $1^3 + 2^3 + 3^3 + \cdots + n^3 = \frac{1}{4}n^2(n + 1)^2$

*13. $1^2 + 3^2 + 5^2 + \cdots + (2n - 1)^2$
    $= \frac{1}{3}n(2n - 1)(2n + 1)$

*14. $1^3 + 3^3 + 5^3 + \cdots + (2n - 1)^3 = n^2(2n^2 - 1)$

15. $1 \cdot 2 + 3 \cdot 4 + 5 \cdot 6 + \cdots + (2n - 1)(2n)$
    $= \frac{1}{3}n(n + 1)(4n - 1)$
16. $1 \cdot 3 + 2 \cdot 4 + 3 \cdot 5 + \cdots + n(n + 2)$
    $= \frac{1}{6}n(n + 1)(2n + 7)$
17. Show that $n^2 - n + 11$ is a prime number for $n = 1, 2, \ldots, 10$ but not for $n = 11$.
18. Consider the proposed equation

$$2 + 4 + 6 + \cdots + 2n \stackrel{?}{=} n^2 + n + 2$$

   a. Under the assumption that the equation is valid for a given positive integer k, prove that it is valid for $k + 1$.
   b. Show that the equation is not valid for $n = 1, 2,$ or 3.

*c. Show that the equation is not valid if n is a multiple of 4, and use parts (a) and (b) to conclude that the equation is false for each positive integer n. (*Hint:* Show that if n is a multiple of 4, then the left side is divisible by 4 but the right side is not.)

In Exercises 19–37, use induction or extended mathematical induction to prove the given inequality or statement for the given values of n.

19. $n \leq 2^{n-1}$ for $n \geq 1$

20. $4^n > 2^n + 3^n$ for $n \geq 2$

21. $n^3 > (n+1)^2$ for $n \geq 3$

***22.** $\left(\frac{3}{2}\right)^{n-1} \geq n$ for $n \geq 5$

23. 2 is a factor of $n^2 + n$ for each positive integer n.

24. 3 is a factor of $n^3 + 2n$ for each positive integer n.

25. 4 is a factor of $5^n - 1$ for each positive integer n. (*Hint:* $5^{k+1} - 1 = 5(5^k - 1) + 4$.)

26. If $x \neq y$, then $x - y$ is a factor of $x^n - y^n$ for any positive integer n. (*Hint:* $x^{k+1} - y^{k+1} = x(x^k - y^k) + y^k(x - y)$.)

27. If $x \neq -y$, then $x + y$ is a factor of $x^{2n+1} + y^{2n+1}$ for any nonnegative integer n. (*Hint:*

$$x^{2k+3} + y^{2k+3} = (x^{2k+2} + y^{2k+2})(x + y) - xy(x^{2k+1} + y^{2k+1}).)$$

28. If $x > 1$, then $x^n > 1$ for any positive integer n.

29. The product of an even number of negative numbers is positive.

30. The product of an odd number of negative numbers is negative.

31. If x is any number and m and n are positive integers, then $(x^m)^n = x^{mn}$. (*Hint:* Let m be any fixed positive integer and use mathematical induction on n.)

32. For any positive number x and any integer $n \geq 2$, $(1 + x)^n > 1 + nx$.

33. If x is any real number and n any positive integer, then

$$(1 - x)(1 + x)(1 + x^2)(1 + x^4)\cdots(1 + x^{2^n}) = 1 - x^{2^{n+1}}$$

34. For any integer $n \geq 2$ and any real number x,

$$\cos x \cdot \cos 2x \cdot \cos 4x \cdots \cos(2^{n-1}x) = \frac{\sin 2^n x}{2^n \sin x}$$

35. For $n \geq 2$ and any real numbers $a_1, a_2, \ldots, a_n$,
a. $e^{a_1 + a_2 + \cdots + a_n} = e^{a_1}e^{a_2}\cdots e^{a_n}$
b. $\ln(a_1 a_2 \cdots a_n) = \ln a_1 + \ln a_2 + \cdots + \ln a_n$

***36.** Let $n \geq 3$. Suppose n distinct points are given, no three of which lie on a single line. Then the total number of lines joining any two of the given points is $\frac{1}{2}n(n-1)$.

***37.** The number of diagonals of a convex polygon of n sides is $\frac{1}{2}n(n-3)$.

38. Use mathematical induction to show that De Moivre's Theorem is valid: For any positive integer n any real numbers r and θ,

$$[r(\cos\theta + i\sin\theta)]^n = r^n(\cos n\theta + i\sin n\theta)$$

39.** A game called the ***Tower of Hanoi consists of three pegs and an arbitrary number of rings of distinct diameters. At the start of the game all the rings are on one peg, arranged according to size, with the smallest on top and the largest on bottom (Figure 11.2). The object of the game is to transfer all the rings to a different peg so that the rings have the same order on the new peg as on the old. Under the rules of the game, only one ring may be moved at a time, and no ring may be placed on top of a smaller ring. All three pegs may be used. Use mathematical induction to show that if there are n rings, then it is possible to complete the game in $2^n - 1$ moves.

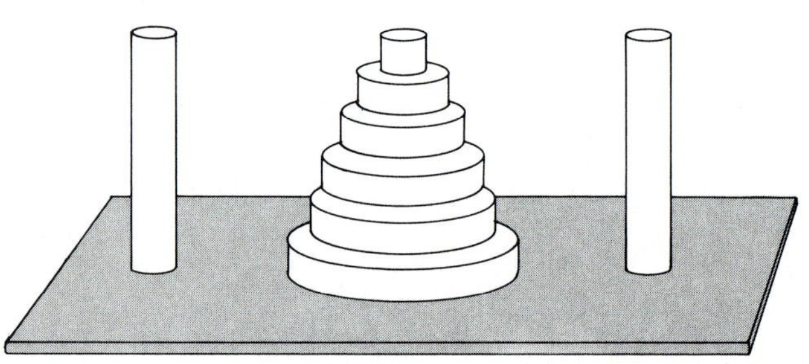

FIGURE 11.2

11.2 SEQUENCES

Earlier we studied functions whose domains were intervals or combinations of intervals. Now we will introduce functions whose domains consist of sets of integers. Throughout the remainder of this chapter the letters m and n will denote integers.

DEFINITION 11.1 A ***sequence*** is a function whose domain is the collection of all integers greater than or equal to a given integer m (usually 0 or 1).

For example, the function f defined by

$$f(n) = n^2 \quad \text{for} \quad n \geq 0 \tag{1}$$

is a sequence, since its domain consists of all nonnegative integers. For a few specific values of f we find that

$$f(0) = 0^2 = 0, \qquad f(4) = 4^2 = 16, \quad \text{and} \quad f(19) = 19^2 = 361$$

In this particular case, the range of f consists of all integers that are squares of integers.

Another example of a sequence is given by

$$g(n) = (-1)^n \quad \text{for} \quad n \geq 5 \tag{2}$$

The range of g consists of -1 and 1, because $(-1)^n = 1$ or -1, depending on whether n is even or odd. A sequence whose range does not consist solely of integers is given by

$$h(n) = \frac{1}{n} \quad \text{for} \quad n \geq 1 \tag{3}$$

The range of h consists of the numbers $1, \frac{1}{2}, \frac{1}{3}, \ldots$.

Let f represent an arbitrary sequence, and let the value assigned to any integer n in the domain be written a_n (read as "a sub n"). Then a formula for f is

$$f(n) = a_n \quad \text{for} \quad n \geq m \tag{4}$$

The numbers $a_m, a_{m+1}, a_{m+2}, \ldots$, completely determine the sequence f, and because the values of the sequence can be listed in this way, we normally suppress the expression $f(n)$ appearing in (4) and write $\{a_n\}_{n=m}^{\infty}$ to represent the sequence in (4). Thus the sequences given in (1)–(3) would be given in the form

$$\{n^2\}_{n=0}^{\infty}, \qquad \{(-1)^n\}_{n=5}^{\infty}, \quad \text{and} \quad \left\{\frac{1}{n}\right\}_{n=1}^{\infty}$$

In order to get a feeling for the numbers in a sequence, we frequently write out the first few terms of the sequence. For example, we would write

$$0, 1, 4, 9, 16, \ldots \quad \text{for} \quad \{n^2\}_{n=0}^{\infty}$$

$$-1, 1, -1, 1, -1, 1, \ldots \quad \text{for} \quad \{(-1)^n\}_{n=5}^{\infty}$$

$$1, \frac{1}{2}, \frac{1}{3}, \frac{1}{4}, \ldots \quad \text{for} \quad \left\{\frac{1}{n}\right\}_{n=1}^{\infty}$$

The numbers $a_m, a_{m+1}, a_{m+2}, \ldots$ are the ***terms*** of the sequence $\{a_n\}_{n=m}^{\infty}$, with a_m the first term, a_{m+1} the second term, and so on. The numbers m, $m+1, m+2, \ldots$ are the ***indexes*** of the sequence, and m is the ***initial index***; after all, a_m is the first term one sees when listing them in order. For $\{1/n\}_{n=1}^{\infty}$ we have

1st term *2nd term* *3rd term* *4th term*

$$1, \frac{1}{2}, \frac{1}{3}, \frac{1}{4}, \ldots$$

EXAMPLE 1. Find the numerical values of the first, second, third, and seventh terms of the sequence $\{a_n\}_{n=m}^{\infty}$, where

a. $a_n = \dfrac{1}{2^n}$ for $n \geq 1$

b. $a_n = 2n - 4$ for $n \geq 0$

c. $a_n = 2n - 4$ for $n \geq 3$

Solution.

a. For this sequence, $m = 1$, so the terms we seek are a_1, a_2, a_3, and a_7:

$$a_1 = \frac{1}{2^1} = \frac{1}{2} \qquad a_3 = \frac{1}{2^3} = \frac{1}{8}$$

$$a_2 = \frac{1}{2^2} = \frac{1}{4} \qquad a_7 = \frac{1}{2^7} = \frac{1}{128}$$

b. Here $m = 0$, so the terms we seek are a_0, a_1, a_2, and a_6:

$$a_0 = 2 \cdot 0 - 4 = -4 \qquad a_2 = 2 \cdot 2 - 4 = 0$$

$$a_1 = 2 \cdot 1 - 4 = -2 \qquad a_6 = 2 \cdot 6 - 4 = 8$$

c. In this case $m = 3$, so the terms we seek are a_3, a_4, a_5, and a_9:

$$a_3 = 2 \cdot 3 - 4 = 2 \qquad a_5 = 2 \cdot 5 - 4 = 6$$

$$a_4 = 2 \cdot 4 - 4 = 4 \qquad a_9 = 2 \cdot 9 - 4 = 14 \quad \square$$

EXAMPLE 2. Find the specified terms of the following sequences.

a. the tenth term, where $a_n = \dfrac{1}{(n+1)(n+2)}$ for $n \geq 1$

b. the eighty-fourth term, where $a_n = \pi$ for $n \geq 3$

Solution.

a. Since $m = 1$, the tenth term is a_{10}. We find that

$$a_{10} = \frac{1}{(10+1)(10+2)} = \frac{1}{11 \cdot 12} = \frac{1}{132}$$

b. Since $m = 3$, the eighty-fourth term is a_{86}, and since each term is π, it follows that

$$a_{86} = \pi \quad \square$$

The sequence given in part (b) of Example 2 has only one value, namely π. We say that $\{a_n\}_{n=m}^{\infty}$ is a ***constant sequence*** if it has but one value, that is, if there is a unique number c such that

$$a_n = c \quad \text{for} \quad n \geq m$$

Thus the sequence in part (b) of Example 2 is a constant sequence.

Not all sequences have terms all of which are as easy to calculate as those in Examples 1 and 2 are. A sequence $\{a_n\}_{n=m}^{\infty}$ is given ***recursively*** (or inductively) if the initial term a_m is given, and if we can determine a_{n+1} by knowing $a_m, a_{m+1}, \ldots, a_n$.

EXAMPLE 3. Find the first four terms of the sequences described below.

a. Initial term is a_1, $a_1 = 3$, and $a_{n+1} = a_n - 2$ for $n \geq 1$

b. Initial term is a_0, $a_0 = 1$, and $a_{n+1} = \frac{1}{3}a_n$ for $n \geq 0$

Solution.

a. $a_1 = 3$ $\qquad a_3 = a_2 - 2 = 1 - 2 = -1$

$a_2 = a_1 - 2 = 3 - 2 = 1$ $\qquad a_4 = a_3 - 2 = -1 - 2 = -3$

b. $a_0 = 1$ $\qquad a_2 = \dfrac{1}{3}a_1 = \dfrac{1}{3} \cdot \dfrac{1}{3} = \dfrac{1}{9}$

$a_1 = \dfrac{1}{3}a_0 = \dfrac{1}{3} \cdot 1 = \dfrac{1}{3}$ $\qquad a_3 = \dfrac{1}{3}a_2 = \dfrac{1}{3} \cdot \dfrac{1}{9} = \dfrac{1}{27}$ $\quad \square$

By the method used in the solution, you should be able to calculate the value of, say, the 10th term, or the 100th term of each sequence, and you should be able to guess a formula for the numerical value of the nth term (see Exercise 67).

An important sequence that can be defined recursively is given by

$$a_0 = 1, \quad \text{and} \quad a_{n+1} = (n+1)a_n \quad \text{for} \quad n \geq 0 \tag{5}$$

The value of a_n is called ***n factorial*** and is denoted by $n!$. The relation $a_{n+1} = (n+1)a_n$ can be written in factorial form as

$$\boxed{(n+1)! = (n+1)n!}$$

The first few terms of $n!$ are given by

$$\begin{aligned}
0! &= 1 \\
1! &= 1 \cdot 0! = 1 \\
2! &= 2 \cdot 1! = 2 \cdot 1 = 2 \\
3! &= 3 \cdot 2! = 3 \cdot 2 \cdot 1 = 6 \\
4! &= 4 \cdot 3! = 4 \cdot 3 \cdot 2 \cdot 1 = 24 \\
5! &= 5 \cdot 4! = 5 \cdot 4 \cdot 3 \cdot 2 \cdot 1 = 120 \\
6! &= 6 \cdot 5! = 6 \cdot 5 \cdot 4 \cdot 3 \cdot 2 \cdot 1 = 720 \\
7! &= 7 \cdot 6! = 7 \cdot 6 \cdot 5 \cdot 4 \cdot 3 \cdot 2 \cdot 1 = 5040
\end{aligned}$$

In general, if $n \geq 1$, then $n!$ is the product of the first n positive integers:

$$\boxed{n! = n(n-1)(n-2)\cdots 3 \cdot 2 \cdot 1} \tag{6}$$

CAUTION: Formula (6) does not give a value for 0!; the right side does not even make sense for $n = 0$. However, we defined 0! by the formula $0! = 1$ above.

When the terms in a sequence are not easily computed, we sometimes approximate their values.

EXAMPLE 4. Calculate the first four terms of the sequence $\left\{\left(1 + \frac{1}{n}\right)^n\right\}_{n=1}^{\infty}$.

Solution. If $a_n = \left(1 + \dfrac{1}{n}\right)^n$ for $n \geq 1$, then

$$\begin{aligned}
a_1 &= \left(1 + \frac{1}{1}\right)^1 = 2 \\
a_2 &= \left(1 + \frac{1}{2}\right)^2 = \left(\frac{3}{2}\right)^2 = \frac{9}{4} = 2.25 \\
a_3 &= \left(1 + \frac{1}{3}\right)^3 = \left(\frac{4}{3}\right)^3 = \frac{64}{27} \approx 2.3703704 \\
a_4 &= \left(1 + \frac{1}{4}\right)^4 = \left(\frac{5}{4}\right)^4 = \frac{625}{256} \approx 2.4414063 \quad \square
\end{aligned}$$

To calculate higher terms of the sequence in Example 4, say a_{10} or a_{1000}, requires ever more patience, or help from a calculator. For instance, by calculator we find that

$$a_{10} = \left(1 + \frac{1}{10}\right)^{10} \approx 2.5937425$$

and

$$a_{1000} = \left(1 + \frac{1}{1000}\right)^{1000} \approx 2.7169239$$

It turns out that as n increases, the value of $(1 + 1/n)^n$ gets nearer and nearer to e. In fact, one can define e to be the value to which the numbers $(1 + 1/n)^n$ tend as n increases without bound.

Not all sequences can be represented by neat numerical formulas. Indeed, if we call a nontrivial factor of an integer n any factor other than 1, -1, n, and $-n$ and if for $n \geq 1$ we let

$$a_n = \text{the number of nontrivial factors of } n$$

then the sequence $\{a_n\}_{n=1}^{\infty}$ begins as follows:

$$0, 0, 0, 2, 0, 4, \ldots$$

However there is no simple formula that tells immediately the numerical value of a_n for each positive integer n.

It is time to mention one of the most famous sequences in the mathematical literature, the Fibonacci sequence, named after the thirteenth-century mathematician Fibonacci (whose real name was Leonardo de Pisa). He posed and solved the following problem.

Suppose a rabbit colony starts out with 1 pair of adult rabbits. Assume that each pair of adult rabbits produces a pair of offspring (one male and one female) every month, and assume that rabbits produce their first offspring at the age of 2 months and continue to live forever. The problem is to determine how many pairs of adult rabbits are there at the end of n months, for any positive integer n.

To help in the solution of this problem, let us call rabbits less than 1 month old babies, those between 1 and 2 months old adolescents, and all older rabbits adults. From one month to the next, babies become adolescents, adolescents become adults, adults remain adults, and two new babies are born to each pair of adults. At the end of successive months we have:

Month	*Pairs of Adults*	*Pairs of Adolescents*	*Pairs of Babies*
1	1	0	1
2	1	1	1
3	2	1	2
4	3	2	3
5	5	3	5
6	8	5	8
7	13	8	13
8	21	13	21

If for $n \geq 1$ we let

$$a_n = \text{the number of pairs of adults at the end of } n \text{ months}$$

then the sequence $\{a_n\}_{n=1}^{\infty}$ is called the ***Fibonacci sequence***. We can read the first few terms of the Fibonacci sequence from our table above:

$$a_1 = 1, \qquad a_2 = 1, \qquad a_3 = 2, \qquad a_4 = 3, \qquad a_5 = 5$$

In general,

$$a_{n+1} = a_n + a_{n-1} \quad \text{for} \quad n \geq 2$$

This sequence turns up not only in higher mathematics but in nature. For example, numbers in the Fibonacci sequence appear in the study of population genetics and in phyllotaxis (the arrangement of leaves along a stem of a plant).

Graphs of Sequences

The graph of a sequence is not as revealing as graphs of the kinds of functions we described in earlier chapters, because the graph of a sequence is just a collection of isolated points, one on each vertical line $x = n$ for $n \geq m$. The following example illustrates one such graph.

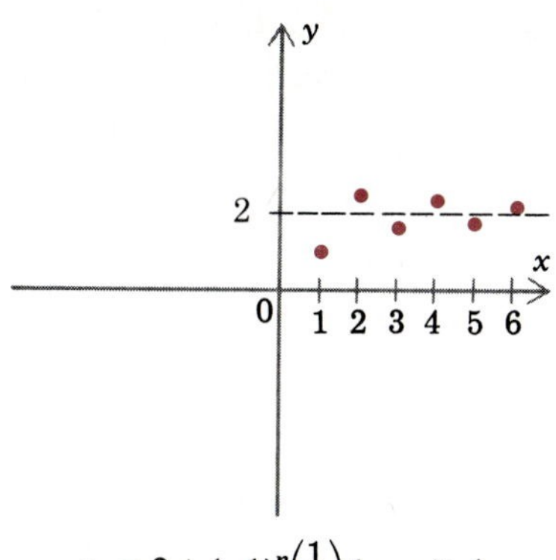

$a_n = 2 + (-1)^n\left(\frac{1}{n}\right)$ for $n \geq 1$

FIGURE 11.3

EXAMPLE 5. Sketch the graph of the sequence

$$\left\{2 + (-1)^n\left(\frac{1}{n}\right)\right\}_{n=1}^{\infty}$$

Solution. If $a_n = 2 + (-1)^n\left(\frac{1}{n}\right)$ for $n \geq 1$, then

$$a_1 = 1, \quad a_2 = \frac{5}{2}, \quad a_3 = \frac{5}{3}, \quad a_4 = \frac{9}{4}, \quad a_5 = \frac{9}{5}, \quad a_6 = \frac{13}{6}$$

The graph, which is shown in Figure 11.3, suggests that a_n gets closer and closer to 2 as n increases. □

Partial Sums of Sequences

Upon occasion it is important to find the sum of the first n terms of a sequence. For any sequence $\{a_n\}_{n=1}^{\infty}$ the expression $\Sigma_{j=1}^{n} a_j$ represents the sum of the first n terms of the sequence and is called the ***nth partial sum*** of the sequence. Thus

$$\sum_{j=1}^{n} a_j = a_1 + a_2 + a_3 + \cdots + a_n$$

The symbol Σ is a capital Greek sigma, which we use to represent a sum.

EXAMPLE 6. Find the following partial sums.

a. $\sum_{j=1}^{5} j$ b. $\sum_{j=1}^{6} j^2$ c. $\sum_{j=1}^{4} \frac{1}{j!}$

Solution.

a. $$\sum_{j=1}^{5} j = 1 + 2 + 3 + 4 + 5 = 15$$

b. $$\sum_{j=1}^{6} j^2 = 1^1 + 2^2 + 3^2 + 4^2 + 5^2 + 6^2 = 91$$

c. $$\sum_{j=1}^{4} \frac{1}{j!} = \frac{1}{1!} + \frac{1}{2!} + \frac{1}{3!} + \frac{1}{4!} = \frac{41}{24}$$ □

EXAMPLE 7. Find the nth partial sum $\Sigma_{j=1}^{n} j$ of the sequence $\{n\}_{n=1}^{\infty}$.

Solution. By Example 1 of Section 11.1,

$$\sum_{j=1}^{n} j = 1 + 2 + 3 + \cdots + n = \frac{1}{2}n(n+1)$$ □

If the initial index of a sequence is not 1, then the nth partial sum is changed accordingly. For example, the nth partial sum of the sequence $\{a_n\}_{n=0}^{\infty}$, which has initial index 0, is $\Sigma_{j=0}^{n-1} a_j$. Thus the fourth partial sum of $\{1/2^n\}_{n=0}^{\infty}$ is

$$\sum_{j=0}^{3} \frac{1}{2^j} = \frac{1}{2^0} + \frac{1}{2^1} + \frac{1}{2^2} + \frac{1}{2^3} = 1 + \frac{1}{2} + \frac{1}{4} + \frac{1}{8} = \frac{15}{8}$$

EXERCISES 11.2

In Exercises 1–6, write the given sequence in the form $\{a_n\}_{n=1}^{\infty}$.

1. $\frac{1}{2}, \frac{1}{3}, \frac{1}{4}, \frac{1}{5}, \cdots$

2. $\frac{1}{2}, \frac{2}{3}, \frac{3}{4}, \frac{4}{5}, \cdots$

3. $-1, 1, -1, 1, \cdots$

4. $0, \dfrac{\ln 2}{2}, \dfrac{\ln 3}{3}, \dfrac{\ln 4}{4}, \cdots$

5. $\frac{8}{9}, \frac{16}{27}, \frac{32}{81}, \frac{64}{243}, \cdots$

6. $1, \frac{1}{2}, \frac{1}{6}, \frac{1}{24}, \frac{1}{120}, \cdots$

In Exercises 7–16, find the first four terms of the sequence $\{a_n\}_{n=1}^{\infty}$ with the given nth term.

7. $a_n = \frac{1}{2}\sqrt{3}$

8. $a_n = 3n - 1$

9. $a_n = 2^n$

10. $a_n = \dfrac{1}{3^{n+1}}$

11. $a_n = (-1)^{n+1}$

12. $a_n = (-1)^{n+1}\dfrac{1}{2^n}$

13. $a_n = \dfrac{1}{n(n+1)}$

14. $a_n = n^n$

15. $a_n = \ln n$

16. $a_n = \dfrac{n-1}{n!}$

In Exercises 17–26, find the first five terms of the sequence defined recursively.

17. $a_1 = 2; a_n = 2a_{n-1}$

18. $a_1 = -1; a_n = 3a_{n-1}$

19. $a_1 = 3; a_n = 4 + a_{n-1}$

20. $a_1 = 3; a_n = 4 - a_{n-1}$

21. $a_1 = 32; a_n = \frac{1}{2}a_{n-1}$

22. $a_1 = -243; a_n = -\frac{1}{3}a_{n-1}$

23. $a_1 = 1; a_n = na_{n-1}$

24. $a_1 = 28; a_n = \dfrac{1}{a_{n-1}}$

25. $a_1 = 1; a_n = \dfrac{na_{n-1}}{(n-1)!}$

26. $a_1 = 2^{1/12}; a_n = a_{n-1}^n$

In Exercises 27–32, sketch the graph of the sequence with the given nth term.

27. $a_n = \dfrac{2}{n}$

28. $a_n = 1 + \dfrac{1}{2^n}$

29. $a_n = (-1)^n$

30. $a_n = n(-1)^n$

31. $a_n = 2(-1)^{n-1}$

32. $a_n = \dfrac{n}{n+1}$

In Exercises 33–42, find the indicated partial sum.

33. $\sum_{j=1}^{10} 1$

34. $\sum_{j=1}^{6} (-3)$

35. $\sum_{j=1}^{9} (-1)^j$

36. $\sum_{j=1}^{100} (-1)^j$

37. $\sum_{j=1}^{5} [2 + (-1)^j]$

38. $\sum_{j=1}^{4} \dfrac{1}{j}$

39. $\sum_{j=1}^{3} \dfrac{1}{2^j}$

40. $\sum_{j=1}^{6} \dfrac{1}{2^j}$

41. $\sum_{j=1}^{10} \left(\dfrac{1}{j+1} - \dfrac{1}{j}\right)$

42. $\sum_{j=1}^{100} \left(\dfrac{1}{j+1} - \dfrac{1}{j}\right)$

In Exercises 43–48, express the given sum in sigma notation.

43. $1 + 2 + 3 + 4 + 5 + 6 + 7$

44. $1^2 + 2^2 + 3^2 + 4^2 + 5^2 + 6^2$

45. $1^1 + 2^2 + 3^3 + 4^4 + 5^5 + 6^6 + 7^7 + 8^8 + 9^9$

46. $\frac{1}{3} + \frac{2}{4} + \frac{3}{5} + \frac{4}{6} + \frac{5}{7}$

47. $1 + 3 + 5 + 7 + 9 + 11 + 13 + 15$

48. $1 + \frac{1}{1} + 2 + \frac{1}{2} + 3 + \frac{1}{3} + 4 + \frac{1}{4}$

In Exercises 49–56, use a calculator to approximate the indicated values of a_n.

49. $a_n = \dfrac{\ln n}{n^2}$; a_{10}, a_{100}

50. $a_n = \dfrac{2n - 19}{n^2 + 17}$; a_{39}, a_{500}

51. $a_n = \dfrac{n!}{n^n}$; a_5, a_{15}

52. $a_n = e^{-n^2}$; a_2, a_4

53. $a_n = \dfrac{\sin n}{n}$; a_{200}, a_{4000}

54. $a_n = \sqrt[n]{100}$; a_{10}, a_{200}

55. $a_n = \sqrt[n]{n}$; a_{100}, a_{1000}

56. $a_n = \sqrt[n]{n!}$; a_{10}, a_{20}

In Exercises 57–60, use a calculator to approximate the value of the given partial sum.

57. $\displaystyle\sum_{j=1}^{10} \frac{1}{j}$ **58.** $\displaystyle\sum_{j=1}^{20} \frac{1}{j}$ **59.** $\displaystyle\sum_{j=1}^{15} \frac{1}{j^2}$ **60.** $\displaystyle\sum_{j=1}^{12} \frac{1}{j!}$

61. Find the eighth partial sum of the sequence of prime numbers, whose first five terms are 2, 3, 5, 7, 11 . . .

62. Use mathematical induction to prove that

$$\sum_{j=1}^{n} \frac{1}{2^j} = 1 - \frac{1}{2^n} \quad \text{for } n \geq 1$$

63. Use mathematical induction to prove that the following equations are valid for $n \geq 1$.

a. $\displaystyle\sum_{j=1}^{2n} (-1)^j = 0$ b. $\displaystyle\sum_{j=1}^{2n-1} (-1)^j = -1$

64. Use mathematical induction to prove that if $\{a_n\}_{n=1}^{\infty}$ is defined recursively by $a_1 = 1$ and $a_n = na_{n-1}$ for $n \geq 2$, then $a_n = n!$ for $n \geq 1$.

65. Let $a_n =$ the nth digit after the decimal point in the decimal expansion of π. Write a_5 and a_8.

66. Let $a_n =$ the nth digit after the decimal point in the decimal expansion of $\frac{1}{7}$. Write a_4, a_{10}, and a_{100}.

67. a. Let $\{a_n\}_{n=1}^{\infty}$ be the sequence defined in Example 3(a). Write out the first several terms of the sequence until you see a pattern emerge. Then guess a formula that expresses a_n in terms of n, and prove it by induction.

b. Repeat part (a) for the sequence $\{a_n\}_{n=0}^{\infty}$ defined in Example 3(b).

68. Let $a_n = \dfrac{(1 + \sqrt{5})^n - (1 - \sqrt{5})^n}{2^n\sqrt{5}}$ for $n \geq 1$.

a. Show that $a_1 = 1$ and $a_2 = 1$.

*b. Show that if $n \geq 1$, then $a_{n+2} = a_{n+1} + a_n$, so that $\{a_n\}_{n=1}^{\infty}$ is the Fibonacci sequence. (*Hint:* Evaluate $a_{n+1} + a_n$ by adding separately the parts containing powers of $(1 + \sqrt{5})$ and the parts containing powers of $(1 - \sqrt{5})$, factoring out common factors in the process. Use the fact that $6 + 2\sqrt{5} = (1 + \sqrt{5})^2$ and $6 - 2\sqrt{5} = (1 - \sqrt{5})^2$.)

c. Use parts (a) and (b) to conclude that a_n is an integer for each positive integer n.

69. Let $\{a_n\}_{n=1}^{\infty}$ be the Fibonacci sequence, defined by $a_1 = 1$, $a_2 = 1$, and $a_{n+1} = a_n + a_{n-1}$ for $n \geq 2$.

a. Show that $a_n = a_{n+2} - a_{n+1}$ for $n \geq 1$.

b. Using (a) and mathematical induction, show that $\sum_{j=1}^{n} a_j = a_{n+2} - 1$ for $n \geq 1$. (*Hint:* In (a) replace n by $n + 1$.)

70. Express the following in terms of factorials:

a. $2 \cdot 4 \cdot 6 \cdots (2n)$ b. $\dfrac{1 \cdot 3 \cdot 5 \cdots (2n - 1)}{2 \cdot 4 \cdot 6 \cdots (2n)}$

11.3 ARITHMETIC SEQUENCES

Suppose we deposit \$100 in a savings account paying 6% simple interest. This means that after the first year the balance is \$106, and after each succeeding year the balance increases by \$6. If a_n denotes the number of dollars in the account after n years, we have

$$a_1 = 106, \qquad a_2 = 112, \qquad a_3 = 118, \qquad a_4 = 124$$

and in general,

$$a_n = 106 + \overbrace{6 + \cdots + 6}^{n-1}$$

which can be written more simply as

$$a_n = 106 + (n-1)6 \tag{1}$$

The sequence $\{a_n\}_{n=1}^{\infty}$ given in (1) has the property that each term after the initial term is obtained from the preceding one by adding the same fixed number, namely, 6. Such a sequence is called an arithmetic sequence (or arithmetic progression).

DEFINITION 11.2 An ***arithmetic sequence*** is a sequence such that the difference between any term and the succeeding term is constant.

By definition, a sequence $\{a_n\}_{n=1}^{\infty}$ is an arithmetic sequence if and only if there is a number d such that

$$a_{n+1} - a_n = d \quad \text{for} \quad n \geq 1$$

or equivalently,

$$\boxed{a_{n+1} = a_n + d \quad \text{for} \quad n \geq 1} \tag{2}$$

The number d is called the ***common difference*** of the sequence. It follows from the discussion before Definition 11.2 that for the sequence given by (1), the common difference $d = 6$.

EXAMPLE 1. Show that the sequences listed below are arithmetic sequences, and find their common differences.

a. $\{2n + 3\}_{n=1}^{\infty}$ b. 3, 2, 1, 0, . . .

Solution.

a. If $a_n = 2n + 3$ for $n \geq 1$, then

$$\begin{aligned} a_{n+1} - a_n &= [2(n+1) + 3] - (2n + 3) \\ &= 2n + 5 - 2n - 3 \\ &= 2 \end{aligned}$$

Therefore the common difference is 2, so the sequence is an arithmetic sequence.

b. In this case, notice that $a_n = 4 - n$ for $n \geq 1$. Then for $n \geq 1$ we have

$$\begin{aligned} a_{n+1} - a_n &= [4 - (n + 1)] - (4 - n) \\ &= 3 - n - 4 + n \\ &= -1 \end{aligned}$$

Consequently the common difference is -1, and this sequence is also an arithmetic sequence. □

If one knows the initial term of an arithmetic sequence, which we normally designate by a_1, and the common difference d, then the value of a_n can be calculated by the formula

$$\boxed{a_n = a_1 + (n - 1)d \quad \text{for} \quad n \geq 1} \tag{3}$$

The formula can be proved by induction (see Exercise 52).

EXAMPLE 2. Find a formula for the nth term of the arithmetic sequence whose initial term a_1 and common difference d are given below.

a. $a_1 = 2,\ d = 5$ b. $a_1 = \frac{2}{3},\ d = -\frac{4}{3}$ c. $a_1 = 2,\ d = 0$

Solution. In each case we use (3):

a. $a_n = 2 + (n - 1)5 = -3 + 5n$

b. $a_n = \frac{2}{3} + (n - 1)(-\frac{4}{3}) = 2 - \frac{4}{3}n$

c. $a_n = 2 + (n - 1)0 = 2$ □

Partial Sums of Arithmetic Sequences

Theorem 11.3, which we present next, includes two convenient formulas for computing partial sums of arithmetic sequences.

THEOREM 11.3 *Let $\{a_n\}_{n=1}^{\infty}$ be an arithmetic sequence. The nth partial sum of the sequence is given by either of the following two formulas:*

$$\begin{aligned} \sum_{j=1}^{n} a_j &= a_1 + (a_1 + d) + (a_1 + 2d) + \cdots + (a_1 + (n - 1)d) \\ &= na_1 + \frac{n(n - 1)}{2} \cdot d \end{aligned} \tag{4}$$

$$\begin{aligned} \sum_{j=1}^{n} a_j &= a_1 + (a_1 + d) + (a_1 + 2d) + \cdots + (a_1 + (n - 1)d) \\ &= \frac{n}{2}(a_1 + a_n) \end{aligned} \tag{5}$$

EXAMPLE 3. Find the sum of the first eight terms of the arithmetic sequence that begins 3, 7, ...

Solution. For the given sequence, $a_1 = 3$ and $a_2 = 7$, so that

$$d = a_2 - a_1 = 7 - 3 = 4$$

Thus by (4) with $n = 8$, the sum S is given by

$$S = 8(3) + \frac{8 \cdot 7}{2} 4 = 24 + 112 = 136 \quad \square$$

EXERCISES 11.3

In Exercises 1–6, show that the sequence is arithmetic, and find the common difference.

1. $\{2 + 3n\}_{n=1}^{\infty}$
2. $\{4 - \pi n\}_{n=1}^{\infty}$
3. $\{5 + \frac{1}{2}n\}_{n=1}^{\infty}$
4. $\left\{\frac{3 + 2n}{4}\right\}_{n=1}^{\infty}$
5. $\{\ln 2^n\}_{n=1}^{\infty}$
6. $\{e^{1 + \ln 3n}\}_{n=1}^{\infty}$

In Exercises 7–14, determine whether the given sequence is arithmetic.

7. 1, 3, 6, 10, . . .
8. 1, −1, 1, −1, . . .
9. 3, 6, 9, 12, . . .
10. −2, 3, 8, 13, . . .
11. $\{2n + 1\}_{n=1}^{\infty}$
12. $\{3 - 2n^2\}_{n=1}^{\infty}$
13. $\{2 - 3 \ln n\}_{n=1}^{\infty}$
14. $\{\ln (3^n)\}_{n=1}^{\infty}$

In Exercises 15–22, find the fourth term and the nth term of the arithmetic sequence whose initial term a_1 and common difference d are given.

15. $a_1 = 2, d = 3$
16. $a_1 = \frac{3}{4}, d = -1$
17. $a_1 = 2\pi, d = 0$
18. $a_1 = \sqrt{2}, d = 2\sqrt{2}$
19. $a_1 = -3, d = \frac{2}{3}$
20. $a_1 = -3, d = -\frac{2}{3}$
21. $a_1 = e, d = 2$
22. $a_1 = 0, d = -\frac{1}{2}$

In Exercises 23–28, find the seventh term and the nth term of the arithmetic sequences whose first two terms are given.

23. 4, 7, . . .
24. 3, 2, . . .
25. $0, -\frac{3}{2}, \ldots$
26. $\sqrt{3}, 3\sqrt{3}, \ldots$
27. ln 2, ln 4, . . .
28. $\log_2 3, \log_2 1, \ldots$

In Exercises 29–34, find a formula for the nth term of the arithmetic sequence $\{a_n\}_{n=1}^{\infty}$ that satisfies the given conditions.

29. $a_2 = 5, a_4 = 3$
30. $a_4 = 3, d = 5$
31. $a_3 = 6, a_6 = 1$
32. $a_9 = 16, d = 2$
33. $a_4 = 3, a_6 - a_1 = 5$
34. $a_6 = -5, a_{10} - a_2 = -12$

In Exercises 35–42, find $\Sigma_{j=1}^{n} a_j$ for the arithmetic sequence that satisfies the given conditions.

35. $a_1 = 3, d = 2, n = 7$
36. $a_1 = 1, d = -3, n = 8$
37. $a_1 = -2, d = 1, n = 12$
38. $a_3 = -\sqrt{2}, d = \sqrt{2}, n = 18$
39. $a_2 = \ln 4, d = \ln 2, n = 20$
40. $a_1 = 2, a_6 = -18, n = 6$
41. $a_1 = -3\pi, a_{10} = 21\pi/2, n = 10$
42. $a_2 = 1, a_6 = -7, n = 7$
43. In the arithmetic sequence that begins −3, 4, . . . , is 277 a term? Explain why or why not.
44. In the arithmetic sequence that begins 2, 5, . . . , is 1000 a term? Explain why or why not.
45. The sum of the second and the sixth terms of a given arithmetic sequence is 2, and the sum of the fifth and the ninth terms is −10. Find a formula for the nth term of the sequence.
46. Find the sum of all positive even integers less than 100.
47. Find the sum of all positive odd integers less than 100.
48. The first three terms of an arithmetic sequence have the form x, $2x + 1$, and $4x - 1$ for a suitable real number x. Determine x and the nth term a_n of the sequence.
49. The sum of the first five terms of an arithmetic sequence is 55, and the fifth term is 15. Determine the nth term a_n of the sequence.

50. The sum of the first six terms of an arithmetic sequence is 24, and the common difference is 24. Determine the nth term a_n of the sequence.

51. Let a and b be real numbers. Show that $(a^2 - 2ab - b^2)^2$, $(a^2 + b^2)^2$, and $(a^2 + 2ab - b^2)^2$ form the first three terms of an arithmetic sequence, and find the common difference.

52. Prove formula (3) by induction.

53. The number of times each hour that a chime clock strikes is equal to the number of the hour. How many times does the clock strike in a 24-hour day?

54. During a 12-week summer vacation a student saves \$10 the first week, \$15 the second week, \$20 the third week, and so on. How much money does the student save during the summer?

55. A boy wishes to save for 8 weeks in order to buy a bicycle for \$66. If he saves \$3 the first week, and wishes to increase his savings by a constant amount each week, how much must the weekly increase be in order that \$66 be saved after 8 weeks?

56. A bricklayer plans to build a brick structure on which to mount a sign. If the first layer of bricks is to contain 30 bricks, the second 29, and third 28, and so on with the top layer containing 9 bricks, how many bricks will the bricklayer need?

***57.** The Great Pyramid of Cheops was originally about 480 feet tall and had a square base approximately 754 feet on a side. Each layer was approximately 3 feet tall and succeeding layers indented 2.35 feet (see Figure 11.4).

The Great Pyramid of Cheops

FIGURE 11.4

a. Approximate the total area of the vertical surfaces on the four sides of the pyramid.
b. Without using sequences, determine the total area of the horizontal surfaces on the four sides of the pyramid.

58. According to legend, during the last weeks of his life the French mathematician Abraham de Moivre noticed that he needed a quarter-hour more sleep each night than on the preceding night, and he predicted that as soon as he needed a 24-hour sleep he would die. If all this were true, how much sleep did he need after the 12-hour sleep and before death?

11.4 GEOMETRIC SEQUENCES

If instead of successive terms of a sequence having the same difference, the successive terms have the same ratio, the sequence is called a geometric sequence (or geometric progression).

DEFINITION 11.4 A ***geometric sequence*** (or ***geometric progression***) is a sequence such that the quotient of any term with its predecessor is the same nonzero number, called the ***common ratio*** of the geometric sequence.

For example, the sequences

$$1, 2, 4, 8, 16, \ldots \quad \text{and} \quad \{3^{-n}\}_{n=0}^{\infty}$$

are geometric. For the first of these sequences the quotient of any term with its predecessor is 2, and for the second it is $\frac{1}{3}$.

If $\{a_n\}_{n=1}^{\infty}$ is a geometric sequence and r its common ratio, then for any $n \geq 1$,

$$\boxed{\frac{a_{n+1}}{a_n} = r} \tag{1}$$

This means that

$$a_2 = a_1 r$$
$$a_3 = a_2 r = (a_1 r)r = a_1 r^2$$
$$a_4 = a_3 r = (a_1 r^2)r = a_1 r^3$$

and in general,

$$\boxed{a_n = a_1 r^{n-1}} \tag{2}$$

a formula that can be proved inductively (see Exercise 70).

EXAMPLE 1. Find the sixth term and the common ratio of the geometric sequence $\{a_n\}_{n=1}^{\infty}$ that begins 2, 6, ...

Solution. We are given that $a_1 = 2$ and $a_2 = 6$. Therefore

$$r = \frac{a_2}{a_1} = \frac{6}{2} = 3$$

so by (2),

$$a_n = a_1 r^{n-1} = 2 \cdot 3^{n-1}$$

Taking $n = 6$, we find that

$$a_6 = 2(3^5) = 2(243) = 486 \quad \square$$

EXAMPLE 2. Find a formula for the nth term of the geometric sequence $\{a_n\}_{n=1}^{\infty}$ such that

a. $a_1 = 4, r = 2$ b. $a_1 = 1, r = \frac{1}{5}$

c. $a_1 = -1, r = -\frac{1}{3}$

Solution. In each case we use (2):

a. $a_n = 4(2)^{n-1} = 2^{n+1}$

b. $a_n = 1\left(\frac{1}{5}\right)^{n-1} = \frac{1}{5^{n-1}}$

c. $a_n = (-1)\left(-\frac{1}{3}\right)^{n-1} = \frac{-1}{(-3)^{n-1}} = \frac{(-1)^n}{3^{n-1}} \quad \square$

EXAMPLE 3. Suppose that the third and sixth terms of a geometric sequence $\{a_n\}_{n=1}^{\infty}$ are $\frac{9}{2}$ and $-\frac{243}{16}$, respectively. Find a formula for the nth term of the sequence.

Solution. Once we know a_1 and r, the nth term is given by (2). For r we use (2) and find that

$$\frac{a_6}{a_3} = \frac{a_1 r^5}{a_1 r^2} = r^3$$

so by hypothesis,

$$r^3 = \frac{a_6}{a_3} = \frac{-243/16}{9/2} = -\frac{27}{8}$$

This means that $r = -\frac{3}{2}$. A second application of (2) tells us that

$$a_1 = \frac{1}{r^2} \cdot a_3 = \frac{1}{(-3/2)^2} \cdot \frac{9}{2} = \frac{4}{9} \cdot \frac{9}{2} = 2$$

It follows from this information and from (2) that

$$a_n = 2\left(-\frac{3}{2}\right)^{n-1} \quad \text{for} \quad n \geq 1 \quad \square$$

Partial sums of a geometric sequence can be expressed by a convenient formula. Before we present the formula, we observe that when written out, a partial sum of a geometric sequence has the form

$$\sum_{j=1}^{n} ar^{j-1} = ar^0 + ar^1 + ar^2 + \cdots + ar^{n-1}$$

or more simply,

$$\sum_{j=1}^{n} ar^{j-1} = a + ar + ar^2 + \cdots + ar^{n-1} \tag{3}$$

THEOREM 11.5 *Let $\{ar^{n-1}\}_{n=1}^{\infty}$ be a geometric sequence with $r \neq 0$ and $r \neq 1$. Then the nth partial sum of the sequence is given by*

$$\sum_{j=1}^{n} ar^{j-1} = a \cdot \frac{1 - r^n}{1 - r} \tag{4}$$

Proof. By (3), along with Example 2 in Section 11.1 (with x replaced by r), we conclude that

$$\begin{aligned}\sum_{j=1}^{n} ar^{j-1} &= a + ar + ar^2 + \cdots + ar^{n-1} \\ &= a(1 + r + r^2 + \cdots + r^{n-1}) \\ &= a \cdot \frac{1 - r^n}{1 - r} \quad \blacksquare\end{aligned}$$

EXAMPLE 4. Find the sum S of the first 10 terms of the geometric sequence $\frac{1}{4}, \frac{1}{8}, \ldots$.

Solution. Applying (1) with $n = 1$, we find that

$$r = \frac{a_2}{a_1} = \frac{\frac{1}{8}}{\frac{1}{4}} = \frac{1}{2}$$

By (4) with $a = \frac{1}{4}$, $r = \frac{1}{2}$, and $n = 10$, we have

$$S = \frac{1}{4} \cdot \frac{1 - \left(\frac{1}{2}\right)^{10}}{1 - \frac{1}{2}} = \frac{1}{2}\left(1 - \frac{1}{1024}\right) = \frac{1023}{2048} \quad \square$$

Next we consider the sequence $\{(\frac{1}{2})^n\}_{n=1}^{\infty}$. The terms start out

$$\frac{1}{2}, \frac{1}{4}, \frac{1}{8}, \frac{1}{16}, \frac{1}{32}, \frac{1}{64}, \ldots$$

Since each term in the sequence is half its predecessor, the value $(\frac{1}{2})^n$ of the nth term can be made as small as we like by taking n large enough. We express this by saying that $(\frac{1}{2})^n$ approaches 0 as n increases without bound. More generally if $|r| < 1$, then r^n approaches 0 as n increases without bound. It follows that

$$\text{if } |r| < 1, \quad \text{then} \quad a \cdot \frac{1 - r^n}{1 - r} \text{ approaches } \frac{a}{1 - r}$$

as n increases without bound. Thus Theorem 11.5 implies that as n increases without bound, the nth partial sum of the geometric sequence $\{ar^{n-1}\}_{n=1}^{\infty}$ approaches the number $a/(1 - r)$. It is therefore natural to define the sum of all the numbers in the geometric sequence $\{ar^{n-1}\}_{n=1}^{\infty}$ to be $a/(1 - r)$. We denote this sum by $\Sigma_{j=1}^{\infty} ar^{j-1}$, an expression we call a ***geometric series.*** Thus

$$\sum_{j=1}^{\infty} ar^{j-1} = a + ar + ar^2 + ar^3 + \cdots = \frac{a}{1 - r} \quad \text{for} \quad |r| < 1 \tag{5}$$

EXAMPLE 5. Find the numerical value of the following geometric series.

a. $\displaystyle\sum_{j=1}^{\infty} \left(\frac{1}{2}\right)^{j-1}$ b. $\displaystyle\sum_{j=1}^{\infty} 4\left(-\frac{2}{3}\right)^{j-1}$ c. $\displaystyle\sum_{j=1}^{\infty} (-2)\left(\frac{3}{5}\right)^{j}$

Solution.

a. By (5) with $a = 1$ and $r = \frac{1}{2}$, we have

$$\sum_{j=1}^{\infty} \left(\frac{1}{2}\right)^{j-1} = \frac{1}{1 - \frac{1}{2}} = 2$$

b. By (5) with $a = 4$ and $r = -\frac{2}{3}$, we have

$$\sum_{j=1}^{\infty} 4\left(-\frac{2}{3}\right)^{j-1} = \frac{4}{1-(-\frac{2}{3})} = \frac{12}{5}$$

c. Notice that the geometric series is not in the form $\Sigma_{j=1}^{\infty} ar^{j-1}$, so in order to apply (5), we alter the series as follows:

$$\sum_{j=1}^{\infty} (-2)\left(\frac{3}{5}\right)^{j} = \sum_{j=1}^{\infty} (-2)\left(\frac{3}{5}\right)\left(\frac{3}{5}\right)^{j-1} = \sum_{j=1}^{\infty} \left(-\frac{6}{5}\right)\left(\frac{3}{5}\right)^{j-1}$$

Now (5) applies, with $a = -\frac{6}{5}$ and $r = \frac{3}{5}$, and we obtain

$$\sum_{j=1}^{\infty} (-2)\left(\frac{3}{5}\right)^{j} = \sum_{j=1}^{\infty} \left(-\frac{6}{5}\right)\left(\frac{3}{5}\right)^{j-1} = \frac{-\frac{6}{5}}{1-\frac{3}{5}} = -3 \quad \square$$

Repeating Decimals

It is well known that the repeating decimal 0.3333 . . . can be identified with the fraction $\frac{1}{3}$, that is,

$$\frac{1}{3} = 0.3333\ldots \tag{6}$$

The way we arrive at this formula is as follows. First, by 0.3333 . . . we mean the sum

$$0.3 + 0.03 + 0.003 + 0.0003 + \cdots$$

that is,

$$\frac{3}{10} + \frac{3}{100} + \frac{3}{1000} + \frac{3}{10{,}000} + \cdots$$

or equivalently,

$$\frac{3}{10} + \frac{3}{10}\left(\frac{1}{10}\right) + \frac{3}{10}\left(\frac{1}{10}\right)^2 + \cdots$$

This is the geometric series whose 1st term is $\frac{3}{10}$ and whose common ratio is $\frac{1}{10}$. In sum notation the series is

$$\sum_{j=1}^{\infty} \frac{3}{10}\left(\frac{1}{10}\right)^{j-1}$$

By (5) with $a = \frac{3}{10}$ and $r = \frac{1}{10}$, we find that

$$\sum_{j=1}^{\infty} \frac{3}{10}\left(\frac{1}{10}\right)^{j-1} = \frac{\frac{3}{10}}{1-\frac{1}{10}} = \frac{3}{10}\cdot\frac{10}{9} = \frac{3}{9} = \frac{1}{3}$$

This completes our proof of (6).

In a similar way we can identify any repeating decimal with a suitable rational number.

EXAMPLE 6. Write 0.535353 . . . as a fraction.

Solution. Notice first that

$$0.535353\ldots = \frac{53}{100} + \frac{53}{100}\left(\frac{1}{100}\right) + \frac{53}{100}\left(\frac{1}{100}\right)^2 + \cdots$$

which is the geometric series whose first term is $\frac{53}{100}$ and whose common ratio is $\frac{1}{100}$. In sum notation the series is

$$\sum_{j=1}^{\infty} \frac{53}{100}\left(\frac{1}{100}\right)^{j-1}$$

By (5) with $a = \dfrac{53}{100}$ and $r = \dfrac{1}{100}$, we have

$$\sum_{j=1}^{\infty} \frac{53}{100}\left(\frac{1}{100}\right)^{j-1} = \frac{\frac{53}{100}}{1 - \frac{1}{100}} = \frac{53}{100} \cdot \frac{100}{99} = \frac{53}{99}$$

Therefore

$$0.535353\ldots = \frac{53}{99} \quad \square$$

EXAMPLE 7. Write 0.4298298298 . . . as a fraction.

Solution. First we separate the repeating part from the rest:

$$0.4298298298 = 0.4 + 0.0298298298\ldots$$

Next we observe that

$$0.0298298298\ldots = \frac{298}{10{,}000} + \frac{298}{10{,}000}\left(\frac{1}{1000}\right) + \frac{298}{10{,}000}\left(\frac{1}{1000}\right)^2 + \cdots$$

which is the geometric series whose first term is $\frac{298}{10{,}000}$ and whose common ratio is $\frac{1}{1000}$. In sum notation the series is

$$\sum_{j=1}^{\infty} \frac{298}{10{,}000}\left(\frac{1}{1000}\right)^{j-1}$$

By (5) with $a = \frac{298}{10{,}000}$ and $r = \frac{1}{1000}$, we have

$$\sum_{j=1}^{\infty} \frac{298}{10{,}000}\left(\frac{1}{1000}\right)^{j-1} = \frac{\frac{298}{10{,}000}}{1 - \frac{1}{1000}} = \frac{298}{10{,}000} \cdot \frac{1000}{999} = \frac{298}{9990}$$

Consequently

$$0.4298298298\ldots = 0.4 + 0.0298298298\ldots$$
$$= \frac{4}{10} + \frac{298}{9990} = \frac{3996}{9990} + \frac{298}{9990} = \frac{4294}{9990} = \frac{2147}{4995} \quad \square$$

EXERCISES 11.4

In Exercises 1–6, show that the sequence is geometric, and find the common ratio.

1. $\{4^n\}_{n=1}^{\infty}$
2. $\left\{\left(\frac{5}{2}\right)^n\right\}_{n=1}^{\infty}$
3. $\{(-0.3)^{n-1}\}_{n=1}^{\infty}$
4. $\{(\sqrt{2})^{n+1}\}_{n=1}^{\infty}$
5. $\{4^{2n}\}_{n=1}^{\infty}$
6. $\{3^{n/2}\}_{n=1}^{\infty}$

In Exercises 7–14, determine whether the given sequence is geometric.

7. $-1, -2, -4, -8, \ldots$
8. $-1, 2, -4, 8, \ldots$
9. $1, 3, 5, 7, \ldots$
10. $0, 3, 6, 9, \ldots$
11. $\{-3(-2)^{n+1}\}_{n=1}^{\infty}$
12. $\{e^n\}_{n=1}^{\infty}$
13. $\{5n^n\}_{n=1}^{\infty}$
14. $\{15 + 2^n\}_{n=1}^{\infty}$

In Exercises 15–22, find the fifth term and the nth term of the geometric sequence whose initial term a_1 and common ratio r are given.

15. $a_1 = 4, r = 2$
16. $a_1 = 6, r = -\frac{1}{2}$
17. $a_1 = 3, r = \frac{1}{6}$
18. $a_1 = \frac{1}{3}, r = \frac{3}{2}$
19. $a_1 = 10, r = 3$
20. $a_1 = 0.5, r = \frac{1}{10}$
21. $a_1 = c, r = 5$
22. $a_1 = c, r = c^2$

In Exercises 23–30, find the sixth term and the nth term of the geometric sequence $\{a_n\}_{n=1}^{\infty}$ whose first two terms are given.

23. $\frac{1}{3}, 1, \ldots$
24. $5, 20, \ldots$
25. $6, 3, \ldots$
26. $1, -2, \ldots$
27. $256, -64, \ldots$
28. $2, 2^2, \ldots$
29. $\frac{1}{6}, -\frac{1}{18}, \ldots$
30. $\ln 10, \ln 100, \ldots$

In Exercises 31–36, find a formula for the nth term of the geometric sequence $\{a_n\}_{n=1}^{\infty}$ that satisfies the given conditions.

31. $a_2 = 6, a_3 = 9$
32. $a_1 = 4, a_4 = -\frac{1}{16}$
33. $a_3 = 250, a_4 = 1250$
34. $a_2 = 192, a_7 = 6$
35. $a_3 = -\frac{4}{5}, r = \frac{2}{5}$
36. $a_8 = 1024, r = 2$

In Exercises 37–40, find the nth partial sum (for the designated value of n) of the geometric sequence $\{a_n\}_{n=1}^{\infty}$ that satisfies the given conditions.

37. $a_1 = 9, r = 3, n = 3$
38. $a_1 = 5, r = 2, n = 5$
39. $a_2 = 8, r = 2, n = 4$
40. $a_3 = -8, r = -\frac{1}{2}, n = 5$

In Exercises 41–48, find the given partial sum.

41. $\sum_{j=1}^{5} 3^{j-1}$
42. $\sum_{j=1}^{4} \left(\frac{3}{2}\right)^{j-1}$
43. $\sum_{j=1}^{7} \left(-\frac{1}{2}\right)^{j-1}$
44. $\sum_{j=1}^{7} \left(-\frac{1}{2}\right)^{j}$
45. $\sum_{j=1}^{5} \frac{4}{3^j}$
46. $\sum_{j=1}^{5} 8\left(\frac{3}{2}\right)^{j}$
47. $\sum_{j=1}^{4} 2\left(-\frac{3}{4}\right)^{j}$
48. $\sum_{j=1}^{3} 3\left(\frac{4}{5}\right)^{j+1}$

In Exercises 49–56, find the numerical value of the geometric series.

49. $\sum_{j=1}^{\infty} \left(\frac{1}{3}\right)^{j-1}$
50. $\sum_{j=1}^{\infty} \left(-\frac{1}{2}\right)^{j-1}$
51. $\sum_{j=1}^{\infty} 2\left(\frac{1}{4}\right)^{j-1}$
52. $\sum_{j=1}^{\infty} \frac{1}{3}\left(-\frac{3}{4}\right)^{j-1}$
53. $\sum_{j=1}^{\infty} 6\left(\frac{2}{3}\right)^{j}$
54. $\sum_{j=1}^{\infty} \frac{1}{2}\left(-\frac{4}{5}\right)^{j+1}$
55. $\sum_{j=1}^{\infty} 2(0.3)^{j}$
56. $\sum_{j=1}^{\infty} \frac{1}{3}(0.62)^{j-1}$

In Exercises 57–64, write the given repeating decimal as a fraction.

57. 0.777 . . .

58. 0.141414 . . .

59. 0.959595 . . .

60. 0.0959595 . . .

61. 5.432432432 . . .

62. 1.697697697 . . .

63. 3.1474747 . . .

64. 2.6858585 . . .

65. If $\{a_n\}_{n=1}^{\infty}$ is a geometric sequence such that $a_5/a_3 = 16$ and $a_2 = 1$, find a formula for a_n.

66. If the product of the second and sixth terms of a geometric sequence is 1 and the common ratio is 3, find a formula for a_n.

67. If 3 and 108 are two terms of a geometric sequence separated by exactly one positive term, find its value.

68. The first two terms of a geometric sequence are $\sqrt[3]{2}$ and $\sqrt{2}$. Find the third and fourth terms.

69. Show that there are two distinct geometric sequences having second term 12 and fourth term 27. Find a formula for the *n*th term of each sequence.

70. Prove (2) by using (1) and mathematical induction.

71. Let $\{a_n\}_{n=1}^{\infty}$ be a geometric sequence with positive terms and common ratio r. Prove that $\{\ln a_n\}_{n=1}^{\infty}$ is an arithmetic sequence, and find the common difference d. (*Hint:* Use (2).)

72. Starting with your parents, how many ancestors do you have going back 8 generations?

73. The enrollment at a certain college doubled every 5 years from 1960 to 1980. If the enrollment was 8000 in 1980, what was the enrollment in 1960?

***74.** When a tennis ball is dropped, it is supposed to bounce back to approximately 55% of its original height. Use a calculator to determine the distance the ball travels during its initial ten bounces (down and up) if it is dropped from a height of 6 feet.

11.5 PERMUTATIONS AND COMBINATIONS

Suppose that a 7-member committee sets out to pick 3 of its members to be officers. The first member picked will be chairperson, the next vice-chairperson, and the third secretary. The following two questions typify the kind of question we will discuss in this section.

Question A: In how many different ways can the set of officers be formed from the committee?

Question B: How many different sets of officers can be formed if we disregard which officer has what title?

As we set out to answer these questions, let us list the 7 committee members as a, b, c, d, e, f, and g. Question A is equivalent to asking in how

many ways we can write down an ordered triplet consisting of 3 of the letters a through g. For example,

$$(a, b, c), \qquad (d, f, e), \qquad (g, e, b), \quad \text{and} \quad (b, e, g)$$

are four such triplets. Here we let the first letter of each triplet denote the chairperson, the second letter the vice-chairperson, and the third letter the secretary. Each of the triplets listed above is considered to be distinct from the others with respect to Question A (even though the last two contain exactly the same letters), because the order of the letters in the triplet is taken into account.

By contrast, Question B is equivalent to asking in how many ways we can write down a triplet of 3 of the letters a through g without regard to the order of the letters. In this case,

$$(b, e, g), \qquad (b, g, e), \qquad (e, b, g), \qquad (e, g, b), \qquad (g, b, e), \quad \text{and} \quad (g, e, b)$$

are identified, because they contain exactly the same letters and because order is irrelevant for Question B. In this case our sole concern is which 3 people are selected.

It follows from the discussion above that the answers to Questions A and B are different. Before we set out to find the numerical answers to the questions, let us put the questions into a more general setting:

i. How many distinct ordered sets of k objects can be chosen from a collection of n objects?
ii. How many distinct sets of k objects may be chosen from a collection of n objects without regard to the order of the objects?

We apply the terms "permutations" and "combinations" to the sets in (i) and (ii), respectively. These terms are defined formally now.

DEFINITION 11.6 Let k and n be integers with n positive and $0 \le k \le n$. A ***permutation*** of n objects taken k at a time is an ordered set of k objects from the set of n objects. A ***combination*** of n objects taken k at a time is a set of k objects from the set of n objects (without regard to order).

The difference between a permutation and a combination is that order is all-important with respect to a permutation but is irrelevant with respect to a combination.

A Counting Principle

Before proceeding further in analyzing permutations and combinations, we discuss a method of counting sets of objects. To illustrate the method, let us count the number of full names that can be formed by using "Barbara" or "Sharon" as the first name and "Lee," "Lopez," or "Smith" as the last name. We may regard the selection of such a full name as a 2-stage process, and we display the possibilities in the following diagram, called a ***tree diagram***:

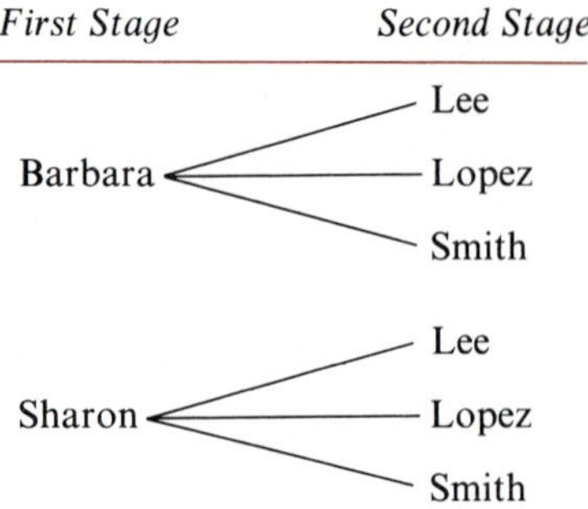

We see that there are 6 possible full names. Indeed, there are 2 possible first names and 3 possible last names, making a total of $2 \times 3 = 6$ possible full names. Thus the total number of possible names is the product of the number of choices at the first stage and at the second stage.

Now let us count the number of full names that can be formed by using "David" or "Mark" as the first name, "Charles" or "Christopher" as the middle name, and "Green," "Jansen," or "Mantle" as the last name. Here we regard the selection of such a full name as a 3-stage process, displayed in the following tree diagram:

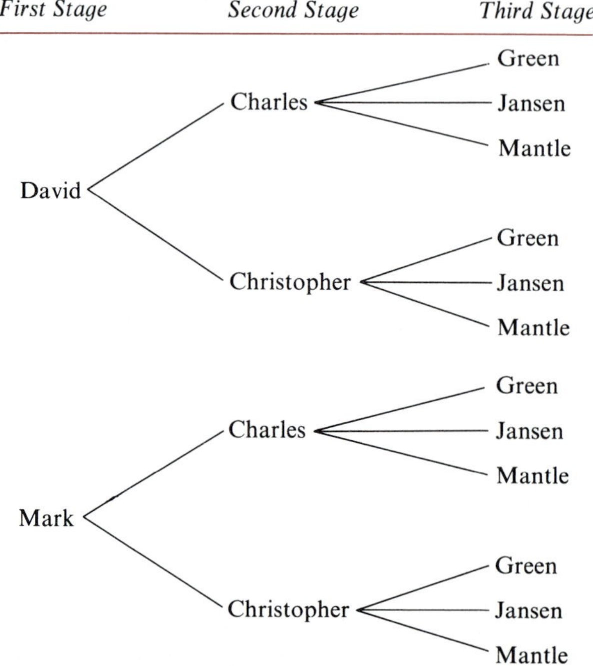

This time we see that there are 12 possible full names. After all, there are 2 possible first names, 2 possible middle names, and 3 possible last names, making a total of $2 \times 2 \times 3 = 12$ possible full names. Thus the total number of possible full names is the product of the number of choices at each of the 3 stages.

In general, if the items to be counted may be selected in k stages, and if we know the number of choices at each of the k stages, then the total number of items is the product of the number of choices at each stage. We now state this result formally.

Counting Principle

Suppose that k objects are to be selected. Assume that the first object can be selected in n_1 distinct ways, and after the first object has been selected the second object can be selected in n_2 distinct ways, and so on, with the kth object selectable in any of n_k distinct ways after the first $k - 1$ objects have been selected. Then the total number of distinct ways of selecting the k objects is the product

$$n_1 n_2 n_3 \ldots n_k$$

In the following example we illustrate the use of the Counting Principle.

EXAMPLE 1. How many 7-digit telephone numbers are there if neither 0 nor 1 can be used as the first or second digit?

Solution. For each of the first and second digits of the telephone number, there are 8 possible choices: 2, 3, 4, 5, 6, 7, 8, and 9. Therefore

$$n_1 = n_2 = 8$$

For each of the remaining digits of the telephone number there are 10 possibilities: 0, 1, 2, 3, 4, 5, 6, 7, 8, and 9. Therefore

$$n_3 = n_4 = n_5 = n_6 = n_7 = 10$$

By the Counting Principle, the total number N of permissible 7-digit telephone numbers is given by

$$\begin{aligned} N &= n_1 n_2 n_3 n_4 n_5 n_6 n_7 = 8 \cdot 8 \cdot 10 \cdot 10 \cdot 10 \cdot 10 \cdot 10 \\ &= 64 \times 10^5 = 6{,}400{,}000 \quad \square \end{aligned}$$

Permutations

The number of permutations of n objects taken k at a time is often written symbolically as ${}_nP_k$ (or sometimes P_k^n or $P(n, k)$). In this notation the number of permutations of 7 objects taken 3 at a time is ${}_7P_3$.

Suppose now that we wish to determine the numerical value of ${}_nP_k$, where n is any positive integer and k is any integer with $0 \leq k \leq n$. We define

${}_nP_0 = 1$. For $k \geq 1$ we consider the process of forming a permutation of n objects taken k at a time as a k-stage process. In the first stage we select the first object, with n possible choices for this object. In the second stage we select the second object, with only $n - 1$ choices for this object, since the object selected first is no longer available. Similarly, there are only $n - 2$ choices for the third object, since the objects selected first and second are no longer available. Thus we have the following setup:

Object	*Number of choices*
First	n
Second	$n - 1$
Third	$n - 2$
⋮	⋮
kth	$n - (k - 1)$

Therefore by the Counting Principle, it follows that the number ${}_nP_k$ of ways of choosing k objects from the total collection of n objects is given by the product of the possible choices just listed. That is,

$$ {}_nP_k = n(n-1)(n-2)\cdots[n-(k-2)][n-(k-1)] \tag{1} $$

We are now in a position to answer question A. Indeed, by (1),

$$ {}_7P_3 = 7 \cdot 6 \cdot 5 = 210 $$

This tells us that there are 210 ways of choosing a chairman, vice-chairman, and secretary from the 7-member committee. This also answers the question posed in the introduction to Chapter 11.

EXAMPLE 2. Suppose 9 horses are set to run the Kentucky Derby. How many ways are there of the horses occupying the first three places in the order of finish?

Solution. Since we are interested in determining the number of ordered sets of 3 horses selected from 9 horses (corresponding to the order in which the first 3 horses cross the finish line), we seek ${}_9P_3$. By (1) with $n = 9$ and $k = 3$, we have

$$ {}_9P_3 = 9 \cdot 8 \cdot 7 = 504 \quad \square $$

If we let $k = n$, then (1) becomes

$$ {}_nP_n = n(n-1)(n-2)\cdots 2 \cdot 1 = n! $$

so that the number ${}_nP_n$ of permutations of n objects taken n at a time is $n!$. But notice that a permutation of n objects taken n at a time corresponds to an ordering of the n objects. Thus we have the following result:

There are $n!$ ways of ordering a set of n objects.

We can obtain a simpler version of (1) by noticing that

$$\begin{aligned} n(n-1)(n-2)\cdots[n-(k-1)] &= n(n-1)(n-2)\cdots[n-(k-1)]\frac{(n-k)[n-(k+1)]\cdots 2\cdot 1}{(n-k)[n-(k+1)]\cdots 2\cdot 1} \\ &= \frac{n(n-1)(n-2)\cdots[n-(k-1)](n-k)[n-(k+1)]\cdots 2\cdot 1}{(n-k)[n-(k+1)]\cdots 2\cdot 1} \\ &= \frac{n!}{(n-k)!} \end{aligned}$$

Consequently

$$_nP_k = \frac{n!}{(n-k)!} \tag{2}$$

EXAMPLE 3. Calculate $_6P_4$ by means of (2).

Solution. By (2),

$$_6P_4 = \frac{6!}{(6-4)!} = \frac{6!}{2!} = \frac{720}{2} = 360 \quad \square$$

Next, suppose that we have 5 marbles, including 3 indistinguishable black ones and 2 indistinguishable white ones. Let us denote the black marbles by B_1, B_2, and B_3, and the white marbles by W_1 and W_2. Then the two ordered sets

$$B_1B_2B_3W_1W_2 \quad \text{and} \quad B_2B_3B_1W_2W_1$$

would be indistinguishable, since both have 3 black marbles followed by 2 white marbles. Thus of the 5! ordered sets of the 5 marbles, many are indistinguishable from one another. It can be proved that there are only

$$\frac{5!}{3!2!}$$

or 10, distinguishable ordered sets of the 5 marbles, rather than the 5!, or 120, that would exist if all 5 marbles were different.

More generally, suppose that there are n objects, but n_1 of them are of one type and are indistinguishable, n_2 are of a second type and are indistinguishable, and so on, with a final kth type of n_k indistinguishable objects. Then it can be proved that the number of distinguishable ordered sets that can be formed from the n objects is

$$\boxed{\frac{n!}{n_1!\, n_2! \cdots n_k!}} \tag{3}$$

EXAMPLE 4. Suppose 6 indistinguishable nickels, 4 indistinguishable dimes, and 2 indistinguishable quarters are drawn out of a bowl, one by one. How many distinguishable ordered sets can be formed from the coins?

Solution. There are 12 coins in all, so using (3) with $n = 12$, $n_1 = 6$, $n_2 = 4$, and $n_3 = 2$, we find that there are

$$\frac{12!}{6!4!2!} = 13{,}860$$

distinguishable ordered sets of the 12 coins. □

Combinations

With permutations we were able to determine in how many ways 3 officers could be selected from 7 committee members. Now we will study the number of sets of 3 officers that can be chosen from the 7 committee members, without regard to which officer has what office. This will involve combinations.

The number of combinations of n objects taken k at a time is written symbolically as ${}_nC_k$. We will derive a formula for ${}_nC_k$ from formula (2) for ${}_nP_k$.

To that end we observe that the number ${}_nP_k$ of permutations of n objects taken k at a time takes into account the order in which the k objects are selected. By contrast, for the number ${}_nC_k$ of combinations the order in which the k objects are selected is irrelevant, so that each distinct combination counted in ${}_nC_k$ corresponds to $k!$ distinct permutations in ${}_nP_k$ (the $k!$ ways the k elements can be ordered). Therefore

$${}_nC_k = \frac{{}_nP_k}{k!}$$

Together with (2), this means that

$$\boxed{{}_nC_k = \frac{n!}{k!(n-k)!}} \tag{4}$$

From (4) we conclude that

$${}_7C_3 = \frac{7!}{3!(7-3)!} = \frac{7!}{3!4!} = \frac{7 \cdot 6 \cdot 5 \cdot 4!}{3 \cdot 2 \cdot 1 \cdot 4!} = \frac{7 \cdot 6 \cdot 5}{3 \cdot 2} = 35$$

Since the number of sets of 3 officers that can be chosen from a 7-member committee is ${}_7C_3$, it follows that there are 35 such sets of officers possible. This then answers Question B posed at the beginning of the section.

EXAMPLE 5. Determine the number of 5-card hands that can be dealt from a 52-card deck.

Solution. Since the order in which cards are dealt into the hand is immaterial to the composition of the hand, what we seek is ${}_{52}C_5$, and by (4) with $n = 52$ and $k = 5$, we have

$$\begin{aligned} {}_{52}C_5 &= \frac{52!}{5!(52-5)!} = \frac{52!}{5!47!} \\ &= \frac{52 \cdot 51 \cdot 50 \cdot 49 \cdot 48(47!)}{5 \cdot 4 \cdot 3 \cdot 2 \cdot 1(47!)} = \frac{52 \cdot 51 \cdot 50 \cdot 49 \cdot 48}{5 \cdot 4 \cdot 3 \cdot 2} = 2{,}598{,}960 \quad \square \end{aligned}$$

As we will see in the following section, when the nth power $(x + y)^n$ of the binomial $x + y$ is written out, the coefficients of the terms that appear in the formula are equal to ${}_nC_k$ for various values of k. But in that context they are normally written in the alternative notation $\binom{n}{k}$ and are called ***binomial coefficients.*** From (4) we find that

$$\boxed{\binom{n}{k} = \frac{n!}{k!(n-k)!}} \tag{5}$$

EXAMPLE 6. Find the numerical value of $\binom{8}{5}$.

Solution. By (5),

$$\binom{8}{5} = \frac{8!}{5!(8-5)!} = \frac{8 \cdot 7 \cdot 6(5!)}{(5!)3 \cdot 2 \cdot 1} = \frac{8 \cdot 7 \cdot 6}{3 \cdot 2} = 56 \quad \square$$

Since $0! = 1$ by definition, it follows from (5) that

$$\binom{n}{0} = \frac{n!}{0!(n-0)!} = \frac{n!}{0!n!} = \frac{n!}{n!} = 1$$

and

$$\binom{n}{n} = \frac{n!}{n!(n-n)!} = \frac{n!}{n!0!} = \frac{n!}{n!} = 1$$

Similarly, since $1! = 1$, we have

$$\binom{n}{1} = \frac{n!}{1!(n-1)!} = \frac{n(n-1)!}{(n-1)!} = n$$

and likewise,

$$\binom{n}{n-1} = n$$

EXERCISES 11.5

In Exercises 1–24, calculate the given number.

1. The number of permutations of 5 objects taken 3 at a time.
2. The number of permutations of 6 objects taken 2 at a time.
3. The number of permutations of 7 objects taken 5 at a time.
4. The number of combinations of 5 objects taken 3 at a time.
5. The number of combinations of 6 objects taken 2 at a time.
6. The number of combinations of 7 objects taken 5 at a time.

7. ${}_4P_1$ 8. ${}_4P_3$ 9. ${}_5P_5$

10. ${}_7P_3$ 11. ${}_{12}P_3$ 12. ${}_{100}P_2$

13. ${}_4C_2$ 14. ${}_6C_3$ 15. ${}_6C_6$

16. ${}_5C_3$ 17. ${}_5C_2$ 18. ${}_8C_5$

19. $\binom{2}{1}$ 20. $\binom{4}{3}$ 21. $\binom{6}{4}$

22. $\binom{7}{4}$ 23. $\binom{17}{15}$ 24. $\binom{40}{3}$

25. Let n be an integer greater than 1.
   a. Find a formula for $\binom{n}{n-2}$.
   b. Show that $\sum_{j=1}^{n} j = \binom{n+1}{n-1}$.

26. Show that ${}_{2n+1}C_n = {}_{2n+1}C_{n+1}$ for any positive integer n.

27. Let n and j be integers with $0 \le j \le n$. Show that

$$\binom{n}{j} = \binom{n}{n-j}$$

28. Show that $({}_{12}C_5)({}_7C_4) = ({}_{12}C_4)({}_8C_5)$.

29. A student is eligible for 3 mathematics courses and 4 physics courses. In how many ways can the student choose one mathematics course and one physics course?

30. A department store has 4 entrances on the first floor, 3 escalators from the first floor to the second floor, and 2 escalators from the second floor to the third floor. In how many ways can a customer enter the store and ascend to the third floor by escalator?

31. a. How many distinct license plate numbers are there that have 2 letters (neither of which may be I or O) followed by 3 digits?
   b. If in part (a) the letters must be distinct and so must the digits, how many distinct license plate numbers are there?

32. At a sandwich bar there are 3 kinds of bread, 4 kinds of cold cuts, 2 kinds of cheese, and 3 kinds of dressing. How many different kinds of sandwiches can be prepared with one kind each of bread, cold cuts, cheese, and dressing?

33. A corporation president must choose a vice-president and a treasurer from among 10 young executives.
   a. In how many ways can the president choose two executives to fill the positions (without regard to which executive gets which position)?
   b. In how many ways can the president choose two executives to fill the positions (considering which executive gets which position)?

34. Find the number of permutations of 4 letters selected from the word "rainbow."

35. A classrom contains 20 seats; in how many ways can a teacher assign seats to 15 students? Leave your answer in factorial form.

36. How many sets of three volunteers are possible from a group of 10 people?

37. In how many ways can 5 patients be given appointments with a doctor if there are 7 appointments available?

38. A doctor must reschedule 4 of 7 appointments. In how many ways can the 4 appointments be chosen?

39. On Halloween a trick-or-treater is offered any 3 of 5 different kinds of candy bars. In how many ways can the choices be made?

40. Find an expression for the number of distinct 13-card bridge hands that can be dealt from a deck of 52 cards. Do not carry out the arithmetic operations.

41. A basketball coach has 2 centers, 5 guards, and 4 forwards. How many different teams consisting of 1 center, 2 guards, and 2 forwards can the coach choose? (*Hint:* Consider the selection a 3-stage process.)

42. A person receives a list of five names from a computerized dating service. In how many different orders can the person date the five people listed?

43. A basketball team has 12 members.

a. In how many ways can the 12 members be introduced one by one before a game?

b. In how many ways can they be introduced if the five preselected starting players are to be introduced first?

44. A person has 5 quarters, 3 dimes, and 4 nickels. In how many ways can the person choose 2 quarters, 1 dime, and 2 nickels to pay for a 70¢ purchase? (Assume that the coins of each denomination are distinguishable.)

45. A baker has 5 vanilla, 3 chocolate, and 4 lemon layers of cake. In how many distinguishable ways can the baker stack the layers to make a cake with 12 layers?

46. Find the number of (possibly meaningless) 5-letter words with exactly 1 vowel. (*Hint:* Assume that there are 5 vowels and 21 consonants in the alphabet. Consider the formation of the word as a six-stage process, the first stage of which involves selecting the position in the word that the vowel is to occupy.)

47. One student claims to another, "I'll bet there are at least 10 times as many distinguishable ordered sets that can be formed from the letters in "Mississippi" as there are distinguishable ordered sets that can be formed from the letters in "Tennessee." Is the student correct?

48. How many 9-digit Social Security numbers are there containing four 5's, two 6's, and three 9's?

11.6 THE BINOMIAL THEOREM

In Section 0.5 we wrote out the following formulas for $(x + y)^2$ and $(x + y)^3$:

$$(x + y)^2 = x^2 + 2xy + y^2 \tag{1}$$

$$(x + y)^3 = x^3 + 3x^2y + 3xy^2 + y^3 \tag{2}$$

Now we will derive a formula for $(x + y)^n$ for any positive integer n.

In preparation for the formula of $(x + y)^n$, let us analyze the right sides of the equations in (1) and (2). Notice that each of their terms has the form (coefficient) x^iy^j for suitable integers i and j. For the terms in (1), $i + j = 2$, whereas for the terms in (2), $i + j = 3$.

Now let us consider the expansion of $(x + y)^4$. Notice that

$$\begin{aligned}(x + y)^4 &= (x + y)(x + y)(x + y)(x + y)\\ &= (xx + xy + yx + yy)(xx + xy + yx + yy)\\ &= xxxx + xxxy + xxyx + xxyy + xyxx\\ &\quad + xyxy + xyyx + xyyy + yxxx + yxxy\\ &\quad + yxyx + yxyy + yyxx + yyxy + yyyx + yyyy\end{aligned}$$

Because $xy = yx$, this can be condensed to

$$\begin{aligned}(x + y)^4 = x^4y^0 &+ x^3y^1 + x^3y^1 + x^2y^2 + x^3y^1\\ &+ x^2y^2 + x^2y^2 + x^1y^3 + x^3y^1 + x^2y^2\\ &+ x^2y^2 + x^1y^3 + x^2y^2 + x^1y^3 + x^1y^3 + x^0y^4\end{aligned} \tag{3}$$

Observe that the term x^4y^0 appears only once on the right side of (3) because it arises by choosing x from each of the four expressions $x + y$ occurring in $(x + y)^4$ and never choosing y. Analogously, x^3y appears four times in (3) because we can pick exactly one y from any of the four expressions $x + y$ occurring in $(x + y)^4$. In general, the coefficient of x^iy^j (with $i + j = 4$) is equal to the number of ways of choosing j y's from the four expressions $x + y$ (the remaining choices being x's). Thus the coefficient of x^iy^j is $\binom{4}{j}$, and consequently (3) can be rewritten as

$$(x + y)^4 = \binom{4}{0}x^4y^0 + \binom{4}{1}x^3y^1 + \binom{4}{2}x^2y^2 + \binom{4}{3}x^1y^3 + \binom{4}{4}x^0y^4$$
$$= x^4 + 4x^3y + 6x^2y^2 + 4xy^3 + y^4$$

In general, if n is a positive integer and we expand $(x + y)^n$ by writing

$$(x + y)^n = \overbrace{(x + y)(x + y)\cdots(x + y)}^{n \text{ expressions } (x + y)} \tag{4}$$

and then multiplying out, the result is a sum of terms of the form (coefficient)x^iy^j, where $i + j = n$. The coefficient of x^iy^j is equal to the number of ways of choosing j y's from the n expressions $x + y$ (the remaining $n - j$ choices being x's) in the parentheses on the right side of (4). Thus that coefficient is $\binom{n}{j}$. The resulting formula for $(x + y)^n$, which we state below, is known as the Binomial Theorem. It was one of the first theorems of Isaac Newton, proved in 1665, yet it was discovered nearly 600 years earlier by the great Persian poet and mathematician Omar Khayyám (1044?–1123?).

THEOREM 11.7
The Binomial Theorem

Let x and y be any real numbers and let n be any positive integer. Then

$$(x + y)^n = \binom{n}{0}x^n + \binom{n}{1}x^{n-1}y + \binom{n}{2}x^{n-2}y^2 + \cdots + \binom{n}{n-1}xy^{n-1} + \binom{n}{n}y^n \tag{5}$$

or more succinctly,

$$(x + y)^n = \sum_{j=0}^{n} \binom{n}{j} x^{n-j}y^j \tag{6}$$

The right side of (5) or (6) is known as the ***binomial expansion*** of the binomial $(x + y)^n$. The Binomial Theorem is also valid if x or y is complex.

EXAMPLE 1. Find the binomial expansion of $(x + y)^6$.

Solution. By (5) with $n = 6$, we have

$$(x+y)^6 = \binom{6}{0}x^6 + \binom{6}{1}x^5y + \binom{6}{2}x^4y^2 + \binom{6}{3}x^3y^3 + \binom{6}{4}x^2y^4 + \binom{6}{5}xy^5 + \binom{6}{6}y^6$$

$$= x^6 + 6x^5y + 15x^4y^2 + 20x^3y^3 + 15x^2y^4 + 6xy^5 + y^6 \quad \square$$

EXAMPLE 2. Find the binomial expansion of $(2r^2 - 3s)^5$.

Solution. By (5) with $n = 5$, and with x replaced by $2r^2$ and y replaced by $-3s$, we have

$$(2r^2 - 3s)^5 = \binom{5}{0}(2r^2)^5 + \binom{5}{1}(2r^2)^4(-3s) + \binom{5}{2}(2r^2)^3(-3s)^2 + \binom{5}{3}(2r^2)^2(-3s)^3 + \binom{5}{4}(2r^2)(-3s)^4 + \binom{5}{5}(-3s)^5$$

$$= 32r^{10} + 5(16r^8)(-3s) + 10(8r^6)(9s^2) + 10(4r^4)(-27s^3) + 5(2r^2)(81s^4) + (-243s^5)$$

$$= 32r^{10} - 240r^8s + 720r^6s^2 - 1080r^4s^3 + 810r^2s^4 - 243s^5 \quad \square$$

Sometimes it is necessary to isolate the kth term in the binomial expansion of $(x + y)^n$. By (6) the kth term is $\binom{n}{k-1}x^{n-k+1}y^{k-1}$.

EXAMPLE 3. Write the sixth term in the binomial expansion of $\left(\frac{1}{x} + y\right)^{12}$.

Solution. By our observation above, with $n = 12$ and $k = 5$, the sixth term is

$$\binom{12}{5}\left(\frac{1}{x}\right)^{12-5} y^5$$

Since

$$\binom{12}{5} = \frac{12!}{7!5!} = \frac{12 \cdot 11 \cdot 10 \cdot 9 \cdot 8(7!)}{(7!)5 \cdot 4 \cdot 3 \cdot 2 \cdot 1} = \frac{12 \cdot 11 \cdot 10 \cdot 9 \cdot 8}{5 \cdot 4 \cdot 3 \cdot 2} = 792$$

and

$$\left(\frac{1}{x}\right)^{12-5} = \left(\frac{1}{x}\right)^7 = x^{-7}$$

the sixth term can be written more simply as $792x^{-7}y^5$. $\square$

EXAMPLE 4. Let a be a positive number and n a positive integer. Use the Binomial Theorem to prove that $(1 + a)^n > 1 + na$.

Solution. By (5), with x replaced by 1 and y by a, we have

$$\begin{aligned}(1+a)^n &= \binom{n}{0}1^n + \binom{n}{1}1^{n-1}a + \binom{n}{2}1^{n-2}a^2 + \cdots + \binom{n}{n-1}a^{n-1} + \binom{n}{n}a^n \\ &= 1 + na + \binom{n}{2}a^2 + \cdots + \binom{n}{n-1}a^{n-1} + \binom{n}{n}a^n\end{aligned}$$

Since a is positive by hypothesis, all terms on the right side are positive, so it follows that $(1+a)^n$ is greater than the sum, $1+na$, of the first two terms. □

Pascal's Triangle

If we write down the values of the binomial coefficients of $(x+y)^n$ for $n = 0, 1, 2, 3, 4$, and 5 in the following fashion, we obtain a triangular array:

Coefficients	*Binomial*
1	$(x+y)^0$
1 1	$(x+y)^1$
1 2 1	$(x+y)^2$
1 3 3 1	$(x+y)^3$
1 4 6 4 1	$(x+y)^4$
1 5 10 10 5 1	$(x+y)^5$

The triangular array that arises from writing out the binomial coefficients of $(x+y)^n$ for $n = 0, 1, 2, 3, \ldots$ is called ***Pascal's triangle***, after the French mathematician Blaise Pascal (1623–1662), who wrote a treatise on properties of the triangular array.* Notice that the borders of the array consist of 1's, and any number in the array not on the border is the sum of the two closest numbers in the preceding row.

EXERCISES 11.6

In Exercises 1–10, use the Binomial Theorem to expand the given power of the given binomial.

1. $(x+y)^5$ **2.** $(x-y)^3$ **3.** $(x+2)^4$

4. $(2x-1)^6$ **5.** $(x+y)^7$ **6.** $(2x-\frac{1}{2})^5$

7. $\left(a-\dfrac{1}{a}\right)^5$ **8.** $(z^2-1)^7$

9. $\left(\sqrt{x}+\dfrac{1}{\sqrt{x}}\right)^8$ **10.** $(x^3-x^{-3})^6$

In Exercises 11–14, use the Binomial Theorem to calculate the value of the given expression.

11. $(\sqrt{2}+1)^4$ **12.** $(\sqrt{2}+1)^4+(\sqrt{2}-1)^4$

13. $(2-i)^5$ **14.** $(2-i)^5-(2+i)^5$

* The triangular array is not actually due to Pascal, and in fact the array appeared even as early as 1303 in a work by a Chinese mathematician, Chu Shi Kei. Nevertheless, Pascal's name is attached to it because of his treatise on the triangular array.

15. Find the fourth term of $(x + y)^8$.

16. Find the seventh term of $(2x - 3)^{10}$.

17. Find the coefficient of x^6y^9 in the binomial expansion of $(2x^2 - y^3)^6$.

18. Find the coefficient of x^4 in the binomial expansion of $(1 - 2x)^7$.

19. Find the constant term in the binomial expansion of $\left(x - \frac{1}{x}\right)^{10}$.

20. a. Fill in the seventh row of Pascal's Triangle. (*Hint:* See Example 1.)
b. Fill in the eighth row of Pascal's Triangle.

21. Use the first three terms of the expansion of $(1 - 0.02)^7$ to approximate $(0.98)^7$.

22. Without the use of a calculator, prove that $(1.012)^{30} > 1.36$.

23. In the binomial expansion of a certain binomial, the fifth term appears before simplification as

$$\frac{12 \cdot 11 \cdot 10 \cdot 9}{4 \cdot 3 \cdot 2 \cdot 1}\left(-\frac{1}{3}x\right)^8 (y^3)^4$$

a. Write the original binomial.
b. Determine the seventh term in the binomial expansion.

24. The coefficients of the sixth and eleventh terms in the binomial expansion of $(x + y)^n$ are the same. Determine the value of n.

25. a. Show that the sum of the coefficients in the binomial expansion of $(x + y)^n$ is 2^n. (*Hint:* Let $x = y = 1$.)
b. Show that the sum of the coefficients in the binomial expansion of $(x - y)^n$ is 0.

26. Show that the sum of the numbers in the nth row of Pascal's Triangle is twice the sum of the preceding row. (*Hint:* Use Exercise 25(a).)

KEY TERMS

mathematical induction
sequence
term of a sequence
factorial
partial sum
arithmetic sequence
common difference
geometric sequence
common ratio
geometric series
repeating decimal
permutation
combination
binomial expansion
binomial coefficient
Pascal's triangle

KEY FORMULAS

$$\sum_{j=1}^{n} a_j = na_1 + \frac{n(n-1)}{2} \cdot d = \frac{n}{2}(a_1 + a_n)$$

$$\sum_{j=1}^{n} ar^{j-1} = a\frac{1-r^n}{1-r}$$

$$\sum_{j=1}^{\infty} ar^{j-1} = \frac{a}{1-r} \quad \text{for } |r| < 1$$

$$n! = n(n-1)(n-2) \cdots 2 \cdot 1$$

$${}_nP_k = \frac{n!}{(n-k)!}$$

$${}_nC_k = \frac{n!}{k!(n-k)!} = \binom{n}{k}$$

$$(x + y)^n = \sum_{j=0}^{n} \binom{n}{j} x^{n-j}y^j$$

KEY THEOREMS AND LAWS

Axiom of Mathematical Induction
Axiom of Extended Mathematical Induction
Binomial Theorem

REVIEW EXERCISES

In Exercises 1–4, use mathematical induction to prove the given formula for all positive integers n.

1. $1 + 3 + 5 + \cdots + (2n - 1) = n^2$

2. $2 + 5 + 8 + \cdots + (3n - 1) = \frac{1}{2}n(3n + 1)$

3. $1 + 5 + 5^2 + \cdots + 5^{n-1} = \frac{1}{4}(5^n - 1)$

4. $\dfrac{1}{1 \cdot 4} + \dfrac{1}{4 \cdot 7} + \dfrac{1}{7 \cdot 10} + \cdots + \dfrac{1}{(3n - 2)(3n + 1)} = \dfrac{n}{3n + 1}$

5. Prove by induction that 3 is a factor of $n^3 + 2n$ for every integer $n \geq 0$.

***6.** Let

$$a_n = \sqrt{2}^{\sqrt{2}^{\left(\sqrt{2}^{\left(\cdot^{\cdot^{\cdot(\sqrt{2})}}\right)}\right)}}$$

Show that $a_n < 2$ for each $n \geq 1$. (*Hint:* Let $a_1 = \sqrt{2}$ and $a_{n+1} = (\sqrt{2})^{a_n}$ for $n \geq 1$. Use mathematical induction to prove that $a_n < 2$ for $n \geq 1$.)

7. For $n \geq 1$ let

$$a_n = \left(1 + \frac{1}{1}\right)\left(1 + \frac{1}{2}\right)\left(1 + \frac{1}{3}\right) \cdots \left(1 + \frac{1}{n}\right)$$

a. Use the fact that for each positive integer k,

$$1 + \frac{1}{k} = \frac{k + 1}{k}$$

to rewrite the formula for a_n. Then cancel where possible to discover a formula that shows that a_n is an integer.

b. Use mathematical induction to prove the formula you obtained in part (a).

8. Use mathematical induction to prove that $4^n > n^2$ for every integer $n \geq 0$.

In Exercises 9–12, find the first five terms of the sequence $\{a_n\}_{n=1}^{\infty}$.

9. $a_n = 4n + (-1)^n$

10. $a_n = 2^{n-3} + 2^{3-n}$

11. $a_1 = 3;\ a_{n+1} = -2a_n + 4$

12. $a_1 = 0;\ a_{n+1} = 2^{a_n}$

In Exercises 13–20, find the indicated partial sum.

13. $\sum_{j=1}^{5} (-2)$

14. $\sum_{j=1}^{8} (-1)^{j+1}$

15. $\sum_{j=1}^{4} \left(-\frac{4}{j^2}\right)$

16. $\sum_{j=1}^{10} j$

17. $\sum_{j=1}^{16} (-5 + 7j)$

18. $\sum_{j=2}^{20} \left(\frac{3}{4} + \frac{j}{2}\right)$

19. $\sum_{j=1}^{4} 2\left(\frac{1}{3}\right)^{j-1}$

20. $\sum_{j=1}^{7} (-2)^j$

21. Find the nth term a_n of the arithmetic sequence whose first term is 5 and whose common difference is $\frac{3}{4}$.

22. Find the nth term a_n of the arithmetic sequence whose second term is 7 and whose common difference is -3.

23. Find the nth term a_n of the arithmetic sequence whose fifth term is 13 and whose ninth term is 5.

24. Find the nth term a_n of the geometric sequence whose first term is 2 and whose common ratio is $\frac{1}{3}$.

25. Find the nth term a_n of the geometric sequence whose fifth term is $\frac{1}{4}$ and whose common ratio is $-\frac{1}{2}$.

26. Find the nth term a_n of the geometric sequence whose third term is -5 and whose sixth term is 625.

In Exercises 27–30, find the numerical value of the geometric series.

27. $\sum_{j=1}^{\infty} 3\left(-\frac{4}{7}\right)^{j-1}$ **28.** $\sum_{j=1}^{\infty} -\left(\frac{99}{100}\right)^{j-1}$

29. $\sum_{j=1}^{\infty} -6\left(\frac{11}{23}\right)^{j}$ **30.** $\sum_{j=1}^{\infty} (0.8)^{j+1}$

31. Express 0.151515 . . . as a fraction.

32. Express 3.062062062 . . . as a fraction.

In Exercises 33–42, calculate the given number.

33. The number of permutations of 6 objects taken 3 at a time.

34. The number of permutations of 8 objects taken 6 at a time.

35. The number of combinations of 4 objects taken 2 at a time.

36. The number of combinations of 8 objects taken 5 at a time.

37. ${}_5P_2$ **38.** ${}_{50}P_3$ **39.** ${}_7C_5$

40. ${}_{10}C_7$ **41.** $\binom{8}{4}$ **42.** $\binom{15}{12}$

In Exercises 43–46, use the Binomial Theorem to expand the given power of the binomial.

43. $(x + y)^8$ **44.** $(x - y)^6$

45. $(-3x + 4)^3$ **46.** $(2x^2 - y)^5$

47. Use the Binomial Theorem to calculate
a. $(\sqrt{2} - \sqrt{3})^4 + (\sqrt{2} + \sqrt{3})^4$
b. $(1 - 4i)^3 - (1 + 4i)^3$

48. Find the fifth term of $(\pi x - y)^6$.

49. Find the coefficient of the third term in the binomial expansion of $(x + 2y)^{16}$.

50. Suppose a university receives \$1,000,000 in contributions in 1991, and each year thereafter the amount of contribution increases by \$100,000. How much will the university have received in contributions by the end of 2000?

51. Which is more valuable—a gift consisting of one dollar each day for 100 days, or a gift consisting of 10¢ the first day, 20¢ the second day, 40¢ the third day, and so on for 10 days?

52. Suppose a king grants a peasant one grain of wheat the first day, 2 grains of wheat the second day, 4 grains of wheat the third day, and so on for 30 days. Use a calculator to compute the number of grains of wheat the peasant would receive.

53. How many 7-digit telephone numbers are there that neither begin nor end in 0?

54. A student has 5 mathematics books, 3 science books, and 2 history books. In how many ways can the books be placed on a shelf in a bookcase if all the books are distinct and the mathematics books are to be together, the science books together, and the history books together?

55. a. How many 2-card blackjack hands can be dealt from a 52-card deck?
b. How many blackjack hands are there consisting of an ace and another card that is either a king, queen, jack, or ten? (*Hint:* There are 4 each of aces, kings, queens, jacks, and tens in a 52-card deck.)

56. A club having 15 members needs 4 officers: president, vice-president, secretary, and treasurer.
a. In how many ways can 4 members be selected as officers, without regard to who has which office?
b. In how many ways can 4 members be selected to fill the offices?
c. If the president can be treasurer as well but the other officers must be distinct, in how many ways can the officers be selected?

57. A swimming club has 4 blue, 4 red, and 2 green sun umbrellas. In how many distinguishable ways can the sun umbrellas be distributed to 10 tables arranged in a straight line?

ANSWERS TO ODD-NUMBERED EXERCISES

SECTION 0.1

1. $6a^2 + 5a - 4$ **3.** $\frac{1}{2}a^2 + \frac{15}{4}a - 2$ **5.** $27 + 10\sqrt{2}$ **7.** $4a^2 + 12ab + 9b^2$
9. $\frac{10}{7}$ **11.** $\frac{7}{15}$ **13.** -2 **15.** l.c.d.: 24; $\frac{7}{24}$ **17.** l.c.d.: 45; $\frac{46}{45}$

19. l.c.d.: 180; $-\frac{7}{180}$ **21.** $-\dfrac{4a}{a^2 - 1}$ **23.** $\dfrac{2b - 3a + 4}{ab}$

25. $(a + 1)(b + 1) = ab + a + b + 1$ **27.** $a - (b + c) = a - b - c$

29. $(a + b)^3 = a^3 + 3a^2b + 3ab^2 + b^3$ **31.** $\dfrac{1}{a + b}$ remains $\dfrac{1}{a + b}$ **33.** 82.81

35. 0.03417635 **37.** 0.0220247086
39. The decimal expansion neither terminates nor repeats.
43. a. $a = 0, b = 1, c = 1$ (other solutions possible)
b. $a = 1, b = 2, c = 2$ (other solutions possible)
53. 26.46

SECTION 0.2

1. 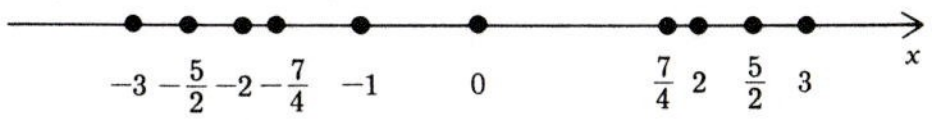

3. $x \ge \sqrt{2}$ **5.** $4x \ge 8$ **7.** $-1 < 6 - r < 1$ **9.** $z < 0$ **11.** $|x - 2| < 0.01$
13. $\sqrt{2} > 1$ **15.** $(-2)^2 > 3$ **17.** $\frac{22}{7} > \pi$ **19.** 1 **21.** 6 **23.** 8.5 **25.** 16
27. 0 **29.** $-1 - \sqrt{2}$ **31.** 2 **33.** x^2 **35.** $x^2 + 1$ **37.** $a - 4$ **39.** $a - b$
41. 0 **43.** $\frac{22}{7}$ is larger; $\frac{223}{71}$ is smaller **45.** $a \ge 0$

SECTION 0.3

1. $\frac{1}{4}$ **3.** 8 **5.** $\frac{1}{1000}$ **7.** 4^5 **9.** $-(2^9)$ **11.** 10^{-2} **13.** 4^7 **15.** $-\frac{1}{7}$
17. 4^{30} **19.** a^{12} **21.** y^{-8} **23.** b^8 **25.** $-c^{10}$ **27.** $x^8y^{12}z^4$ **29.** $r^{16}s^{14}$

31. t^2 **33.** $\dfrac{1}{ab}$ **35.** $\dfrac{a^3b^3}{(a+b)^3}$ **37.** Negative

39. 4.832×10^2 **41.** 1.009×10^0 (or 1.009) **43.** 9.999×10^{-1}
45. 2.7168876×10^7 **47.** 1.7659446×10^{-1} **49.** 3×10^{-7} **51.** 1.817×10^9
53. $8.74457025 \times 10^{-2}$ **55.** $7.806637807 \times 10^{-3}$ **57.** $A = \pi r^2$ **59.** $V = \frac{4}{3}\pi r^3$
61. $V = \frac{1}{3}\pi r^2 h$ **65.** 10 **67.** 5,980,000,000,000,000,000,000,000
69. Approximately 10^{22} **71.** Approximately 9.2529×10^7 miles
73. a. Approximately 5.8697136×10^{12} miles
b. Approximately 8.699 light years

SECTION 0.4

1. $\frac{1}{5}$ **3.** $\frac{1}{12}$ **5.** $6\sqrt{2}$ **7.** 10^7 **9.** $\frac{5}{7}$ **11.** $2x^3y^{-2}\sqrt{6}$ **13.** $\dfrac{8\sqrt{2}}{r\sqrt{r}}$

15. $\dfrac{1}{c^{12}d^4\sqrt{d}}$ **17.** $(4a - 7b)^2$ **19.** $2ab^2$ **21.** $\sqrt{a} + \sqrt{b}$ **23.** 3 **25.** 0.05

27. 0.2 **29.** ab^2 **31.** $-\dfrac{1}{5xy^2\sqrt[3]{z}}$ **33.** t **35.** $|a|^3b^2$ **37.** $a^2|b|$ **39.** 32

41. $\frac{1}{25}$ **43.** $\frac{125}{27}$ **45.** x **47.** $216a^{9/2}b^6$ **49.** $x^2y^4z^{16/3}$ **51.** $\dfrac{z^{9/4}}{8}$

53. $p^{1/3}q^{1/3}$ **55.** $\dfrac{z^2}{64(x-y)^3}$ **57.** $b^{13/6}$ **59.** Approximately 1.741101127

61. Approximately 1.464591888 **63.** Approximately 0.8992180438

67. $r = \sqrt{\dfrac{S}{4\pi}}$ **73.** Approximately $9.662614165 \times 10^{-28}$ grams

75. Approximately 3152.608104 **77.** The new pressure $p_1 = \dfrac{p}{2^{1.4}}$

SECTION 0.5

1. 3; 1; 10 **3.** $7x^2 - 6$ **5.** $-x^3 + x^2 + 2x + 4$ **7.** $8x^2 - 22x + 15$
9. $7x^2 - 2x - 3$ **11.** $12x - 15$ **13.** $-s^3 + 10s^2 - 32s + 32$ **15.** $2y^6 + 16$
17. $4x^2 + xy + 2y^2$ **19.** $4x^4 - 4x^2y^3 + y^6$ **21.** $8x^3 - 36x^2y + 54xy^2 - 27y^3$
23. $2xh + h^2$ **25.** $x^3 - y^3$ **27.** $x^2 + 2xy + y^2 - z^2$ **29.** $u + 2u^{1/2}v^{1/2} + v$

31. $\dfrac{1}{r^2} - \dfrac{2}{rs} + \dfrac{1}{s^2}$ **35.** $x = \frac{17}{9}, y = \frac{26}{27}$

39.

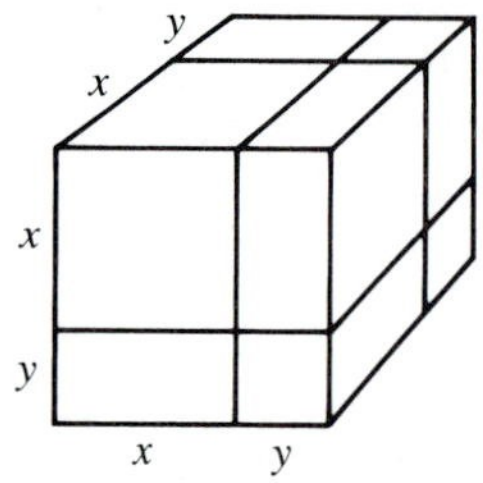

41. They are the same.

SECTION 0.6

1. $(x+6)(x+2)$ **3.** $(x-6)(x-1)$ **5.** $2(t-4)(t+1)$ **7.** $(7-b)(3-b)$
9. $(a-7)^2$ **11.** $(x+\frac{11}{2})^2$ **13.** $(4-3z)(4+3z)$ **15.** $(2x+1)(x+3)$
17. $(x-3)(-x+2)$ or $(x-2)(-x+3)$ **19.** $(7x-4)(x+3)$ **21.** $(x+y)^2$
23. $(x-2y)(x+2y)$ **25.** $(x-y)^2(x+y)^2$ **27.** $(5x+y)(x-3y)$
29. $x^6(x-1)$ **31.** $(3x+5)(x+1)x$ **33.** $(x-3)^2(x+3)^2$ **35.** $(x+1)^2(x-1)$
37. $(3x+2y+4)(3x+2y-3)$ **39.** $(x-5-2y)(x-5+2y)$
41. $(x+1)(x^2-x+1)$ **43.** $(x-2)(x^2+2x+4)$ **45.** $(2x-1)(4x^2+2x+1)$
47. $(2x-1)(16x^4+8x^3+4x^2+2x+1)$ **49.** Yes **51.** No **55.** -1
57. -2

SECTION 0.7

1. $\dfrac{1}{x+4}$ **3.** $\dfrac{3y+2}{y-\frac{1}{3}}$ **5.** $\dfrac{(s+3)(s+2)(s-2)}{s-4}$ **7.** $\dfrac{x+y}{x^2+xy+y^2}$

9. $(x+1)(x-2)$ **11.** $y(y-2)$ **13.** $\dfrac{(z+3)^3}{(z-2)(z+2)(z-1)}$ **15.** $\dfrac{(x-y)^2}{x+y}$

17. $\dfrac{3x-1}{x(x-1)}$ **19.** $\dfrac{3y-4}{(y-3)(y+3)}$ **21.** $\dfrac{2(z^2+1)}{(z-1)(z+1)}$ **23.** $\dfrac{5}{t-3}$ **25.** $\dfrac{v-1}{v+2}$

27. $\dfrac{b^2}{a^2+b^2}$ **29.** $\dfrac{qr+pr+pq}{pqr}$ **31.** $-\dfrac{1}{x(x+h)}$ **33.** $\dfrac{x-y}{x+y}$ **35.** $\dfrac{1+x^2}{2+x}$

37. x **39.** $\dfrac{y}{x+y}$ **41.** $\dfrac{(x-y)(x^2+xy+y^2)}{x+y}$ **43.** $\frac{1}{3}\sqrt{3}$ **45.** $\dfrac{\sqrt{x}+9}{x-81}$

47. $\dfrac{\sqrt{x}+\sqrt{y}}{x-y}$ **49.** $\dfrac{1}{\sqrt{x+3}+\sqrt{3}}$ **51.** $\dfrac{x-y}{x+2\sqrt{xy}+y}$ **53.** $\dfrac{pq}{p+q}$

CHAPTER 0 REVIEW

1. $\frac{1}{27}$ **3.** 16 **5.** 0.3 **7.** $-\frac{2}{3}$ **9.** $a-\sqrt{5}\geq 0$ **11.** 4.3 **13.** 1.59×10^5
15. $\frac{1}{8}a^{-2}$ **17.** $6\sqrt{2}$ **19.** $\frac{1}{15}$ **21.** $x^{15/2}y^{-3}z^9$ **23.** 1
25. $7.276362762\times 10^{-26}$ **27.** 0.1284520595 **29.** $\frac{1}{4}a^2-\frac{2}{3}a+\frac{4}{9}$
31. $x-3x^{1/3}+3x^{-1/3}-x^{-1}$ **33.** $8x^6+60x^4y+150x^2y^2+125y^3$
35. Negative **37.** $(x-9)(x+3)$ **39.** $(t-10)^2$ **41.** $(4x-1)(3x-2)$
43. $(y-1)^3(y-3)(y+1)$ **45.** $(z^2+2)(z-1)(z+1)$ **47.** $(x+1)(x^2+1)$
49. $(x+7y)(x-5y)$ **51.** $(4x^2+1)(2x-1)(2x+1)$ **53.** -14 **55.** $\frac{28}{97}$

57. $\dfrac{x-1}{x+2}$ **59.** $\dfrac{(2x-1)(x-2)}{(x+4)(3x-1)}$ **61.** 1 **63.** $\dfrac{x^2+1}{(x-1)^2(x+1)}$ **65.** $\dfrac{x+2}{x-1}$

67. $\dfrac{x^2+y^2}{(x-y)(x+y)}$ **69.** $\dfrac{x-1}{x+1}$ **71.** $\sqrt{5}+2$ **73.** $S=2s^2+4sh$

75. $A=\frac{1}{4}\sqrt{3}s^2$ **77.** Approximately 0.04811 **79.** 1 and -1
83. b. i. $a=4, b=3, c=5$
ii. $a=6, b=8, c=10$
iii. $a=12, b=5, c=13$

SECTION 1.1

1. 7 **3.** $\frac{5}{11}$ **5.** $-\frac{2.6}{9.1}$ **7.** $\frac{1}{2}$ **9.** -5 **11.** -1 **13.** $-\frac{7}{3}$ **15.** $\frac{3}{2}$ **17.** 1, 7
19. $-1, 6$ **21.** 0, 36 **23.** $x=3y-\frac{1}{2}$ **25.** $x=-\frac{7}{3}y+\frac{1}{15}$ **27.** $x=-6y-8$
29. $y=\frac{1}{2}x-\frac{7}{2}$ **31.** -2 **33.** 39.37 **35.** 91 **37.** 5 and 9

39. 47, 49, and 51 **41.** 38 nickels and 26 dimes **43.** \$25.50 **45.** \$4,250
47. \$3,200 in the 7% account, and \$2,800 in the 8% account
49. length: 38 feet; width: 15 feet **51.** 21 feet by 7 feet **53.** 5 inches

55. $p = \dfrac{fq}{q - f}$ **57.** 55,160,000 square meters **59.** $\frac{5}{6}$ hour **61.** 3 miles

63. length: 250 meters; speed: 25 meters per second **65.** 12 minutes
67. $\frac{36}{23}$ hours **69.** 3 quarts **71.** 1.6 gallons **73.** 9 gallons **75.** 84 years old

SECTION 1.2

1. -2 **3.** 25 **5.** $\frac{2}{3}$ **7.** $\sqrt{2}$ **9.** $-\frac{1}{4}$ **11.** $-4, -1$ **13.** $2, -5$ **15.** $1, \frac{1}{2}$
17. $1 + \sqrt{7}, 1 - \sqrt{7}$ **19.** $5 + 3\sqrt{3}, 5 - 3\sqrt{3}$ **21.** $-1 + \frac{1}{2}\sqrt{10}, -1 - \frac{1}{2}\sqrt{10}$
23. $\frac{3}{2} + \frac{1}{2}\sqrt{5}, \frac{3}{2} - \frac{1}{2}\sqrt{5}$ **25.** None **27.** $1 + \sqrt{5}, 1 - \sqrt{5}$
29. $\frac{1}{2} + \frac{1}{2}\sqrt{7}, \frac{1}{2} - \frac{1}{2}\sqrt{7}$ **31.** $1 + \frac{1}{3}\sqrt{6}, 1 - \frac{1}{3}\sqrt{6}$ **33.** $-\frac{1}{2}, -\frac{1}{3}$
35. $\frac{2}{3}(\sqrt{3} + \sqrt{6}), \frac{2}{3}(\sqrt{3} - \sqrt{6})$ **37.** $-\frac{1}{2}\sqrt{2} + 1, -\frac{1}{2}\sqrt{2} - 1$ **39.** None
41. $\frac{3}{2} + \frac{1}{2}\sqrt{29}, \frac{3}{2} - \frac{1}{2}\sqrt{29}$ **43.** $-2 + 2\sqrt{3}, -2 - 2\sqrt{3}$
45. $3 + \sqrt{14}, 3 - \sqrt{14}$ **47.** $6, -6$ **49.** 0 **51.** $2, -2$
53. $6.794954775, -1.128288108$ **55.** $1.433642043, -1.019629304$
57. root: $-\frac{3}{2}$; $c \doteq \frac{27}{2}$ **59.** $\frac{8}{9}$ **61.** 2.47

SECTION 1.3

1. $\frac{3}{2}\sqrt{5}$ seconds (approximately 3.35 seconds) **3.** 2 seconds **5.** 1 second
7. $\sqrt{61}$ inches **9.** $-\frac{1}{2} + \frac{1}{2}\sqrt{199}$ meters, $\frac{1}{2} + \frac{1}{2}\sqrt{199}$ meters **11.** 3000 feet
13. 40 miles per hour **15.** 300 and 400 miles per hour **17.** 170 miles per hour
19. 40 miles per hour in midday, 24 miles per hour in rush hour **21.** 32, 33
23. $(0, 2\sqrt{5})$ and $(0, -2\sqrt{5})$ **25.** 9 inches by 7 inches **27.** A 10-foot strip
29. 28 inches and 16 inches **31.** 4 inches
33. a. \$3 and \$10; 70 umbrellas and 0 umbrellas, respectively
b. \$5; 50 umbrellas
35. 8 feet

SECTION 1.4

1. $2 + 8i$ **3.** $2 + 2i$ **5.** $2 + \frac{1}{6}i$ **7.** $7\sqrt{2} - 6\sqrt{5}i$ **9.** $10 + 6i$ **11.** $11 + 3i$
13. $-5 + \frac{5}{2}i$ **15.** $\frac{13}{36}$ **17.** $-1 + 2\sqrt{2} + (2 + 2\sqrt{2})i$ **19.** $2i$ **21.** $2\sqrt{2}i$
23. i **25.** $-2 + 2\sqrt{3}i$ **27.** $-2i, 2i$ **29.** $-2\sqrt{3}i, 2\sqrt{3}i$ **31.** $1 + 2i, 1 - 2i$
33. $\frac{1}{2} + \frac{1}{2}i, \frac{1}{2} - \frac{1}{2}i$ **35.** $\frac{1}{2} + \frac{1}{2}\sqrt{3}i, \frac{1}{2} - \frac{1}{2}\sqrt{3}i$ **37.** $1 + \sqrt{2}i, 1 - \sqrt{2}i$
39. $\frac{3}{2} + \frac{1}{2}\sqrt{11}i, \frac{3}{2} - \frac{1}{2}\sqrt{11}i$ **41.** $\frac{1}{3} + \frac{1}{3}\sqrt{5}i, \frac{1}{3} - \frac{1}{3}\sqrt{5}i$
43. $-1 \pm \sqrt{\pi - 1}i$; $-1 \pm 1.463i$ **45.** $2 - 3i$ **47.** $-\frac{1}{2} - \frac{1}{3}i$ **49.** $-9i$
51. $2 - 3i$

SECTION 1.5

1. $-11, -4$ **3.** $7, -4$ **5.** $2, -2, \sqrt{2}, -\sqrt{2}$ **7.** $3, -3$ **9.** $\sqrt[3]{28}, -1$
11. 16 **13.** 81 **15.** 8, 1 **17.** 16 **19.** $\sqrt{5}, -\sqrt{5}$ **21.** $7, -2$ **23.** $\frac{1}{3}$

25. $\sqrt{3}, -\sqrt{3}$ **27.** 5 **29.** 1 **31.** 1 **33.** $\frac{3}{2}$ **35.** $\dfrac{\sqrt[3]{10}}{3}$

37. $2^{1/n}$ (and $-2^{1/n}$ if n is even) **39.** -1 **41.** $\sqrt{3}, -\sqrt{3}$ **43.** $\sqrt[3]{23}$
45. $0, -1$ **47.** $0, 1, -1$ **49.** $2, -2, 4$ **51.** $3, -3$ **53.** 1 **55.** $0, 2, -2, 3$
57. $-1, 1, -i, i$ **59.** $-i, i$ **61.** $-2, 2, -i, i$ **63.** $1, -\frac{1}{2} - \frac{1}{2}\sqrt{3}i, -\frac{1}{2} + \frac{1}{2}\sqrt{3}i$

65. a. -1 b. -1 **67.** $a = \sqrt{\left(\dfrac{Q}{E}\right)^{2/3} - 1}$ **69.** $r = \sqrt[4]{\dfrac{8\eta l V}{\pi p}}$

71. a. $100^{1/5}$ b. $d = c^3$ c. $c \approx 2.512$; $d \approx 15.849$

SECTION 1.6

1. Open, bounded **3.** Closed, unbounded **5.** Half-open, bounded
7. Closed, bounded **9.** $(-4, 3]$ **11.** $(-1.1, -0.9)$ **13.** $[-1, 1)$
15. $(-\infty, \frac{1}{4}]$ **17.** $(-\infty, \frac{5}{2})$ **19.** $(6, \infty)$ **21.** $(-\infty, 7]$ **23.** $(-\infty, -\frac{3}{2}]$
25. $(-5, -2)$ **27.** $[-3, -1)$ **29.** $(1.99, 2.01)$ **31.** $(a, a + d)$
35. Between 20 and 120, inclusive **37.** From \$0 to \$5.50 inclusive
39. From \$0 to \$2 inclusive

SECTION 1.7

1. $(-\infty, 1), [2, \infty)$ **3.** $[-3, -1]$ **5.** $(-\infty, 3), (3, \infty)$ **7.** $(-2, 2)$
9. $[1, 3]$ **11.** $(-\infty, -3), (5, \infty)$ **13.** $[-9, 0]$ **15.** $(-\infty, -5), (4, \infty)$
17. $(-2, \frac{3}{2}]$ **19.** $(-1, 0), (1, \infty)$ **21.** $(1, 2), (3, \infty)$ **23.** $(7, \infty)$ **25.** $(\frac{1}{2}, 2]$
27. $(-\infty, 0), (1, \infty)$ **29.** $(-\frac{1}{2} - \frac{1}{2}\sqrt{5}, -\frac{1}{2} + \frac{1}{2}\sqrt{5})$
31. $(-1 - \frac{1}{2}\sqrt{2}, -1 + \frac{1}{2}\sqrt{2})$ **33.** $(-4, 4)$ **35.** $[-\frac{1}{5}, \frac{1}{5}]$
37. $(-\infty, -\frac{9}{2}], [\frac{9}{2}, \infty)$ **39.** $(2, 8)$ **41.** $[-6, 0]$ **43.** $(-\infty, 6), (8, \infty)$
45. $(-\infty, \frac{5}{3}], [\frac{7}{3}, \infty)$ **47.** $(-1, 2)$ **49.** $(-\infty, \frac{2}{3}), (\frac{2}{3}, \infty)$ **51.** $(-\sqrt{2}, 0), (0, \sqrt{2})$
53. $(-6, 6)$ **55.** $(3, 4], [6, 7)$ **57.** $[-\frac{7}{6}, -\frac{5}{6}], [-\frac{1}{2}, -\frac{1}{6})$
59. a. $(-2\sqrt{5}, 2\sqrt{5})$ b. $(-\infty, -2\sqrt{5}), (2\sqrt{5}, \infty)$
61. $(-\infty, -2), [-1, \infty)$ **65.** a. No b. Yes **67.** During the first six seconds
69. Between 3 and 4 seconds **73.** Between 0 and 2 inches

CHAPTER 1 REVIEW

1. 26 **3.** 3 **5.** $-\frac{2}{19}$ **7.** $\frac{1}{2}, -\frac{5}{3}$ **9.** $1, \frac{7}{3}$ **11.** $\sqrt{3}, -\sqrt{3}$ **13.** None
15. $\frac{1}{4} + \frac{1}{4}\sqrt{17}, \frac{1}{4} - \frac{1}{4}\sqrt{17}$ **17.** $3, -3$ **19.** $-3 + \sqrt{6}, -3 - \sqrt{6}$
21. $\sqrt{2}, -\sqrt{2}$ **23.** $\frac{1}{2}, 1$ **25.** $-4 - 2\sqrt{5}$ **27.** $0, -3 + 3\sqrt{2}, -3 - 3\sqrt{2}$
29. $0, \frac{1}{2}, 1, -1$ **31.** $2, -2, 2\sqrt{2}, -2\sqrt{2}$ **33.** $2\sqrt{3}, -2\sqrt{3}, 2\sqrt{5}, -2\sqrt{5}$
35. $-4 + 6i$ **37.** 10 **39.** $2\sqrt{3}i$ **41.** $-3\sqrt{3}i, 3\sqrt{3}i$
43. $-\frac{1}{2} - \frac{1}{6}\sqrt{3}i, -\frac{1}{2} + \frac{1}{6}\sqrt{3}i$ **45.** $-4 + 3i$ **47.** $(-\infty, \frac{5}{2}]$ **49.** $(-\infty, \frac{9}{2})$
51. $[-\frac{1}{2}, \infty)$ **53.** $(-\infty, 4), (5, \infty)$ **55.** $(-2, -\frac{1}{3})$ **57.** $[-\frac{1}{2}, \frac{2}{5})$
59. $(-\infty, -\frac{7}{6}], (3, \infty)$ **61.** $(-\infty, -\sqrt{2}], [\sqrt{2}, \infty)$ **63.** $[-4, 9]$ **65.** -1
67. $(3, 8)$ **69.** \$66.78 **71.** 4 **73.** 13 miles **75.** $\frac{9}{2}$ hours
77. Between 1 and 2 seconds **79.** 4 **81.** \$10,000 **83.** $4\sqrt{2}$ feet **85.** $46\frac{2}{3}\%$

87. a. 7.5 b. $\frac{4}{3}$ **89.** $r = \dfrac{A_n - P}{Pn}$ **91.** $x = \dfrac{Qy}{17{,}860y - 1.798Q}$

93. 0.90 seconds

SECTION 2.1

1. 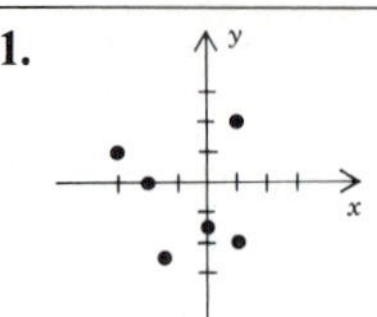 **3.** 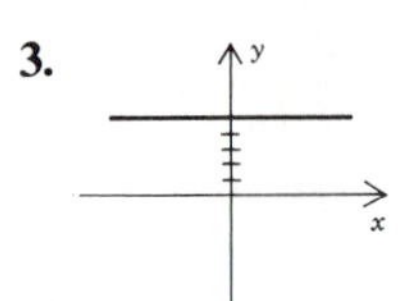 **5.** 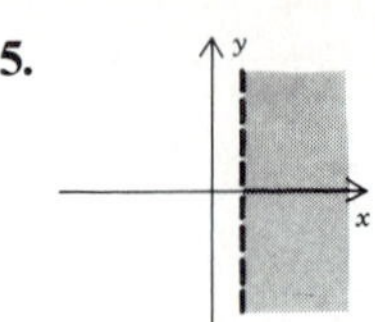

7. x and y axes

9. 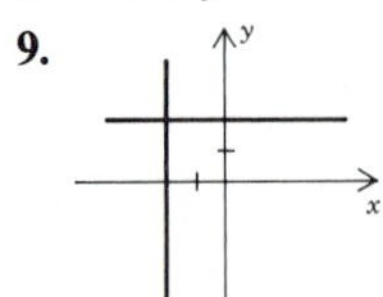 **11.** 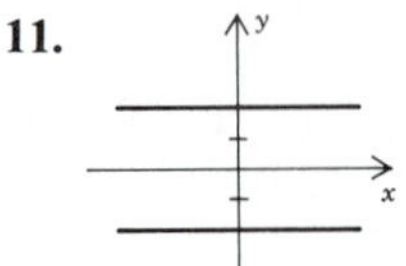 **13.** 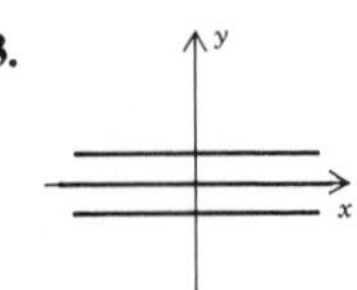

15. 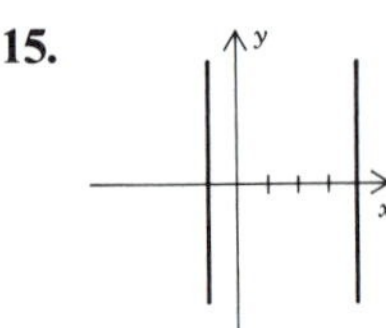 **17.** 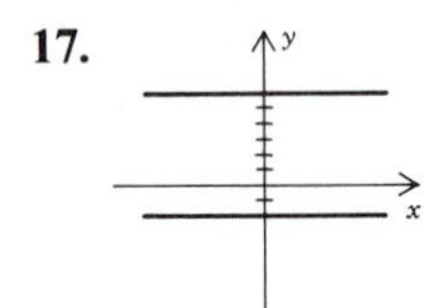 **19.** 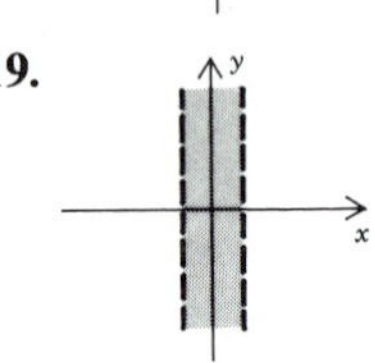

21. Second and fourth quadrants, not including axes **23.** 2 **25.** 13 **27.** $2\sqrt{10}$ **29.** 13 **31.** $\sqrt{29-2\sqrt{2}}$ **33.** $5a$ **35.** (6, 6) **37.** (0, 7) **39.** (0.7, 2.85) **41.** $(0, a)$ **43.** $(4+\sqrt{3}, 0)$ and $(4-\sqrt{3}, 0)$ **45.** $(\frac{1}{3}a+\frac{2}{3}c, \frac{1}{3}b+\frac{2}{3}d)$ **51.** $90\sqrt{2}$ feet (approximately 127.28 feet)

SECTION 2.2

1. 3 **3.** $\frac{2}{3}$ **5.** 2.378 **7.** $y-(-4)=0(x-3)$ (or $y+4=0$) **9.** $y-0=\frac{2}{3}(x-3)$ **11.** $y-\frac{3}{2}=-1(x-0)$ **13.** $y-1=2x$ **15.** $y+2=-4(x+1)$ **17.** $x=2$ **19.** $y=0x-2$ (or $y=-2$) **21.** $y=-\frac{4}{5}x-1$ **23.** $y=-\sqrt{3}x+\frac{1}{2}$ **25.** $m=0; b=3$ **27.** $m=3; b=-\frac{1}{2}$ **29.** $m=-\frac{2}{3}; b=2$ **31.** $m=3; b=0$ **33.** Perpendicular **35.** Perpendicular **37.** Neither **39.** Parallel **41.** Neither **43.** Perpendicular

45. b. $y-y_1=\dfrac{y_2-y_1}{x_2-x_1}(x-x_1)$ **47.** a. $\frac{1}{3}$ b. -5 c. $-\frac{3}{4}$

49. a. $\dfrac{x}{-1}+\dfrac{y}{2}=1$ b. $\dfrac{x}{3}+\dfrac{y}{\frac{1}{2}}=1$ c. $\dfrac{x}{0.7}+\dfrac{y}{0.3}=1$ d. $\dfrac{x}{-2}+\dfrac{y}{-3}=1$

51. a, c, f **53.** $-\frac{4}{3}$ **55.** $f(x)=-2x-6$ **57.** $f(x)=4x+4$ **59.** $y-5=-\frac{4}{3}(x+2)$ **61.** $(-1, 2)$

SECTION 2.3

1. $-14; -14; -14$ **3.** $4; 8; t^4-3t^2+4$ **5.** 1.797752809, 3.809523810 **7.** $2; \sqrt{7}; |x|$ **9.** $-2.311653282, -158.1023022$ **11.** 5; 0 **13.** 11; 3 **15.** $\frac{4}{5}; \frac{5}{6}$

17. $-2; -6$ **19.** $\frac{2}{3}; \dfrac{3\sqrt[3]{2}}{3\sqrt[3]{2}-1}\approx 1.35974288$ **21.** All real numbers

23. All real numbers except 0, $\frac{5}{2}$ **25.** All real numbers
27. All real numbers except 1, -6 **29.** All real numbers
31. All negative numbers **33.** All numbers in $(-\infty, -2]$ and $[2, \infty)$
35. All numbers in $(-\infty, 0]$ and $(1, \infty)$ **37.** All real numbers
39. All numbers in $[1, \infty)$ **41.** All real numbers except 0 **43.** $4x + 2h$

45. $\dfrac{-1}{x(x+h)}$ **47.** c **49.** No

55. a. 107.9080743 b. -2.346718531 c. 6218.891838
57. a. 2.87923601 b. 2.185812841 c. 2.971266923
59. a. $C = 2\pi r$ for $r \geq 0$ b. $C = \pi d$ for $d \geq 0$

61. a. $C = 2.54I$ for $I \geq 0$ b. $I = \dfrac{1}{2.54}C$ for $C \geq 0$

63. $L(x) = \frac{1}{3}x$ for $x > 0$ **65.** a. $L = 15 + \dfrac{1}{750}(T - 10)$ b. 85°C

SECTION 2.4

1. $p = 4s$ **3.** $S = 6s^2$ **5.** $E = hv$ **7.** $R = cv^2$

9. a. $A = \dfrac{\sqrt{3}}{4}s^2$ b. 6 centimeters c. 3.8 centimeters

11. a. $l_f = \dfrac{1250}{381}l_m$ b. $l_y = cl_m$, where $c = \dfrac{1250}{1143}$

13. Approximately 2.498×10^{19} ergs
15. a. $K = \frac{1}{2}mv^2$ b. Velocity of lighter one is $\sqrt{3}$ times as great

17. $\lambda = \dfrac{c}{v}$ **19.** $F = \dfrac{cq_1q_2}{r^2}$

SECTION 2.5

1. y intercept: 3; x intercept: -3

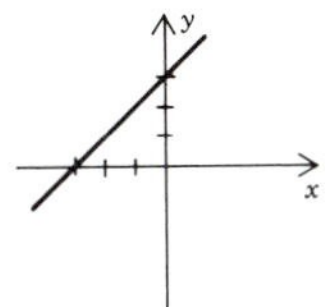

3. y intercept: 1; no x intercept

5. y intercept: 1; x intercept: -1

7. y intercept: -2; x intercepts: -2, 2

9. y intercept: 2; x intercept: 2

11. y intercept: -1; x intercept: 1

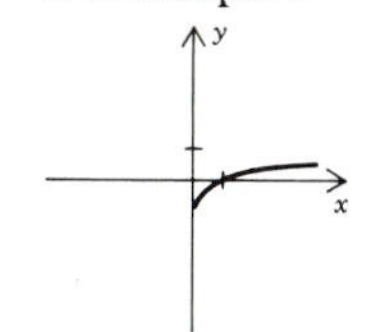

13. no y intercept; x intercept: 1

15. no y intercept; x intercept: $\frac{1}{3}$

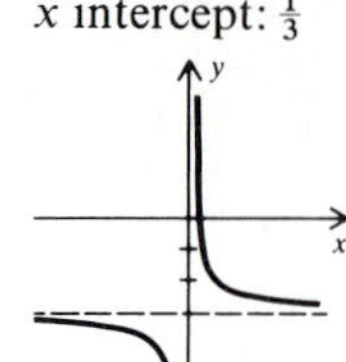

17. y intercept: 0; x intercept: 0

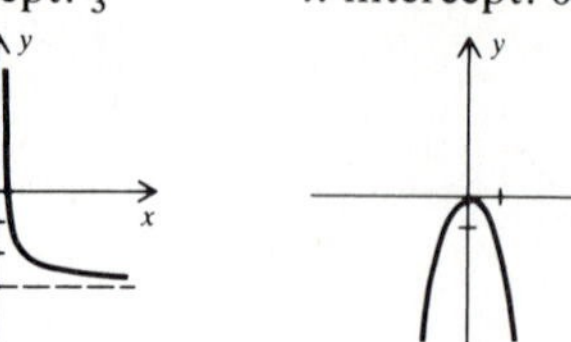

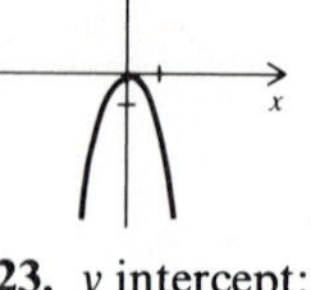

19. y intercept: 2; x intercepts: $-2-\sqrt{2}$ and $-2+\sqrt{2}$

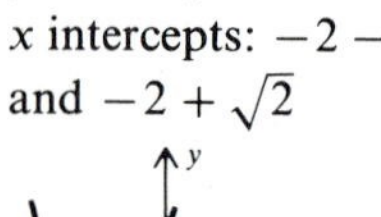

21. y intercept: 0; x intercept: 0

23. y intercept: 0; x intercept: 0

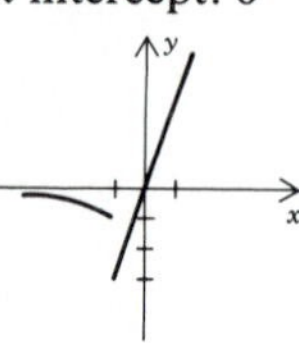

25. y intercept: 0; x intercepts: 0, 4 **27.** y intercept: 0; x intercepts: -3, 0
29. y intercept: $\frac{1}{3}$; x intercepts: -1 **31.** y axis **33.** Neither **35.** y axis
37. y axis **39.** Neither **41.** Odd **43.** Neither **45.** Neither **47.** Even
49. a, c, g, h **51.**

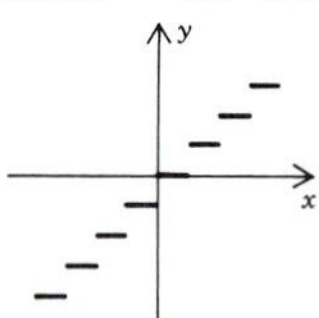

SECTION 2.6

1. $f(x)=x-1, g(x)=2x$ **3.** $f(x)=x^4, g(x)=\dfrac{3}{x}$ **5.** $f(t)=\sqrt{t}, g(t)=\dfrac{1}{2t}$

7. Domain: all real numbers; rule: $(g \circ f)(x)=2x^2-3x+2$

9. Domain: $(0, \infty)$; rule: $(g \circ f)(x)=\sqrt{\dfrac{1}{2x}}$

11. Domain: $[1, \infty)$; rule: $(g \circ f)(x)=\sqrt{x-1}+1$

13. Domain: all real numbers except $0, \frac{1}{2}, -\frac{1}{2}$; rule: $(g \circ f)(x)=\dfrac{4x^2}{1-4x^2}$

15. Domain: all real numbers except 0; rule: $(g \circ f)(x)=x$

17. Domain: all real numbers except -3 and -1; rule: $(g \circ f)(x)=-\dfrac{4x+2}{x+3}$

19. Domain: $(-\infty, -\sqrt{3}]$ and $[\sqrt{3}, \infty)$; rule: $(g \circ f)(x)=\sqrt{x^2-3}$
25. $(9, 10]$ **27.** m **29.** $V(r(t))=\frac{9}{2}\pi t^6$

SECTION 2.7

1. $f^{-1}(x)=\frac{1}{2}x$ **3.** $f^{-1}(x)=3-2x$ **5.** $f^{-1}(x)=x^{1/3}$
7. $f^{-1}(x)=[\frac{1}{3}(x+5)]^{1/3}$ **9.** $f^{-1}(x)=(1-x)^{1/5}$

11. $f^{-1}(x) = x^2$ for $x \geq 0$ **13.** $f^{-1}(x) = x^2 + 3$ for $x \geq 0$

15. $f^{-1}(x) = \dfrac{1}{x^2}$ for $x > 0$ **17.** $f^{-1}(x) = 2 - \dfrac{1}{2x^3}$ **19.** $f^{-1}(x) = \dfrac{x+1}{1-x}$

21. $f^{-1}(x) = \left(\dfrac{3x+1}{2-x}\right)^{1/3}$

23.

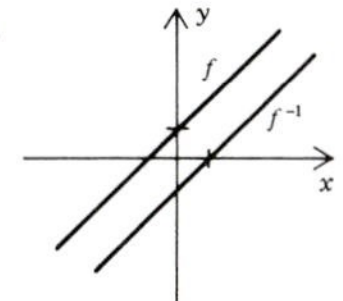

25.

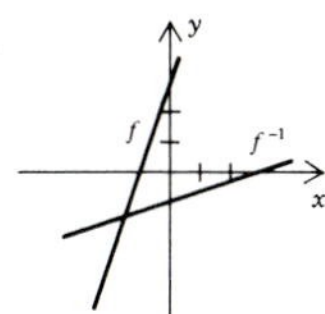

27.

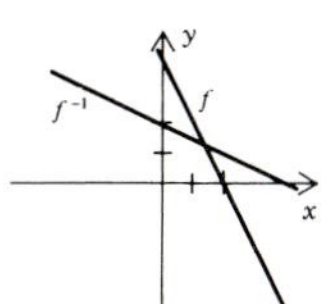

29. $f = f^{-1}$ **31.**

33. Increasing **35.** Neither **37.** Increasing **39.** Neither

43. a. $s = \dfrac{2\sqrt{A}}{\sqrt[4]{3}}$ b. $\dfrac{8}{\sqrt[4]{3}}$

SECTION 2.8

1. $a = \frac{1}{2}$: They increase toward 1.
$a = 5$: They decrease toward 1.
$a = 30$: They decrease toward 1.
$a = 1000$: They decrease toward 1.

3. $a = 0$: They increase toward $\frac{1}{2} - \frac{1}{4}\sqrt{2}$.
$a = .5$: They increase toward $\frac{1}{2} - \frac{1}{4}\sqrt{2}$.

5.

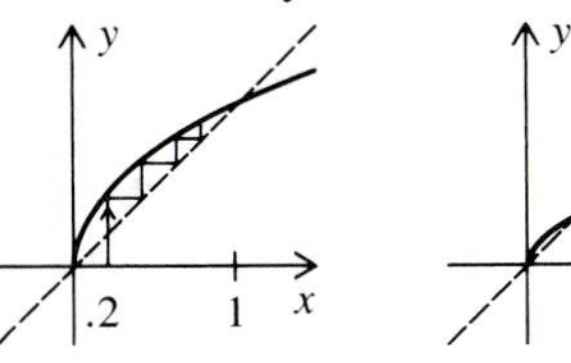

7.

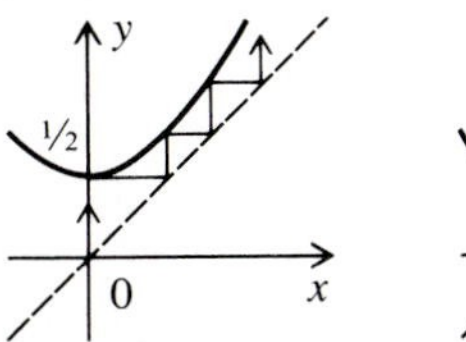

9. 0, 1 **11.** They approach $\frac{1}{2}$.

13. They eventually oscillate between approximately .50088, .87499, .38281, and .82694.

15. After the first iterate, all iterates of .5 are 0. The iterates of .3 and .8 seem to be random.

17. They oscillate between a, $\dfrac{1}{1-a}$ and $1 - \dfrac{1}{a}$.

CHAPTER 2 REVIEW

1. a. $3\sqrt{5}$ b. $\frac{1}{2}\sqrt{82}$ **3.** All real numbers except -3
5. All real numbers except $-2+\sqrt{11}$ and $-2-\sqrt{11}$
7. All numbers in $[\frac{2}{3}, \infty)$ **9.** All numbers in $(-\infty, -5]$ and $[5, \infty)$
13. $-\frac{1}{3}$ **15.** a. All numbers in $(-\infty, -2]$ and $[2, \infty)$

17. y intercept: -5;
x intercept: $\frac{5}{2}$; no symmetry

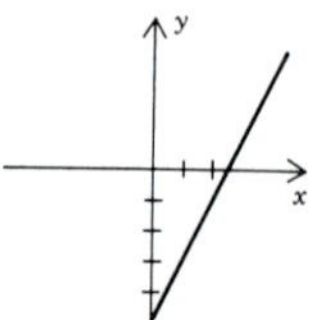

19. y intercept: -6;
symmetry with respect to y axis
x intercepts: $\sqrt{6}$ and $-\sqrt{6}$

21. y intercept: 2;
x intercepts: 2 and -2;
symmetry with respect to y axis

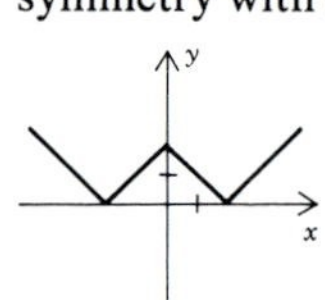

23. y intercept: 0; x intercept: 0;
no symmetry

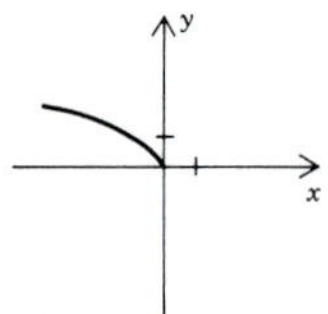

25. No intercepts;
symmetry with respect to origin

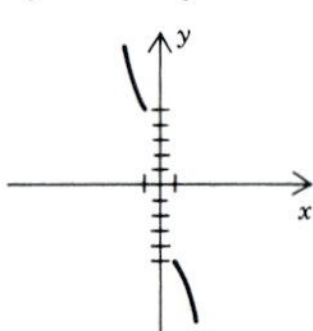

27. $(f \circ g)(x) = 1/x^2$ for all x except 0
$(g \circ f)(x) = 125/x^2$ for all x except 0
29. $(f \circ g)(x) = (2x+3)^{3/2}$ for $x \geq -\frac{3}{2}$
$(g \circ f)(x) = \sqrt{2x^3+3}$ for all $x \geq -(\frac{3}{2})^{1/3}$
31. $(f \circ g)(x) = \sqrt{x^2-4}$ for all x in $(-\infty, -2]$ and $[2, \infty)$
$(g \circ f)(x) = \sqrt{x^2-4}$ for all x in $(-\infty, -2]$ and $[2, \infty)$

35. No **37.** $f^{-1}(x) = \dfrac{1}{x} - 3$ **39.** $f^{-1}(x) = \dfrac{2x+1}{4-3x}$

41. For $a = \frac{1}{2}$, they tend to 1. For $a = 2$, they tend to 1. For $a = -3$, they tend to -1.
43. They oscillate between two numbers approximately equal to 0.4794268623 and 0.8236032618.
45. $2\sqrt{a+b}$ **47.** $(7+2\sqrt{5}, 2+\sqrt{5})$ and $(7-2\sqrt{5}, 2-\sqrt{5})$
49. $(x-\frac{1}{2})^2 + (y+\frac{1}{4})^2 = \frac{67}{16}$ **51.** $y+3 = -\frac{1}{2}(x+1)$ **53.** $6y - 8x = 7$

55. No **61.** a. $L = \dfrac{cI}{D^2}$ b. Multiplied by 9

SECTION 3.1

1. Vertex: (0, 0); y intercept: 0; x intercept: 0

3. Vertex: (0, 1); y intercept: 1

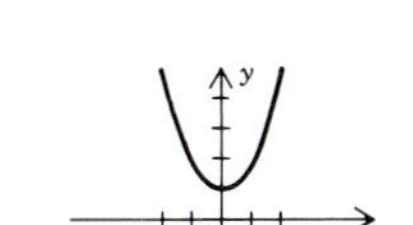

5. Vertex: (0, 8); y intercept: 8; x intercepts: 2 and -2

7. Vertex: (1, 0); y intercept: $-\frac{3}{2}$; x intercept: 1

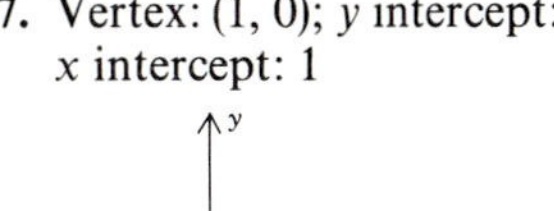

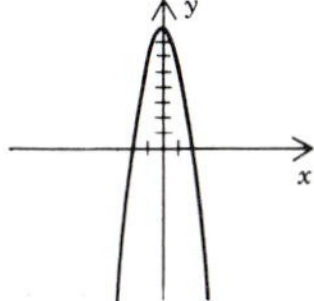

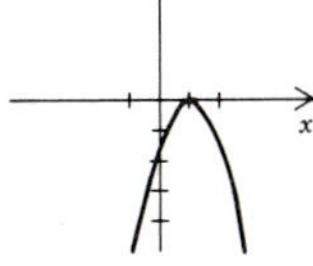

9. Vertex: $\left(-\dfrac{1}{3}, -\dfrac{1}{3}\right)$; y intercept: $-\dfrac{1}{6}$; x intercepts: $-\dfrac{1}{3} + \dfrac{\sqrt{2}}{3}$ and $-\dfrac{1}{3} - \dfrac{\sqrt{2}}{3}$

11. Vertex: $(-1, -1)$; y intercept: 0; x intercepts: 0 and -2

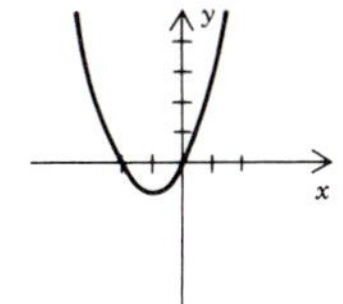

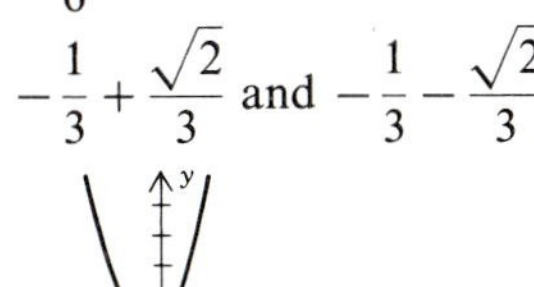

13. Vertex: (1, 1); y intercept: 2

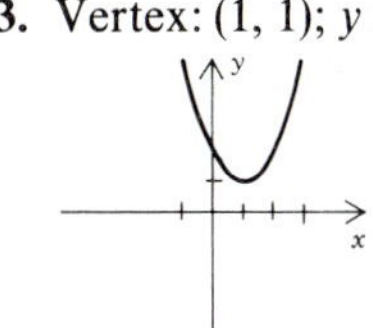

15. Vertex: (3, 4); y intercept: -5; x intercepts: 5 and 1

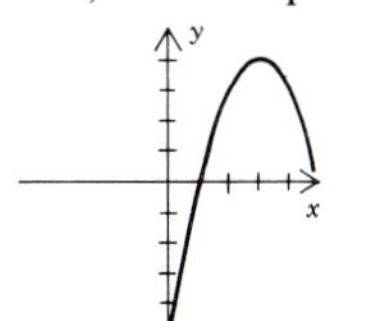

17. Vertex: $(-\frac{1}{3}, \frac{4}{3})$; y intercept: 1; x intercepts: $\frac{1}{3}$ and -1

19. Vertex: $(-\frac{1}{4}, -\frac{9}{8})$; y intercept: -1; x intercepts: $\frac{1}{2}$ and -1

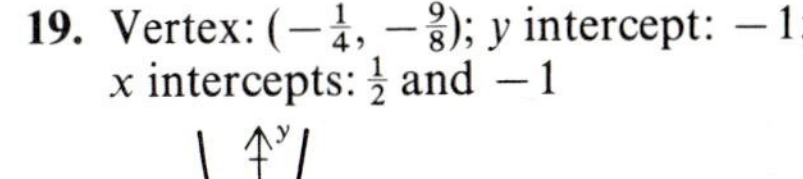

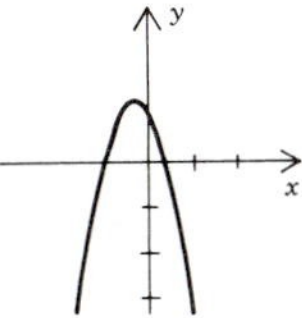

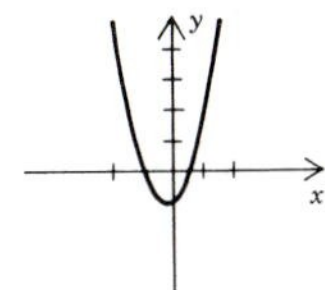

21. $y = \frac{1}{2}(x - 1)^2$ 23. $y = -(x - 1)^2 + 1$ 25. a. $g(x) = -2x^2 + 4$ b. $g(x) = -2x^2$ c. $g(x) = -2(x - \frac{1}{2})^2 + 3$ d. $g(x) = -2(x + 4)^2 + 3$
27. 3 29. 8 33. 12 feet

SECTION 3.2

1. Minimum value: 5 **3.** Minimum value: -20 **5.** Maximum value: 1 **7.** Minimum value: -28 **9.** Minimum value: $-\frac{9}{4}$ **11.** 1 mile by $\frac{1}{2}$ mile **13.** In half **15.** 3 inches each **17.** 4 by 4 **19.** -8 and -8 **21.** 8 and 24 **23.** $(2, -2)$ **25.** $(\frac{1}{2}\sqrt{2}, \frac{1}{2})$ and $(-\frac{1}{2}\sqrt{2}, \frac{1}{2})$ **27.** \$2.50 **29.** 100 feet; 1.5 seconds

SECTION 3.3

1. y intercept: 1; x intercept: -1

3. y intercept: $\frac{1}{4}$; x intercept: $-\frac{1}{2}$

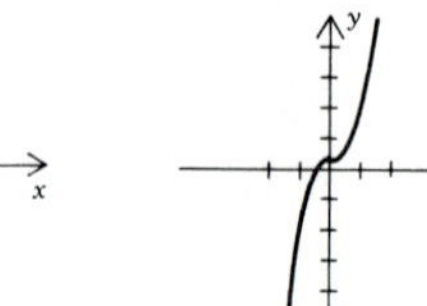

5. y intercept: -2; x intercept: -1

7. y intercept: 1; x intercept: -1

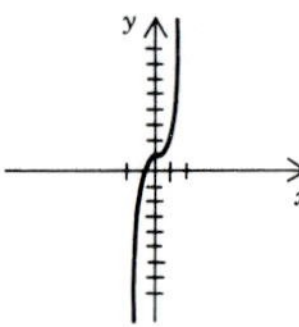

9. y intercept: 2; x intercepts: -1, 1, and 2

11. y intercept: 0; x intercepts: -2, -1, and 0

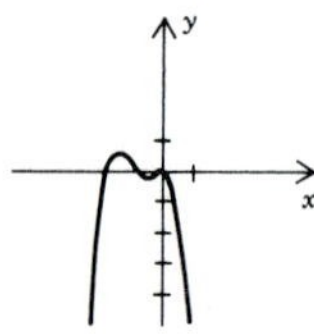

13. y intercept: 0; x intercepts: 0 and 1

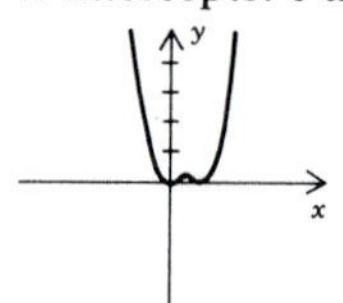

15. y intercept: -1; x intercepts: 1 and -1

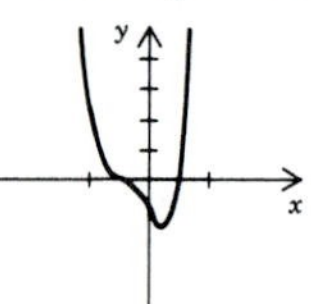

17. y intercept: 0; x intercepts: -1, 0, and 2

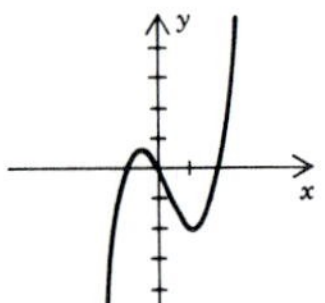

19. y intercept: 0; x intercepts: -1, 0, and 1

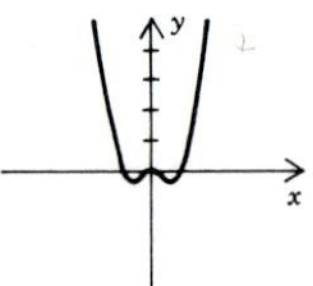

21. y intercept: 0; x intercepts: 0 and 1

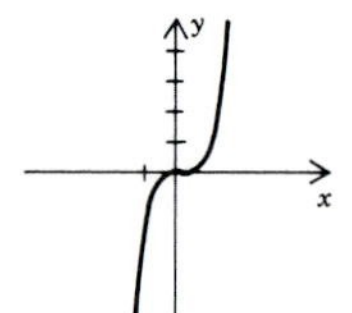

23. y intercept: 4; x intercepts: -2 -1, 1, and 2

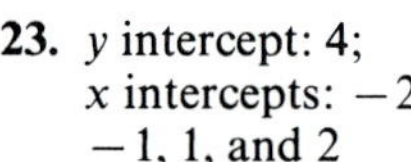

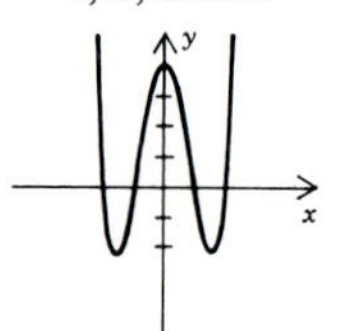

31. $f(x) = (x + 3)(x + 2)(x - 1)$

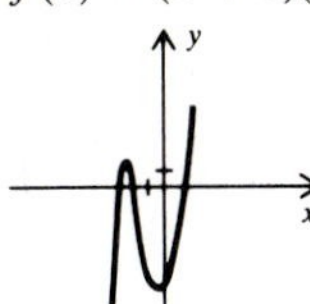

33. $f(x) = 3(x + 1)(x - \frac{1}{3})(x - 2)$

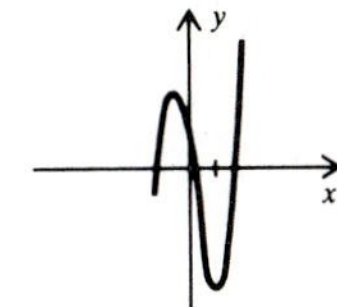

35. $f(x) = 2(x + 2)(x + 1)(x - \frac{3}{2})$

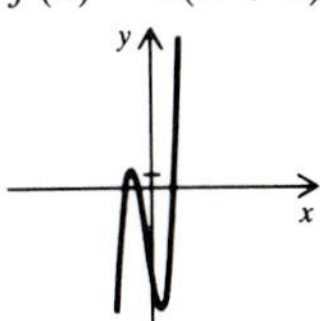

37. 2 **39.** -9

SECTION 3.4

1. No intercepts; vertical asymptote: $x = 0$; horizontal asymptote: $y = 0$

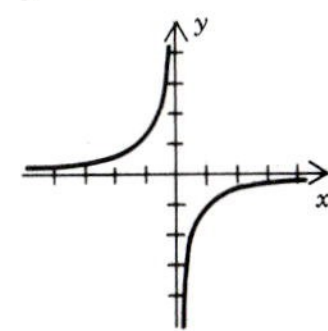

3. y intercept: 2; no x intercepts; vertical asymptote: $x = 2$; horizontal asymptote: $y = 0$

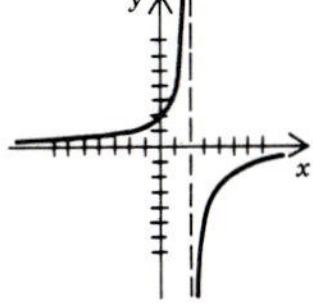

5. y intercept: -1; no x intercepts; vertical asymptote: $x = -1$; horizontal asymptote: $y = 0$

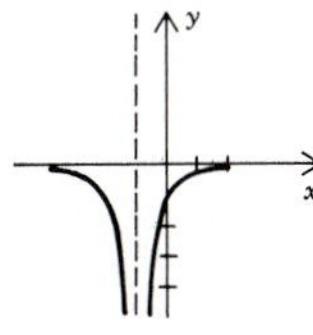

7. y intercept: -4; no x intercepts; vertical asymptote: $x = -\frac{1}{2}$; horizontal asymptote: $y = 0$

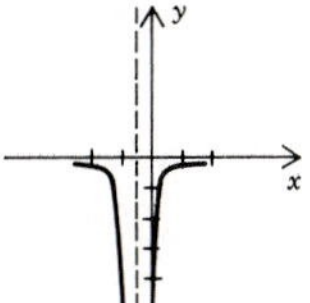

9. y intercept: 0; x intercept: 0; vertical asymptote: $x = 2$; horizontal asymptote: $y = 1$

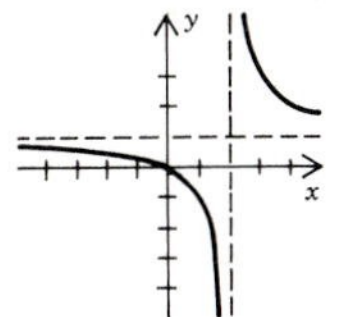

11. y intercept: 0; x intercept: 0; vertical asymptotes: $x = 2$ and $x = -2$; horizontal asymptote: $y = 0$

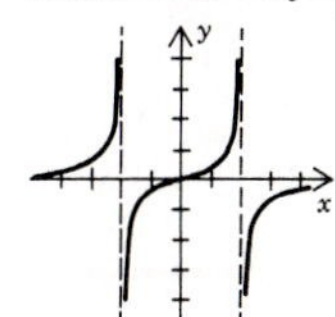

13. y intercept: -1; no x intercepts; vertical asymptotes: $x = 1$ and $x = -2$; horizontal asymptote: $y = 0$

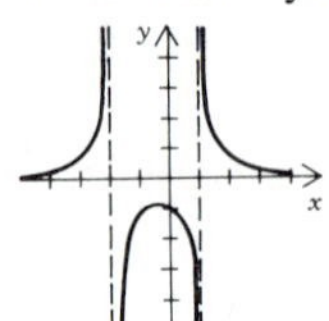

15. y intercept: 0; x intercept: 0; vertical asymptotes: $x = -3$ and $x = -1$; horizontal asymptote: $y = 0$

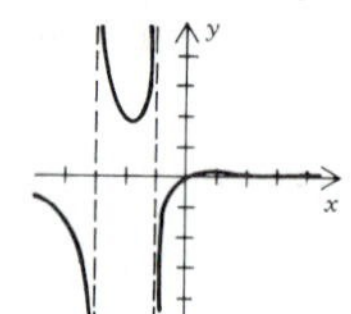

17. y intercept: 0; x intercept: 0; vertical asymptotes: $x = 2$ and $x = -\frac{1}{2}$; horizontal asymptote: $y = \frac{1}{2}$

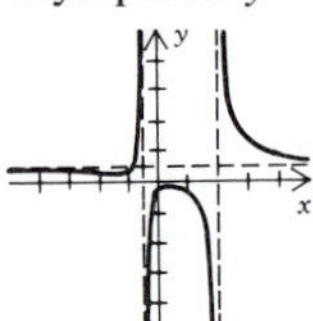

19. No y intercept; x intercept: $-\frac{1}{2}$; vertical asymptote: $x = 0$; horizontal asymptote: $y = 4$

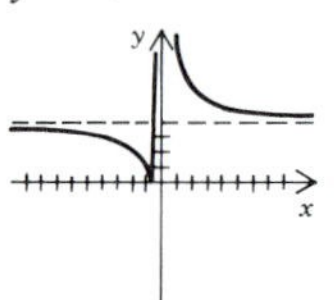

21. y intercept: $-\frac{1}{4}$; t intercept: $-\frac{1}{3}$; no vertical asymptotes; horizontal asymptote: $y = 0$

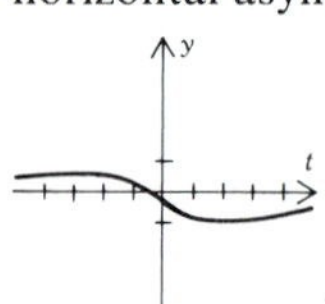

23. y intercept: $\frac{3}{4}$; x intercept: -1; vertical asymptote: $x = -4$; horizontal asymptote: $y = 0$

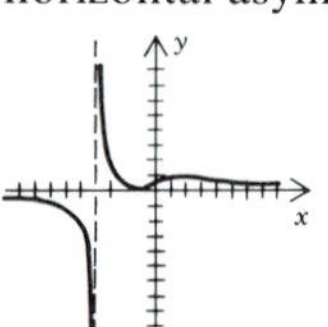

SECTION 3.5

1. $x^2 + y^2 = 9$ **3.** $(x + 1)^2 + (y - 4)^2 = 4$ **5.** $x^2 + y^2 = 17$
7. $(x + 2)^2 + (y - 3)^2 = 25$

9. y intercept: 3; symmetry symmetry with respect to y axis

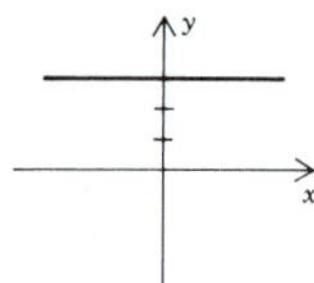

11. y intercept: 0; x intercept: 0; symmetry with respect to x axis

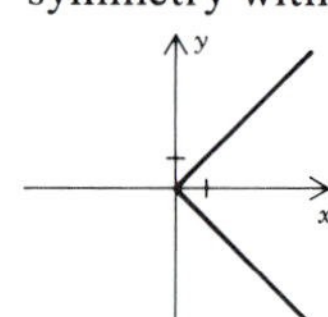

13. x intercept: 1; symmetry with respect to x axis

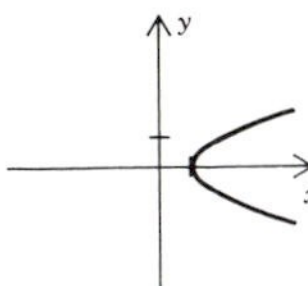

15. y intercept: 0; x intercept: 0; symmetry with respect to origin

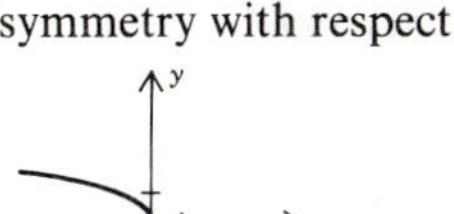

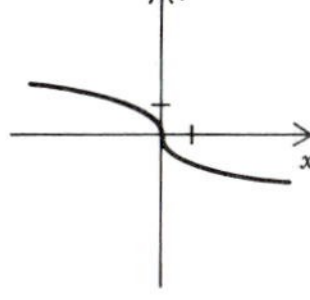

17. Circle: y intercepts and x intercepts: 2 and -2; all symmetries

19. y intercept: 0; x intercept: 0; no symmetry

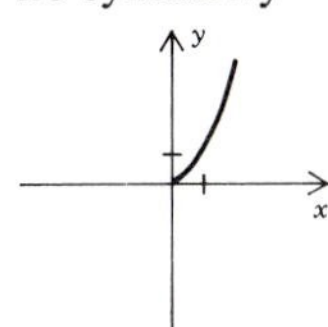

21. y intercepts: $\frac{1}{2}$ and $-\frac{1}{2}$; x intercepts: 1 and -1 **23.** x intercept: 3

25. x intercepts: 1 and -1 **27.** y intercepts: $\sqrt{5}$ and $-\sqrt{5}$; x intercept: 5

29. y intercepts: 3 and -2 **31.** y intercepts: 1 and -3; x intercepts: 1 and 3

SECTION 3.6

1. Vertex: (0, 0); axis: $y = 0$

3. Vertex: $(-5, 1)$; axis: $y = 1$

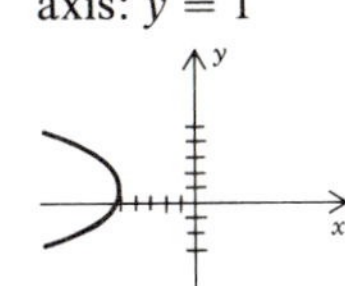

5. Vertex: $(-\frac{1}{4}, -\frac{1}{2})$; axis: $y = -\frac{1}{2}$

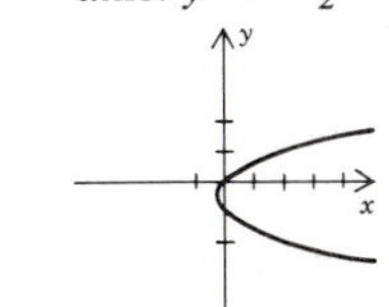

7. Vertex: $(\frac{1}{2}, \frac{1}{2})$; axis: $y = \frac{1}{2}$

9. Vertex: (3, 2); axis: $x = 3$

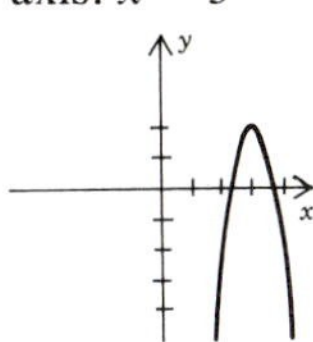

11. Vertex: $(-1, 3)$; axis: $y = 3$

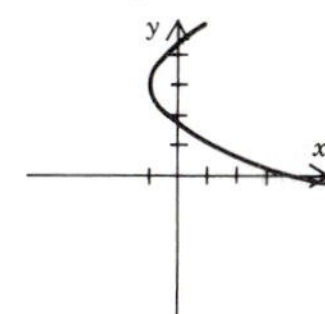

13. Vertex: $(-2, -3)$; axis: $x = -2$

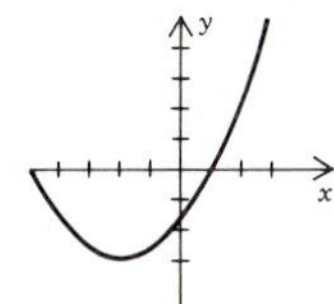

15. $y = \frac{5}{4}x^2$ **17.** $y - 5 = \frac{1}{2}(x + 2)^2$ **19.** $y - 4 = (x - 3)^2$
21. The lines $x - 3y + 4 = 0$ and $x + 3y - 2 = 0$ **23.** The point $(-\frac{1}{2}, \frac{1}{3})$

25. -15 **27.** $(3, -\frac{1}{2})$ and $(3, 1)$ **29.** $y = \dfrac{316}{(1750)^2}x^2$

SECTION 3.7

1. Center: (0, 0); major axis between (−4, 0) and (4, 0); minor axis between (0, −2) and (0, 2); vertices: (−4, 0), (4, 0)

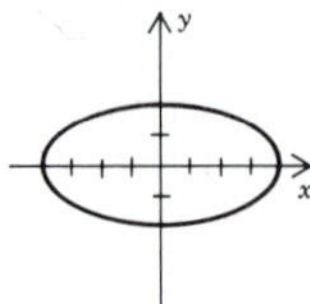

3. Circle: center (0, 0), radius $\sqrt{3}$

5. Center: (0, 0); major axis between (0, −3) and (0, 3); minor axis between (−2, 0) and (2, 0); vertices: (0, −3), (0, 3)

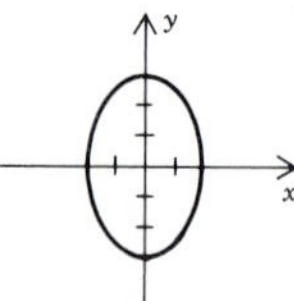

7. Center: (4, −3); major axis between (4, −7) and (4, 1); minor axis between (1, −3) and (7, −3); vertices: (4, −7), (4, 1)

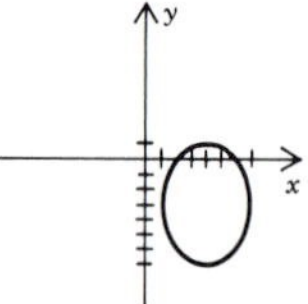

9. Center: (−2, −1); major axis between (−6, −1) and (2, −1); minor axis between (−2, −3) and (−2, 1); vertices: (−6, −1), (2, −1)

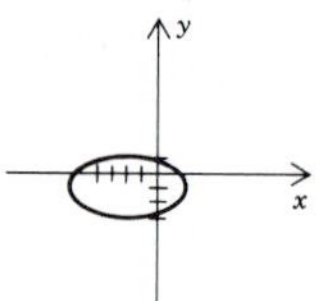

11. Center: (1, −2); major axis between (1, −6) and (1, 2); minor axis between (−1, −2) and (3, −2); vertices: (1, −6), (1, 2)

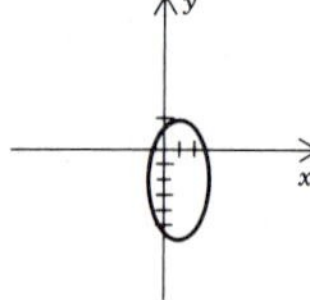

13. Center: (−3, 3); major axis between (−3, −2) and (−3, 8); minor axis between (−7, 3) and (1, 3); vertices: (−3, −2), (−3, 8)

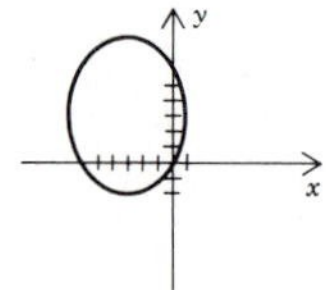

15. $\dfrac{x^2}{4} + \dfrac{y^2}{9} = 1$ **17.** $\dfrac{x^2}{4} + \dfrac{y^2}{16} = 1$ **19.** $\dfrac{(x-3)^2}{49} + \dfrac{(y-2)^2}{25} = 1$

21. $\dfrac{(x+2)^2}{1} + \dfrac{(y+5)^2}{4} = 1$ **23.** $4(x+\frac{3}{2})^2 + \dfrac{4(y-\frac{1}{2})^2}{25} = 1$

25. $\dfrac{4x^2}{9} + \dfrac{(y-4)^2}{1} = 1$ and $\dfrac{x^2}{1} + \dfrac{4(y-4)^2}{9} = 1$ **29.** $4 + \frac{3}{2}\sqrt{3}$ and $4 - \frac{3}{2}\sqrt{3}$

31. b. $(\frac{24}{5}, -\frac{12}{5})$ **33.** a. $\frac{1}{2}\sqrt{3}$ b. $\frac{1}{2}\sqrt{3}$ c. $\frac{1}{3}\sqrt{5}$ d. 0

SECTION 3.8

1. Center: (0, 0); vertices: (−1, 0), (1, 0); asymptotes: $y = x$, $y = -x$

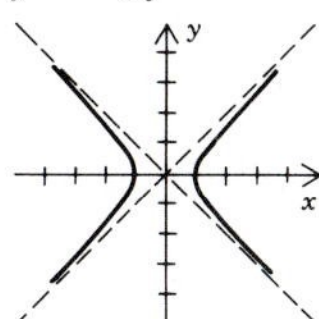

3. Center: (0, 0); vertices: (−3, 0), (3, 0); asymptotes: $y = \frac{2}{3}x$, $y = -\frac{2}{3}x$

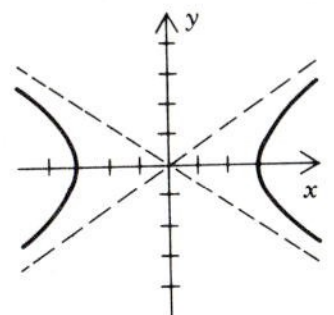

5. Center: (0, 0); vertices: (0, −2), (0, 2); asymptotes: $y = \frac{2}{5}x$, $y = -\frac{2}{5}x$

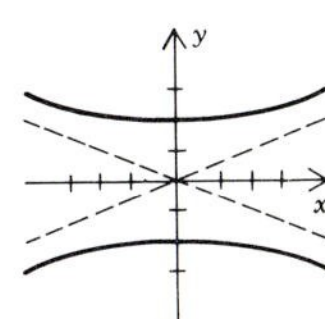

7. Center: (−2, −1); vertices: (−6, −1), (2, −1); asymptotes: $y + 1 = \frac{5}{4}(x+2)$, $y + 1 = -\frac{5}{4}(x+2)$

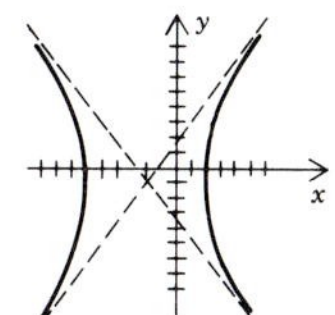

9. Center: (0, −1); vertices: (0, −2), (0, 0); asymptotes: $y + 1 = \frac{1}{2}x$, $y + 1 = -\frac{1}{2}x$

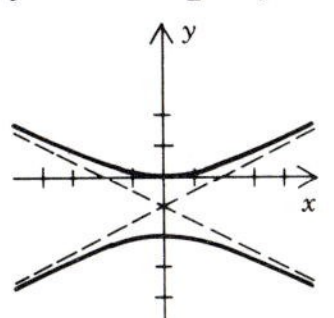

11. Center: (3, 0); vertices: (3, −3), (3, 3); asymptotes: $y = \frac{3}{4}(x-3)$, $y = -\frac{3}{4}(x-3)$

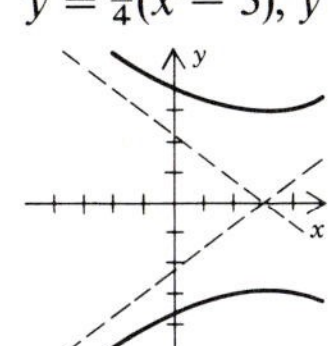

13. Center: (1, −3); vertices: (−1, −3), (3, −3); asymptotes: $y + 3 = \frac{3}{2}(x-1)$, $y + 3 = -\frac{3}{2}(x-1)$

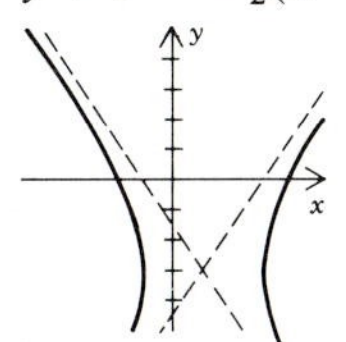

15. $\frac{x^2}{4} - \frac{y^2}{36} = 1$ **17.** $\frac{y^2}{16} - \frac{11x^2}{256} = 1$ **19.** $\frac{(x-2)^2}{4} - \frac{(y+3)^2}{4} = 1$
21. $\frac{1}{2}\sqrt{3}$ and $-\frac{1}{2}\sqrt{3}$ **23.** b. $(\frac{13}{6}, -\frac{5}{4})$ **25.** a. $\sqrt{5}$ b. $\frac{1}{2}\sqrt{5}$ c. $\frac{1}{2}\sqrt{13}$
d. $\sqrt{2}$ **27.** Ellipse; center: (0, 0) **29.** Parabola; vertex: $(-\frac{3}{2}, \frac{1}{2})$
31. Ellipse; center: $(-4, 3)$ **33.** Ellipse; center: $(-3, 0)$
35. Hyperbola; center: $(2, -2)$ **37.** Hyperbola; center: (3, 4)

SECTION 3.9

1. $-\frac{11}{16}$ **3.** $-\frac{23}{16}$ **5.** $1, \frac{9}{16}$ **7.** f has no negative values. **9.** 0.6825658
11. 0.518914525 **13.** -0.6611842 **15.** $(-\infty, 1.46052635)$
17. $(-\infty, 0.36842105), (0.539473685, \infty)$ **19.** $(-\infty, -0.2331414)$ **21.** $c \le 1.56$

CHAPTER 3 REVIEW

1. Vertex: $(0, -4)$; axis: $x = 0$; y intercept: -4; x intercepts: 4 and -4

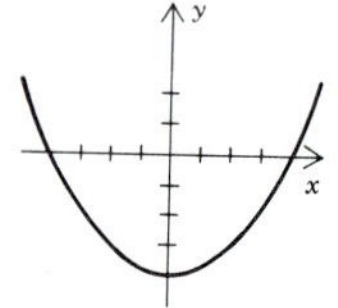

3. Vertex: $(-2, -3)$; axis: $x = -2$; y intercept: -7; no x intercepts

5. y intercept: 2; x intercept: $2^{1/3}$

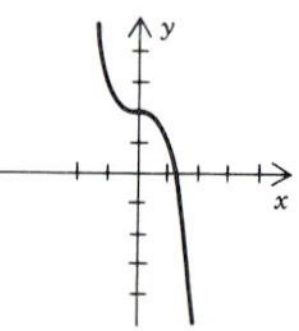

7. y intercept: 2; no x intercepts

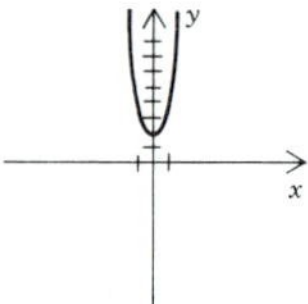

9. y intercept: 0; x intercepts: 0 and 2

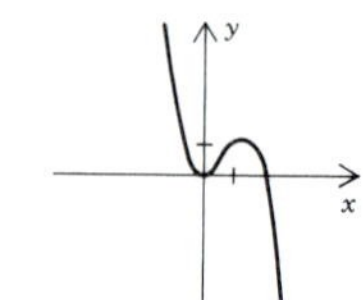

11. y intercept: 9; x intercepts: -3, -1, 1, and 3

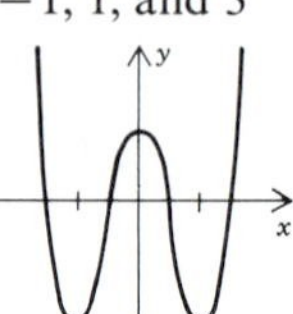

13. y intercept: -1; x intercept: -1; vertical asymptote: $x = 1$; horizontal asymptote: $y = 1$

15. No y intercept; x intercept: -2; vertical asymptotes: $x = 0$ and $x = -4$; horizontal asymptote: $y = 0$

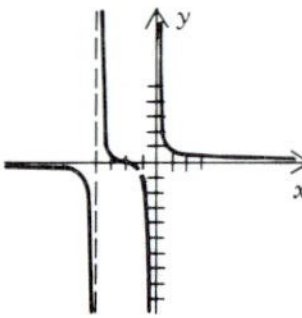

17. $f(x) = 4(x-1)(x-\frac{3}{2})^2$

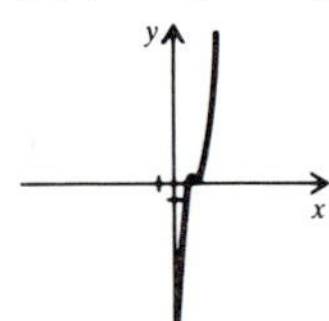

19. $-\frac{7}{16}$ **21.** 0.81907895

23. y intercept: $\sqrt{3}$; x intercept: $-\sqrt{2}$; no symmetry

25. y intercepts: 3 and -5; x intercepts: 3 and 5; no symmetry

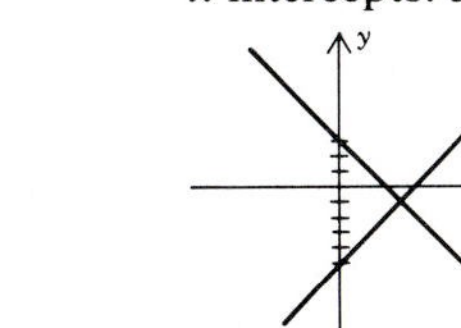

27. y intercepts: $-3 + \sqrt{3}$ and $-3 - \sqrt{3}$; no symmetry

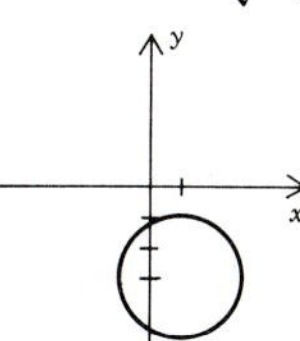

29. y intercepts: $\sqrt{3}$ and $-\sqrt{3}$; x intercept: -3; symmetry with respect to x axis

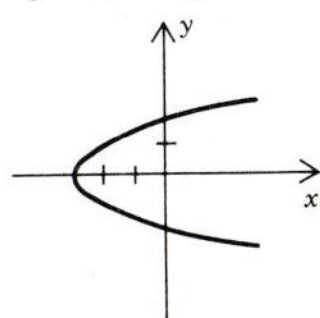

31. Circle: center $(-1, 2)$, radius $\sqrt{3}$

33. Center: $(\frac{1}{2}, -\frac{1}{2})$; vertices: $(-\frac{3}{2}, -\frac{1}{2})$ and $(\frac{5}{2}, -\frac{1}{2})$; asymptotes: $y + \frac{1}{2} = x - \frac{1}{2}$ and $y + \frac{1}{2} = -(x - \frac{1}{2})$

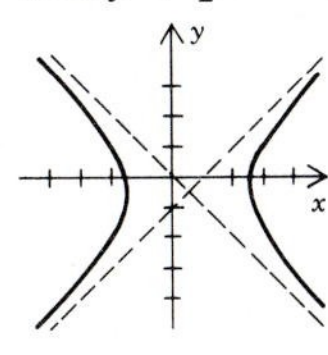

35. Center: $(-3, -4)$; vertices: $(-3, -4 - 2\sqrt{3})$, $(-3, -4 + 2\sqrt{3})$;

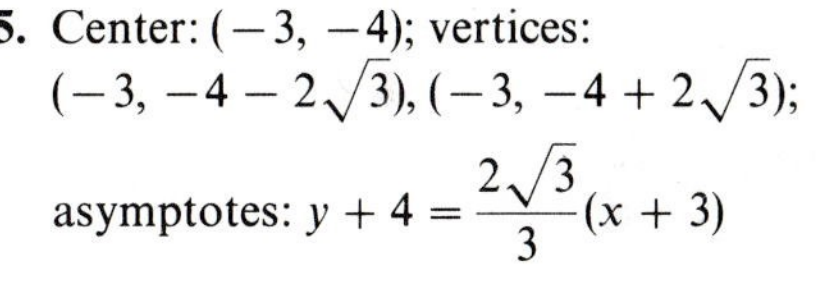

asymptotes: $y + 4 = \dfrac{2\sqrt{3}}{3}(x + 3)$

and $y + 4 = -\dfrac{2\sqrt{3}}{3}(x + 3)$

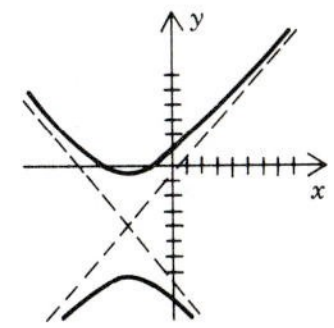

37. $-\frac{1}{3}$ **39.** $-9, -1$ **41.** \$350, and 350 people **43.** 4096 feet

45. For the square: $\dfrac{8}{\pi + 4}$ feet; for the circle: $\dfrac{2\pi}{\pi + 4}$ feet

47. Length: $1 - \frac{1}{2}\sqrt{2}$; width: $\frac{1}{4}\sqrt{2} + \frac{1}{2}$

CUMULATIVE REVIEW I: CHAPTERS 0–3

1. $-\frac{3}{4}\sqrt[3]{3}$ **3.** $x^{39/4}y$ **5.** $(x-3)^2(2x-3)^2(4x^2-27)$ **7.** $\dfrac{4-2x^3}{3x^3+x}$

9. $\dfrac{12x}{(x+2)^2(x-2)}$ **11.** $\frac{5}{4}, \frac{1}{4}$ **13.** $-2 \pm \frac{1}{2}\sqrt{26}$ **15.** $-\sqrt{3}i, \sqrt{3}i$

17. All numbers in $(-\infty, -1]$ and $[1, \frac{5}{4}]$ **19.** b. $y + \frac{1}{3} = -\frac{3}{2}(x+1)$

21. a. $[-2, 2]$ b. $\sqrt{2}$ c. $[0, 2]$ **23.** $f^{-1}(x) = \dfrac{3-x}{x+2}$

25.

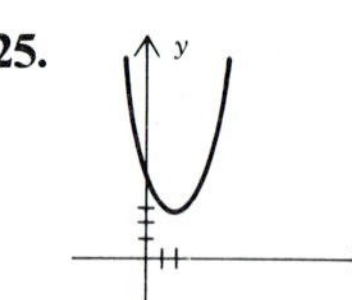

27.

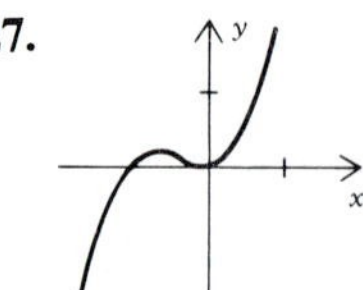

29.

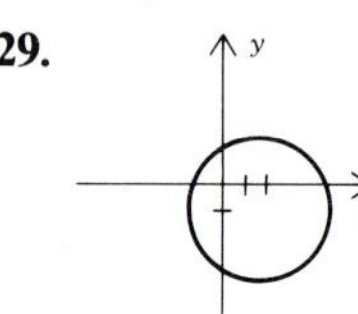

31.

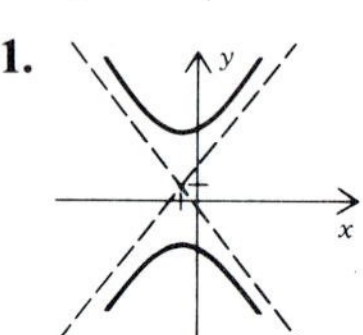

33. $\dfrac{(x+1)^2}{4} + \dfrac{(y+2)^2}{9} = 1$ **35.** $a = 2b$ **37.** c. $\frac{1}{7} + \frac{1}{91}$ **39.** $C = ct^{3/2}$

41. $V = (24-2x)(16-2x)x$ **43.** $\frac{1}{75}$ cup **45.** 6 hours

47. a. $40\sqrt{6}$ feet per second b. $\frac{5}{2}\sqrt{6}$ seconds

49. a. $[0, 26]$ b. 5 feet

SECTION 4.1

1. **3.**

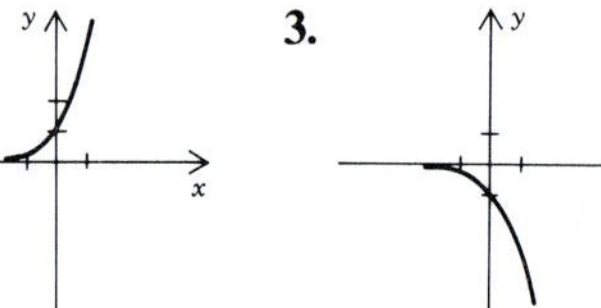

5. The line $y = 1$ **7.**

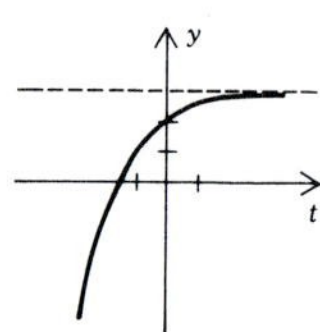

9. **11.**

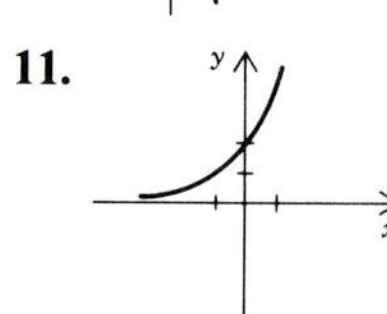

13.

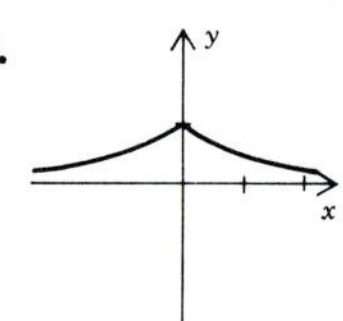

15. π^2 **17.** $5^{\sqrt{6}}$ **19.** $\frac{1}{4}$ **21.** 3^{π} **23.** 2 **25.** 7 **27.** $\frac{1}{3}$ **29.** $\frac{1}{4}$ **31.** $\frac{1}{64}$

33. 1.404947591 **35.** 0.3011942119 **37.** 1.648721271 **39.** 11 **41.** 22

43. 0 and $\frac{12}{5}$ **45.** $-2\sqrt{2}-1$ **47.** $\sqrt{5}$ and $-\sqrt{5}$

49.

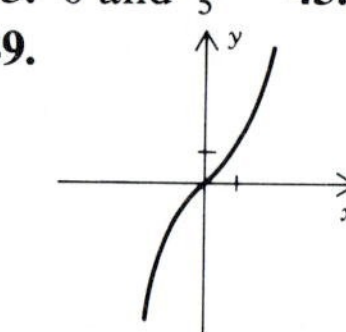

(a)

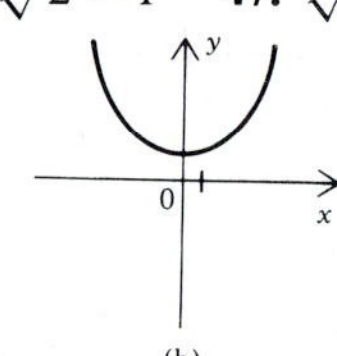

(b)

51. $e^{(e^e)}$ **53.** b. 6 **57.** a. Approximately 6.76×10^{-6} atmospheres
b. Approximately 3.07×10^{-3} atmospheres
c. Approximately 5.99×10^{-1} atmospheres
d. Approximately 1.12 atmospheres

59. Approximately $7.521206186 \times 10^{-4}$ coulombs

SECTION 4.2

1. $\log_{10} 10 = 1$ **3.** $\log_{1/2} 32 = -5$ **5.** $\log_x 16 = -4$ **7.** $10^3 = 1000$
9. $4^{2.5} = 32$ **11.** $e^2 = x$ **13.** 3 **15.** 4 **17.** -4 **19.** 0 **21.** $\frac{3}{2}$ **23.** -1
25. $\frac{1}{3}$ **27.** -2 **29.** -1 **31.** -4 **33.** $\frac{5}{2}$ **35.** $\frac{4}{3}$ **37.** 6 **39.** 4 **41.** 5
43. 1 **45.** 16 **47.** 2 **49.** $\frac{1}{16}$ **51.** 1 **53.** 12 **55.** 67 **57.** 4 **59.** $\frac{2}{3}\sqrt{3}$
61. 11, -11 **63.** 8 **65.** 4 **67.** 5 **69.** 2, -2 **71.** 2 **73.** 9 **75.** $\frac{1}{3}$
77. $\frac{1}{4}$

79.

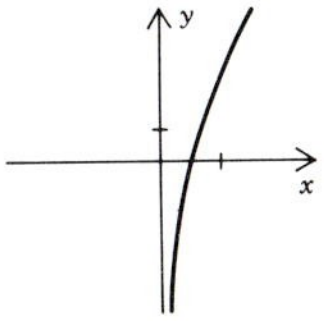

y
x

81.

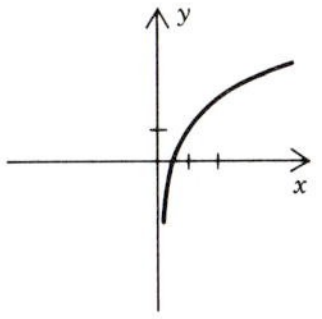

y
x

83.

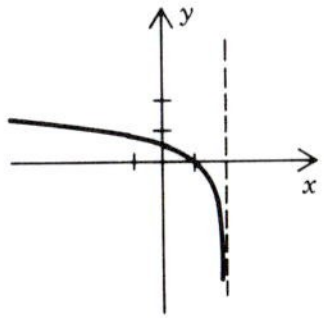

y
x

85.

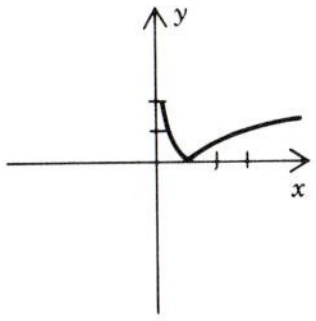

y
x

87. No, since $\log_{10} \frac{1}{6} < 0$ **89.** The display alternates between two numbers.

SECTION 4.3

1. 15 **3.** $\frac{7}{18}$ **5.** 1.6021 **7.** -0.6021 **9.** 5.7318 **11.** 27.318
13. 0.6989700043 **15.** -2.07109231 **17.** -3.970291914 **19.** $\log_2 5$

21. $\log_2 \dfrac{x^6}{y}$ **23.** $\log_{10} \dfrac{x^6}{y}$ **25.** $\log_a x + \log_a y - 3 \log_a z$

27. $3 \log_a z + \frac{1}{2} \log_a x + \frac{1}{2} \log_a y$ **29.** $\frac{1}{2} \log_a x + \frac{5}{2} \log_a y - \frac{3}{2} \log_a z$
31. 4 **33.** $\frac{1}{2}$ **35.** 3 **37.** 2.321928095 **39.** 1.367102966
41. -0.5634737703 **43.** -9.385735739
53. b. The graph of $\log_3 x$ lies between the x axis and the graph of $\log_2 x$.

SECTION 4.4

1. $\dfrac{\log 5}{\log 2}$ **3.** $\dfrac{1}{2}\left(\dfrac{\log 51}{\log 5}\right)$ **5.** $\dfrac{\log 5}{\log 4} - 2$ **7.** $\dfrac{3 \log 5 + 2 \log 3}{5 \log 3 - 2 \log 5}$ **9.** $0, \dfrac{\log 3}{\log 2}$

11. $-2, \dfrac{\log 4}{\log 5} + 2$ **15.** 10^{-7} watts per square meter **17.** 20 decibels

19. $10^{0.06} \approx 1.148153621$ **21.** $10^{8.5} \approx 316{,}227{,}766$ **23.** $10^{-1.4} \approx 0.0398107171$
25. a. 0 b. 3 c. 4 **27.** $\log 2 \approx 0.3$ **29.** Highly acidic **31.** Wheat

SECTION 4.5

1. $\frac{1}{3} \ln 4$ **3.** $\dfrac{1}{1.2} \ln 2$ **5.** 3 **7.** 0, ln 2 **9.** 0 **13.** a. $f(t) = 1000e^{(\ln 1.537)t}$

b. $\dfrac{\ln 2}{\ln 1.537} \approx 1.61$ weeks **15.** No **17.** At 32.3°C **19.** 5.223 billion

23. Approximately 20.35% **25.** Yes **27.** Approximately 79%
29. Approximately 1.2% **31.** Approximately 65.25 million years
33. a. 29.92 inches of mercury b. Approximately 24.50 inches of mercury
c. Approximately 17.28 inches of mercury **35.** No

SECTION 4.6

1. a. \$1276.28 b. \$1283.36 c. \$1284.00
3. a. \$23,673.64 b. \$24,513.57 c. \$24,593.30
5. \$759.41 **7.** The \$1500 deposit
9. a. Approximately 11.55 years b. Approximately 7.70 years
11. Approximately 166.58%
13. 1¢ (\$849,465.05 in a regular year, \$849,465.06 in a leap year)
15. a. Approximately \$2,018,408,628 b. Approximately \$3,173,350,575
17. b. Approximately 7.76% c. $r = k[(E + 1)^{1/k} - 1]$
19. a. \$3286.39 b. \$65,727.80 **21.** \$32,686.13 **23.** Approximately 15.83%

CHAPTER 4 REVIEW

1.
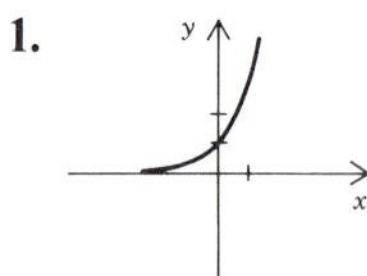

3.
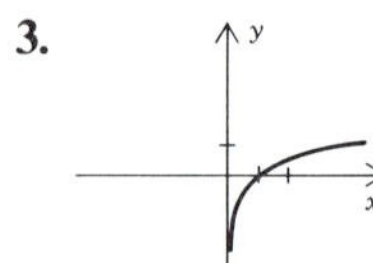

5. 2 **7.** $6 + \ln 6$ **9.** $-\frac{1}{2} + \frac{1}{2}\sqrt{5}, -\frac{1}{2} - \frac{1}{2}\sqrt{5}$ **11.** $\frac{1}{8}$ **13.** $\frac{7}{5}$

15. $-\dfrac{\log 7}{\log 2}$ **17.** $0, \left(\dfrac{\log 2}{\log 3}\right)^{1/3}$ **19.** $\frac{1}{5}\ln 4.37$ **21.** -4 **23.** 5

25. $x = \dfrac{e^{2y} + 1}{2e^y} = \dfrac{1}{2}(e^y + e^{-y})$ **29.** 6 **31.** 100

33. a. $\dfrac{\ln 2}{\ln 11070 - \ln 50} \approx 0.128$ years (or 47 days)

b. Approximately 0.193 years (or 70 days)
35. Approximately 0.9701 gram **37.** Approximately 2130 years old in 1970
39. a. \$4317.85 b. \$4439.28 c. \$4450.69 **41.** \$3093.92

SECTION 5.1

1.
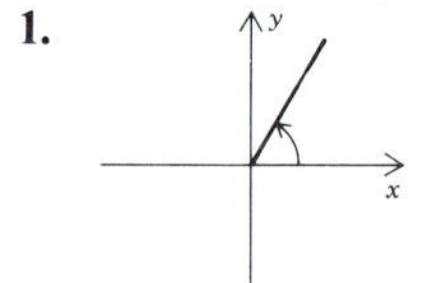

3.
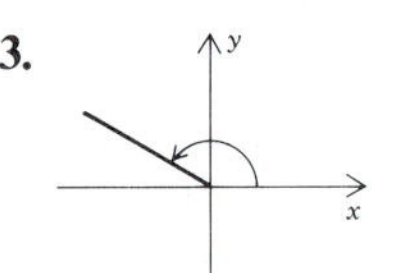

5.
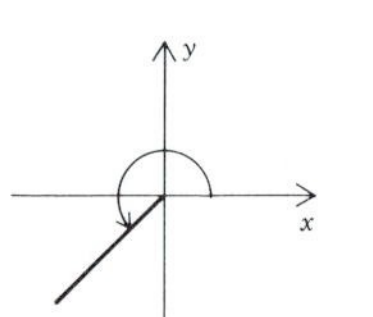

7.
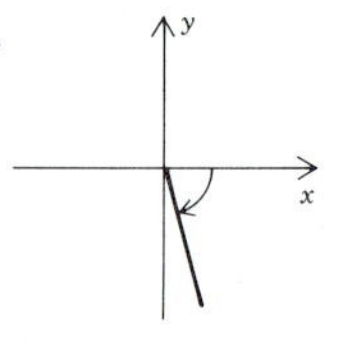

9.
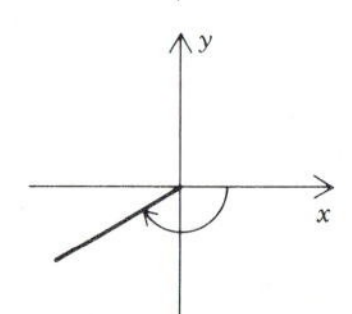

11.

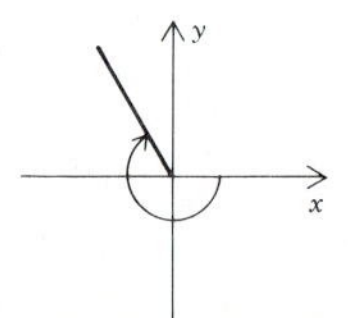

13.

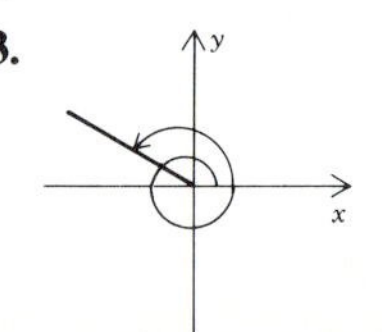

15.

17.

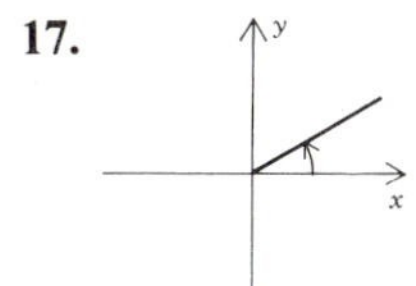

19.

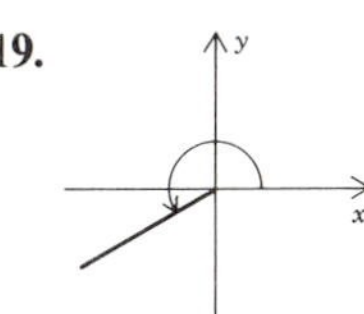

21.

23.

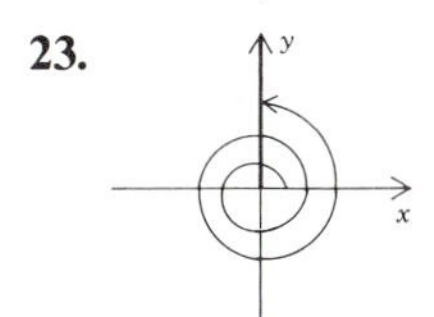

25.

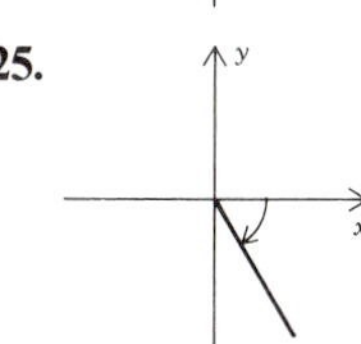

27.

29.

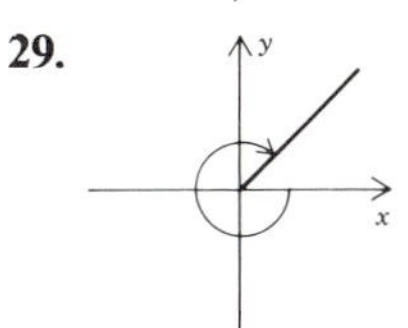

31.

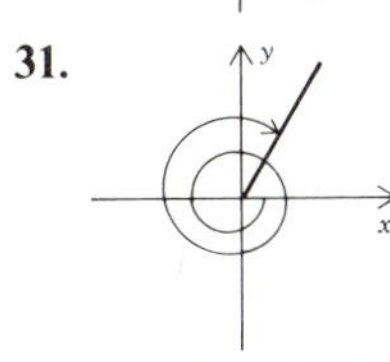

33. 0 **35.** $-\pi/4$ **37.** $13\pi/6$ **39.** $-\pi$ **41.** $49\pi/6$ **43.** $-2\pi/5$ **45.** $\pi/40$ **47.** 405° **49.** 1080° **51.** 1980° **53.** −585° **55.** 70° **57.** $(324/\pi)°$ **59.** Second **61.** Third **63.** Fourth **65.** Second **67.** Fourth **69.** $2\pi/3$ radians: b, c, h, i; $4\pi/3$ radians: a, d, e, f, g, j **71.** a. $\pi/8$ b. $\pi/24$ c. $\pi/2$ **73.** 11.25°, $\pi/16$ **75.** a. $\pi/12$ b. $\pi/720$ c. $\pi/43{,}200$
77. a. Approximately 1.15 miles b. Approximately 2408.66 miles
79. Approximately 1151.57 miles

SECTION 5.2

1. $\sqrt{2} - \frac{3}{2}\sqrt{3}$ **3.** 0 **5.** $2 + \frac{2}{3}\sqrt{3}$ **7.** *a*, *d*, and *h*; *e* and *g*
9. $\sin\theta = \frac{1}{37}\sqrt{37}$; $\cos\theta = \frac{6}{37}\sqrt{37}$; $\tan\theta = \frac{1}{6}$; $\cot\theta = 6$; $\sec\theta = \sqrt{37}/6$; $\csc\theta = \sqrt{37}$ **11.** $\sin\theta = \frac{3}{5}$; $\cos\theta = \frac{4}{5}$; $\tan\theta = \frac{3}{4}$; $\cot\theta = \frac{4}{3}$; $\sec\theta = \frac{5}{4}$; $\csc\theta = \frac{5}{3}$ **13.** $\sin\theta = \frac{12}{13}$; $\cos\theta = \frac{5}{13}$; $\tan\theta = \frac{12}{5}$; $\cot\theta = \frac{5}{12}$; $\sec\theta = \frac{13}{5}$; $\csc\theta = \frac{13}{12}$ **15.** $\sin\theta = \frac{1}{2}\sqrt{2}$; $\cos\theta = \frac{1}{2}\sqrt{2}$; $\tan\theta = 1$; $\cot\theta = 1$; $\sec\theta = \sqrt{2}$; $\csc\theta = \sqrt{2}$ **17.** $\sin\theta = \frac{24}{25}$; $\cos\theta = \frac{7}{25}$; $\tan\theta = \frac{24}{7}$; $\cot\theta = \frac{7}{24}$; $\sec\theta = \frac{25}{7}$; $\csc\theta = \frac{25}{24}$ **19.** $\sin\theta = \frac{1}{6}$; $\cos\theta = \frac{1}{6}\sqrt{35}$; $\tan\theta = \frac{1}{35}\sqrt{35}$; $\cot\theta = \sqrt{35}$; $\sec\theta = \frac{6}{35}\sqrt{35}$; $\csc\theta = 6$ **21.** $\cos\theta = \sqrt{21}/5$; $\tan\theta = \frac{2}{21}\sqrt{21}$; $\cot\theta = \sqrt{21}/2$; $\sec\theta = \frac{5}{21}\sqrt{21}$; $\csc\theta = \frac{5}{2}$ **23.** $\sin\theta = \sqrt{7}/4$; $\tan\theta = \sqrt{7}/3$; $\cot\theta = \frac{3}{7}\sqrt{7}$; $\sec\theta = \frac{4}{3}$; $\csc\theta = \frac{4}{7}\sqrt{7}$ **25.** $\sin\theta = \frac{2}{5}\sqrt{5}$; $\cos\theta = \frac{1}{5}\sqrt{5}$; $\cot\theta = \frac{1}{2}$; $\sec\theta = \sqrt{5}$; $\csc\theta = \sqrt{5}/2$ **27.** $\sin\theta = \frac{10}{269}\sqrt{269}$; $\cos\theta = \frac{13}{269}\sqrt{269}$; $\tan\theta = \frac{10}{13}$; $\sec\theta = \frac{1}{13}\sqrt{269}$; $\csc\theta = \frac{1}{10}\sqrt{269}$
29. $\sin\theta = \sqrt{21}/11$; $\cos\theta = \frac{10}{11}$; $\tan\theta = \sqrt{21}/10$; $\cot\theta = \frac{10}{21}\sqrt{21}$; $\csc\theta = \frac{11}{21}\sqrt{21}$
31. $\sin\theta = \frac{3}{5}$; $\cos\theta = \frac{4}{5}$; $\tan\theta = \frac{3}{4}$; $\cot\theta = \frac{4}{3}$; $\sec\theta = \frac{5}{4}$ **33.** 0.139173101 **35.** 0.8910065242 **37.** 0.2493280028 **39.** 1.150368407 **41.** 1.000152328 **43.** 1.236067978 **45.** 0.6202354913 **47.** 0.9604558336 **49.** 1.556663862 **51.** 0.8938696821 **53.** 57.27404894 **55.** 1.000000111 **57.** a. 0.0998334166 b. $9.999999983 \times 10^{-5}$ c. 0.000001 d. 0.9950041653 e. 0.9999995 f. 1
59. h. 3.14159265359 **61.** $(\frac{\pi}{6}, \frac{\pi}{2})$ **71.** a. 1 b. 4
73. Approximately 29.78 kilometers per second
75. Approximately 370.82 foot-pounds
77. a. Approximately 51.2 miles per hour
b. It is increased from approximately 51.2 to approximately 56.2 miles per hour.

SECTION 5.3

1. $a \approx 1.035, b \approx 3.864, \beta = 75°$ **3.** $b \approx 2.384, c \approx 3.111, \beta = 50°$
5. $a \approx 4.037, c \approx 5.152, \alpha = 51.6°$
7. $a \approx 8, \alpha \approx 28.07° (\approx 28° 4'), \beta \approx 61.93° (\approx 61° 56')$
9. $b = 1, \alpha \approx 71.57° (\approx 71° 34'), \beta \approx 18.43° (\approx 18° 26')$
11. $c \approx 0.5207, \alpha \approx 50.19° (\approx 50° 12'), \beta \approx 39.81° (\approx 39° 48')$
13. Approximately 38.39 feet **15.** Approximately 48.59° ($\approx$48° 35′)
17. Approximately 4.67 miles **19.** Approximately 77.65 feet
21. Approximately 362.0 feet **23.** Approximately 1.30° ($\approx$1° 18′)
25. Approximately 482 feet

SECTION 5.4

1. sine, cosine, tangent, secant **3.** sine, cosine, cotangent, cosecant
5. sine, cosine, cotangent, cosecant **7.** sine, cosine, cotangent, cosecant
9. 30° **11.** $\pi/4$ **13.** $\sin 135° = \sqrt{2}/2$, $\cos 135° = -\sqrt{2}/2$, $\tan 135° = -1$, $\cot 135° = -1$, $\sec 135° = -\sqrt{2}$, $\csc 135° = \sqrt{2}$ **15.** $\sin 5\pi/3 = -\sqrt{3}/2$, $\cos 5\pi/3 = \frac{1}{2}$, $\tan 5\pi/3 = -\sqrt{3}$, $\cot 5\pi/3 = -\sqrt{3}/3$, $\sec 5\pi/3 = 2$, $\csc 5\pi/3 = -\frac{2}{3}\sqrt{3}$ **17.** $\sin(-120°) = -\sqrt{3}/2$, $\cos(-120°) = -\frac{1}{2}$, $\tan(-120°) = \sqrt{3}$, $\cot(-120°) = \sqrt{3}/3$, $\sec(-120°) = -2$, $\csc(-120°) = -2\sqrt{3}/3$ **19.** $\sin(-3\pi/4) = -\sqrt{2}/2$, $\cos(-3\pi/4) = -\sqrt{2}/2$, $\tan(-3\pi/4) = 1$, $\cot(-3\pi/4) = 1$, $\sec(-3\pi/4) = -\sqrt{2}$, $\csc(-3\pi/4) = -\sqrt{2}$
21. $\sin 750° = \frac{1}{2}$, $\cos 750° = \sqrt{3}/2$, $\tan 750° = \sqrt{3}/3$, $\cot 750° = \sqrt{3}$, $\sec 750° = \frac{2}{3}\sqrt{3}$, $\csc 750° = 2$ **23.** $\sin(41\pi/6) = \frac{1}{2}$, $\cos(41\pi/6) = -\sqrt{3}/2$, $\tan(41\pi/6) = -\sqrt{3}/3$, $\cot(41\pi/6) = -\sqrt{3}$, $\sec(41\pi/6) = -2\sqrt{3}/3$, $\csc(41\pi/6) = 2$ **25.** $\cos\theta = -\frac{4}{5}$, $\tan\theta = \frac{3}{4}$, $\cot\theta = \frac{4}{3}$, $\sec\theta = -\frac{5}{4}$, $\csc\theta = -\frac{5}{3}$ **27.** $\sin\theta = -\sqrt{6}/3$, $\cos\theta = -\sqrt{3}/3$, $\cot\theta = \sqrt{2}/2$, $\sec\theta = -\sqrt{3}$, $\csc\theta = -\sqrt{6}/2$ **29.** $\sin\theta = \sqrt{6}/3$, $\cos\theta = -\sqrt{3}/3$, $\tan\theta = -\sqrt{2}$, $\cot\theta = -\sqrt{2}/2$, $\csc\theta = \sqrt{6}/2$ **31.** False **33.** False
35. True **37.** True **39.** $\sqrt{2(1 - \cos s \cot t - \sin s \sin t)}$

SECTION 5.5

1.
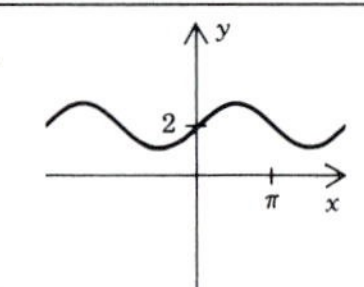

3.
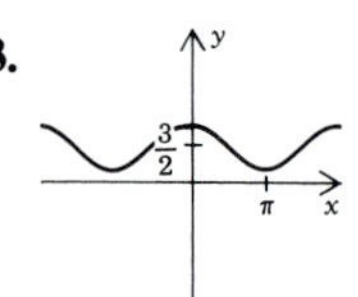

5.
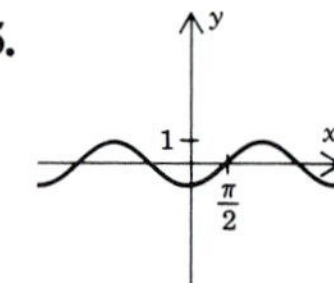

7.
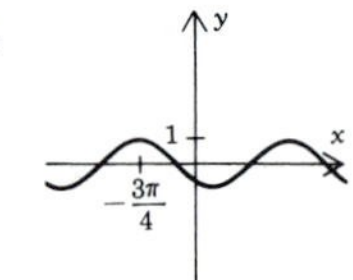

9.
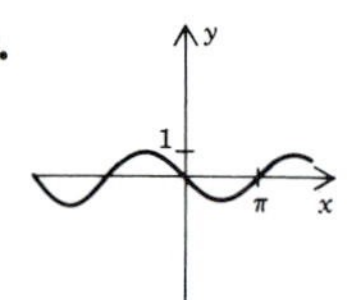

11.
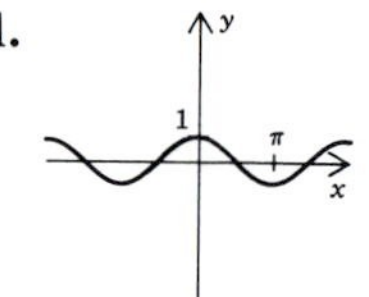

13.
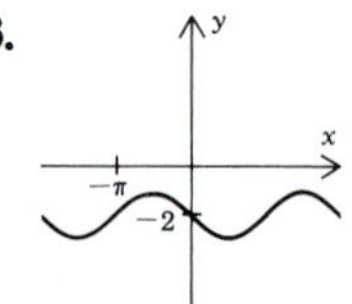

15.
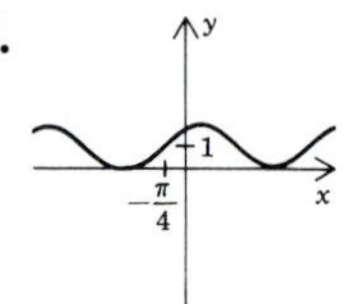

17.
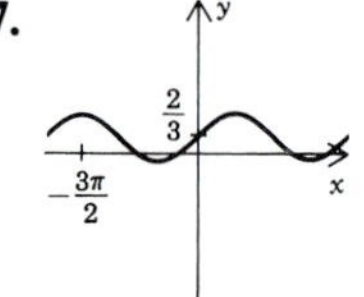

19. 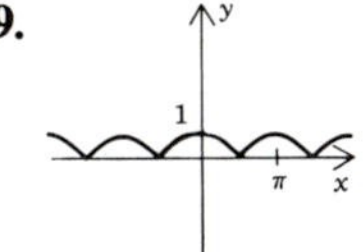

21. 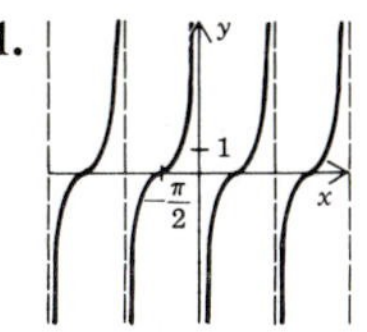

23. 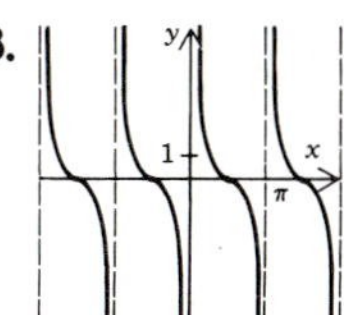

25. 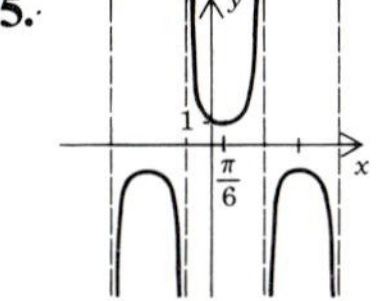

27. 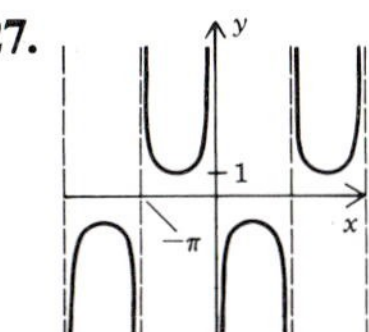

29. 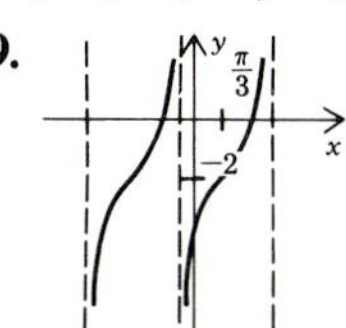

31. 0.736842105

SECTION 5.6

1. 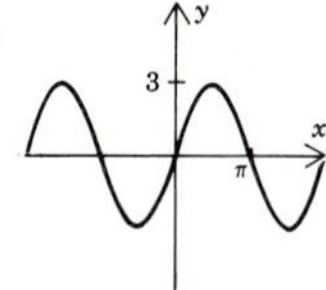

3. 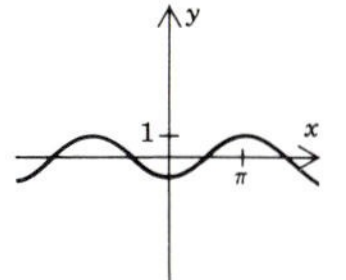

5. 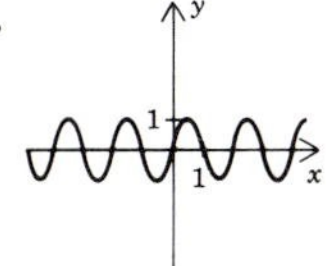

7. 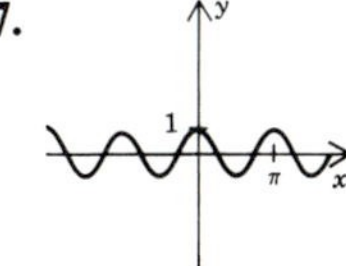

9. 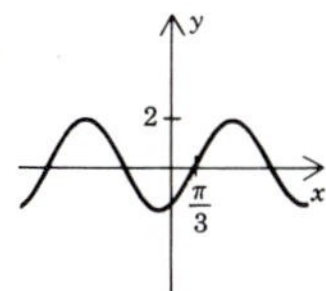

11. 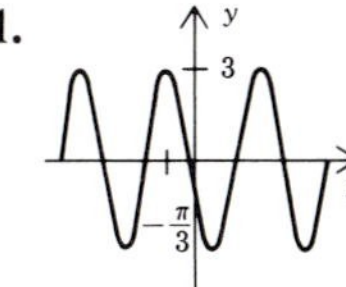

13. $4 \sin 3(x + \pi)$ **15.** $\frac{3}{2} \sin 2(x + \frac{\pi}{3})$

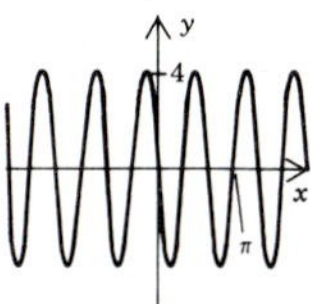

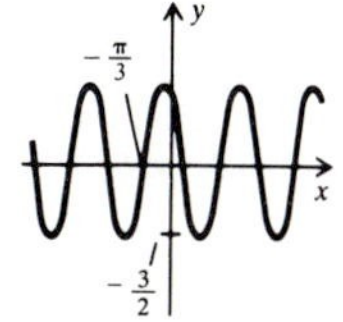

SECTION 5.7

1. $-\pi/2$ **3.** $\pi/2$ **5.** 0 **7.** $\pi/2$ **9.** $3\pi/4$ **11.** 0 **13.** $2\pi/3$ **15.** $\pi/2$
17. $-\pi/2$ **19.** 0.304692654 **21.** 1.000359217 **23.** 0.1095595268
25. 0.816316143 **27.** 1.141020895 **29.** −0.2144048647 **31.** $\pi/6$ **33.** $-\pi/3$
35. $2\pi/3$ **37.** $3\pi/4$ **39.** $\pi/13$ **41.** $5\pi/6$ **43.** $\pi/6$ **45.** 0 **47.** $\frac{1}{4}$
49. $-\frac{1}{2}$ **51.** $\sqrt{2}/2$ **53.** −1 **55.** $\frac{3}{5}$ **57.** $-\frac{4}{3}$ **59.** $3\sqrt{2}/4$ **61.** $\pi/2$
63. $\pi/6$

71. 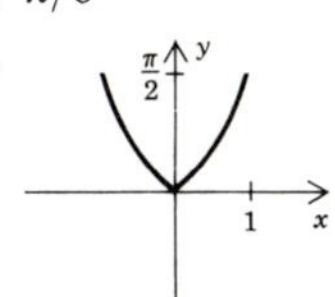

73. 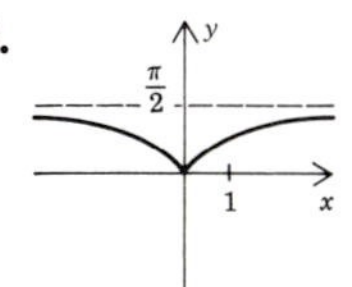

79. b. 1×10^{-13}

CHAPTER 5 REVIEW

1. **3.**

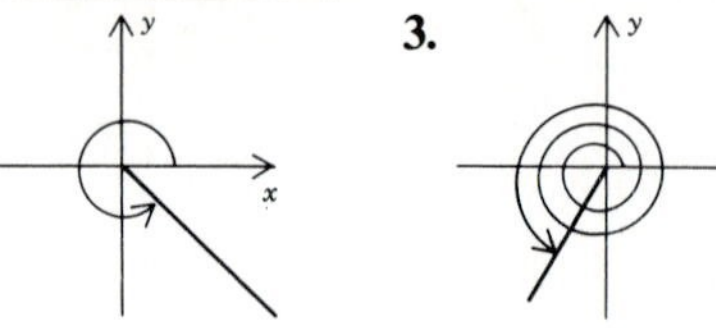

5.

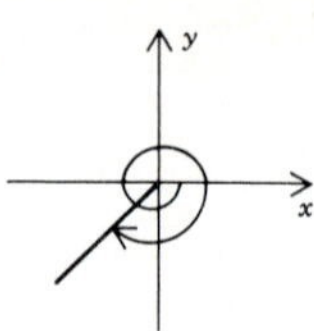

7. $-\sqrt{2}/2$ **9.** $-\sqrt{3}/2$ **11.** -1 **13.** $\sqrt{3}/3$ **15.** $2\sqrt{3}/3$ **17.** $2\sqrt{3}/3$
19. $-\frac{1}{2}$ **21.** $\sqrt{3}/3$
23. $\sin\theta = -\frac{2}{5}$, $\cos\theta = -\sqrt{21}/5$, $\tan\theta = 2\sqrt{21}/21$, $\cot\theta = \sqrt{21}/2$, $\sec\theta = -5\sqrt{21}/21$, $\csc\theta = -\frac{5}{2}$
25. $\sin\theta = 8\sqrt{65}/65$, $\cos\theta = -\sqrt{65}/65$, $\tan\theta = -8$, $\cot\theta = -\frac{1}{8}$, $\sec\theta = -\sqrt{65}$, $\csc\theta = \sqrt{65}/8$
27. $\sin\theta = \sqrt{15}/4$, $\cos\theta = \frac{1}{4}$, $\tan\theta = \sqrt{15}$, $\cot\theta = \sqrt{15}/15$, $\sec\theta = 4$, $\csc\theta = 4\sqrt{15}/15$
29. 0.2475627214 **31.** 0.6743023876 **33.** 0.1429146854
35. $a \approx 0.5290$, $c \approx 3.046$, $\beta = 80°$
37. $c \approx 8.360$, $\alpha \approx 53.27°$ ($\approx 53°\ 16'$), $\beta \approx 36.73°$ ($\approx 36°\ 44'$)
39. $\sin\theta = 5\sqrt{29}/29$, $\cos\theta = 2\sqrt{29}/29$, $\tan\theta = \frac{5}{2}$, $\cot\theta = \frac{2}{5}$, $\sec\theta = \sqrt{29}/2$, $\csc\theta = \sqrt{29}/5$
41.

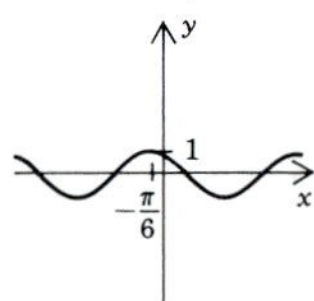

43. $\pi/3$ **45.** $\pi/6$ **47.** $\pi/3$ **49.** $-\pi/4$ **51.** $\pi/2$ **53.** $\pi/4$ **55.** $-2\sqrt{5}/5$
57. a. 1 b. 0 c. 1 d. 0 **63.** 136π inches **65.** Approximately 5874 feet
67. Approximately 2678 feet **69.** 24,500 miles

SECTION 6.1

57. $\pi/4$ **59.** $\pi/4$ **61.** $-\pi/4$ **63.** Graph of $2 \sec x$

SECTION 6.2

1. $\frac{1}{4}(\sqrt{6} - \sqrt{2})$ **3.** $\frac{1}{4}(\sqrt{2} - \sqrt{6})$ **5.** $\frac{1}{4}(\sqrt{6} - \sqrt{2})$ **7.** $\frac{1}{4}(\sqrt{2} - \sqrt{6})$
9. $\dfrac{\sqrt{3} - 1}{\sqrt{3} + 1}$ **11.** $\dfrac{\sqrt{3} + 1}{\sqrt{3} - 1}$ **13.** $-\frac{1}{8}(\sqrt{15} + \sqrt{3})$ **15.** $\frac{1}{20}(1 - 6\sqrt{10})$
17. $\frac{1}{12}(1 - 2\sqrt{30})$ **55.** $x = \pi/2$, $y = \pi/2$ **57.** $x = \pi/6$, $y = \pi/6$
59. Graph of $\sqrt{2}\sin(x + \pi/4)$ **61.** Graph of $4\sqrt{3}\sin(x + \pi/3)$
63. Graph of $\cos 2x$

SECTION 6.3

1. $\sin 2x = -4\sqrt{2}/9$, $\cos 2x = \frac{7}{9}$, $\tan 2x = -4\sqrt{2}/7$
3. $\sin 2x = -2\sqrt{2}/3$, $\cos 2x = -\frac{1}{3}$, $\tan 2x = 2\sqrt{2}$
5. $\sin 2x = -4\sqrt{6}/25$, $\cos 2x = -\frac{23}{25}$, $\tan 2x = 4\sqrt{6}/23$
7. $\sin\frac{1}{2}x = \frac{1}{4}\sqrt{8 + 3\sqrt{7}}$, $\cos\frac{1}{2}x = \frac{1}{4}\sqrt{8 - 3\sqrt{7}}$, $\tan\frac{1}{2}x = \dfrac{1}{8 - 3\sqrt{7}}$

9. $\sin \frac{1}{2}x = \sqrt{\frac{5+2\sqrt{5}}{10}}$, $\cos \frac{1}{2}x = -\sqrt{\frac{5-2\sqrt{5}}{10}}$, $\tan \frac{1}{2}x = \frac{1}{2-\sqrt{5}}$

11. $\sin \frac{1}{2}x = \sqrt{6}/3$, $\cos \frac{1}{2}x = -\sqrt{3}/3$, $\tan \frac{1}{2}x = -\sqrt{2}$ **13.** $\frac{1}{2}\sqrt{2-\sqrt{2}}$

15. $-\frac{1}{2}\sqrt{2-\sqrt{3}}$ **17.** $-\frac{1}{2}\sqrt{2-\sqrt{2}}$ **19.** $\frac{1}{2}\sqrt{2-\sqrt{3}}$ **21.** $\frac{\sqrt{2}}{\sqrt{2}-2}$

23. $-\frac{1}{2+\sqrt{3}}$ **25.** a. $\frac{1}{2}\sqrt{2-\sqrt{2+\sqrt{2}}}$ b. $\frac{1}{2}\sqrt{2+\sqrt{2+\sqrt{2}}}$

c. $\frac{\sqrt{2-\sqrt{2+\sqrt{2}}}}{\sqrt{2+\sqrt{2+\sqrt{2}}}}$, or $\frac{\sqrt{2-\sqrt{2}}}{2+\sqrt{2+\sqrt{2}}}$

53. Graph of $\frac{1}{2}\sin 4x$ **55.** Graph of $2 \tan x$

SECTION 6.4

1. $\frac{1}{2}(\cos 2x - \cos 6x)$ **3.** $\frac{1}{2}(\cos 4x + \cos 2x)$ **5.** $\frac{1}{2}(\cos 2x + \cos 4x/3)$
7. $\frac{1}{2}(\sin 10x - \sin 4x)$ **9.** $2 \sin 3x \cos x$ **11.** $2(\sin 3x/4)(\cos 5x/4)$
13. $2 \cos x \cos x/2$ **15.** $-2 \sin 5x \sin 2x$

SECTION 6.5

1. $\frac{\pi}{2} + 2n\pi$ **3.** $\frac{2\pi}{3} + 2n\pi, \frac{4\pi}{3} + 2n\pi$ **5.** $\frac{\pi}{3} + n\pi$ **7.** $\frac{\pi}{6} + n\pi$

9. $\frac{\pi}{3} + 2n\pi, \frac{2\pi}{3} + 2n\pi$ **11.** $\frac{\pi}{18} + \frac{2n\pi}{3}, \frac{5\pi}{18} + \frac{2n\pi}{3}$ **13.** $\frac{5\pi}{4} + 5n\pi$

15. $\frac{\pi}{24} + \frac{n\pi}{3}, -\frac{\pi}{24} + \frac{n\pi}{3}$ **17.** $\frac{\pi}{3} + 2n\pi, \frac{5\pi}{3} + 2n\pi$ **19.** $\frac{\pi}{3} + n\pi$

21. $\frac{5\pi}{6} + 2n\pi, \frac{7\pi}{6} + 2n\pi$ **23.** $n\pi, \frac{\pi}{3} + 2n\pi, -\frac{\pi}{3} + 2n\pi$ **25.** $-\frac{\pi}{4} + n\pi$

27. $\frac{\pi}{2} + n\pi$ **29.** $\frac{\pi}{4} + 2n\pi, \frac{3\pi}{4} + n\pi$ **31.** $\frac{3\pi}{2} + 2n\pi$

33. $\frac{2\pi}{3} + 2n\pi, \frac{4\pi}{3} + 2n\pi, 2n\pi$ **35.** $\frac{2\pi}{3} + 2n\pi, \frac{4\pi}{3} + 2n\pi$

37. $\frac{\pi}{6} + n\pi, -\frac{\pi}{3} + n\pi$ **39.** $\frac{\pi}{4} + \frac{n\pi}{2}$

41. $-\frac{\pi}{6} + 2n\pi, \frac{7\pi}{6} + 2n\pi, \frac{\pi}{2} + 2n\pi$ **43.** $\frac{\pi}{3} + 2n\pi, -\frac{\pi}{3} + 2n\pi$

45. 30°, 150°, 199° 28′, 340° 32′ **47.** 78° 28′, 281° 32′, 109° 28′, 250° 32′
49. 54° 44′, 305° 16′, 125° 16′, 234° 44′

CHAPTER 6 REVIEW

33. $\frac{1}{2}\sqrt{2+\sqrt{2}}$ **35.** $\frac{4}{\sqrt{6}-\sqrt{2}}$, or $\frac{2}{\sqrt{2-\sqrt{3}}}$ **37.** $-3\sqrt{3}/14$

39. $-\sqrt{15}/8$ **43.** a. No b. Yes **45.** a. $\frac{1}{2}\sqrt{2+\sqrt{3}}$

47. Graph of $2 \sin(x + \pi/6)$ **49.** $\frac{5\pi}{6} + 2n\pi, \frac{7\pi}{6} + 2n\pi$ **51.** $n\pi$

53. $\frac{\pi}{6} + 2n\pi, \frac{5\pi}{6} + 2n\pi, \frac{3\pi}{2} + 2n\pi$ **55.** $\frac{\pi}{6} + 2n\pi, \frac{5\pi}{6} + 2n\pi, \frac{3\pi}{2} + 2n\pi$

57. $\frac{\pi}{3} + n\pi, -\frac{\pi}{3} + n\pi$

SECTION 7.1

1. $\gamma = 65°, a \approx 8.458, c \approx 8.851$ **3.** $\alpha = 29°, a \approx 3.430, b \approx 3.430$
5. $\alpha = 123°, b \approx 3.334, c \approx 2.329$ **7.** $\beta = 142.9°, a \approx 7.024, c \approx 4.004$
9. $\beta = 91° 27', a \approx 1.366, b \approx 2.967$
11. Two triangles: $\beta \approx 59.29°$ ($\approx 59° 17'$), $\gamma = 83.71°$ ($\approx 83° 43'$), $c \approx 2.312$; $\beta \approx 120.71°$ ($\approx 120° 43'$), $\gamma = 22.29°$ ($\approx 22° 17'$), $c \approx 0.8822$
13. No triangle **15.** $\beta = 90°, \gamma = 60°, c = 3\sqrt{3} \approx 5.196$
17. $\alpha \approx 44.77°$ ($\approx 44° 46'$), $\gamma \approx 60.33°$ ($\approx 60° 20'$), $a \approx 3.647$
19. Two triangles: $\alpha \approx 119.21°$ ($\approx 119° 12'$), $\beta \approx 33.99°$ ($\approx 33° 59'$), $a \approx 4.840$; $\alpha \approx 7.19°$ ($\approx 7° 12'$), $\beta \approx 146.01°$ ($\approx 146° 1'$), $a \approx 0.6942$
21. $\beta \approx 35.37°$ ($\approx 35° 22'$), $\gamma \approx 33.88°$ ($\approx 33° 53'$), $c \approx 2.504$
23. The guard at A is closer, by approximately 69.26 yards.
25. Approximately 0.9493 miles **27.** Approximately 6.158 miles

SECTION 7.2

1. $\alpha \approx 41.75°$ ($\approx 41° 45'$), $\beta \approx 50.98°$ ($\approx 50° 59'$), $\gamma \approx 87.27°$ ($\approx 87° 16'$)
3. $\alpha \approx 21.79°$ ($\approx 21° 47'$), $\beta \approx 38.21°$ ($\approx 38° 13'$), $\gamma \approx 120.0°$
5. $\alpha \approx 38.27°$ ($\approx 38° 16'$), $\beta \approx 56.60°$ ($\approx 56° 36'$), $\gamma \approx 85.13°$ ($\approx 85° 8'$)
7. $a \approx 1.564$, $\beta \approx 40.99°$ ($\approx 40° 59'$), $\gamma \approx 119.01°$ ($\approx 119° 1'$)
9. $b \approx 0.6498$, $\alpha \approx 92.96°$ ($\approx 92° 57'$), $\gamma \approx 75.34°$ ($\approx 75° 20'$)
11. $c \approx 10.97$, $\alpha \approx 4.75°$ ($\approx 4° 45'$), $\beta \approx 3.55°$ ($\approx 3° 33'$)
13. $a \approx 4.726$, $\beta \approx 12.64°$ ($\approx 12° 38'$), $\gamma \approx 114.68°$ ($\approx 114° 41'$)
15. Two triangles: $a \approx 5.519$, $\alpha \approx 55.65°$ ($\approx 55° 39'$), $\gamma \approx 65.85°$ ($\approx 65° 51'$); $a \approx 0.8552$, $\alpha \approx 7.35°$ ($\approx 7° 21'$), $\gamma \approx 114.15°$ ($\approx 114° 9'$)
17. $b \approx 5.945$, $\alpha \approx 35.90°$ ($\approx 35° 54'$), $\beta \approx 44.20°$ ($\approx 44° 12'$)
19. $b \approx 3.382$, $\alpha \approx 22.16°$ ($\approx 22° 10'$), $\gamma \approx 120.44°$ ($\approx 120° 26'$)
21. $\alpha \approx 61.21°$ ($\approx 61° 13'$), $\beta \approx 87.96°$ ($\approx 87° 58'$), $\gamma \approx 30.83°$ ($\approx 30° 50'$)
23. $a \approx 0.2084$, $\alpha \approx 19.61°$ ($\approx 19° 37'$), $\gamma \approx 9.27°$ ($\approx 9° 16'$)
25. $500\sqrt{3}$ feet (≈ 866 feet) **27.** Approximately 23.18 miles
29. Approximately 44.42° ($\approx 44° 25'$), 57.12° ($\approx 57° 7'$), and 78.46° ($\approx 78° 28'$)
31. Approximately 4.62 feet
33. The angle at (0, 0) $\approx 26.57°$ ($\approx 26° 34'$); the angle at (3, 0) $= 45°$; the angle at (2, 1) $\approx 108.43°$ ($\approx 108° 26'$).
35. Approximately 19.91 feet and 24.57 feet
37. a. $\frac{1}{c}(2.333)$ b. Approximately $\frac{1}{c}(2.367)$

SECTION 7.3

1. $\mathscr{A} = 3\sqrt{2} \approx 4.243$ **3.** $\mathscr{A} \approx 2.431$ **5.** $\mathscr{A} \approx 10$ **7.** $\mathscr{A} \approx 3.093$
9. $\mathscr{A} = 4\sqrt{5} \approx 8.944$ **11.** $\mathscr{A} = 1/\sqrt{2} \approx 0.7071$ **13.** $\mathscr{A} = \sqrt{3} \approx 1.732$
15. $\mathscr{A} \approx 4.081$

SECTION 7.4

1. $(0, 0)$ **3.** $(-5, 0)$ **5.** $(0, 3)$ **7.** $(-2.3, 0)$ **9.** $(\frac{3}{2}, 3\sqrt{3}/2)$
11. $(\frac{1}{2}, -\sqrt{3}/2)$ **13.** $(-2, 0)$ **15.** $(2, 2\sqrt{3})$
17. Approximately $(1.080604612, 1.68294197)$
19. $(5, 2n\pi)$ and $(-5, \pi + 2n\pi)$ for any integer n
21. $(5, \pi + 2n\pi)$ and $(-5, 2n\pi)$ for any integer n

23. $\left(2, \frac{\pi}{4} + 2n\pi\right)$ and $\left(-2, \frac{5\pi}{4} + 2n\pi\right)$ for any integer n

25. $\left(2, \frac{\pi}{6} + 2n\pi\right)$ and $\left(-2, \frac{7\pi}{6} + 2n\pi\right)$ for any integer n

27. $\left(2\sqrt{2}, \frac{2\pi}{3} + 2n\pi\right)$ and $\left(-2\sqrt{2}, \frac{5\pi}{3} + 2n\pi\right)$ for any integer n

29. Approximately $(20.59126028, 0.5070985044 + 2n\pi)$ and $(-20.59126028, 3.648691158 + 2n\pi)$ for any integer n

31.
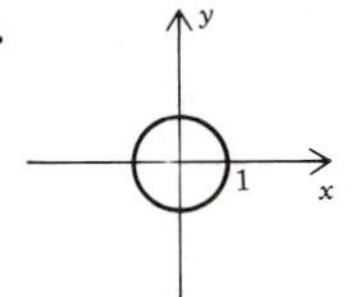

33.
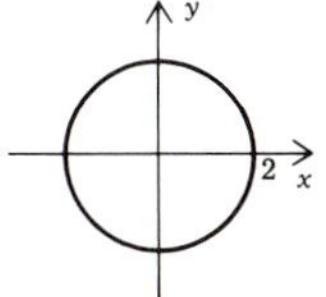

35.
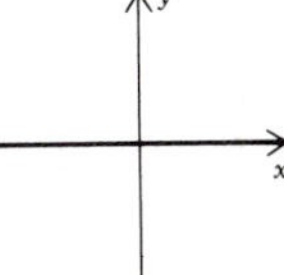

37.
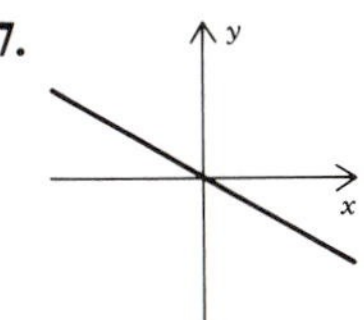

39.
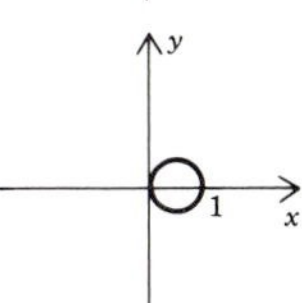

41.
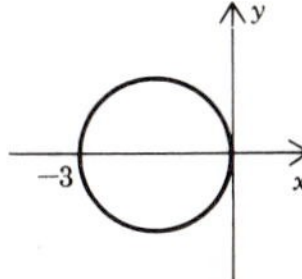

43.
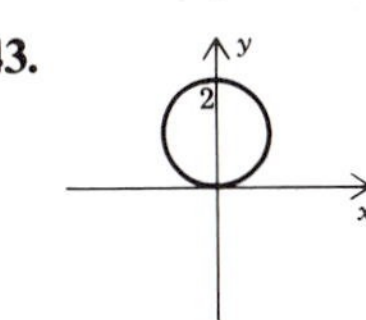

45.
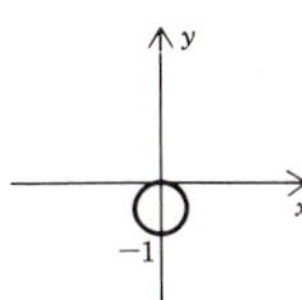

47.
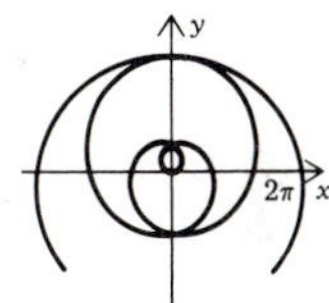

49. The graph of $r = 1 + \cos\theta$ is obtained by rotating the graph of $r = 1 + \sin\theta$ clockwise through an angle of $\pi/2$ radians about the origin.
51. The graph of $r = \frac{1}{2} + \sin\theta$ is obtained by rotating the graph of $r = \frac{1}{2} - \cos\theta$ clockwise through an angle of $\pi/2$ radians about the origin.

SECTION 7.5

1. $\mathbf{a} + \mathbf{b} = (1, 5)$, $\mathbf{a} - \mathbf{b} = (5, -3)$, $2\mathbf{a} - 3\mathbf{b} = (12, -10)$, $-\frac{5}{2}\mathbf{a} = (-\frac{15}{2}, -\frac{5}{2})$
3. $\mathbf{a} + \mathbf{b} = (-3, 3)$, $\mathbf{a} - \mathbf{b} = (-9, -9)$, $2\mathbf{a} - 3\mathbf{b} = (-21, -24)$, $-\frac{5}{2}\mathbf{a} = (15, \frac{15}{2})$
5. $\mathbf{a} + \mathbf{b} = (-2, 3)$, $\mathbf{a} - \mathbf{b} = (-2, -3)$, $2\mathbf{a} - 3\mathbf{b} = (-4, -9)$, $-\frac{5}{2}\mathbf{a} = (5, 0)$
7. $4\mathbf{a} - \mathbf{b} = 25\mathbf{i} + 10\mathbf{j}$, $3\mathbf{a} + 2\mathbf{b} = 16\mathbf{i} + 2\mathbf{j}$
9. $4\mathbf{a} - \mathbf{b} = 5\mathbf{i} + 6\mathbf{j}$, $3\mathbf{a} + 2\mathbf{b} = -10\mathbf{i} + 21\mathbf{j}$
11. $4\mathbf{a} - \mathbf{b} = 3\mathbf{i} - 5\mathbf{j}$, $3\mathbf{a} + 2\mathbf{b} = 5\mathbf{i} - \mathbf{j}$ **19.** $\sqrt{10}$ **21.** $\sqrt{5}$ **23.** 2 **25.** No

27. No **29.** Yes **31.** $\sqrt{2}\left(\cos\frac{5\pi}{4}\mathbf{i} + \sin\frac{5\pi}{4}\mathbf{j}\right)$ **33.** $2\left(\cos\frac{\pi}{3}\mathbf{i} + \sin\frac{\pi}{3}\mathbf{j}\right)$

35. $\pi\sqrt{2}\left(\cos\frac{3\pi}{4}\mathbf{i} + \sin\frac{3\pi}{4}\mathbf{j}\right)$ **39.** $\mathbf{a} = -\frac{3}{4}\mathbf{i} - \mathbf{j}$, $\mathbf{b} = -\frac{3}{2}\mathbf{i} - 2\mathbf{j}$ **41.** -11

43. 0 **45.** $300\sqrt{3} \approx 519.6$ foot-pounds
47. Velocity vector: $(150\sqrt{3} + 30)\mathbf{i} + 150\mathbf{j}$;
ground speed: $30\sqrt{101 + 10\sqrt{3}} \approx 326.3$ miles per hour
Actual course: approximately $27.37°$ ($\approx 27°\ 22'$) north of east

CHAPTER 7 REVIEW

1. $\alpha \approx 75.52°$ ($\approx 75°\ 31'$), $\beta \approx 75.52°$ ($\approx 75°\ 31'$), $\gamma \approx 28.96°$ ($\approx 28°\ 58'$)
3. $b \approx 46.61$, $c \approx 59.9$, $\gamma = 97.4°$ ($= 97°\ 24'$)
5. $c \approx 28.65$, $\beta \approx 10.84°$ ($\approx 10°\ 50'$), $\gamma \approx 155.71°$ ($\approx 155°\ 43'$)
7. $c \approx 0.07401$, $\alpha \approx 40.27°$, ($\approx 40°\ 16'$), $\gamma \approx 15.41°$ ($\approx 15°\ 25'$)
9. Two triangles: $c \approx 2.716$, $\alpha \approx 84.48°$ ($\approx 84°\ 29'$), $\gamma \approx 15.22°$ ($\approx 15°\ 13'$);
$c \approx 0.7548$, $\alpha \approx 95.52°$ ($\approx 95°\ 31'$), $\gamma \approx 4.18°$ ($\approx 4°\ 11'$)
11. $\alpha \approx 10.96°$ ($\approx 10°\ 58'$), $\beta \approx 130.48°$ ($\approx 130°\ 29'$), $\gamma \approx 38.56°$ ($\approx 38°\ 34'$)
13. $a \approx 1.809$, $c \approx 2.225$, $\alpha \approx 21.03°$ ($\approx 21°\ 2'$)
15. $\alpha \approx 32.20°$ ($\approx 32°\ 12'$), $\beta = 60°$, $\gamma \approx 87.80°$ ($\approx 87°\ 48'$)
17. $b \approx 10.33$, $\beta \approx 64.24°$ ($\approx 64°\ 14'$), $\gamma \approx 55.06°$ ($\approx 55°\ 4'$)
19. $c = 106$, $\alpha \approx 58.11°$ ($\approx 58°\ 7'$), $\beta \approx 31.89°$ ($\approx 31°\ 53'$)
21. $\mathscr{A} \approx 32.43$ **23.** $\mathscr{A} \approx 1.735$ **25.** $(\frac{3}{4}\sqrt{2}, -\frac{3}{4}\sqrt{2})$ **27.** $(-\frac{1}{2}e, -\frac{1}{2}\sqrt{3}e)$

29. $\left(2\sqrt{5}, -\dfrac{\pi}{3} + 2n\pi\right)$ and $\left(-2\sqrt{5}, \dfrac{2\pi}{3} + 2n\pi\right)$ for any integer n

31. $\left(\pi, \dfrac{\pi}{2} + 2n\pi\right)$ and $\left(-\pi, \dfrac{3\pi}{2} + 2n\pi\right)$ for any integer n

33.

35.

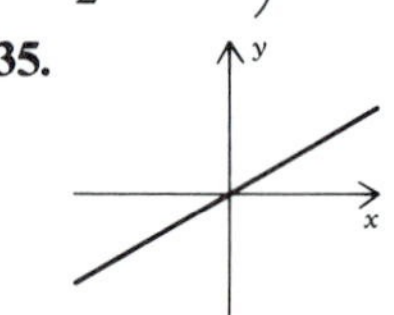

37.

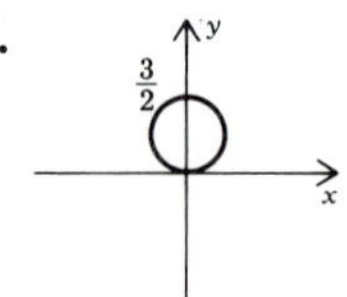

39. $(\frac{14}{3}, -2)$ **41.** $6\mathbf{i} - 8\mathbf{j}$ **43.** $4\sqrt{3}$ **45.** $7\sqrt{2} - \sqrt{3}\pi$ **47.** $\sqrt{3}$

49. $2\sqrt{7}\left(\cos\dfrac{5\pi}{6}\mathbf{i} + \sin\dfrac{5\pi}{6}\mathbf{j}\right)$ **51.** $\dfrac{1}{5}\left(\cos\dfrac{\pi}{3}\mathbf{i} + \sin\dfrac{\pi}{3}\mathbf{j}\right)$

53. Approximately $78.97°$ ($\approx 78°\ 58'$) **55.** Approximately 51.03 feet
57. Approximately $88.83°$ ($\approx 88°\ 50'$) **59.** Approximately $92.82°$ ($\approx 92°\ 49'$)

CUMULATIVE REVIEW II: CHAPTERS 4–7

1. 9 **3.** $-\frac{5}{6}$ **5.** 2, 1 **7.** 25 **9.** $50 \ln \frac{8}{3}$ **11.** $\dfrac{\ln 3 + \ln 2}{\ln 3 - 2 \ln 2}$ **13.** $-2x$
17. $\sqrt{2}$
19.

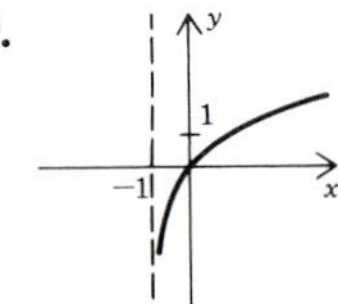

21. $\frac{1}{2}$ **23.** 2 **25.** $\frac{1}{2}\sqrt{2 + \sqrt{3}}$
27. $\sin\theta = -\frac{1}{5}\sqrt{5}$, $\cos\theta = -\frac{2}{5}\sqrt{5}$, $\cot\theta = 2$, $\sec\theta = -\frac{1}{2}\sqrt{5}$, $\csc\theta = -\sqrt{5}$

29. $-\frac{7}{10}\sqrt{2}$ **31.** $-\pi/4$

33.

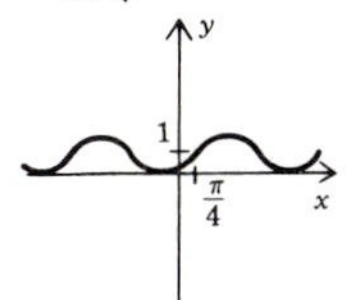

41. $\dfrac{\pi}{2} + n\pi$ for any integer n

43. Two triangles: $c \approx 6.212$, $\beta \approx 53.46°$ ($\approx 53°\ 28'$), $\gamma \approx 86.54°$ ($\approx 86°\ 32'$); $c \approx 1.449$, $\beta \approx 126.54°$ ($\approx 126°\ 32'$), $\gamma \approx 13.46°$ ($\approx 13°\ 28'$)

45. $\alpha \approx 16.52°$ ($\approx 16°\ 31'$), $\beta \approx 45.31°$ ($\approx 45°\ 19'$), $\gamma \approx 118.17°$ ($\approx 118°\ 10'$)

47. $\sqrt{231}$ **49.** $\left(12, -\dfrac{\pi}{3} + 2n\pi\right)$ and $\left(-12, \dfrac{2\pi}{3} + 2n\pi\right)$ for any integer n

51. No **53.** $3\dfrac{\ln 10}{\ln 2} \approx 9.97$ weeks

55. Approximately 11.54 pounds per square inch **57.** a. 200 b. 6000

59. Approximately 101.25 meters

61. Magnitude: $2\sqrt{5}$ pounds; direction: approximately 18.43°, or 18° 26′, south of east

SECTION 8.1

5. $(1, -2)$ **7.** $(3, 1)$ **9.** $\left(a, \dfrac{7a}{5} + \dfrac{6}{5}\right)$ for any number a **11.** $(-1, 4)$

13. $(1.7, 2.1)$ **15.** No solution **17.** No solution **19.** $(1, 3), (-1, -3)$

21. $(4, 2)$ **23.** $(3\sqrt{5}/5, -6\sqrt{5}/5), (-3\sqrt{5}/5, 6\sqrt{5}/5)$ **25.** $(\frac{1}{2}, \frac{1}{4}), (-\frac{1}{3}, \frac{1}{9})$

27. $(\log_2 3, 3)$ **29.** $(a, -b)$

31. a. $\left(\dfrac{s + \sqrt{s^2 - 4}}{2}, \dfrac{2}{s + \sqrt{s^2 - 4}}\right), \left(\dfrac{s - \sqrt{s^2 - 4}}{2}, \dfrac{2}{s - \sqrt{s^2 - 4}}\right)$

33. a. $a > \frac{3}{4}$ b. $a = \frac{3}{4}$ c. $a < \frac{3}{4}$ **35.** $a = -\frac{3}{7}, b = \frac{5}{7}$

37. 12 and 18, or -12 and -18 **39.** $2\sqrt{10}, 2\sqrt{10}$ **41.** 17 feet and 24 feet

43. $\frac{9}{8}$ quarts tap water, $\frac{15}{8}$ quarts heated water.

45. Girl is 12 years old; father is 40 years old.

47. 2 centimeters and 4 centimeters

SECTION 8.2

1. $(3, -1)$ **3.** $(2, 4)$ **5.** $(-2, -3)$ **7.** $\left(a, \dfrac{a}{10} - 1\right)$ for any number a

9. No solution **11.** $\left(a, \dfrac{3a}{2} - \dfrac{7}{4}\right)$ for any number a **13.** $(4a, 2b)$

17. $a = 3, b = -4$ **19.** \$20 and \$30

21. 2 ounces of 24-carat, 3 ounces of 14-carat **23.** 60

SECTION 8.3

5. $(4, 0, -2)$, **7.** $(1, -1, 3)$ **9.** $(a, 0, a)$ for any number a **11.** No solution

13. $(0, 0, 0)$ **15.** $(4, 1, -1)$ **17.** $(3, -1, -2)$ **19.** $(-\frac{1}{2}, \frac{1}{2}, -1)$

21. $(\frac{1}{2}, -\frac{2}{3}, \frac{1}{6})$ **23.** $(1.5, -1, 0.5)$ **25.** $a = 3, b = -2, c = 5$ **27.** 14, 16, 28

29. 8 **31.** 320

SECTION 8.4

1. $\begin{bmatrix} 2 & -4 \\ -5 & 1 \end{bmatrix}, \left[\begin{array}{cc|c} 2 & -4 & 7 \\ -5 & 1 & 6 \end{array}\right]$ **3.** $\begin{bmatrix} 1 & 1 \\ 0 & 1 \end{bmatrix}, \left[\begin{array}{cc|c} 1 & 1 & 2 \\ 0 & 1 & 1 \end{array}\right]$

5. $\begin{bmatrix} \frac{1}{3} & -\frac{1}{4} & \frac{1}{2} \\ \frac{2}{5} & 0 & 1 \\ 1 & -\frac{1}{2} & 0 \end{bmatrix}, \left[\begin{array}{ccc|c} \frac{1}{3} & -\frac{1}{4} & \frac{1}{2} & 1 \\ \frac{2}{5} & 0 & 1 & 3 \\ 1 & -\frac{1}{2} & 0 & 4 \end{array}\right]$

13. $(1, -1)$ **15.** $(\frac{7}{2}, \frac{1}{2})$ **17.** $(0.6, 0.4)$ **19.** $(-2, -1, 3)$
21. $(2, 5, 1)$ **23.** $(-c + 2, 5c - 2, c)$ for any number c

SECTION 8.5

1. $M_{21} = 10, M_{22} = -32, M_{23} = 20, A_{21} = -10, A_{22} = -32, A_{23} = -20$
3. 8 **5.** 0 **7.** -55 **9.** -28 **11.** -14 **13.** 4 **15.** 97 **17.** 1
23. $3, -2$ **29.** a. 5 b. 40

SECTION 8.6

1. $(1, -2)$ **3.** $(3, 1)$ **5.** $(3, -1)$ **7.** $(2, 4)$ **9.** $(-2, -3)$ **11.** $(4, 1, -2)$
13. $(3, -2, 8)$ **15.** $(1, -2, 3, 0)$ **17.** $(0, -1, 5, 3)$

SECTION 8.7

1. $\begin{bmatrix} 5 & -2 & 3 \\ -3 & 5 & 7 \end{bmatrix}$ **3.** $\begin{bmatrix} -6 & 12 \\ 15 & -36 \end{bmatrix}$ **5.** $\begin{bmatrix} -2 & 5 \\ -8 & 5 \end{bmatrix}$ **7.** $\begin{bmatrix} 3 & \frac{3}{2} \\ -4 & 1 \end{bmatrix}$ **9.** $\begin{bmatrix} -5 \\ -3 \end{bmatrix}$

11. $\begin{bmatrix} 6 & -8 \\ -3 & 4 \end{bmatrix}$ **13.** $\begin{bmatrix} 4 & -2 & 5 \\ 18 & -2 & -14 \\ -9 & 18 & 9 \end{bmatrix}$ **15.** $\begin{bmatrix} 1 & 0 & 0 \\ 0 & 1 & 0 \\ 0 & 0 & 1 \end{bmatrix}$ **17.** $[-7]$

19. $AB = \begin{bmatrix} -8 & -6 \\ -7 & -3 \end{bmatrix}, BA = \begin{bmatrix} -11 & 9 \\ 2 & 0 \end{bmatrix}$

21. $AB = \begin{bmatrix} -5 & 5 & 7 \\ 27 & 3 & 1 \\ 2 & 2 & -3 \end{bmatrix}, BA = \begin{bmatrix} 10 & -3 & 9 \\ 1 & -6 & 1 \\ 6 & 13 & -9 \end{bmatrix}$

23. $(AB)C = A(BC) = \begin{bmatrix} -8 & 0 \\ 35 & 9 \end{bmatrix}$ **25.** Inverse: $\begin{bmatrix} 0 & 1 \\ 1 & 0 \end{bmatrix}$ **27.** No inverse

29. Inverse: $\frac{1}{15}\begin{bmatrix} 2 & -1 \\ 3 & 6 \end{bmatrix}$ **31.** $\frac{1}{ab}\begin{bmatrix} b & 0 \\ -c & a \end{bmatrix}$ **33.** Inverse: $\begin{bmatrix} 1 & 0 & 0 \\ 0 & 1 & 0 \\ 0 & 0 & 1 \end{bmatrix}$

35. No inverse **37.** Inverse: $\begin{bmatrix} 2 & -3 & 1 \\ -6 & 9 & -2 \\ -3 & 5 & -1 \end{bmatrix}$

39. Inverse: $\frac{1}{10}\begin{bmatrix} 11 & 18 & 2 \\ 6 & 8 & 2 \\ -7 & -16 & -4 \end{bmatrix}$

41. $(-1, -2)$ **43.** $(-6, 1)$ **45.** $(1, -1, 2)$ **47.** $(\frac{1}{2}, -\frac{1}{4}, \frac{3}{4})$

SECTION 8.8

1. Yes **3.** No **5.** No

7.

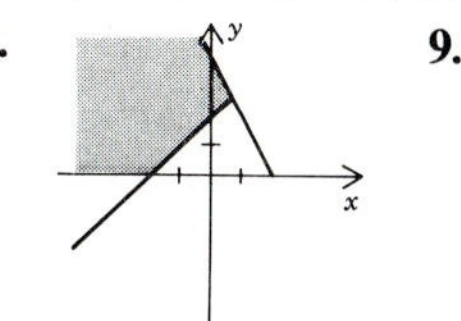

9.

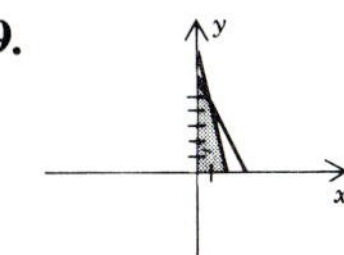

11. (0, 0), (3, 0), (3, 2), (0, 5)

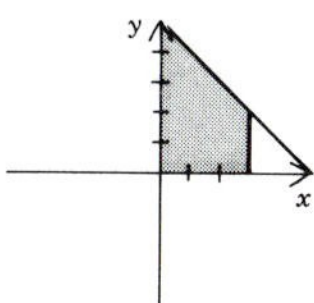

13. (0, 0), (4, 0), (18, 7), (0, 7)

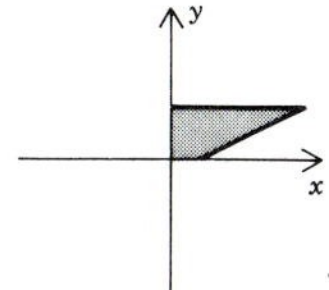

15. $(-\frac{4}{3}, \frac{1}{3})$, (4, 3), (2, 7)

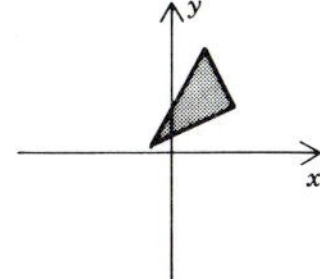

17. (0, 0), (10, 0), (8, 4), (4, 8), (0, 10)

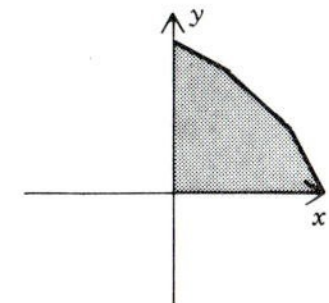

19. (0, −3), (−3, 0)

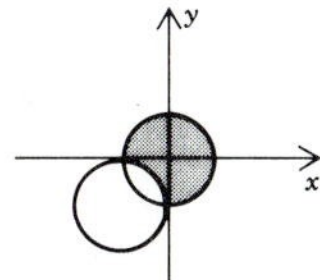

21. $(\frac{1}{2}\sqrt{2}, \frac{1}{2})$, $(-\frac{1}{2}\sqrt{2}, \frac{1}{2})$

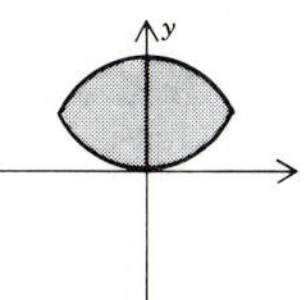

SECTION 8.9

1. Maximum: 5 at (−1, 1); minimum: −14 at (3, −2)
3. Maximum: 4 at (4, 6); minimum: 0 at (0, 0)
5. Maximum: 28 at (−4, 2); minimum: −2 at (2, 2)
7. Maximum: 19 at (6, 21); minimum: 3 at (0, 9)
9. Maximum: 42 at (5, 8); minimum: 0 at (0, 0) **11.** \$70 **13.** \$4000
15. 180 bushels of peaches and \$1260 of profit **17.** \$1200

CHAPTER 8 REVIEW

1. (9, −2) **3.** No solution **5.** (2*b*, *b*) for any number *b* **7.** $(\frac{1}{2}, -1)$, $(\frac{1}{4}, -2)$
9. (1, 2) **11.** (1, −2, 4) **13.** (3*c*, 2*c* − 1, *c*) for any number *c* **15.** (−3, 5, 7)
17. $(0, \frac{2}{5}, -\frac{3}{5})$ **19.** 22 **21.** 132 **23.** −51 **25.** (5, 7) **27.** (−3, 4, 5)

29. $\begin{bmatrix} 0 & -13 & 11 \\ 2 & -14 & 28 \\ 14 & 11 & 0 \end{bmatrix}$ **31.** Inverse: $\frac{1}{8}\begin{bmatrix} 4 & 1 \\ 0 & 2 \end{bmatrix}$ **33.** Inverse: $\frac{1}{2}\begin{bmatrix} 1 & 1 \\ -1 & 1 \end{bmatrix}$

35. Inverse: $\frac{1}{3}\begin{bmatrix} 11 & 1 & -5 \\ -4 & 3 & 3 \\ -6 & 0 & 3 \end{bmatrix}$ **37.** (−7, −10) **39.** (2, 1, 0)

41. $(0, 0), (\frac{3}{4}, 0), (\frac{1}{2}, \frac{1}{3}), (0, \frac{2}{3})$ **43.** $(0, 0), (5, 0), (4, 4), (2, 8), (0, 10)$

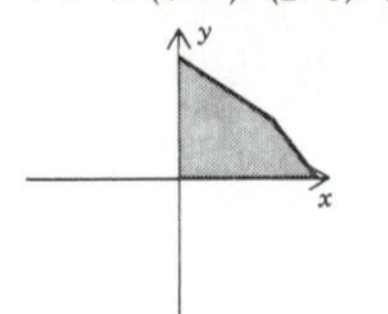

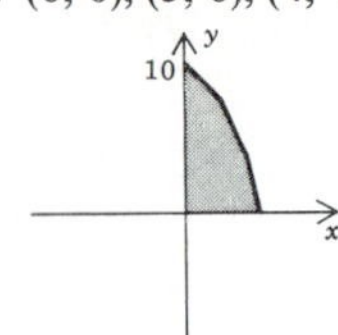

45. Maximum: $\frac{8}{3}$ at $(0, \frac{2}{3})$; minimum: 0 at $(0, 0)$
47. Maximum: 16 at $(4, 4)$; minimum: 0 at $(0, 0)$ **51.** $9, -9$ **53.** \$0.15
55. \$4000 at 8%, \$16,000 at 10% **57.** 300 doors, 400 windows

SECTION 9.1

1. $7 - 5i$ **3.** $6 + 19i$ **5.** 23 **7.** $6 + 2i$ **9.** $5 + \frac{1}{2}i$ **11.** $\frac{1}{2} - \frac{1}{2}i$ **13.** $-i$
15. $\frac{1}{15} - \frac{2}{15}i$ **17.** i **19.** $\frac{3}{5} + \frac{2}{5}i$ **21.** $-\frac{24}{25} + \frac{7}{25}i$ **23.** 169 **25.** $\frac{12}{5} + \frac{4}{5}i$
29. b. $3 + 4i$ **35.** $0, 1, -1, i, -i$

SECTION 9.2

1. 1 **3.** 1 **5.** 5 **7.** $\sqrt{2}$ **9.** $5\sqrt{2}$ **11.** $\sqrt{26}/\sqrt{20}$ **13.** $\sqrt{2}/4$
15. Circle: center 0, radius 1 **17.** Circle: center 0, radius $\sqrt{3}$
19. Circle: center 1, radius 1 **21.** Circle: center $-2 + 4i$, radius $\frac{3}{2}$

23. $\sqrt{2}\left(\cos\frac{\pi}{4} + i\sin\frac{\pi}{4}\right)$ **25.** $\sqrt{2}\left(\cos\frac{3\pi}{4} + i\sin\frac{3\pi}{4}\right)$ **27.** $1(\cos\pi + i\sin\pi)$

29. $7\sqrt{2}\left(\cos\frac{3\pi}{4} + i\sin\frac{3\pi}{4}\right)$ **31.** $1\left(\cos\frac{5\pi}{3} + i\sin\frac{5\pi}{3}\right)$

33. $z_1z_2 = \cos\frac{5\pi}{12} + i\sin\frac{5\pi}{12}, \quad \frac{z_1}{z_2} = \cos\frac{\pi}{12} + i\sin\frac{\pi}{12}$

35. $z_1z_2 = 8\left(\cos\frac{\pi}{8} + i\sin\frac{\pi}{8}\right), \quad \frac{z_1}{z_2} = \frac{1}{2}\left(\cos\frac{\pi}{24} + i\sin\frac{\pi}{24}\right)$

37. $\frac{5\pi}{12}$; $14\sqrt{2}\left(\cos\frac{5\pi}{12} + i\sin\frac{5\pi}{12}\right)$ **39.** $\frac{7\pi}{12}$; $2\sqrt{2}\left(\cos\frac{7\pi}{12} + i\sin\frac{7\pi}{12}\right)$

41. 0; $4\sqrt{2}(\cos 0 + i\sin 0)$ **43.** $\frac{5\pi}{4}$; $\frac{2\sqrt{2}}{3}\left(\cos\frac{5\pi}{4} + i\sin\frac{5\pi}{4}\right)$

49. a. $\pi/2$ plus an argument of z is an argument of iz, and the modulus of z is equal to the modulus of iz.

b. $\pi/4$ plus an argument of z is an argument of $\frac{1}{\sqrt{2}}(1 + i)z$, and the modulus of z is equal to the modulus of $\frac{1}{\sqrt{2}}(1 + i)z$.

53. a. $\log i = \frac{\pi}{2}i$, $\log(-1) = \pi i$, $\log(-i) = \frac{3\pi}{2}i$

SECTION 9.3

1. $64i$ **3.** $-\frac{27}{2} + \frac{27\sqrt{3}}{2}i$ **5.** 16 **7.** $\frac{1}{64} + \frac{\sqrt{3}}{64}i$ **9.** $1000i$

11. $2\left(\cos\frac{\pi}{36} + i\sin\frac{\pi}{36}\right), 2\left(\cos\frac{25\pi}{36} + i\sin\frac{25\pi}{36}\right), 2\left(\cos\frac{49\pi}{36} + i\sin\frac{49\pi}{36}\right)$

13. $10^{1/5}\left[\cos\left(-\frac{\pi}{40}\right) + i\sin\left(-\frac{\pi}{40}\right)\right], 10^{1/5}\left(\cos\frac{3\pi}{8} + i\sin\frac{3\pi}{8}\right),$
$10^{1/5}\left(\cos\frac{31\pi}{40} + i\sin\frac{31\pi}{40}\right), 10^{1/5}\left(\cos\frac{47\pi}{40} + i\sin\frac{47\pi}{40}\right),$
$10^{1/5}\left(\cos\frac{63\pi}{40} + i\sin\frac{63\pi}{40}\right)$

15. $\cos\frac{\pi}{4} + i\sin\frac{\pi}{4}, \cos\frac{7\pi}{12} + i\sin\frac{7\pi}{12}, \cos\frac{11\pi}{12} + i\sin\frac{11\pi}{12},$
$\cos\frac{5\pi}{4} + i\sin\frac{5\pi}{4}, \cos\frac{19\pi}{12} + i\sin\frac{19\pi}{12}, \cos\frac{23\pi}{12} + i\sin\frac{23\pi}{12}$

17. $2^{1/3}\left[\cos\left(-\frac{\pi}{9}\right) + i\sin\left(-\frac{\pi}{9}\right)\right], 2^{1/3}\left(\cos\frac{5\pi}{9} + i\sin\frac{5\pi}{9}\right),$
$2^{1/3}\left(\cos\frac{11\pi}{9} + i\sin\frac{11\pi}{9}\right)$

19. $\frac{3\sqrt{3}}{2} + \frac{3}{2}i, -\frac{3\sqrt{3}}{2} - \frac{3}{2}i$ **21.** $1, -\frac{1}{2} + \frac{\sqrt{3}}{2}i, -\frac{1}{2} - \frac{\sqrt{3}}{2}i$

23. $2\left[\cos\left(\frac{\pi}{4} + \frac{2\pi k}{3}\right) + i\sin\left(\frac{\pi}{4} + \frac{2\pi k}{3}\right)\right], k = 0, 1, 2$

25. $2\left[\cos\left(-\frac{\pi}{6} + \frac{2\pi k}{5}\right) + i\sin\left(-\frac{\pi}{6} + \frac{2\pi k}{5}\right)\right], k = 0, 1, 2, 3, 4$

27. $-\frac{\sqrt{2}}{2} + \frac{\sqrt{2}}{2}i, \frac{\sqrt{2}}{2} - \frac{\sqrt{2}}{2}i$ **29.** $\cos\left(\frac{3\pi}{8} + \frac{\pi k}{2}\right) + i\sin\left(\frac{3\pi}{8} + \frac{\pi k}{2}\right), k = 0, 1, 2, 3$

31. $\cos\frac{\pi k}{4} + i\sin\frac{\pi k}{4}, k = 0, 1, 2, 3, 4, 5, 6, 7$ **33.** $2, -1 + \sqrt{3}i, -1 - \sqrt{3}i$

35. $\sqrt{2}, \frac{\sqrt{2}}{2} + \frac{\sqrt{6}}{2}i, -\frac{\sqrt{2}}{2} + \frac{\sqrt{6}}{2}i, -\sqrt{2}, -\frac{\sqrt{2}}{2} - \frac{\sqrt{6}}{2}i, \frac{\sqrt{2}}{2} - \frac{\sqrt{6}}{2}i$

37. $\frac{\sqrt{3}}{2} + \frac{1}{2}i, -\frac{1}{2} + \frac{\sqrt{3}}{2}i, -\frac{\sqrt{3}}{2} - \frac{1}{2}i, \frac{1}{2} - \frac{\sqrt{3}}{2}i$

39. b. $\sin 2\theta = 2\sin\theta\cos\theta, \cos 2\theta = \cos^2\theta - \sin^2\theta$ **41.** $1 + i, 1 - i, -1 - i$
43. Their arguments differ by an integral multiple of π.

CHAPTER 9 REVIEW

1. $-3 - 4i$ **3.** $-4 - 4i$ **5.** $-\frac{2}{13} + \frac{3}{13}i$ **7.** 1 **9.** 10 **11.** $\sqrt{5}/2$

13. $1 + \sqrt{5}i, 1 - \sqrt{5}i$ **15.** $i, -\frac{\sqrt{3}}{2} - \frac{1}{2}i, \frac{\sqrt{3}}{2} - \frac{1}{2}i$ **17.** $\frac{\pi}{2}$; $\cos\frac{\pi}{2} + i\sin\frac{\pi}{2}$

19. $\frac{5\pi}{6}$; $4\left(\cos\frac{5\pi}{6} + i\sin\frac{5\pi}{6}\right)$ **21.** $\frac{11\pi}{12}$; $2\left(\cos\frac{11\pi}{12} + i\sin\frac{11\pi}{12}\right)$

23. Circle: center 0, radius $\frac{2}{3}$ **25.** Circle: center $1 + i$, radius 3

27. $\sqrt{2}\left[\cos\left(\frac{\pi}{12}+\frac{\pi k}{2}\right)+i\sin\left(\frac{\pi}{12}+\frac{\pi k}{2}\right)\right]$, $k=0, 1, 2, 3$

29. $2^{1/6}\left[\cos\left(\frac{7\pi}{12}+\frac{2\pi k}{3}\right)+i\sin\left(\frac{7\pi}{12}+\frac{2\pi k}{3}\right)\right]$, $k=0, 1, 2$

31. $\cos\frac{2\pi k}{9}+i\sin\frac{2\pi k}{9}$, $k=0, 1, 2, 3, 4, 5, 6, 7, 8$

33. All points on the line $y=x$

SECTION 10.1

1. Quotient: $2x^2+3x-2$; remainder: $-7x+9$
3. Quotient: x^2-1; remainder: 2 **5.** Quotient: $3x-2$; remainder: 1
7. Quotient: $3x-5$; remainder: 12 **9.** Quotient: $2x^2+4x+2$; remainder: 6
11. Quotient: $6x^2-3x+3$; remainder: 0
13. Quotient: $5x^4-4x^3+x^2-3x+31$; remainder: -124
15. Quotient: $x^6+x^5+x^4+x^3+x^2+x+1$; remainder: 0
17. Quotient: $x^4+ix^3-x^2-ix+1-i$; remainder: $2+i$ **31.** $1, \frac{1}{5}$

SECTION 10.2

1. 1 **3.** 21 **5.** $3\sqrt{2}-2$ **7.** a^4+a^2 **9.** $2-6i$ **11.** $(x-2)^3$
13. $(x+2)^2(x-1)$ **15.** $2(x-\frac{1}{2})(x+3)(x-1)$ **17.** $x(x-2)(x+3i)(x-3i)$
19. $2x(x-\frac{1}{2}i)(x-2i)(x-i)$ **29.** x^2-x-6 **31.** x^3-12x^2-13x
33. x^3-3x^2-6x+8 **35.** x^4-5x^2+4 **37.** x^2+25 **39.** $x^2-2\sqrt{2}x+6$
41. $(x+2i)(x-2i)$ **43.** $(x+\sqrt{7}i)(x-\sqrt{7}i)$ **45.** $2(x+i)(x-i)$
47. $x(x+i)(x-i)$ **49.** $-\frac{3}{2}$

SECTION 10.3

1. 1 (multiplicity 2), $-\frac{1}{2}$ (multiplicity 3)
3. 0 (multiplicity 2), $\sqrt{3}i$ (multiplicity 2), $-\sqrt{3}i$ (multiplicity 2)
5. 3 (multiplicity 1), -3 (multiplicity 1), $\sqrt{3}i$ (multiplicity 1), $-\sqrt{3}i$ (multiplicity 1)
7. 0 (multiplicity 2), $-\frac{1}{2}$ (multiplicity 2), $\frac{\sqrt{3}}{3}i$ (multiplicity 3), $-\frac{\sqrt{3}}{3}i$ (multiplicity 3)
9. $x(x-1)(x-2)$ **11.** $(x-2)(x+2)(x-\sqrt{6})(x+\sqrt{6})$ **13.** $(x+3i)^2(x-3i)^2$
15. $-(x-2+3i)(x-2-3i)$ **19.** 0 **21.** $(x-2)^3(x-i)(x+i)$
23. $(x+i)^4(x-2i)$ **25.** $(x-3)^4(x-1+i)$
27. $(x-1-i)^2(x-1+i)^2(x+6)^3$ **29.** 2 **31.** 2

SECTION 10.4

1. $f(x)=2x^2-4x+10$ **3.** $f(x)=-\frac{3}{2}x^2+3x-3$

5. $f(x)=\frac{4}{3}(x^4+2x^3+x^2+2x)$ **7.** $f(x)=\frac{\sqrt{2}}{15}(x^4-1)$

9. $f(x)=7(x^3-6x^2+15x-14)$
11. $f(x)=\frac{1}{4}(x^4-12x^3+62x^2-140x+125)$ **13.** $(x+1)^2(x^2+1)^3$
15. $(x-1-i)(x-1+i)(2x-1)$ **17.** $(x-1)(x+2-3i)(x+2+3i)$
19. $(x-2-3i)(x-2+3i)(\sqrt{2}x+i)(\sqrt{2}x-i)$
21. a. $2(x-2-i)(x-2+i)(x-1+i)(x-1-i)$
b. $2(x^2-4x+5)(x^2-2x+2)$

SECTION 10.5

1. $1, -1, 2, -2$ **3.** $1, -1, 5, -5$ **5.** $1, -1, 2, -2, 4, -4, 8, -8$
7. $1, -1, 2, -2, 3, -3, 4, -4, 6, -6, 8, -8, 12, -12, 24, -24$ **9.** $-1, 2, -5$
11. $1, -4$ **13.** $1, -1$ **15.** $-1, 2, -2, 6$ **17.** $(x+3)(2x-1)$
19. $(x-5)(x-i)(x+i)$ **21.** $(x-2)(x-3)(x-4)$ **23.** $(x+1)(2x-3)(2x+5)$

25. $2(x-1)(x+1)\left(x-\frac{1}{2}+\frac{\sqrt{11}}{2}i\right)\left(x-\frac{1}{2}-\frac{\sqrt{11}}{2}i\right)$

27. $(x-1)(x+2)^2(x-i)(x+i)$ **29.** $x(2x-1)(3x+1)(4x+1)$
49. a. $(-2, -1), (-1, 0), (0, 1), (1, 2)$

SECTION 10.6

1. $-\frac{2}{x}+\frac{3}{x+1}$ **3.** $\frac{3}{4x-2}+\frac{6}{-x+4}$ **5.** $-\frac{2}{2x-1}+\frac{1}{x-4}-\frac{1}{x+3}$

7. $\frac{1/2}{x}+\frac{1/2}{x+1}$ **9.** $-\frac{2}{x}+\frac{2x}{x^2+3}$ **11.** $\frac{3}{2x^2+3}-\frac{4}{5x^2+4}$ **13.** $\frac{1}{x}-\frac{x}{x^2+1}$

15. $\frac{x+\sqrt{2}}{x^2+\sqrt{2}x+1}+\frac{-x+\sqrt{2}}{x^2-\sqrt{2}x+1}$ **17.** $-\frac{3}{x-1}+\frac{2}{x+1}+\frac{2}{(x+1)^2}$

19. $\frac{1}{x^2+2}-\frac{2}{(x^2+2)^2}$ **21.** $-\frac{2}{x}+\frac{1}{x^2}+\frac{4x}{2x^2+1}-\frac{1}{(2x^2+1)^2}$

CHAPTER 10 REVIEW

1. Quotient: x^2+x+2; remainder: $x+3$
3. Quotient: $-x-2$; remainder: -4 **5.** Quotient: x^2-ix+1; remainder: $2i$
7. Quotient: $x^5+(-1-i)x^3+(-2-4i)x^2+(-6-6i)x-14-18i$; remainder: $-39-54i$
13. -7 **15.** 0 **17.** $(x+2)(x-1)(x-3)$ **19.** $3(x+\frac{1}{3})(x-1)^3$

21. $x(x-\sqrt{2})\left(x+\frac{\sqrt{2}}{2}+\frac{\sqrt{2}}{2}i\right)\left(x+\frac{\sqrt{2}}{2}-\frac{\sqrt{2}}{2}i\right)$

23. $x(x-2+2i)(x-2-2i)$ **25.** $(x-2)(x+2)(x-\sqrt{3}i)(x+\sqrt{3}i)$

27. $(x+3)(x+2+\sqrt{3}i)(x+2-\sqrt{3}i)$ **29.** $(x-i)(x-1)(x+1-i)$
31. $2i$ (multiplicity 2), $7i$ (multiplicity 4), $-7i$ (multiplicity 4), 3 (multiplicity 7)

33. 0 (multiplicity 3), $\frac{\sqrt{3}}{3}$ (multiplicity 4), $\frac{1}{3}i$ (multiplicity 2), $-\frac{1}{3}i$ (multiplicity 2)

35. $x(x-3+5i)^3(x-2i)^4$ **37.** 2 **39.** $2x^3+12x-40$
41. $(x-2+i)^2(x-2-i)^2$ **43.** $(x^2-2x+2)(x^2+2x+5)$ **45.** -1
47. $-3, 4$ **49.** $(x-2)(\sqrt{2}x-1)(\sqrt{2}x+1)$ **51.** $x(x+1)(x-2)(x-i)(x+i)$

53. $(2x+5)(4x-1)(2x-1)$ **63.** $-\frac{3}{3x-2}+\frac{1}{x-2}$ **65.** $\frac{1}{x+2}-\frac{1}{x^2+1}$

67. $\frac{3/4}{x+2}-\frac{3/4}{x-2}-\frac{4}{x^2+4}$

SECTION 11.1

There are no answers for this section.

SECTION 11.2

1. $\left\{\frac{1}{n+1}\right\}_{n=1}^{\infty}$ **3.** $\left\{(-1)^n\right\}_{n=1}^{\infty}$ **5.** $\left\{\frac{2^{n+2}}{3^{n+1}}\right\}_{n=1}^{\infty}$ **7.** $\frac{\sqrt{3}}{2}, \frac{\sqrt{3}}{2}, \frac{\sqrt{3}}{2}, \frac{\sqrt{3}}{2}$

9. 2, 4, 8, 16 **11.** 1, −1, 1, −1 **13.** $\frac{1}{2}, \frac{1}{6}, \frac{1}{12}, \frac{1}{20}$ **15.** 0, ln 2, ln 3, ln 4

17. 2, 4, 8, 16, 32 **19.** 3, 7, 11, 15, 19 **21.** 32, 16, 8, 4, 2 **23.** 1, 2, 6, 24, 120
25. 1, 2, 3, 2, $\frac{5}{12}$

27. **29.** **31.**

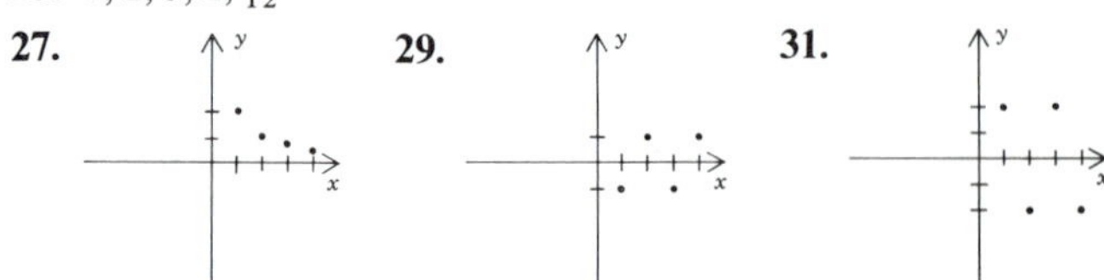

33. 10 **35.** −1 **37.** 9 **39.** $\frac{7}{8}$ **41.** $-\frac{10}{11}$ **43.** $\sum_{j=1}^{7} j$ **45.** $\sum_{j=1}^{9} j^j$

47. $\sum_{j=1}^{8} (2j-1)$ **49.** $a_{10} \approx 2.30258509 \times 10^{-2}$, $a_{100} \approx 4.605170186 \times 10^{-4}$

51. $a_5 = 3.84 \times 10^{-2}$, $a_{15} \approx 2.986281373 \times 10^{-6}$
53. $a_{200} \approx -4.3664865 \times 10^{-3}$, $a_{4000} \approx -1.708759484 \times 10^{-4}$
55. $a_{100} \approx 1.047128548$, $a_{1000} \approx 1.006931669$ **57.** 2.928968254
59. 1.580440283 **61.** 77 **65.** $a_5 = 9$, $a_8 = 5$
67. a. $a_n = 5 - 2n$ for $n \geq 1$ b. $a_n = 1/3^n$ for $n \geq 0$

SECTION 11.3

1. 3 **3.** $\frac{1}{2}$ **5.** ln 2 **7.** No **9.** Yes **11.** Yes **13.** No
15. $a_4 = 11$, $a_n = 2 + 3(n-1) = -1 + 3n$ **17.** $a_4 = 2\pi$, $a_n = 2\pi$

19. $a_4 = -1$, $a_n = -3 + \frac{2}{3}(n-1) = -\frac{11}{3} + \frac{2n}{3}$

21. $a_4 = e + 6$, $a_n = e + 2(n-1) = e - 2 + 2n$

23. $a_7 = 22$, $a_n = 4 + 3(n-1) = 1 + 3n$ **25.** $a_7 = -9$, $a_n = -\frac{3}{2}(n-1) = \frac{3}{2} - \frac{3n}{2}$

27. $a_7 = 7 \ln 2$, $a_n = n \ln 2$ **29.** $a_n = 6 - (n-1) = 7 - n$

31. $a_n = \frac{28}{3} - \frac{5}{3}(n-1) = 11 - \frac{5n}{3}$ **33.** $a_n = n - 1$ **35.** 63 **37.** 42

39. 210 ln 2 **41.** $\frac{75\pi}{2}$ **43.** Yes **45.** $a_n = 9 - 2n$ **47.** 2500

49. $a_n = 7 + 2(n-1) = 5 + 2n$ **51.** Common difference: $4a^3b - 4ab^3$
53. 156 **55.** $1.50 **57.** a. 730,272 square feet b. 568,516 square feet

SECTION 11.4

1. 4 **3.** −0.3 **5.** 16 **7.** Yes **9.** No **11.** Yes **13.** No
15. $a_5 = 64$, $a_n = 2^{n+1}$ **17.** $a_5 = \frac{1}{432}$, $a_n = 3(\frac{1}{6})^{n-1}$ **19.** $a_5 = 810$, $a_n = 10(3^{n-1})$
21. $a_5 = 625c$, $a_n = c(5^{n-1})$ **23.** $a_6 = 81$, $a_n = 3^{n-2}$ **25.** $a_6 = \frac{3}{16}$, $a_n = 6(\frac{1}{2})^{n-1}$
27. $a_6 = -\frac{1}{4}$, $a_n = 256(-\frac{1}{4})^{n-1}$ **29.** $a_6 = -\frac{1}{1458}$, $a_n = \frac{1}{6}(-\frac{1}{3})^{n-1}$
31. $a_n = 4(\frac{3}{2})^{n-1}$ **33.** $a_n = 10(5)^{n-1}$ **35.** $a_n = -5(\frac{2}{5})^{n-1}$ **37.** 117 **39.** 60

41. 121 **43.** $\frac{43}{64}$ **45.** $\frac{484}{243}$ **47.** $-\frac{75}{128}$ **49.** $\frac{3}{2}$ **51.** $\frac{8}{3}$ **53.** 12 **55.** $\frac{6}{7}$
57. $\frac{7}{9}$ **59.** $\frac{95}{99}$ **61.** $\frac{201}{37}$ **63.** $\frac{3116}{990}$ **65.** $a_n = 4^{n-2}$ **67.** 18
69. $8(\frac{3}{2})^{n-1}$ and $-8(-\frac{3}{2})^{n-1}$ **71.** $d = \ln r$ **73.** 500

SECTION 11.5

1. 60 **3.** 2520 **5.** 15 **7.** 4 **9.** 120 **11.** 1320 **13.** 6 **15.** 1 **17.** 10

19. 2 **21.** 15 **23.** 136 **25.** a. $\frac{1}{2}n(n-1)$ **29.** 12

31. a. 576,000 b. 397,440 **33.** a. 45 b. 90 **35.** $\frac{20!}{5!}(\approx 2 \times 10^{16})$

37. 2520 **39.** 10 **41.** 120 **43.** a. 12! (=479,001,600) b. 5! × 7! (=604,800)
45. 27,720 **47.** No

SECTION 11.6

1. $x^5 + 5x^4y + 10x^3y^2 + 10x^2y^3 + 5xy^4 + y^5$ **3.** $x^4 + 8x^3 + 24x^2 + 32x + 16$
5. $x^7 + 7x^6y + 21x^5y^2 + 35x^4y^3 + 35x^3y^4 + 21x^2y^5 + 7xy^6 + y^7$

7. $a^5 - 5a^3 + 10a - \frac{10}{a} + \frac{5}{a^3} - \frac{1}{a^5}$

9. $x^4 + 8x^3 + 28x^2 + 56x + 70 + \frac{56}{x} + \frac{28}{x^2} + \frac{8}{x^3} + \frac{1}{x^4}$ **11.** $17 + 12\sqrt{2}$

13. $-38 - 41i$ **15.** $56x^5y^3$ **17.** -160 **19.** -252 **21.** 0.8684

23. a. $\left(-\frac{1}{3}x + y^3\right)^{12}$ b. $\frac{308}{243}x^6y^{18}$

CHAPTER 11 REVIEW

9. 3, 9, 11, 17, 19 **11.** 3, −2, 8, −12, 28 **13.** −10 **15.** $-\frac{205}{36}$

17. 872 **19.** $\frac{80}{27}$ **21.** $a_n = 5 + \frac{3}{4}(n-1) = \frac{17+3n}{4}$

23. $a_n = 21 - 2(n-1) = 23 - 2n$ **25.** $a_n = 4(-\frac{1}{2})^{n-1}$ **27.** $\frac{21}{11}$ **29.** $-\frac{11}{2}$
31. $\frac{5}{33}$ **33.** 120 **35.** 6 **37.** 20 **39.** 21 **41.** 70
43. $x^8 + 8x^7y + 28x^6y^2 + 56x^5y^3 + 70x^4y^4 + 56x^3y^5 + 28x^2y^6 + 8xy^7 + y^8$
45. $-27x^3 + 108x^2 - 144x + 64$ **47.** a. 98 b. $104i$ **49.** 480
51. The ten-day gift **53.** 8,100,000 **55.** a. 1326 b. 64 **57.** 3150

INDEX

RIGHT TRIANGLES

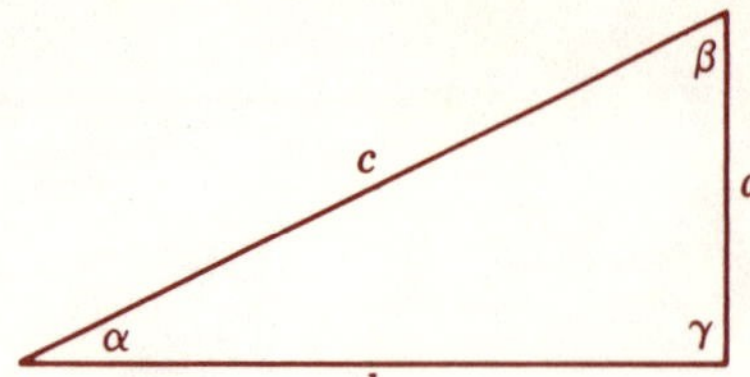

$\sin\alpha = \dfrac{a}{c}$ $\quad\tan\alpha = \dfrac{a}{b}$ $\quad\sec\alpha = \dfrac{c}{b}$

$\cos\alpha = \dfrac{b}{c}$ $\quad\cot\alpha = \dfrac{b}{a}$ $\quad\csc\alpha = \dfrac{c}{a}$

Pythagorean Theorem: $c^2 = a^2 + b^2$

OBLIQUE TRIANGLES

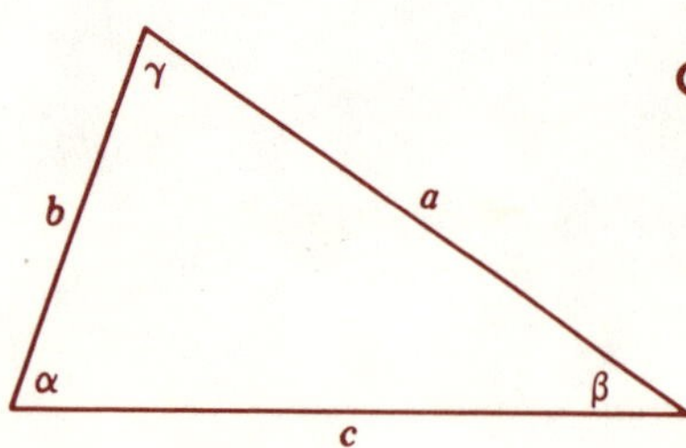

$\alpha + \beta + \gamma = 180°$

Law of Sines: $\dfrac{\sin\alpha}{a} = \dfrac{\sin\beta}{b} = \dfrac{\sin\gamma}{c}$

Law of Cosines: $a^2 = b^2 + c^2 - 2bc\cos\alpha$

TRIGONOMETRIC IDENTITIES

$\tan x = \dfrac{\sin x}{\cos x}$ $\quad\csc x = \dfrac{1}{\sin x}$ $\quad\sin(-x) = -\sin x$ $\quad\cot(-x) = -\cot x$

$\cot x = \dfrac{\cos x}{\sin x}$ $\quad\cot x = \dfrac{1}{\tan x}$ $\quad\cos(-x) = \cos x$ $\quad\sec(-x) = \sec x$

$\sec x = \dfrac{1}{\cos x}$ $\quad\tan(-x) = -\tan x$ $\quad\csc(-x) = -\csc x$

$\sin 2x = 2\sin x\cos x$ $\quad\sin^2 x + \cos^2 x = 1$

$\cos 2x = \cos^2 x - \sin^2 x$ $\quad\tan^2 x + 1 = \sec^2 x$

$\tan 2x = \dfrac{2\tan x}{1 - \tan^2 x}$ $\quad\cot^2 x + 1 = \csc^2 x$

$$\cos(x + y) = \cos x\cos y - \sin x\sin y$$

$$\cos(x - y) = \cos x\cos y + \sin x\sin y$$

$$\sin(x + y) = \sin x\cos y + \cos x\sin y$$

$$\sin(x - y) = \sin x\cos y - \cos x\sin y$$

$$\tan(x + y) = \frac{\tan x + \tan y}{1 - \tan x\tan y}$$

$$\tan(x - y) = \frac{\tan x - \tan y}{1 + \tan x\tan y}$$